Universitext

T0202884

For further volumes:
http://www.springer.com/series/223

Joseph J. Rotman

An Introduction to
Homological Algebra

Second Edition

 Springer

Joseph J. Rotman
Department of Mathematics
University of Illinois at Urbana-Champaign
Urbana IL 61801
USA
rotman@math.uiuc.edu

ISBN: 978-0-387-24527-0   e-ISBN: 978-0-387-68324-9
DOI 10.1007/978-0-387-68324-9

Library of Congress Control Number: 2008936123

Mathematics Subject Classification (2000): 18-01

To the memory of my mother

Rose Wolf Rotman

לזכר אמי מורתי

רוז וולף רוטמן

Contents

Chapter 9   Homology and Groups

Chapter 10   Spectral Sequences

Preface to the Second Edition

Homological Algebra has grown in the nearly three decades since the first edition of this book appeared in 1979. Two books discussing more recent results are Weibel, *An Introduction to Homological Algebra*, 1994, and Gelfand–Manin, *Methods of Homological Algebra*, 2003. In their Foreword, Gelfand and Manin divide the history of Homological Algebra into three periods: the first period ended in the early 1960s, culminating in applications of Homological Algebra to regular local rings. The second period, greatly influenced by the work of A. Grothendieck and J.-P. Serre, continued through the 1980s; it involves abelian categories and sheaf cohomology. The third period, involving derived categories and triangulated categories, is still ongoing. Both of these newer books discuss all three periods (see also Kashiwara–Schapira, *Categories and Sheaves*). The original version of this book discussed the first period only; this new edition remains at the same introductory level, but it now introduces the second period as well. This change makes sense pedagogically, for there has been a change in the mathematics population since 1979; today, virtually all mathematics graduate students have learned something about functors and categories, and so I can now take the categorical viewpoint more seriously.

When I was a graduate student, Homological Algebra was an unpopular subject. The general attitude was that it was a grotesque formalism, boring to learn, and not very useful once one had learned it. Perhaps an algebraic topologist was forced to know this stuff, but surely no one else should waste time on it. The few true believers were viewed as workers at the fringe of mathematics who kept tinkering with their elaborate machine, smoothing out rough patches here and there.

This attitude changed dramatically when J.-P. Serre characterized regular local rings using Homological Algebra (they are the commutative noetherian local rings of "finite global dimension"), for this enabled him to prove that any localization of a regular local ring is itself regular (until then, only special cases of this were known). At the same time, M. Auslander and D. A. Buchsbaum also characterized regular local rings, and they went on to complete work of M. Nagata by using global dimension to prove that every regular local ring is a unique factorization domain. As Grothendieck and Serre revolutionized Algebraic Geometry by introducing schemes and sheaves, resistance to Homological Algebra waned. Today, it is just another standard tool in a mathematician's kit. For more details, we recommend C. A. Weibel's chapter, "History of Homological Algebra," in the book of James, *History of Topology*.

Homological Algebra presents a great pedagogical challenge for authors and for readers. At first glance, its flood of elementary definitions (which often originate in other disciplines) and its space-filling diagrams appear forbidding. To counter this first impression, S. Lang set the following exercise on page 105 of his book, *Algebra*:

> Take any book on homological algebra and prove all the theorems without looking at the proofs given in that book.

Taken literally, the statement of the exercise is absurd. But its spirit is absolutely accurate; the subject only *appears* difficult. However, having recognized the elementary character of much of the early material, one is often tempted to "wave one's hands": to pretend that minutiae always behave well. It should come as no surprise that danger lurks in this attitude. For this reason, I include many details in the beginning, at the risk of boring some readers by so doing (of course, such readers are free to turn the page). My intent is twofold: to allow readers to see that complete proofs can, in fact, be written compactly; to give readers the confidence to believe that they, too, can write such proofs when, later, the lazy author asks them to. However, we must caution the reader; some "obvious" statements are not only false, they may not even make sense. For example, if R is a ring and A and B are left R-modules, then $\mathrm{Hom}_R(A, B)$ may not be an R-module at all; and, if it is a module, it is sometimes a left module and sometimes a right module. Is an alleged function with domain a tensor product well-defined? Is an isomorphism really natural? Does a diagram really commute? After reading the first three chapters, the reader should be able to deal with such matters efficiently.

This book is my attempt to make Homological Algebra lovable, and I believe that this requires the subject be presented in the context of other mathematics. For example, Chapters 2, 3, and 4 form a short course in module theory, investigating the relation between a ring and its projective, injective, and flat modules. Making the subject lovable is my reason for delaying the formal introduction of homology functors until Chapter 6 (although simplicial and singular homology do appear in Chapter 1). Many readers wanting to learn Homological Algebra are familiar with the first properties of Hom and tensor; even so, they should glance at the first chapters, for there may be some unfamiliar items therein. Some category theory appears throughout, but it makes a more brazen appearance in Chapter 5, where we discuss limits, adjoint functors, and sheaves. Although presheaves are introduced in Chapter 1, we do not introduce sheaves until we can observe that they usually form an abelian category. Chapter 6 constructs homology functors, giving the usual fundamental results about long exact sequences, natural connecting homomorphisms, and independence of choices of projective, injective, and flat resolutions used to construct them. Applications of sheaves are most dramatic in the context of Several Complex Variables and in Algebraic Geometry; alas, I say only a few words pointing the reader to appropriate texts, but there is a brief discussion of the Riemann–Roch Theorem over compact Riemann surfaces. Chapters 7, 8, and 9 consider the derived functors of Hom and tensor, with applications to ring theory (via global dimension), cohomology of groups, and division rings.

Learning Homological Algebra is a two-stage affair. First, one must learn the language of Ext and Tor and what it describes. Second, one must be able to compute these things and, often, this involves yet another language, that of spectral sequences. Chapter 10 develops spectral sequences via exact couples, always taking care that bicomplexes and their multiple indices are visible because almost all applications occur in this milieu.

A word about notation. I am usually against spelling reform; if everyone is comfortable with a symbol or an abbreviation, who am I to say otherwise? However, I do use a new symbol to denote the integers mod m because, nowadays, two different symbols are used: $\mathbb{Z}/m\mathbb{Z}$ and $\mathbb{Z}_m$. My quarrel with the first symbol is that it is too complicated to write many times in an argument; my quarrel with the simpler second symbol is that it is ambiguous: when p is a prime, the symbol $\mathbb{Z}_p$ often denotes the p-adic integers and not the integers mod p. Since capital I reminds us of *integers* and since blackboard font is in common use, as in $\mathbb{Z}, \mathbb{Q}, \mathbb{R}, \mathbb{C}$, and $\mathbb{F}_q$, I denote the integers mod m by $\mathbb{I}_m$.

It is a pleasure to thank again those who helped with the first edition. I also thank the mathematicians who helped with this revision: Matthew Ando, Michael Barr, Steven Bradlow, Kenneth S. Brown, Daniel Grayson, Phillip Griffith, William Haboush, Aimo Hinkkanen, Ilya Kapovich, Randy McCarthy, Igor Mineyev, Thomas A. Nevins, Keith Ramsay, Derek Robinson, and Lou van den Dries. I give special thanks to Mirroslav Yotov who not only made many valuable suggestions improving the entire text but who, having seen my original flawed subsection on the Riemann–Roch Theorem, patiently guided my rewriting of it.

Joseph J. Rotman
May 2008
Urbana IL

How to Read This Book

Some exercises are starred; this means that they will be cited somewhere in the book, perhaps in a proof.

One may read this book by starting on page 1, then continuing, page by page, to the end, but a mathematics book cannot be read as one reads a novel. Certainly, this book is not a novel! A reader knowing very little homology (or none at all) should begin on page 1 and then read only the portion of Chapter 1 that is unfamiliar. Homological Algebra developed from Algebraic Topology, and it is best understood if one knows its origins, which are described in Sections 1.1 and 1.3. Section 1.2 introduces categories and functors; at the outset, the reader may view this material as a convenient language, but it is very important for the rest of the text.

After Chapter 1, one could go directly to Chapter 6, Homology, but I don't advise it. It is not necessary to digest all the definitions and constructions in the first five chapters before studying homology, but one should read enough to become familiar with the point of view being developed, returning to read or reread items in earlier chapters when necessary.

I believe that it is wisest to learn homology in a familiar context in which it can be applied. To illustrate, one of the basic constructs in defining homology is that of a *complex*: a sequence of homomorphisms

$$\to C_{n+1} \xrightarrow{d_{n+1}} C_n \xrightarrow{d_n} C_{n-1} \to$$

in which $d_n d_{n+1} = 0$ for all $n \in \mathbb{Z}$. There is no problem digesting such a simple definition, but one might wonder where it comes from and why it is significant. The reader who has seen some Algebraic Topology (as in our Chapter 1) recognizes a geometric reason for considering complexes. But this observation only motivates the singular complex of a topological space. A more perspicacious reason arises in Algebra. Every R-module M is a quotient of a free module; thus, $M \cong F/K$, where F is free and $K \subseteq F$ is the submodule of *relations*; that is, $0 \to K \to F \to M \to 0$ is exact. If X is a basis of F, then $(X \mid K)$ is called a *presentation* of G. Theoretically, $(X \mid K)$ is a complete description of M (to isomorphism) but, in practice, it is difficult to extract information about M from a presentation of it. However, if R is a principal ideal domain, then every submodule of a free module is free, and so K has a basis, say, Y [we also say that $(X \mid Y)$ is a presentation]. For example, the canonical forms for matrices over a field k arise from presentations of certain $k[x]$-modules. For a general ring R, we can iterate the idea of presentations. If $M \cong F/K$, where F is free, then $K \cong F_1/K_1$ for some free module F_1 (thus, K_1 can be thought of as relations among the relations; Hilbert called them *syzygies*). Now $0 \to K_1 \to F_1 \to K \to 0$ is exact; splicing it to the earlier exact sequence gives exactness of

$$0 \to K_1 \to F_1 \xrightarrow{d} F \to M \to 0$$

(where $d\colon F_1 \to F$ is the composite $F_1 \to K \subseteq F$), for $\operatorname{im} d = K = \ker(F \to M)$. Repeat: $K_1 \cong F_2/K_2$ for some free F_2, and continuing the construction above gives an infinitely long exact sequence of free modules and homomorphisms, called a *resolution* of M, which serves as a generalized presentation. A standard theme of Homological Algebra is to replace a module by a resolution of it. Resolutions are exact sequences, and exact sequences are complexes (if $\operatorname{im} d_{n+1} = \ker d_n$, then $d_n d_{n+1} = 0$). Why do we need the extra generality present in the definition of complex? One answer can be seen by returning to Algebraic Topology. We are interested not only in the homology groups of a space, but also in its cohomology groups, and these arise by applying contravariant Hom functors to the singular complex. In Algebra, the problem of classifying group extensions also leads to applying Hom functors to resolutions. Even though resolutions are exact sequences, they become mere complexes after applying Hom. Homological Algebra is a tool that extracts information from such sequences. As the reader now sees, the context is interesting, and it puts flesh on abstract definitions.

1

Introduction

1.1 Simplicial Homology

Homological Algebra is an outgrowth of Algebraic Topology, and so we begin
with a historical discussion of the origins of homology in topology. Let X be
an open set in the plane, and fix points a and b in X. Given a path[1] β in X

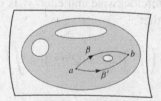

Fig. 1.1 Two paths.

from a to b, and given a pair $P(x, y)$ and $Q(x, y)$ of real-valued, continuously
differentiable functions on X, one wants to evaluate the line integral

$$\int_{\beta} P\,dx + Q\,dy.$$

---

[1]Let $\mathbf{I} = [0, 1]$ be the closed unit interval. A ***path*** β ***in*** X ***from*** a ***to*** b is a continuous
function $\beta \colon \mathbf{I} \to X$ with $\beta(0) = a$ and $\beta(1) = b$; thus, a path is a parametrized curve. A
path β is ***closed at*** a if $\beta(0) = a = \beta(1)$ or, what is the same thing, if β is a continuous
map of the unit circle S^1 into X with $f \colon (1, 0) \mapsto a$.

J.J. Rotman, *An Introduction to Homological Algebra*, Universitext,
DOI 10.1007/978-0-387-68324-9_1, © Springer Science+Business Media LLC 2009

It is wise to regard β as a finite union of paths, for β may be only piece-wise smooth; for example, it may be a polygonal path. For the rest of this discussion, we ignore (necessary) differentiability hypotheses.

A fundamental question asks when the line integral is independent of the path β: is $\int_{\beta'} P\,dx + Q\,dy = \int_{\beta} P\,dx + Q\,dy$ if β' is another path in X from a to b? If γ is the closed path $\gamma = \beta - \beta'$ [that is, γ goes from a to b via β, and then goes back via $-\beta'$ from b to a, where $-\beta'(t) = \beta'(1-t)$], then the integral is independent of the paths β and β' if and only if $\int_{\gamma} P\,dx + Q\,dy = 0$. Suppose there are "bad" points $z_1, \ldots, z_n$ deleted from X (for example, if P or Q has a singularity at some z_i). The line integral along γ is affected by

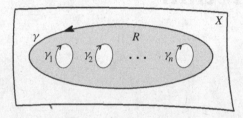

Fig. 1.2 Green's theorem.

whether any of these bad points lies inside γ. In Fig. 1.2, each path γ_i is a simple closed path in X (that is, γ_i is a homeomorphism from the unit circle S^1 to $\operatorname{im}\gamma_i \subseteq \mathbb{R}^2$) containing z_i inside, while all the other z_j are outside γ_i. If γ is oriented counterclockwise and each γ_i is oriented clockwise, then **Green's Theorem** states that

$$\int_{\gamma} P\,dx + Q\,dy + \sum_{i=1}^{n} \left(\int_{\gamma_i} P\,dx + Q\,dy \right) = \iint_{R} \left(\frac{\partial Q}{\partial x} - \frac{\partial P}{\partial y} \right) dx\,dy,$$

where R is the shaded two-dimensional region in Fig. 1.2. One is tempted to write $\sum_{i=1}^{n} \left(\int_{\gamma_i} P\,dx + Q\,dy \right)$ more concisely, as $\int_{\sum_i \gamma_i} P\,dx + Q\,dy$. More-over, instead of mentioning orientations explicitly, we may write sums and differences of paths, where a negative coefficient reverses direction. For sim-ple paths, the notions of "inside" and "outside" make sense.[2] A path may wind around some z_i several times along γ_i, and so it makes sense to write formal $\mathbb{Z}$-linear combinations of paths; that is, we may allow integer coeffi-cients other than ± 1. Recall that if Y is any set, then the **free abelian group**

---

[2]The *Jordan curve theorem* says that if γ is a simple closed path in the plane $\mathbb{R}^2$, then the complement $\mathbb{R}^2 - \operatorname{im}\gamma$ has exactly two components, one of which is bounded. The bounded component is called the *inside* of γ, and the other (necessarily unbounded) component is called the *outside*.

with **basis** Y is an abelian group $G[Y]$ in which each element $g \in G[Y]$ has a unique expression of the form $g = \sum_{y \in Y} m_y y$, where $m_y \in \mathbb{Z}$ and only finitely many $m_y \neq 0$ (see Proposition 2.33). In particular, Green's Theorem involves the free abelian group $G[Y]$ with basis Y being the (huge) set of all paths $\sigma : \mathbf{I} \to X$. Intuitively, elements of $G[Y]$ are unions of finitely many (not necessarily closed) paths σ in X.

Consider those ordered pairs (P, Q) of functions $X \to \mathbb{R}$ satisfying $\partial Q/\partial x = \partial P/\partial y$. The double integral in Green's Theorem vanishes for such function pairs: $\int_{m\gamma + \sum_i m_i \gamma_i} P\, dx + Q\, dy = 0$. An equivalence relation on $G[Y]$ suggests itself. If $\beta = \sum m_i \sigma_i$ and $\beta' = \sum m_i' \sigma_i \in G[Y]$, call β and β' **equivalent** if, for all (P, Q) with $\partial Q/\partial x = \partial P/\partial y$, the values of their line integrals agree:

$$\int_\beta P\, dx + Q\, dy = \int_{\beta'} P\, dx + Q\, dy.$$

The equivalence class of β is called its **homology class**, from the Latin word *homologia* meaning *agreement*. If $\beta - \beta' = \sum m_\sigma \sigma$, where $\bigcup_{m_\sigma \neq 0} \operatorname{im} \sigma$ is the boundary of a two-dimensional region in X, then $\int_{\beta - \beta'} P\, dx + Q\, dy = 0$; that is, $\int_\beta P\, da + Q\, dy = \int_{\beta'} P\, da + Q\, dy$. In short, integration is independent of paths lying in the same homology class.

Homology can be defined without using integration of function pairs. Poincaré recognized that whether a topological space X has different kinds of holes is a kind of connectivity. To illustrate, suppose that X is a *finite simplicial complex*; that is, X can be "triangulated" into finitely many *n-simplexes* for $n \geq 0$, where 0-simplexes are points, say, $v_1, \ldots, v_q$, 1-simplexes are certain edges $[v_i, v_j]$ (with endpoints v_i and v_j), 2-simplexes are certain triangles $[v_i, v_j, v_k]$ (with vertices v_i, v_j, v_k), 3-simplexes are certain tetrahedra $[v_i, v_j, v_k, v_\ell]$, and there are higher-dimensional analogs $[v_{i_0}, \ldots, v_{i_n}]$ for larger n. The question to ask is whether a union of n-simplexes in X that "ought" to be the boundary of some $(n+1)$-simplex actually is such a boundary. For example, when $n = 0$, two points a and b in X ought to be the boundary (endpoints) of a path in X; if, for each pair of points $a, b \in X$, there is a path in X from a to b, then X is called **path connected**; if there is no such path, then X has a 0-dimensional hole. For an example of a one-dimensional hole, let X be the **punctured plane**; that is, X is the plane with the origin deleted. The perimeter of a triangle Δ ought to be the boundary of a 2-simplex, but it is not if Δ contains the origin in its interior; thus, X has a one-dimensional hole. If the interior of X were missing a line segment containing the origin, or even a small disk containing the origin, this hole would still be one-dimensional; we are not considering the size of the hole, but the size of the possible boundary. We must keep our eye on the doughnut and not upon the hole!

The triangle $[a, b, c]$ in Fig. 1.3 has vertices a, b, c and edges $[b, c]$, $[a, c]$, $[a, b]$; its boundary $\partial[a, b, c]$ should be $[b, c] \cup [c, a] \cup [a, c]$. But edges are oriented; think of $[a, c]$ as a path from a to c and $[c, a] = -[a, c]$ as the reverse path from c to a. Thus, the boundary is

$$\partial[a, b, c] = [b, c] \cup -[a, c] \cup [a, b].$$

Assume now that paths can be added and subtracted; that is, view paths as lying in the free abelian group $C_1(X)$ with basis all 1-simplexes. Then

$$\partial[a, b, c] = [b, c] - [a, c] + [a, b].$$

Similarly, define the boundary of $[a, b]$ to be $\partial[a, b] = b - a \in C_0(X)$, the free abelian group with basis all 0-simplexes, and define the boundary of a point to be 0. Note that

$$\partial(\partial[a, b, c]) = \partial([b, c] - [a, c] + [a, b])$$
$$= (c - b) - (c - a) + (b - a) = 0.$$

Fig. 1.3.   The rectangle $\square$.

The rectangle $\square$ with vertices a, b, c, d is the union of two triangles, namely, $[a, b, c] \cup [a, c, d]$; let us check its boundary. If we assume that ∂ is a homomorphism, then

$$\partial(\square) = \partial[a, b, c] \cup \partial[a, c, d]$$
$$= \partial[a, b, c] + \partial[a, c, d]$$
$$= ([b, c] - [a, c] + [a, b]) + ([c, d] - [a, d] + [a, c])$$
$$= [a, b] + [b, c] + [c, d] - [a, d]$$
$$= [a, b] + [b, c] + [c, d] + [d, a].$$

Note that the diagonal $[a, c]$ occurred twice, with different signs, and so it canceled, as it should. We have seen that the formalism suggests the use of signs to describe boundaries as certain alternating sums of edges or points.

Such ideas lead to the following construction. For each $n \geq 0$, consider all formal linear combinations of n-simplexes; that is, form the free abelian group $C_n(X)$ with basis all n-simplexes $[v_{i_0}, \ldots, v_{i_n}]$, and call such linear combinations **simplicial n-chains**; define $C_{-1}(X) = \{0\}$. Some n-chains ought to be boundaries of some union of $(n+1)$-simplexes; call them *simplicial n-cycles*. For example, adding the three edges of a triangle (with appropriate choice of signs) is a 1-cycle, as is the sum of the four outer edges of a rectangle. Certain simplicial *n*-chains actually are boundaries, and these are called *simplicial n-boundaries*. For example, if Δ is a triangle in the punctured plane X, then the alternating sum of the edges of Δ is a 1-cycle; this 1-cycle is a 1-boundary if and only if the origin does not lie in the interior of Δ. Here are the precise definitions.

Definition. Let X be a finite simplicial complex. If $n \geq 1$, define

$$\partial_n : C_n(X) \to C_{n-1}(X)$$

by

$$\partial_n[v_0, \dots, v_n] = \sum_{i=0}^{n} (-1)^i [v_0, \dots, \widehat{v_i}, \dots, v_n]$$

(the notation $\widehat{v_i}$ means that v_i is omitted). Define $\partial_0 : C_0(X) \to C_{-1}(X)$ to be identically zero [since $C_{-1}(X) = \{0\}$, this definition is forced on us]. As every simplicial *n*-chain has a unique expression as a linear combination of simplicial *n*-simplexes, ∂_n *extends by linearity*[3] to a function defined on all of $C_n(X)$. The homomorphisms ∂_n are called *simplicial boundary maps*.

The presence of signs gives the following fundamental result.

Proposition 1.1.   *For all* $n \geq 0$,

$$\partial_{n-1}\partial_n = 0.$$

Proof.   Each term of $\partial_n[x_0, \dots, x_n]$ has the form $(-1)^i[x_0, \dots, \widehat{x_i}, \dots, x_n]$. Hence, $\partial_n[x_0, \dots, x_n] = \sum_i (-1)^i [x_0, \dots, \widehat{x_i}, \dots, x_n]$, and

$$\partial_{n-1}[x_0, \dots, \widehat{x_i}, \dots, x_n] = [\widehat{x_0}, \dots, \widehat{x_i}, \dots, x_n] + \cdots$$
$$+ (-1)^{i-1}[x_0, \dots, \widehat{x_{i-1}}, \widehat{x_i}, \dots, x_n]$$
$$+ (-1)^i[x_0, \dots, \widehat{x_i}, \widehat{x_{i+1}} \dots, x_n] + \cdots$$
$$+ (-1)^{n-1}[x_0, \dots, \widehat{x_i}, \dots, \widehat{x_n}]$$

---

[3] Proposition 2.34 says that if F is a free abelian group with basis Y, and if $f : Y \to G$ is any function with values in an abelian group G, then there exists a unique homomorphism $\widetilde{f} : F \to G$ with $\widetilde{f}(y) = f(y)$ for all $y \in Y$. The map $\widetilde{f}$ is obtained from f by *extending by linearity*: if $u = m_1 y_1 + \cdots + m_p y_p$, then $\widetilde{f}(u) = m_1 f(y_1) + \cdots + m_p f(y_p)$.

(when $k \geq i + 1$, the sign of $[x_0, \ldots, \widehat{x_i}, \ldots, \widehat{x_k}, \ldots, x_n]$ is $(-1)^{k-1}$, for the vertex x_k is the $(k-1)$st term in the list $x_0, \ldots, \widehat{x_i}, \ldots, x_k, \ldots, x_n$). Thus,

$$\partial_{n-1}[x_0, \ldots, \widehat{x_i}, \ldots, x_n] = \sum_{j=0}^{i-1}(-1)^j[x_0, \ldots, \widehat{x_j}, \cdots, \widehat{x_i}, \ldots, x_n]$$
$$+ \sum_{k=i+1}^{n}(-1)^{k-1}[x_0, \ldots, \widehat{x_i}, \ldots, \widehat{x_k}, \ldots, x_n].$$

Now $[x_0, \ldots, \widehat{x_i}, \ldots, \widehat{x_j}, \ldots, x_n]$ occurs twice in $\partial_{n-1}\partial_n[x_0, \ldots, x_n]$: from $\partial_{n-1}[x_0, \ldots, \widehat{x_i}, \ldots, x_n]$ and from $\partial_{n-1}[x_0, \ldots, \widehat{x_j}, \ldots, x_n]$. Therefore, the first term has sign $(-1)^{i+j}$, while the second term has sign $(-1)^{i+j-1}$. Thus, the $(n-2)$-tuples cancel in pairs, and $\partial_{n-1}\partial_n = 0$. •

Definition. For each $n \geq 0$, the subgroup $\ker \partial_n \subseteq C_n(X)$ is denoted by $Z_n(X)$; its elements are called *simplicial n-cycles*. The subgroup $\operatorname{im} \partial_{n+1} \subseteq C_n(X)$ is denoted by $B_n(X)$; its elements are called *simplicial n-boundaries*.

Corollary 1.2. *For all n,*

$$B_n(X) \subseteq Z_n(X).$$

Proof. If $\alpha \in B_n$, then $\alpha = \partial_{n+1}(\beta)$ for some $(n+1)$-chain β. Hence, $\partial_n(\alpha) = \partial_n\partial_{n+1}(\beta) = 0$, so that $\alpha \in \ker \partial_n = Z_n$. •

We have defined a sequence of abelian groups and homomorphisms in which composites of consecutive arrows are 0:

$$\cdots \to C_3(X) \xrightarrow{\partial_3} C_2(X) \xrightarrow{\partial_2} C_1(X) \xrightarrow{\partial_1} C_0(X) \xrightarrow{\partial_0} 0.$$

The interesting group is the quotient group $Z_n(X)/B_n(X)$.

Definition. The nth *simplicial homology group* of a finite simplicial complex X is
$$H_n(X) = Z_n(X)/B_n(X).$$

What survives in the quotient group $Z_n(X)/B_n(X)$ are the n-dimensional holes; that is, those n-cycles that are not n-boundaries; $H_n(X) = \{0\}$ means that X has no n-dimensional holes.[4] For example, if X is the punctured plane,

---

[4]Eventually, homology groups will be defined for mathematical objects other than topological spaces. It is always a good idea to translate $H_n(X) = \{0\}$ into concrete terms, if possible, as some interesting property of X. One can then interpret the elements of $H_n(X)$ as being *obstructions* to whether X enjoys this property.

then $H_1(X) \neq \{0\}$: if $\Delta = [a, b, c]$ is a triangle in X having the origin in its interior, then $\alpha = [b, c] - [a, c] + [a, b]$ is a 1-cycle that is not a boundary,[5] and the coset $\alpha + B_1(X)$ is a nonzero element of $H_1(X)$. Topologists modify this construction in two ways. They introduce homology with *coefficients* in an abelian group G by tensoring the sequence of chain groups by G and then taking homology groups; they also consider *cohomology with coefficients* in G by applying $\mathrm{Hom}(\square, G)$ to the sequence of chain groups and then taking homology groups. Homological Algebra arose in trying to compute and to find relations between homology groups and cohomology groups of spaces.

1.2   Categories and Functors

Let us now pass from the concrete to the abstract. Categories are the context for discussing general properties of systems such as groups, rings, modules, sets, or topological spaces, in tandem with their respective transformations: homomorphisms, functions, or continuous maps.

There are well-known set-theoretic "paradoxes" showing that contradictions arise if we are not careful about how the undefined terms *set* and *element* are used. For example, ***Russell's paradox*** gives a contradiction arising from regarding every collection as a set. Define a *Russell set* to be a set S that is not a member of itself; that is, $S \notin S$, and define R to be the collection of all Russell sets. Either R is a Russell set or it is not a Russell set. If R is a Russell set, then $R \notin R$, by definition. But all Russell sets lie in the collection of all Russell sets, namely, R; that is, $R \in R$, a contradiction. On the other hand, if R is not a Russell set, then R does not lie in the collection of all Russell sets; that is, $R \notin R$. But now R satisfies the criterion for being a Russell set, another contradiction. We conclude that some conditions are needed to determine which collections are allowed to be sets. Such conditions are given in the ***Zermelo–Fraenkel axioms*** for set theory, specifically, by the ***axiom of comprehension***; the collection R is not a set, and this resolves the Russell paradox. Another approach to resolving this paradox involves restrictions on the membership relation: some say that $x \in x$ is not a well-formed formula; others say that $x \in x$ is well-formed, but it is always false.

Let us give a bit more detail. The Zermelo–Fraenkel axioms have primitive terms *class* and $\in$ and rules for constructing classes, as well as for constructing certain special classes, called ***sets***. For example, finite classes and the natural numbers $\mathbb{N}$ are assumed to be sets. A class is called ***small*** if it has a cardinal number, and it is a theorem that a class is a set if and only if it is small; a class that is not a set is called a ***proper class***. For example, $\mathbb{N}$, $\mathbb{Z}$, $\mathbb{Q}$, $\mathbb{R}$,

---

[5]Of course, $\alpha = \partial[a, b, c]$, but $[a, b, c]$ is not a 2-simplex in X because Δ has the origin in its interior. One must do more, however, to prove that $\alpha \notin B_1(X)$.

and $\mathbb{C}$ are sets, the collection of all sets is a proper class, and the collection R of all Russell classes is not even a class. For a more complete discussion, see Mac Lane, *Categories for the Working Mathematician*, pp. 21–24, Douady–Douady, *Algèbre et Théories Galoisiennes*, pp. 24–25, and Herrlich–Strecker, *Category Theory*, Chapter II and the Appendix. We quote Herrlich–Strecker, p. 331.

> There are two important points (in different approaches to Category Theory). ... First, there is no such thing as *the* category **Sets** of all sets. If one approaches set theory from a naive standpoint, inconsistencies will arise, and approaching it from any other standpoint requires an axiom scheme, so that the properties of **Sets** will depend upon the foundation chosen. ... The second point is that (there is) a foundation that allows us to perform all of the categorical-theoretical constructions that at the moment seem desirable. If at some later time different constructions that cannot be performed within this system are needed, then the foundation should be expanded to accommodate them, or perhaps should be replaced entirely. After all, the purpose of foundations is not to arbitrarily restrict inquiry, but to provide a framework wherein one can legitimately perform those constructions and operations that are mathematically interesting and useful, so long as they are not inconsistent within themselves.

We will be rather relaxed about Set Theory. As a practical matter, when an alleged class arises, there are three possibilities: it is a set; it is a proper class; it is not a class at all (one consequence of the axioms is that a proper class is forbidden to be an element of any class). In this book, we will not worry about the possibility that an alleged class is not a class.

Definition. A *category* $\mathcal{C}$ consists of three ingredients: a class obj$(\mathcal{C})$ of *objects*, a *set* of *morphisms* Hom(A, B) for every ordered pair (A, B) of objects, and *composition* Hom$(A, B) \times$ Hom$(B, C) \to$ Hom(A, C), denoted by

$$(f, g) \mapsto gf,$$

for every ordered triple A, B, C of objects. [We often write $f \colon A \to B$ or $A \overset{f}{\to} B$ instead of $f \in$ Hom(A, B).] These ingredients are subject to the following axioms:

(i) the Hom sets are pairwise disjoint[6]; that is, each $f \in$ Hom(A, B) has a unique *domain* A and a unique *target* B;

---

[6]In the unlikely event that some particular candidate for a category does not have disjoint Hom sets, then one can force them to be disjoint by redefining Hom(A, B) as $Hom(A, B) = \{A\} \times$ Hom$(A, B) \times \{B\}$, so that each morphism $f \in$ Hom(A, B) is now relabeled as (A, f, B). If $(A, B) \neq (A', B')$, then $Hom(A, B) \cap Hom(A', B') = \varnothing$.

(ii) for each object A, there is an *identity morphism* $1_A \in \mathrm{Hom}(A, A)$ such that $f1_A = f$ and $1_B f = f$ for all $f \colon A \to B$;

(iii) composition is associative: given morphisms $A \overset{f}{\to} B \overset{g}{\to} C \overset{h}{\to} D$, then

$$h(gf) = (hg)f.$$

A more important notion in this circle of ideas is that of *functor*, which we will define soon. Categories are necessary because they are an essential ingredient in the definition of functor. A similar situation occurs in Linear Algebra: linear transformation is the more important notion, but vector spaces are needed in order to define it.

The following examples will explain certain fine points of the definition of category.

Example 1.3.

(i) **Sets.** The objects in this category are sets (not proper classes), morphisms are functions, and composition is the usual composition of functions.

It is an axiom of set theory that if A and B are sets, then the class $\mathrm{Hom}(A, B)$ of all functions from A to B is also a set. That Hom sets are pairwise disjoint is just a reflection of the definition of equality of functions, which says that two functions are equal if they have the same domains and the same targets (as well as having the same graphs). For example, if $U \subsetneq X$ is a proper subset of a set X, then the inclusion function $U \to X$ is distinct from the identity function 1_U, for they have different targets. If $f \colon A \to B$ and $g \colon C \to D$ are functions, we define their composite $gf \colon A \to D$ if $B = C$. In contrast, in Analysis, one often says gf is defined when $B \subseteq C$. We do not recognize this; for us, gf is not defined, but gif is defined, where $i \colon B \to C$ is the inclusion.

(ii) **Groups.** Objects are groups, morphisms are homomorphisms, and composition is the usual composition (homomorphisms are functions). Part of the verification that **Groups** is a category involves checking that identity functions are homomorphisms and that the composite of two homomorphisms is itself a homomorphism [one needs to know that if $f \in \mathrm{Hom}(A, B)$ and $g \in \mathrm{Hom}(B, C)$, then $gf \in \mathrm{Hom}(A, C)$].

(iii) A partially ordered set X can be regarded as the category whose objects are the elements of X, whose Hom sets are either empty or have only one element:

$$\mathrm{Hom}(x, y) = \begin{cases} \varnothing & \text{if } x \npreceq y, \\ \{\iota_y^x\} & \text{if } x \preceq y \end{cases}$$

(the symbol ι_y^x is the unique element in the Hom set when $x \preceq y$), and whose composition is given by $\iota_z^y \iota_y^x = \iota_z^x$. Note that $1_x = \iota_x^x$, by reflexivity, while composition makes sense because $\preceq$ is transitive. The converse is false: if $\mathcal{C}$ is a category with $|\operatorname{Hom}(x, y)| \leq 1$ for every $x, y \in \operatorname{obj}(\mathcal{C})$, define $x \preceq y$ if $\operatorname{Hom}(x, y) \neq \varnothing$. Then $\mathcal{C}$ may not be partially ordered because $\preceq$ need not be antisymmetric. The two-point category $\bullet \rightrightarrows \bullet$ having only two nonidentity morphisms is such an example that is not partially ordered.

We insisted, in the definition of category, that each $\operatorname{Hom}(A, B)$ be a set, but we did not say it was nonempty. The category X, where X is a partially ordered set, is an example in which this possibility occurs. [Not every Hom set in a category $\mathcal{C}$ can be empty, for $1_A \in \operatorname{Hom}(A, A)$ for every $A \in \operatorname{obj}(\mathcal{C})$.]

(iv) Let X be a topological space, and let $\mathcal{U}$ denote its topology; that is, $\mathcal{U}$ is the family of all the open subsets of X. Then $\mathcal{U}$ is a partially ordered set under ordinary inclusion, and so it is a category as in part (iii). In this case, we can realize the morphism ι_V^U, when $U \subseteq V$, as the inclusion $i_V^U : U \to V$.

(v) View a natural number $n \geq 1$ as the partially ordered set whose elements are $0, 1, \ldots, n-1$ and $0 \leq 1 \leq \cdots \leq n-1$. As in part (iii), there is a category **n** with $\operatorname{obj}(\mathbf{n}) = \{0, 1, \ldots, n-1\}$ and with morphisms $i \to j$ for all $0 \leq i \leq j \leq n-1$.

(vi) Another special case of part (iii) is $\mathbb{Z}$ viewed as a partially ordered set (which we order by reverse inequality, so that $n \preceq n-1$):

$$\cdots \longrightarrow \underset{n+1}{\bullet} \longrightarrow \underset{n}{\bullet} \longrightarrow \underset{n-1}{\bullet} \longrightarrow \cdots .$$

Actually, there are some morphisms we have not drawn: loops at each n, corresponding to identity morphisms $n \to n$, and composites $m \to n$ for all $m > n + 1$; that is, $m \prec n + 1$.

(vii) **Top**. Objects are all topological spaces, morphisms are continuous functions, and composition is the usual composition of functions. In checking that **Top** is a category, one must note that identity functions are continuous and that composites of continuous functions are continuous.

(viii) The category **Sets**$_*$ of all *pointed sets* has as its objects all ordered pairs (X, x_0), where X is a nonempty set and x_0 is a point in X, called the *basepoint*. A morphism $f : (X, x_0) \to (Y, y_0)$ is called a *pointed map*; it is a function $f : X \to Y$ with $f(x_0) = y_0$. Composition is the usual composition of functions.

One defines the category **Top**$_*$ of all *pointed spaces* in a similar way; obj(**Top**$_*$) consists of all ordered pairs (X, x_0), where X is a nonempty topological space and $x_0 \in X$, and morphisms $f : (X, x_0) \to (Y, y_0)$ are continuous functions $f : X \to Y$ with $f(x_0) = y_0$.

(ix) We now define the category **Asc** of *abstract simplicial complexes* and *simplicial maps*.

Definition.   An *abstract simplicial complex* K is a nonempty set Vert(K), called *vertices*, together with a family of nonempty finite subsets $\sigma \subseteq$ Vert(K), called *simplexes*, such that

(a) $\{v\}$ is a simplex for every point $v \in$ Vert(K),

(b) every nonempty subset of a simplex is itself a simplex.

A simplex σ with $|\sigma| = n+1$ is called an *n-simplex*. If K and L are simplicial complexes, then a *simplicial map* is a function $\varphi :$ Vert$(K) \to$ Vert(L) such that, whenever σ is a simplex in K, then $\varphi(\sigma)$ is a simplex in L. [We do not assume that φ is injective; if σ is an n-simplex, then $\varphi(\sigma)$ need not be an n-simplex. For example, a constant function Vert$(K) \to$ Vert(L) is a simplicial map.]

(x) Let $\mathcal{U}$ be an open cover of a topological space X; that is, $\mathcal{U} = (U_i)_{i \in I}$ is an indexed family of open subsets with $X = \bigcup_{i \in I} U_i$. We define the *nerve*[7] $N(\mathcal{U})$ to be the abstract simplicial complex having vertices Vert$(N(\mathcal{U})) = \mathcal{U}$ and simplexes $\{U_{i_0}, U_{i_1}, \ldots, U_{i_n}\} \subseteq \mathcal{U}$ such that $\bigcap_{j=0}^{n} U_{i_j} \neq \varnothing$.

(xi) Recall that a *monoid* is a nonempty set G having an associative binary operation and an identity element e: that is, $ge = g = eg$ for all $g \in G$. For example, every group is a monoid and, if we forget its addition, every ring R is a monoid under multiplication. The following description defines a category $\mathcal{C}(G)$: there is only one object, denoted by $*$, Hom$(*, *) = G$, and composition

$$\text{Hom}(*, *) \times \text{Hom}(*, *) \to \text{Hom}(*, *)$$

---

[7]Nerves first appeared in Lebesgue, "Sur la non applicabilité de deux domaines appartenant ā des espaces de n et $n + p$ dimensions," *Math. Ann.* 70 (1911), 166–168. *Lebesgue's Covering Theorem* states that a separable metric space has dimension $\leq n$ if and only if every finite open cover has a refinement whose nerve has dimension $\leq n$; see Hurewicz-Wallman, *Dimension Theory*, p. 42.

(that is, $G \times G \to G$) is the given multiplication of G. We leave verification of the axioms to the reader.

The category $C(G)$ has an unusual feature. Since $*$ is merely an object, not a set, there are no *functions* $* \to *$, and so morphisms here are not functions.

(xii) A less artificial example in which morphisms are not functions arises in Algebraic Topology. Recall that if $f_0, f_1 \colon X \to Y$ are continuous functions, where X and Y are topological spaces, then f_0 is **homotopic** to f_1, denoted by $f_0 \simeq f_1$, if there exists a continuous function $h \colon \mathbf{I} \times X \to Y$ (where $\mathbf{I} = [0, 1]$ is the closed unit interval) such that $h(0, x) = f_0(x)$ and $h(1, x) = f_1(x)$ for all $x \in X$. One calls h a **homotopy**, and we think of it as deforming f_0 into f_1. A **homotopy equivalence** is a continuous map $f \colon X \to Y$ for which there exists a continuous $g \colon Y \to X$ such that $gf \simeq 1_X$ and $fg \simeq 1_Y$. One can show that homotopy is an equivalence relation on the set of all continuous functions $X \to Y$, and the equivalence class $[f]$ of f is called its **homotopy class**.

The **homotopy category Htp** has as its objects all topological spaces and as its morphisms all homotopy classes of continuous functions (thus, a morphism here is not a function but a certain equivalence class of functions). Composition is defined by $[f][g] = [fg]$ (if $f \simeq f'$ and $g \simeq g'$, then their composites are homotopic: $fg \simeq f'g'$), and identity morphisms are homotopy classes $[1_X]$. ◄

The next examples are more algebraic.

Example 1.4.

(i) **Ab**. Objects are abelian groups, morphisms are homomorphisms, and composition is the usual composition.

(ii) **Rings**. Objects are rings, morphisms are ring homomorphisms, and composition is the usual composition. We assume that all rings R have a unit element 1, but we do not assume that $1 \neq 0$. (Should $1 = 0$, however, the equation $1r = r$ for all $r \in R$ shows that $R = \{0\}$, because $0r = 0$. In this case, we call R the **zero ring**.) We agree, as part of the definition, that $\varphi(1) = 1$ for every ring homomorphism φ. Since the inclusion map $S \to R$ of a subring should be a homomorphism, it follows that the unit element 1 in a subring S must be the same as the unit element 1 in R.

(iii) **ComRings**. Objects are commutative rings, morphisms are ring homomorphisms, and composition is the usual composition. ◄

We now introduce R-modules, a common generalization of abelian groups and of vector spaces, for many applications of Homological Algebra arise in this context. An R-module is just a "vector space over a ring R"; that is, in the definition of vector space, allow the scalars to be in R instead of in a field.

Definition.   A *left R-module*, where R is a ring, is an additive abelian group M having a *scalar multiplication* $R \times M \to M$, denoted by $(r, m) \mapsto rm$, such that, for all $m, m' \in M$ and $r, r' \in R$,

(i) $r(m + m') = rm + rm'$,

(ii) $(r + r')m = rm + r'm$,

(iii) $(rr')m = r(r'm)$,

(iv) $1m = m$.[8]

We often write $_R M$ to denote M being a left R-module.

Example 1.5.

(i) Every vector space over a field k is a left k-module.

(ii) Every abelian group is a left $\mathbb{Z}$-module.

(iii) Every ring R is a left module over itself if we define scalar multiplication $R \times R \to R$ to be the given multiplication of elements of R. More generally, every left ideal in R is a left R-module.

(iv) If S is a subring of a ring R, then R is a left S-module, where scalar multiplication $S \times R \to R$ is just the given multiplication $(s, r) \mapsto sr$. For example, the *center* $Z(R)$ of a ring R is

$$Z(R) = \{a \in R : ar = ra \text{ for all } r \in R\}.$$

It is easy to see that $Z(R)$ is a subring of R and that R is a left $Z(R)$-module. A ring R is commutative if and only if $Z(R) = R$. ◄

---

[8]If we do not assume that $1m = m$ for all $m \in M$, then the abelian group M is a direct sum $M_0 \oplus M_1$, where M_0 is an abelian group in which $rm = 0$ for all $r \in R$ and $m \in M_0$, and M_1 is a left R-module. See Exercise 1.10 on page 34.

Definition.   If M and N are left R-modules, then an R-**homomorphism** (or an R-**map**) is a function $f : M \to N$ such that, for all $m, m' \in M$ and $r \in R$,

(i) $f(m + m') = f(m) + f(m')$,

(ii) $f(rm) = rf(m)$.

An R-**isomorphism** is a bijective R-homomorphism.

Note that the composite of R-homomorphisms is an R-homomorphism and, if f is an R-isomorphism, then its inverse function f^{-1} is also an R-isomorphism.

Example 1.6.

(i) If R is a field, then left R-modules are vector spaces and R-maps are linear transformations.

(ii) Every homomorphism of abelian groups is a $\mathbb{Z}$-map.

(iii) Let M be an R-module, where R is a ring. If $r \in Z(R)$, then **multiplication by r** (or *homothety*) is the function $\mu_r : M \to M$ defined by $m \mapsto rm$. The functions μ_r are R-maps because $r \in Z(R)$: if $a \in R$ and $m \in M$, then

$$\mu_r(am) = r(am) = (ra)m = (ar)m = a(rm) = a\mu_r(m). \quad \blacktriangleleft$$

Definition.   A *right R-module*, where R is a ring, is an additive abelian group M having a *scalar multiplication* $M \times R \to M$, denoted by $(m, r) \mapsto mr$, such that, for all $m, m' \in M$ and $r, r' \in R$,

(i) $(m + m')r = mr + m'r$,

(ii) $m(r + r') = mr + mr'$,

(iii) $m(rr') = (mr)r'$,

(iv) $m = m1$.

We often write M_R to denote M being a right R-module.

Example 1.7.   Every ring R is a right module over itself if we define scalar multiplication $R \times R \to R$ to be the given multiplication of elements of R. More generally, every right ideal I in R is a right R-module, for if $i \in I$ and $r \in R$, then $ir \in I$. $\blacktriangleleft$

Definition. If M and M' are right R-modules, then an **R-homomorphism** (or an **R-map**) is a function $f: M \to M'$ such that, for all $m, m' \in M$ and $r \in R$,

(i) $f(m + m') = f(m) + f(m')$,

(ii) $f(mr) = f(m)r$.

An **R-isomorphism** is a bijective R-homomorphism.

If, in a right R-module M, we had denoted mr by rm, then all the axioms in the definition of left module would hold for M with the exception of axiom (iii): this axiom now reads

$$(rr')m = r'(rm).$$

This remark shows that if R is a commutative ring, then right R-modules are the same thing as left R-modules. Thus, when R is commutative, we usually say **R-module**, dispensing with the adjectives *left* and *right*.

There is a way to treat properties of right and left R-modules at the same time, instead of first discussing left modules and then saying that a similar discussion can be given for right modules. Strictly speaking, a ring R is an ordered triple (R, α, μ), where R is a set, $\alpha: R \times R \to R$ is addition, and $\mu: R \times R \to R$ is multiplication, and these obey certain axioms. Of course, we usually abbreviate the notation and, instead of saying that (R, α, μ) is a ring, we merely say that R is a ring.

Definition. If (R, α, μ) is a ring, then its **opposite ring** R^{op} is (R, α, μ^o), where $\mu^o: R \times R \to R$ is defined by

$$\mu^o(r, r') = \mu(r', r).$$

It is easy to check that R^{op} is a ring. Informally, we have reversed the order of multiplication. It is obvious that $(R^{\mathrm{op}})^{\mathrm{op}} = R$, and that $R^{\mathrm{op}} = R$ if and only if R is commutative. Exercise 1.11 on page 35 says that every right R-module is a left R^{op}-module and every left R-module is a right R^{op}-module.

Definition. The category $_R\mathbf{Mod}$ of all *left R-modules* (where R is a ring) has as its objects all left R-modules, asa its morphisms all R-homomorphisms, and as its composition the usual composition of functions. We denote the sets $\mathrm{Hom}(A, B)$ in $_R\mathbf{Mod}$ by

$$\mathrm{Hom}_R(A, B).$$

If $R = \mathbb{Z}$, then $_{\mathbb{Z}}\mathbf{Mod} = \mathbf{Ab}$, for abelian groups are $\mathbb{Z}$-modules and homomorphisms are $\mathbb{Z}$-maps.

There is also a category of right R-modules.

Definition.   The category $\mathbf{Mod}_R$ of all *right R-modules* (where R is a ring) has as its objects all *right R-modules*, as its morphisms all R-homomorphisms, and as its composition the usual composition. We denote the sets $\text{Hom}(A, B)$ in $\mathbf{Mod}_R$ by

$$\text{Hom}_R(A, B)$$

(we use the same notation for Hom as in $_R\mathbf{Mod}$).

Definition.   A category $\mathcal{S}$ is a *subcategory* of a category $\mathcal{C}$ if

(i) $\text{obj}(\mathcal{S}) \subseteq \text{obj}(\mathcal{C})$,

(ii) $\text{Hom}_{\mathcal{S}}(A, B) \subseteq \text{Hom}_{\mathcal{C}}(A, B)$ for all $A, B \in \text{obj}(\mathcal{S})$, where we denote Hom sets in $\mathcal{S}$ by $\text{Hom}_{\mathcal{S}}(\square, \square)$,

(iii) if $f \in \text{Hom}_{\mathcal{S}}(A, B)$ and $g \in \text{Hom}_{\mathcal{S}}(B, C)$, then the composite $gf \in \text{Hom}_{\mathcal{S}}(A, C)$ is equal to the composite $gf \in \text{Hom}_{\mathcal{C}}(A, C)$,

(iv) if $A \in \text{obj}(\mathcal{S})$, then the identity $1_A \in \text{Hom}_{\mathcal{S}}(A, A)$ is equal to the identity $1_A \in \text{Hom}_{\mathcal{C}}(A, A)$.

A subcategory $\mathcal{S}$ of $\mathcal{C}$ is a *full subcategory* if, for all $A, B \in \text{obj}(\mathcal{S})$, we have $\text{Hom}_{\mathcal{S}}(A, B) = \text{Hom}_{\mathcal{C}}(A, B)$.

For example, **Ab** is a full subcategory of **Groups**. Call a category *discrete* if its only morphisms are identity morphisms. If $\mathcal{S}$ is the discrete category with $\text{obj}(\mathcal{S}) = \text{obj}(\mathbf{Sets})$, then $\mathcal{S}$ is a subcategory of **Sets** that is not a full subcategory. On the other hand, the homotopy category **Htp** is not a subcategory of **Top**, even though $\text{obj}(\mathbf{Htp}) = \text{obj}(\mathbf{Top})$, for morphisms in **Htp** are not continuous functions.

Example 1.8.   If $\mathcal{C}$ is any category and $\mathcal{S} \subseteq \text{obj}(\mathcal{C})$, then the *full subcategory generated by* $\mathcal{S}$, also denoted by $\mathcal{S}$, is the subcategory with $\text{obj}(\mathcal{S}) = \mathcal{S}$ and with $\text{Hom}_{\mathcal{S}}(A, B) = \text{Hom}_{\mathcal{C}}(A, B)$ for all $A, B \in \text{obj}(\mathcal{S})$. For example, we define the category $\mathbf{Top}_2$ to be the full subcategory of **Top** generated by all Hausdorff spaces.   ◄

Functors[9] are homomorphisms of categories.

---

[9]The term *functor* was coined by the philosopher R. Carnap, and S. Mac Lane thought it was the appropriate term in this context.

Definition. If C and D are categories, then a *functor* $T: C \to D$ is a function such that

(i) if $A \in \text{obj}(C)$, then $T(A) \in \text{obj}(D)$,

(ii) if $f: A \to A'$ in C, then $T(f): T(A) \to T(A')$ in D,

(iii) if $A \xrightarrow{f} A' \xrightarrow{g} A''$ in C, then $T(A) \xrightarrow{T(f)} T(A') \xrightarrow{T(g)} T(A'')$ in D and

$$T(gf) = T(g)T(f),$$

(iv) $T(1_A) = 1_{T(A)}$ for every $A \in \text{obj}(C)$.

In Exercise 1.1 on page 33, we see that there is a bijection between objects A and their identity morphism 1_A. Thus, we may regard a category as consisting only of morphisms (in almost all uses of categories, however, it is more natural to think of two sorts of entities: objects and morphisms). Viewing a category in this way shows that the notation $T: C \to D$ for a functor is consistent with standard notation for functions.

Example 1.9.

(i) If S is a subcategory of a category C, then the definition of subcategory may be restated to say that the inclusion $I: S \to C$ is a functor [this is one reason for the presence of Axiom (iv)].

(ii) If C is a category, then the *identity functor* $1_C: C \to C$ is defined by $1_C(A) = A$ for all objects A and $1_C(f) = f$ for all morphisms f.

(iii) If C is a category and $A \in \text{obj}(C)$, then the **Hom** *functor* $T_A: C \to$ **Sets**, usually denoted by $\text{Hom}(A, \square)$, is defined by

$$T_A(B) = \text{Hom}(A, B) \text{ for all } B \in \text{obj}(C),$$

and if $f: B \to B'$ in C, then $T_A(f): \text{Hom}(A, B) \to \text{Hom}(A, B')$ is given by

$$T_A(f): h \mapsto fh.$$

We call $T_A(f) = \text{Hom}(A, f)$ the *induced map*, and we denote it by f_*; thus,

$$f_*: h \mapsto fh.$$

Because of the importance of this example, we will verify the parts of the definition in detail. First, the very definition of category says that $\text{Hom}(A, B)$ is a set. Note that the composite fh makes sense:

$$A \underset{h}{\overset{fh}{\rightrightarrows}} B \xrightarrow{f} B'.$$

Suppose now that $g: B' \to B''$. Let us compare the functions

$$(gf)_*, g_* f_*: \operatorname{Hom}(A, B) \to \operatorname{Hom}(A, B'').$$

If $h \in \operatorname{Hom}(A, B)$, i.e., if $h: A \to B$, then

$$(gf)_*: h \mapsto (gf)h;$$

on the other hand, associativity of composition gives

$$g_* f_*: h \mapsto fh \mapsto g(fh) = (gf)h,$$

as desired. Finally, if f is the identity map $1_B: B \to B$, then

$$(1_B)_*: h \mapsto 1_B h = h$$

for all $h \in \operatorname{Hom}(A, B)$, so that $(1_B)_* = 1_{\operatorname{Hom}(A,B)}$.

(iv) A functor $T: \mathbb{Z} \to \mathcal{C}$, where $\mathbb{Z}$ is the category obtained from $\mathbb{Z}$ viewed as a partially ordered set [as in Example 1.3(vi)], is a sequence

$$\cdots \to C_{n+1} \to C_n \to C_{n-1} \to \cdots.$$

(v) Define the *forgetful functor* $U:$ **Groups** $\to$ **Sets** as follows: $U(G)$ is the underlying set of a group G and $U(f)$ is a homomorphism f regarded as a mere function. Strictly speaking, a group is an ordered pair (G, μ) [where G is its (underlying) set and $\mu: G \times G \to G$ is its operation], and $U((G, \mu)) = G$; the functor U "forgets" the operation and remembers only the set. There are many variants. For example, a ring is an ordered triple (R, α, μ) [where $\alpha: R \times R \to R$ is addition and $\mu: R \times R \to R$ is multiplication], and there are forgetful functors $U':$ **Rings** $\to$ **Ab** with $U'(R, \alpha, \mu) = (R, \alpha)$, the *additive group* of R, and $U'':$ **Rings** $\to$ **Sets** with $U''(R, \alpha, \mu) = R$, the underlying set. ◄

We can draw pictures in a category.

Definition.   A *diagram* in a category $\mathcal{C}$ is a functor $D: \mathcal{D} \to \mathcal{C}$, where $\mathcal{D}$ is a *small category*; that is, $\operatorname{obj}(\mathcal{D})$ is a set.

Let us see that this formal definition captures the intuitive idea of a diagram. We think of an abstract diagram as a *directed multigraph*; that is, as a set V of *vertices* and, for each ordered pair $(u, v) \in V \times V$, a (possibly empty) set $\operatorname{arr}(u, v)$ of *arrows* from u to v. A diagram in a category $\mathcal{C}$ should be a multigraph each of whose vertices is labeled by an object of $\mathcal{C}$ and each of whose arrows is labeled by a morphism of $\mathcal{C}$. But this is precisely the image of a functor $D: \mathcal{D} \to \mathcal{C}$: if u and v label vertices, then $u = Da$ and $v = Db$, where $a, b \in \operatorname{obj}(\mathcal{D})$, and $\operatorname{arr}(u, v) = \{Df: Da \to Db \mid f \in \operatorname{Hom}_{\mathcal{D}}(a, b)\}$. That each $a \in \operatorname{obj}(\mathcal{D})$ has an identity morphism says that there is a loop 1_u at each vertex $u = Da$. In drawing a diagram, we usually omit these loops; also, we usually omit morphisms that arise as composites of other morphisms.

Definition. A *path* in a category C is a functor $P\colon \mathbf{n+1} \to C$, where $\mathbf{n+1}$ is the category arising from the partially ordered set $0 \le 1 \le \cdots \le n$, as in Example 1.3(v). Thus, P is a sequence $P_0, P_1, \ldots, P_n$, where $P_i \in \mathrm{obj}(C)$ for all i. A *labeled path* is a path in which the morphisms $f_i\colon P_i \to P_{i+1}$ are displayed:

$$P_0 \xrightarrow{f_0} P_1 \xrightarrow{f_1} P_2 \to \cdots \to P_{n-1} \xrightarrow{f_{n-1}} P_n.$$

A path is *simple* if $P_0, P_1, \ldots, P_{n-1}$ are distinct (we allow $P_n = P_0$).

A diagram in a category *commutes* if, for each pair of vertices A and B, the composites $f_{n-1} \cdots f_1 f_0$ of the labels on any two simple labeled paths from A to B are equal.

The triangular diagram (arising from a category D with three objects and four nonidentity morphisms) commutes if $gf = h$ and $kf = h$, and the square diagram (arising from a category D with four objects and four nonidentity morphisms) commutes if $gf = f'g'$. The term *commutes* in this context arises from this last example.

A second type of functor reverses arrows.

Definition. A *contravariant functor* $T\colon C \to D$, where C and D are categories, is a function such that

(i) if $C \in \mathrm{obj}(C)$, then $T(C) \in \mathrm{obj}(D)$,

(ii) if $f\colon C \to C'$ in C, then $T(f)\colon T(C') \to T(C)$ in D (note the reversal of arrows),

(iii) if $C \xrightarrow{f} C' \xrightarrow{g} C''$ in C, then $T(C'') \xrightarrow{T(g)} T(C') \xrightarrow{T(f)} T(C)$ in D and

$$T(gf) = T(f)T(g),$$

(iv) $T(1_A) = 1_{T(A)}$ for every $A \in \mathrm{obj}(C)$.

To distinguish them from contravariant functors, the functors defined earlier are called *covariant functors*.

Example 1.10. If $\mathcal{C}$ is a category and $B \in \mathrm{obj}(\mathcal{C})$, then the ***contravariant* Hom *functor*** $T^B \colon \mathcal{C} \to \mathbf{Sets}$, usually denoted by $\mathrm{Hom}(\square, B)$, is defined, for all $C \in \mathrm{obj}(\mathcal{C})$, by

$$T^B(C) = \mathrm{Hom}(C, B),$$

and if $f \colon C \to C'$ in $\mathcal{C}$, then $T^B(f) \colon \mathrm{Hom}(C', B) \to \mathrm{Hom}(C, B)$ is given by

$$T^B(f) \colon h \mapsto hf.$$

We also call $T^B(f) = \mathrm{Hom}(f, B)$ the ***induced map***, and we denote it by f^*; thus,

$$f^* \colon h \mapsto hf.$$

Because of the importance of this example, we verify the axioms, showing that $\mathrm{Hom}(\square, B)$ is a (contravariant) functor. Note that the composite hf makes sense:

$$C \underset{f}{\overset{hf}{\rightrightarrows}} C' \overset{}{\underset{h}{\rightrightarrows}} B.$$

Given homomorphisms

$$C \overset{f}{\to} C' \overset{g}{\to} C'',$$

let us compare the functions

$$(gf)^*, f^*g^* \colon \mathrm{Hom}(C'', B) \to \mathrm{Hom}(C, B).$$

If $h \in \mathrm{Hom}(C'', B)$, i.e., if $h \colon C'' \to B$, then

$$(gf)^* \colon h \mapsto h(gf);$$

on the other hand,

$$f^*g^* \colon h \mapsto hg \mapsto (hg)f = h(gf) = (hg)f,$$

as desired. Finally, if f is the identity map $1_C \colon C \to C$, then

$$(1_C)^* \colon h \mapsto h1_C = h$$

for all $h \in \mathrm{Hom}(C, B)$, so that $(1_C)^* = 1_{\mathrm{Hom}(C,B)}$. ◀

Example 1.11. Here is a special case of a contravariant Hom functor. Recall that a ***linear functional*** on a vector space V over a field k is a linear transformation $\varphi \colon V \to k$ [remember that k is a (one-dimensional) vector space over itself]. For example, if $V = \{\text{continuous } f \colon [0, 1] \to \mathbb{R}\}$, then integration $f \mapsto \int_0^1 f(t)\,dt$ is a linear functional on V. If V is a vector space over a field k, then its ***dual space*** is $V^* = \mathrm{Hom}_k(V, k)$, the set of all the linear functionals on V. Now V^* is a k-module if we define $af \colon V \to k$ (for $f \in V^*$ and

$a \in k$) by $af \colon v \mapsto a[f(v)]$; that is, V^* is a vector space over k. Moreover, if $f \colon V \to W$ is a linear transformation, then the induced map $f^* \colon W^* \to V^*$ is also a linear transformation. (By Exercise 2.9 on page 66, if A is a matrix of f, then the transpose A^T is a matrix for f^*.) The **dual space functor** is $\mathrm{Hom}_k(\square, k) \colon {}_k\mathbf{Mod} \to {}_k\mathbf{Mod}$. ◄

Example 1.12.   Recall Example 1.3(iii): a partially ordered set X can be viewed as a category, where $x \preceq x'$ in X if and only if $\mathrm{Hom}_X(x, x') \neq \varnothing$; that is, $\mathrm{Hom}_X(x, x') = \{\iota^x_{x'}\}$. If Y is a partially ordered set and $T \colon X \to Y$ is a covariant functor, then $T(\iota^x_{x'}) = \iota^{Tx}_{Tx'}$; that is, $Tx \preceq Tx'$ in Y. In other words, a covariant functor is an **order-preserving** function: if $x \preceq x'$, then $Tx \preceq Tx'$. Similarly, if $T \colon X \to Y$ is a contravariant functor, then T is **order-reversing**: if $x \preceq x'$, then $Tx \succeq Tx'$. ◄

Example 1.13.   A functor $T \colon \mathcal{C} \to \mathcal{D}$ is **faithful** if, for all $A, B \in \mathrm{obj}(\mathcal{C})$, the functions $\mathrm{Hom}_{\mathcal{C}}(A, B) \to \mathrm{Hom}_{\mathcal{D}}(TA, TB)$ given by $f \mapsto Tf$ are injections. A category $\mathcal{C}$ is **concrete** if there is a faithful functor $U \colon \mathcal{C} \to \mathbf{Sets}$. Informally, a concrete category is one whose morphisms may be regarded as functions. The homotopy category **Htp** is not concrete, but the proof of this fact is not obvious. ◄

Example 1.14.   If U is an open subset of a topological space X, define

$$\mathcal{P}(U) = \{\text{continuous } f \colon U \to \mathbb{R}\}.$$

It is easy to see that $\mathcal{P}(U)$ is a commutative ring under pointwise operations: if $f, g \in \mathcal{P}(U)$ and $x \in U$, then

$$f + g \colon x \mapsto f(x) + g(x) \quad \text{and} \quad fg \colon x \to f(x)g(x).$$

If V is an open set containing U, then restriction $f \mapsto f|U$ is a ring homomorphism $\mathrm{res}^V_U \colon \mathcal{P}(V) \to \mathcal{P}(U)$.

  We generalize this construction. As in Example 1.3(iii), the topology $\mathcal{U}$ of X is a category: $\mathrm{obj}(\mathcal{U}) = \mathcal{U}$ and, if $U, V \in \mathcal{U}$, then

$$\mathrm{Hom}_{\mathcal{U}}(U, V) = \begin{cases} \varnothing & \text{if } U \not\subseteq V, \\ \{i^U_V\} & \text{if } U \subseteq V, \end{cases}$$

where $i^U_V \colon U \to V$ is the inclusion.

Definition.   If $\mathcal{U}$ is the topology of a topological space X and $\mathcal{C}$ is a category, then a **presheaf over** X is a contravariant functor $\mathcal{P} \colon \mathcal{U} \to \mathcal{C}$. (This definition is sometimes generalized by defining a *presheaf* over an arbitrary category $\mathcal{A}$ to be a contravariant functor $\mathcal{A} \to \mathcal{C}$.)

The construction at the beginning of this example gives a presheaf $\mathcal{P}$ of commutative rings over X. We have already defined $\mathcal{P}(U)$ for each open set U in X. If $U \subseteq V$, then the restriction map $\mathrm{res}_U^V : \mathcal{P}(V) \to \mathcal{P}(U)$ is defined by $f \mapsto f i_V^U = f|U$. It is routine to check that $\mathcal{P}$ is a contravariant functor (arrows are reversed). ◄

Just as opposite rings can be used to convert right modules to left modules, *opposite categories* can convert contravariant functors to covariant functors.

Definition. If $\mathcal{C}$ is a category, define its *opposite category* $\mathcal{C}^{\mathrm{op}}$ to be the category with $\mathrm{obj}(\mathcal{C}^{\mathrm{op}}) = \mathrm{obj}(\mathcal{C})$, with morphisms $\mathrm{Hom}_{\mathcal{C}^{\mathrm{op}}}(A, B) = \mathrm{Hom}_{\mathcal{C}}(B, A)$ (we may write morphisms in $\mathcal{C}^{\mathrm{op}}$ as f^{op}, where f is a morphism in $\mathcal{C}$), and with composition the reverse of that in $\mathcal{C}$; that is, $g^{\mathrm{op}} f^{\mathrm{op}} = (fg)^{\mathrm{op}}$.

We illustrate composition in $\mathcal{C}^{\mathrm{op}}$: a diagram $C \xrightarrow{f^{\mathrm{op}}} B \xrightarrow{g^{\mathrm{op}}} A$ in $\mathcal{C}^{\mathrm{op}}$ corresponds to the diagram $A \xrightarrow{g} B \xrightarrow{f} C$ in $\mathcal{C}$.

Opposite categories are difficult to visualize. In **Sets**^{op}, for example, the set $\mathrm{Hom}_{\mathbf{Sets}^{\mathrm{op}}}(X, \varnothing)$, for any set X, has exactly one element, namely, i^{op}, where i is the inclusion $\varnothing \to X$ in $\mathrm{Hom}_{\mathbf{Sets}}(\varnothing, X)$. But $i^{\mathrm{op}} : X \to \varnothing$ cannot be a function, for there are no functions from a nonempty set X to $\varnothing$.

It is easy to show that a contravariant functor $T : \mathcal{A} \to \mathcal{C}$ is the same thing as a (covariant) functor $S : \mathcal{A}^{\mathrm{op}} \to \mathcal{C}$ (see Exercise 1.4 on page 33). Recall that a diagram in a category $\mathcal{C}$ is a covariant functor $D : \mathcal{D} \to \mathcal{C}$, where $\mathcal{D}$ is a small category. The *opposite diagram* is $D^{\mathrm{op}} : \mathcal{D}^{\mathrm{op}} \to \mathcal{C}$, which is just the diagram in $\mathcal{A}$ obtained by reversing the direction of all arrows.

Here is how to translate *isomorphism* into categorical language.

Definition. A morphism $f : A \to B$ in a category $\mathcal{C}$ is an *isomorphism* if there exists a morphism $g : B \to A$ in $\mathcal{C}$ with

$$gf = 1_A \quad \text{and} \quad fg = 1_B.$$

The morphism g is called the *inverse* of f.

Exercise 1.1 on page 33 says that an isomorphism has a unique inverse.

Identity morphisms in a category are always isomorphisms. In X, where X is a partially ordered set, the only isomorphisms are identities; in $\mathcal{C}(G)$, where G is a group [see Example 1.3(xi)], every morphism is an isomorphism. In **Sets**, isomorphisms are bijections; in **Groups**, $_R\mathbf{Mod}$, **Rings**, or **ComRings**, isomorphisms are isomorphisms in the usual sense; in **Top**, isomorphisms are homeomorphisms. A *homotopy equivalence* is a continuous map $f : X \to Y$ for which there exists a continuous $g : Y \to X$ such that $gf \simeq 1_X$ and $fg \simeq 1_Y$. In the homotopy category, isomorphisms are homotopy classes of homotopy equivalences. We say that X and Y have the *same* *homotopy type* if they are isomorphic in **Htp**; that is, if there is a homotopy equivalence between them.

The following result is useful, even though it is very easy to prove. See Corollary 1.26 for an interesting application of this simple result.

Proposition 1.15. *Let $T: \mathcal{C} \to \mathcal{D}$ be a functor of either variance. If f is an isomorphism in $\mathcal{C}$, then $T(f)$ is an isomorphism in $\mathcal{D}$.*

Proof. If g is the inverse of f, apply the functor T to the equations $gf = 1$ and $fg = 1$.   •

How could we prove this result when $\mathcal{C} = \mathbf{Ab}$ if an isomorphism is viewed as a homomorphism that is an injection and a surjection? This proposition illustrates, admittedly at a low level, one reason why it is useful to give categorical definitions: functors can recognize definitions phrased solely in terms of objects, morphisms, and diagrams.

Just as homomorphisms compare algebraic objects and functors compare categories, *natural transformations* compare functors.

Definition. Let $S, T: \mathcal{A} \to \mathcal{B}$ be covariant functors. A ***natural transformation*** $\tau: S \to T$ is a one-parameter family of morphisms in $\mathcal{B}$,

$$\tau = (\tau_A: SA \to TA)_{A \in \mathrm{obj}(\mathcal{A})},$$

making the following diagram commute for all $f: A \to A'$ in $\mathcal{A}$:

$$
\begin{array}{ccc}
SA & \xrightarrow{\ \tau_A\ } & TA \\
{\scriptstyle Sf}\downarrow & & \downarrow{\scriptstyle Tf} \\
SA' & \xrightarrow[\ \tau_{A'}\ ]{} & TA'.
\end{array}
$$

Natural transformations between contravariant functors are defined similarly (replace $\mathcal{A}$ by $\mathcal{A}^{\mathrm{op}}$). A ***natural isomorphism*** is a natural transformation τ for which each τ_A is an isomorphism.

Natural transformations can be composed. If $\tau: S \to T$ and $\sigma: T \to U$ are natural transformations, where $S, T, U: \mathcal{A} \to \mathcal{B}$ are functors of the same variance, then define $\sigma\tau: S \to U$ by

$$(\sigma\tau)_A = \sigma_A \tau_A$$

for all $A \in \mathrm{obj}(\mathcal{A})$. It is easy to check that $\sigma\tau$ is a natural transformation (see Exercise 1.15 on page 35).

For any functor $S: \mathcal{A} \to \mathcal{B}$, define the ***identity natural transformation*** $\omega_S: S \to S$ by setting $(\omega_S)_A: SA \to SA$ to be the identity morphism 1_{SA}. The reader may check, using Exercise 1.15, that a natural transformation $\tau: S \to T$ is a natural isomorphism if and only if there is a natural transformation $\sigma: T \to S$ with $\sigma\tau = \omega_S$ and $\tau\sigma = \omega_T$.

Example 1.16.

(i) Let k be a field and let $\mathcal{V} = {}_k\mathbf{Mod}$ be the category of all vector spaces over k. As in Example 1.11, if $V \in \mathrm{obj}(\mathcal{V})$, then its dual space $V^* = \mathrm{Hom}_k(V, k)$ is the vector space of all linear functionals on V. If $f \in V^*$ and $v \in V$, denote $f(v)$ by (f, v). Of course, we are accustomed to fixing f and letting v vary, thereby describing f as $(f, \square)$. On the other hand, if we fix v and let f vary, then $(\square, v)$ assigns a value in k to every $f \in V^*$; that is, if $(\square, v)$ is denoted by v^e, then $v^e \colon V^* \to k$ is the evaluation function defined by $v^e(f) = (f, v) = f(v)$. In fact, v^e is a linear functional on V^*: the definitions of addition and scalar multiplication in V^* give $v^e(f + g) = (f + g, v) = f(v) + g(v) = v^e(f) + v^e(g)$ and, if $a \in k$, then $v^e(af) = (af, v) = a[f(v)] = av^e(f)$. Thus, $v^e \in (V^*)^*$ (which is usually denoted by V^{**}).

For each $V \in \mathrm{obj}(\mathcal{V})$, define $\tau_V \colon V \to V^{**}$ by $v \mapsto v^e = (\square, v)$. We have two (covariant) functors $\mathcal{V} \to \mathcal{V}$: the identity functor $1_{\mathcal{V}}$ and the double dual $\square^{**} = \mathrm{Hom}_k(\mathrm{Hom}_k(\square, k), k)$, and we claim that $\tau \colon 1_{\mathcal{V}} \to \square^{**}$ is a natural transformation. The reader may show that each $\tau_V \colon V \to V^{**}$ is linear; let us show commutativity of the diagram

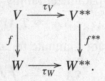

If $f \colon V \to W$, then the induced map $f^* \colon W^* \to V^*$ is given by $g \mapsto gf$ (the dual space functor is contravariant!); similarly, $f^{**} \colon V^{**} \to W^{**}$ is given by $h \mapsto hf^*$. Now take $v \in V$. Going clockwise, $v \mapsto v^e \mapsto v^e f^*$; going counterclockwise, $v \mapsto f(v) \mapsto (fv)^e$. To see that these elements of W^{**} are the same, evaluate each on $h \in W^*$:

$$v^e f^*(h) = v^e(hf) = (hf)v \quad \text{and} \quad (fv)^e(h) = h(f(v)).$$

It is not difficult to see that each τ_V is an injection, but τ may not be a natural isomorphism. If V is infinite-dimensional, then $\dim(V^*) > \dim(V)$; hence, $\dim(V^{**}) > \dim(V)$, and there is no isomorphism $V \to V^{**}$. On the other hand, if $\dim(V) < \infty$, a standard result of Linear Algebra (Exercise 2.9 on page 66) shows that $\dim(V^*) = \dim(V)$, and so the injection $\tau_V \colon V \to V^{**}$ must be an isomorphism. Thus, if $\mathcal{S} \subseteq \mathcal{V}$ is the subcategory of all finite-dimensional vector spaces over k, then $\tau | \mathcal{S}$ is a natural isomorphism.

We remark that there is no natural transformation from the identity functor to the dual space functor, for these two functors have different variances.

(ii) Choose a one-point set $P = \{p\}$. We claim that $\mathrm{Hom}(P, \square)\colon \mathbf{Sets} \to \mathbf{Sets}$ is naturally isomorphic to the identity functor on $\mathbf{Sets}$. If X is a set, define $\tau_X\colon \mathrm{Hom}(P, X) \to X$ by $f \mapsto f(p)$. Each τ_X is a bijection, as is easily seen, and we now show that τ is a natural transformation. Let X and Y be sets, and let $h\colon X \to Y$; we must show that the following diagram commutes:

$$
\begin{array}{ccc}
\mathrm{Hom}(P, X) & \xrightarrow{\ h_*\ } & \mathrm{Hom}(P, Y) \\
{\scriptstyle \tau_X}\downarrow & & \downarrow{\scriptstyle \tau_Y} \\
X & \xrightarrow[\ h\ ]{} & Y,
\end{array}
$$

where $h_*\colon f \mapsto hf$. Going clockwise, $f \mapsto hf \mapsto hf(p)$, while going counterclockwise, $f \mapsto f(p) \mapsto h(f(p))$. ◀

Notation. If $F, G\colon \mathcal{A} \to \mathcal{B}$ are functors of the same variance, then

$$
\mathrm{Nat}(F, G) = \{\text{natural transformations } F \to G\}.
$$

The notation $\mathrm{Nat}(F, G)$ should be accepted in light of our remarks on Set Theory on page 8. In general, $\mathrm{Nat}(F, G)$ may not be a class and, even if it is a class, it may be a proper class (see Exercise 1.19 on page 36). The next theorem shows that $\mathrm{Nat}(F, G)$ is a set when $F = \mathrm{Hom}_{\mathcal{C}}(A, \square)$.

Theorem 1.17 (Yoneda Lemma). *Let $\mathcal{C}$ be a category, let $A \in \mathrm{obj}(\mathcal{C})$, and let $G\colon \mathcal{C} \to \mathbf{Sets}$ be a covariant functor. Then there is a bijection*

$$
y\colon \mathrm{Nat}(\mathrm{Hom}_{\mathcal{C}}(A, \square), G) \to G(A)
$$

given by $y\colon \tau \mapsto \tau_A(1_A)$.

Proof. If $\tau\colon \mathrm{Hom}_{\mathcal{C}}(A, \square) \to G$ is a natural transformation, then $y(\tau) = \tau_A(1_A)$ lies in the set $G(A)$, for $\tau_A\colon \mathrm{Hom}_{\mathcal{C}}(A, A) \to G(A)$. Thus, y is a well-defined function.

For each $B \in \mathrm{obj}(\mathcal{C})$ and $\varphi \in \mathrm{Hom}_{\mathcal{C}}(A, B)$, there is a commutative diagram

$$
\begin{array}{ccc}
\mathrm{Hom}_{\mathcal{C}}(A, A) & \xrightarrow{\ \tau_A\ } & GA \\
{\scriptstyle \varphi_*}\downarrow & & \downarrow{\scriptstyle G\varphi} \\
\mathrm{Hom}_{\mathcal{C}}(A, B) & \xrightarrow[\ \tau_B\ ]{} & GB,
\end{array}
$$

so that

$$
(G\varphi)\tau_A(1_A) = \tau_B \varphi_*(1_A) = \tau_B(\varphi 1_A) = \tau_B(\varphi).
$$

If σ: $\mathrm{Hom}_{\mathcal{C}}(A, \square) \to G$ is another natural transformation, then $\sigma_B(\varphi) = (G\varphi)\sigma_A(1_A)$. Hence, if $\sigma_A(1_A) = \tau_A(1_A)$, then $\sigma_B = \tau_B$ for all $B \in \mathrm{obj}(\mathcal{C})$ and, hence, $\sigma = \tau$. Therefore, y is an injection.

To see that y is a surjection, take $x \in G(A)$. For $B \in \mathrm{obj}(\mathcal{C})$ and $\psi \in \mathrm{Hom}_{\mathcal{C}}(A, B)$, define

$$\tau_B(\psi) = (G\psi)(x)$$

[note that $G\psi \colon GA \to GB$, so that $(G\psi)(x) \in GB$]. We claim that τ is a natural transformation; that is, if $\theta \colon B \to C$ is a morphism in $\mathcal{C}$, then the following diagram commutes:

$$
\begin{array}{ccc}
\mathrm{Hom}_{\mathcal{C}}(A, B) & \xrightarrow{\ \tau_B\ } & GB \\
{\scriptstyle \theta_*}\big\downarrow & & \big\downarrow{\scriptstyle G\theta} \\
\mathrm{Hom}_{\mathcal{C}}(A, C) & \xrightarrow[\ \tau_C\ ]{} & GC.
\end{array}
$$

Going clockwise, we have $(G\theta)\tau_B(\psi) = G\theta G\psi(x)$; going counterclockwise, we have $\tau_C\theta_*(\psi) = \tau_C(\theta\psi) = G(\theta\psi)(x)$. Since G is a functor, however, $G(\theta\psi) = G\theta G\psi$; thus, τ is a natural transformation. Now $y(\tau) = \tau_A(1_A) = G(1_A)(x) = x$, and so y is a bijection.   •

Definition.   A covariant functor $F \colon \mathcal{C} \to \mathbf{Sets}$ is *representable* if there exists $A \in \mathrm{obj}(\mathcal{C})$ with $F \cong \mathrm{Hom}_{\mathcal{C}}(A, \square)$.

Theorem 5.50 characterizes all representable functors $_R\mathbf{Mod} \to \mathbf{Ab}$. The most interesting part of the next corollary is part (iii), which says that if F is representable, then the object A is essentially unique.

Corollary 1.18.   *Let $\mathcal{C}$ be a category and let $A, B \in \mathrm{obj}(\mathcal{C})$.*

(i) *If τ: $\mathrm{Hom}_{\mathcal{C}}(A, \square) \to \mathrm{Hom}_{\mathcal{C}}(B, \square)$ is a natural transformation, then for all $C \in \mathrm{obj}(\mathcal{C})$, we have $\tau_C = \psi^*$, where $\psi = \tau_A(1_A) \colon B \to A$ and ψ^* is the induced map $\mathrm{Hom}_{\mathcal{C}}(A, C) \to \mathrm{Hom}_{\mathcal{C}}(B, C)$ given by $\varphi \mapsto \varphi\psi$. Moreover, the morphism ψ is unique: if $\tau_C = \theta^*$, then $\theta = \psi$.*

(ii) *Let $\mathrm{Hom}_{\mathcal{C}}(A, \square) \xrightarrow{\tau} \mathrm{Hom}_{\mathcal{C}}(B, \square) \xrightarrow{\sigma} Hom_{\mathcal{C}}(B', \square)$ be natural transformations. If $\sigma_C = \eta^*$ and $\tau_C = \psi^*$ for all $C \in \mathrm{obj}(\mathcal{C})$, then*

$$(\sigma\tau)_C = (\psi\eta)^*.$$

(iii) *If $\mathrm{Hom}_{\mathcal{C}}(A, \square)$ and $\mathrm{Hom}_{\mathcal{C}}(B, \square)$ are naturally isomorphic functors, then $A \cong B$. (The converse is also true; see Exercise 1.16 on page 36.)*

Proof.

(i) If we denote $\tau_A(1_A) \in \mathrm{Hom}_\mathcal{C}(B, A)$ by ψ, then the Yoneda Lemma says, for all $C \in \mathrm{obj}(\mathcal{C})$ and all $\varphi \in \mathrm{Hom}_\mathcal{C}(A, C)$, that $\tau_C(\varphi) = \varphi_*(\psi)$. But $\varphi_*(\psi) = \varphi\psi = \psi^*(\varphi)$. The uniqueness assertion follows from injectivity of the Yoneda function y.

(ii) By part (i), there are unique morphisms $\psi \in \mathrm{Hom}_\mathcal{C}(B, A)$ and $\eta \in \mathrm{Hom}_\mathcal{C}(B', B)$ with

$$\tau_C(\varphi) = \psi^*(\varphi) \quad \text{and} \quad \sigma_C(\varphi') = \eta^*(\varphi')$$

for all $\varphi \in \mathrm{Hom}_\mathcal{C}(A, C)$ and $\varphi' \in \mathrm{Hom}_\mathcal{C}(B, C)$. By definition, $(\sigma\tau)_C = \sigma_C\tau_C$, and so

$$(\sigma\tau)_C(\varphi) = \sigma_C(\psi^*(\varphi)) = \eta^*\psi^*(\varphi) = (\psi\eta)^*(\varphi).$$

(iii) If $\tau \colon \mathrm{Hom}_\mathcal{C}(A, \square) \to \mathrm{Hom}_\mathcal{C}(B, \square)$ is a natural isomorphism, then there is a natural isomorphism $\sigma \colon \mathrm{Hom}_\mathcal{C}(B, \square) \to \mathrm{Hom}_\mathcal{C}(A, \square)$ with $\sigma\tau = \omega_{\mathrm{Hom}_\mathcal{C}(A, \square)}$ and $\tau\sigma = \omega_{\mathrm{Hom}_\mathcal{C}(B, \square)}$. By part (i), there are morphisms $\psi \colon B \to A$ and $\eta \colon A \to B$ with $\tau_C = \psi^*$ and $\sigma_C = \eta^*$ for all $C \in \mathrm{obj}(\mathcal{C})$. By part (ii), we have $\tau\sigma = \psi^*\eta^* = (\eta\psi)^* = 1_B^*$ and $\sigma\tau = (\psi\eta)^* = 1_A^*$. The uniqueness in part (i) now gives $\psi\eta = 1_A$ and $\eta\psi = 1_B$, so that $\eta \colon A \to B$ is an isomorphism. $\quad\bullet$

Example 1.19.

(i) Informally, if $\mathcal{A}$ and $\mathcal{B}$ are categories, the ***functor category*** $\mathcal{B}^\mathcal{A}$ has as its objects all covariant functors $\mathcal{A} \to \mathcal{B}$ and as its morphisms all natural transformations. Each functor $F \colon \mathcal{A} \to \mathcal{B}$ has an identity natural transformation $\omega_F \colon F \to F$, and a composite of natural transformations is itself a natural transformation (see Exercise 1.15 on page 35). But there is a set-theoretic problem here: $\mathrm{Nat}(F, G)$ need not be a set. Recall that a category $\mathcal{A}$ is *small* if the class $\mathrm{obj}(\mathcal{A})$ is a set; in this case, $\mathrm{Nat}(F, G)$ is a set, and so the formal definition of the functor category $\mathcal{B}^\mathcal{A}$ requires $\mathcal{A}$ to be a small category (see Exercise 1.19 on page 36). Note that all contravariant functors and natural transformations form the category $\mathcal{B}^{\mathcal{A}^{\mathrm{op}}}$, which is essentially the same as $(\mathcal{B}^{\mathrm{op}})^\mathcal{A}$.

(ii) A diagram in a category $\mathcal{C}$ is a (covariant) functor $D \colon \mathcal{D} \to \mathcal{C}$, where $\mathcal{D}$ is a small category. Thus, $D \in \mathcal{C}^\mathcal{D}$.

(iii) As in Example 1.3(vi), view $\mathbb{Z}$ as a partially ordered set under reverse inequality. A functor in $\mathbf{Ab}^\mathbb{Z}$ is essentially a sequence (we have not drawn identities and composites)

$$\cdots \to A_{n+1} \xrightarrow{d_{n+1}} A_n \xrightarrow{d_n} A_{n-1} \to \cdots.$$

A natural transformation between two such functors is a sequence of maps $(\ldots, f_{n+1}, f_n, f_{n-1}, \ldots)$ making the following diagram commute.

$$
\begin{array}{ccccccc}
\cdots \longrightarrow & A_{n+1} & \xrightarrow{d_{n+1}} & A_n & \xrightarrow{d_n} & A_{n-1} & \longrightarrow \cdots \\
& \downarrow{f_{n+1}} & & \downarrow{f_n} & & \downarrow{f_{n-1}} & \\
\cdots \longrightarrow & A'_{n+1} & \xrightarrow{d'_{n+1}} & A'_n & \xrightarrow{d'_n} & A'_{n-1} & \longrightarrow \cdots
\end{array}
$$

A *complex* (or *chain complex*) is such a functor with the property that $d_n d_{n+1} = 0$ for all n, and a natural transformation between complexes is usually called a *chain map*. Define **Comp** to be the full subcategory of $\mathbf{Ab}^{\mathbb{Z}}$ generated by all complexes.

(iv) Recall (Example 1.14): a *presheaf* $\mathcal{P}$ over a topological space X is a contravariant functor $\mathcal{P} \colon \mathcal{U} \to \mathbf{Ab}$, where the topology $\mathcal{U}$ on X is viewed as a (small) category. If $\mathcal{G} \colon \mathcal{U} \to \mathbf{Ab}$ is another presheaf over X, then a natural transformation $\tau \colon \mathcal{P} \to \mathcal{G}$ is a one-parameter family of morphisms $\tau_U \colon \mathcal{P}(U) \to \mathcal{G}(U)$, where $U \in \mathcal{U}$, making the following diagram commute for all $U \subseteq V$.

$$
\begin{array}{ccc}
\mathcal{P}(V) & \xrightarrow{\tau_V} & \mathcal{G}(V) \\
\downarrow{\mathcal{P}(f)} & & \downarrow{\mathcal{G}(f)} \\
\mathcal{P}(U) & \xrightarrow{\tau_U} & \mathcal{G}(U)
\end{array}
$$

Presheaves over a space X comprise the functor category $\mathbf{Ab}^{\mathcal{U}^{\mathrm{op}}}$, which we denote by $\mathbf{pSh}(X, \mathbf{Ab})$. ◄

Corollary 1.20 (Yoneda Imbedding). *If $\mathcal{C}$ is a small category, then there is a functor $Y \colon \mathcal{C}^{\mathrm{op}} \to \mathbf{Sets}^{\mathcal{C}}$ that is injective on objects and whose image is a full subcategory of $\mathbf{Sets}^{\mathcal{C}}$.*

Proof. Define Y on objects by $Y(A) = \mathrm{Hom}_{\mathcal{C}}(A, \Box)$. If $A \neq A'$, then pairwise disjointness of Hom sets gives $\mathrm{Hom}_{\mathcal{C}}(A, \Box) \neq \mathrm{Hom}_{\mathcal{C}}(A', \Box)$; that is, $Y(A) \neq Y(A')$, and so Y is injective on objects. If $\psi \colon B \to A$ is a morphism in $\mathcal{C}$, then there is a natural transformation $Y(\psi) \colon \mathrm{Hom}_{\mathcal{C}}(A, \Box) \to \mathrm{Hom}_{\mathcal{C}}(B, \Box)$ with $Y(\psi)_C = \psi^*$ for all $C \in \mathrm{obj}(\mathcal{C})$, by Corollary 1.18(i). Now Corollary 1.18(ii) gives $Y(\psi\eta) = Y(\eta)Y(\psi)$, and so Y is a contravariant functor. Finally, surjectivity of the Yoneda function y in Theorem 1.17 shows that every natural transformation $\mathrm{Hom}_{\mathcal{C}}(A, \Box) \to \mathrm{Hom}_{\mathcal{C}}(B, \Box)$ arises as $Y(\psi)$ for some ψ. Therefore, the image of Y is a full subcategory of the functor category $\mathbf{Sets}^{\mathcal{C}}$. •

We paraphrase the Yoneda imbedding by saying that every small category is a full subcategory of a category of presheaves.

1.3 Singular Homology

In the first section, we defined homology groups $H_n(X)$ for every finite simplicial complex X; we are now going to generalize this construction so that it applies to all topological spaces. Once this is done, we shall see that each H_n is actually a functor **Top** $\to$ **Ab**.[10] The reader will see that the construction has two parts: a topological half and an algebraic half.

Definition. Recall that ***Hilbert space*** is the set $\mathcal{H}$ of all sequences (x_i), where $x_i \in \mathbb{R}$ for all $i \geq 0$, such that $\sum_{i=0}^{\infty} x_i^2 < \infty$. ***Euclidean space*** $\mathbb{R}^n$ is the subset of $\mathcal{H}$ consisting of all sequences of the form $(x_0, \ldots, x_{n-1}, 0, \ldots)$ with $x_i = 0$ for all $i \geq n$.

We begin by generalizing the notion of n-simplex, where a 0-*simplex* is a point, a 1-*simplex* is a line segment, a 2-*simplex* is a triangle (with interior), a 3-*simplex* is a (solid) tetrahedron, and so forth. Here is the precise definition.

Definition. The ***standard n-simplex*** is the set of all (***convex***) ***combinations***

$$\Delta^n = [e_0, \ldots, e_n] = \left\{ t_0 e_0 + \cdots + t_n e_n : t_i \geq 0 \text{ and } \sum_{i=0}^{n} t_i = 1 \right\},$$

where e_i denotes the sequence in $\mathcal{H}$ having 1 in the ith coordinate and 0 everywhere else. We may also write $t_0 e_0 + \cdots + t_n e_n$ as the vector $(t_0, \ldots, t_n)$ in $\mathbb{R}^{n+1} \subseteq \mathcal{H}$. The ***$i$th vertex*** of Δ^n is e_i; the ***jth faces*** of Δ^n, for $0 \leq j \leq n$, are the convex combinations of j of its vertices.

If X is a topological space, then a ***singular n-simplex in*** X is a continuous map $\sigma : \Delta^n \to X$, where Δ^n is the standard n-simplex.

Definition. If X is a topological space, define $S_{-1}(X) = \{0\}$ and, for each $n \geq 0$, define $S_n(X)$ to be the free abelian group with basis the set of all singular n-simplexes in X. The elements of $S_n(X)$ are called ***singular n-chains***.

The boundary of a singular n-simplex σ ought to be

$$\partial_n(\sigma) = \sum_{i=0}^{n} (-1)^i \sigma \big| [e_0, \ldots, \widehat{e_i}, \ldots, e_n].$$

However, this is not an $(n-1)$-chain, because $[e_0, \ldots, \widehat{e_i}, \ldots, e_n]$ is not the *standard* $(n-1)$-simplex, and so the restriction $\sigma | [e_0, \ldots, \widehat{e_i}, \ldots, e_n]$ is not a singular $(n-1)$-simplex. We remedy this by introducing *face maps*.

---

[10]Simplicial homology H_n is also functorial, but defining $H_n(f)$ for a simplicial map f is more complicated, needing the *Simplicial Approximation Theorem*.

Definition.   Define the ith *face map* $\epsilon_i^n \colon \Delta^{n-1} \to \Delta^n$, where $0 \le i \le n$, by putting 0 in the ith coordinate and preserving the ordering of the other coordinates: the points of $[e_0, \ldots, e_{n-1}]$ are convex combinations $(t_0, \ldots, t_n) = t_0 e_0 + \cdots + t_{n-1} e_{n-1}$, and so

$$\epsilon_i^n \colon (t_0, \ldots, t_{n-1}) \mapsto \begin{cases} (0, t_0, \ldots, t_{n-1}) \text{ if } i = 0, \\ (t_0, \ldots, t_{i-1}, 0, t_i, \ldots, t_{n-1}) \text{ if } i > 0. \end{cases}$$

The superscript indicates that the target of ϵ_i^n is Δ^n. For example, there are three face maps $\epsilon_i^2 \colon \Delta^1 \to \Delta^2$: $\epsilon_0^2 \colon (t_0, t_1) \mapsto (0, t_0, t_1)$; $\epsilon_1^2 \colon (t_0, t_1) \mapsto (t_0, 0, t_1)$; and $\epsilon_2^2 \colon (t_0, t_1) \mapsto (t_0, t_1, 0)$. The images of these face maps are the 1-faces of the triangle $[e_0, e_1, e_2]$.

The following identities hold for face maps.

Lemma 1.21.   *If $0 \le j < i \le n - 1$, then the face maps satisfy*

$$\epsilon_i^n \epsilon_j^{n-1} = \epsilon_j^n \epsilon_{i-1}^{n-1} \colon \Delta^{n-2} \to \Delta^n.$$

Proof.   The straightforward calculations are left to the reader.   •

We can now define boundary maps.

Definition.   Let X be a topological space. If σ is a 0-simplex in X, define $\partial_0(\sigma) = 0$. If $n \ge 1$ and $\sigma \colon \Delta^n \to X$ is an n-simplex, define

$$\partial_n(\sigma) = \sum_{i=0}^{n} (-1)^i \sigma \epsilon_i^n.$$

Define the *singular boundary map* $\partial_n \colon S_n(X) \to S_{n-1}(X)$ by extending by linearity.

Proposition 1.22.   *For all $n \ge 1$,*

$$\partial_{n-1} \partial_n = 0.$$

Proof.   We mimic the proof of Proposition 1.1. For any n-simplex σ,

$$\partial_{n-1} \partial_n(\sigma) = \partial_{n-1}\left(\sum_i (-1)^i \sigma \epsilon_i^n \right)$$

$$= \sum_i (-1)^i \partial_{n-1}(\sigma \epsilon_i^n)$$

$$= \sum_i (-1)^i \sum_j (-1)^j \sigma \epsilon_i^n \epsilon_j^{n-1}$$

$$= \sum_{j<i} (-1)^{i+j} \sigma \epsilon_i^n \epsilon_j^{n-1} + \sum_{j \ge i} (-1)^{i+j} \sigma \epsilon_i^n \epsilon_j^{n-1}.$$

By Lemma 1.21, we have $\sigma \epsilon_i^n \epsilon_j^{n-1} = \sigma \epsilon_j^n \epsilon_{i-1}^{n-1}$ if $j < i$. The term $\sigma \epsilon_i^n \epsilon_j^{n-1}$ occurs in the first sum (over all $j < i$) with sign $(-1)^{i+j}$, and the term $\sigma \epsilon_j^n \epsilon_{i-1}^{n-1}$ occurs in the second sum (the first index j is now the larger index), and with opposite sign $(-1)^{j+i-1}$. Thus, all terms in $\partial \partial (\sigma)$ cancel, and $\partial \partial = 0$. •

We can now define singular cycles and singular boundaries.

Definition. For each $n \geq 0$, the group of *singular n-cycles* is $Z_n(X) = \ker \partial_n$, and the group of *singular n-boundaries* is $B_n(X) = \operatorname{im} \partial_{n+1}$.

Corollary 1.23. $B_n(X) \subseteq Z_n(X)$ *for all n.*

Proof. If $z \in B_n(X) = \operatorname{im} \partial_{n+1}$, then $z = \partial_{n+1} c$ for some $c \in C_{n+1}$, and $\partial_n z = \partial_n \partial_{n+1} c = 0$. •

Definition. The *n*th *singular homology group* of a topological space X is

$$H_n(X) = Z_n(X)/B_n(X).$$

We are now going to show that each H_n is a functor. If $f : X \to Y$ is a continuous map and $\sigma : \Delta^n \to X$ is an n-simplex in X, then the composite $f\sigma : \Delta^n \to Y$ is an n-simplex in Y, for a composite of continuous functions is continuous. Hence, $f\sigma \in S_n(Y)$, and we define the *chain map* $f_\# : S_n(X) \to S_n(Y)$ by

$$f_\# : \sum_\sigma m_\sigma \sigma \mapsto \sum_\sigma m_\sigma f\sigma.$$

It is usually reckless to be careless with notation, but the next lemma shows that one can sometimes do so without causing harm (moreover, it is easier to follow an argument when notational clutter is absent). Strictly speaking, notation for a chain map $f_\#$ should display n, X, and Y, while boundary maps $\partial_n : S_n(X) \to S_{n-1}(X)$ obviously depend on X.

Lemma 1.24. *If $f : X \to Y$ is a continuous map, then $\partial_n f_\# = f_\# \partial_n$; that is, there is a commutative diagram*

$$
\begin{array}{ccc}
S_n(X) & \xrightarrow{\partial_n} & S_{n-1}(X) \\
{\scriptstyle f_\#} \downarrow & & \downarrow {\scriptstyle f_\#} \\
S_n(Y) & \xrightarrow{\partial_n} & S_{n-1}(Y).
\end{array}
$$

Proof. It suffices to evaluate each composite on a basis element σ of $S_n(X)$. Now

$$f_\# \partial \sigma = f_\# \left(\sum (-1)^i \sigma \epsilon_i \right) = \sum (-1)^i f_\#(\sigma \epsilon_i) = \sum (-1)^i f(\sigma \epsilon_i).$$

On the other hand,

$$\partial f_{\#}(\sigma) = \partial(f\sigma) = \sum (-1)^i (f\sigma)\epsilon_i.$$

These are equal, by associativity of composition.   •

Theorem 1.25. *For every $n \geq 0$, singular homology H_n: **Top** $\to$ **Ab** is a functor.*

Proof. Having already defined H_n on objects by $H_n(X) = Z_n(X)/B_n(X)$, we need only define it on morphisms. If $f\colon X \to Y$ is a continuous map, define $H_n(f)\colon H_n(X) \to H_n(Y)$ by

$$H_n(f)\colon \operatorname{cls} z_n \mapsto \operatorname{cls} f_{\#}z_n,$$

where z_n is an n-cycle in X and $\operatorname{cls} z_n = z_n + B_n(X)$. In more detail, z_n is a linear combination of simplexes σ_i in X, and $H_n(f)$ sends $\operatorname{cls} z_n$ into the corresponding linear combination of cosets of simplexes $f\sigma_i$ in Y.

We claim that $f_{\#}z_n$ is an n-cycle in Y. If z_n is a cycle, then Lemma 1.24 gives $\partial f_{\#}z_n = f_{\#}\partial z_n = 0$. Thus, $f_{\#}(Z_n(X)) \subseteq Z_n(Y)$. But we also have $f_{\#}(B_n(X)) \subseteq B_n(Y)$, for if $\partial u \in B_n(X)$, then Lemma 1.24 gives $f_{\#}\partial u = \partial f_{\#}u \in B_n(Y)$. It follows that $H_n(f)$ is a well-defined function. We let the reader prove that $H_n(f)$ is a homomorphism, that $H_n(1_X) = 1_{H_n(X)}$, and that if $g\colon Y \to Y'$ is a continous map, then $H_n(gf) = H_n(g)H_n(f)$.   •

It is true that if $f_0, f_1\colon X \to Y$ are homotopic maps, then $H_n(f_0) = H_n(f_1)$ for all $n \geq 0$ (Spanier, *Algebraic Topology*, p. 175). It follows that the homology functors are actually defined on the homotopy category **Htp** (recall that morphisms in **Htp** are homotopy classes $[f]$ of continuous maps), and we may now define $H_n([f]) = H_n(f)$.

Corollary 1.26. *If X and Y are topological spaces having the same homotopy type, then $H_n(X) \cong H_n(Y)$ for all $n \geq 0$.*

Proof. An isomorphism in **Htp** is a homotopy class $[f]$ of a homotopy equivalence f. By Proposition 1.15, any functor takes an isomorphism into an isomorphism.   •

A topological space X is called **contractible** if $1_X \simeq c$, where $c\colon X \to X$ is a constant map $[c(x) = x_0 \in X$ for all $x \in X]$. For example, euclidean space $\mathbb{R}^n$ is contractible. It is easy to see that contractible spaces have the same homotopy type as a one-point set, and so their homology groups are easily computable: $H_0(X) = \mathbb{Z}$ and $H_n(X) = \{0\}$ for $n \geq 1$.

Define a functor $S\colon$ **Top** $\to$ **Comp**, the category of complexes defined in Example 1.19(iii): to each topological space X, assign the complex

$$\mathbf{S}_{\bullet}(X) = \cdots \to S_{n+1} \xrightarrow{\partial_{n+1}} S_n \xrightarrow{\partial_n} S_{n-1} \to \cdots;$$

to each continuous map $X \rightarrow Y$, assign the chain map $f_\# \colon \mathbf{S}_\bullet(X) \rightarrow \mathbf{S}_\bullet(Y)$. The nth homology functor is the composite $\mathbf{Top} \rightarrow \mathbf{Comp} \rightarrow \mathbf{Ab}$, where $\mathbf{Comp} \rightarrow \mathbf{Ab}$ is defined by $\mathbf{S}_\bullet(X) \mapsto H_n(X)$ and $f_\# \mapsto H_n(f)$. Thus, in a very precise way, we see that homology has a topological half and an algebraic half, for the functor $\mathbf{Top} \rightarrow \mathbf{Comp}$ involves the topological notions of spaces and continuous maps, while the functor $\mathbf{Comp} \rightarrow \mathbf{Ab}$ involves only algebra. Homological Algebra is the study of this algebraic half.

Exercises

*1.1   **(i)**   Prove, in every category $\mathcal{C}$, that each object $A \in \mathcal{C}$ has a unique identity morphism.

(ii)   If f is an isomorphism in a category, prove that its inverse is unique.

1.2   **(i)**   Prove that there is a functor $F \colon \mathbf{ComRings} \rightarrow \mathbf{ComRings}$ defined on objects by $F \colon R \mapsto R[x]$ and on morphisms $\varphi \colon R \rightarrow S$ by $F\varphi \colon r_0 + r_1 x + \cdots + r_n x^n \mapsto \varphi(r_0) + \varphi(r_1)x + \cdots + \varphi(r_n)x^n$.

(ii)   Prove that there is a functor on $\mathbf{Dom}$, the category of all (integral) domains, defined on objects by $R \mapsto \mathrm{Frac}(R)$, and on morphisms $f \colon R \rightarrow S$ by $r/1 \mapsto f(r)/1$.

1.3   Let $\mathcal{A} \overset{S}{\longrightarrow} \mathcal{B} \overset{T}{\longrightarrow} \mathcal{C}$ be functors. If the variances of S and T are the same, prove that the composite $TS \colon \mathcal{A} \rightarrow \mathcal{C}$ is a covariant functor; if the variances of S and T are different, prove that TS is a contravariant functor

*1.4   If $T \colon \mathcal{A} \rightarrow \mathcal{B}$ is a functor, define $T^{\mathrm{op}} \colon \mathcal{A}^{\mathrm{op}} \rightarrow \mathcal{B}$ by $T^{\mathrm{op}}(A) = T(A)$ for all $A \in \mathrm{obj}(\mathcal{A})$ and $T^{\mathrm{op}}(f^{\mathrm{op}}) = T(f)$ for all morphisms f in $\mathcal{A}$. Prove that T^{op} is a functor having variance opposite to the variance of T.

1.5   **(i)**   If X is a set and k is a field, define the vector space k^X to be the set of all functions $X \rightarrow k$ under pointwise operations. Prove that there is a functor $G \colon \mathbf{Sets} \rightarrow {}_k\mathbf{Mod}$ with $G(X) = k^X$.

(ii)   Define $U \colon {}_k\mathbf{Mod} \rightarrow \mathbf{Sets}$ to be the forgetful functor [see Example 1.9(v)]. What are the composites $GU \colon {}_k\mathbf{Mod} \rightarrow {}_k\mathbf{Mod}$ and $UG \colon \mathbf{Sets} \rightarrow \mathbf{Sets}$?

*1.6   **(i)**   If X is a set, define FX to be the free group having basis X; that is, the elements of FX are reduced words on the alphabet X and multiplication is juxtaposition followed by cancellation. If $\varphi \colon X \rightarrow Y$ is a function, prove that there is a unique homomorphism $F\varphi\colon FX \to FY$ such that $(F\varphi)|X = \varphi$.

(ii) Prove that the $F\colon$ **Sets** $\to$ **Groups** is a functor (F is called the *free functor*).

*1.7 (i) Define $\mathcal{C}$ to have objects all ordered pairs (G, H), where G is a group and H is a normal subgroup of G, and to have morphisms $\varphi_*\colon (G, H) \to (G_1, H_1)$, where $\varphi\colon G \to G_1$ is a homomorphism with $\varphi(H) \subseteq H_1$. Prove that $\mathcal{C}$ is a category if composition in $\mathcal{C}$ is defined to be ordinary composition.

(ii) Construct a functor $Q\colon \mathcal{C} \to$ **Groups** with $Q(G, H) = G/H$.

(iii) Prove that there is a functor **Groups** $\to$ **Ab** taking each group G to G/G', where G' is its commutator subgroup.

1.8 If X is a topological space, define $C(X)$ to be its ring of continuous real-valued functions, $C(X) = \{f\colon X \to \mathbb{R} : f \text{ is continuous}\}$, under pointwise operations: $f + g\colon x \mapsto f(x) + g(x)$ and $fg\colon x \mapsto f(x)g(x)$. Prove that there is a contravariant functor $T\colon$ **Top** $\to$ **ComRings** with $T(X) = C(X)$. [A theorem of Gelfand and Kolmogoroff (see Dugundji, *Topology*, p. 289) says that if X and Y are compact Hausdorff spaces and the rings $C(X)$ and $C(Y)$ are isomorphic, then the spaces X and Y are homeomorphic.]

1.9 Let X be a set and let $\mathcal{B}(X)$ be the *Boolean ring* whose elements are the subsets of X, whose multiplication is intersection, and whose addition is symmetric difference: if $A, B \subseteq X$, then $AB = A \cap B$, $A + B = (A - B) \cup (B - A)$, and $-A = A$. You may assume that $\mathcal{B}(X)$ is a commutative ring under these operations in which $\varnothing$ is the zero element and X is the 1 element.

(i) Prove that $\mathcal{B}(X)$ has only one unit (recall that an element $u \in R$ is a *unit* if there is $v \in R$ with $uv = 1 = vu$).

(ii) If $Y \subsetneq X$ is a proper subset of X, prove that $\mathcal{B}(Y)$ is *not* a subring of $\mathcal{B}(X)$.

(iii) Prove that a nonempty subset $I \subseteq \mathcal{B}(X)$ is an ideal if and only if $A \in I$ implies that every subset of A also lies in I. In particular, the principal ideal (A) generated by a subset A is the family of all the subsets of A.

(iv) Prove that every maximal ideal M in $\mathcal{B}(X)$ is a principal ideal of the form $(X - \{x\})$ for some $x \in X$.

(v) Prove that every prime ideal in $\mathcal{B}(X)$ is a maximal ideal.

*1.10 Let R be a ring. Call an (additive) abelian group M an *almost left R-module* if there is a function $R \times M \to M$ satisfying all the axioms of a left R-module except axiom (iv): we do not assume that $1m = m$ for all $m \in M$. Prove that $M = M_1 \oplus M_0$ (direct sum of abelian groups), where $M_1 = \{m \in M : 1m = m\}$ and $M_0 = \{m \in M : rm = 0 \text{ for all } r \in R\}$ are subgroups of M that are almost left R-modules; in fact, M_1 is a left R-module.

*1.11 Prove that every right R-module is a left R^{op}-module, and that every left R-module is a right R^{op}-module.

1.12 If R and A are rings, an ***anti-homomorphism*** $\varphi : R \to A$ is an additive function for which $\varphi(rr') = \varphi(r')\varphi(r)$ for all $r, r' \in R$.

    (i) Prove that R and A are anti-isomorphic if and only if $A \cong R^{\mathrm{op}}$.

    (ii) Prove that transposition $B \mapsto B^T$ is an anti-isomorphism of a matrix ring $\mathrm{Mat}_n(R)$ with itself, where R is a commutative ring. (If R is not commutative, then $B \mapsto B^T$ is an isomorphism $[\mathrm{Mat}_n(R)]^{\mathrm{op}} \cong \mathrm{Mat}_n(R^{\mathrm{op}})$.)

*1.13 An R-map $f : M \to M$, where M is a left R-module, is called an ***endomomorphism***.

    (i) Prove that $\mathrm{End}_R(M) = \{f : M \to M : f \text{ is an } R\text{-map}\}$ is a ring (under pointwise addition and composition as multiplication) and that M is a left $\mathrm{End}_R(M)$-module. We call $\mathrm{End}_R(M)$ the ***endomorphism ring*** of M.

    (ii) If a ring R is regarded as a left R-module, prove that there is an isomorphism $\mathrm{End}_R(R) \to R^{\mathrm{op}}$ of rings.

1.14   **(i)** Give an example of topological spaces X, Y and an injective continuous map $i : X \to Y$ whose induced map $H_n(i) : H_n(X) \to H_n(Y)$ is not injective for some $n \geq 0$.

    Hint. You may assume that $H_1(S^1) \cong \mathbb{Z}$.

    (ii) Give an example of a subspace $A \subseteq X$ of a topological space X and a continuous map $f : X \to Y$ such that $H_n(f) \neq 0$ for some $n \geq 0$ and $H_n(f|A) = 0$.

*1.15 Let $F, G : \mathcal{A} \to \mathcal{B}$ and $F', G' : \mathcal{B} \to \mathcal{C}$ be functors of the same variance, and let $\tau : F \to G$ and $\tau' : F' \to G'$ be natural transformations.

    (i) Prove that their composite $\tau'\tau$ is a natural transformation $F'F \to G'G$ where, for each $A \in \mathrm{obj}(\mathcal{A})$, we define

$$(\tau'\tau)_A = \tau'_{FA}\tau_A : F'F(A) \to G'G(A).$$

    (ii) If $\tau : F \to G$ is a natural isomorphism, define $\sigma_C : FC \to GC$ for all $C \in \mathrm{obj}(\mathcal{A})$ by $\sigma_C = \tau_C^{-1}$. Prove that σ is a natural transformation $G \to F$.

*1.16 Let $\mathcal{C}$ be a category and let $A, B \in \text{obj}(\mathcal{C})$. Prove the converse of Corollary 1.18: if $A \cong B$, then $\text{Hom}_{\mathcal{C}}(A, \square)$ and $\text{Hom}_{\mathcal{C}}(B, \square)$ are naturally isomorphic functors.

Hint. If $\alpha: A \to B$ is an isomorphism, define $\tau_C = (\alpha^{-1})^*$.

1.17   **(i)**   Let $\mathcal{A}$ be the category with $\text{obj}(\mathcal{A}) = \{A, B, C, D\}$ and morphisms $\text{Hom}_{\mathcal{A}}(A, B) = \{f\}$, $\text{Hom}_{\mathcal{A}}(C, D) = \{g\}$, and four identities. Define $F: \mathcal{A} \to \textbf{Sets}$ by $F(A) = \{1\}$, $F(B) = \{2\} = F(C)$, and $F(D) = \{3\}$. Prove that F is a functor but that im F is not a subcategory of **Sets**.

Hint. The composite $Fg \circ Ff$, which is defined in **Sets**, does not lie in im F.

(ii)   Prove that if $F: \mathcal{A} \to \mathcal{B}$ is a functor with $F | \text{obj}(\mathcal{A})$ an injection, then im F is a subcategory of $\mathcal{B}$.

1.18 Let $\mathcal{A}$ and $\mathcal{B}$ be categories. Prove that $\mathcal{A} \times \mathcal{B}$ is a category, where $\text{obj}(\mathcal{A} \times \mathcal{B}) = \text{obj}(\mathcal{A}) \times \text{obj}(\mathcal{B})$, where $\text{Hom}_{\mathcal{A} \times \mathcal{B}}((A_1, B_1), (A_2, B_2))$ consists of all $(f, g) \in \text{Hom}_{\mathcal{A}}(A_1, A_2) \times \text{Hom}_{\mathcal{B}}(B_1, B_2)$, and where composition is $(f', g')(f, g) = (f'f, g'g)$.

*1.19   **(i)**   If $\mathcal{A}$ is a small category and $F, G: \mathcal{A} \to \mathcal{B}$ are functors of the same variance, prove that $\text{Nat}(F, G)$ is a set (not a proper class).

(ii)   Give an example of categories $\mathcal{C}, \mathcal{D}$ and functors $S, T: \mathcal{C} \to \mathcal{D}$ such that $\text{Nat}(S, T)$ is a proper class. [As discussed on page 8, do not worry whether $\text{Nat}(S, T)$ is a class.]

Hint. Let $\mathcal{S}$ be a discrete subcategory of **Sets**, and consider $\text{Nat}(T, T)$, where $T: \mathcal{S} \to \textbf{Sets}$ is the inclusion functor.

1.20 Show that **Cat** is a category, where $\text{obj}(\textbf{Cat})$ is the class of all *small* categories, where $\text{Hom}_{\textbf{Cat}}(\mathcal{A}, \mathcal{B}) = \mathcal{B}^{\mathcal{A}}$, and where composition is the usual composition of functors. [We assume that categories here are small in order that $\text{Hom}_{\textbf{Cat}}(\mathcal{A}, \mathcal{B})$ be a set.]

2

Hom and Tensor

The most important functors studied in Homological Algebra are Hom, tensor product, and functors derived from them. We begin by describing certain constructs in $_R\mathbf{Mod}$, such as sums, products, and exact sequences, and we will then apply Hom functors to them. Tensor products will then be introduced, and we will apply these functors to the constructs in $_R\mathbf{Mod}$ as well. There is an intimate relationship between Hom and tensor—they form an *adjoint pair* of functors.

2.1 Modules

Many properties of vector spaces and of abelian groups are also enjoyed by modules. We assume that much of this section is familiar to most readers, and so our account is written to refresh one's memory. All rings R in this book are assumed to have an *identity element* 1 (or *unit*) (where $r1 = r = 1r$ for all $r \in R$). We do not insist that $1 \neq 0$; however, should $1 = 0$, then R is the *zero ring* having only one element. If $f : R \to S$ is a ring homomorphism, we assume that $f(1) = 1$; that is, $f(1)$ is the identity element of S.

We can view modules as a tool for studying rings. If M is an abelian group, then we saw, in Exercise 1.13 on page 35, that

$$\mathrm{End}_{\mathbb{Z}}(M) = \{\text{homomorphisms } f : M \to M\}$$

is a ring under pointwise addition $[f + g : m \mapsto f(m) + g(m)]$ and composition as multiplication. A *representation* of a ring R is a ring homomorphism $\varphi : R \to \mathrm{End}_{\mathbb{Z}}(M)$ for some abelian group M.

J.J. Rotman, *An Introduction to Homological Algebra*, Universitext, DOI 10.1007/978-0-387-68324-9_2, © Springer Science+Business Media LLC 2009

Proposition 2.1.   *Let R be a ring and let M an abelian group. If $\varphi: R \to$ End(M) is a representation, define $\sigma: R \times M \to M$ by $\sigma(r, m) = \varphi_r(m)$, where we write $\varphi(r) = \varphi_r$; then σ is a scalar multiplication making M into a left R-module. Conversely, if M is a left R-module, then the function $\psi: R \to$ End(M), given by $\psi(r): m \mapsto rm$, is a representation.*

Proof.   The proof is straightforward.   •

Example 2.2.   Let G be a finite[1] group and let k be a commutative ring. The **group ring** is the set of all functions $\alpha: G \to k$ made into a ring with pointwise operations: for all $x \in G$,

$$\alpha + \beta: x \mapsto \alpha(x) + \beta(x) \quad \text{and} \quad \alpha\beta: x \mapsto \alpha(x)\beta(x).$$

If $y \in G$, the function δ_y, defined by

$$\delta_y(x) = \begin{cases} 1 & \text{if } x = y, \\ 0 & \text{if } x \neq y, \end{cases}$$

is usually denoted by y. It is easy to check that kG is a k-module and that each $\gamma \in kG$ has a unique expression

$$\gamma = \sum_{y \in G} a_y y,$$

where $a_y \in k$. In this notation, elements of G multiply as they do in G; in particular, the identity element 1 in G is also the unit in kG. Multiplication in kG is called **convolution**, and a formula for it is

$$\left(\sum_x a_x x \right)\left(\sum_y b_y y \right) = \sum_{x,y} a_x b_y xy = \sum_z \left(\sum_x a_x b_{x^{-1}z} \right) z.$$

Recall that if G is a group and k is a commutative ring, then a k-**representation** of G is a function $\sigma: G \to \text{Mat}_n(k)$ with

$$\sigma(xy) = \sigma(x)\sigma(y),$$
$$\sigma(1) = I, \quad \text{the identity matrix.}$$

For all $x \in G$, we have $\sigma(x)$ nonsingular, for $I = \sigma(1) = \sigma(xx^{-1}) = \sigma(x)\sigma(x^{-1})$. It follows that σ is a group homomorphism $G \to \text{GL}(n, k)$, the multiplicative group of all nonsingular $n \times n$ matrices over k. It is easy to see that σ extends to a ring homomorphism $\widetilde{\sigma}: kG \to \text{Mat}_n(k)$ by

$$\widetilde{\sigma}\left(\sum_y a_y y \right) = \sum_y a_y \sigma(y).$$

---

[1]This construction can be done for infinite groups G as well; the elements of kG are the functions $\alpha: G \to k$ for which $\alpha(y) = 0$ for all but a finite number of $y \in G$.

Thus, $\tilde{\sigma}: kG \to \mathrm{End}(k^n)$, so that $\tilde{\sigma}$ is a representation of the group ring kG. By Proposition 2.1, the group representation σ corresponds to a left kG-module. ◄

Lemma 2.3. *If $A, B \in \mathrm{obj}(_R\mathbf{Mod})$ [or if $A, B \in \mathrm{obj}(\mathbf{Mod}_R)$], then the set $\mathrm{Hom}_R(A, B)$ is an abelian group. Moreover, if $p: A' \to A$ and $q: B \to B'$ are R-maps, then*

$$(f + g)p = fp + gp \quad and \quad q(f + g) = qf + qg.$$

Proof. One easily checks that $f + g$ is an R-map; thus, $f + g \in \mathrm{Hom}_R(A, B)$ and addition is an operation on $\mathrm{Hom}_R(A, B)$. The zero in $\mathrm{Hom}_R(A, B)$ is the constant map $a \mapsto 0$, and the inverse of $f: A \to B$ is $-f: a \mapsto -[f(a)]$. It is routine to see that addition is associative, and so $\mathrm{Hom}_R(A, B)$ is an abelian group. The last equations are checked by evaluating each on $a \in A$. •

Proposition 2.4. *Let R be a ring, and let A, B, B' be left R-modules.*

(i) *$\mathrm{Hom}_R(A, \square)$ is an additive functor $_R\mathbf{Mod} \to \mathbf{Ab}$.*

(ii) *If A is a left R-module, then $\mathrm{Hom}_R(A, B)$ is a $Z(R)$-module, where $Z(R)$ is the center of R, if we define*

$$rf: a \mapsto f(ra)$$

for $r \in Z(R)$ and $f: A \to B$. If $q: B \to B'$ is an R-map, then the induced map $q_*: \mathrm{Hom}_R(A, B) \to \mathrm{Hom}_R(A, B')$ is a $Z(R)$-map, and $\mathrm{Hom}_R(A, \square)$ takes values in $_{Z(R)}\mathbf{Mod}$. In particular, if R is commutative, then $\mathrm{Hom}_R(A, \square)$ is a functor $_R\mathbf{Mod} \to {}_R\mathbf{Mod}$.

Proof.

(i) Lemma 2.3 shows that $\mathrm{Hom}_R(A, B)$ is an abelian group and, for all $q: B \to B'$, that $q(f + g) = qf + qg$; that is, $q_*(f + g) = q_*(f) + q_*(g)$. Hence, q_* is a homomorphism. Since $\mathrm{Hom}_R(A, \square)$ preserves identities and composition, it is an additive functor with values in $\mathbf{Ab}$.

(ii) We show that if $r \in Z(R)$, then rf, as defined in the statement, is a $Z(R)$-map. If $s \in R$, then $rs = sr$ and

$$(rf)(sa) = f(r(sa)) = f((rs)a) = (rs)fa = (sr)fa = s[rf](a).$$

It follows that $\mathrm{Hom}_R(A, B)$ is a $Z(R)$-module. If $q: B \to B'$ is an R-map, we show that the induced map $q_*: f \mapsto qf$ is a $Z(R)$-map. Now q_* is additive, by part (i). We check that $q_*(rf) = rq_*(f)$, where $r \in Z(R)$; that is, $q(rf) = (rq)f$. But $q(rf): a \mapsto q(f(ra))$, while

$(rq)f: a \mapsto (rq)(f(a)) = q(rf(a)) = qf(ra)$, because q is a $Z(R)$-map and f is an R-map. Therefore, q_* is a $Z(R)$-map. The last statement is true because R is commutative if and only if $R = Z(R)$.   •

We have generalized the familiar fact that if V and W are vector spaces over a field k, then $\text{Hom}_k(V, W)$ is also a vector space over k.

If R is a ring and B is a left R-module, then the contravariant Hom functor $\text{Hom}_R(\square, B): \mathbf{Mod}_R \to \mathbf{Sets}$ also has more structure.

Proposition 2.5.   *Let R be a ring, and let A, A', B be left R-modules.*

(i) $\text{Hom}_R(\square, B)$ *is a contravariant functor $_R\mathbf{Mod} \to \mathbf{Ab}$.*

(ii) *If B is a left R-module, then $\text{Hom}_R(A, B)$ is a $Z(R)$-module, where $Z(R)$ is the center of R, if we define*

$$rf: a \mapsto f(ra)$$

for $r \in Z(R)$ and $f: A \to B$. If $p: A \to A'$ is an R-map, then the induced map $p^*: \text{Hom}_R(A', B) \to \text{Hom}_R(A, B)$ is a $Z(R)$-map, and $\text{Hom}(\square, B)$ takes values in $_{Z(R)}\mathbf{Mod}$. In particular, if R is commutative, then $\text{Hom}_R(\square, B)$ is a contravariant functor $_R\mathbf{Mod} \to {}_R\mathbf{Mod}$.

Proof.   Similar to the proof of Proposition 2.4.   •

Example 2.6.   As in Example 1.11, the dual space $V^* = \text{Hom}_k(V, k)$ of a vector space V over a field k is also a vector space over k.   ◀

Definition.   A functor $T: {}_R\mathbf{Mod} \to \mathbf{Ab}$ of either variance is called an ***additive functor*** if, for every pair of R-maps $f, g: A \to B$, we have

$$T(f + g) = T(f) + T(g).$$

We have just seen that Hom functors $_R\mathbf{Mod} \to \mathbf{Ab}$ of either variance are additive functors. If $T: \mathcal{C} \to \mathcal{D}$ is a covariant functor between categories $\mathcal{C}$ and $\mathcal{D}$, then there are functions

$$T_{AB}: \text{Hom}_{\mathcal{C}}(A, B) \to \text{Hom}_{\mathcal{D}}(TA, TB),$$

namely, $h \mapsto T(h)$. If $T: {}_R\mathbf{Mod} \to \mathbf{Ab}$ is an additive functor, then each T_{AB} is a homomorphism of abelian groups; the analogous statement for contravariant functors is also true.

Proposition 2.7. *Let* $T: {}_R\mathbf{Mod} \to \mathbf{Ab}$ *be an additive functor of either variance.*

(i) *If* $0: A \to B$ *is the **zero map**, that is, the map* $a \mapsto 0$ *for all* $a \in A$, *then* $T(0) = 0$.

(ii) $T(\{0\}) = \{0\}$.

Proof.

(i) Since T is additive, the function T_{AB} between Hom sets is a homomorphism, and so it preserves identity elements; that is, $T(0) = 0$.

(ii) If A is a left R-module, then $0 = 1_A$ if and only if $A = \{0\}$ [sufficiency is obvious; for necessity, if $1_A = 0$, then for all $a \in A$, we have $a = 1_A(a) = 0(a) = 0$, and so $A = \{0\}$]. By part (i), we have $T(1_{\{0\}}) = T(0) = 0$, and so $T(\{0\}) = \{0\}$. •

We now show that many constructions made for abelian groups and for vector spaces can be generalized to left modules over any ring. A *submodule S* is a left R-module contained in a larger left R-module M such that if $s, s' \in S$ and $r \in R$, then $s + s'$ and rs have the same meaning in S as in M.

Definition. If M is a left R-module, then a *submodule* N of M, denoted by $N \subseteq M$, is an additive subgroup N of M closed under scalar multiplication: $rn \in N$ whenever $n \in N$ and $r \in R$. A similar definition holds for right modules.

Example 2.8.

(i) A submodule of a $\mathbb{Z}$-module (i.e., of an abelian group) is a subgroup, and a submodule of a vector space is a subspace.

(ii) Both $\{0\}$ and M are submodules of a module M. A *proper submodule* of M is a submodule $N \subseteq M$ with $N \neq M$. In this case, we may write $N \subsetneqq M$.

(iii) If a ring R is viewed as a left module over itself, then a submodule of R is a left ideal; I is a proper submodule when it is a proper left ideal. Similarly, if R is viewed as a right module over itself, then its submodules are its right ideals.

(iv) If M is an R-module and $r \in R$, where R is a commutative ring, then

$$rM = \{rm : m \in M\}$$

is a submodule of M. Here is a generalization. If J is an ideal in R and M is an R-module, then

$$JM = \left\{ \sum_i j_i m_i : j_i \in J \text{ and } m_i \in M \right\}$$

is a submodule of M.

(v) If S and T are submodules of a left module M, then

$$S + T = \{s + t : s \in S \text{ and } t \in T\}$$

is a submodule of M that contains S and T.

(vi) If $(S_i)_{i \in I}$ is a family of submodules of a left R-module M, then $\bigcap_{i \in I} S_i$ is a submodule of M.

(vii) A left R-module S is *cyclic* if there exists $s \in S$ with $S = \{rs : r \in R\}$. If M is an R-module and $m \in M$, then the *cyclic submodule generated by* m, denoted by $\langle m \rangle$, is

$$\langle m \rangle = \{rm : r \in R\}.$$

More generally, if X is a subset of an R-module M, then

$$\langle X \rangle = \left\{ \sum_{\text{finite}} r_i x_i : r_i \in R \text{ and } x_i \in X \right\},$$

the set of all *R-linear combinations* of elements in X. We call $\langle X \rangle$ the *submodule generated by* X. Exercise 2.10 on page 66 states that $\langle X \rangle = \bigcap_{X \subseteq S} S$. ◄

Definition.   A left R-module M is *finitely generated* if M is generated by a finite set; that is, if there is a finite subset $X = \{x_1, \ldots, x_n\}$ with $M = \langle X \rangle$.

For example, a vector space V over a field k is a finitely generated k-module if and only if V is finite-dimensional.

Definition.   If N is a submodule of a left R-module M, then the *quotient module* is the quotient group M/N (remember that M is an abelian group and N is a subgroup) equipped with the scalar multiplication

$$r(m + N) = rm + N.$$

The *natural map* $\pi : M \to M/N$, given by $m \mapsto m + N$, is easily seen to be an R-map.

Scalar multiplication in the definition of quotient module is well-defined: if $m + N = m' + N$, then $m - m' \in N$. Hence, $r(m - m') \in N$ (because N is a submodule), $rm - rm' \in N$, and $rm + N = rm' + N$.

Example 2.9. If $N \subseteq M$ is merely an additive subgroup of M but not a submodule, then the abelian group M/N is not an R-module. For example, let V be a vector space over a field k. If $a \in k$ and $v \in V$, then $av = 0$ if and only if $a = 0$ or $v = 0$ [if $a \neq 0$, then $0 = a^{-1}(av) = (a^{-1}a)v = v$]. Now $\mathbb{Q}$ is a vector space over itself, but $\mathbb{Q}/\mathbb{Z}$ is not a vector space over $\mathbb{Q}$ [we have $2(\frac{1}{2} + \mathbb{Z}) = \mathbb{Z}$ in $\mathbb{Q}/\mathbb{Z}$, and neither factor is zero]. ◀

Example 2.10.

(i) Recall that an additive subgroup $J \subseteq R$ of a ring R is a *two-sided ideal* if $x \in J$ and $r \in R$ imply $rx \in J$ and $xr \in J$. If $R = \text{Mat}_2(k)$, the ring of all 2×2 matrices over a field k, then $I = \left\{ \left[\begin{smallmatrix} a & 0 \\ b & 0 \end{smallmatrix} \right] : a, b \in k \right\}$ is a left ideal and $I' = \left\{ \left[\begin{smallmatrix} a & b \\ 0 & 0 \end{smallmatrix} \right] : a, b \in k \right\}$ is a right ideal, but neither is a two-sided ideal.

(ii) If J is a left (or right) ideal in R, then R/J is a left (or right) R-module. If J is a two-sided ideal, then R/J is a ring with multiplication

$$(r + J)(s + J) = rs + J.$$

This multiplication is well-defined, for if $r + J = r' + J$ and $s + J = s' + J$, then $rs + J = r's' + J$, because

$$rs - r's' = rs - r's + r's - r's' = (r - r')s + r'(s - s') \in J. \quad ◀$$

We continue extending definitions from abelian groups and vector spaces to modules.

Definition. If $f: M \to N$ is an R-map between left R-modules, then

> ***kernel*** $f = \ker f = \{m \in M : f(m) = 0\}$,
> ***image*** $f = \text{im } f = \{n \in N : \text{there exists } m \in M \text{ with } n = f(m)\}$,
> ***cokernel*** $f = \text{coker } f = N/\text{im } f$.

It is routine to check that $\ker f$ is a submodule of M and that $\text{im } f$ is a submodule of N.

Theorem 2.11 (First Isomorphism Theorem). *If* $f: M \to N$ *is an R-map of left R-modules, then there is an R-isomorphism*

$$\varphi: M/\ker f \to \text{im } f$$

given by

$$\varphi: m + \ker f \mapsto f(m).$$

Proof. If we view M and N only as abelian groups, then the first isomorphism theorem for groups says that $\varphi: M/\ker f \to \operatorname{im} f$ is a well-defined

$$
\begin{array}{ccc}
M & \xrightarrow{\ \ f\ \ } & N \\
{\scriptstyle\text{nat}}\downarrow & \nearrow & \uparrow {\scriptstyle\text{inc}} \\
M/\ker f & \dashrightarrow[\varphi]{} & \operatorname{im} f
\end{array}
$$

isomorphism of abelian groups. But φ is an R-map: if $r \in R$ and $m \in M$, then $\varphi(r(m + N)) = \varphi(rm + N) = f(rm)$; since f is an R-map, however, $f(rm) = rf(m) = r\varphi(m + N)$, as desired. •

The second and third isomorphism theorems are corollaries of the first.

Theorem 2.12 (Second Isomorphism Theorem). *If S and T are submodules of a left R-module M, then there is an R-isomorphism*

$$S/(S \cap T) \to (S + T)/T.$$

Proof. If $\pi: M \to M/T$ is the natural map, then $\ker \pi = T$; define $f = \pi|S$, so that $f: S \to M/T$. Now

$$\ker f = S \cap T \quad \text{and} \quad \operatorname{im} f = (S + T)/T,$$

for $(S + T)/T$ consists of all those cosets in M/T having a representative in S. The first isomorphism theorem now applies. •

Definition. If $T \subseteq S \subseteq M$ is a tower of submodules of a left R-module M, then *enlargement of coset* $e: M/T \to M/S$ is defined by

$$e: m + T \mapsto m + S$$

(e is well-defined, for if $m + T = m' + T$, then $m - m' \in T \subseteq S$ and $m + S = m' + S$).

Theorem 2.13 (Third Isomorphism Theorem). *If $T \subseteq S \subseteq M$ is a tower of submodules of a left R-module M, then enlargement of coset $e: M/T \to M/S$ induces an R-isomorphism*

$$(M/T)/(S/T) \to M/S.$$

Proof. The reader may check that $\ker e = S/T$ and $\operatorname{im} e = M/S$, so that the first isomorphism theorem applies at once. •

If $f: M \to N$ is a map of left R-modules and $S \subseteq N$, then the reader may check that $f^{-1}(S) = \{m \in M: f(m) \in S\}$ is a submodule of M containing $f^{-1}(\{0\}) = \ker f$.

Theorem 2.14 (Correspondence Theorem). *If T is a submodule of a left R-module M, then $\varphi \colon S \mapsto S/T$ is a bijection:*

$$\varphi \colon \{\text{intermediate submodules } T \subseteq S \subseteq M\} \to \{\text{submodules of } M/T\}.$$

Moreover, $T \subseteq S \subseteq S'$ in M if and only if $S/T \subseteq S'/T$ in M/T.

Proof. Since every module is an additive abelian group, every submodule is a subgroup, and so the usual correspondence theorem for groups shows that φ is an injection that preserves inclusions: $S \subseteq S'$ in M if and only if $S/T \subseteq S'/T$ in M/T. Moreover, φ is surjective: if $S^* \subseteq M/T$, then there is a unique submodule $S \supseteq T$ with $S^* = S/T$. The remainder of this proof is a repetition of the usual proof for groups, checking only that images and inverse images of submodules are submodules.   •

The correspondence theorem is usually invoked tacitly: a submodule S^* of M/T is equal to $S^* = S/T$ for some unique intermediate submodule S.

Here is a ring-theoretic version.

Theorem 2.15 (Correspondence Theorem for Rings). *If I is a two-sided ideal of a ring R, then $\varphi \colon J \mapsto J/I$ is a bijection:*

$$\varphi \colon \{\text{intermediate left ideals } I \subseteq J \subseteq R\} \to \{\text{left ideals of } R/I\}.$$

Moreover, $I \subseteq J \subseteq J'$ in R if and only if $J/I \subseteq J'/I$ in R/I.

Proof. The reader may supply a variant of the proof of Theorem 2.14.   •

Proposition 2.16. *A left R-module M is cyclic if and only if $M \cong R/I$ for some left ideal I.*

Proof. If M is cyclic, then $M = \langle m \rangle$ for some $m \in M$. Define $f \colon R \to M$ by $f(r) = rm$. Now f is surjective, since M is cyclic, and its kernel is a submodule of R; that is, $\ker f$ is a left ideal I. The first isomorphism theorem gives $R/I \cong M$.

Conversely, R/I is cyclic with generator $m = 1 + I$.   •

Definition. A left R-module M is **simple** (or **irreducible**) if $M \neq \{0\}$ and M has no proper nonzero submodules; that is, $\{0\}$ and M are the only submodules of M.

Corollary 2.17. *A left R-module M is simple if and only if $M \cong R/I$, where I is a maximal left ideal.*

Proof. This follows from the correspondence theorem.   •

For example, an abelian group G is simple if and only if G is cyclic of order p for some prime p. The existence of maximal left ideals guarantees the existence of simple modules.

Definition. A finite or infinite sequence of R-maps and left R-modules

$$\cdots \to M_{n+1} \xrightarrow{f_{n+1}} M_n \xrightarrow{f_n} M_{n-1} \to \cdots$$

is called an ***exact sequence***[2] if im $f_{n+1} = \ker f_n$ for all n.

Observe that there is no need to label arrows[3] $0 \xrightarrow{f} A$ or $B \xrightarrow{g} 0$: in either case, there is a unique map, namely, $f : 0 \mapsto 0$ or the constant homomorphism $g(b) = 0$ for all $b \in B$. Here are some simple consequences of a sequence of homomorphisms being exact.

Proposition 2.18.

(i) *A sequence* $0 \to A \xrightarrow{f} B$ *is exact if and only if* f *is injective.*

(ii) *A sequence* $B \xrightarrow{g} C \to 0$ *is exact if and only if* g *is surjective.*

(iii) *A sequence* $0 \to A \xrightarrow{h} B \to 0$ *is exact if and only if* h *is an isomorphism.*

Proof.

(i) The image of $0 \to A$ is $\{0\}$, so that exactness gives $\ker f = \{0\}$, and so f is injective. Conversely, given $f : A \to B$, there is an exact sequence $\ker f \xrightarrow{i} A \xrightarrow{f} B$, where i is the inclusion. If f is injective, then $\ker f = \{0\}$.

(ii) The kernel of $C \to 0$ is C, so that exactness gives im $g = C$, and so g is surjective. Conversely, given $g : B \to C$, there is an exact sequence $B \xrightarrow{g} C \xrightarrow{\pi} C / \text{im } g$, where π is the natural map. If g is surjective, then $C = \text{im } g$ and $C / \text{im } g = \{0\}$.

(iii) Part (i) shows that h is injective if and only if $0 \to A \xrightarrow{h} B$ is exact, and part (ii) shows that h is surjective if and only if $A \xrightarrow{h} B \to 0$ is exact. Therefore, h is an isomorphism if and only if the sequence $0 \to A \xrightarrow{h} B \to 0$ is exact. •

---

[2]This terminology comes from Advanced Calculus, where a differential form ω is called ***closed*** if $d\omega = 0$ and is called ***exact*** if $\omega = dh$ for some function h. The term *exact sequence* was coined by the algebraic topologist W. Hurewicz. It is interesting to look at the wonderful book by Hurewicz and Wallman, *Dimension Theory*, which was written just before this coinage. Many results there would have been much simpler to state had the term *exact sequence* been available.

[3]We usually write 0 instead of $\{0\}$ in sequences and diagrams.

Definition.   A *short exact sequence* is an exact sequence of the form

$$0 \to A \xrightarrow{f} B \xrightarrow{g} C \to 0.$$

We also call this short exact sequence an *extension* of A by C.

Some authors call this an extension of C by A; some authors say that the middle module B is an extension.

The next proposition restates the first and third isomorphism theorems in terms of exact sequences.

Proposition 2.19.

(i) *If* $0 \to A \xrightarrow{f} B \xrightarrow{g} C \to 0$ *is a short exact sequence, then*

$$A \cong \operatorname{im} f \quad and \quad B/\operatorname{im} f \cong C.$$

(ii) *If* $T \subseteq S \subseteq M$ *is a tower of submodules, then there is an exact sequence*

$$0 \to S/T \to M/S \to M/T \to 0.$$

Proof.

(i) Since f is injective, changing its target gives an isomorphism $A \to \operatorname{im} f$. The first isomorphism theorem gives $B/\ker g \cong \operatorname{im} g$. By exactness, however, $\ker g = \operatorname{im} f$ and $\operatorname{im} g = C$; therefore, $B/\operatorname{im} f \cong C$.

(ii) This is just a restatement of the third isomorphism theorem. Define $f: S/T \to M/T$ to be the inclusion and $g: M/T \to M/S$ to be enlargement of coset: $g: m + T \mapsto m + S$. As in the proof of Theorem 2.13, g is surjective, and $\ker g = S/T = \operatorname{im} f$.   •

In the special case when A is a submodule of B and $f: A \to B$ is the inclusion, then exactness of $0 \to A \xrightarrow{f} B \xrightarrow{g} C \to 0$ gives $B/A \cong C$.

The familiar notions of direct sum of vector spaces and direct sum of abelian groups extend to modules. Recall that if S and T are abelian groups, then their *external direct sum* $S \boxplus T$ is the abelian group whose underlying set is the cartesian product and whose binary operation is pointwise addition. If S and T are subgroups of an abelian group such that $S + T = G$ and $S \cap T = \{0\}$, then $G = S \oplus T$ is their *internal direct sum*. Both versions give isomorphic abelian groups. The external-internal viewpoints persist for modules as well.

Definition.   If S and T are left R-modules, where R is a ring, then their *external direct sum*, denoted by $S \boxplus T$, is the cartesian product with coordinate-wise operations:

$$(s, t) + (s', t') = (s + s', t + t'),$$
$$r(s, t) = (rs, rt),$$

where $s, s' \in S, t, t' \in T$, and $r \in R$.

Proposition 2.20.   *The following statements are equivalent for left R-modules $M, S,$ and T.*

(i) $S \boxplus T \cong M$.

(ii) *There exist injective R-maps $i: S \to M$ and $j: T \to M$ such that*

$$M = \operatorname{im} i + \operatorname{im} j \quad and \quad \operatorname{im} i \cap \operatorname{im} j = \{0\}.$$

(iii) *There exist R-maps $i: S \to M$ and $j: T \to M$ such that, for every $m \in M$, there are unique $s \in S$ and $t \in T$ with $m = is + jt$.*

(iv) *There are R-maps $i: S \to M$, $j: T \to M$, called **projections**, and R-maps $p: M \to S$, $q: M \to T$, called **injections**, such that*

$$pi = 1_S, \quad qj = 1_T, \quad pj = 0, \quad qi = 0, \quad and \quad ip + jq = 1_M.$$

(v) *The map $\psi: M \to S \boxplus T$, given by $m \mapsto (pm, qm)$, is an isomorphism.*

Remark.   The equations $pi = 1_S$ and $qj = 1_T$ show that the maps i and j must be injective (so that $\operatorname{im} i \cong S$ and $\operatorname{im} j \cong T$) and the maps p and q must be surjective.   ◄

Proof.

(i) $\Rightarrow$ (ii). Let $\varphi: S \boxplus T \to M$ be an isomorphism. Define $\sigma: S \to S \boxplus T$ by $s \mapsto (s, 0)$ and $\tau: T \to S \boxplus T$ by $t \mapsto (0, t)$. Clearly, σ and τ are injective R-maps, and so their composites $i = \varphi\sigma: S \to M$ and $j = \varphi\tau: T \to M$ are also injections.

If $m \in M$, then φ surjective implies that there exist $s \in S$ and $t \in T$ with

$$m = \varphi(s, t) = \varphi(s, 0) + \varphi(0, t) = is + jt \in \operatorname{im} i + \operatorname{im} j.$$

Finally, if $x \in \operatorname{im} i \cap \operatorname{im} j$, then $x = \varphi\sigma(s) = \varphi(s, 0)$ and $x = \varphi\tau(t) = \varphi(0, t)$. Since φ is injective, $(s, 0) = (0, t)$, so that $s = 0$ and $x = \varphi(s, 0) = 0$.

(ii) $\Rightarrow$ (iii). Since $M = \operatorname{im} i + \operatorname{im} j$, each $m \in M$ has an expression $m = is + jt$ with $s \in S$ and $t \in T$, so that only uniqueness need be proved. If also $m = is' + jt'$, then $i(s - s') = j(t' - t) \in \operatorname{im} i \cap \operatorname{im} j = \{0\}$. Therefore, $i(s - s') = 0$ and $j(t - t') = 0$. Since i and j are injections, we have $s = s'$ and $t = t'$.

(iii) $\Rightarrow$ (iv). If $m \in M$, then there are unique $s \in S$ and $t \in T$ with $m = is + jt$. The functions p and q, given by $p(m) = s$ and $q(m) = t$, are thus well-defined. It is routine to check that p and q are R-maps and that the first four equations in the statement hold (they follow from the definitions of p and q). For the last equation, if $m \in M$, then $m = is + jt$, and $ip(m) + jq(m) = is + jt = m$.

(iv) $\Rightarrow$ (v). Define $\varphi \colon S \boxplus T \to M$ by $\varphi(s, t) = is + jt$. It is easy to check that φ is an R-map. Now φ is surjective: if $m \in M$, then $ip + jq = 1_M$ gives $m = ipm + jqm = \varphi(pm, qm)$. To see that ψ is injective, suppose that $\varphi(s, t) = 0$; that is, $is = -jt$. Then $s = pis = -pjt = 0$ and $-t = -qjt = qis = 0$. Therefore, φ is an isomorphism, and its inverse is $m \mapsto (pm, qm)$.

(v) $\Rightarrow$ (i). Obvious. $\bullet$

Corollary 2.21.   *If $T \colon {}_R\mathbf{Mod} \to \mathbf{Ab}$ is an additive functor of either variance, then*

$$T(A \boxplus B) \cong T(A) \boxplus T(B).$$

In particular, if T is covariant, then $x \mapsto (T(p)x, T(q)x)$ is an isomorphism, where $p \colon A \boxplus B \to A$ and $q \colon A \boxplus B \to B$ are the projections.

Proof.   By Proposition 2.7, an additive functor preserves the equations in Proposition 2.20(iv), and the displayed isomorphism is that given in the proof of (iv) $\Rightarrow$ (i) of the proposition. $\bullet$

Internal direct sum is the most important instance of a module isomorphic to a direct sum.

Definition.   If S and T are submodules of a left R-module M, then M is their ***internal direct sum*** if each $m \in M$ has a unique expression of the form $m = s + t$, where $s \in S$ and $t \in T$. We denote an internal direct sum by

$$M = S \oplus T.$$

Notice that we use equality here and not isomorphism.

Here are restatements of Proposition 2.20 and Corollary 2.21 for internal direct sums.

Corollary 2.22.

(i) *Let M be a left R-module having submodules S and T. Then $M = S \oplus T$ if and only if $S + T = M$ and $S \cap T = \{0\}$. Thus, $S \oplus T \cong S \boxplus T$.*

(ii) *If $T: {}_R\mathbf{Mod} \to \mathbf{Ab}$ is an additive functor of either variance, then $T(A \oplus B) \cong T(A) \oplus T(B)$. In particular, if T is covariant and $x \in T(A \oplus B)$, then $\psi: x \mapsto (T(p)x, T(q)x)$ is an isomorphism, where $p: A \oplus B \to A$ and $q: A \oplus B \to B$ are the projections.*

Proof. This follows at once from the equivalence of parts (ii) and (iii) of Proposition 2.20 by taking i and j to be inclusions. The second statement follows from Corollary 2.21. •

We now forsake the notation $S \boxplus T$, and we write as the mathematical world writes: either version of direct sum is denoted by $S \oplus T$.

Definition. A submodule S of a left R-module M is a ***direct summand*** of M if there exists a submodule T of M with $M = S \oplus T$. The submodule T is called a ***complement*** of S.

Complements of a direct summand S of M are not unique. For example, let V be a two-dimensional vector space over a field k, and let a, b be a basis. For any $\alpha \in k$, the one-dimensional subspace $\langle \alpha a + b \rangle$ is a complement of $\langle a \rangle$. On the other hand, all complements of S are isomorphic (to M/S).

The next corollary relates direct summands to a special type of homomorphism.

Definition. A submodule S of a left R-module M is a ***retract*** of M if there exists an R-map $\rho: M \to S$, called a ***retraction***, with $\rho(s) = s$ for all $s \in S$.

Equivalently, ρ is a retraction if and only if $\rho i = 1_S$, where $i: S \to M$ is the inclusion.

Corollary 2.23. *A submodule S of a left R-module M is a direct summand if and only if there exists a retraction $\rho: M \to S$.*

Proof. In this case, we let $i: S \to M$ be the inclusion. We show that $M = S \oplus T$, where $T = \ker \rho$. If $m \in M$, then $m = (m - \rho m) + \rho m$. Plainly, $\rho m \in \operatorname{im} \rho = S$. On the other hand, $\rho(m - \rho m) = \rho m - \rho\rho m = 0$, because $\rho m \in S$ and so $\rho\rho m = \rho m$. Therefore, $M = S + T$.

If $m \in S$, then $\rho m = m$; if $m \in T = \ker \rho$, then $\rho m = 0$. Hence, if $m \in S \cap T$, then $m = 0$. Therefore, $S \cap T = \{0\}$, and $M = S \oplus T$.

For the converse, if $M = S \oplus T$, then each $m \in M$ has a unique expression of the form $m = s + t$, where $s \in S$ and $t \in T$, and it is easy to check that $\rho: M \to S$, defined by $\rho: s + t \mapsto s$, is a retraction $M \to S$. •

Corollary 2.24.

(i) *If $M = S \oplus T$ and $S \subseteq N \subseteq M$, then $N = S \oplus (N \cap T)$.*

(ii) *If $M = S \oplus T$ and $S' \subseteq S$, then $M/S' = S/S' \oplus (T + S')/S'$.*

Proof.

(i) Let $\rho : M \to S$ be the retraction $s + t \mapsto s$. Since $S \subseteq N$, the restriction $\rho | N : N \to S$ is a retraction with $\ker(\rho | N) = N \cap T$.

(ii) The map $\overline{\rho} : M/S' \to S/S'$ is a retraction with $\ker \overline{\rho} = T + S'$. •

The direct sum constructions can be extended to finitely many submodules. There are external and internal versions, and we temporarily revive the $\boxplus$ notation.

Definition. Given left R-modules $S_1, \ldots, S_n$, their (*external*) *direct sum* $S_1 \boxplus \cdots \boxplus S_n$ is the left R-module whose underlying set is the cartesian product and whose operations are

$$(s_1, \ldots, s_n) + (s'_1, \ldots, s'_n) = (s_1 + s'_1, \ldots, s_n + s'_n),$$
$$r(s_1, \ldots, s_n) = (rs_1, \ldots, rs_n).$$

Let M be a left R-module, and let $S_1, \ldots, S_n$ be submodules of M. Define M to be their (*internal*) *direct sum*

$$M = S_1 \oplus \cdots \oplus S_n$$

if each $m \in M$ has a unique expression of the form $m = s_1 + \cdots + s_n$, where $s_i \in S_i$ for all $i = 1, \ldots, n$. We also write the internal direct sum as $\bigoplus_{i=1}^{n} S_i$.

The reader can prove that both external and internal versions, when the latter is defined, are isomorphic: $S_1 \boxplus \cdots \boxplus S_n \cong S_1 \oplus \cdots \oplus S_n$. We shall no longer use the adjectives *external* and *internal*, and we shall no longer use the $\boxplus$ notation.

If $S_1, \ldots, S_n$ are submodules of a left R-module M, let

$$S_1 + \cdots + S_n$$

be the submodule generated by the S_i; that is, $S_1 + \cdots + S_n$ is the set of all elements $m \in M$ having a (not necessarily unique) expression of the form $m = s_1 + \cdots + s_n$ with $s_i \in S_i$ for all i. When is $S_1 + \cdots + S_n$ equal to their direct sum? A common mistake is to say that it suffices to assume that $S_i \cap S_j = \{0\}$ for all $i \neq j$, but the next example shows that this is not enough.

Example 2.25. Let V be a two-dimensional vector space over a field K, and let x, y be a basis. The vector space V is a k-module, and it is a direct sum: $V = \langle x \rangle \oplus \langle y \rangle$, where $\langle x \rangle$ is the one-dimensional subspace spanned by x. Now

$$\langle x + y \rangle \cap \langle x \rangle = \{0\} = \langle x + y \rangle \cap \langle y \rangle,$$

but we do not have $V = \langle x + y \rangle \oplus \langle x \rangle \oplus \langle y \rangle$, because 0 has two expressions:

$$0 = 0 + 0 + 0 \quad \text{and} \quad 0 = (x + y) - x - y. \quad \blacktriangleleft$$

Proposition 2.26. *Let* $M = S_1 + \cdots + S_n$, *where the* S_i *are submodules. Then* $M = S_1 \oplus \cdots \oplus S_n$ *if and only if, for each* i,

$$S_i \cap \langle S_1 + \cdots + \widehat{S_i} + \cdots + S_n \rangle = \{0\},$$

where $\widehat{S_i}$ *means that the term* S_i *is omitted from the sum.*

Proof. If $M = S_1 \oplus \cdots \oplus S_n$ and $x \in S_i \cap (S_1 + \cdots + \widehat{S_i} + \cdots + S_n)$, then $x = s_i \in S_i$ and $s_i = \sum_{j \neq i} s_j$, where $s_j \in S_j$. Unless all the $s_j = 0$, the element 0 has two distinct expressions: $0 = -s_i + \sum_{j \neq i} s_j$ and $0 = 0 + 0 + \cdots + 0$. Therefore, all $s_j = 0$ and $x = s_i = 0$.

We prove the converse by induction on $n \geq 2$. The base step is Corollary 2.22(i). For the inductive step, define $T = S_1 + \cdots + S_n$, so that $M = T \oplus S_{n+1}$. If $a \in M$, then a has a unique expression of the form $a = t + s_{n+1}$, where $t \in T$ and $s_{n+1} \in S_{n+1}$ (by the base step). But the inductive hypothesis says that t has a unique expression of the form $t = s_1 + \cdots + s_n$, where $s_i \in S_i$ for all $i \leq n$, as desired. $\bullet$

Example 2.27. If V is an n-dimensional vector space over a field k and $v_1, \ldots, v_n$ is a basis, then $V = \langle v_1 \rangle \oplus \cdots \oplus \langle v_n \rangle$, for each vector $v \in V$ has a unique expression $v = \sum \alpha_i v_i$ with $\alpha_i v_i \in \langle v_i \rangle$. Thus, V is a direct sum of n one-dimensional vector spaces if and only if $\dim(V) = n$. $\blacktriangleleft$

Direct sums can be described in terms of exact sequences.

Definition. A short exact sequence

$$0 \to A \xrightarrow{i} B \xrightarrow{p} C \to 0$$

is *split* if there exists a map $j \colon C \to B$ with $pj = 1_C$.

Note that jp is a retraction $B \to \operatorname{im} j$.

Proposition 2.28. *If an exact sequence*

$$0 \to A \xrightarrow{i} B \xrightarrow{p} C \to 0$$

is split, then $B \cong A \oplus C$.

Remark.   See Exercise 2.8 on page 65.   ◄

Proof.   We show that $B = \operatorname{im} i \oplus \operatorname{im} j$, where $j \colon C \to B$ satisfies $pj = 1_C$. If $b \in B$, then $pb \in C$ and $b - jpb \in \ker p$, for $p(b - jpb) = pb - pj(pb) = 0$ because $pj = 1_C$. By exactness, there is $a \in A$ with $ia = b - jpb$. It follows that $B = \operatorname{im} i + \operatorname{im} j$. It remains to prove $\operatorname{im} i \cap \operatorname{im} j = \{0\}$. If $ia = x = jc$, then $px = pia = 0$, because $pi = 0$, whereas $px = pjc = c$, because $pj = 1_C$. Therefore, $x = jc = 0$, and so $B \cong A \oplus C$.   •

We shall see, in Example 2.29, that the converse of Proposition 2.28 is not true.

There are (at least) two ways to extend the notion of direct sum of modules from finitely many summands to infinitely many summands.

Definition.   Let R be a ring and let $(A_i)_{i \in I}$ be an indexed family of left R-modules. The ***direct product*** $\prod_{i \in I} A_i$ is the cartesian product [i.e., the set of all I-tuples (a_i) whose ith coordinate a_i lies in A_i for all i] with coordinate-wise addition and scalar multiplication:

$$(a_i) + (b_i) = (a_i + b_i) \quad \text{and} \quad r(a_i) = (ra_i),$$

where $r \in R$ and $a_i, b_i \in A_i$ for all i. In particular, if all A_i are equal, say, $A_i = A$ for all $i \in I$, then we may write A^I instead of $\prod_{i \in I} A_i$.

The ***direct sum***, denoted by $\bigoplus_{i \in I} A_i$, is the submodule of $\prod_{i \in I} A_i$ consisting of all (a_i) having only finitely many nonzero coordinates.

If $B = \prod_{i \in I} A_i$, then the jth ***projection*** (for $j \in I$) is the map $p_j \colon B \to A_j$ defined by $(a_i) \mapsto a_j$. The jth ***injection*** (for $j \in I$) is the map $a_i \mapsto (e_i) \in B$, where $e_i = 0$ if $i \neq j$ and $e_j = a_j$.

We can be more precise. An ***I-tuple*** is a function $\varphi \colon I \to \bigcup_i A_i$ with $\varphi(i) \in A_i$ for all $i \in I$. Thus, the direct product consists of all I-tuples, while the direct sum consists of all those I-tuples having finite ***support***, where $\operatorname{supp}(\varphi) = \{i \in I : \varphi(i) \neq \{0\}\}$. Another way to say that φ has finite support is to say that ***almost all*** the coordinates of (a_i) are zero; that is, only finitely many a_i are nonzero.

If $a_i \in A_i$, let $\mu_i a_i$ be the I-tuple in $\prod_i A_i$ whose ith coordinate is a_i and whose other coordinates are 0. Recall that two functions $f, g \colon X \to Y$ are equal if and only if $f(x) = g(x)$ for all $x \in X$; thus, two vectors are equal if and only if they have the same coordinates. It follows that each $m \in \bigoplus_{i \in I} A_i$ has a unique expression of the form $m = \sum_{i \in I} \mu_i a_i$, where $a_i \in A_i$ and almost all $a_i = 0$.

Note that if the index set I is finite, then $\prod_{i \in I} A_i = \bigoplus_{i \in I} A_i$. On the other hand, when I is infinite and infinitely many $A_i \neq 0$, then the direct sum is a proper submodule of the direct product. An infinite direct product is almost never isomorphic to an infinite direct sum. For example, if k is a field and $A_i = k$ for all $i \in \mathbb{N}$, then both $\prod A_i$ and $\bigoplus A_i$ are vector spaces over k; the product has uncountable dimension while the sum has countable dimension.

Example 2.29. There are exact sequences $0 \to S \to S \oplus T \to T \to 0$ that are not split.

Let $A = \langle a \rangle$ and $B = \langle b \rangle$ be cyclic groups of order 2 and 4, respectively. If $i: A \to B$ is defined by $i(a) = 2b$ and $p: B \to A$ is defined by $p(b) = a$, then $0 \to A \xrightarrow{i} B \xrightarrow{p} A \to 0$ is an exact sequence that is not split (because $\mathbb{I}_4 \not\cong \mathbb{I}_2 \oplus \mathbb{I}_2$). By Exercise 2.5 on page 65, for any abelian group M, there is an exact sequence

$$0 \to A \xrightarrow{i'} B \oplus M \xrightarrow{p'} A \oplus M \to 0, \qquad (1)$$

where $i'(a) = (ia, 0)$ and $p'(b, m) = (pb, m)$, and this sequence does not split either. If we choose M to be the direct sum of infinitely many copies of $A \oplus B$, then $A \oplus M \cong M \cong B \oplus M$. The middle group in extension (1) is now isomorphic to the direct sum of the two outer groups. ◀

The next theorem says that covariant Hom functors preserve direct products; the following theorem says that contravariant Hom functors convert direct sums to direct products.

Theorem 2.30. *Let R be a ring, let A be a left R-module, and let $(B_i)_{i \in I}$ be a family of left R-modules.*

(i) *There is a $Z(R)$-isomorphism*

$$\varphi: \operatorname{Hom}_R\left(A, \prod_{i \in I} B_i\right) \to \prod_{i \in I} \operatorname{Hom}_R(A, B_i)$$

with $\varphi: f \mapsto (p_i f)$, where the p_i are the projections of the direct product $\prod_{i \in I} B_i$. If R is commutative, then φ is an R-isomorphism.

(ii) *The isomorphism φ is natural: if $(C_j)_{j \in J}$ is a family of left R-modules and, for each $i \in I$, there exist $j \in J$ and an R-map $\sigma_{ij}: B_i \to C_j$, then there is a commutative diagram*

$$
\begin{array}{ccc}
\operatorname{Hom}_R\left(A, \prod_{i \in I} B_i\right) & \xrightarrow{\sigma_*} & \operatorname{Hom}_R\left(A, \prod_{j \in J} C_j\right) \\
\varphi \downarrow & & \downarrow \varphi \\
\prod_{i \in I} \operatorname{Hom}_R(A, B_i) & \xrightarrow{\tilde{\sigma}} & \prod_{j \in J} \operatorname{Hom}_R(A, C_j),
\end{array}
$$

where $\sigma: \prod_i B_i \to \prod_j C_j$ is given by $(b_i) \mapsto (\sigma_{ij} b_i)$, and $\tilde{\sigma}: (g_i) \mapsto (\sigma_{ij} g_i)$.

Proof.

(i) To see that φ is surjective, let $(f_i) \in \prod \operatorname{Hom}_R(A, B_i)$; then $f_i : A \to B_i$ for every i.

Define an R-map $\theta : A \to \prod B_i$ by $\theta(a) = (f_i(a))$; it is easy to see that $\varphi(\theta) = (p_i\theta) = (f_i)$, and so φ is surjective.

To see that φ is injective, let $f, f' \in \operatorname{Hom}_R(A, \prod B_i)$. Now if $a \in A$, then $f(a) = (b_i)$ and $(p_i f)(a) = b_i$; similarly, $f'(a) = (b_i')$ and $(p_i f')(a) = b_i'$. If $\varphi(f) = \varphi(f')$, then $(p_i f) = (p_i f')$, and so $p_i f = p_i f'$ for all i. Thus, for all i and all $a \in A$, we have $b_i = b_i'$; that is, $f(a) = f'(a)$, and $f = f'$.

To see that φ is a $Z(R)$-map, note, for each i and each $r \in Z(R)$, that $p_i rf = rp_i f$; therefore,

$$\varphi : rf \mapsto (p_i rf) = (rp_i f) = r(p_i f) = r\varphi(f).$$

(ii) Going clockwise, $f \mapsto \sigma_*(f) = \sigma f \mapsto (q_j \sigma f)$, where q_j is the jth projection $\prod_j C_j \to C_j$; going counterclockwise, $f \mapsto (p_i f) \mapsto (q_j \tilde{\sigma} f)$. To see that these are equal, evaluate each at $a \in A$. Note that if $fa = (b_i) \in \prod_i B_i$, then $p_i fa = b_i$. Hence, $q_j \sigma fa = \sigma_{ij} fa = \sigma_{ij} b_i$. On the other hand, $[q_j \tilde{\sigma} f]a = q_j(\sigma_{ij} f)a = q_j(\sigma_{ij} b_i) = \sigma_{ij} b_i$. $\bullet$

Theorem 2.31. *Let R be a ring, let B be a left R-module, and let $(A_i)_{i \in I}$ be a family of left R-modules.*

(i) *There is a $Z(R)$-isomorphism*

$$\psi : \operatorname{Hom}_R\left(\bigoplus_{i \in I} A_i, B\right) \to \prod_{i \in I} \operatorname{Hom}_R(A_i, B),$$

with $\psi : f \mapsto (f\alpha_i)$, where the α_i are the injections into the direct sum $\bigoplus_{i \in I} A_i$. If R is commutative, then φ is an R-isomorphism.

(ii) *The isomorphism ψ is natural: if $(D_j)_{j \in J}$ is a family of left R-modules and, for each $j \in J$, there exist $i \in I$ and an R-map $\tau_{ji} : D_j \to A_i$, then there is a commutative diagram*

$$\begin{array}{ccc} \operatorname{Hom}_R(\bigoplus_{j \in J} D_j, B) & \xleftarrow{\ \tau^* \ } & \operatorname{Hom}_R(\bigoplus_{i \in I} A_i, B) \\ \psi \downarrow & & \downarrow \psi \\ \prod_{j \in J} \operatorname{Hom}_R(D_j, B) & \xleftarrow{\ \tilde{\tau} \ } & \prod_{i \in I} \operatorname{Hom}_R(A_i, B), \end{array}$$

where $\tau : \prod_j D_J \to \prod_i A_i$ *is given by* $(d_j) \mapsto (\tau_{ji} d_j)$, *and* $\widehat{\tau} : (h_j) \mapsto (h_j \tau_{ji})$.

Proof.   This proof, similar to that of Theorem 2.30, is left to the reader.   •

There are examples showing that there are no other isomorphisms involving Hom, $\bigoplus$, and $\prod$ (see Exercise 2.25 on page 68).

Here is a new proof of Corollary 2.22(ii).

Corollary 2.32.   *If A, A', B, and B' are left R-modules, then there are $Z(R)$-isomorphisms*

$$\operatorname{Hom}_R(A, B \oplus B') \cong \operatorname{Hom}_R(A, B) \oplus \operatorname{Hom}_R(A, B')$$

and

$$\operatorname{Hom}_R(A \oplus A', B) \cong \operatorname{Hom}_R(A, B) \oplus \operatorname{Hom}_R(A', B).$$

If R is commutative, these are R-isomorphisms.

Proof.   When the index set is finite, the direct sum and the direct product of modules are equal.   •

The simplest modules are *free modules*.

Definition.   A left R-module F is a *free left R-module* if F is isomorphic to a direct sum of copies of R: that is, there is a (possibly infinite) index set B with $F = \bigoplus_{b \in B} R_b$, where $R_b = \langle b \rangle \cong R$ for all $b \in B$. We call B a *basis* of F.

By the definition of direct sum, each $m \in F$ has a unique expression of the form

$$m = \sum_{b \in B} r_b b,$$

where $r_b \in R$ and almost all $r_b = 0$. It follows that $F = \langle B \rangle$.

A free $\mathbb{Z}$-module is called a *free abelian group*. Every ring R, when considered as a left module over itself, is itself a free R-module.

In the first chapter, we defined the singular chain groups $S_n(X)$ of a topological space X as the free abelian group with basis all singular n-simplexes in X. We now prove that such huge abelian groups exist.

Proposition 2.33.   *Let R be a ring. Given any set B, there exists a free left R-module F with basis B.*

Proof. The set of all functions $R^B = \{\varphi : B \to R\}$ is a left R-module where, for all $b \in B$ and $r \in R$, we define $\varphi + \psi : b \mapsto \varphi(b) + \psi(b)$ and $r\varphi : b \mapsto r[\varphi(b)]$. In vector notation, a function φ is written as the B-tuple whose bth coordinate is $\varphi(b)$. In particular, the function μ_b, defined by

$$\mu_b(b') = \begin{cases} 1 & \text{if } b' = b, \\ 0 & \text{if } b' \neq b, \end{cases}$$

is the B-tuple whose bth coordinate is 1 and whose other coordinates are all 0. If we denote μ_b by b, then R^B is the direct product $\prod_{b \in B} \langle b \rangle$. Now $R \cong \langle b \rangle$ via the map $r \mapsto r\mu_b$, and so the submodule F of R^B generated by B is a direct sum of copies of R; that is, $F = \bigoplus \langle b \rangle$ is a free left R-module with basis B. •

A basis of a free module has a strong resemblance to a basis of a vector space. If k is a field, then every vector space V over k has a basis, in the sense of Linear Algebra. It is easy to see that the two notions of basis coincide in this case (see Example 2.27); moreover, a vector space V is a finitely generated free k-module if and only if it is finite-dimensional. The theorem of Linear Algebra that linear transformations are described by matrices can be rephrased to say that if $v_1, \ldots, v_n$ is a basis of a vector space V and if $w_1, \ldots, w_n$ is a list (possibly with repetitions) of vectors in a vector space W, then there exists a unique linear transformation $T : V \to W$ with $T(v_i) = w_i$ for all i. Since T has the formula

$$T(a_1 v_1 + \cdots + a_n v_n) = a_1 w_1 + \cdots + a_n w_n,$$

one says that T arises by *extending by linearity*. This idea can be used for free R-modules.

Proposition 2.34 (Extending by Linearity). *Let R be a ring and let F be the free left R-module with basis X. If M is any left R-module and if $f : X \to M$ is any function, then there exists a unique R-map $\widetilde{f} : F \to M$ with $\widetilde{f}\mu = f$, where $\mu : X \to F$ is the inclusion; that is, $\widetilde{f}(x) = f(x)$ for all $x \in X$, so that $\widetilde{f}$ extends f.*

$$\begin{array}{ccc} F & & \\ \mu \uparrow & \searrow \widetilde{f} & \\ X & \xrightarrow{f} & M \end{array}$$

Proof. Every element $v \in F$ has a unique expression of the form

$$v = \sum_{x \in X} r_x x,$$

where $r_x \in R$ and almost all $r_x = 0$; it follows that there is a well-defined function $\tilde{f}: F \to M$ given by $\tilde{f}(v) = \sum_{x \in X} r_x f(x)$. Obviously, $\tilde{f}$ extends f. If $s \in R$, then $sv = \sum sr_x x$; if $v' = \sum r_x' x$, then $v + v' = \sum (r_x + r_x')x$. The formula for $\tilde{f}$ shows that it is an R-map. Finally, $\tilde{f}$ is the unique R-map extending f: since $F = \langle X \rangle$, Exercise 2.3 on page 64 shows that two R-maps agreeing on a generating set must be equal.   •

Arbitrary modules can be described in terms of free modules.

Theorem 2.35.   *Every left R-module M is a quotient of a free left R-module F. Moreover, M is finitely generated if and only if F can be chosen to be finitely generated.*

Proof.   Choose a generating set X of M, and let F be the free module with basis $\{u_x : x \in X\}$. By Proposition 2.34, there is an R-map $g: F \to M$ with $g(\acute{u}_x) = x$ for all $x \in X$. Now g is a surjection, for im g is a submodule of M containing X, and so $F / \ker g \cong M$.

If M is finitely generated, then there is a finite generating set X, and the free module F just constructed is finitely generated. The converse is obvious, for any image of a finitely generated module is itself finitely generated   •

If F is a free left R-module, then we would like to know that there is an analog of dimension; that is, the number of elements in a basis of F is an invariant. The next proposition shows that this is so when R is commutative. However, there are noncommutative rings R for which $R \cong R \oplus R$ as left R-modules; that is, R is a free left R-module having bases of different sizes.

Example 2.36.   Let V be an infinite-dimensional vector space over a field k, so that there is a k-isomorphism $\theta: V \to V \oplus V$. Define projections $p, q: V \oplus V \to V$ by $p: (v, w) \mapsto v$ and $q: (v, w) \mapsto w$. Let $R = \text{End}_k(V)$ be the ring of all k-linear transformations $f: V \to V$ (with composition as multiplication). Now apply $\text{Hom}_k(V, \square)$ to obtain a k-isomorphism

$$\theta_*: \text{Hom}_k(V, V) \to \text{Hom}_k(V, V \oplus V),$$

namely, $\theta_*: g \mapsto \theta_*(g) = \theta g$ for $g \in \text{Hom}_k(V, V)$. Let

$$\psi: \text{Hom}_k(V, V \oplus V) \to \text{Hom}_k(V, V) \oplus \text{Hom}_k(V, V)$$

be given by $f \mapsto (pf, qf)$ [ψ is the isomorphism of Corollary 2.22(ii) with $T = \text{Hom}_k(V, \square)$]. Consider the k-isomorphism $\psi\theta_*: R \to R \oplus R$. As usual, R is a right R-module via right multiplication, and $R \oplus R$ is a right R-module via $(f, g)h = (fh, gh)$ for $f, g, h \in R$. We show that $\psi\theta_*$ is an

R-isomorphism. If $f, h \in R$, then

$$
\begin{aligned}
(\psi\theta_*)(fh) &= \psi(\theta_*[fh]) \\
&= \psi(\theta fh) \\
&= (p\theta fh, q\theta fh) \\
&= (p\theta f, q\theta f)h \\
&= (\psi\theta_*)(f)h.
\end{aligned}
$$

Therefore, $\psi\theta_*$ is an R-isomorphism, $R \cong R \oplus R$ as right R-modules. (Of course, replacing R by R^{op} gives a similar example for left modules; this amounts to writing composites fg as gf.) ◀

Proposition 2.37. *Let R be a nonzero commutative ring.*

(i) *Any two bases of a free R-module F have the same cardinality.*

(ii) *Free R-modules F and F' are isomorphic if and only if there are bases of each having the same cardinality.*

(iii) *If m and n are natural numbers, then $R^m \cong R^n$ if and only if $m = n$.*

Proof.

(i) Choose a maximal ideal I in R (which exists, by Zorn's lemma). If X is a basis of the free R-module F, then Exercise 2.12 on page 66 shows that the set of cosets $\{v + IF : v \in X\}$ is a basis of the vector space F/IF over the field R/I. If Y is another basis of F, then the same argument gives $\{u + IF : u \in Y\}$ a basis of F/IF. But any two bases of a vector space have the same size (which is the dimension of the space), and so X and Y have the same cardinality.

(ii) Let X be a basis of F, let X' be a basis of F', and let $\gamma : X \to X'$ be a bijection (which exists by hypothesis). Composing γ with the inclusion $X' \to F'$, we may assume that $\gamma : X \to F'$. By Proposition 2.34, there is a unique R-map $\varphi : F \to F'$ extending γ. Similarly, we may regard $\gamma^{-1} : X' \to X$ as a function $X' \to F$, and there is a unique $\psi : F' \to F$ extending γ^{-1}. Finally, both $\psi\varphi$ and 1_F extend 1_X, so that $\psi\varphi = 1_F$. Similarly, $\psi\varphi = 1_{F'}$, and so $\varphi : F \to F'$ is an isomorphism.

Conversely, suppose that $\varphi : F \to F'$ is an isomorphism. If $\{v_i : i \in I\}$ is a basis of F, then it is easy to see that $\{\varphi(v_i) : i \in I\}$ is a basis of F'. But any two bases of the free module F' have the same cardinality, by part (i). Hence, bases of F and of F' have the same cardinality.

(iii) If $m = n$, then $R^m \cong R^n$ is obvious. Conversely, if $R^m \cong R^n$, part (ii) applies. •

Definition. A ring R has **IBN** (*invariant basis number*) if $R^m \cong R^n$ as left R-modules implies $m = n$. If R has IBN, then the number of elements in a basis of a free left R-module F is called the **rank** of F and is denoted by rank(F).

If a ring R has IBN, then it is also true that $R^m \cong R^n$ as *right* R-modules implies $m = n$; see Exercise 2.37 on page 97. If R has IBN and F is a finitely generated free left R-module, then every two bases of F have the same number of elements, for if $x_1, \ldots, x_n$ is a basis of F, then $F \cong R^n$. Thus, rank(F) is well-defined for rings with IBN. Free modules having an infinite basis are considered in Exercise 2.26 on page 69.

Proposition 2.37(i) shows that every nonzero commutative ring R has IBN. Division rings have IBN: if Δ is a division ring and F is a finitely generated free left Δ-module, then any two bases of F have the same number of elements (see Rotman, *Advanced Modern Algebra*, p. 537). The proof above that commutative rings have IBN generalizes to any noncommutative ring R having a two-sided ideal I for which R/I is a division ring [every *local ring* is such a ring (see Proposition 4.56(iii)]. We shall see, in Theorem 3.24, that all left noetherian rings also have IBN. On the other hand, Example 2.36 shows that if $R = \mathrm{End}_k(V)$, where V is an infinite-dimensional vector space over a field k, then R does not have IBN.

Corollary 2.22(ii) says that the Hom functors $_R\mathbf{Mod} \to \mathbf{Ab}$ preserve direct sums of modules: $\mathrm{Hom}_R(X, A \oplus C) \cong \mathrm{Hom}_R(X, A) \oplus \mathrm{Hom}_R(X, C)$. If we regard such a direct sum as a split short exact sequence, then we may rephrase the corollary by saying that if $0 \to A \xrightarrow{i} B \xrightarrow{p} C \to 0$ is a split short exact sequence, then so is

$$0 \to \mathrm{Hom}_R(X, A) \xrightarrow{i_*} \mathrm{Hom}_R(X, B) \xrightarrow{p_*} \mathrm{Hom}_R(X, C) \to 0.$$

This leads us to a more general question: if $0 \to A \xrightarrow{i} B \xrightarrow{p} C \to 0$ is any short exact sequence, not necessarily split, is

$$0 \to \mathrm{Hom}_R(X, A) \xrightarrow{i_*} \mathrm{Hom}_R(X, B) \xrightarrow{p_*} \mathrm{Hom}_R(X, C) \to 0$$

also an exact sequence? Here is the answer (there is no misprint in the statement of the theorem: "$\to 0$" should not appear at the end of the sequences, and we shall discuss this point after the proof).

Theorem 2.38 (Left Exactness). *If* $0 \to A \xrightarrow{i} B \xrightarrow{p} C$ *is an exact sequence of left R-modules, and if X is a left R-module, then there is an exact sequence of $Z(R)$-modules*

$$0 \to \mathrm{Hom}_R(X, A) \xrightarrow{i_*} \mathrm{Hom}_R(X, B) \xrightarrow{p_*} \mathrm{Hom}_R(X, C).$$

If R is commutative, then the latter sequence is an exact sequence of R-modules and R-maps.

Proof. That $\text{Hom}_R(X, A)$ is a $Z(R)$-module follows from Proposition 2.4.

(i) $\ker i_* = \{0\}$.

If $f \in \ker i_*$, then $f: X \to A$ and $i_*(f) = 0$; that is, $if(x) = 0$ for all $x \in X$. Since i is injective, $f(x) = 0$ for all $x \in X$, and so $f = 0$.

(ii) $\text{im}\, i_* \subseteq \ker p_*$.

If $g \in \text{im}\, i_*$, then $g: X \to B$ and there is some $f: X \to A$ with $g = i_*(f) = if$. But $p_*(g) = pg = pif = 0$ because exactness of the original sequence, namely, $\text{im}\, i = \ker p$, implies $pi = 0$.

(iii) $\ker p_* \subseteq \text{im}\, i_*$.

If $g \in \ker p_*$, then $g: X \to B$ and $p_*(g) = pg = 0$. Hence, $pg(x) = 0$ for all $x \in X$, so that $g(x) \in \ker p = \text{im}\, i$. Thus, $g(x) = i(a)$ for some $a \in A$; since i is injective, this element a is unique. Hence, the function $f: X \to A$, given by $f(x) = a$ if $g(x) = i(a)$, is well-defined. It is easy to check that $f \in \text{Hom}_R(X, A)$; that is, f is an R-homomorphism. Since $g(x + x') = g(x) + g(x') = i(a) + i(a') = i(a + a')$, we have $f(x + x') = a + a' = f(x) + f(x')$. A similar argument shows that $f(rx) = rf(x)$ for all $r \in R$. But, $i_*(f) = if$ and $if(x) = i(a) = g(x)$ for all $x \in X$; that is, $i_*(f) = g$, and so $g \in \text{im}\, i_*$. •

Example 2.39. Even if the map $p: B \to C$ in the original exact sequence is surjective, the functored sequence need not end with "$\to 0$"; that is, the induced map $p_*: \text{Hom}_R(X, B) \to \text{Hom}_R(X, C)$ may fail to be surjective.

The abelian group $\mathbb{Q}/\mathbb{Z}$ consists of cosets $q + \mathbb{Z}$ for $q \in \mathbb{Q}$, and its element $x = \frac{1}{2} + \mathbb{Z}$ has order 2 ($x \neq 0$ and $2x = 0$). It follows that $\text{Hom}_{\mathbb{Z}}(\mathbb{I}_2, \mathbb{Q}/\mathbb{Z}) \neq \{0\}$, for it contains the nonzero homomorphism $[1] \mapsto \frac{1}{2} + \mathbb{Z}$.

Apply the functor $\text{Hom}_{\mathbb{Z}}(\mathbb{I}_2, \square)$ to

$$0 \to \mathbb{Z} \xrightarrow{i} \mathbb{Q} \xrightarrow{p} \mathbb{Q}/\mathbb{Z} \to 0,$$

where i is the inclusion and p is the natural map. We have just seen that

$$\text{Hom}_{\mathbb{Z}}(\mathbb{I}_2, \mathbb{Q}/\mathbb{Z}) \neq \{0\}.$$

On the other hand, $\text{Hom}_{\mathbb{Z}}(\mathbb{I}_2, \mathbb{Q}) = \{0\}$ because $\mathbb{Q}$ has no (nonzero) elements of finite order. Therefore, the induced map $p_*: \text{Hom}_{\mathbb{Z}}(\mathbb{I}_2, \mathbb{Q}) \to \text{Hom}_{\mathbb{Z}}(\mathbb{I}_2, \mathbb{Q}/\mathbb{Z})$ cannot be surjective. ◄

Definition.   A covariant functor $T: {}_R\mathbf{Mod} \to \mathbf{Ab}$ is called *left exact* if exactness of

$$0 \to A \xrightarrow{i} B \xrightarrow{p} C$$

implies exactness of abelian groups

$$0 \to T(A) \xrightarrow{T(i)} T(B) \xrightarrow{T(p)} T(C).$$

Thus, Theorem 2.38 shows that the covariant functors $\mathrm{Hom}_R(X, \square)$ are left exact functors.

There is an analogous result for contravariant Hom functors.

Theorem 2.40 (Left Exactness).   *If*

$$A \xrightarrow{i} B \xrightarrow{p} C \to 0$$

is an exact sequence of left R-modules, and if Y is a left R-module, then there is an exact sequence of $Z(R)$-modules

$$0 \to \mathrm{Hom}_R(C, Y) \xrightarrow{p^*} \mathrm{Hom}_R(B, Y) \xrightarrow{i^*} \mathrm{Hom}_R(A, Y).$$

If R is commutative, then the latter sequence is an exact sequence of R-modules and R-maps.

Proof.   That $\mathrm{Hom}_R(A, Y)$ is a $Z(R)$-module follows from Proposition 2.5.

(i) $\ker p^* = \{0\}$.

If $h \in \ker p^*$, then $h: C \to Y$ and $0 = p^*(h) = hp$. Thus, $h(p(b)) = 0$ for all $b \in B$, so that $h(c) = 0$ for all $c \in \mathrm{im}\, p$. Since p is surjective, $\mathrm{im}\, p = C$, and $h = 0$.

(ii) $\mathrm{im}\, p^* \subseteq \ker i^*$.

If $g \in \mathrm{Hom}_R(C, Y)$, then $i^* p^*(g) = (pi)^*(g) = 0$, because exactness of the original sequence, namely, $\mathrm{im}\, i = \ker p$, implies $pi = 0$.

(iii) $\ker i^* \subseteq \mathrm{im}\, p^*$.

If $g \in \ker i^*$, then $g: B \to Y$ and $i^*(g) = gi = 0$. If $c \in C$, then $c = p(b)$ for some $b \in B$, because p is surjective. Define $f: C \to Y$ by $f(c) = g(b)$ if $c = p(b)$. Note that f is well-defined: if $p(b) = p(b')$, then $b - b' \in \ker p = \mathrm{im}\, i$, so that $b - b' = i(a)$ for some $a \in A$. Hence, $g(b) - g(b') = g(b - b') = gi(a) = 0$, because $gi = 0$. The reader may check that f is an R-map. Finally,

$$p^*(f) = fp = g,$$

because if $c = p(b)$, then $g(b) = f(c) = f(p(b))$. Therefore, $g \in \mathrm{im}\, p^*$.   •

Example 2.41.   Even if the map $i: A \to B$ in the original exact sequence is assumed to be injective, the functored sequence need not end with "$\to 0$"; that is, the induced map $i^*: \operatorname{Hom}_R(B, Y) \to \operatorname{Hom}_R(A, Y)$ may fail to be surjective.

We claim that $\operatorname{Hom}_{\mathbb{Z}}(\mathbb{Q}, \mathbb{Z}) = \{0\}$. Let $f: \mathbb{Q} \to \mathbb{Z}$, and let $f(a/b) = m \in \mathbb{Z}$. For all $n > 0$,

$$nf(a/nb) = f(na/nb) = f(a/b) = m.$$

Thus, m is divisible by every positive integer n [for $f(a/nb) \in \mathbb{Z}$], and this forces $m = 0$. Therefore, $f = 0$.

If we apply the functor $\operatorname{Hom}_{\mathbb{Z}}(\square, \mathbb{Z})$ to the short exact sequence

$$0 \to \mathbb{Z} \xrightarrow{i} \mathbb{Q} \xrightarrow{p} \mathbb{Q}/\mathbb{Z} \to 0,$$

where i is the inclusion and p is the natural map, then the induced map

$$i^*: \operatorname{Hom}_{\mathbb{Z}}(\mathbb{Q}, \mathbb{Z}) \to \operatorname{Hom}_{\mathbb{Z}}(\mathbb{Z}, \mathbb{Z})$$

cannot be surjective, for $\operatorname{Hom}_{\mathbb{Z}}(\mathbb{Q}, \mathbb{Z}) = \{0\}$ while $\operatorname{Hom}_{\mathbb{Z}}(\mathbb{Z}, \mathbb{Z}) \neq \{0\}$ because it contains $1_{\mathbb{Z}}$. ◄

Definition.   A contravariant functor $T: {}_R\mathbf{Mod} \to \mathbf{Ab}$ is called *left exact* if exactness of

$$A \xrightarrow{i} B \xrightarrow{p} C \to 0$$

implies exactness of

$$0 \to T(C) \xrightarrow{T(p)} T(B) \xrightarrow{T(i)} T(A).$$

Thus, Theorem 2.40 shows that the contravariant functors $\operatorname{Hom}_R(\square, Y)$ are left exact functors.[4]

There is a converse of Theorem 2.40 (a similar statement for covariant Hom functors is true but not very interesting; see Exercise 2.13 on page 66).

Proposition 2.42.   *Let $i: B' \to B$ and $p: B \to B''$ be R-maps, where R is a ring. If, for every left R-module M,*

$$0 \to \operatorname{Hom}_R(B'', M) \xrightarrow{p^*} \operatorname{Hom}_R(B, M) \xrightarrow{i^*} \operatorname{Hom}_R(B', M)$$

is an exact sequence of abelian groups, then

$$B' \xrightarrow{i} B \xrightarrow{p} B'' \to 0$$

is an exact sequence of left R-modules.

---

[4]These functors are called *left exact* because the functored sequence has "$0 \to$" on the left-hand side.

Proof.

(i) p is surjective.

Let $M = B'' / \operatorname{im} p$ and let $f : B'' \to B'' / \operatorname{im} p$ be the natural map, so that $f \in \operatorname{Hom}(B'', M)$. Then $p^*(f) = fp = 0$, so that $f = 0$, because p^* is injective. Therefore, $B'' / \operatorname{im} p = 0$, and p is surjective.

(ii) $\operatorname{im} i \subseteq \ker p$.

Since $i^* p^* = 0$, we have $0 = (pi)^*$. Hence, if $M = B''$ and $g = 1_{B''}$, so that $g \in \operatorname{Hom}(B'', M)$, then $0 = (pi)^* g = gpi = pi$, and so $\operatorname{im} i \subseteq \ker p$.

(iii) $\ker p \subseteq \operatorname{im} i$.

Now choose $M = B / \operatorname{im} i$ and let $h : B \to M$ be the natural map, so that $h \in \operatorname{Hom}(B, M)$. Clearly, $i^* h = hi = 0$, so that exactness of the Hom sequence gives an element $h' \in \operatorname{Hom}_R(B'', M)$ with $p^*(h') = h'p = h$. We have $\operatorname{im} i \subseteq \ker p$, by part (ii); hence, if $\operatorname{im} i \neq \ker p$, there is an element $b \in B$ with $b \notin \operatorname{im} i$ and $b \in \ker p$. Thus, $hb \neq 0$ and $pb = 0$, which gives the contradiction $hb = h'pb = 0$. •

The single condition that $i^* : \operatorname{Hom}_R(B, M) \to \operatorname{Hom}_R(B', M)$ be surjective is much stronger than the hypotheses of Proposition 2.42 (see Exercise 2.20 on page 68).

Exercises

Unless we say otherwise, all modules in these exercises are left R-modules.

2.1 Let R and S be rings, and let $\varphi : R \to S$ be a ring homomorphism. If M is a left S-module, prove that M is also a left R-module if we define
$$rm = \varphi(r)m,$$
for all $r \in R$ and $m \in M$.

2.2 Give an example of a left R-module $M = S \oplus T$ having a submodule N such that $N \neq (N \cap S) \oplus (N \cap T)$.

*2.3 Let $f, g : M \to N$ be R-maps between left R-modules. If $M = \langle X \rangle$ and $f|X = g|X$, prove that $f = g$.

*2.4 Let $(M_i)_{i \in I}$ be a (possibly infinite) family of left R-modules and, for each i, let N_i be a submodule of M_i. Prove that
$$\left(\bigoplus_i M_i \right) \Big/ \left(\bigoplus_i N_i \right) \cong \bigoplus_i (M_i / N_i).$$

*2.5 Let $0 \to A \to B \to C \to 0$ be a short exact sequence of left R-modules. If M is any left R-module, prove that there are exact sequences

$$0 \to A \oplus M \to B \oplus M \to C \to 0$$

and

$$0 \to A \to B \oplus M \to C \oplus M \to 0.$$

*2.6   (i)   Let $\to A_{n+1} \xrightarrow{d_{n+1}} A_n \xrightarrow{d_n} A_{n-1} \to$ be an exact sequence, and let $\operatorname{im} d_{n+1} = K_n = \ker d_n$ for all n. Prove that

$$0 \to K_n \xrightarrow{i_n} A_n \xrightarrow{d'_n} K_{n-1} \to 0$$

is an exact sequence for all n, where i_n is the inclusion and d'_n is obtained from d_n by changing its target. We say that the original sequence has been *factored* into these short exact sequences.

  (ii)   Let

$$\to A_1 \xrightarrow{f_1} A_0 \xrightarrow{f_0} K \to 0$$

and

$$0 \to K \xrightarrow{g_0} B_0 \xrightarrow{g_1} B_1 \to$$

be exact sequences. Prove that

$$\to A_1 \xrightarrow{f_1} A_0 \xrightarrow{g_0 f_0} B_0 \xrightarrow{g_1} B_1 \to$$

is an exact sequence. We say that the original two sequences have been *spliced* to form the new exact sequence.

*2.7 Use left exactness of Hom to prove that if G is an abelian group, then $\operatorname{Hom}_{\mathbb{Z}}(\mathbb{I}_n, G) \cong G[n]$, where $G[n] = \{g \in G : ng = 0\}$.

*2.8   (i)   Prove that a short exact sequence in $_R\mathbf{Mod}$,

$$0 \to A \xrightarrow{i} B \xrightarrow{p} C \to 0,$$

splits if and only if there exists $q \colon B \to A$ with $qi = 1_A$. (Note that q is a retraction $B \to \operatorname{im} i$.)

  (ii)   A sequence $A \xrightarrow{i} B \xrightarrow{p} C$ in **Groups** is *exact* if $\operatorname{im} i = \ker p$; an exact sequence

$$1 \to A \xrightarrow{i} B \xrightarrow{p} C \to 1$$

in **Groups** is *split* if there is a homomorphism $j \colon C \to B$ with $pj = 1_C$. Prove that $1 \to A_3 \to S_3 \to \mathbb{I}_2 \to 1$ is a split exact sequence. In contrast to part (i), show, in a split exact sequence in **Groups**, that there may not be a homomorphism $q \colon B \to A$ with $qi = 1_A$.

*2.9 (i) Let $v_1, \ldots, v_n$ be a basis of a vector space V over a field k. Let $v_i^*: V \to k$ be the evaluation $V^* \to k$ defined by $v_i^* = (\Box, v_i)$ (see Example 1.16). Prove that $v_1^*, \ldots, v_n^*$ is a basis of V^* (it is called the **dual basis** of $v_1, \ldots, v_n$).

Hint. Use Corollary 2.22(ii) and Example 2.27.

(ii) Let $f: V \to V$ be a linear transformation, and let A be the matrix of f with respect to a basis $v_1, \ldots, v_n$ of V; that is, the ith column of A consists of the coordinates of $f(v_i)$ with respect to the given basis $v_1, \ldots, v_n$. Prove that the matrix of the induced map $f^*: V^* \to V^*$ with respect to the dual basis is the transpose A^T of A.

*2.10 If X is a subset of a left R-module M, prove that $\langle X \rangle$, the submodule of M generated by X, is equal to $\bigcap S$, where the intersection ranges over all those submodules S of M that contain X.

*2.11 Prove that if $f: M \to N$ is an R-map and K is a submodule of a left R-module M with $K \subseteq \ker f$, then f induces an R-map $\widehat{f}: M/K \to N$ by $\widehat{f}: m + K \mapsto f(m)$.

*2.12 (i) Let R be a commutative ring and let J be an ideal in R. Recall Example 2.8(iv): if M is an R-module, then JM is a submodule of M. Prove that M/JM is an R/J-module if we define scalar multiplication:

$$(r + J)(m + JM) = rm + JM.$$

Conclude that if $JM = \{0\}$, then M itself is an R/J-module. In particular, if J is a maximal ideal in R and $JM = \{0\}$, then M is a vector space over R/J.

(ii) Let I be a maximal ideal in a commutative ring R. If X is a basis of a free R-module F, prove that F/IF is a vector space over R/I and that $\{$cosets $x + IF : x \in X\}$ is a basis.

*2.13 Let M be a left R-module.

(i) Prove that the map $\varphi_M: \operatorname{Hom}_R(R, M) \to M$, given by $\varphi_M: f \mapsto f(1)$, is an R-isomorphism.

Hint. Make the abelian group $\operatorname{Hom}_R(R, M)$ into a left R-module by defining rf (for $f: R \to M$ and $r \in R$) by $rf: s \mapsto f(sr)$ for all $s \in R$.

(ii) If $g: M \to N$, prove that the following diagram commutes:

$$
\begin{array}{ccc}
\operatorname{Hom}_R(R, M) & \xrightarrow{\;\varphi_M\;} & M \\
{\scriptstyle g_*}\downarrow & & \downarrow{\scriptstyle g} \\
\operatorname{Hom}_R(R, N) & \xrightarrow[\;\varphi_N\;]{} & N.
\end{array}
$$

Conclude that $\varphi = (\varphi_M)_{M \in \mathrm{obj}(_R\mathbf{Mod})}$ is a natural isomorphism from $\mathrm{Hom}_R(R, \square)$ to the identity functor on $_R\mathbf{Mod}$. [Compare with Example 1.16(ii).]

2.14 Let $A \xrightarrow{f} B \xrightarrow{g} C$ be a sequence of module maps. Prove that $gf = 0$ if and only if $\mathrm{im}\, f \subseteq \ker g$. Give an example of such a sequence that is not exact.

*2.15   **(i)**   Prove that $f: M \to N$ is surjective if and only if $\mathrm{coker}\, f = \{0\}$.

       (ii) If $f: M \to N$ is a map, prove that there is an exact sequence

$$0 \to \ker f \to M \xrightarrow{f} N \to \mathrm{coker}\, f \to 0.$$

*2.16   **(i)**   If $0 \to M \to 0$ is an exact sequence, prove that $M = \{0\}$.

       (ii) If $A \xrightarrow{f} B \xrightarrow{g} C \xrightarrow{h} D$ is an exact sequence, prove that f is surjective if and only if h is injective.

      (iii) Let $A \xrightarrow{\alpha} B \xrightarrow{\beta} C \xrightarrow{\gamma} D \xrightarrow{\delta} E$ be exact. If α and δ are isomorphisms, prove that $C = \{0\}$.

*2.17 If $A \xrightarrow{f} B \xrightarrow{g} C \xrightarrow{h} D \xrightarrow{k} E$ is exact, prove that there is an exact sequence

$$0 \to \mathrm{coker}\, f \xrightarrow{\alpha} C \xrightarrow{\beta} \ker k \to 0,$$

where $\alpha: b + \mathrm{im}\, f \mapsto gb$ and $\beta: c \mapsto hc$.

*2.18 Let $0 \to A \xrightarrow{i} B \xrightarrow{p} C \to 0$ be a short exact sequence.

       (i)   Assume that $A = \langle X \rangle$ and $C = \langle Y \rangle$. For each $y \in Y$, choose $y' \in B$ with $p(y') = y$. Prove that

$$B = \langle i(X) \cup \{y': y \in Y\} \rangle.$$

      (ii) Prove that if both A and C are finitely generated, then B is finitely generated. More precisely, prove that if A can be generated by m elements and C can be generated by n elements, then B can be generated by $m + n$ elements.

*2.19 Let R be a ring, let A and B be left R-modules, and let $r \in Z(R)$.

       (i)   If $\mu_r: B \to B$ is multiplication by r, prove that the induced map $(\mu_r)_*: \mathrm{Hom}_R(A, B) \to \mathrm{Hom}_R(A, B)$ is also multiplication by r.

      (ii) If $m_r: A \to A$ is multiplication by r, prove that the induced map $(m_r)^*: \mathrm{Hom}_R(A, B) \to \mathrm{Hom}_R(A, B)$ is also multiplication by r.

*2.20 Suppose one assumes, in the hypothesis of Proposition 2.42, that the induced map $i^*\colon \operatorname{Hom}_R(B, M) \to \operatorname{Hom}_R(B', M)$ is surjective for every M. Prove that $0 \to B' \xrightarrow{i} B \xrightarrow{p} B'' \to 0$ is a split short exact sequence.

*2.21 If $T\colon \mathbf{Ab} \to \mathbf{Ab}$ is an additive functor, prove, for every abelian group G, that the function $\operatorname{End}(G) \to \operatorname{End}(TG)$, given by $f \mapsto Tf$, is a ring homomorphism.

*2.22 (i) Prove that $\operatorname{Hom}_{\mathbb{Z}}(\mathbb{Q}, C) = \{0\}$ for every cyclic group C.

(ii) Let R be a commutative ring. If M is an R-module such that $\operatorname{Hom}_R(M, R/I) = \{0\}$ for every nonzero ideal I, prove that $\operatorname{im} f \subseteq \bigcap I$ for every R-map $f\colon M \to R$, where the intersection is over all nonzero ideals I in R.

(iii) Let R be a domain and suppose that M is an R-module with $\operatorname{Hom}_R(M, R/I) = \{0\}$ for all nonzero ideals I in R. Prove that $\operatorname{Hom}_R(M, R) = \{0\}$.

Hint. Every $r \in \bigcap_{I \neq 0} I$ is nilpotent.

2.23 Generalize Proposition 2.26. Let $(S_i)_{i \in I}$ be a family of submodules of a left R-module M. If $M = \langle \bigcup_{i \in I} S_i \rangle$, then the following conditions are equivalent.

(i) $M = \bigoplus_{i \in I} S_i$.

(ii) Every $a \in M$ has a unique expression of the form $a = s_{i_1} + \cdots + s_{i_n}$, where $s_{i_j} \in S_{i_j}$.

(iii) $S_i \cap \langle \bigcup_{j \neq i} S_j \rangle = \{0\}$ for each $i \in I$.

*2.24 (i) Prove that any family of R-maps $(f_j\colon U_j \to V_j)_{j \in J}$ can be assembled into an R-map $\varphi\colon \bigoplus_j U_j \to \bigoplus_j V_j$, namely, $\varphi\colon (u_j) \mapsto (f_j(u_j))$.

(ii) Prove that φ is an injection if and only if each f_j is an injection.

*2.25 (i) If $Z_i \cong \mathbb{Z}$ for all i, prove that

$$\operatorname{Hom}_{\mathbb{Z}}\left(\prod_{i=1}^{\infty} Z_i, \mathbb{Z}\right) \not\cong \prod_{i=1}^{\infty} \operatorname{Hom}_{\mathbb{Z}}(Z_i, \mathbb{Z}).$$

Hint. A theorem of J. Łos and, independently, of E. C. Zeeman (see Fuchs, *Infinite Abelian Groups* II, Section 94) says that

$$\operatorname{Hom}_{\mathbb{Z}}\left(\prod_{i=1}^{\infty} Z_i, \mathbb{Z}\right) \cong \bigoplus_{i=1}^{\infty} \operatorname{Hom}_{\mathbb{Z}}(Z_i, \mathbb{Z}) \cong \bigoplus_{i=1}^{\infty} Z_i.$$

(ii) Let p be a prime and let B_n be a cyclic group of order p^n, where n is a positive integer. If $A = \bigoplus_{n=1}^{\infty} B_n$, prove that

$$\operatorname{Hom}_k\left(A, \bigoplus_{n=1}^{\infty} B_n\right) \not\cong \bigoplus_{n=1}^{\infty} \operatorname{Hom}_k(A, B_n).$$

Hint. Prove that $\operatorname{Hom}(A, A)$ has an element of infinite order, while every element in $\bigoplus_{n=1}^{\infty} \operatorname{Hom}_k(A, B_n)$ has finite order.

(iii) Prove that $\operatorname{Hom}_{\mathbb{Z}}(\prod_{n \geq 2} \mathbb{I}_n, \mathbb{Q}) \not\cong \prod_{n \geq 2} \operatorname{Hom}_{\mathbb{Z}}(\mathbb{I}_n, \mathbb{Q})$.

*2.26 Let R be a ring with IBN.

(i) If R^{∞} is a free left R-module having an infinite basis, prove that $R \oplus R^{\infty} \cong R^{\infty}$.

(ii) Prove that $R^{\infty} \not\cong R^n$ for any $n \in \mathbb{N}$.

(iii) If X is a set, denote the free left R-module $\bigoplus_{x \in X} Rx$ by $R^{(X)}$. Let X and Y be sets, and let $R^{(X)} \cong R^{(Y)}$. If X is infinite, prove that Y is infinite and that $|X| = |Y|$; that is, X and Y have the same cardinal.

Hint. Since X is a basis of $R^{(X)}$, each $u \in R^{(X)}$ has a unique expression $u = \sum_{x \in X} r_x x$; define

$$\operatorname{Supp}(u) = \{x \in X : r_x \neq 0\}.$$

Given a basis B of $R^{(X)}$ and a finite subset $W \subseteq X$, prove that there are only finitely many elements $b \in B$ with $\operatorname{Supp}(b) \subseteq W$. Conclude that $|B| = \operatorname{Fin}(X)$, where $\operatorname{Fin}(X)$ is the family of all the finite subsets of X. Finally, using the fact that $|\operatorname{Fin}(X)| = |X|$ when X is infinite, conclude that $R^{(X)} \cong R^{(Y)}$ implies $|X| = |Y|$.

2.2 Tensor Products

One of the most compelling reasons to introduce tensor products comes from Algebraic Topology. The homology groups of a space are interesting (for example, computing the homology groups of spheres enables us to prove the Jordan Curve Theorem), and the homology groups of the cartesian product $X \times Y$ of two topological spaces are computed (by the *Künneth formula*) in terms of the tensor product of the homology groups of the factors X and Y.

Here is a second important use of tensor products. We saw, in Example 2.2, that if k is a field, then every k-representation $\varphi \colon H \to \operatorname{Mat}_n(k)$ of a group H to $n \times n$ matrices makes the vector space k^n into a left kH-module;

conversely, every such module gives a representation of H. If H is a subgroup of a group G, can we obtain a k-representation of G from a k-representation of H; that is, can we construct a kG-module from a kH-module? Now kH is a subring of kG; can we "adjoin more scalars" to form a kG-module from the kH-module? Tensor products will give a very simple construction, *induced modules*, which does exactly this.

More generally, if S is a subring of a ring R and M is a left S-module, can we adjoin more scalars to form a left R-module M' that contains M? If a left S-module M is generated by a set X (so that each $m \in M$ has an expression of the form $m = \sum_i s_i x_i$ for $s_i \in S$ and $x_i \in X$), can we define a left R-module M' containing M as the set of all expressions of the form $\sum_i r_i x_i$ for $r_i \in R$? Recall that if V is a vector space over a field k and $qv = 0$ in V, where $q \in k$ and $v \in V$, then either $q = 0$ or $v = 0$. Now suppose that $M = \langle a \rangle$ is a cyclic $\mathbb{Z}$-module (abelian group) of order 2; if M could be imbedded in a $\mathbb{Q}$-module (i.e., a vector space V over $\mathbb{Q}$), then $2a = 0$ in V and yet neither factor is 0. Thus, our goal of extending scalars has merit, but we cannot be so cavalier about its solution. We must consider two problems: given a left S-module M, can we extend scalars to obtain a left R-module M' (always); if we can extend scalars, does M imbed in M' (sometimes).

Definition.   Let R be a ring, let A_R be a right R-module, let $_R B$ be a left R-module, and let G be an (additive) abelian group. A function $f : A \times B \to G$ is called *R-biadditive* if, for all $a, a' \in A$, $b, b' \in B$, and $r \in R$, we have

$$f(a + a', b) = f(a, b) + f(a', b),$$
$$f(a, b + b') = f(a, b) + f(a, b'),$$
$$f(ar, b) = f(a, rb).$$

If R is *commutative* and A, B, and M are R-modules, then a function $f : A \times B \to M$ is called *R-bilinear* if f is R-biadditive and also

$$f(ar, b) = f(a, rb) = rf(a, b)$$

[$rf(a, b)$ makes sense here because $f(a, b)$ now lies in the R-module M].

Example 2.43.

(i) If R is a ring, then its multiplication $\mu : R \times R \to R$ is R-biadditive; the first two axioms are the right and left distributive laws, while the third axiom is associativity:

$$\mu(ar, b) = (ar)b = a(rb) = \mu(a, rb).$$

If R is a commutative ring, then μ is R-bilinear, for $(ar)b = a(rb) = r(ab)$.

(ii) If we regard a left R-module $_RM$ as its underlying abelian group, then the scalar multiplication $\sigma: R \times M \to M$ is $\mathbb{Z}$-bilinear.

(iii) If R is commutative and M and N are R-modules, then $\text{Hom}_R(M, N)$ is an R-module if we define $rf: M \to N$ by $rf: m \mapsto f(rm)$, where $f \in \text{Hom}_R(M, N)$ and $r \in R$. With this definition, we can now see that *evaluation* $e: M \times \text{Hom}_R(M, N) \to N$, given by $(m, f) \mapsto f(m)$, is R-bilinear. The dual space V^* of a vector space V over a field k is a special case of this construction: evaluation $V \times V^* \to k$ is k-bilinear.

(iv) The *Pontrjagin dual* of an abelian group G is defined to be $G^* = \text{Hom}_{\mathbb{Z}}(G, \mathbb{R}/\mathbb{Z})$, and evaluation $G \times G^* \to \mathbb{R}/\mathbb{Z}$ is $\mathbb{Z}$-bilinear (see Exercise 3.19 on page 130). ◄

Tensor product converts biadditive functions into linear ones.

Definition. Given a ring R and modules A_R and $_RB$, then their *tensor product* is an abelian group $A \otimes_R B$ and an R-biadditive function

$$h: A \times B \to A \otimes_R B$$

such that, for every abelian group G and every R-biadditive $f: A \times B \to G$, there exists a unique $\mathbb{Z}$-homomorphism $\tilde{f}: A \otimes_R B \to G$ making the following diagram commute.

Proposition 2.44. *If U and $A \otimes_R B$ are tensor products of A_R and $_RB$ over R, then $A \otimes_R B \cong U$.*

Proof. Assume that $\eta: A \times B \to U$ is an R-biadditive function such that, for every abelian group G and every R-biadditive $f: A \times B \to G$, there exists a unique $\mathbb{Z}$-homomorphism $f': A \otimes_R B \to G$ making the following diagram commute.

Setting $G = A \otimes_R B$ and $f = h$, there is a homomorphism $h': U \to A \otimes_R B$ with $h'\eta = h$. Similarly, setting $G = U$ and $f = \eta$ in the diagram defining $A \otimes_R B$, there is a homomorphism $\tilde{\eta}: A \otimes_R B \to U$ with $\tilde{\eta}h = \eta$.

Consider the following new diagram.

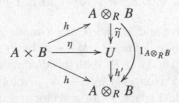

Now $h'\widetilde{\eta}$ makes the big triangle with vertices $A \times B$, $A \otimes_R B$, and $A \otimes_R B$ commute. But the identity $1_{A \otimes_R B}$ also makes this diagram commute. By the uniqueness of the completing arrow (in the definition of tensor product), we have $h'\widetilde{\eta} = 1_{A \otimes_R B}$. A similar argument shows that $\widetilde{\eta}h' = 1_U$. Hence, $\widetilde{\eta} \colon A \otimes_R B \to U$ is an isomorphism. •

Tensor product has been defined as a solution to a **universal mapping problem**; it is an abelian group that admits a unique map making many diagrams commute (a precise definition can be found in Mac Lane, *Categories for the Working Mathematician*, Chapter III). There are many universal mapping problems, and the proof of Proposition 2.44 is a paradigm proving that solutions, if they exist, are unique to isomorphism (we will give a second paradigm in Chapter 5).

Proposition 2.45.   *If R is a ring and A_R and $_RB$ are modules, then their tensor product exists.*

Proof.   Let F be the free abelian group with basis $A \times B$; that is, F is free on all ordered pairs (a, b), where $a \in A$ and $b \in B$. Define S to be the subgroup of F generated by all elements of the following three types:

$$(a, b + b') - (a, b) - (a, b');$$

$$(a + a', b) - (a, b) - (a', b);$$

$$(ar, b) - (a, rb).$$

Define $A \otimes_R B = F/S$, denote the coset $(a, b) + S$ by $a \otimes b$, and define

$$h \colon A \times B \to A \otimes_R B \quad \text{by} \quad h \colon (a, b) \mapsto a \otimes b$$

(thus, h is the restriction of the natural map $F \to F/S$). We have the following identities in $A \otimes_R B$:

$$a \otimes (b + b') = a \otimes b + a \otimes b';$$
$$(a + a') \otimes b = a \otimes b + a' \otimes b;$$
$$ar \otimes b = a \otimes rb.$$

It is now obvious that h is R-biadditive.

Consider the following diagram, where G is an abelian group, f is R-biadditive, and $i: A \times B \to F$ is the inclusion.

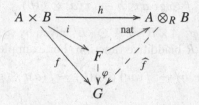

Since F is free abelian with basis $A \times B$, Proposition 2.34 says that there exists a homomorphism $\varphi: F \to G$ with $\varphi(a, b) = f(a, b)$ for all (a, b). Now $S \subseteq \ker \varphi$ because f is R-biadditive, and so Exercise 2.11 on page 66 says that φ induces a map $\widehat{f}: A \otimes_R B = F/S \to G$ by

$$\widehat{f}(a \otimes b) = \widehat{f}((a, b) + S) = \varphi(a, b) = f(a, b).$$

This equation may be rewritten as $\widehat{f}h = f$; that is, the diagram commutes. Finally, $\widehat{f}$ is unique because $A \otimes_R B$ is generated by the set of all $a \otimes b$'s, and Exercise 2.3 on page 64 says that two homomorphisms agreeing on a set of generators are equal. •

Remark. Since $A \otimes_R B$ is generated by the elements of the form $a \otimes b$, every $u \in A \otimes_R B$ has the form

$$u = \sum_i a_i \otimes b_i.$$

This expression for u is not unique; for example, there are expressions

$$0 = a \otimes (b + b') - a \otimes b - a \otimes b',$$
$$0 = (a + a') \otimes b - a \otimes b - a' \otimes b,$$
$$0 = ar \otimes b - a \otimes rb.$$

Therefore, given some abelian group G, we must be suspicious of a *definition* of a map $u: A \otimes_R B \to G$ given by specifying u on the generators $a \otimes b$; such a "function" u may not be well-defined, because elements have many expressions in terms of these generators. In essence, u is defined only on F, the free abelian group with basis $A \times B$, and we must still show that $u(S) = \{0\}$, because $A \otimes_R B = F/S$. The simplest (and safest!) procedure is to define an R-biadditive function on $A \times B$; it will yield a (well-defined) homomorphism. We illustrate this procedure in the next proof. ◄

Proposition 2.46. *Let* $f: A_R \to A'_R$ *and* $g: {}_R B \to {}_R B'$ *be maps of right R-modules and left R-modules, respectively. Then there is a unique $\mathbb{Z}$-map, denoted by $f \otimes g: A \otimes_R B \to A' \otimes_R B'$, with*

$$f \otimes g: a \otimes b \mapsto f(a) \otimes g(b).$$

Proof. The function $\varphi: A \times B \to A' \otimes_R B'$, given by $(a, b) \mapsto f(a) \otimes g(b)$, is easily seen to be an R-biadditive function. For example,

$$\varphi: (ar, b) \mapsto f(ar) \otimes g(b) = f(a)r \otimes g(b)$$

and

$$\varphi: (a, rb) \mapsto f(a) \otimes g(rb) = f(a) \otimes rg(b);$$

these are equal because of the identity $a'r \otimes b' = a' \otimes rb'$ in $A' \otimes_R B'$. The biadditive function φ yields a unique homomorphism $A \otimes_R B \to A' \otimes_R B'$ taking $a \otimes b \mapsto f(a) \otimes g(b)$.   •

Corollary 2.47. *Given maps of right R-modules, $A \xrightarrow{f} A' \xrightarrow{f'} A''$, and maps of left R-modules, $B \xrightarrow{g} B' \xrightarrow{g'} B''$,*

$$(f' \otimes g')(f \otimes g) = f'f \otimes g'g.$$

Proof. Both maps take $a \otimes b \mapsto f'f(a) \otimes g'g(b)$, and so the uniqueness of such a homomorphism gives the desired equation.   •

Theorem 2.48. *Given A_R, there is an additive functor $F_A: {}_R\mathbf{Mod} \to \mathbf{Ab}$, defined by*

$$F_A(B) = A \otimes_R B \quad and \quad F_A(g) = 1_A \otimes g,$$

where $g: B \to B'$ is a map of left R-modules.

Similarly, given ${}_R B$, there is an additive functor $G_B: \mathbf{Mod}_R \to \mathbf{Ab}$, defined by

$$G_B(A) = A \otimes_R B \quad and \quad G_B(f) = f \otimes 1_B,$$

where $f: A \to A'$ is a map of right R-modules.

Proof. First, note that F_A preserves identities: $F_A(1_B) = 1_A \otimes 1_B$ is the identity $1_{A \otimes B}$, because it fixes every generator $a \otimes b$. Second, F_A preserves composition:

$$F_A(g'g) = 1_A \otimes g'g = (1_A \otimes g')(1_A \otimes g) = F_A(g')F_A(g),$$

by Corollary 2.47. Therefore, F_A is a functor.

To see that F_A is additive, we must show that $F_A(g+h) = F_A(g)+F_A(h)$, where $g, h: B \to B'$; that is, $1_A \otimes (g + h) = 1_A \otimes g + 1_A \otimes h$. This is also easy, for both these maps send $a \otimes b \mapsto a \otimes g(b) + a \otimes h(b)$.   •

Notation. We denote the functor F_A by $A \otimes_R \square$, and we denote the functor G_B by $\square \otimes_R B$.

Corollary 2.49. *If $f: M \to M'$ and $g: N \to N'$ are, respectively, isomorphisms of right and left R-modules, then $f \otimes g: M \otimes_R N \to M' \otimes_R N'$ is an isomorphism of abelian groups.*

Proof. Now $f \otimes 1_N$ is the value of the functor $\square \otimes_R N$ on the isomorphism f, and hence $f \otimes 1_N$ is an isomorphism; similarly, $1_M \otimes g$ is an isomorphism. By Corollary 2.47, we have $f \otimes g = (f \otimes 1_N)(1_M \otimes g)$. Therefore, $f \otimes g$ is an isomorphism, being the composite of isomorphisms. •

Before continuing with properties of tensor products, we pause to discuss a technical point. In general, the tensor product of two modules is only an abelian group; is it ever a module? If so, do the tensor product functors then take values in a module category, not merely in **Ab**; that is, is $1 \otimes f$ then a map of modules? The notion of *bimodule* usually answers such questions.

Definition. Let R and S be rings and let M be an abelian group. Then M is an (R, S)-*bimodule*, denoted by $_RM_S$, if M is a left R-module and a right S-module, and the two scalar multiplications are related by an associative law:

$$r(ms) = (rm)s$$

for all $r \in R$, $m \in M$, and $s \in S$.

If M is an (R, S)-bimodule, it is permissible to write rms with no parentheses, for the definition of bimodule says that the two possible associations agree.

Example 2.50.

(i) Every ring R is an (R, R)-bimodule; the extra identity is just the associativity of multiplication in R. More generally, if $S \subseteq R$ is a subring, then R is an (R, S)-bimodule.

(ii) Every two-sided ideal in a ring R is an (R, R)-bimodule.

(iii) If M is a left R-module (i.e., if $M = {}_RM$), then M is an $(R, \mathbb{Z})$-bimodule; that is, $M = {}_RM_{\mathbb{Z}}$. Similarly, a right R-module N is a bimodule $_{\mathbb{Z}}N_R$.

(iv) If R is commutative, then every left (or right) R-module is an (R, R)-bimodule. In more detail, if $M = {}_RM$, define a new scalar multiplication $M \times R \to M$ by $(m, r) \mapsto rm$. To see that M is a right R-module, we must show that $m(rr') = (mr)r'$, that is, $(rr')m = r'(rm)$, and this is so because $rr' = r'r$. Finally, M is an (R, R)-bimodule because both $r(mr')$ and $(rm)r'$ are equal to $(rr')m$.

(v) We can make any left kG-module M into a right kG-module by defining $mg = g^{-1}m$ for every $m \in M$ and every g in the group G. Even though M is both a left and right kG-module, it is usually not a (kG, kG)-bimodule because the required associativity formula may not hold. In more detail, let $g, h \in G$ and $m \in M$. Now $g(mh) = g(h^{-1}m) = (gh^{-1})m$; on the other hand, $(gm)h = h^{-1}(gm) = (h^{-1}g)m$. To see that these can be different, take $M = kG$, $m = 1$, and g and h noncommuting elements of G.   ◄

The next proposition uses tensor product to extend scalars.

Proposition 2.51 (Extending Scalars).   *Let S be a subring of a ring R.*

(i) *Given a bimodule $_R A_S$ and a left module $_S B$, then the tensor product $A \otimes_S B$ is a left R-module, where*

$$r(a \otimes b) = (ra) \otimes b.$$

Similarly, given A_S and $_S B_R$, the tensor product $A \otimes_S B$ is a right R-module, where $(a \otimes b)r = a \otimes (br)$.

(ii) *The ring R is an (R, S)-bimodule and, if M is a left S-module, then $R \otimes_S M$ is a left R-module.*

Proof.

(i) For fixed $r \in R$, the multiplication $\mu_r : A \to A$, defined by $a \mapsto ra$, is an S-map, for A being a bimodule gives

$$\mu_r(as) = r(as) = (ra)s = \mu_r(a)s.$$

If $F = \square \otimes_S B : \mathbf{Mod}_S \to \mathbf{Ab}$, then $F(\mu_r) : A \otimes_S B \to A \otimes_S B$ is a (well-defined) $\mathbb{Z}$-homomorphism. Thus, $F(\mu_r) = \mu_r \otimes 1_B : a \otimes b \mapsto (ra) \otimes b$, and so the formula in the statement of the lemma makes sense. It is now straightforward to check that the module axioms do hold for $A \otimes_S B$.

(ii) Example 2.50(i) shows that R can be viewed as an (R, S)-bimodule, and so part (i) applies.   •

For example, if V and W are vector spaces over a field k, then their tensor product $V \otimes_k W$ is also a vector space over k.

Example 2.52. If H is a subgroup of a group G, then a representation of H gives a left kH-module B. Now $kH \subseteq kG$ is a subring, so that kG is a (kG, kH)-bimodule. Therefore, Proposition 2.51(ii) shows that $kG \otimes_{kH} B$ is a left kG-module. The corresponding representation of G is called the ***induced representation***. ◄

We see that proving properties of tensor product is often a matter of showing that obvious maps are, indeed, well-defined functions.

Corollary 2.53.

(i) *Given a bimodule $_S A_R$, the functor $A \otimes_R \square \colon {}_R\mathbf{Mod} \to \mathbf{Ab}$ actually takes values in $_S\mathbf{Mod}$.*

(ii) *If R is a ring, then $A \otimes_R B$ is a $Z(R)$-module, where*

$$r(a \otimes b) = (ra) \otimes b = a \otimes rb$$

for all $r \in Z(R)$, $a \in A$, and $b \in B$.

(iii) *If R is a ring, $r \in Z(R)$, and $\mu_r \colon B \to B$ is multiplication by r, then*

$$1_A \otimes \mu_r \colon A \otimes_R B \to A \otimes_R B$$

is also multiplication by r.

Proof.

(i) By Proposition 2.51, $A \otimes_R B$ is a left S-module, where $s(a \otimes b) = (sa) \otimes b$, and so it suffices to show that if $g \colon B \to B'$ is a map of left R-modules, then $1_A \otimes g$ is an S-map. But

$$\begin{aligned}
(1_A \otimes g)[s(a \otimes b)] &= (1_A \otimes g)[(sa) \otimes b] \\
&= (sa) \otimes gb \\
&= s(a \otimes gb) \qquad \text{by Proposition 2.51} \\
&= s(1_A \otimes g)(a \otimes b).
\end{aligned}$$

(ii) Since the center $Z(R)$ is commutative, we may regard A and B as $(Z(R), Z(R))$-bimodules by defining $ar = ra$ and $br = rb$ for all $r \in Z(R)$, $a \in A$, and $b \in B$. Proposition 2.51(i) now gives

$$r(a \otimes b) = (ra) \otimes b = (ar) \otimes b = a \otimes rb.$$

(iii) This statement merely sees the last equation $a \otimes rb = r(a \otimes b)$ from a different viewpoint:

$$(1_A \otimes \mu_r)(a \otimes b) = a \otimes rb = r(a \otimes b). \qquad \bullet$$

The next technical result complements Proposition 2.51: when one of the modules is a bimodule, then Hom also has extra structure. The reader will frequently refer back to this.

Proposition 2.54.   *Let R and S be rings.*

  (i) *Given $_RA_S$ and $_RB$, then $\mathrm{Hom}_R(A, B)$ is a left S-module, where $sf: a \mapsto f(as)$, and $\mathrm{Hom}_R(A, \square)$ is a functor $_R\mathbf{Mod} \to {}_S\mathbf{Mod}$.*

  (ii) *Given $_RA_S$ and B_S, then $\mathrm{Hom}_S(A, B)$ is a right R-module, where $fr: a \mapsto f(ra)$, and $\mathrm{Hom}_S(A, \square)$ is a functor $\mathbf{Mod}_S \to \mathbf{Mod}_R$.*

  (iii) *Given $_SB_R$ and A_R, then $\mathrm{Hom}_R(A, B)$ is a left S-module, where $sf: a \mapsto s[f(a)]$, and $\mathrm{Hom}_R(\square, B)$ is a functor $\mathbf{Mod}_R \to {}_S\mathbf{Mod}$.*

  (iv) *Given $_SB_R$ and $_SA$, then $\mathrm{Hom}_S(A, B)$ is a right R-module, where $fr: a \mapsto f(a)r$, and $\mathrm{Hom}_S(A, \square)$ is a functor $_S\mathbf{Mod} \to \mathbf{Mod}_R$.*

Proof.   All parts are routine.   ●

Remark.   Let $f: A \to B$ be an R-map. Suppose we write fa [instead of $f(a)$] when A is a right R-module and af [instead of $(a)f$] when A is a left R-module (that is, write the function symbol f on the side opposite the scalar action). With this notation, each of the four parts of Proposition 2.54 is an associative law. For example, in part (i) with both A and B left R-modules, writing sf for $s \in S$, we have $a(sf) = (as)f$. Similarly, in part (ii), we define fr, for $r \in R$ so that $(fr)a = f(ra)$.   ◄

We have made some progress in our original problem: given a left S-module M, where S is a subring of a ring R, we can create a left R-module from M by *extending scalars*; that is, Proposition 2.51 shows that $R \otimes_S M$ is a left R-module. However, we still ask, among other things, whether a left S-module M can be imbedded in $R \otimes_S M$. More generally, let $A' \subseteq A$ be right R-modules and let $i: A' \to A$ be the inclusion; if B is a left R-module, is $i \otimes 1_B: A' \otimes_R B \to A \otimes_R B$ an injection? Example 2.64 gives a negative answer, and investigating $\ker i \otimes 1_B$ was one of the first tasks of Homological Algebra. The best way to attack this problem is to continue studying properties of tensor functors.

We have defined R-biadditive functions for arbitrary, possibly noncommutative, rings R, whereas we have defined R-bilinear functions only for commutative rings. Recall that if R is commutative and A, B, and M are R-modules, then a function $f: A \times B \to M$ is *R-bilinear* if f is R-biadditive and $f(ar, b) = f(a, rb) = rf(a, b)$. Tensor product was defined as the solution of a certain universal mapping problem involving R-biadditive functions; we show now, when R is commutative, that tensor product $A \otimes_R B$ also solves the universal mapping problem for R-bilinear functions.

Proposition 2.55. *If R is a commutative ring and A, B are R-modules, then $A \otimes_R B$ is an R-module, the function $h: A \times B \to A \otimes_R B$ is R-bilinear, and, for every R-module M and every R-bilinear function $g: A \times B \to M$, there exists a unique R-homomorphism $\widehat{g}: A \otimes_R B \to M$ making the following diagram commute.*

$$
\begin{array}{ccc}
A \times B & \xrightarrow{\quad h \quad} & A \otimes_R B \\
 & \searrow_{g} \quad \swarrow_{\widehat{g}} & \\
 & M &
\end{array}
$$

Proof. By uniqueness, as in the proof of Proposition 2.44, it suffices to show that $A \otimes_R B$ is a solution if we define $h(a, b) = a \otimes b$; note that h is also R-bilinear, thanks to Corollary 2.53. Since g is R-bilinear, it is R-biadditive, and so there does exist a $\mathbb{Z}$-homomorphism $\widehat{g}: A \otimes_R B \to M$ with $\widehat{g}(a \otimes b) = g(a, b)$ for all $(a, b) \in A \times B$. We need only show that $\widehat{g}$ is an R-map. If $u \in R$,

$$
\begin{aligned}
\widehat{g}(u(a \otimes b)) &= \widehat{g}((ua) \otimes b) \\
&= g(ua, b) \\
&= ug(a, b), \qquad \text{for } g \text{ is R-bilinear} \\
&= u\widehat{g}(a \otimes b). \quad \bullet
\end{aligned}
$$

The tensor functors obey certain commutativity and associativity laws that have no analogs for the Hom functors.

Proposition 2.56 (Commutativity).

(i) *If R is a ring and M_R, $_R N$ are modules, then there is a $\mathbb{Z}$-isomorphism*

$$\tau: M \otimes_R N \to N \otimes_{R^{\mathrm{op}}} M$$

with $\tau: m \otimes n \mapsto n \otimes m$. The map τ is natural in the sense that the following diagram commutes:

$$
\begin{array}{ccc}
M \otimes_R N & \xrightarrow{\quad \tau \quad} & N \otimes_{R^{\mathrm{op}}} M \\
{\scriptstyle f \otimes g}\downarrow & & \downarrow{\scriptstyle g \otimes f} \\
M' \otimes_R N' & \xrightarrow{\quad \tau \quad} & N' \otimes_{R^{\mathrm{op}}} M'.
\end{array}
$$

(ii) *If R is a commutative ring and M and N are R-modules, then τ is an R-isomorphism.*

Proof. Consider the diagram

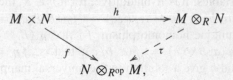

$$
\begin{array}{ccc}
M \times N & \xrightarrow{\quad h \quad} & M \otimes_R N \\
 & \searrow_{f} \quad \swarrow_{\tau} & \\
 & N \otimes_{R^{\mathrm{op}}} M, &
\end{array}
$$

where $f(m, n) = n \otimes m$. It is easy to see that f is R-biadditive, and so there is a unique $\mathbb{Z}$-map $\tau : M \otimes_R N \to N \otimes_{R^{op}} M$ with $\tau : m \otimes n \mapsto n \otimes m$. A similar diagram, interchanging the roles of $M \otimes_R N$ and $N \otimes_{R^{op}} M$, gives a $\mathbb{Z}$-map in the reverse direction taking $n \otimes m \mapsto m \otimes n$. Both composites of these maps are obviously identity maps, and so τ is an isomorphism. The proof of naturality is routine, as is the proof of (ii). •

Proposition 2.57 (Associativity). *Given $A_R, {}_R B_S,$ and ${}_S C$, there is an isomorphism*

$$\theta : A \otimes_R (B \otimes_S C) \cong (A \otimes_R B) \otimes_S C$$

given by

$$a \otimes (b \otimes c) \mapsto (a \otimes b) \otimes c.$$

Proof. Define a **triadditive** function $h : A \times B \times C \to G$, where G is an abelian group, to be a function satisfying

$$h(ar, b, c) = h(a, rb, c) \quad \text{and} \quad h(a, bs, c) = h(a, b, sc)$$

for all $r \in R$ and $s \in S$, and that is additive in each of the three variables when we fix the other two [e.g., $h(a + a', b, c) = h(a, b, c) + h(a', b, c)$]. Consider the univeral mapping problem described by the diagram

$$A \times B \times C \xrightarrow{\quad h \quad} T(A, B, C)$$

with f and $\tilde{f}$ mapping to G.

Here, an abelian group $T(A, B, C)$ and a triadditive function h are given once for all, while G is any abelian group, f is triadditive, and $\tilde{f}$ is the unique $\mathbb{Z}$-homomorphism making the diagram commute.

We show that $A \otimes_R (B \otimes_S C)$ is a solution to this universal mapping problem. Define a triadditive function $h : A \times B \times C \to A \otimes_R (B \otimes_S C)$ by $h : (a, b, c) \mapsto a \otimes (b \otimes c)$. Let $f : A \times B \times C \to G$ be triadditive, where G is some abelian group. If $a \in A$, the function $f_a : B \times C \to G$, defined by $(b, c) \mapsto f(a, b, c)$, is S-biadditive, and so it gives a unique homomorphism $\tilde{f}_a : B \otimes_S C \to G$ taking $b \otimes c \mapsto f(a, b, c)$. If $a, a' \in A$, then $\tilde{f}_{a+a'}(b \otimes c) = f(a + a', b, c) = f(a, b, c) + f(a', b, c) = \tilde{f}_a(b \otimes c) + \tilde{f}_{a'}(b \otimes c)$. It follows that the function $\varphi : A \times (B \otimes_S C) \to G$, defined by $\varphi(a, b \otimes c) = \tilde{f}_a(b \otimes c)$, is additive in both variables. It is R-biadditive, for if $r \in R$, then $\varphi(ar, b \otimes c) = \tilde{f}_{ar}(b \otimes c) = f(ar, b, c) = f(a, rb, c) = \tilde{f}_a(rb \otimes c) = \varphi(a, r(b \otimes c))$. Hence, there is a unique homomorphism $\tilde{f} : A \otimes_R (B \otimes_S C) \to G$ with $a \otimes (b \otimes c) \mapsto \varphi(a, b \otimes c) = f(a, b, c)$; that is, $\tilde{f}\eta = f$. Therefore, $A \otimes_R (B \otimes_S C)$ and h give a solution to the universal mapping problem.

In a similar way, we can prove that $(A \otimes_R B) \otimes_S C$ and the triadditive function $(a, b, c) \mapsto a \otimes b \otimes c$ is another solution. Uniqueness of solutions to universal mapping problems, as in the proof of Proposition 2.44, shows that there is an isomorphism $A \otimes_R (B \otimes_S C) \to (A \otimes_R B) \otimes_S C$ with $a \otimes (b \otimes c) \mapsto (a \otimes b) \otimes c$. •

The reader can construct another proof of associativity in the spirit of our proof of the existence of $A \otimes_R B$ as a quotient of a free abelian group; see Exercise 2.30 on page 94.

A set A with an associative binary operation satisfies *generalized associativity* if for all $n \geq 3$ and $a_i \in A$ for all i, every product $a_1 \cdots a_n$ needs no parentheses to be well-defined. Generalized associativity for tensor product does hold: every product $A_1 \otimes \cdots \otimes A_n$ needs no parentheses to be well-defined (see Exercise 2.30 on page 94); however, it does not follow from the fact just cited because equality $A \otimes_R (B \otimes_S C) = (A \otimes_R B) \otimes_S C$ was not proved; Proposition 2.57 only says that these two groups are isomorphic. (See Mac Lane, *Categories for the Working Mathematician*, Section VII 3).

Recall Exercise 2.13 on page 66: for any left R-module M, for any $f \in \text{Hom}_R(R, M)$, and for any $r, s \in R$, the map $rf : s \mapsto f(sr)$ defines a natural isomorphism from $\text{Hom}_R(R, \Box) \to M$ to the identity functor on $_R\mathbf{Mod}$. Here is the analog for tensor products.

Proposition 2.58.   *There is a natural R-isomorphism*

$$\varphi_M : R \otimes_R M \to M,$$

for every left R-module M, where $\varphi_M : r \otimes m \mapsto rm$.

Proof.   The function $R \times M \to M$, given by $(r, m) \mapsto rm$, is R-biadditive, and so there is an R-homomorphism $\varphi : R \otimes_R M \to M$ with $r \otimes m \mapsto rm$ [we are using the fact that R is an (R, R)-bimodule]. To see that φ is an R-isomorphism, it suffices to find a $\mathbb{Z}$-homomorphism $f : M \to R \otimes_R M$ with φf and $f\varphi$ identity maps (for it is now only a question of whether the *function* φ is a bijection). Such a $\mathbb{Z}$-map is given by $f : m \mapsto 1 \otimes m$.

Naturality is proved by showing commutativity of the following diagram.

$$
\begin{array}{ccc}
R \otimes_R M & \xrightarrow{\varphi_M} & M \\
{\scriptstyle 1 \otimes f} \downarrow & & \downarrow {\scriptstyle f} \\
R \otimes_R N & \xrightarrow[\varphi_N]{} & N
\end{array}
$$

It suffices to check the maps on generators $r \otimes m$. Going clockwise, $r \otimes m \mapsto rm \mapsto f(rm)$; going counterclockwise, $r \otimes m \mapsto r \otimes f(m) \mapsto rf(m)$. These are equal because f is an R-map.   •

Definition. Let k be a commutative ring. Then a ring R is a k-*algebra* if R is a k-module satisfying

$$a(rs) = (ar)s = r(as)$$

for all $a \in k$ and $r, s \in R$.

Consider the important special case in which k is isomorphic to a subring k' of R. In this case, the defining identity with $s = 1$ is $ar = ra$ for all $r \in R$; that is, $k' \subseteq Z(R)$, the center of R. This explains why we assume, in the definition of k-algebra, that k is commutative.

Example 2.59.

   (i) If R is a k-algebra, then $R[x]$ is also a k-algebra.

  (ii) If k is a commutative ring, then the ring of matrices $R = \mathrm{Mat}_n(k)$ is a k-algebra.

 (iii) If k is a field and R is a finite-dimensional k-algebra, then every left or right ideal I in R is a subspace of R, so that $\dim(I) \leq \dim(R)$. A basis of I generates I as a k-module; a fortiori, it generates I as an R-module, and so I is finitely generated.

 (iv) If k is a commutative ring and G is a (multiplicative) group, then the group ring kG is a k-algebra. ◄

The tensor product of two k-algebras is itself a k-algebra.

Proposition 2.60. *If k is a commutative ring and A and B are k-algebras, then the tensor product $A \otimes_k B$ is a k-algebra if we define*

$$(a \otimes b)(a' \otimes b') = aa' \otimes bb'.$$

Proof. First, $A \otimes_k B$ is a k-module, by Theorem 2.48. Let $\mu: A \times A \to A$ and $\nu: B \times B \to B$ be the given multiplications on the algebras A and B, respectively. We must show there is a multiplication on $A \otimes_k B$ as in the statement; that is, there is a k-bilinear function $\lambda: (A \otimes_k B) \times (A \otimes_k B) \to A \otimes_k B$ with $\lambda: (a \otimes b, a' \otimes b') \mapsto aa' \otimes bb'$. Such a function λ exists because it is the composite

$$(A \otimes_k B) \times (A \otimes_k B) \to (A \otimes_k B) \otimes (A \otimes_k B) \tag{1}$$

$$\to [(A \otimes_k B) \otimes_k A] \otimes_k B \tag{2}$$

$$\to [A \otimes_k (B \otimes_k A)] \otimes_k B \tag{3}$$

$$\to [A \otimes_k (A \otimes_k B)] \otimes_k B \tag{4}$$

$$\to [(A \otimes_k A) \otimes_k B] \otimes_k B \tag{5}$$

$$\to (A \otimes_k A) \otimes (B \otimes_k B) \tag{6}$$

$$\to A \otimes_k B : \tag{7}$$

map (1) is $(a \otimes b, a' \otimes b') \mapsto a \otimes b \otimes a' \otimes b'$; maps (2) and (3) are associativity; map (4) is $1 \otimes \tau \otimes 1$, where $\tau: B \otimes_k A \to A \otimes_k B$ takes $b \otimes a \mapsto a \otimes b$; maps (5) and (6) are associativity; the last map is $\mu \otimes \nu$. Each of these maps is well-defined (see my book *Advanced Modern Algebra*, Section 9.6). The reader may now verify that the k-module $A \otimes_k B$ is a k-algebra.   ●

Bimodules can be viewed as left modules over a suitable ring.

Corollary 2.61.   *Let R and S be k-algebras, where k is a commutative ring. Every (R, S)-bimodule M is a left $R \otimes_k S^{op}$-module, where*

$$(r \otimes s)m = rms.$$

Proof.   The function $R \times S^{op} \times M \to M$, given by $(r, s, m) \mapsto rms$, is k-trilinear, and this can be used to prove that $(r \otimes s)m = rms$ is well-defined. Let us write $s * s'$ for the product in S^{op}; that is, $s * s' = s's$. The only axiom that is not obvious is axiom (iii) in the definition of module: if $a, a' \in R \otimes_k S^{op}$, then $(aa')m = a(a'm)$, and it is enough to check that this is true for generators $a = r \otimes s$ and $a' = r' \otimes s'$ of $R \otimes_k S^{op}$. But

$$
\begin{aligned}
[(r \otimes s)(r' \otimes s')]m &= [rr' \otimes s * s']m \\
&= (rr')m(s * s') \\
&= (rr')m(s's) \\
&= r(r'ms')s.
\end{aligned}
$$

On the other hand,

$$(r \otimes s)[(r' \otimes s')m] = (r \otimes s)[r'(ms')] = r(r'ms')s.   ●$$

Definition.   If A is a k-algebra, where k is a commutative ring, then its *enveloping algebra* is

$$A^e = A \otimes_k A^{op}.$$

Corollary 2.62.   *If A is a k-algebra, where k is a commutative ring, then A is a left A^e-module whose submodules are the two-sided ideals. If A is a simple k-algebra, then A is a simple A^e-module.*

Proof.   Since a k-algebra A is an (A, A)-bimodule, it is a left A^e-module.   ●

We now present properties of tensor products that will help us compute them. First, we give a result about Hom, and then we give the analogous result for tensor. Corollary 2.22(ii) says that any additive functor $T: {}_R\mathbf{Mod} \to \mathbf{Ab}$ preserves direct sums of modules: $T(A \oplus C) \cong T(A) \oplus T(C)$. If we regard such a direct sum as a split short exact sequence, then we may specialize the corollary by taking $T = X \otimes_R \square$ and saying that if

$$0 \to A \xrightarrow{i} B \xrightarrow{p} C \to 0$$

is a split short exact sequence, then so is

$$0 \to X \otimes_R A \xrightarrow{1 \otimes i} X \otimes_R B \xrightarrow{1 \otimes p} X \otimes_R C \to 0.$$

This leads us to a more general question: if

$$0 \to A \xrightarrow{i} B \xrightarrow{p} C \to 0$$

is any short exact sequence, not necessarily split, is

$$0 \to X \otimes_R A \xrightarrow{1 \otimes i} X \otimes_R B \xrightarrow{1 \otimes p} X \otimes_R C \to 0$$

also an exact sequence? Here is the answer (there is no misprint in the statement of the theorem: "$0 \to$" should not appear at the beginning of the sequences, and we shall discuss this point after the proof).

Theorem 2.63 (Right Exactness).[5] *Let A be a right R-module, and let*

$$B' \xrightarrow{i} B \xrightarrow{p} B'' \to 0$$

be an exact sequence of left R-modules. Then

$$A \otimes_R B' \xrightarrow{1 \otimes i} A \otimes_R B \xrightarrow{1 \otimes p} A \otimes_R B'' \to 0$$

is an exact sequence of abelian groups.

Remark.   We will give a nicer proof of this theorem once we prove the Adjoint Isomorphism (see Proposition 2.78). ◀

Proof.   There are three things to check.

(i) $\text{im}(1 \otimes i) \subseteq \ker(1 \otimes p)$.

It suffices to prove that the composite is 0; but

$$(1 \otimes p)(1 \otimes i) = 1 \otimes pi = 1 \otimes 0 = 0.$$

---

[5]These functors are called *right exact* because the functored sequence has "$\to 0$" on the right-hand side.

(ii) $\ker(1 \otimes p) \subseteq \operatorname{im}(1 \otimes i)$.

Let $E = \operatorname{im}(1 \otimes i)$. By part (i), $E \subseteq \ker(1 \otimes p)$, and so $1 \otimes p$ induces a map $\widehat{p} \colon (A \otimes B)/E \to A \otimes B''$ with

$$\widehat{p} \colon a \otimes b + E \mapsto a \otimes pb,$$

where $a \in A$ and $b \in B$. Now if $\pi \colon A \otimes B \to (A \otimes B)/E$ is the natural map, then

$$\widehat{p}\pi = 1 \otimes p,$$

for both send $a \otimes b \mapsto a \otimes pb$.

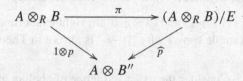

Suppose we show that $\widehat{p}$ is an isomorphism. Then

$$\ker(1 \otimes p) = \ker \widehat{p}\pi = \ker \pi = E = \operatorname{im}(1 \otimes i),$$

and we are done. To see that $\widehat{p}$ is, indeed, an isomorphism, we construct its inverse $A \otimes B'' \to (A \otimes B)/E$. Define

$$f \colon A \times B'' \to (A \otimes B)/E$$

as follows. If $b'' \in B''$, there is $b \in B$ with $pb = b''$, because p is surjective; let

$$f \colon (a, b'') \mapsto a \otimes b.$$

Now f is well-defined: if $pb_1 = b''$, then $p(b - b_1) = 0$ and $b - b_1 \in \ker p = \operatorname{im} i$. Thus, there is $b' \in B'$ with $ib' = b - b_1$, and hence $a \otimes (b - b_1) = a \otimes ib' \in \operatorname{im}(1 \otimes i) = E$. Clearly, f is R-biadditive, and so the definition of tensor product gives a homomorphism $\widehat{f} \colon A \otimes B'' \to (A \otimes B)/E$ with $\widehat{f}(a \otimes b'') = a \otimes b + E$. The reader may check that $\widehat{f}$ is the inverse of $\widehat{p}$, as desired.

(iii) $1 \otimes p$ is surjective.

If $\sum_i a_i \otimes b_i'' \in A \otimes B''$, then there exist $b_i \in B$ with $pb_i = b_i''$ for all i, for p is surjective. But

$$1 \otimes p \colon \sum a_i \otimes b_i \mapsto \sum a_i \otimes pb_i = \sum a_i \otimes b_i''. \quad \bullet$$

A similar statement holds for the functor $\square \otimes_R B$: if B is a left R-module and $A' \xrightarrow{i} A \xrightarrow{p} A'' \to 0$ is a short exact sequence of right R-modules, then the sequence

$$A' \otimes_R B \xrightarrow{i \otimes 1} A \otimes_R B \xrightarrow{p \otimes 1} A'' \otimes_R B \to 0$$

is exact.

Definition.   A (covariant) functor $T : {}_R\mathbf{Mod} \to \mathbf{Ab}$ is called *right exact* if exactness of a sequence of left R-modules

$$B' \xrightarrow{i} B \xrightarrow{p} B'' \to 0$$

implies exactness of the sequence

$$T(B') \xrightarrow{T(i)} T(B) \xrightarrow{T(p)} T(B'') \to 0.$$

There is a similar definition for covariant functors $\mathbf{Mod}_R \to \mathbf{Ab}$.

The functors $A \otimes_R \square$ and $\square \otimes_R B$ are right exact functors.

The next example shows why "$0 \to$" is absent in Theorem 2.63.

Example 2.64.   Consider the exact sequence of abelian groups

$$0 \to \mathbb{Z} \xrightarrow{i} \mathbb{Q} \to \mathbb{Q}/\mathbb{Z} \to 0,$$

where i is the inclusion. By right exactness, there is an exact sequence

$$\mathbb{I}_2 \otimes \mathbb{Z} \xrightarrow{1 \otimes i} \mathbb{I}_2 \otimes \mathbb{Q} \to \mathbb{I}_2 \otimes (\mathbb{Q}/\mathbb{Z}) \to 0$$

(we abbreviate $\otimes_{\mathbb{Z}}$ to $\otimes$ here). Now $\mathbb{I}_2 \otimes \mathbb{Z} \cong \mathbb{I}_2$, by Proposition 2.58. On the other hand, if $a \otimes q$ is a generator of $\mathbb{I}_2 \otimes \mathbb{Q}$, then

$$a \otimes q = a \otimes (2q/2) = 2a \otimes (q/2) = 0 \otimes (q/2) = 0.$$

Therefore, $\mathbb{I}_2 \otimes \mathbb{Q} = \{0\}$, and so $1 \otimes i$ cannot be an injection.

Thus, tensor product may not preserve injections: if $i : B' \to B$ is injective, the map $1_X \otimes i$ may have a nonzero kernel. We will determine $\ker 1_X \otimes i$ in general when we study the functor Tor.   ◄

The next theorem says that tensor product preserves arbitrary direct sums; compare it with Theorems 2.30 and 2.31.

Theorem 2.65.   *Let A_R be a right module, and let $({}_R B_i)_{i \in I}$ be a family of left R-modules.*

(i) *There is a $Z(R)$-isomorphism*

$$\tau : A \otimes_R \bigoplus_{i \in I} B_i \to \bigoplus_{i \in I} (A \otimes_R B_i)$$

with $\tau : a \otimes (b_i) \mapsto (a \otimes b_i)$. Moreover, if R is commutative, then τ is an R-isomorphism.

(ii) *The isomorphism τ is natural: if $(C_j)_{j \in J}$ is a family of left R-modules and, for each $i \in I$, there exist $j \in J$ and an R-map $\sigma_{ij} : B_i \to C_j$, then there is a commutative diagram*

$$
\begin{array}{ccc}
A \otimes_R \bigoplus_{i \in I} B_i & \xrightarrow{\ 1 \otimes \sigma\ } & A \otimes_R \bigoplus_{j \in J} C_j \\
\tau \downarrow & & \downarrow \tau \\
\bigoplus_{i \in I} (A \otimes_R B_i) & \xrightarrow[\ \widetilde{\sigma}\ ]{} & \bigoplus_{j \in J} (A \otimes_R C_j),
\end{array}
$$

where $\sigma : (b_i) \mapsto (\sigma_{ij} b_i)$ and $\widetilde{\sigma} : (a \otimes b_i) \mapsto (a \otimes \sigma_{ij} b_i)$.

Proof.

(i) Since the function $f : A \times \left(\bigoplus_i B_i \right) \to \bigoplus_i (A \otimes_R B_i)$, given by $f : (a, (b_i)) \mapsto (a \otimes b_i)$, is R-biadditive, there is a $\mathbb{Z}$-homomorphism

$$
\tau : A \otimes_R \left(\bigoplus_i B_i \right) \to \bigoplus_i (A \otimes_R B_i)
$$

with $\tau : a \otimes (b_i) \mapsto (a \otimes b_i)$. Now $A \otimes_R \left(\bigoplus_{i \in I} B_i \right)$ and $\bigoplus_{i \in I} (A \otimes_R B_i)$ are $Z(R)$-modules, and τ is a $Z(R)$-map (for τ is the function given by the universal mapping problem in Proposition 2.55).

To see that τ is an isomorphism, we give its inverse. Denote the injection $B_k \to \bigoplus_i B_i$ by λ_k [where $\lambda_k b_k \in \bigoplus_i B_i$ has kth coordinate b_k and all other coordinates 0], so that $1_A \otimes \lambda_k : A \otimes_R B_k \to A \otimes_R \left(\bigoplus_i B_i \right)$. That direct sum is the coproduct in ${}_R\mathbf{Mod}$ gives a homomorphism $\theta : \bigoplus_i (A \otimes_R B_i) \to A \otimes_R \left(\bigoplus_i B_i \right)$ with $\theta : (a \otimes b_i) \mapsto a \otimes \sum_i \lambda_i b_i$. It is now routine to check that θ is the inverse of τ, so that τ is an isomorphism.

(ii) Going clockwise, $a \otimes (b_i) \mapsto a \otimes (\sigma_{ij} b_i) \mapsto (a \otimes \sigma_{ij} b_i)$; going counterclockwise, $a \otimes (b_i) \mapsto (a \otimes b_i) \mapsto (a \otimes \sigma_{ij} b_i)$. •

We shall see, in Example 3.52, that tensor product may not commute with direct products.

Example 2.66. Let V and W be vector spaces over a field k. Now W is a free k-module; say, $W = \bigoplus_{i \in I} \langle w_i \rangle$, where $\{w_i : i \in I\}$ is a basis of W. Therefore, $V \otimes_k W \cong \bigoplus_{i \in I} V \otimes_k \langle w_i \rangle$. Similarly, $V = \bigoplus_{j \in J} \langle v_j \rangle$, where $\{v_j : j \in J\}$ is a basis of V and, for each i, $V \otimes_k \langle w_i \rangle \cong \bigoplus_{j \in J} \langle v_j \rangle \otimes_k \langle w_i \rangle$. But the one-dimensional vector spaces $\langle v_j \rangle$ and $\langle w_i \rangle$ are isomorphic to k, and Proposition 2.58 gives $\langle v_j \rangle \otimes_k \langle w_i \rangle \cong \langle v_j \otimes w_i \rangle$. Hence, $V \otimes_k W$ is a vector space over k having $\{v_j \otimes w_i : i \in I \text{ and } j \in J\}$ as a basis. In case both V and W are finite-dimensional, we have

$$
\dim(V \otimes_k W) = \dim(V) \dim(W). \quad \blacktriangleleft
$$

Example 2.67. We now show that there may exist elements in a tensor product $V \otimes_k V$ that cannot be written in the form $u \otimes w$ for $u, w \in V$.

Let v_1, v_2 be a basis of a two-dimensional vector space V over a field k. As in Example 2.66, a basis for $V \otimes_k V$ is

$$v_1 \otimes v_1, \quad v_1 \otimes v_2, \quad v_2 \otimes v_1, \quad v_2 \otimes v_2.$$

We claim that there do not exist $u, w \in V$ with $v_1 \otimes v_2 + v_2 \otimes v_1 = u \otimes w$. Otherwise, write u and w in terms of v_1 and v_2:

$$
\begin{aligned}
v_1 \otimes v_2 + v_2 \otimes v_1 &= u \otimes w \\
&= (av_1 + bv_2) \otimes (cv_1 + dv_2) \\
&= acv_1 \otimes v_1 + adv_1 \otimes v_2 + bcv_2 \otimes v_1 + bdv_2 \otimes v_2.
\end{aligned}
$$

By linear independence of the basis, $ac = 0 = bd$ and $ad = 1 = bc$. The first equation gives $a = 0$ or $c = 0$, and either possibility, when substituted into the second equation, gives $0 = 1$. ◀

Proposition 2.68. *If R is a ring, $r \in Z(R)$, and M is a left R-module, then*

$$R/(r) \otimes_R M \cong M/rM.$$

In particular, for every abelian group B, we have $\mathbb{I}_n \otimes_{\mathbb{Z}} B \cong B/nB$.

Proof. There is an exact sequence

$$R \xrightarrow{\mu_r} R \xrightarrow{p} R^* \to 0,$$

where $R^* = R/(r)$ and μ_r is multiplication by r. Since $\square \otimes_r M$ is right exact, there is an exact sequence

$$R \otimes_R M \xrightarrow{\mu_r \otimes 1} R \otimes_R M \xrightarrow{p \otimes 1} R^* \otimes_R M \to 0.$$

Consider the diagram

$$
\begin{array}{ccccccc}
R \otimes_R M & \xrightarrow{\mu_r \otimes 1} & R \otimes_R M & \xrightarrow{p \otimes 1} & R^* \otimes_R M & \longrightarrow & 0 \\
\theta \downarrow & & \downarrow \theta & & & & \\
M & \xrightarrow[\mu_r]{} & M & \xrightarrow[\pi]{} & M/rM & \longrightarrow & 0,
\end{array}
$$

where $\theta \colon R \otimes_R M \to M$ is the isomorphism of Proposition 2.58, namely, $\theta \colon a \otimes m \mapsto am$, where $a \in R$ and $m \in M$. This diagram commutes, for both composites take $a \otimes m \mapsto ram$. Proposition 2.70 applies to this diagram, yielding $R^* \otimes_R M \cong M/rM$. •

Example 2.69. Exercise 2.29 on page 94 shows that there is an isomorphism of abelian groups $\mathbb{I}_m \otimes \mathbb{I}_n \cong \mathbb{I}_d$, where $d = (m, n)$. It follows that if $(m, n) = 1$, then $\mathbb{I}_m \otimes \mathbb{I}_n = \{0\}$. Of course, this tensor product is still $\{0\}$ if we regard $\mathbb{I}_m$ and $\mathbb{I}_n$ as $\mathbb{Z}$-algebras. In this case, the tensor product is the zero ring. Had we insisted, in the definition of ring, that $1 \neq 0$, then the tensor product of rings would not always be defined.   ◄

Proposition 2.70.   *Given a commutative diagram with exact rows,*

$$
\begin{array}{ccccccc}
A' & \xrightarrow{\ i\ } & A & \xrightarrow{\ p\ } & A'' & \longrightarrow & 0 \\
{\scriptstyle f}\downarrow & & {\scriptstyle g}\downarrow & & \vdots\,{\scriptstyle h} & & \\
B' & \xrightarrow[\ j\ ]{} & B & \xrightarrow[\ q\ ]{} & B'' & \longrightarrow & 0,
\end{array}
$$

there exists a unique map $h \colon A'' \to B''$ making the augmented diagram commute. Moreover, h is an isomorphism if f and g are isomorphisms.

Proof. If $a'' \in A''$, then there is $a \in A$ with $p(a) = a''$ because p is surjective. Define $h(a'') = qg(a)$. Of course, we must show that h is well-defined; that is, if $u \in A$ satifies $p(u) = a''$, then $qg(u) = qg(a)$. Since $p(a) = p(u)$, we have $p(a - u) = 0$, so that $a - u \in \ker p = \operatorname{im} i$, by exactness. Hence, $a - u = i(a')$, for some $a' \in A'$. Thus,

$$
qg(a - u) = qgi(a') = qjf(a') = 0,
$$

because $qj = 0$. Therefore, h is well-defined. If $h' \colon A'' \to B''$ satisfies $h'p = qg$ and if $a'' \in A''$, choose $a \in A$ with $pa = a''$. Then $h'pa = h'a'' = qga = ha''$, and so h is unique.

To see that the map h is an isomorphism, we construct its inverse. As in the first paragraph, there is a map h' making the following diagram commute:

$$
\begin{array}{ccccccc}
B' & \xrightarrow{\ j\ } & B & \xrightarrow{\ q\ } & B'' & \longrightarrow & 0 \\
{\scriptstyle f^{-1}}\downarrow & & {\scriptstyle g^{-1}}\downarrow & & \vdots\,{\scriptstyle h'} & & \\
A' & \xrightarrow[\ i\ ]{} & A & \xrightarrow[\ p\ ]{} & A'' & \longrightarrow & 0.
\end{array}
$$

We claim that $h' = h^{-1}$. Now $h'q = pg^{-1}$. Hence,

$$
h'hp = h'qg = pg^{-1}g = p;
$$

since p is surjective, we have $h'h = 1_{A''}$. A similar calculation shows that the other composite hh' is also the identity. Therefore, h is an isomorphism.   ●

The proof of the last proposition is an example of **diagram chasing**. Such proofs appear long, but they are, in truth, quite mechanical. We choose an element and, at each step, there are only two things to do with it: either push it along an arrow or lift it (i.e., choose an inverse image) back along another arrow. The next proposition is also proved in this way.

Proposition 2.71.   *Given a commutative diagram with exact rows,*

$$
\begin{array}{ccccccc}
0 & \longrightarrow & A' & \xrightarrow{\ i\ } & A & \xrightarrow{\ p\ } & A'' \\
 & & \downarrow{\scriptstyle f} & & \downarrow{\scriptstyle g} & & \downarrow{\scriptstyle h} \\
0 & \longrightarrow & B' & \xrightarrow{\ j\ } & B & \xrightarrow{\ q\ } & B'',
\end{array}
$$

there exists a unique map $f : A' \to B'$ making the augmented diagram commute. Moreover, f is an isomorphism if g and h are isomorphisms.

Proof.   A diagram chase.   •

Who would think that a lemma about 10 modules and 13 homomorphisms could be of any interest?

Proposition 2.72 (Five Lemma).   *Consider a commutative diagram with exact rows.*

$$
\begin{array}{ccccccccc}
A_1 & \longrightarrow & A_2 & \longrightarrow & A_3 & \longrightarrow & A_4 & \longrightarrow & A_5 \\
\downarrow{\scriptstyle h_1} & & \downarrow{\scriptstyle h_2} & & \downarrow{\scriptstyle h_3} & & \downarrow{\scriptstyle h_4} & & \downarrow{\scriptstyle h_5} \\
B_1 & \longrightarrow & B_2 & \longrightarrow & B_3 & \longrightarrow & B_4 & \longrightarrow & B_5
\end{array}
$$

(i) *If h_2 and h_4 are surjective and h_5 is injective, then h_3 is surjective.*

(ii) *If h_2 and h_4 are injective and h_1 is surjective, then h_3 is injective.*

(iii) *If h_1, h_2, h_4, and h_5 are isomorphisms, then h_3 is an isomorphism.*

Proof.   A diagram chase.   •

We have already seen, in Example 2.69, that a tensor product of two nonzero abelian groups can be zero. Here is another instance of this.

Definition.   An abelian group D is called **divisible** if, for each $d \in D$ and every nonzero natural number n, there exists $d' \in D$ with $d = nd'$.

It is easy to see that $\mathbb{Q}$ is a divisible abelian group. Moreover, every direct sum and every direct product of divisible groups is divisible. Thus, $\mathbb{R}$ is divisible, for it is a vector space over $\mathbb{Q}$ and, hence, it is a direct sum of copies of $\mathbb{Q}$ (for it has a basis); similarly, $\mathbb{C}$ is divisible. Every quotient of a divisible group is divisible. Thus, $\mathbb{R}/\mathbb{Z}$ and $\mathbb{Q}/\mathbb{Z}$ are divisible abelian groups (note that $\mathbb{R}/\mathbb{Z} \cong S^1$, the unit circle in $\mathbb{C}$, via $x + \mathbb{Z} \mapsto e^{2\pi i x}$).

Proposition 2.73. *If T is an abelian group with every element of finite order and if D is a divisible abelian group, then $T \otimes_{\mathbb{Z}} D = \{0\}$.*

Proof. It suffices to show that each generator $t \otimes d$, where $t \in T$ and $d \in D$, is 0 in $T \otimes_{\mathbb{Z}} D$. Since t has finite order, there is a nonzero integer n with $nt = 0$. As D is divisible, there exists $d' \in D$ with $d = nd'$. Hence,

$$t \otimes d = t \otimes nd' = nt \otimes d' = 0 \otimes d' = 0. \quad \bullet$$

We now understand why we cannot make a finite abelian group G into a nonzero $\mathbb{Q}$-module, for $G \otimes_{\mathbb{Z}} \mathbb{Q} = \{0\}$.

Corollary 2.74. *If D is a nonzero divisible abelian group with every element of finite order (e.g., $D = \mathbb{Q}/\mathbb{Z}$), then there is no multiplication $D \times D \to D$ making D a ring.*

Proof. Suppose that there is a multiplication $\mu: D \times D \to D$ making D a ring. If 1 is the identity, we have $1 \neq 0$, lest D be the zero ring, which has only one element. Since multiplication in a ring is $\mathbb{Z}$-bilinear, there is a homomorphism $\widetilde{\mu}: D \otimes_{\mathbb{Z}} D \to D$ with $\widetilde{\mu}(d \otimes d') = \mu(d, d')$ for all $d, d' \in D$. In particular, if $d \neq 0$, then $\widetilde{\mu}(d \otimes 1) = \mu(d, 1) = d \neq 0$. But $D \otimes_{\mathbb{Z}} D = \{0\}$, by Proposition 2.73, so that $\widetilde{\mu}(d \otimes 1) = 0$. This contradiction shows that no multiplication μ on D exists. $\bullet$

2.2.1 Adjoint Isomorphisms

There is a remarkable relationship between Hom and $\otimes$. The key idea is that a function of two variables, say, $f : A \times B \to C$, can be viewed as a one-parameter family of functions of one variable: if we fix $a \in A$, then define $f_a: B \to C$ by $b \mapsto f(a, b)$. Recall Proposition 2.51: if R and S are rings and A_R and $_R B_S$ are modules, then $A \otimes_R B$ is a right S-module, where $(a \otimes b)s = a \otimes (bs)$. Furthermore, if C_S is a module, then $\mathrm{Hom}_S(B, C)$ is a right R-module, where $(fr)(b) = f(rb)$; thus $\mathrm{Hom}_R(A, \mathrm{Hom}_S(B, C))$ makes sense, for it consists of R-maps between right R-modules. Finally, if $F \in \mathrm{Hom}_R(A, \mathrm{Hom}_S(B, C))$, we denote its value on $a \in A$ by F_a, so that $F_a: B \to C$, defined by $F_a: b \mapsto F(a)(b)$, is a one-parameter family of functions. There are two versions of the adjoint isomorphism, arising from two ways in which B can be a bimodule (either $_R B_S$ or $_S B_R$).

Theorem 2.75 (Adjoint Isomorphism, First Version).   *Given modules A_R, $_R B_S$, and C_S, where R and S are rings, there is a natural isomorphism:*

$$\tau_{A,B,C} \colon \operatorname{Hom}_S(A \otimes_R B, C) \to \operatorname{Hom}_R(A, \operatorname{Hom}_S(B, C)),$$

namely, for $f \colon A \otimes_R B \to C$, $a \in A$, and $b \in B$,

$$\tau_{A,B,C} \colon f \mapsto \tau(f), \text{ where } \tau(f)_a \colon b \mapsto f(a \otimes b).$$

Remark.   In more detail, fixing any two of A, B, C, each $\tau_{A,B,C}$ is a natural isomorphism:

$$\operatorname{Hom}_S(\square \otimes_R B, C) \to \operatorname{Hom}_R(\square, \operatorname{Hom}_S(B, C)),$$

$$\operatorname{Hom}_S(A \otimes_R \square, C) \to \operatorname{Hom}_R(A, \operatorname{Hom}_S(\square, C)),$$

$$\operatorname{Hom}_S(A \otimes_R B, \square) \to \operatorname{Hom}_R(A, \operatorname{Hom}_S(B, \square)).$$

For example, if $f \colon A \to A'$, there is a commutative diagram

$$
\begin{array}{ccc}
\operatorname{Hom}_S(A' \otimes_R B, C) & \xrightarrow{\;\tau_{A',B,C}\;} & \operatorname{Hom}_R(A', \operatorname{Hom}_S(B, C)) \\
{\scriptstyle (f \otimes 1_B)^*}\downarrow & & \downarrow{\scriptstyle f^*} \\
\operatorname{Hom}_S(A \otimes_R B, C) & \xrightarrow[\;\tau_{A,B,C}\;]{} & \operatorname{Hom}_R(A, \operatorname{Hom}_S(B, C)).
\end{array}
$$
◄

Proof.   To prove that $\tau = \tau_{A,B,C}$ is a $\mathbb{Z}$-map, let $f, g \colon A \otimes_R B \to C$. The definition of $f + g$ gives, for all $a \in A$,

$$
\begin{aligned}
\tau(f + g)_a \colon b &\mapsto (f + g)(a \otimes b) \\
&= f(a \otimes b) + g(a \otimes b) \\
&= \tau(f)_a(b) + \tau(g)_a(b).
\end{aligned}
$$

Therefore, $\tau(f + g) = \tau(f) + \tau(g)$.

Next, τ is injective. If $\tau(f)_a = 0$ for all $a \in A$, then $0 = \tau(f)_a(b) = f(a \otimes b)$ for all $a \in A$ and $b \in B$. Therefore, $f = 0$ because it vanishes on every generator of $A \otimes_R B$.

We now show that τ is surjective. If $F \colon A \to \operatorname{Hom}_S(B, C)$ is an R-map, define $\varphi \colon A \times B \to C$ by $\varphi(a, b) = F_a(b)$. Now consider the diagram

$$
\begin{array}{ccc}
A \times B & \xrightarrow{\;\;h\;\;} & A \otimes_R B \\
& {\scriptstyle \varphi}\searrow \quad \swarrow{\scriptstyle \widetilde{\varphi}} & \\
& C. &
\end{array}
$$

It is straightforward to check that φ is R-biadditive, and so there exists a $\mathbb{Z}$-homomorphism $\widetilde{\varphi} \colon A \otimes_R B \to C$ with $\widetilde{\varphi}(a \otimes b) = \varphi(a, b) = F_a(b)$ for all $a \in A$ and $b \in B$. Therefore, $F = \tau(\widetilde{\varphi})$ and τ is surjective.

The reader can check that the maps τ are natural.   •

If $B = {}_S B_R$ is an (S, R)-bimodule, there is a variant of Theorem 2.75 (whose proof is left to the reader).

Theorem 2.76 (Adjoint Isomorphism, Second Version).   *Given modules* $_RA$, $_SB_R$, *and* $_SC$, *where R and S are rings, there is a natural isomorphism:*

$$\tau'_{A,B,C}\colon \operatorname{Hom}_S(B \otimes_R A, C) \to \operatorname{Hom}_R(A, \operatorname{Hom}_S(B, C)),$$

namely, for $f\colon B \otimes_R A \to C$, $a \in A$, *and* $b \in B$,

$$\tau'_{A,B,C}\colon f \mapsto \tau'(f), \text{ where } \tau'(f)_a\colon b \mapsto f(b \otimes a).$$

Corollary 2.77.

(i) *Given modules* $_RB_S$ *and* C_S, *the functors* $\operatorname{Hom}_R(\square, \operatorname{Hom}_S(B, C))$ *and* $\operatorname{Hom}_S(\square \otimes_S B, C)\colon \mathbf{Mod}_R \to \mathbf{Ab}$, *are naturally isomorphic.*

(ii) *Given modules* $_SB_R$, *and* $_SC$, *the functors* $\operatorname{Hom}_R(\square, \operatorname{Hom}_S(B, C))$ *and* $\operatorname{Hom}_S(B \otimes_S \square, C)\colon {}_R\mathbf{Mod} \to \mathbf{Ab}$ *are naturally isomorphic.*

Proof.   If B and C are fixed, then the maps $\tau_{A,B,C}$ and $\tau'_{A,B,C}$ form natural isomorphisms.   •

As promised earlier, here is another proof of Theorem 2.63, the right exactness of tensor product.

Proposition 2.78 (Right Exactness).   *Let* A_R *be a right R-module, and let*

$$B' \xrightarrow{i} B \xrightarrow{p} B'' \to 0$$

be an exact sequence of left R-modules. Then

$$A \otimes_R B' \xrightarrow{1_A \otimes i} A \otimes_R B \xrightarrow{1_A \otimes p} A \otimes_R B'' \to 0$$

is an exact sequence of abelian groups.

Proof.   Regard a left R-module B as an $(R, \mathbb{Z})$-bimodule, and note, for any abelian group C, that $\operatorname{Hom}_{\mathbb{Z}}(B, C)$ is a right R-module, by Proposition 2.54. In light of Proposition 2.42, it suffices to prove that the top row of the following diagram is exact for every C:

$$0 \to \operatorname{Hom}_{\mathbb{Z}}(A \otimes_R B'', C) \to \operatorname{Hom}_{\mathbb{Z}}(A \otimes_R B, C) \to \operatorname{Hom}_{\mathbb{Z}}(A \otimes_R B', C)$$

$$\tau''_{A,C}\downarrow \qquad\qquad\qquad \downarrow\tau_{A;C} \qquad\qquad\qquad \downarrow\tau'_{A,C}$$

$$0 \longrightarrow \operatorname{Hom}_R(A, H'') \longrightarrow \operatorname{Hom}_R(A, H) \longrightarrow \operatorname{Hom}_R(A, H'),$$

where $H'' = \operatorname{Hom}_{\mathbb{Z}}(B'', C)$, $H = \operatorname{Hom}_{\mathbb{Z}}(B, C)$, and $H' = \operatorname{Hom}_{\mathbb{Z}}(B', C)$. By the Adjoint Isomorphism, the vertical maps are isomorphisms and the diagram commutes. The bottom row is exact, for it arises from the given exact sequence $B' \to B \to B'' \to 0$ by first applying the left exact (contravariant) functor $\operatorname{Hom}_{\mathbb{Z}}(\square, C)$, and then applying the left exact (covariant) functor $\operatorname{Hom}_R(A, \square)$. Exactness of the top row now follows from Exercise 2.31 on page 95.   •

Exercises

2.27 Let V and W be finite-dimensional vector spaces over a field F, say, and let $v_1, \ldots, v_m$ and $w_1, \ldots, w_n$ be bases of V and W, respectively. Let $S \colon V \to V$ be a linear transformation having matrix $A = [a_{ij}]$, and let $T \colon W \to W$ be a linear transformation having matrix $B = [b_{k\ell}]$. Show that the matrix of $S \otimes T \colon V \otimes_k W \to V \otimes_k W$, with respect to a suitable listing of the vectors $v_i \otimes w_j$, is the $nm \times nm$ matrix K, which we write in block form:

$$A \otimes B = \begin{bmatrix} a_{11}B & a_{12}B & \cdots & a_{1m}B \\ a_{21}B & a_{22}B & \cdots & a_{2m}B \\ \vdots & \vdots & \vdots & \vdots \\ a_{m1}B & a_{m2}B & \cdots & a_{mm}B \end{bmatrix}.$$

Remark. The matrix $A \otimes B$ is called the **Kronecker product** of the matrices A and B. ◄

2.28 Let R be a domain with $Q = \text{Frac}(R)$, its field of fractions. If A is an R-module, prove that every element in $Q \otimes_R A$ has the form $q \otimes a$ for $q \in Q$ and $a \in A$ (instead of $\sum_i q_i \otimes a_i$). (Compare this result with Example 2.67.)

*2.29   **(i)** Let p be a prime, and let p, q be relatively prime. Prove that if A is a p-primary group and $a \in A$, then there exists $x \in A$ with $qx = a$.

(ii) If D is a finite cyclic group of order m, prove that D/nD is a cyclic group of order $d = (m, n)$.

(iii) Let m and n be positive integers, and let $d = (m, n)$. Prove that there is an isomorphism of abelian groups

$$\mathbb{I}_m \otimes \mathbb{I}_n \cong \mathbb{I}_d.$$

(iv) Let G and H be finitely generated abelian groups, so that

$$G = A_1 \oplus \cdots \oplus A_n \quad \text{and} \quad H = B_1 \oplus \cdots \oplus B_m,$$

where A_i and B_j are cyclic groups. Compute $G \otimes_\mathbb{Z} H$ explicitly.

Hint. $G \otimes_\mathbb{Z} H \cong \sum_{i,j} A_i \otimes_\mathbb{Z} B_j$. If A_i or B_j is infinite cyclic, use Proposition 2.58; if both are finite, use part (ii).

*2.30   **(i)** Given $A_R, {}_R B_S$, and ${}_S C$, define $T(A, B, C) = F/N$, where F is the free abelian group on all ordered triples $(a, b, c) \in A \times B \times C$, and N is the subgroup generated by all

$$(ar, b, c) - (a, rb, c),$$

$$(a, bs, c) - (a, b, sc),$$

$$(a + a', b, c) - (a, b, c) - (a', b, c),$$

$$(a, b + b', c) - (a, b, c) - (a, b', c),$$

$$(a, b, c + c') - (a, b, c) - (a, b, c').$$

Define $h: A \times B \times C \to T(A, B, C)$ by $h: (a, b, c) \mapsto a \otimes b \otimes c$, where $a \otimes b \otimes c = (a, b, c) + N$. Prove that this construction gives a solution to the universal mapping problem for triadditive functions.

(ii) Let R be a commutative ring and let $A_1, \ldots, A_n$, M be R-modules, where $n \geq 2$. An **R-*multilinear function*** is a function $h: A_1 \times \cdots \times A_n \to M$ if h is additive in each variable (when we fix the other $n - 1$ variables), and $f(a_1, \ldots, ra_i, \ldots, a_n) = rf(a_1, \ldots, a_i, \ldots, a_n)$ for all i and all $r \in R$. Let F be the free R-module with basis $A_1 \times \cdots \times A_n$, and define $N \subseteq F$ to be the submodule generated by all the elements of the form

$$(a_1, \ldots, ra_i, \ldots, a_n) - r(a_1, \ldots, a_i, \ldots, a_n)$$

and

$$(\ldots, a_i + a'_i, \ldots) - (\ldots, a_i, \ldots) - (\ldots, a'_i, \ldots).$$

Define $T(A_1, \ldots, A_n) = F/N$ and $h: A_1 \times \cdots \times A_n \to T(A_1, \ldots, A_n)$ by $(a_1, \ldots, a_n) \mapsto (a_1, \ldots, a_n) + N$. Prove that h is R-multilinear, and that h and $T(A_1, \ldots, A_n)$ solve the univeral mapping problem for R-multilinear functions.

(iii) Let R be a commutative ring and prove generalized associativity for tensor products of R-modules.

Hint. Prove that any association of $A_1 \otimes \cdots \otimes A_n$ is also a solution to the universal mapping problem.

*2.31 Assume that the following diagram commutes, and that the vertical arrows are isomorphisms.

$$
\begin{array}{ccccccccc}
0 & \longrightarrow & A' & \longrightarrow & A & \longrightarrow & A'' & \longrightarrow & 0 \\
& & \downarrow & & \downarrow & & \downarrow & & \\
0 & \longrightarrow & B' & \longrightarrow & B & \longrightarrow & B'' & \longrightarrow & 0
\end{array}
$$

Prove that the bottom row is exact if and only if the top row is exact.

*2.32 (**3 × 3 *Lemma***) Consider the following commutative diagram in $_R$**Mod** having exact columns.

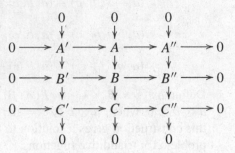

If the bottom two rows are exact, prove that the top row is exact; if the top two rows are exact, prove that the bottom row is exact.

*2.33 Consider the following commutative diagram in $_R$**Mod** having exact rows and columns.

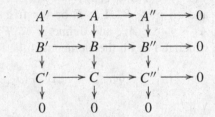

If $A'' \to B''$ and $B' \to B$ are injections, prove that $C' \to C$ is an injection. Similarly, if $C' \to C$ and $A \to B$ are injections, then $A'' \to B''$ is an injection. Conclude that if the last column and the second row are short exact sequences, then the third row is a short exact sequence and, similarly, if the bottom row and the second column are short exact sequences, then the third column is a short exact sequence.

2.34 Give an example of a commutative diagram with exact rows and vertical maps h_1, h_2, h_4, h_5 isomorphisms

$$
\begin{array}{ccccccccc}
A_1 & \longrightarrow & A_2 & \longrightarrow & A_3 & \longrightarrow & A_4 & \longrightarrow & A_5 \\
{\scriptstyle h_1}\downarrow & & {\scriptstyle h_2}\downarrow & & & & \downarrow{\scriptstyle h_4} & & \downarrow{\scriptstyle h_5} \\
B_1 & \longrightarrow & B_2 & \longrightarrow & B_3 & \longrightarrow & B_4 & \longrightarrow & B_5
\end{array}
$$

for which there does not exist a map $h_3 \colon A_3 \to B_3$ making the diagram commute.

*2.35 If $\mathcal{A}$, $\mathcal{B}$, and $\mathcal{C}$ are categories, then a ***bifunctor*** $T \colon \mathcal{A} \times \mathcal{B} \to \mathcal{C}$ assigns, to each ordered pair of objects (A, B), where $A \in \mathrm{ob}(\mathcal{A})$ and $B \in \mathrm{ob}(\mathcal{B})$, an object $T(A, B) \in \mathrm{ob}(\mathcal{C})$, and to each ordered pair of morphisms $f: A \to A'$ in $\mathcal{A}$ and $g: B \to B'$ in $\mathcal{B}$, a morphism $T(f, g): T(A, B) \to T(A', B')$, such that

(a) fixing either variable is a functor; for example, if $A \in \mathrm{ob}(\mathcal{A})$, then $T_A = T(A, \square): \mathcal{B} \to \mathcal{C}$ is a functor, where $T_A(B) = T(A, B)$ and $T_A(g) = T(1_A, g)$,

(b) the following diagram commutes:

$$\begin{array}{ccc} T(A, B) & \xrightarrow{T(1_A,g)} & T(A, B') \\ {\scriptstyle T(f,1_B)}\downarrow & {\scriptstyle T(f,g)}\searrow & \downarrow{\scriptstyle T(f,1_{B'})} \\ T(A', B) & \xrightarrow[T(1_{A'},g)]{} & T(A', B'). \end{array}$$

  (i)   Prove that $\otimes: \mathbf{Mod}_R \times {}_R\mathbf{Mod} \to \mathbf{Ab}$ is a bifunctor.
  (ii) Prove that Hom: ${}_R\mathbf{Mod} \times {}_R\mathbf{Mod} \to \mathbf{Ab}$ is a bifunctor if we modify the definition of bifunctor to allow contravariance in one variable.

*2.36 Let R be a commutative ring, and let F be a free R-module.
  (i)   If $\mathfrak{m}$ is a maximal ideal in R, prove that $(R/\mathfrak{m}) \otimes_R F$ and $F/\mathfrak{m}F$ are isomorphic as vector spaces over $R/\mathfrak{m}$.
  (ii) Prove that $\mathrm{rank}(F) = \dim((R/\mathfrak{m}) \otimes_R F)$.
  (iii) If R is a domain with fraction field Q, prove that $\mathrm{rank}(F) = \dim(Q \otimes_R F)$.

*2.37 Assume that a ring R has IBN; that is, if $R^m \cong R^n$ as left R-modules, then $m = n$. Prove that if $R^m \cong R^n$ as right R-modules, then $m = n$.
  Hint. If $R^m \cong R^n$ as right R-modules, apply $\mathrm{Hom}_R(\square, R)$, using Proposition 2.54(iii).

*2.38 Let R be a domain and let A be an R-module.
  (i)   Prove that if the multiplication $\mu_r: A \to A$ is an injection for all $r \neq 0$, then A is **torsion-free**; that is, there are no nonzero $a \in A$ and $r \in R$ with $ra = 0$.
  (ii) Prove that if the multiplication $\mu_r: A \to A$ is a surjection for all $r \neq 0$, then A is divisible.
  (iii) Prove that if the multiplication $\mu_r: A \to A$ is an isomorphism for all $r \neq 0$, then A is a vector space over Q, where $Q = \mathrm{Frac}(R)$.
        Hint. A module A is a vector space over Q if and only if it is torsion-free and divisible.
  (iv) If either C or A is a vector space over Q, prove that both $C \otimes_R A$ and $\mathrm{Hom}_R(C, A)$ are also vector spaces over Q.

3

Special Modules

There are special modules that make Hom and tensor functors exact; namely, projectives, injectives, and flats.

3.1 Projective Modules

The functors $\operatorname{Hom}_R(X, \square)$ and $\operatorname{Hom}_R(\square, Y)$ almost preserve short exact sequences; they are left exact functors. Similarly, the functors $\square \otimes_R Y$ and $X \otimes_R \square$ almost preserve short exact sequences; they are right exact functors. Are there any functors that do preserve short exact sequences?

Definition. A covariant functor $T : {}_R\mathbf{Mod} \to \mathbf{Ab}$ is an *exact functor* if, for every exact sequence

$$0 \to A \xrightarrow{i} B \xrightarrow{p} C \to 0,$$

the sequence

$$0 \to T(A) \xrightarrow{T(i)} T(B) \xrightarrow{T(p)} T(C) \to 0$$

is also exact. A contravariant functor $T : {}_R\mathbf{Mod} \to \mathbf{Ab}$ is an *exact functor* if there is always exactness of

$$0 \to T(C) \xrightarrow{T(p)} T(B) \xrightarrow{T(i)} T(A) \to 0.$$

In Theorem 2.35, we saw that every left module is a quotient of a free left module. Here is a property of free modules that does not mention bases.

J.J. Rotman, *An Introduction to Homological Algebra*, Universitext, DOI 10.1007/978-0-387-68324-9_3, © Springer Science+Business Media LLC 2009

Theorem 3.1. *Let F be a free left R-module. If $p: A \to A''$ is surjective, then for every $h: F \to A''$, there exists an R-homomorphism g making the following diagram commute*:

$$
\begin{array}{ccc}
 & & F \\
 & {}^{g}\swarrow & \downarrow h \\
A & \xrightarrow{p} A'' & \longrightarrow 0.
\end{array}
$$

Proof. Let B be a basis of F. For each $b \in B$, the element $h(b) \in A''$ has the form $h(b) = p(a_b)$ for some $a_b \in A$, because p is surjective; by the Axiom of Choice, there is a function $u: B \to A$ with $u(b) = a_b$ for all $b \in B$. Proposition 2.34 gives an R-homomorphism $g: F \to A$ with

$$g(b) = a_b \text{ for all } b \in B.$$

Now $pg(b) = p(a_b) = h(b)$, so that pg agrees with h on the basis B; since $\langle B \rangle = F$, we have $pg = h$. •

Definition. A *lifting* of a map $h: C \to A''$ is a map $g: C \to A$ with $pg = h$.

$$
\begin{array}{ccc}
 & & C \\
 & {}^{g}\swarrow & \downarrow h \\
A & \xrightarrow{p} A'' &
\end{array}
$$

That g is a lifting of h says that $h = p_*(g)$, where p_* is the induced map $\text{Hom}_R(C, A) \to \text{Hom}_R(C, A'')$.

If C is any, not necessarily free, module, then a lifting g of h, should one exist, need not be unique. Exactness of

$$0 \to \ker p \xrightarrow{i} A \xrightarrow{p} A'',$$

where i is the inclusion, gives $pi = 0$. Any other lifting has the form $g + if$ for $f: C \to \ker p$; this follows from exactness of

$$0 \to \text{Hom}(C, \ker p) \xrightarrow{i_*} \text{Hom}(C, A) \xrightarrow{p_*} \text{Hom}(C, A''),$$

for any two liftings of h differ by a map $if \in \text{im } i_* = \ker p_*$.

We promote this (basis-free) property of free modules to a definition.

Definition. A left R-module P is *projective* if, whenever p is surjective and h is any map, there exists a lifting g; that is, there exists a map g making the following diagram commute:

$$
\begin{array}{ccc}
 & & P \\
 & {}^{g}\swarrow & \downarrow h \\
A & \xrightarrow{p} A'' & \longrightarrow 0.
\end{array}
$$

Theorem 3.1 says that every free left R-module is projective. Is every projective R-module free? The answer to this question depends on the ring R. Note that if projective R-modules happen to be free, then free modules are characterized without mentioning bases.

Let us now see that projective modules arise in a natural way. We know that the Hom functors are left exact; that is, for any module P, applying $\mathrm{Hom}_R(P, \square)$ to an exact sequence

$$0 \to A' \xrightarrow{i} A \xrightarrow{p} A''$$

gives an exact sequence

$$0 \to \mathrm{Hom}_R(P, A') \xrightarrow{i_*} \mathrm{Hom}_R(P, A) \xrightarrow{p_*} \mathrm{Hom}_R(P, A'').$$

Proposition 3.2. *A left R-module P is projective if and only if $\mathrm{Hom}_R(P, \square)$ is an exact functor.*

Remark.   Since $\mathrm{Hom}_R(P, \square)$ is a left exact functor, the thrust of the proposition is that p_* is surjective whenever p is surjective. ◀

Proof.   If P is projective, then given $h \colon P \to A''$, there exists a lifting $g \colon P \to A$ with $pg = h$. Thus, if $h \in \mathrm{Hom}_R(P, A'')$, then $h = pg = p_*(g) \in \mathrm{im}\, p_*$, and so p_* is surjective. Hence, $\mathrm{Hom}(P, \square)$ is an exact functor.

For the converse, assume that $\mathrm{Hom}(P, \square)$ is an exact functor, so that p_* is surjective: if $h \in \mathrm{Hom}_R(P, A'')$, there exists $g \in \mathrm{Hom}_R(P, A)$ with $h = p_*(g) = pg$. This says that given p and h, there exists a lifting g making the diagram commute; that is, P is projective. •

Proposition 3.3. *A left R-module P is projective if and only if every short exact sequence $0 \to A \xrightarrow{i} B \xrightarrow{p} P \to 0$ splits.*

Proof.   If P is projective, then there exists $j \colon P \to B$ with $1_P = p_*(j) = pj$; that is, P is a retract of B. Corollary 2.23 now gives the result.

$$B \xrightarrow[p]{} P \longrightarrow 0.$$

Conversely, assume that every short exact sequence ending with P splits. Consider the diagram

$$B \xrightarrow[p]{} C \longrightarrow 0$$

with p surjective. Let F be a free left R-module for which there exists a surjective $h\colon F \to P$ (by Theorem 2.35), and consider the augmented diagram

$$
\begin{array}{ccc}
F & \xrightarrow{\ \ h\ \ } & P \\
\scriptstyle g_0 \downarrow & \xleftarrow{\ \ j\ \ } & \downarrow \scriptstyle f \\
B & \xrightarrow{\ \ p\ \ } & C \longrightarrow 0.
\end{array}
$$

By hypothesis, there is a map $j\colon P \to F$ with $hj = 1_P$. Since F is free, there is a map $g_0\colon F \to B$ with $pg_0 = fh$. If we define $g = g_0 j$, then $pg = pg_0 j = fhj = f$. Therefore, P is projective.   •

We restate half this proposition without mentioning the word *exact*.

Corollary 3.4.   *Let A be a submodule of a left R-module B. If B/A is projective, then A has a complement; that is, there is a submodule C of B with $C \cong B/A$ and $B = A \oplus C$.*

The next result gives a concrete characterization of projective modules.

Theorem 3.5.

   (i) *A left R-module P is projective if and only if P is a direct summand of a free left R-module.*

   (ii) *A finitely generated left R-module P is projective if and only if P is a direct summand of R^n for some n.*

Proof.

   (i) Assume that P is projective. By Theorem 2.35, every module is a quotient of a free module. Thus, there are a free module F and a surjection $g\colon F \to P$, and so there is an exact sequence

$$
0 \to \ker g \to F \xrightarrow{\ g\ } P \to 0.
$$

Proposition 3.3 now shows that P is a direct summand of F.

Suppose that P is a direct summand of a free module F, so there are maps $q\colon F \to P$ and $j\colon P \to F$ with $qj = 1_P$. Now consider the diagram

$$
\begin{array}{ccc}
F & \xrightarrow{\ \ q\ \ } & P \\
\scriptstyle h \downarrow & \xleftarrow{\ \ j\ \ } & \downarrow \scriptstyle f \\
B & \xrightarrow{\ \ p\ \ } & C \longrightarrow 0,
\end{array}
$$

where p is surjective. The composite fq is a map $F \to C$; since F is free, it is projective, and so there is a map $h: F \to B$ with $ph = fq$. Define $g: P \to B$ by $g = hj$. It remains to prove that $pg = f$. But

$$pg = phj = fqj = f1_P = f.$$

(ii) Sufficiency follows from part (i). For necessity, let $P = \langle a_1, \ldots, a_n \rangle$. If $\{x_1, \ldots, x_n\}$ is a basis of R^n, define $\varphi: R^n \to P$ by $x_i \mapsto a_i$; the map φ is surjective because P is generated by $a_1, \ldots, a_n$. But $R^n / \ker \varphi \cong P$, so that projectivity shows that P is a direct summand of R^n. •

Corollary 3.6.

(i) *Every direct summand of a projective module is itself projective.*

(ii) *Every direct sum of projective modules is projective.*

Proof.

(i) We can use the proof in the second paragraph of Theorem 3.5. Alternatively, we can use the statement of Theorem 3.5 along with the simple observation that if A is a direct summand of B and B is a direct summand of C, then A is a direct summand of C.

(ii) Let $(P_i)_{i \in I}$ be a family of projective modules. For each i, there exists a free module F_i with $F_i = P_i \oplus Q_i$ for some $Q_i \subseteq F_i$. But $\bigoplus_i F_i$ is free (a basis being the union of the bases of the F_i), and

$$\bigoplus_i F_i = \bigoplus_i (P_i \oplus Q_i) = \bigoplus_i P_i \oplus \bigoplus_i Q_i. \quad •$$

We can now give an example of a commutative ring R and a projective R-module that is not free.

Example 3.7. The ring $R = \mathbb{I}_6$ is the direct sum of two ideals:

$$\mathbb{I}_6 = J \oplus I,$$

where

$$J = \{[0], [2], [4]\} \cong \mathbb{I}_3 \quad \text{and} \quad I = \{[0], [3]\} \cong \mathbb{I}_2.$$

Now $\mathbb{I}_6$ is a free module over itself, and so J and I, being direct summands of a free module, are projective $\mathbb{I}_6$-modules. Neither J nor I can be free, however. After all, a (finitely generated) free $\mathbb{I}_6$-module F is a direct sum of, say, n copies of $\mathbb{I}_6$, and so F has 6^n elements. Therefore, J is too small to be free, for it has only three elements. ◄

The next very general result allows us to focus on countably generated projectives. Let us first consider an ascending sequence of submodules of a module P,

$$\{0\} = P_0 \subseteq P_1 \subseteq P_2 \subseteq \cdots,$$

with $P = \bigcup_{n \geq 0} P_n$. Suppose that each P_n is a direct summand of P_{n+1}; that is, there are complementary submodules X_n with $P_{n+1} = P_n \oplus X_n$ for all n. By induction, we have $P_{n+1} = X_1 \oplus \cdots \oplus X_n$ (since $P_0 = \{0\}$, we have $P_1 = X_1$). We claim that $P = \bigoplus_n X_n$. Now $P = \langle X_n : n \geq 0 \rangle$, because $P = \bigcup_{n \geq 0} P_n$ and $P_{n+1} = \langle X_1, \ldots, X_n \rangle$. To see that the X_n generate their direct sum, suppose that $x_{n_1} + \cdots + x_{n_m} = 0$ is a shortest equation with $x_{n_i} \in X_{n_i}$ and $n_1 < \cdots < n_m$. Then $-x_{n_m} \in X_{n_m} \cap X_{n_{m-1}} = \{0\}$ (for the Xs are ascending), and this gives a shorter equation.

The same reasoning applies to an ascending transfinite sequence of submodules (P_α) indexed by some well-ordered set (which may as well consist of ordinals). The reader may prove that if $P = \bigcup_\alpha P_\alpha$, if each P_α is a direct summand of $P_{\alpha+1}$, and if $P_\alpha = \bigcup_{\beta < \alpha} P_\beta$ for every limit ordinal α, then P is isomorphic to the direct sum $\bigoplus_\alpha (P_{\alpha+1}/P_\alpha)$. (We have essentially treated the limit ordinal case in the previous paragraph.)

Proposition 3.8 (Kaplansky).   *If R is a ring and $P \oplus Q = \bigoplus_{i \in I} M_i$, where I is any (possibly uncountable) index set and each M_i is a countably generated left R-module, then P is a direct sum of countably generated left R-modules.*

Proof.   Write $M = \bigoplus_{i \in I} M_i$. We are going to construct an ascending family $(S_\alpha)_{\alpha \in A}$ of submodules with $M = \bigcup_\alpha S_\alpha$, where A is a well-ordered set, such that

 (i) if α is a limit ordinal, then $S_\alpha = \bigcup_{\beta < \alpha} S_\beta$,

 (ii) $S_{\alpha+1}/S_\alpha$ is countably generated,

 (iii) each $S_\alpha = \bigoplus_{j \in J_\alpha} M_j$ for some $J_\alpha \subseteq A$,

 (iv) $S_\alpha = P_\alpha \oplus Q_\alpha$, where $P_\alpha = S_\alpha \cap P$ and $Q_\alpha = S_\alpha \cap Q$.

Before giving the construction, let us see that such a family can be used to prove the proposition. Now P_α is a direct summand of S_α, by (iv), and S_α is a direct summand of M, by (iii), so that P_α is a direct summand of M. Apply Corollary 2.24(i) to $P_\alpha \subseteq P_{\alpha+1} \subseteq M$ to see that P_α is a direct summand of $P_{\alpha+1}$. Now

$$S_{\alpha+1}/S_\alpha = (P_{\alpha+1} \oplus Q_{\alpha+1})/(P_\alpha \oplus Q_\alpha) = (P_{\alpha+1}/P_\alpha) \oplus (Q_{\alpha+1}/Q_\alpha),$$

by Exercise 2.4 on page 64. It follows that $P_{\alpha+1}/P_\alpha$ is an image of $S_{\alpha+1}/S_\alpha$, which is countably generated, by (ii); hence, $P_{\alpha+1}/P_\alpha$ is countably generated. If α is a limit ordinal, then

$$P_\alpha = S_\alpha \cap P = \left(\bigcup_{\beta < \alpha} S_\beta \right) \cap P = \bigcup_{\beta < \alpha} (S_\beta \cap P) = \bigcup_{\beta < \alpha} P_\beta.$$

As in the preamble to this proposition, we have $P \cong \bigoplus_\alpha \left(P_{\alpha+1}/P_\alpha \right)$. This is what we want, for we have already noted that each $P_{\alpha+1}/P_\alpha$ is countably generated.

Let us construct (S_α). Set $S_0 = \{0\}$. Let $\alpha > 0$, and assume that S_β has been constructed for each $\beta < \alpha$; we must construct S_α. If α is a limit ordinal, define $S_\alpha = \bigcup_{\beta < \alpha} S_\beta$. Let $\alpha = \beta + 1$; we may assume some M_j is not contained in S_β (otherwise, $S_\beta = M$ and we are done). Choose a countable generating set of M_j, say, $x_{11}, x_{12}, x_{13}, \ldots$ (the reason for the double subscript will soon be apparent). As any element of $M = P \oplus Q$, there is an expression $x_{11} = p + q$ with $p \in P, q \in Q$. Now each of p and q has only finitely many nonzero coordinates in the decomposition $M = \bigoplus_i M_i$. The finitely many M_i corresponding to these coordinates generate a countably generated submodule of M; let $x_{21}, x_{22}, x_{23}, \ldots$ be a countable set of generators of it. Next, repeat this procedure for x_{12}, getting a new countable set $x_{31}, x_{32}, x_{33}, \ldots$. We have constructed the first three rows of an infinite matrix. Proceed in this fashion, pursuing the elements along successive diagonals in the order $x_{11}, x_{12}, x_{21}, x_{13}, x_{22}, x_{31}, \ldots$. Let I_β be the set of all the coordinates arising from all the x_{ij} in the infinite matrix, define $J_\alpha = J_\beta \cup I_\beta$, and define $S_\alpha = S_{\beta+1}$ to be the submodule of M generated by S_β and all the xs. Note that $M_j \subseteq S_\alpha$. The reader may check that the family of these S_α satisfies (i) through (iv). •

Corollary 3.9. *Let R be a ring.*

(i) *Every projective left R-module P is a direct sum of countably generated projective left R-modules.*

(ii) *If every countably generated projective left R-module is free, then every projective left R-module is free.*

Proof. By Theorem 3.5(i), P is a direct summand of a free module. Since the ring R is a countably generated left R-module (it is even cyclic), Proposition 3.8 gives $P \cong \bigoplus_{j \in J} P_j$, where each P_j is countably generated, and each P_j is projective, by Corollary 3.6(i). The second statement follows immediately from the first. •

Classifying projective R-modules is a problem very much dependent on the ring R. If R is a PID (principal ideal domain: a domain in which every ideal is principal), then every submodule of a free module is itself free (Corollary 4.15). It follows from Theorem 3.5 that every projective R-module is free in this case. A much harder result is that if $R = k[x_1, \ldots, x_n]$ is the polynomial ring in n variables over a field k, then every projective R-module is also free. This question was raised by J.-P. Serre, and it was proved, independently, by D. Quillen and by A. Suslin (see Theorem 4.100).

There are domains having projective modules that are not free. For example, the ring of all the algebraic integers in an *algebraic number field* K (that is, K is a field extension of $\mathbb{Q}$ of finite degree) is an example of a ***Dedekind ring***. There are many equivalent definitions of Dedekind rings, one of which is that they are domains in which every ideal is a projective module. There are Dedekind rings, even rings of integers in algebraic number fields, that are not PIDs, and any nonprincipal ideal in a Dedekind ring is a projective module that is not free.

Here is another characterization of projective modules. Note that if A is a free left R-module with basis $\{a_i : i \in I\} \subseteq A$, then each $x \in A$ has a unique expression $x = \sum_{i \in I} r_i a_i$, and the coordinate functions $\varphi_i : x \mapsto r_i$ are R-maps $\varphi_i : A \to R$.

Proposition 3.10 (Projective Basis). *A left R-module A is projective if and only if there exist elements $(a_i \in A)_{i \in I}$ and R-maps $(\varphi_i : A \to R)_{i \in I}$ such that*

(i) *for each $x \in A$, almost all $\varphi_i(x) = 0$,*

(ii) *for each $x \in A$, we have $x = \sum_{i \in I} (\varphi_i x) a_i$.*

Moreover, A is generated by $\{a_i : i \in I\} \subseteq A$ in this case.

Proof. Assume that A is projective. There are a free left R-module F and a surjective R-map $\psi : F \to A$; by Proposition 3.3, projectivity of A gives an R-map $\varphi : A \to F$ with $\psi\varphi = 1_A$. Let $\{e_i : i \in I\}$ be a basis of F, and define $a_i = \psi(e_i)$. Now if $x \in A$, then there is a unique expression $\varphi(x) = \sum_i r_i e_i$, where $r_i \in R$ and almost all $r_i = 0$. Define $\varphi_i : A \to R$ by $\varphi_i(x) = r_i$. Of course, given x, we have $\varphi_i(x) = 0$ for almost all i. Finally,

$$x = \psi\varphi(x) = \psi\left(\sum r_i e_i\right)$$
$$= \sum r_i \psi(e_i) = \sum (\varphi_i x)\psi(e_i) = \sum (\varphi_i x) a_i.$$

Since ψ is surjective, A is generated by $\{a_i = \psi(e_i) : i \in I\}$.

Conversely, given $(a_i \in A)_{i \in I}$ and a family of R-maps $(\varphi_i : A \to R)_{i \in I}$ as in the statement, define F to be the free left R-module with basis $\{e_i : i \in I\}$, and define an R-map $\psi : F \to A$ by $\psi : e_i \mapsto a_i$. It suffices to find an R-map $\varphi : A \to F$ with $\psi\varphi = 1_A$, for then A is (isomorphic to) a retract (i.e., A is a direct summand of F), and hence A is projective. Define φ by $\varphi(x) = \sum_i (\varphi_i x) e_i$, for $x \in A$. The sum is finite, by condition (i), and so φ is well-defined. By condition (ii),

$$\psi\varphi(x) = \psi \sum (\varphi_i x) e_i = \sum (\varphi_i x) \psi(e_i) = \sum (\varphi_i x) a_i = x;$$

that is, $\psi\varphi = 1_A$.   •

Definition.   If A is a left R-module, then a family $(a_i \in A)_{i \in I}$ and a family of R-maps $(\varphi_i : A \to R)_{i \in I}$ satisfying the condition in Proposition 3.10 are called a ***projective basis***.

Using sheaves, R. Bkouche, "Pureté, molesse, et paracompacité," *C. R. Acad. Sci. Paris*, Sér. A 270 (1970), 1653–1655, proved the following theorem. Let X be a locally compact Hausdorff space, let $C(X)$ be the ring of all continuous real-valued functions on X, and let J be the ideal in $C(X)$ consisting of all such functions having compact support. Then X is a paracompact space if and only if J is a projective $C(X)$-module. An elementary proof of Bkouche's theorem, using projective bases, is due to R. L. Finney and J. Rotman, "Paracompactness of locally compact Hausdorff spaces," *Mich. Math. J.* 17 (1970), 359–361.

Definition.   Let $X = \{x_i : i \in I\}$ be a basis of a free left R-module F, and let $Y = \{\sum_i r_{ji} x_i : j \in J\}$ be a subset of F. If K is the submodule of F generated by Y, then we say that a module $M \cong F/K$ has ***generators*** X and ***relations*** Y.[1] We also say that the ordered pair $(X|Y)$ is a ***presentation*** of M. An R-module M is ***finitely presented*** if there is an exact sequence

$$R^m \to R^n \to M \to 0,$$

where $m, n \in \mathbb{N}$.

Thus, a module M is *finitely generated* if it has a presentation $(X|Y)$ in which X is finite, while M is *finitely presented* if it has a presentation $(X|Y)$ in which both X and Y are finite. Example 3.14 displays a finitely generated module that is not finitely presented.

Proposition 3.11.   *Every finitely generated projective left R-module P is finitely presented.*

Proof.   Let $P = \langle a_1, \ldots, a_n \rangle$, and let F be the free left R-module with basis $\{x_1, \ldots, x_n\}$. Define $\varphi : F \to P$ by $\varphi : x_j \mapsto a_j$, so there is an exact sequence

$$0 \to \ker \varphi \to F \xrightarrow{\varphi} P \to 0.$$

---

[1] A module is called *free* because it has no entangling relations.

This sequence splits, because P is projective, so that $F \cong P \oplus \ker \varphi$. Now $\ker \varphi$ is finitely generated, for it is a direct summand, hence an image, of the finitely generated module F. Therefore, P is finitely presented. •

If M is a finitely presented left R-module, then there is a short exact sequence

$$0 \to K \to F \to M \to 0,$$

where F is free and both K and F are finitely generated. Equivalently, M is finitely presented if there is an exact sequence

$$F' \to F \to M \to 0,$$

where both F' and F are finitely generated free modules (just map a finitely generated free module F' onto K). Note that the second exact sequence does not begin with "$0 \to$."

Every finitely presented module is finitely generated, but we will soon see that the converse may be false. We begin by comparing two presentations of a module (we generalize a bit by replacing free modules by projectives).

Proposition 3.12 (Schanuel's Lemma). *Given exact sequences*

$$0 \to K \xrightarrow{i} P \xrightarrow{\pi} M \to 0$$

and

$$0 \to K' \xrightarrow{i'} P' \xrightarrow{\pi'} M \to 0,$$

where P and P' are projective, then there is an isomorphism

$$K \oplus P' \cong K' \oplus P.$$

Proof. Consider the diagram with exact rows

$$
\begin{array}{ccccccccc}
0 & \longrightarrow & K & \xrightarrow{\ i\ } & P & \xrightarrow{\ \pi\ } & M & \longrightarrow & 0 \\
 & & \downarrow{\alpha} & & \downarrow{\beta} & & \downarrow{1_M} & & \\
0 & \longrightarrow & K' & \xrightarrow[\ i'\ ]{} & P' & \xrightarrow[\ \pi'\ ]{} & M & \longrightarrow & 0.
\end{array}
$$

Since P is projective, there is a map $\beta \colon P \to P'$ with $\pi'\beta = \pi$; that is, the right square in the diagram commutes. A diagram chase, Proposition 2.71, shows that there is a map $\alpha \colon K \to K'$ making the other square commute. This commutative diagram with exact rows gives an exact sequence

$$0 \to K \xrightarrow{\theta} P \oplus K' \xrightarrow{\psi} P' \to 0,$$

where $\theta \colon x \mapsto (ix, \alpha x)$ and $\psi \colon (u, x') \mapsto \beta u - i'x'$, for $x \in K$, $u \in P$, and $x' \in K'$. Exactness of this sequence is a straightforward calculation that is left to the reader; this sequence splits because P' is projective. •

Corollary 3.13. *If M is finitely presented and*

$$0 \to K \to F \to M \to 0$$

is an exact sequence, where F is a finitely generated free module, then K is finitely generated.

Proof. Since M is finitely presented, there is an exact sequence

$$0 \to K' \to F' \to M \to 0$$

with F' free and with both F' and K' finitely generated. By Schanuel's Lemma, $K \oplus F' \cong K' \oplus F$. Now $K' \oplus F$ is finitely generated because both summands are, so that the left side is also finitely generated. But K, being a summand, is also a homomorphic image of $K \oplus F'$, and hence it is finitely generated. •

We can now give an example of a finitely generated module that is not finitely presented.

Example 3.14. Let R be a commutative ring containing an ideal I that is not finitely generated (see Exercise 3.7 on page 114 for an example). We claim that the R-module $M = R/I$ is finitely generated but not finitely presented. Of course, M is finitely generated; it is even cyclic. If M were finitely presented, then there would be an exact sequence $0 \to K \to F \to M \to 0$ with F free and both K and F finitely generated. Comparing this with the exact sequence $0 \to I \to R \to M \to 0$, as in Corollary 3.13, gives I finitely generated, a contradiction. Therefore, M is not finitely presented. ◄

Finitely generated modules are the most important modules, and they are intimately related to a chain condition.

Definition. A left R-module M (over some ring R) has **ACC** (*ascending chain condition*) if every ascending chain of submodules

$$S_1 \subseteq S_2 \subseteq S_3 \subseteq \cdots$$

stops; that is, there is an integer n with $S_n = S_{n+1} = S_{n+2} = \cdots$.

Proposition 3.15. *The following three conditions are equivalent for a left R-module M.*

(i) *M has ACC on submodules.*

(ii) *M satisfies the **maximum condition**: every nonempty family $\mathcal{F}$ of submodules has a maximal element; that is, there is some $S_0 \in \mathcal{F}$ for which there is no $S \in \mathcal{F}$ with $S_0 \subsetneq S$.*

(iii) *Every submodule of M is finitely generated.*

Proof.

(i) $\Rightarrow$ (ii) Let $\mathcal{F}$ be a family of submodules of M, and assume that $\mathcal{F}$ has no maximal element. Choose $S_1 \in \mathcal{F}$. Since S_1 is not a maximal element, there is $S_2 \in \mathcal{F}$ with $S_1 \subsetneqq S_2$. Now S_2 is not a maximal element in $\mathcal{F}$, and so there is $S_3 \in \mathcal{F}$ with $S_2 \subsetneqq S_3$. Continuing in this way, we can construct an ascending chain of submodules that does not stop, contradicting ACC.

(ii) $\Rightarrow$ (iii) Let S be a submodule of M, and define $\mathcal{F}$ to be the family of all the finitely generated submodules contained in S; of course, $\mathcal{F} \neq \varnothing$ (for $\{0\} \in \mathcal{F}$). By hypothesis, there exists a maximal element $S^* \in \mathcal{F}$. Now $S^* \subseteq S$ because $S^* \in \mathcal{F}$. If S^* is a proper submodule of S, then there is $s \in S$ with $s \notin S^*$. The submodule $S^{**} = \langle S^*, s \rangle \subseteq S$ is finitely generated, and so $S^{**} \in \mathcal{F}$; but $S^* \subsetneqq S^{**}$, contradicting the maximality of S^*. Therefore, $S^* = S$, and so S is finitely generated.

(iii) $\Rightarrow$ (i) Assume that every submodule of M is finitely generated, and let $S_1 \subseteq S_2 \subseteq \cdots$ be an ascending chain of submodules. It is easy to see that the ascending union $S^* = \bigcup_{n \geq 1} S_n$ is a submodule. By hypothesis, S^* is finitely generated, say, $S^* = \langle s_1, \ldots, s_q \rangle$. Now s_i got into S^* by being in S_{n_i} for some n_i. If N is the largest n_i, then $S_{n_i} \subseteq S_N$ for all i; hence, $s_i \in S_N$ for all i, and $S^* = \langle s_1, \ldots, s_q \rangle \subseteq S_N$. If $n \geq N$, then $S^* \subseteq S_N \subseteq S_n \subseteq S^*$; therefore, $S_n = S^*$, the chain stops, and M has ACC.   •

Corollary 3.16.   *The following conditions are equivalent for a ring R.*

(i) *R has ACC on left ideals.*

(ii) *R satisfies the maximum condition: every nonempty family $\mathcal{F}$ of left ideals in R has a maximal element.*

(iii) *Every left ideal in R is finitely generated.*

Proof.   This is the special case of the proposition when $M = {}_R R$.   •

Definition.   A ring R is **left noetherian** if it satisfies any of the equivalent conditions in Corollary 3.16.

Of course, every PID is noetherian. We will soon prove the Hilbert Basis Theorem, which says that if R is left noetherian, then so is $R[x]$ (where we assume the indeterminate x commutes with the coefficients in R). A ring R

is called **right noetherian** if every right ideal is finitely generated. Obviously, every commutative left noetherian ring is right noetherian, and one omits the adjective *left*, calling them **noetherian**. If k is a field, then every finite-dimensional k-algebra R is both left noetherian and right noetherian, for every left or right ideal is a vector space over k, and so every strictly increasing chain of left (or right) ideals has length $\leq \dim_k(R)$. An example of a ring that is noetherian on one side only is given in Exercise 3.8 on page 114.

Corollary 3.17. *Every quotient ring of a left noetherian ring R is left noetherian.*

Proof. Let I be a two-sided ideal in R, so that R/I is a ring. If J is a left ideal in R/I, then $J' = v^{-1}(J)$ is a left ideal in R, where $v: R \to R/I$ is the natural map. Since R is left noetherian, J' is finitely generated; say, $J' = (r_1, \dots, r_n)$. Hence, $J = v(J')$ is generated by $v(r_1), \dots, v(r_n)$. Thus, every left ideal in R/I is finitely generated, and so R/I is left noetherian.   •

Let $\mathbb{Z}\langle x, y \rangle$ be the ring of all polynomials with integer coefficients in noncommuting indeterminates x and y. If (y^2, yx) is the two-sided ideal generated by y^2, yx, then Dieudonné showed that $R = \mathbb{Z}\langle x, y \rangle / (y^2, yx)$ is left noetherian but not right noetherian (see Lam, *A First Course in Noncommutative Rings*, p. 23). It follows from Corollary 3.17 that $\mathbb{Z}\langle x, y \rangle$ is not right noetherian.

Proposition 3.18.

(i) *If R is left noetherian, then every submodule of a finitely generated left R-module M is itself finitely generated.*

(ii) *If R is a PID and an R-module M can be generated by n elements, then every submodule of M can be generated by n or fewer elements.*

Remark. Part (ii) is not true more generally. For example, $R = \mathbb{Q}[x, y]$ is not a PID, and so there is some ideal I that is not principal. Thus, R has one generator while its submodule I cannot be generated by one element.   ◀

Proof.

(i) The proof is by induction on $n \geq 1$, where $M = \langle x_1, \dots, x_n \rangle$. If $n = 1$, then M is cyclic, and so Proposition 2.16 gives $M \cong R/I$ for some left ideal I. If $S \subseteq M$, then the correspondence theorem for rings, Theorem 2.15, gives a left ideal J with $I \subseteq J \subseteq R$ and $S \cong J/I$. But R is left noetherian, so that J, and hence J/I, is finitely generated.

If $n \geq 1$ and $M = \langle x_1, \ldots, x_n, x_{n+1} \rangle$, consider the exact sequence

$$0 \to M' \xrightarrow{i} M \xrightarrow{p} M'' \to 0,$$

where $M' = \langle x_1, \ldots, x_n \rangle$, $M'' = M/M'$, i is the inclusion, and p is the natural map. Note that M'' is cyclic, being generated by $x_{n+1} + M'$. If $S \subseteq M$ is a submodule, there is an exact sequence

$$0 \to S \cap M' \to S \to S/(S \cap M') \to 0.$$

Now $S \cap M' \subseteq M'$, and hence it is finitely generated, by the inductive hypothesis. Furthermore, $S/(S \cap M') \cong (S + M')/M' \subseteq M/M'$, so that $S/(S \cap M')$ is finitely generated, by the base step. Using Exercise 2.18 on page 67, we conclude that S is finitely generated.

(ii) We prove the statement by induction on $n \geq 1$. If M is cyclic, then $M \cong R/I$; if $S \subseteq M$, then $S \cong J/I$ for some ideal J in R containing I. Since R is a PID, J is principal, and so J/I is cyclic.

For the inductive step, we refer to the exact sequence

$$0 \to S \cap M' \to S \to S/(S \cap M') \to 0$$

in part (i), where $M = \langle x_1, \ldots, x_n, x_{n+1} \rangle$ and $M' = \langle x_1, \ldots, x_n \rangle$. By the inductive hypothesis, $S \cap M'$ can be generated by n or fewer elements, while the base step shows that $S/(S \cap M')$ is cyclic. Exercise 2.18 on page 67 shows that S can be generated by $n + 1$ or fewer elements.

●

Corollary 3.19. *If R is a left noetherian ring, then every finitely generated left R-module is finitely presented.*

Proof. If M is a finitely generated R-module, then there are a finitely generated free left R-module F and a surjection $\varphi \colon F \to M$. Since R is left noetherian, Proposition 3.18 says that every submodule of F is finitely generated. In particular, $\ker \varphi$ is finitely generated, and so M is finitely presented.

●

In 1890, Hilbert proved the famous Hilbert Basis Theorem, showing that every ideal in $\mathbb{C}[x_1, \ldots, x_n]$ is finitely generated. As we shall see, the proof is nonconstructive in the sense that it does not give an explicit set of generators of an ideal (nowadays, this is often possible using Gröbner bases). It is reported that when P. Gordan, one of the leading algebraists of the time, first saw Hilbert's proof, he said, "Das ist nicht Mathematik. Das ist Theologie!" ("This is not mathematics. This is theology!"). On the other hand, Gordan said, in 1899 when he published a simplified proof of Hilbert's theorem, "I have convinced myself that theology also has its merits."

Lemma 3.20. *A ring R is left noetherian if and only if, for every sequence $a_1, \ldots, a_n, \ldots$ of elements in R, there exists $m \geq 1$ and $r_1, \ldots, r_m \in R$ with $a_{m+1} = r_1 a_1 + \cdots + r_m a_m$.*

Proof. Assume that R is left noetherian and that $a_1, \ldots, a_n, \ldots$ is a sequence of elements in R. If I_n is the left ideal generated by $a_1, \ldots, a_n$, then there is an ascending chain of left ideals, $I_1 \subseteq I_2 \subseteq \cdots$. By the ACC, there exists $m \geq 2$ with $I_m = I_{m+1}$. Therefore, $a_{m+1} \in I_{m+1} = I_m$, and so there are $r_i \in R$ with $a_{m+1} = r_1 a_1 + \cdots + r_m a_m$.

Conversely, suppose that R satisfies the condition on sequences of elements. If R is not left noetherian, then there is an ascending chain of left ideals $I_1 \subseteq I_2 \subseteq \cdots$ that does not stop. Deleting any repetitions if necessary, we may assume that $I_n \subsetneq I_{n+1}$ for all n. For each n, choose $a_{n+1} \in I_{n+1}$ with $a_{n+1} \notin I_n$. By hypothesis, there exist m and $r_i \in R$ for $i \leq m$ with $a_{m+1} = \sum_{i \leq m} r_i a_i \in I_m$. This contradiction implies that R is left noetherian. •

Notation. If R is a ring, not necessarily commutative, then $R[x]$ denotes the polynomial ring in which the indeterminate x commutes with every element in R.

Theorem 3.21 (Hilbert Basis Theorem).   *If R is a left noetherian ring, then $R[x]$ is also left noetherian.*

Proof. (**Sarges**) Assume that I is a left ideal in $R[x]$ that is not finitely generated; of course, $I \neq \{0\}$. Define $f_0(x)$ to be a polynomial in I of minimal degree and define, inductively, $f_{n+1}(x)$ to be a polynomial of minimal degree in $I - (f_0, \ldots, f_n)$. Note that $f_n(x)$ exists for all $n \geq 0$; if $I - (f_0, \ldots, f_n)$ were empty, then I would be finitely generated. It is clear that

$$\deg(f_0) \leq \deg(f_1) \leq \deg(f_2) \leq \cdots .$$

Let a_n denote the leading coefficient of $f_n(x)$. Since R is left noetherian, Lemma 3.20 applies to give an integer m with $a_{m+1} \in (a_0, \ldots, a_m)$; that is, there are $r_i \in R$ with $a_{m+1} = r_0 a_0 + \cdots + r_m a_m$. Define

$$f^*(x) = f_{m+1}(x) - \sum_{i=0}^{m} x^{d_{m+1} - d_i} r_i f_i(x),$$

where $d_i = \deg(f_i)$. Now $f^*(x) \in I - (f_0(x), \ldots, f_m(x))$; otherwise, $f_{m+1}(x) \in (f_0(x), \ldots, f_m(x))$. It suffices to see that $\deg(f^*) < \deg(f_{m+1})$, for this contradicts $f_{m+1}(x)$ having minimal degree among polynomials in I

that are not in $(f_0, \ldots, f_m)$. If $f_i(x) = a_i x^{d_i} +$ lower terms, then

$$f^*(x) = f_{m+1}(x) - \sum_{i=0}^{m} x^{d_{m+1} - d_i} r_i f_i(x)$$

$$= (a_{m+1} x^{d_{m+1}} + \text{lower terms}) - \sum_{i=0}^{m} x^{d_{m+1} - d_i} r_i (a_i x^{d_i} + \text{lower terms}).$$

The leading term being subtracted is thus $\sum_{i=0}^{m} r_i a_i x^{d_{m+1}} = a_{m+1} x^{d_{m+1}}$. •

Corollary 3.22.

  (i) *If k is a field, then $k[x_1, \ldots, x_n]$ is noetherian.*

  (ii) *The ring $\mathbb{Z}[x_1, \ldots, x_n]$ is noetherian.*

Proof.   The proofs are by induction on $n \geq 1$. •

We are now going to show that every left noetherian ring has IBN.

Lemma 3.23.   *Let R be a ring and let A be a left R-module having ACC. If $\varphi \colon A \to A$ is surjective, then φ is an isomorphism.*

Proof.   For all $n \geq 0$, define $K_n = \ker \varphi^n$; note that $\varphi^0 = 1_A$, so that $K_0 = \{0\}$. Now $K_n \subseteq K_{n+1}$, for if $\varphi^n(x) = 0$, then $\varphi^{n+1}(x) = 0$. Thus, there is an ascending sequence of submodules

$$K_0 \subseteq K_1 \subseteq K_2 \subseteq \cdots .$$

Since A has ACC, this sequence stops; let t be the smallest integer such that $K_t = K_{t+1} = K_{t+2} = \cdots$. We claim that $t = 0$, which will prove the result. Otherwise, $t \geq 1$, and there is $x \in K_t$ with $x \notin K_{t-1}$; that is, $\varphi^t(x) = 0$ and $\varphi^{t-1}(x) \neq 0$. Since φ is surjective, there is $a \in A$ with $x = \varphi(a)$. Hence, $0 = \varphi^t(x) = \varphi^{t+1}(a)$, so that $a \in K_{t+1} = K_t$. Therefore, $0 = \varphi^t(a) = \varphi^{t-1}(\varphi(a)) = \varphi^{t-1}(x)$, a contradiction. Thus, φ is an injection, and hence it is an isomorphism. •

Theorem 3.24.   *If R is a left noetherian ring, then R has IBN.*

Proof.   Let A be a free left R-module, and assume that $A \cong R^m \cong R^n$, where $m \geq n$. If $m > n$, then there is a surjection $\varphi \colon A \to A$ having a nonzero kernel (just project an m-tuple onto its first n coordinates). Now A is obviously finitely generated, so that A has the ACC, by Proposition 3.18. Therefore, φ is an isomorphism, by Lemma 3.23, a contradiction. •

Exercises

3.1 Let M be a free R-module, where R is a domain. Prove that if $rm = 0$, where $r \in R$ and $m \in M$, then either $r = 0$ or $m = 0$. (This is false if R is not a domain.)

*3.2 Let R be a ring and let S be a nonzero submodule of a free right R-module F. Prove that if $a \in R$ is not a right zero-divisor[2], then $Sa \neq \{0\}$.

3.3 Define projectivity in **Groups**, and prove that a group G is projective if and only if G is a free group.
Hint. Recall the Nielsen–Schreier Theorem: Every subgroup of a free group is free.

*3.4   (i)   (**Pontrjagin**) If A is a countable torsion-free abelian group each of whose subgroups S of finite rank is free abelian, prove that A is free abelian (the **rank** of an abelian group S is defined as $\dim_{\mathbb{Q}}(\mathbb{Q} \otimes_{\mathbb{Z}} S)$; cf. Exercise 2.36 on page 97).
Hint. See the discussion on page 103.

   (ii)   Prove that every subgroup of finite rank in $\mathbb{Z}^{\mathbb{N}}$ (the product of countably many copies of $\mathbb{Z}$) is free abelian.

   (iii)   Prove that every countable subgroup of $\mathbb{Z}^{\mathbb{N}}$ is free. (In Theorem 4.17, we will see that $\mathbb{Z}^{\mathbb{N}}$ itself is not free.)

*3.5 (**Eilenberg**) Prove that every projective left R-module P has a free complement; that is, there exists a free left R-module F such that $P \oplus F$ is free.
Hint. If $P \oplus Q$ is free, consider $Q \oplus P \oplus Q \oplus P \oplus \cdots$.

3.6 Let k be a commutative ring, and let P and Q be projective k-modules. Prove that $P \otimes_k Q$ is a projective k-module.

*3.7   (i)   Prove that $R = C(\mathbb{R})$, the ring of all real-valued functions on $\mathbb{R}$ under pointwise operations, is not noetherian.

   (ii)   Recall that $f : \mathbb{R} \to \mathbb{R}$ is a C^{∞}-function if $\partial^n f / \partial x^n$ exists and is continuous for all n. Prove that $R = C^{\infty}(\mathbb{R})$, the ring of all C^{∞}-functions on $\mathbb{R}$ under pointwise operations, is not noetherian.

   (iii)   If k is a commutative ring, prove that $k[X]$, the polynomial ring in infinitely many indeterminates X, is not noetherian.

*3.8 (**Small**) Let R be the ring of all 2×2 matrices $\left[\begin{smallmatrix} a & 0 \\ b & c \end{smallmatrix}\right]$ with $a \in \mathbb{Z}$ and $b, c \in \mathbb{Q}$ is a ring. Schematically, we can describe R as $\left[\begin{smallmatrix} \mathbb{Z} & 0 \\ \mathbb{Q} & \mathbb{Q} \end{smallmatrix}\right]$. Prove that R is left noetherian, but that R is not right noetherian.

---

[2]An element $a \in R$ is a **zero-divisor** if $a \neq 0$ and there exists a nonzero $b \in R$ with $ab = 0$ or $ba = 0$. More precisely, a is a **right zero-divisor** if there is a nonzero b with $ba = 0$; that is, multiplication $r \mapsto ra$ is not an injection $R \to R$.

*3.9 Let V be a vector space over a field k.
  - **(i)**   Prove that V is a free k-module.
  - **(ii)** Prove that a subset B of V is a basis of V considered as a vector space if and only if B is a basis of V considered as a free k-module.

3.10
  - **(i)**   If R is a domain and I and J are nonzero ideals in R, prove that $I \cap J \neq \{0\}$.
  - **(ii)** Let R be a domain and let I be an ideal in R that is a free R-module; prove that I is a principal ideal.

*3.11 Prove that $\mathrm{Hom}_R(P, R) \neq \{0\}$ if P is a nonzero projective left R-module.

3.12 If P is a finitely generated left R-module, prove that P is projective if and only if $1_P \in \mathrm{im}\,\nu$, where $\nu\colon \mathrm{Hom}_R(P, R) \otimes_R P \to \mathrm{Hom}_R(P, P)$ is defined, for all $x \in P$, by $f \otimes x \mapsto \tilde{f}$, where $\tilde{f}\colon y \mapsto f(y)x$.
Hint. Use a projective basis.

*3.13 Let R be a commutative ring, and let A and B be finitely generated R-modules.
  - **(i)**   Prove that $A \otimes_R B$ is a finitely generated R-module.
  - **(ii)** If R is noetherian, prove that $\mathrm{Hom}_R(A, B)$ is a finitely generated R-module.
  - **(iii)** Give an example showing that $\mathrm{Hom}_R(A, B)$ may not be finitely generated if R is not noetherian.

    Hint (Griffith). Let V be an infinite-dimensional vector space over $\mathbb{F}_p$, and let $R = \mathbb{Z} \oplus V$, where $(m, v)(m', v') = (mm', mv' + m'v)$. Then V is an ideal in R that is not finitely generated, and if $A = (R/V)/p(R/V)$, then $\mathrm{Hom}_R(A, R) \cong V$ as R-modules.

3.2   Injective Modules

There is another type of module that turns out to be interesting.

Definition.   A left R-module E is *injective* if, whenever i is an injection, a dashed arrow exists making the following diagram commute.

$$
\begin{array}{ccc}
 & E & \\
 {\scriptstyle f}\uparrow & \nwarrow {\scriptstyle g} & \\
0 \longrightarrow & A \underset{i}{\longrightarrow} & B
\end{array}
$$

In words, every homomorphism from a submodule into E can always be extended to a homomorphism from the big module into E.

Proposition 3.25.   *A left R-module E is injective if and only if $\operatorname{Hom}_R(\square, E)$ is an exact functor.*

Proof.   If

$$0 \to A \xrightarrow{i} B \xrightarrow{p} C \to 0$$

is a short exact sequence, we must prove exactness of

$$0 \to \operatorname{Hom}_R(C, E) \xrightarrow{p^*} \operatorname{Hom}_R(B, E) \xrightarrow{i^*} \operatorname{Hom}_R(A, E) \to 0.$$

Since $\operatorname{Hom}_R(\square, E)$ is a left exact contravariant functor, the thrust of the proposition is that the induced map i^* is surjective whenever i is injective. If $f \in \operatorname{Hom}_R(A, E)$, there exists $g \in \operatorname{Hom}_R(B, E)$ with $f = i^*(g) = gi$; that is, the appropriate diagram commutes, showing that E is an injective module.

For the converse, if E is injective, then given $f : A \to E$, there exists $g : B \to E$ with $gi = f$. Thus, if $f \in \operatorname{Hom}_R(A, E)$, then $f = gi = i^*(g) \in \operatorname{im} i^*$, and so the induced map i^* is surjective. Therefore, $\operatorname{Hom}(\square, E)$ is an exact functor.   •

Compare the next result to Proposition 3.3.

Proposition 3.26.   *If a left R-module E is injective, then every short exact sequence $0 \to E \xrightarrow{i} B \xrightarrow{p} C \to 0$ splits.*

Proof.

$$
\begin{array}{ccc}
 & E & \\
 & {\scriptstyle 1_E}\!\uparrow\;\nwarrow{\scriptstyle q} & \\
0 \longrightarrow & E \dashrightarrow[\;i\;] & B
\end{array}
$$

Since E is injective, there exists $q : B \to E$ making the diagram commute; that is, $qi = 1_E$. Exercise 2.8 on page 65 now gives the result.   •

The converse of Proposition 3.26 is also true; it is Proposition 3.40.

This proposition can be restated without using the word *exact*.

Corollary 3.27.   *If an injective module E is a submodule of a module M, then E is a direct summand of M: there is a complement S with $M = E \oplus S$.*

Proposition 3.28.

(i) *If $(E_k)_{k \in K}$ is a family of injective left R-modules, then $\prod_{k \in K} E_k$ is also an injective left R-module.*

(ii) *Every direct summand of an injective left R-module E is injective.*

Proof.

(i) Consider the diagram in which $E = \prod E_k$.

$$
\begin{array}{ccc}
 & E & \\
f \nwarrow \uparrow & & \searrow g \\
0 \longrightarrow A & \underset{i}{\longrightarrow} & B
\end{array}
$$

Let $p_k : E \to E_k$ be the kth projection, so that $p_k f : A \to E_k$. Since E_k is an injective module, there is $g_k : B \to E_k$ with $g_k i = p_k f$. Now define $g : B \to E$ by $g : b \mapsto (g_k(b))$. The map g does extend f, for if $b = ia$, then

$$g(ia) = (g_k(ia)) = (p_k f a) = fa,$$

because $x = (p_k x)$ for every x in the product.

(ii) Assume that $E = E_1 \oplus E_2$, let $i : E_1 \to E$ be the inclusion, and let $p : E \to E_1$ be the projection (so that $pi = 1_{E_1}$).

$$
\begin{array}{ccc}
E_1 \underset{p}{\overset{i}{\rightleftarrows}} & E_1 \oplus E_2 & \\
f \uparrow & \uparrow \\
0 \longrightarrow B & \underset{j}{\longrightarrow} & C
\end{array}
$$

The reader should be able to complete the proof using the diagram as a guide. ●

Corollary 3.29. *A finite direct sum of injective left R-modules is injective.*

Proof. The direct sum of finitely many modules coincides with the direct product. ●

An infinite direct sum of injective left R-modules need not be injective; indeed, we shall see that all such direct sums are injective if and only if R is left noetherian (see Proposition 3.31 and Theorem 3.39).

The zero module $\{0\}$ is injective, but there are no obvious examples of nonzero injective left R-modules (analogous to free modules as examples of projective modules). Nevertheless, we are going to see that injective modules do exist in abundance. We begin with an important result of R. Baer.

Theorem 3.30 (Baer Criterion).   *A left R-module E is injective if and only if every R-map $f: I \to E$, where I is an ideal in R, can be extended to R.*

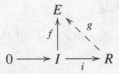

Proof.   Since any left ideal I is a submodule of R, the existence of an extension g of f is just a special case of the definition of injectivity of E.

Suppose we have the diagram

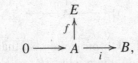

where A is a submodule of a left R-module B. For notational convenience, let us assume that i is the inclusion [this assumption amounts to permitting us to write a instead of $i(a)$ whenever $a \in A$]. We are going to use Zorn's lemma. Let X be the set of all ordered pairs (A', g'), where $A \subseteq A' \subseteq B$ and $g': A' \to E$ extends f; that is, $g'|A = f$. Note that $X \neq \varnothing$ because $(A, f) \in X$. Partially order X by defining

$$(A', g') \preceq (A'', g'')$$

to mean $A' \subseteq A''$ and g'' extends g'. The reader may supply the argument that chains in X have upper bounds in X; hence, Zorn's lemma applies, and there exists a maximal element (A_0, g_0) in X. If $A_0 = B$, we are done, and so we may assume that there is some $b \in B$ with $b \notin A_0$.

Define

$$I = \{r \in R : rb \in A_0\}.$$

It is easy to see that I is a left ideal in R. Define $h: I \to E$ by

$$h(r) = g_0(rb).$$

By hypothesis, there is a map $h^*: R \to E$ extending h. Finally, define $A_1 = A_0 + \langle b \rangle$ and $g_1: A_1 \to E$ by

$$g_1(a_0 + rb) = g_0(a_0) + rh^*(1),$$

where $a_0 \in A_0$ and $r \in R$.

Let us show that g_1 is well-defined. If $a_0 + rb = a_0' + r'b$, then $(r - r')b = a_0' - a_0 \in A_0$; it follows that $r - r' \in I$. Therefore, $g_0((r - r')b)$ and $h(r - r')$ are defined, and we have

$$g_0(a_0' - a_0) = g_0((r - r')b) = h(r - r') = h^*(r - r') = (r - r')h^*(1).$$

Thus, $g_0(a_0') - g_0(a_0) = rh^*(1) - r'h^*(1)$ and $g_0(a_0') + r'h^*(1) = g_0(a_0) + rh^*(1)$, as desired. Clearly, $g_1(a_0) = g_0(a_0)$ for all $a_0 \in A_0$, so that the map g_1 extends g_0. We conclude that $(A_0, g_0) \prec (A_1, g_1)$, contradicting the maximality of (A_0, g_0). Therefore, $A_0 = B$, the map g_0 is a lifting of f, and E is injective. •

Proposition 3.31. *If R is a left noetherian ring and $(E_k)_{k \in K}$ is a family of injective left R-modules, then $\bigoplus_{k \in K} E_i$ is an injective left R-module.*

Proof. By the Baer Criterion, it suffices to complete the diagram

$$
\begin{array}{ccc}
& & \bigoplus_{k \in K} E_k \\
& f \nearrow & \\
0 \longrightarrow I & \xrightarrow{\ i\ } & R,
\end{array}
$$

where I is a left ideal in R. If $x \in \bigoplus_k E_k$, then $x = (e_k)$, where $e_k \in E_k$; define $\mathrm{Supp}(x) = \{k \in K : e_k \neq 0\}$. Since R is left noetherian, I is finitely generated, say, $I = (a_1, \ldots, a_n)$. As any element in $\bigoplus_{k \in K} E_k$, each fa_j, for $j = 1, \ldots, n$, has finite support $\mathrm{Supp}(fa_j) \subseteq K$. Thus, $S = \bigcup_{j=1}^n \mathrm{Supp}(fa_j)$ is a finite set, and so im $f \subseteq \bigoplus_{\ell \in S} E_\ell$; by Corollary 3.29, this finite direct sum is injective. Hence, there is an R-map $g' : R \to \bigoplus_{\ell \in S} E_\ell$ extending f. Composing g' with the inclusion of $\bigoplus_{\ell \in S} E_\ell$ into $\bigoplus_{k \in K} E_k$ completes the given diagram. •

Theorem 3.39 will show that the converse of Proposition 3.31 is true; if every direct sum of injective left R-modules is injective, then R is left noetherian.

We generalize the definition of divisible abelian groups to divisible R-modules.

Definition. Let M be an R-module over a domain R. If $r \in R$ and $m \in M$, then we say that m is ***divisible by*** r if there is some $m' \in M$ with $m = rm'$. We say that M is a ***divisible module*** if each $m \in M$ is divisible by every nonzero $r \in R$.

If R is a domain, $r \in R$, and M is an R-module, then the function $\varphi_r : M \to M$, defined by $\varphi_r : m \mapsto rm$, is an R-map. It is clear that M is a divisible module if and only if φ_r is surjective for every $r \neq 0$.

Remark. One can define divisible left R-modules for every ring R. There are several different definitions in the literature, but we prefer that given by Lam, *Lectures on Modules and Rings*, p. 70. If $r \in R$, define

$$\mathrm{ann}_\ell(r) = \{a \in R : ar = 0\}.$$

Suppose that M is a left R-module and $m \in M$ is divisble by r; i.e., $m = rm'$ for some $m' \in M$. Note that if $ar = 0$, then $am = arm' = 0$. Define M to be **divisible** if this necessary condition is sufficient: if every $m \in M$ is divisible by r whenever $\text{ann}_\ell(r) \subseteq \text{ann}(m)$; that is, whenever $ar = 0$ implies $am = 0$. (This generalizes the definition when R is a domain, for if $r \neq 0$, then $\text{ann}_\ell(r) = \{0\}$ and, of course, $0m = 0$.) A left R-module is divisible if and only if every R-map $f \colon Rr \to M$ extends to an R-map $R \to M$ (Lam, Ibid., Proposition 3.17). ◄

Example 3.32. Let R be a domain.

(i) $\text{Frac}(R)$ is a divisible R-module.

(ii) Direct sums and direct products of divisible R-modules are divisible. It follows that every vector space over $\text{Frac}(R)$ is a divisible R-module.

(iii) Every quotient of a divisible R-module is divisible. It follows that every direct summand of a divisible R-module is divisible, for direct summands are quotients (in fact, they are retracts). ◄

Lemma 3.33. *If R is a domain, then every injective R-module E is a divisible module.*

Proof. Assume that E is injective. Let $e \in E$ and let $r_0 \in R$ be nonzero; we must find $x \in E$ with $e = r_0 x$. Define $f \colon Rr_0 \to E$ by $f(rr_0) = re$ (note that f is well-defined because R is a domain: $rr_0 = r'r_0$ implies $r = r'$). Since E is injective, there exists $h \colon R \to E$ extending f. In particular,

$$e = f(r_0) = h(r_0) = r_0 h(1),$$

so that $x = h(1)$ is the element in E required by the definition of divisible. •

Remark. Lemma 3.33 is true for left R-modules over any ring R if one uses Lam's definition of divisible left R-modules. ◄

The converse of Lemma 3.33 is true for some domains, but it is false in general. For example, Theorem 4.24 shows that if a domain R is not a Dedekind ring, then there exists a divisible R-module that is not injective.

We can now give some examples of injective modules.

Proposition 3.34. *Let R be a domain and let $Q = \text{Frac}(R)$.*

(i) *Q is an injective R-module.*

(ii) *Every vector space E over Q is an injective R-module.*

Proof.

(i) By Baer's Criterion, it suffices to extend an R-map $f\colon I \to Q$, where I is an ideal in R, to all of R. Note first that if $a, b \in I$ are nonzero, then $af(b) = f(ab) = bf(a)$, so that

$$f(a)/a = f(b)/b \text{ in } Q \text{ for all nonzero } a, b \in I;$$

let $c \in Q$ denote their common value (note how I being an ideal is needed to define c: the product ab must be defined, and either factor can be taken outside the parentheses). Define $g\colon R \to Q$ by

$$g(r) = rc$$

for all $r \in R$. It is obvious that g is an R-map. To see that g extends f, suppose that $a \in I$; then

$$g(a) = ac = af(a)/a = f(a).$$

It now follows from Baer's Criterion that Q is an injective R-module.

(ii) If R were noetherian, then direct sums of injective R-modules would be injective, by Proposition 3.31, and the result would follow at once from part (i). Since we are not assuming that R is noetherian, however, we must proceed differently; fortunately, a variation of the proof of part (i) works.

Assume that $f\colon I \to E$ is an R-map from an ideal I to E. For each nonzero $a \in I$, divisibility of E provides $e_a \in E$ with $f(a) = ae_a$. We claim that $e_a = e_b$ for all $a, b \in I$. Since f is an R-map, we have $f(ab) = af(b) = a(be_b)$; similarly, $f(ba) = b(ae_a)$. But R is commutative, so that $ab = ba$ and $abe_a = abe_b$; that is, $ab(e_a - e_b) = 0$. Since E is a vector space over Q and $ab \neq 0$, we have $e_a = e_b$. Define $\widetilde{f}\colon R \to E$ by $\widetilde{f}(r) = rf(1) = re_a$ (for some choice of nonzero $a \in I$). Then $\widetilde{f}$ is an R-map extending f, and so E is injective. •

Remark. If R is a domain with $Q = \mathrm{Frac}(R)$, then the proof of Proposition 3.34(ii) shows that every torsion-free divisible R-module E is injective (recall that an R-module M is *torsion-free* if both $r \in R$ and $m \in M$ nonzero implies $rm \neq 0$). However, this observation does not give a more general result, for every torsion-free divisible R-module is a vector space over Q. ◄

Corollary 3.35. *Let R be a principal ideal domain.*

(i) *An R-module E is injective if and only if it is divisible.*

(ii) *Every quotient of an injective R-module E is itself injective.*

Remark.   This corollary is true for rings R in which every left ideal is principal; the proof uses the notion of divisible left R-module.   ◄

Proof.

(i) We use Baer's Criterion, Theorem 3.30. Assume that $f: I \to E$ is an R-map, where I is a nonzero ideal; by hypothesis, $I = Ra$ for some nonzero $a \in I$. Since E is divisible, there is $e \in E$ with $f(a) = ae$. Define $h: R \to E$ by $h(s) = se$ for all $s \in R$. It is easy to check that h is an R-map; moreover, h extends f, for if $s = ra \in I$, then $h(s) = h(ra) = rae = rf(a) = f(ra)$. Therefore, E is injective.

(ii) Since E is injective, it is divisible; hence, if $M \subseteq E$ is any submodule, then E/M is divisible. By part (i), E/M is injective.   •

Theorem 2.35 says that every module is a quotient of a projective module (actually, it is a stronger result: every module is a quotient of a free module). The next result is the "dual" result for $\mathbb{Z}$-modules: every abelian group can be imbedded in an injective abelian group.

Corollary 3.36.   *Every abelian group M can be imbedded as a subgroup of some injective abelian group.*

Proof.   By Theorem 2.35, there is a free abelian group $F = \bigoplus_i \mathbb{Z}_i$ with $M = F/K$ for some $K \subseteq F$. Now

$$M = F/K = \left(\bigoplus_i \mathbb{Z}_i\right)/K \subseteq \left(\bigoplus_i \mathbb{Q}_i\right)/K,$$

where we have merely imbedded each copy $\mathbb{Z}_i$ of $\mathbb{Z}$ into a copy $\mathbb{Q}_i$ of $\mathbb{Q}$. But Example 3.32 gives each $\mathbb{Q}_i$ divisible, hence gives $\bigoplus_i \mathbb{Q}_i$ divisible, and hence gives divisibility of the quotient $(\bigoplus_i \mathbb{Q}_i)/K$. By Corollary 3.35, $(\bigoplus_i \mathbb{Q}_i)/K$ is injective.   •

Lemma 3.37.   *If D is a divisible abelian group, then $\mathrm{Hom}_{\mathbb{Z}}(R, D)$ is an injective left R-module.*

Proof.   First of all, $\mathrm{Hom}_{\mathbb{Z}}(R, D)$ is a left R-module, by Proposition 2.54: if $f: R \to D$ and $a \in R$, define $(af)(r) = f(ra)$ for all $r \in R$. To prove that $\mathrm{Hom}_{\mathbb{Z}}(R, D)$ is injective, we show that $\mathrm{Hom}_R(\square, \mathrm{Hom}_{\mathbb{Z}}(R, D))$ is an exact functor. By Corollary 2.77, essentially the adjoint isomorphism, this functor is naturally isomorphic to $\mathrm{Hom}_{\mathbb{Z}}(R \otimes_R \square, D)$, which is the composite $\mathrm{Hom}_{\mathbb{Z}}(\square, D) \circ (R \otimes_R \square)$. Since D is $\mathbb{Z}$-injective, by Corollary 3.35, and $R \otimes_R \square$ is naturally isomorphic to the identity functor on $_R\mathbf{Mod}$, both of these functors are exact, and so their composite is also exact.   •

Theorem 3.38. *For every ring R, every left R-module M can be imbedded as a submodule of an injective left R-module.*

Proof. Regard M as an abelian group, and define $\varphi \colon M \to \text{Hom}_{\mathbb{Z}}(R, M)$ by $m \mapsto \varphi_m$, where $\varphi_m(r) = rm$; it is easy to see that φ is a $\mathbb{Z}$-homomorphism, and we now show it is an injection. If $\varphi_m = \varphi_{m'}$, then $rm = \varphi_m(r) = \varphi_{m'}(r) = rm'$ for all $r \in R$; in particular, this is true for $r = 1$, and so $m = m'$.

By Corollary 3.36, there exist a injective abelian group D and an injective $\mathbb{Z}$-homomorphism $i \colon M \to D$. Left exactness of Hom gives an injection $i_* \colon \text{Hom}_{\mathbb{Z}}(R, M) \to \text{Hom}_{\mathbb{Z}}(R, D)$, and so the composite $i_*\varphi$ is an injective $\mathbb{Z}$-map. It remains to show that $i_*\varphi$ is an R-map; that is, if $a \in R$ and $m \in M$, then $(i_*\varphi)(am) = a[(i_*\varphi)(m)]$. Now $(i_*\varphi) \colon am \mapsto i\varphi_{am}$, where $i\varphi_{am} \colon r \mapsto r(am)$ [the function i merely views the element $r(am) \in M$ as an element of D]. On the other hand, $a[(i_*\varphi)(m)] = a(i_*\varphi_m)$, where $a(i_*\varphi_m)(r) = (i_*\varphi_m)(ra)$ [this is the definition of the left module structure on $\text{Hom}_{\mathbb{Z}}(R, D)$]. Hence, $(i_*\varphi_m)(ra) = i(ra)m = (ra)m$, as desired.   ●

After Proposition 3.54, we will use *character modules* to give another proof of Theorem 3.38.

We have seen, in Proposition 3.31, that if R is a left noetherian ring, then every direct sum of injective left R-modules is injective; we now prove the converse.

Theorem 3.39 (Bass–Papp). *If R is a ring for which every direct sum of injective left R-modules is an injective module, then R is left noetherian.*

Proof. We show that if R is not left noetherian, then there are a left ideal I and an R-map from I to a sum of injectives that cannot be extended to R. Since R is not left noetherian, Corollary 3.16 gives a strictly ascending chain of left ideals $I_1 \subsetneq I_2 \subsetneq \cdots$; let $I = \bigcup I_n$. We note that $I/I_n \neq \{0\}$ for all n. By Theorem 3.38, we may imbed I/I_n in an injective left R-module E_n; we claim that $E = \bigoplus_n E_n$ is not injective.

Let $\pi_n \colon I \to I/I_n$ be the natural map. For each $a \in I$, note that $\pi_n(a) = 0$ for large n (because $a \in I_n$ for some n), and so the R-map $f \colon I \to \prod(I/I_n)$, defined by

$$f \colon a \mapsto (\pi_n(a)),$$

does have its image in $\bigoplus_n(I/I_n)$; that is, for each $a \in I$, almost all of the coordinates of $f(a)$ are 0. Composing with the inclusion $\bigoplus(I/I_n) \to \bigoplus E_n = E$, we may regard f as a map $I \to E$. If there is an R-map $g \colon R \to E$ extending f, then $g(1)$ is defined; say, $g(1) = (e_n)$. Choose an index m and choose $a_m \in I$ with $a_m \notin I_m$; since $a_m \notin I_m$, we have $\pi_m(a_m) \neq 0$, and so $g(a_m) = f(a_m)$ has nonzero mth coordinate $\pi_m(a_m)$.

But $g(a_m) = a_m g(1) = a_m(e_n) = (a_m e_n)$, so that $\pi_m(a_m) = a_m e_m$. It follows that $e_m \neq 0$ for all m, and this contradicts $g(1)$ lying in the direct *sum* $E = \bigoplus E_n$. •

Here is the converse of Proposition 3.26.

Proposition 3.40.   *A left R-module E is injective if and only if every short exact sequence* $0 \to E \to B \to C \to 0$ *splits.*

Proof.   Necessity has already been proved in Proposition 3.26, and so we only prove sufficiency. By Theorem 3.38, there is an exact sequence $0 \to E \to M \to M'' \to 0$ with M injective. By hypothesis, this sequence splits, and so $M \cong E \oplus M''$. It follows from Proposition 3.28(ii) that E is injective, for it is a direct summand of an injective module.   •

We can improve this last result by showing that it suffices to consider only those short exact sequences $0 \to A \to B \to C \to 0$ with C cyclic.

Lemma 3.41.   *The diagram with exact row*

$$0 \longrightarrow A \xrightarrow{\alpha} B \xrightarrow{\beta} C \longrightarrow 0$$
$$\gamma \downarrow$$
$$E$$

can be completed to a commutative diagram with exact rows:

$$
\begin{array}{ccccccccc}
0 & \longrightarrow & A & \xrightarrow{\alpha} & B & \xrightarrow{\beta} & C & \longrightarrow & 0 \\
 & & \gamma \downarrow & & \gamma' \downarrow & & \downarrow 1_C & & \\
0 & \longrightarrow & E & \xrightarrow{\alpha'} & P & \xrightarrow{\beta'} & C & \longrightarrow & 0.
\end{array}
$$

Proof.   Our first guess is to define $P = E \oplus B$, $\alpha' : e \mapsto (e, 0)$, and $\gamma' : b \mapsto (0, b)$, but the first square does not commute because $(\gamma a, 0) \neq (0, \alpha a)$. Let

$$S = \{(\gamma a, -\alpha a) : a \in A\},$$

and note that S is a submodule of $E \oplus B$. Make new definitions:

$$P = (E \oplus B)/S, \quad \alpha' : e \mapsto (e, 0) + S, \quad \text{and} \quad \gamma' : b \mapsto (0, b) + S.$$

Now define $\beta' : P \to C$ by $(e, b) + S \mapsto \beta b$. The reader may check that β' is well-defined and that the diagram commutes.

It remains to prove that the bottom row is exact. If $\alpha'(e) = (e, 0) + S$ is zero, then $(e, 0) = (\gamma a, -\alpha a)$ for some $a \in A$. Hence, $0 = -\alpha a$, so that $a = 0$ because α is an injection; thus, $0 = \gamma a = e$, and so α' is an injection.

Now $\operatorname{im}\alpha' \subseteq \ker\beta'$ because $\beta'\alpha' = 0$. Let us prove the reverse inclusion $\ker\beta' \subseteq \operatorname{im}\alpha'$. If $(e, b) + S \in \ker\beta'$, then $\beta b = 0$ and $b \in \ker\beta = \operatorname{im}\alpha$; that is, $b = \alpha a$ for some $a \in A$. Hence, $(e, b) + S = (e, \alpha a) + S = (e + \gamma a, 0) + S \in \operatorname{im}\alpha'$. Finally, β' is surjective, for if $c \in C$, then surjectivity of β gives $c = \beta b$ for some $b \in B$, and so $c = \beta'((0, b) + S)$. $\bullet$

The construction of the first square in the diagram is called a *pushout*, and we will meet it again in Chapter 5.

Proposition 3.42. *A left R-module E is injective if and only if every short exact sequence* $0 \to E \overset{i}{\to} B \overset{p}{\to} C \to 0$ *in which C is cyclic splits.*

Remark. A proof of this proposition using Ext is in Lemma 8.15. ◀

Proof. If E is injective, then Proposition 3.40 says that the sequence splits for every (not necessarily cyclic) module C.

For the converse, let I be a left ideal of R and let $f : I \to E$ be an R-map. By Lemma 3.41, there is a commutative diagram with exact rows

$$
\begin{array}{ccccccccc}
0 & \longrightarrow & I & \overset{i}{\longrightarrow} & R & \longrightarrow & R/I & \longrightarrow & 0 \\
& & f \downarrow & & f' \downarrow & & \downarrow 1 & & \\
0 & \longrightarrow & E & \underset{\alpha'}{\longrightarrow} & P & \longrightarrow & R/I & \longrightarrow & 0.
\end{array}
$$

Since R/I is cyclic, our hypothesis is that the bottom row splits. Thus, there is a map $q : P \to E$ with $q\alpha' = 1_E$. Now define $g : R \to E$ by $g = qf'$. It is easy to see that $gi = f$, and the Baer criterion shows that E is injective. $\bullet$

Theorem 3.38 can be improved, for there is a smallest injective module containing any given module, called its *injective envelope*.

Definition. Let M and E be left R-modules. Then E is an ***essential extension*** of M if there is an injective R-map $\alpha : M \to E$ with $S \cap \alpha(M) \neq \{0\}$ for every nonzero submodule $S \subseteq E$. If also $\alpha(M) \subsetneq E$, then E is called a ***proper essential extension*** of M.

The additive group $\mathbb{Q}$ is an essential extension of $\mathbb{Z}$; indeed, every intermediate subgroup G (i.e., $\mathbb{Z} \subseteq G \subseteq \mathbb{Q}$) is an essential extension of $\mathbb{Z}$.

Proposition 3.43. *A left R-module M is injective if and only if M has no proper essential extensions.*

Proof. Let M be an injective module. If E is an essential extension of M, then there is an injection $\alpha : M \to E$ with $\alpha(M) \neq E$ with $S \cap \alpha(M) \neq \{0\}$ for every nonzero submodule S of E. Since M, and hence $\alpha(M)$, is injective,

Corollary 3.27 says that $\alpha(M)$ is a direct summand of E; that is, there is a submodule $S \subseteq E$ with $E = \alpha(M) \oplus S$, and so $S \cap \alpha(M) = \{0\}$. But S is nonzero, because E is a proper extension. This contradiction shows that M has no proper essential extensions.

Conversely, assume that M has no proper essential extensions. By Theorem 3.38, there exist an injective left R-module E and an injection $i: M \to E$. If E is an essential extension of $i(M)$, then i is an isomorphism and we are done. Otherwise, there exists a nonzero submodule $S \subseteq E$ with $S \cap i(M) = \{0\}$. Using Zorn's lemma, there exists a submodule $N \subseteq E$ maximal such that $S \subseteq N$ and $N \cap i(M) = \{0\}$. If $\pi: E \to E/N$ is the natural map, then $\ker \pi \cap i(M) = N \cap i(M) = \{0\}$, so that $\pi | i(M)$ is an injective R-map. Now Exercise 3.23 on page 130 says that π must also be an injective R-map; that is, $N = \ker \pi = \{0\}$. But $S \subseteq N$ is nonzero; this contradiction shows that E must be an essential extension of $i(M)$, and this completes the proof.   •

Lemma 3.44.   *Given a left R-module M, the following conditions are equivalent for a module $E \supseteq M$.*

(i) *E is a maximal essential extension of M; that is, no proper extension of E is an essential extension of M.*

(ii) *E is an injective module and E is an essential extension of M.*

(iii) *E is an injective module and there is no proper injective intermediate submodule E'; that is, there is no injective E' with $M \subseteq E' \subsetneq E$.*

Proof.   We may assume that M is not injective, for all three statements are equivalent in this case, by Proposition 3.43.

(i) $\Rightarrow$ (ii) Since being an essential extension is a transitive relation, by Exercise 3.20 on page 130, it follows that E has no proper essential extensions. Proposition 3.43 says that E is an injective module.

(ii) $\Rightarrow$ (iii) Assume that E is an injective essential extension of M. If there exists an injective module E' with $M \subseteq E' \subsetneq E$, then E' is a direct summand: $E = E' \oplus E''$, where $E'' \neq \{0\}$. But $M \cap E'' \subseteq E' \cap E'' = \{0\}$, and this contradicts E being an essential extension of M.

(iii) $\Rightarrow$ (i) We show that E is a maximal essential extension of M. Let $\mathcal{F}$ be the family of all submodules $S \subseteq E$ that are essential extensions of M. Now $\mathcal{F} \neq \varnothing$, for $M \in \mathcal{F}$. Partially order $\mathcal{F}$ by inclusion. By Exercise 3.22 on page 130, chains in $\mathcal{F}$ have upper bounds, and so Zorn's lemma says that $\mathcal{F}$ has a maximal element, say, E'. Now E' is an essential extension of M that has no proper essential extension

$N \subseteq E$. Can E' have an essential extension elsewhere? If there is a left R-module N that is an essential extension of E', then consider the diagram

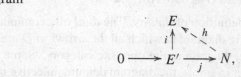

where i and j are inclusions. A map h exists with $hj = i$, because E is injective, so that $hj = h|E$ is an injection. Since N is an essential extension of E, we have h injective, by Exercise 3.23 on page 130. Therefore, $h(N)$ is an essential extension of E' in $\mathcal{F}$. By maximality of E', we have $E' = h(N)$; that is, E' has no proper essential extensions. By Proposition 3.43, E' is injective. But our hypothesis says that there are no injective intermediate submodules, so that either $E' = M$ or $E' = E$. If $E' = M$, then M is injective, which has been considered at the beginning of the proof; if $E' = E$, then E is a maximal essential extension of M. •

Definition. If M is a left R-module, then a left R-module E containing M is an ***injective envelope*** of M, denoted by $\text{Env}(M)$, if any of the equivalent conditions in Lemma 3.44 hold.

Theorem 3.45 (Eckmann–Schöpf). *Let M be a left R-module.*

(i) *There exists an injective envelope* $\text{Env}(M)$ *of M.*

(ii) *If E and E' are injective envelopes of M, then there exists an R-iso-morphism* $\varphi: E \to E'$ *that fixes M pointwise.*

Proof.

(i) Let Z be an injective left R-module containing M, and construct a maximal essential extension of M, as in the proof of (iii) $\Rightarrow$ (i) in Lemma 3.44.

(ii) Let $i: M \to E$ and $j: M \to E'$ be inclusions. Since E' is injective, there exists an R-map $\varphi: E \to E'$ with $\varphi i = j$. Thus, $\varphi|M = \varphi i$ is an injection, and so φ is an injection, by Exercise 3.23 on page 130. Therefore, $\varphi(E) \subseteq E'$ is an intermediate injective submodule, so that $\varphi(E) = E'$, and φ is an isomorphism fixing M pointwise. •

Remark. One might think that injective envelope is functorial, but this is not so. Example 5.20 shows that there is no additive functor $T \colon \mathbf{Ab} \to \mathbf{Ab}$ with $T(G) = \mathrm{Env}(G)$ for all $G \in \mathrm{obj}(\mathbf{Ab})$. ◄

Here is an informal definition of *duality*. The **dual** of a commutative diagram D is the commutative diagram in which all the arrows in D are reversed; that is, the corresponding diagram in the opposite category. Some terms are defined with diagrams; for example, the diagram defining injective modules is the *dual* of the diagram defining projective modules. (We will discuss duality more in Chapter 5 when we will see, in $_R\mathbf{Mod}$, that kernel and cokernel are dual, that direct sum and direct product are dual, and that injective and surjective morphisms are dual; moreover, exact sequence and direct summand are each self-dual.) Informally, the **dual statement** of a statement (see Mac Lane, *Categories for the Working Mathematician*, pp. 31–32, for a formal definition) is the new statement in which every noun, adjective, and diagram is replaced by its dual (when defined). The dual of a theorem may also be a theorem; if so, there are two possibilities. The proofs may be dual: for example, the proof that every short exact sequence $0 \to A \to B \to C \to 0$ with C projective splits is dual to the proof of Proposition 3.26: every short exact sequence $0 \to A \to B \to C \to 0$ with A injective splits. However, proofs may not be dual: for example, the proof that every module is a quotient of a projective module is not dual to the proof that every module is a submodule of an injective module. It is also possible that a theorem that holds in $_R\mathbf{Mod}$ for every ring R may have a false dual. Every left R-module has an injective envelope, but the dual statement is false in some module categories: Example 4.61 shows that a *projective cover* (the dual of an injective envelope) of a $\mathbb{Z}$-module may not exist. Writing a module as a quotient of a free module is the essence of describing it by generators and relations. We may now think of Theorem 3.38 as dualizing this idea.

Exercises

*3.14 Prove the dual of Schanuel's Lemma. Given exact sequences

$$0 \to M \xrightarrow{i} E \xrightarrow{p} Q \to 0 \quad \text{and} \quad 0 \to M \xrightarrow{i'} E' \xrightarrow{p'} Q' \to 0,$$

where E and E' are injective, then there is an isomorphism

$$Q \oplus E' \cong Q' \oplus E.$$

*3.15 (Schanuel) Let B be a left R-module over some ring R, and consider two exact sequences

$$0 \to K \to P_n \to P_{n-1} \to \cdots \to P_1 \to P_0 \to B \to 0$$

and

$$0 \to K' \to Q_n \to \mathbb{Q}_{n-1} \to \cdots \to Q_1 \to Q_0 \to B \to 0,$$

where the Ps and Qs are projective. Prove that

$$K \oplus Q_n \oplus P_{n-1} \oplus \cdots \cong K' \oplus P_n \oplus Q_{n-1} \oplus \cdots.$$

*3.16 Let R be a ring with IBN.

    (i) If $0 \to F_n \to \cdots \to F_0 \to 0$ is an exact sequence with each F_i a finitely generated free R-module, prove that $\sum_{i=0}^{n}(-1)^i \operatorname{rank}(F_i) = 0$.

    (ii) Let $0 \to F_n \to \cdots \to F_0 \to M \to 0$ and $0 \to F'_m \to \cdots \to F'_0 \to M \to 0$ be exact sequences of left R-modules, where each F_i and F'_j is finitely generated and free. Prove that

$$\sum_{i=0}^{n}(-1)^i \operatorname{rank}(F_i) = \sum_{j=0}^{m}(-1)^j \operatorname{rank}(F'_j).$$

The common integer value is called the **Euler characteristic** of M and is denoted by $\chi(M)$.

Hint. Use Exercise 3.15.

    (iii) Let $0 \to M' \to M \to M'' \to 0$ be an exact sequence of finitely generated left R-modules. If two of the modules have an Euler characteristic, prove that the third does also, and

$$\chi(M) = \chi(M') + \chi(M'').$$

Hint. Use Corollary 3.13.

3.17   **(i)** Prove that every vector space over a field k is an injective k-module.

    (ii) Prove that if $0 \to U \to V \to W \to 0$ is an exact sequence of vector spaces, then the corresponding sequence of dual spaces $0 \to W^* \to V^* \to U^* \to 0$ is also exact.

3.18   **(i)** Prove that if a domain R is an injective R-module, then R is a field.

    (ii) Prove that $\mathbb{I}_6$ is simultaneously an injective and a projective module over itself.

    (iii) Let R be a domain that is not a field, and let M be an R-module that is both injective and projective. Prove that $M = \{0\}$.

Hint. Use Exercises 2.22 on page 68 and 3.11 on page 115.

*3.19 (**Pontrjagin Duality**) If G is a (discrete) abelian group, its **Pontrjagin dual** is the group

$$G^* = \mathrm{Hom}_{\mathbb{Z}}(G, \mathbb{R}/\mathbb{Z}).$$

(More generally, the Pontrjagin dual of a locally compact abelian topological group G consists of all the *continuous* homomorphisms from G into the *circle group* $S^1 \cong \mathbb{R}/\mathbb{Z}$.)

(i)   If G is an abelian group and $a \in G$ is nonzero, prove that there is a homomorphism $f : G \to \mathbb{R}/\mathbb{Z}$ with $f(a) \neq 0$.

(ii)   Prove that $\mathbb{R}/\mathbb{Z}$ is an injective abelian group.

(iii)   Prove that if $0 \to A \to G \to B \to 0$ is an exact sequence of abelian groups, then so is $0 \to B^* \to G^* \to A^* \to 0$.

(iv)   If G is a finite abelian group, then $G^* \cong \mathrm{Hom}_{\mathbb{Z}}(G, \mathbb{Q}/\mathbb{Z})$.

(v)   If G is a finite abelian group, prove that $G^* \cong G$.

(vi)   Prove that every quotient group G/H of a finite abelian group G is isomorphic to a subgroup of G.

Remark.   The analogous statement for nonabelian groups is false: if **Q** is the group of quaternions, then $\mathbf{Q}/Z(\mathbf{Q}) \cong$ **V**, where $Z(\mathbf{Q})$ is the center of **Q** and **V** is the four-group. But **Q** has only one element of order 2 while **V** has three elements of order 2, so that **V** is not isomorphic to a subgroup of **Q**. Part (vi) is also false for infinite abelian groups: since $\mathbb{Z}$ has no element of order 2, it has no subgroup isomorphic to $\mathbb{Z}/2\mathbb{Z} = \mathbb{I}_2$. ◄

*3.20   Being an essential extension is transitive. Let $M \subseteq E \subseteq E_1$ be submodules of a left R-module E_1. If E is an essential extension of M and E_1 is an essential extension of E, prove that E_1 is an essential extension of M.

*3.21   **(i)**   Let $M \subseteq E$ be left R-modules. Prove that E is an essential extension of M if and only if, for every nonzero $e \in E$, there is $r \in R$ with $re \in M$ and $re \neq 0$.

(ii)   Let $M \subseteq E$ be left R-modules, and let $\mathcal{S}$ be a chain of intermediate submodules; that is, $M \subseteq S \subseteq E$ for all $S \in \mathcal{S}$ and, if $S, S' \in \mathcal{S}$, either $S \subseteq S'$ or $S' \subseteq S$. If each $S \in \mathcal{S}$ is an essential extension of M, use part (i) to prove that $\bigcup_{S \in \mathcal{S}} S$ is an essential extension of M.

*3.22   Let $M \subseteq E' \subseteq E$ be left R-modules. If both E' and E are essential extensions of M, prove that E is an essential extension of E'.

*3.23   Let E be an essential extension of a left R-module M. If $\varphi : E \to N$ is an R-map with $\varphi|M$ injective, prove that φ is injective.

Hint. Consider $M \cap \ker \varphi$.

3.24 If R is a domain, prove that $\text{Frac}(R) = \text{Env}(R)$, its injective envelope.

*3.25 Recall that every abelian group G having no elements of infinite order has a ***primary decomposition***: $G = \bigoplus_p G_p$, where p is a prime and $G_p = \{g \in G : \text{order } g \text{ is some power of } p\}$. In particular, the p-primary component of $G = \mathbb{Q}/\mathbb{Z}$ is called the ***Prüfer group***; it is denoted by $\mathbb{Z}(p^\infty)$.

  (i) Prove that $\mathbb{Z}(p^\infty)$ is an injective abelian group.

  (ii) Prove that the injective envelope $\text{Env}(\mathbb{I}_{p^n})$ is $\mathbb{Z}(p^\infty)$.

*3.26 (i) If A is the abelian group with the presentation

$$A = (a_n, n \geq 0 \mid pa_0 = 0, \, pa_{n+1} = a_n),$$

prove that $A \cong \mathbb{Z}(p^\infty)$.

  (ii) Give an example of two injective submodules of a module whose intersection is not injective.

  Hint. Define $E = A \oplus \{0\}$ and $E' = \langle \{(a_{n+1}, a_n) : n \geq 0\} \rangle$ in $A \oplus A$.

3.3 Flat Modules

The next type of module arises from tensor products in the same way that projective and injective modules arise from Hom.

Definition.   If R is a ring, then a right R-module A is **flat**[3] if $A \otimes_R \square$ is an exact functor; that is, whenever

$$0 \to B' \xrightarrow{i} B \xrightarrow{p} B'' \to 0$$

is an exact sequence of left R-modules, then

$$0 \to A \otimes_R B' \xrightarrow{1_A \otimes i} A \otimes_R B \xrightarrow{1_A \otimes p} A \otimes_R B'' \to 0$$

is an exact sequence of abelian groups. Flatness of left R-modules is defined similarly.

Because the functors $A \otimes_R \square : {}_R\mathbf{Mod} \to \mathbf{Ab}$ are right exact, we see that a right R-module A is flat if and only if, whenever $i : B' \to B$ is an injection, then $1_A \otimes i : A \otimes_R B' \to A \otimes_R B$ is also an injection.

---

[3] This term arose as the translation into algebra of a geometric property of varieties.

Proposition 3.46.   *Let R be an arbitrary ring.*

(i) *The right R-module R is a flat right R-module.*

(ii) *A direct sum $\bigoplus_j M_j$ of right R-modules is flat if and only each M_j is flat.*

(iii) *Every projective right R-module P is flat.*

Proof.

(i) Consider the commutative diagram

$$
\begin{array}{ccc}
A & \xrightarrow{\;\;i\;\;} & B \\
\sigma\downarrow & & \downarrow\tau \\
R\otimes_R A & \xrightarrow[1_R\otimes i]{} & R\otimes_R B,
\end{array}
$$

where $i\colon A \to B$ is an injection, $\sigma\colon a \mapsto 1\otimes a$, and $\tau\colon b \mapsto 1\otimes b$. Now both σ and τ are isomorphisms, by Proposition 2.58, and so $1_R \otimes i = \tau i \sigma^{-1}$ is an injection. Therefore, R is a flat module over itself.

(ii) By Exercise 2.24 on page 68, any family of R-maps $(f_j\colon U_j \to V_j)_{j\in J}$ can be assembled into an R-map $\varphi\colon \bigoplus_j U_j \to \bigoplus_j V_j$, namely,

$$\varphi\colon (u_j) \mapsto (f_j(u_j));$$

it is easy to see that φ is an injection if and only if each f_j is an injection. Let $i\colon A \to B$ be an injection. There is a commutative diagram

$$
\begin{array}{ccc}
(\bigoplus_j M_j)\otimes_R A & \xrightarrow{\;1\otimes i\;} & (\bigoplus_j M_j)\otimes_R B \\
\downarrow & & \downarrow \\
\bigoplus_j(M_j\otimes_R A) & \xrightarrow[\varphi]{} & \bigoplus_j(M_j\otimes_R B),
\end{array}
$$

where $\varphi\colon (m_j\otimes a) \mapsto (m_j\otimes ia)$, 1 is the identity map on $\bigoplus_j M_j$, and the downward maps are the isomorphisms of Proposition 2.65. By our initial observation, $1\otimes i$ is an injection if and only if each $1_{M_j}\otimes i$ is an injection; this says that $\bigoplus_j M_j$ is flat if and only if each M_j is flat.

(iii) Combining the first two parts, we see that a free right R-module, being a direct sum of copies of R, must be flat. Moreover, since a module is projective if and only if it is a direct summand of a free module, part (ii) shows that projective modules are always flat.   •

The next results will help us recognize whether a given module is flat.

Lemma 3.47. *Let* $0 \to A \xrightarrow{i} B$ *be an exact sequence of left R-modules, and let M be a right R-module. If* $u \in \ker(1_M \otimes i)$, *then there are a finitely generated submodule* $N \subseteq M$ *and an element* $u' \in N \otimes_R A$ *such that*

(i) $u' \in \ker(1_N \otimes i)$,

(ii) $u = (\kappa \otimes 1_A)(u')$, *where* $\kappa \colon N \to M$ *is the inclusion.*

Proof. Let $u = \sum_j m_j \otimes a_j \in \ker(1_M \otimes i)$, where $m_j \in M$ and $a_j \in A$. There is an equation in $M \otimes_R B$,

$$0 = (1_M \otimes i)(u) = \sum_{j=1}^n m_j \otimes ia_j.$$

Let F be the free abelian group with basis $M \times B$, and let S be the subgroup of F consisting of the relations of $F/S \cong M \otimes_R B$ (as in the construction of the tensor product in Proposition 2.45); thus, S is generated by all elements in F of the form

$$(m, b + b') - (m, b) - (m, b'),$$
$$(m + m', b) - (m, b) - (m', b),$$
$$(mr, b) - (m, rb).$$

Let $0 \to S \to F \xrightarrow{\nu} M \otimes_R B \to 0$, where $\nu \colon (m, b) \mapsto m \otimes b$. Since $(1_M \otimes i)(u) = \sum_j m_j \otimes ia_j = 0$ in $M \otimes_R B$, we have $\sum_j m_j \otimes ia_j = \sum_k \nu(m'_k, b'_k) \in \nu(S)$, where $m'_k \in M$ and $b'_k \in B$. Define N to be the submodule of M generated by $m_1, \dots, m_n$ together with the (finite number of) first coordinates m'_k. Of course, N is a finitely generated submodule of M. If we define $u' = \sum_j m_j \otimes a_j$ in $N \otimes_R A$, then $(\kappa \otimes 1_A)(u') = \sum \kappa(m_j) \otimes b_j = u$. Finally, $(1_N \otimes i)(u') = 0$, for we have taken care that all the relations making $(1_N \otimes i)(u) = 0$ are present in $N \otimes_R B$ (identify $N \otimes_R B$ with F'/S', where F' is free with basis $N \times B$). •

Proposition 3.48. *If every finitely generated submodule of a right R-module M is flat, then M is flat.*

Proof. It suffices to prove that exactness of $0 \to A \xrightarrow{i} B$ gives exactness of $0 \to M \otimes_R A \xrightarrow{1_M \otimes i} M \otimes_R B$. If $u \in \ker(1_M \otimes i)$, then the lemma provides a finitely generated submodule $N \subseteq M$ and an element $u' \in N \otimes_R A$ with $u' \in \ker(1_N \otimes i)$ and $u = (\kappa \otimes 1_A)(u')$, where $\kappa \colon N \to M$ is the inclusion. Now $1_N \otimes i$ is injective, by hypothesis, so that $u' = 0$; moreover, $u = (\kappa \otimes 1_A)(u') = 0$. Therefore, $1_M \otimes i$ is an injection and M is flat. •

Definition.   If R is a domain and M is an R-module, then its ***torsion***[4] ***submodule*** is

$$tM = \{ m \in M : rm = 0 \text{ for some nonzero } r \in R \}.$$

We say that M is a ***torsion module*** if $tM = M$; we say that M is ***torsion-free*** if $tM = \{0\}$.

Were R not a domain, then tM might not be a submodule. If $m, m' \in tM$, then there are nonzero $r, r' \in R$ with $rm = 0 = r'm'$. Now $rr'(m - m') = 0$, but it is possible that $rr' = 0$, and so we cannot conclude that $m - m' \in tM$.

For every R-module M over a domain R, there is an exact sequence

$$0 \to tM \to M \to M/tM \to 0.$$

It is easy to check that M/tM is torsion-free, and so every R-module is an extension of a torsion module by a torsion-free module. Moreover, every submodule of a torsion-free module is itself torsion-free.

Proposition 3.49.   *If R is a domain and A is a flat R-module, then A is torsion-free.*

Proof.   Let $Q = \mathrm{Frac}(R)$. Since A is flat, the functor $\square \otimes_R A$ is exact. Hence, exactness of $0 \to R \to Q$ gives exactness of $0 \to R \otimes_R A \to Q \otimes_R A$. But $R \otimes_R A \cong A$ and $Q \otimes_R A$ is torsion-free (it is a vector space over Q).   •

Corollary 3.50.   *If R is a PID, then every torsion-free R-module B is flat.*

Proof.   If R is a PID, then the Fundamental Theorem says that every finitely generated R-module M is a direct sum of cyclic modules. In particular, if M is also torsion-free, then it is a direct sum of copies of R; that is, M is a free module. Thus, every finitely generated submodule M of B is flat, by Proposition 3.46, and so B is flat, by Proposition 3.48.   •

Corollary 3.51.   *An R-module A over a PID R is flat if and only if A is torsion-free.*

Proof.   This follows from Proposition 3.49 and Corollary 3.50.   •

---

[4]There are generalizations of *torsion* to rings R that are not domains. For example, call an element m in a left R-module M a ***torsion element*** if there is a nonzero $r \in R$, not a zero-divisor, with $rm = 0$. Call M ***torsion-free*** if it has no torsion elements. Another generalization involves *torsion theories*. If $\mathcal{A}$ is an abelian category (defined in Chapter 5), then a ***torsion theory*** is an ordered pair $(\mathbf{T}, \mathbf{F})$ of subclasses of $\mathrm{obj}(\mathcal{A})$ that is maximal with the property that $\mathrm{Hom}_{\mathcal{A}}(T, F) = 0$ for all $T \in \mathbf{T}$ and $F \in \mathbf{F}$ (see Rowen, *Ring Theory* I).

Example 3.52. We show that tensor product may not commute with direct products:

$$\mathbb{Q} \otimes_{\mathbb{Z}} \prod_{n \geq 2} \mathbb{I}_n \ncong \prod_{n \geq 2} (\mathbb{Q} \otimes_{\mathbb{Z}} \mathbb{I}_n).$$

The right side is $\{0\}$ because $\mathbb{Q} \otimes_{\mathbb{Z}} \mathbb{I}_n = \{0\}$ for all n, by Proposition 2.73. On the other hand, $\prod_{n \geq 2} \mathbb{I}_n$ contains an element of infinite order: if $\mathbb{I}_n = \langle a_n \rangle$, then there is no positive integer m with $0 = m(a_n) = (ma_n)$; hence, there is an exact sequence $0 \to \mathbb{Z} \to \prod_{n \geq 2} \mathbb{I}_n$. Since $\mathbb{Q}$ is flat, by Corollary 3.50, there is exactness of $0 \to \mathbb{Q} \otimes_{\mathbb{Z}} \mathbb{Z} \to \mathbb{Q} \otimes_{\mathbb{Z}} \prod_{n \geq 2} \mathbb{I}_n$. But $\mathbb{Q} \otimes_{\mathbb{Z}} \mathbb{Z} \cong \mathbb{Q}$, and so $\mathbb{Q} \otimes_{\mathbb{Z}} \prod_{n \geq 2} \mathbb{I}_n \neq \{0\}$. ◄

We are now going to give a connection between flat modules and injective modules (see Exercise 3.19 on page 130).

Definition. If B is a right R-module, define its *character module*[5] B^* as the left R-module

$$B^* = \operatorname{Hom}_{\mathbb{Z}}(B, \mathbb{Q}/\mathbb{Z}).$$

Recall Proposition 2.54: $B^* = \operatorname{Hom}_{\mathbb{Z}}(B, \mathbb{Q}/Z)$ is a left R-module if one defines rf, for $r \in R$ and $f : B \to \mathbb{Q}/\mathbb{Z}$, by $rf : b \mapsto f(br)$. We now improve Proposition 2.42: if $i : A' \to A$ and $p : A \to A''$ are maps and $0 \to \operatorname{Hom}(A'', B) \xrightarrow{p^*} \operatorname{Hom}(A, B) \xrightarrow{i^*} \operatorname{Hom}(A', B)$ is an exact sequence for every module B, then $A' \xrightarrow{i} A \xrightarrow{p} A'' \to 0$ is an exact sequence.

Lemma 3.53. *A sequence of right R-modules*

$$A \xrightarrow{\alpha} B \xrightarrow{\beta} C$$

is exact if and only if the sequence of character modules

$$C^* \xrightarrow{\beta^*} B^* \xrightarrow{\alpha^*} A^*$$

is exact.

Remark. Note the special cases $A = \{0\}$ and $C = \{0\}$. ◄

Proof. If the original sequence is exact, then so is the sequence of character modules, for the contravariant functor $\operatorname{Hom}_{\mathbb{Z}}(\square, \mathbb{Q}/\mathbb{Z})$ is exact, because $\mathbb{Q}/\mathbb{Z}$ is an injective $\mathbb{Z}$-module, by Corollary 3.35.

We prove the converse.

[5]If B is a $\mathbb{Z}$-module, then its character module coincides with its Pontrjagin dual (see Exercise 3.19 on page 130).

$\operatorname{im}\alpha \subseteq \ker\beta$. If $x \in A$ and $\alpha x \notin \ker\beta$, then $\beta\alpha(x) \neq 0$. By Exercise 3.19(i) on page 130, there is a map $f\colon C \to \mathbb{Q}/\mathbb{Z}$ with $f\beta\alpha(x) \neq 0$. Thus, $f \in C^*$ and $f\beta\alpha \neq 0$, which contradicts the hypothesis that $\alpha^*\beta^* = 0$.

$\ker\beta \subseteq \operatorname{im}\alpha$. If $y \in \ker\beta$ and $y \notin \operatorname{im}\alpha$, then $y + \operatorname{im}\alpha$ is a nonzero element of $B/\operatorname{im}\alpha$. Therefore, there is a map $g\colon B/\operatorname{im}\alpha \to \mathbb{Q}/\mathbb{Z}$ with $g(y + \operatorname{im}\alpha) \neq 0$, by Exercise 3.19(i). If $\nu\colon B \to B/\operatorname{im}\alpha$ is the natural map, define $g' = g\nu \in B^*$; note that $g'(y) \neq 0$, for $g'(y) = g\nu(y) = g(y + \operatorname{im}\alpha)$. Now $g'(\operatorname{im}\alpha) = \{0\}$, so that $0 = g'\alpha = \alpha^*(g')$ and $g' \in \ker\alpha^* = \operatorname{im}\beta^*$. Thus, $g' = \beta^*(h)$ for some $h \in C^*$; that is, $g' = h\beta$. Hence, $g'(y) = h\beta(y)$, which is a contradiction, for $g'(y) \neq 0$, while $h\beta(y) = 0$, because $y \in \ker\beta$. •

Proposition 3.54 (Lambek). *A right R-module B is flat if and only if its character module B^* is an injective left R-module.*

Proof. The functors $\operatorname{Hom}_R(\square, \operatorname{Hom}_{\mathbb{Z}}(B, \mathbb{Q}/\mathbb{Z})) = \operatorname{Hom}_R(\square, B^*)$ and $\operatorname{Hom}_{\mathbb{Z}}(\square, \mathbb{Q}/\mathbb{Z})) \circ (B \otimes_R \square)$ are naturally isomorphic, by Corollary 2.77. If B is flat, then each of the functors in the composite is exact, for $\mathbb{Q}/\mathbb{Z}$ is $\mathbb{Z}$-injective; hence, $\operatorname{Hom}_R(\square, B^*)$ is exact and B^* is injective.

Conversely, assume that B^* is an injective left R-module and $A' \to A$ is an injection between left R-modules A' and A. Since $\operatorname{Hom}_R(A, B^*) = \operatorname{Hom}_R(A, \operatorname{Hom}_{\mathbb{Z}}(B, \mathbb{Q}/\mathbb{Z}))$, the (second version of the) adjoint isomorphism, Theorem 2.76, gives a commutative diagram in which the vertical maps are isomorphisms.

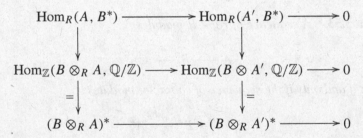

Exactness of the top row now gives exactness of the bottom row. The sequence $0 \to B \otimes_R A' \to B \otimes_R A$ is exact, by Lemma 3.53, and this gives B flat. •

We now sketch another proof of Theorem 3.38: every left R-module M can be imbedded in an injective left R-module. The character module $M^* = \operatorname{Hom}_{\mathbb{Z}}(M, \mathbb{Q}/\mathbb{Z})$ is a right R-module, M^{**} is a left R-module, and the R-map $\varphi\colon M \to M^{**}$, given by $m \mapsto \varphi_m$ [where $\varphi_m(f) = f(m)$ for all $f \in M^*$], is an injection. If F is a free (hence flat) right R-module and $F \to M^* \to 0$ is exact, there is an exact sequence of left R-modules $0 \to M^{**} \to F^{**}$. The composite $M \to M^{**} \to F^{**}$ is an injective R-map of left R-modules, and F^{**} is an injective left R-module, by Proposition 3.54.

Lemma 3.55. *Given modules $({}_R X, {}_R Y_S, Z_S)$, where R and S are rings.*

(i) *There is a natural transformation in X, Y, and Z,*

$$\tau = \tau_{X,Y,Z}: \operatorname{Hom}_S(Y, Z) \otimes_R X \to \operatorname{Hom}_S(\operatorname{Hom}_R(X, Y), Z),$$

that is an isomorphism whenever X is a finitely generated free left R-module.

(ii) *If X is finitely presented and $Z = \mathbb{Q}/\mathbb{Z}$, then*

$$\tau: Y^* \otimes_R X \to \operatorname{Hom}_R(X, Y)^*$$

is an isomorphism.

Proof.

(i) Note that both $\operatorname{Hom}_S(Y, Z)$ and $\operatorname{Hom}_R(X, Y)$ make sense, for Y is a bimodule. If $f \in \operatorname{Hom}_S(Y, Z)$ and $x \in X$, define $\tau_{X,Y,Z}(f \otimes x)$ to be the S-map $\operatorname{Hom}_R(X, Y) \to Z$ given by

$$\tau_{X,Y,Z}(f \otimes x): g \mapsto f(g(x)).$$

It is straightforward to check that $\tau_{X,Y,Z}$ is a homomorphism natural in X, that $\tau_{X,Y,Z}$ is an isomorphism when $X = R$, and, more generally, that $\tau_{X,Y,Z}$ is an isomorphism when X is a finitely generated free left R-module.

(ii) Consider the following diagram, where $F' \to F \to X \to 0$ is an exact sequence with both F' and F finitely generated free modules, and $\tau_\square = \tau_{\square, Y, \mathbb{Q}/\mathbb{Z}}$.

$$
\begin{array}{ccccccc}
Y^* \otimes_R F' & \longrightarrow & Y^* \otimes_R F & \longrightarrow & Y^* \otimes_R X & \longrightarrow & 0 \\
\downarrow{\scriptstyle \tau_{F'}} & & \downarrow{\scriptstyle \tau_F} & & \downarrow{\scriptstyle \tau_X} & & \\
\operatorname{Hom}_R(F', Y)^* & \longrightarrow & \operatorname{Hom}_R(F, Y)^* & \longrightarrow & \operatorname{Hom}_R(X, Y)^* & \longrightarrow & 0
\end{array}
$$

By the naturality in part (i), this diagram commutes [the middle term is $Y^* \otimes_R F = \operatorname{Hom}_{\mathbb{Z}}(Y, \mathbb{Q}/\mathbb{Z}) \otimes_R F$] and the first two vertical maps are isomorphisms. The top row is exact, because $Y^* \otimes_R \square$ is right exact. The bottom row is also exact, because $\operatorname{Hom}_R(\square, Y)^*$ is the composite of the contravariant functors $\operatorname{Hom}_R(\square, Y)$, which is left exact, and $* = \operatorname{Hom}_{\mathbb{Z}}(\square, \mathbb{Q}/\mathbb{Z})$, which is exact. Proposition 2.70 now shows that the third vertical arrow, $\tau_X: Y^* \otimes_R X \to \operatorname{Hom}_R(X, Y)^*$, is an isomorphism.[6] •

---

[6]Proposition 2.72, the Five Lemma, can also be used to prove that τ_X is an isomorphism: just add $\to 0 \to 0$ at the end of each row, and draw downward arrows $0 \to 0$ (which are isomorphisms!).

Theorem 3.56.   *A finitely presented left R-module B is flat if and only if it is projective.*

Proof.   All projective modules are flat, by Proposition 3.46, and so only the converse is significant. Since B is finitely presented, there is an exact sequence

$$F' \to F \to B \to 0,$$

where both F' and F are finitely generated free left R-modules. We begin by showing, for every left R-module Y [which is necessarily an $(R, \mathbb{Z})$-bimodule], that the map $\tau_B = \tau_{B,Y,\mathbb{Q}/\mathbb{Z}} \colon Y^* \otimes_R B \to \mathrm{Hom}_R(B, Y)^*$ of Lemma 3.55 is an isomorphism.

To prove that B is projective, it suffices to prove that $\mathrm{Hom}(B, \square)$ preserves surjections: that is, exactness of $A \to A'' \to 0$ implies exactness of $\mathrm{Hom}(B, A) \to \mathrm{Hom}(B, A'') \to 0$ is exact. By Lemma 3.53, it suffices to show that $0 \to \mathrm{Hom}(B, A'')^* \to \mathrm{Hom}(B, A)^*$ is exact. Consider the diagram

$$
\begin{array}{ccccc}
0 & \longrightarrow & A''^* \otimes_R B & \longrightarrow & A^* \otimes_R B \\
& & \tau \downarrow & & \downarrow \tau \\
0 & \longrightarrow & \mathrm{Hom}(B, A'')^* & \longrightarrow & \mathrm{Hom}(B, A)^*.
\end{array}
$$

Naturality of τ gives commutativity of the diagram, while the vertical maps τ are isomorphisms, by Lemma 3.55(ii), for B is finitely presented. Since $A \to A'' \to 0$ is exact, $0 \to A''^* \to A^*$ is exact, and so the top row is exact, because B is flat. It follows that the bottom row is also exact; that is, $0 \to \mathrm{Hom}(B, A'')^* \to \mathrm{Hom}(B, A'')^*$ is exact, which is what we were to show. Therefore, B is projective.   •

Corollary 3.57.   *If R is left noetherian, then a finitely generated left R-module B is flat if and only if it is projective.*

Proof.   This follows from the theorem once we recall Proposition 3.19: every finitely generated left R-module over a left noetherian ring R is finitely presented.   •

If A is a right R-module and I is a left ideal in R, then

$$AI = \Big\{ \sum_j a_j r_j : a_j \in A \text{ and } r_j \in I \Big\}.$$

Proposition 3.58.   *The following three statements are equivalent for a right R-module A.*

(i)   *A is flat.*

(ii)   *The sequence $0 \to A \otimes_R I \xrightarrow{1_A \otimes i} A \otimes_R R$ is exact for every left ideal I, where $i \colon I \to R$ is the inclusion.*

(iii) *The sequence* $0 \to A \otimes_R J \xrightarrow{1_A \otimes j} A \otimes_R R$ *is exact for every finitely generated left ideal* J, *where* $j : J \to R$ *is the inclusion.*

Remark.   This proposition, mutatis mutandis, also characterizes flat left R-modules.   ◀

Proof.

(i) $\Rightarrow$ (ii) If A is flat, then the sequence $0 \to A \otimes_R I \to A \otimes_R R$ is exact for every left R-module I.

(ii) $\Rightarrow$ (iii) This is obvious.

(iii) $\Rightarrow$ (i) Let I be a left ideal in R. By hypothesis, $0 \to A \otimes_R J \to A \otimes_R R$ is exact for every finitely generated left ideal $J \subseteq I$, and so Lemma 3.47 (for left modules instead of right modules) says that $0 \to A \otimes_R I \xrightarrow{1 \otimes i} A \otimes_R R$ is exact. There is thus an exact sequence of character modules: $(A \otimes_R R)^* \to (A \otimes_R I)^* \to 0$ and, as in the proof of Proposition 3.54, the adjoint isomorphism gives exactness of $\operatorname{Hom}_R(R, A^*) \to \operatorname{Hom}_R(I, A^*) \to 0$. This says that every map from any ideal I to A^* extends to a map $R \to A^*$. Thus, A^* is injective, by the Baer Criterion, and so A is flat, by Proposition 3.54.   ●

Corollary 3.59.   *If A is a flat right R-module and I is a left ideal, then the $\mathbb{Z}$-map $\theta_A : A \otimes_R I \to AI$, given by $a \otimes i \mapsto ai$, is an isomorphism.*

Proof.   Let $\kappa : I \to R$ be the inclusion, and let $\varphi_A : A \otimes_R R \to A$ be the isomorphism $a \otimes r \mapsto ar$ of Proposition 2.58. The composite

$$\varphi_A(1_A \otimes \kappa) : A \otimes_R I \to A \otimes_R R \to A$$

is given by $a \otimes i \mapsto ai \in R$, and its image is AI. Now $1_A \otimes \kappa$ is an injection, because A is flat, and so composing it with the isomomorphism φ_A is an injection. Therefore, the composite $\theta_A(1_A \otimes \kappa)$ is an injection, so that $\theta_A : a \otimes i \mapsto ai$ is an isomorphism.   ●

A quotient of a flat module need not be flat; after all, free modules are flat, and every module is a quotient of a free module.

Proposition 3.60.   *Let $0 \to K \to F \xrightarrow{\varphi} A \to 0$ be an exact sequence of right R-modules in which F is flat. Then A is a flat module if and only if $K \cap FI = KI$ for every finitely generated left ideal I.*

Proof. We give a preliminary discussion before proving the lemma. For every left ideal I, right exactness of $\square \otimes_R I$ gives exactness of

$$K \otimes_R I \to F \otimes_R I \xrightarrow{\varphi \otimes 1} A \otimes_R I \to 0.$$

By Corollary 3.59, there is an isomorphism $\theta_F : F \otimes_R I \to FI$ with $f \otimes i \mapsto fi$; of course, $\theta_K : K \otimes_R I \to KI$ is a surjection. The following diagram commutes, where inc is the inclusion and nat is the natural map.

$$
\begin{array}{ccccccc}
K \otimes_R I & \longrightarrow & F \otimes_R I & \xrightarrow{\varphi \otimes 1} & A \otimes_R I & \longrightarrow & 0 \\
\downarrow{\scriptstyle \theta_K} & & \downarrow{\scriptstyle \theta_F} & & \vert{\scriptstyle \gamma} & & \\
& & & & \downarrow & & \\
KI & \xrightarrow[\text{inc}]{} & FI & \xrightarrow[\text{nat}]{} & FI/KI & \longrightarrow & 0
\end{array}
$$

By Proposition 2.70, there exists a map $\gamma : A \otimes_R I \to FI/KI$, given by $\varphi f \otimes i \mapsto fi + KI$, where $f \in F$ and $i \in I$; since θ_K is a surjection and θ_F is an isomorphism, the map γ is an isomorphism. Now

$$\varphi(FI) = \left\{ \varphi\left(\sum_j f_j i_j\right) : f_j \in F, i_j \in I \right\} = \left\{ \sum_j (\varphi f_j) i_j \right\} = AI.$$

Therefore, the first isomorphism theorem provides an isomorphism

$$\delta : FI/(FI \cap K) \to \varphi(FI) = AI,$$

namely, $fi + (FI \cap K) \mapsto \varphi(fi)$. We assemble these maps to obtain the composite σ:

$$FI/KI \xrightarrow{\gamma^{-1}} A \otimes_R I \xrightarrow{\theta_A} AI \xrightarrow{\delta^{-1}} FI/(FI \cap K).$$

Explicitly, $\sigma : fi + KI \mapsto fi + (FI \cap K)$. But $KI \subseteq FI \cap K$, so that σ is the enlargement of coset map of the third isomorphism theorem and, hence, $\ker \sigma = (FI \cap K)/KI$. Therefore, σ is an isomorphism if and only if $KI = FI \cap K$. Moreover, since the flanking maps γ^{-1} and δ^{-1} are isomorphisms, σ is an isomorphism if and only if θ_A is an isomorphism.

If A is flat, then Lemma 3.59 says that θ_A is an isomorphism. Therefore, σ is an isomorphism and $KI = FI \cap K$. Conversely, if $KI = FI \cap K$ for every finitely generated left ideal I, then θ_A is an isomorphism, and Proposition 3.58 says that A is flat. $\bullet$

Here are some more characterizations of flatness.

Lemma 3.61. *Let $0 \to K \to F \to A \to 0$ be an exact sequence of right R-modules, where F is free with basis $\{x_j : j \in J\}$. For each $v \in F$, define $I(v)$ to be the left ideal in R generated by the "coordinates" $r_1, \ldots, r_t \in R$ of v, where $v = x_{j_1} r_1 + \cdots + x_{j_t} r_t$. Then A is flat if and only if $v \in KI(v)$ for every $v \in K$.*

Proof. If A is flat and $v \in K$, then $v \in K \cap FI(v) = KI(v)$, by Proposition 3.60.

Conversely, let I be any left ideal, and let $v \in K \cap FI$. Then $I(v) \subseteq I$, so the hypothesis gives $v \in KI(v) \subseteq KI$. Hence, $K \cap FI \subseteq KI$. As the reverse inclusion always holds, Proposition 3.60 says that A is flat. •

Theorem 3.62 (Villamayor). *Let $0 \to K \to F \to A \to 0$ be an exact sequence of right R-modules, where F is free. The following statements are equivalent.*

(i) *A is flat.*

(ii) *For every $v \in K$, there is an R-map $\theta : F \to K$ with $\theta(v) = v$.*

(iii) *For every $v_1, \ldots, v_n \in K$, there is an R-map $\theta : F \to K$ with $\theta(v_i) = v_i$ for all i.*

Proof.

(i) $\Rightarrow$ (ii) Assume that A is flat. Choose a basis $\{x_j : j \in J\}$ of F. If $v \in K$, then $I(v)$ is the left ideal generated by $r_1, \ldots, r_t$, where $v = x_{j_1}r_1 + \cdots + x_{j_t}r_t$. By Lemma 3.61, $v \in KI(v)$, and so $v = \sum k_p s_p$, where $k_p \in K$ and $s_p \in I(v)$. Hence, $s_p = \sum u_{pi}r_i$, where $u_{pi} \in R$. Rewrite: $v = \sum k_i' r_i$, where $k_i' = \sum k_p u_{pi} \in K$, and define $\theta : F \to K$ by $\theta(x_{j_i}) = k_i'$ and $\theta(x_j) = 0$ for all other basis elements x_j. Clearly, $\theta(v) = v$.

(ii) $\Rightarrow$ (i) Let $v \in K$, and let $\theta : F \to K$ be a map with $\theta(v) = v$. Choose a basis $\{x_j : j \in J\}$ of F, and write $v = x_{j_1}r_1 + \cdots + x_{j_t}r_t$. Then $v = \theta(v) = \theta(x_{j_1})r_1 + \cdots + \theta(x_{j_t})r_t \in KI(v)$. Hence, A is flat, by Lemma 3.61.

Since (iii) obviously implies (ii), it only remains to prove (ii) $\Rightarrow$ (iii). The proof is by induction on n. The base step is our hypothesis (ii). Let $v_1, \ldots, v_n \in K$, where $n \geq 2$. There is a map $\theta_n : F \to K$ with $\theta_n(v_n) = v_n$. By induction, there is a map $\theta' : F \to K$ with $\theta'[v_i - \theta_n(v_i)] = v_i - \theta_n(v_i)$ for all $i = 1, \ldots, n-1$. Now define $\theta : F \to K$ by

$$\theta(u) = \theta_n(u) + \theta'[u - \theta_n(u)]$$

for all $u \in F$. It is routine to see that $\theta(v_i) = v_i$ for all i. •

Here is a variant of Theorem 3.56.

Theorem 3.63. *A finitely generated right R-module B is projective if and only if it is finitely presented and flat.*

Proof. Every projective module is flat, by Proposition 3.46; if it is also finitely generated, then it is finitely presented, by Proposition 3.11.

Let $0 \to K \to F \to B \to 0$ be an exact sequence of right R-modules, where K, F are finitely generated and F is free. If $K = \langle v_1, \ldots, v_n \rangle$, then Theorem 3.62 gives $\theta \colon F \to K$ with $\theta(v_i) = v_i$ for all i (because A is flat). Therefore, K is a retraction of F, and hence it is a direct summand: $F \cong K \oplus B$. Therefore, B is projective.   •

We have seen that direct sums of injective left R-modules are injective if and only if R is left noetherian. We are going to characterize those rings R for which direct products of flat left R-modules are flat. However, we will not complete the proof of this until Chapter 7.

Definition.   A ring R is called *left coherent* if every finitely generated left ideal is finitely presented.

Example 3.64.

   (i) Every left noetherian ring is left coherent.

  (ii) If k is a field, then the polynomial ring $R = k[X]$ in infinitely many indeterminates X is coherent but not noetherian.

 (iii) There are left coherent rings that are not right coherent (see Lam, *Lectures on Modules and Rings*, p. 138).

 (iv) A *left semihereditary* ring is a ring all of whose finitely generated left ideals are projective. Theorem 4.32 shows that all such rings are left coherent.   ◄

Lemma 3.65.   *Let R be a ring and let A be a right R-module. Then the following statements are equivalent.*

   (i) *A is flat.*

  (ii) *Whenever $\sum_{j=1}^{n} a_j r_{ji} = 0$, where $a_j \in A$, $r_{ji} \in R$, and $i = 1, \ldots, d$, there exist $a_q' \in A$, for $q = 1, \ldots, m$, and $s_{qj} \in R$ with $\sum_j s_{qj} r_{ji} = 0$ for all q, i and $\sum_{q=1}^{m} a_q' s_{qj} = a_j$ for all j.*

 (iii) *Whenever $\sum_{j=1}^{n} a_j r_j = 0$, where $a_j \in A$ and $r_j \in R$, there exist $a_q' \in A$, for $q = 1, \ldots, m$, and $s_{qj} \in R$ with $\sum_j s_{qj} r_j = 0$ for all q and $\sum_{q=1}^{m} a_q' s_{qj} = a_j$ for all j.*

Proof.

(i) $\Rightarrow$ (ii). Consider a short exact sequence $0 \to K \xrightarrow{\text{inc}} F \xrightarrow{\varphi} A \to 0$, where F is a free right R-module, $K = \ker \varphi$, and inc is the inclusion. Choose $y_i, \ldots, y_m \in F$ with $\varphi(y_j) = a_j$. Define $u_i = \sum_j y_j r_{ji}$ for $i = 1, \ldots, d$; note that $u_i \in K$, for $\varphi(u_i) = \varphi(\sum_j y_j r_{ji}) = \sum_j a_j r_{ji} = 0$. Since A is flat, Theorem 3.62 gives an R-map $\theta : F \to K$ with $\theta(u_i) = u_i$ for all i. Let X be a basis of F, and write

$$y_j - \theta(y_j) = \sum_q x_q s_{qj},$$

where $x_q \in X$ and $s_{qj} \in R$. Define $a'_q = \varphi(x_q)$; now

$$a_j = \varphi(y_j) = \varphi(y_j - \theta(y_j))$$

because $\operatorname{im} \theta \subseteq K = \ker \varphi$, and

$$a_j = \varphi(y_j - \theta(y_j)) = \varphi\left(\sum_q x_q s_{qj}\right) = \sum_q a'_q s_{qj}.$$

Finally,

$$\begin{aligned}
0 = u_i - \theta(u_i) &= \sum_j y_j r_{ji} - \theta\left(\sum_j y_j r_{ji}\right) \\
&= \sum_j (y_j - \theta(y_j)) r_{ji} \\
&= \sum_j \left(\sum_q x_q s_{qj}\right) r_{ji} \\
&= \sum_q x_q \left(\sum_j s_{qj} r_{ji}\right).
\end{aligned}$$

Since the x_q are part of the basis X of F, we have $0 = \sum_j s_{qj} r_{ji}$ for all q, i, as desired.

(ii) $\Rightarrow$ (iii). This is the special case of (ii) with $d = 1$.

(iii) $\Rightarrow$ (i). We prove that A is flat using Proposition 3.58: if I is a left ideal, then the map $1_A \otimes \lambda : A \otimes_R I \to A \otimes_R R$ is injective, where $\lambda : I \to R$ is the inclusion. If $\sum_j a_j \otimes r_j \in \ker(1_A \otimes \lambda)$, where $a_j \in A$ and $r_j \in I$, then $\sum_j a_j \otimes r_j = 0$ in $A \otimes_R R$; hence, $\sum_j a_j r_j = 0$ in A (because $A \otimes_R R \cong A$ via $a \otimes r \mapsto ar$). By hypothesis, there exist

$a'_1, \ldots, a'_m \in A$ and $s_{ij} \in R$ with $\sum_j s_{ij} r_j = 0$ and $\sum_i a'_i s_{ij} = a_j$. Substituting,

$$\sum_j a_j \otimes r_j = \sum_j \left(\sum_i a'_i s_{ij} \right) \otimes r_j = \sum_i \left(a'_i \otimes \sum_j s_{ij} r_j \right) = 0.$$

Therefore, $1_A \otimes \lambda$ is injective, and so A is flat.   •

Theorem 3.66 (Chase).   *The following are equivalent for a ring R.*

(i) *For every set X, the right R-module R^X (the direct product of $|X|$ copies of R) is flat.*

(ii) *Every finitely generated submodule of a free left R-module is finitely presented.*

(iii) *R is left coherent.*

Proof.

(i) $\Rightarrow$ (ii) Let B be a finitely generated submodule of a free left R-module G; say, B has generators $b_1, \ldots, b_n$. Since each b_j involves only finitely many elements of a basis of G, we may assume that $G = R^d$, so that its elements are d-tuples. Thus, $b_j = (r_{j1}, \ldots, r_{jd})$ for $j = 1, \ldots, n$. If F is the free left R-module with basis $\{x_1, \ldots, x_n\}$ and if $\varphi : F \to B$ is defined by $\varphi(x_j) = b_j$, then there is an exact sequence

$$0 \to K \to F \xrightarrow{\varphi} B \to 0,$$

where $K = \ker \varphi$. We must show that B is finitely presented, and so it suffices to prove that K is finitely generated. Each $k \in K$ has a unique expression

$$k = a_1(k) x_1 + \cdots + a_n(k) x_n,$$

where $a_j(k) \in R$ is the jth coordinate of $k \in K \subseteq F$. Let us view R^K as a right R-module. For $j = 1, \ldots, n$, define

$$a_j = (a_j(k)) \in R^K.$$

The kth row in the $|K| \times d$ matrix below displays the coordinates of $k \in K$ with respect to the basis $x_1, \ldots, x_d$ of G, while the jth column is $a_j \in R^K$.

$$
\begin{array}{cccc}
\vdots & \vdots & \vdots & \vdots \\
a_1(k) & a_2(k) & \cdots & a_d(k) \\
a_1(k') & a_2(k') & \cdots & a_d(k') \\
\vdots & \vdots & \vdots & \vdots
\end{array}
$$

If $k \in K$, we have $0 = \varphi(k) = a_1(k)b_1 + \cdots + a_n(k)b_n$ in B; that is, $\sum_j a_j(k)b_j = 0$ for all $k \in K$. View these last equations in $B \subseteq R^d$:

$$0 = \sum_j a_j(k)(r_{j1}, \ldots, r_{jd}) = \left(\sum_j a_j(k)r_{j1}, \ldots, \sum_j a_j(k)r_{jd} \right).$$

Thus, all coordinates are 0, and $\sum_j a_j(k)r_{ji} = 0$ for all $i = 1, \ldots, d$; that is, $\sum_j a_j r_{ji} = 0$ for all i. Since R^K is flat, Lemma 3.65 gives $a'_q \in R^K$, for $q = 1, \ldots, m$, and $s_{qj} \in R$, such that $\sum_j s_{qj}r_{ji} = 0$ for all q, i and $\sum_q a'_q s_{qj} = a_j$ for all q. Define

$$z_q = \sum_j s_{qj} x_j \in F.$$

The first set of equations gives $z_q \in K$ for all q:

$$\begin{aligned}
\varphi(z_q) &= \sum_j s_{qj} b_j \\
&= \sum_j s_{qj}(r_{j1}, \ldots, r_{jd}) \\
&= \left(\sum_j s_{qj} r_{j1}, \ldots, \sum_j s_{qj} r_{jd} \right) = 0.
\end{aligned}$$

To prove that K is finitely generated, it is enough to show that $K = \langle z_1, \ldots, z_m \rangle$. Define $a'_q \in R^K$ by

$$a'_q = (a'_q(k)).$$

Rewrite the set of equations $\sum_q a'_q s_{qj} = a_j$ as

$$a_j = (a_j(k)) = \sum_q (a'_q(k)) s_{qj} = \left(\sum_q a'_q(k) s_{qj} \right).$$

Hence $a_j(k) = \sum_q a'_q(k) s_{qj}$. If $k \in K$, then

$$k = \sum_j a_j(k) x_j = \sum_j \left(\sum_q a'_q(k) s_{qj} \right) x_j = \sum_q a'_q(k) z_q;$$

that is, K is finitely generated.

(ii) $\Rightarrow$ (iii) This is a special case, for every finitely generated left ideal is a submodule of the free module R.

(iii) $\Rightarrow$ (i) We use the criterion of Lemma 3.65(iii) to prove that $A = R^Y$ is a flat right R-module. Suppose that

$$\sum_{j=1}^{n} a_j r_j = 0$$

for $a_j \in R^Y$ and $r_j \in R$. Write $a_j = \big(a_j(y)\big)$. Let I be the left ideal generated by $r_1, \ldots, r_n$, let F be the free left R-module with basis $x_1, \ldots, x_n$, define $\varphi \colon F \to I$ by $\varphi(x_j) = r_j$, and consider the exact sequence $0 \to K \to F \xrightarrow{\varphi} I \to 0$, where $K = \ker \varphi$. By hypothesis, K is finitely generated, say, $K = \langle z_1, \ldots, z_t \rangle$. Since $K \subseteq F$, there are equations $z_i = \sum_j s_{ij} x_j$, where $s_{ij} \in R$. For each $y \in Y$, define

$$u(y) = a_1(y) x_1 + \cdots + a_n(y) x_n \in F.$$

Now $\varphi(u(y)) = \sum_j a_j(y) r_j = 0$, for this is the yth coordinate of the original equation $\sum_{j=1}^{n} a_j r_j = 0$. Thus, for each $y \in Y$, we have $u(y) \in K = \langle z_1, \ldots, z_t \rangle$, and so there are $b_i(y) \in R$ with

$$u(y) = \sum_i b_i(y) z_i = \sum_j \Big(\sum_i b_i(y) s_{ij}\Big) x_j.$$

Since $x_1, \ldots, x_n$ is a basis of F, we may equate coordinates: $a_j(y) = \sum_i b_i(y) s_{ij}$ for all j, y. Define $b_i \in R^Y$ by $b_i = \big(b_i(y)\big)$; then $a_j = \sum_i b_i s_{ij}$ for all j. Finally, $\sum_j s_{ij} r_j = \varphi(z_i) = 0$. Therefore, $A = R^Y$ is flat. •

Remark.   Each of the statements in the theorem is equivalent to every direct product of flat right R-modules being flat. As we mentioned earlier, it is more convenient to prove this using the functor Tor (see Theorem 7.9). ◄

3.3.1 Purity

Let us consider loss of exactness from a different viewpoint. We have blamed A if $0 \to B' \to B$ is exact but $0 \to A \otimes_R B' \to A \otimes_R B$ is not exact. Tensoring by "good" modules preserves exactness (and we have called them flat). Perhaps, however, the fault is not in our modules but in our sequences.

Definition.   An exact sequence $0 \to B' \xrightarrow{\lambda} B \to B'' \to 0$ of left R-modules is **pure exact** if, for every right R-module A, we have exactness of $0 \to A \otimes_R B' \xrightarrow{1_A \otimes \lambda} A \otimes_R B \to A \otimes_R B'' \to 0$. We say that $\lambda B' \subseteq B$ is a **pure submodule** in this case.

Every split short exact sequence is pure exact, but the next result shows that the converse is false.

Proposition 3.67. *A left R-module B'' is flat if and only if every exact sequence $0 \to B' \to B \to B'' \to 0$ of left R-modules is pure exact.*

Proof. Let $\to A' \to A \to A'' \to 0$ be an exact sequence of right R-modules. Since tensor is a bifunctor, right exact in each variable, there is a commutative diagram with exact rows and columns.

$$
\begin{array}{ccccccc}
A' \otimes_R B' & \longrightarrow & A' \otimes_R B & \longrightarrow & A' \otimes_R B'' & \longrightarrow & 0 \\
\downarrow & & \downarrow & & \downarrow & & \\
A \otimes_R B' & \longrightarrow & A \otimes_R B & \longrightarrow & A \otimes_R B'' & \longrightarrow & 0 \\
\downarrow & & \downarrow & & \downarrow & & \\
A'' \otimes_R B' & \longrightarrow & A'' \otimes_R B & \longrightarrow & A'' \otimes_R B'' & \longrightarrow & 0 \\
\downarrow & & \downarrow & & \downarrow & & \\
0 & & 0 & & 0 & &
\end{array}
$$

Specialize the diagram so that A is a free right R-module and B'' is a flat left R-module; this forces both the third column and the second row to be short exact sequences. By Exercise 2.33 on page 96, the bottom row is a short exact sequence for every A''; that is, $0 \to B' \to B \to B'' \to 0$ is pure exact.

Conversely, assume that every exact sequence $0 \to B' \to B \to B'' \to 0$ is pure exact; in particular, there is such a sequence with B free. Now take any short exact sequence $0 \to A' \to A \to A'' \to 0$ of right R-modules and form the 3×3 diagram as above (now the middle row is a short exact sequence because B is free and, hence, flat; moreover, all the rows are short exact sequences). Purity says that the bottom row is a short exact sequence, and Exercise 2.33 says that the last column is a short exact sequence; that is, B'' is flat. $\bullet$

See Corollary 7.3 for another proof of Proposition 3.67 using Tor.

By Corollary 3.51, an abelian group D is flat if and only if it is torsion-free. Thus, the sequence $0 \to tG \to G \to G/tG \to 0$ of abelian groups is always pure exact even though it may not split (see Exercise 3.31 on page 151).

Here is a variant of Lemma 3.65.

Lemma 3.68. *Let A be a finitely presented right R-module with generators $a_1, \ldots, a_n$ and relations $\sum_j a_j r_{ji}$, where $i = 1, \ldots, m$. If B is a left R-module with*

$$\sum_{j=1}^{n} a_j \otimes b_j = 0 \quad \text{in } A \otimes_R B,$$

then there exist elements $h_i \in B$ with $b_j = \sum_i r_{ji} h_i$ for all j.

Proof. Let us make the statement precise. We assume that F is a free right R-module with basis $\{x_1, \ldots, x_n\}$, that $\varphi : F \to A$ is defined by $\varphi x_j = a_j$, that there is an exact sequence $0 \to K \xrightarrow{\lambda} F \xrightarrow{\varphi} A \to 0$ with $K = \ker \varphi$ generated by $\sum_j x_j r_{ji}$, where $\lambda \colon K \to F$ is the inclusion. Tensoring by B gives exactness of

$$K \otimes_R B \xrightarrow{\lambda \otimes 1} F \otimes_R B \xrightarrow{\varphi \otimes 1} A \otimes_R B \to 0.$$

By hypothesis, $\sum_j x_j \otimes b_j \in \ker(\varphi \otimes 1) = \operatorname{im}(\lambda \otimes 1)$. But every element of $K \otimes_R B$ has an expression of the form $\sum_{ij} x_j r_{ji} \otimes h_i$, where $h_i \in B$. In particular,

$$\sum_j x_j \otimes b_j = (\lambda \otimes 1) \sum_{ij} x_j r_{ji} \otimes h_i = \sum_j x_j \otimes \left(\sum_i r_{ji} h_i \right).$$

Since F is free on the x_js, every element of $F \otimes_R B = \bigoplus (x_j R) \otimes_R B$ has a unique expression of the form $\sum_j x_j \otimes \beta_j$, where $\beta_j \in B$. It follows that $b_j = \sum_i r_{ji} h_i$ for all j. •

Theorem 3.69 (Cohn).   *Let $\lambda \colon B' \to B$ be an injection of left R-modules. Then $\lambda B'$ is a pure submodule of B if and only if, given any commutative diagram with F_0, F_1 finitely generated free left R-modules, there is a map $F_0 \to B'$ making the upper triangle[7] commute.*

Remark.   The diagrammatic condition can be restated: if $b'_1, \ldots, b'_n \in B'$ satisfy equations $\lambda b'_j = \sum_i r_{ji} b_i$ for each j, where $b_1, \ldots, b_m \in B$ and $r_{ji} \in R$, then there exist $h'_i \in B'$ with $b'_j = \sum_i r_{ji} h'_i$ for all j. ◄

Proof.   Assume that $\lambda B'$ is a pure submodule of B. Let $b'_1, \ldots, b'_n \in B'$, and assume there are equations $\lambda b'_j = \sum_i r_{ji} b_i$ for all j, where $i = 1, \ldots, m$. Let F be the free right R-module with basis $\{x_1, \ldots, x_n\}$, and define $A = F/K$, where $K \subseteq F$ is generated by the m elements $\sum_j x_j r_{ji}$. Obviously, A is a finitely presented module generated by $\{a_i = x_i + K : i = 1, \ldots, m\}$. In $A \otimes_R B$, we have

$$\sum_j a_j \otimes \lambda b'_j = \sum_j a_j \otimes \left(\sum_i r_{ji} b_i \right) = \sum_i \left(\sum_j a_j r_{ji} \otimes b_i \right) = 0.$$

---

[7]If $F_1 \to F_0$ were surjective, then commutativity of the square and of the upper triangle would imply commutativity of the lower triangle.

Purity says that $1 \otimes \lambda$ is injective, so that $\sum_j a_j \otimes b'_j = 0$ in $A \otimes_R B'$. By Lemma 3.68, there are elements $h'_i \in B'$ with $b'_j = \sum_i r_{ji} h'_i$ for all j.

For the converse, we must show that $1 \otimes \lambda \colon A \otimes B' \to A \otimes B$ is an injection for every A. By Exercise 3.42 on page 152, we may assume that A is finitely presented, say, with generators $a_i, \ldots, a_n$ and relations $\sum_j a_j r_{ji}$, $i = 1, \ldots, m$. A typical element of $A \otimes_R B'$ can be written as $\sum_j a_j \otimes b'_j$ for $b'_j \in B'$. If $(1 \times \lambda) \sum_j a_j \otimes b'_j = 0$ in $A \otimes_R B$, is $\sum_j a_j \otimes b'_j = 0$ in $A \otimes_R B'$? By Lemma 3.42, there are elements $h_i \in B$ with $\lambda b'_j = \sum_i r_{ji} h_i$ for all j. By hypothesis, there are elements $h'_i \in B'$ with $b'_j = \sum_i r_{ji} h'_i$ for all j. Therefore,

$$\sum_j a_j \otimes b'_j = \sum_j a_j \otimes \left(\sum_i a_j r_{ji} \right) = \sum_i \left(\sum_j a_j r_{ji} \right) \otimes h'_i = 0 \quad \text{in } A \otimes_R B'.$$

Hence, $1 \otimes \lambda$ is an injection, and so $\lambda B'$ is a pure submodule of B. $\quad \bullet$

Lemma 3.70. *Let* $0 \to B' \xrightarrow{\ i\ } B \xrightarrow{\ p\ } B'' \to 0$ *be a pure exact sequence, where i is the inclusion. If M is a finitely presented left R-module, then* $p_* \colon \operatorname{Hom}_R(M, B) \to \operatorname{Hom}_R(M, B'')$ *is surjective.*

Proof. Since M is finitely presented, there is an exact sequence

$$R^m \xrightarrow{\ f\ } R^n \xrightarrow{\ g\ } M \to 0.$$

If $\varphi \in \operatorname{Hom}_R(M, B'')$, we construct the commutative diagram with exact rows:

$$
\begin{array}{ccccccccc}
R^m & \xrightarrow{\ f\ } & R^n & \xrightarrow{\ g\ } & M & \longrightarrow & 0 \\
{\scriptstyle\sigma}\downarrow & {\scriptstyle\eta}\swarrow & {\scriptstyle\tau}\downarrow & {\scriptstyle\psi}\swarrow & {\scriptstyle\varphi}\downarrow & & \\
0 & \longrightarrow & B' & \xrightarrow{\ i\ } & B & \xrightarrow{\ p\ } & B'' & \longrightarrow & 0.
\end{array}
$$

Since R^n is free, hence projective, the map $\varphi g \colon R^n \to B''$ can be lifted to $\tau \colon R^n \to B$; that is, a map τ exists making the square on the right commute. Now $p\tau f = \varphi g f = 0$, so that $\operatorname{im} \tau f \subseteq \ker p = \operatorname{im} i = B'$. Thus, if we define $\sigma = \tau f$, then the first square commutes. By Theorem 3.69, pure exactness gives a map $\eta \colon R^n \to B'$ with $\eta f = \sigma$. Now $i\eta f = i\sigma = \tau f$, so that $(\tau - i\eta) f = 0$. Hence, if we define $\tau' = \tau - i\eta \colon R^n \to B$, then $\operatorname{im} f \subseteq \ker \tau'$; thus, τ' induces a map $\psi \colon M \to B$ with $\tau' = \psi g$ (for $M \cong R^n / \operatorname{im} f$). But $p\psi g = p\tau' = p(\tau - i\eta) = p\tau = \varphi g$. Since g is surjective, we conclude that $p\psi = \varphi$. $\quad \bullet$

Proposition 3.71. *Let* $0 \to B' \to B \xrightarrow{p} B'' \to 0$ *be an exact sequence of left R-modules.*

(i) *If B'' is finitely presented, then this sequence is pure exact if and only if it is split.*

(ii) *If R is left noetherian and B'' is finitely generated, then the sequence is pure exact if and only if it splits.*

Proof.

(i) If we define $M = B''$ and $\varphi = 1_{B''}$, then Lemma 3.70 provides a map $\psi : B'' \to B$ with $p\psi = 1_{B''}$.

(ii) Every finitely generated module over a noetherian ring is finitely presented. •

In the theory of Abelian Groups, one calls a subgroup S of a group G a *pure subgroup* if $S \cap nG = nS$ for all $n \in \mathbb{Z}$. We now show that this notion of pure subgroup coincides with that of pure $\mathbb{Z}$-submodule.

Corollary 3.72. *Let* $0 \to S \xrightarrow{\lambda} G \to G/S \to 0$ *be an exact sequence of abelian groups, where λ is the inclusion. This sequence is pure exact if and only if S is pure in the sense of abelian groups; that is, $S \cap nG = nS$ for all $n \in \mathbb{Z}$.*

Proof.   Necessity is the special case of Theorem 3.69 with $n = 1 = m$.

For the converse, it suffices to prove that $1_A \otimes \lambda : A \otimes_{\mathbb{Z}} S \to A \otimes_{\mathbb{Z}} G$ is an injection for every abelian group A. Suppose that $A = \langle a \rangle$ is cyclic. If A is infinite cyclic, then $A \cong \mathbb{Z}$ is flat, and $1_A \otimes \lambda$ is injective. Thus, we may assume that A has a presentation $A = (a \mid qa)$ for some $q > 0$. Now a typical element $u \in A \otimes_{\mathbb{Z}} G$ is $\sum_j k_j a \otimes g_j$, where $k_j \in \mathbb{Z}$ and $g_j \in G$. But

$$u = \sum_j k_j a \otimes g_j = \sum_j (a \otimes k_j g_j) = a \otimes \left(\sum_j k_j g_j \right);$$

that is, $u = a \otimes g$ for some $g \in G$. If $a \otimes s \in A \otimes_{\mathbb{Z}} S$ lies in $\ker(1_A \otimes \lambda)$, then $a \otimes s = 0$ in $A \otimes_{\mathbb{Z}} G$, and Lemma 3.68 (with $n = 1 = m$) gives $h \in G$ with $s = qh \in S \cap qG$. But $S \cap qG = qS$, by hypothesis, so that there is $s' \in S$ with $s = qs'$. Hence, $a \otimes s = a \otimes qs' = aq \otimes s' = 0$ in $A \otimes_{\mathbb{Z}} S$, and $1_A \otimes \lambda$ is injective in this case as well.

If A is finitely generated, then it is a direct sum of cyclic groups. Since tensor product commutes with direct sums, it follows easily that $1_A \otimes \lambda$ is injective in this case. The result for general A follows from Lemma 3.47. •

Thus, if a subgroup S of an abelian group G satisfies $S \cap nG = nS$ for all $n \in \mathbb{Z}$, then the sequence $0 \to A \otimes_{\mathbb{Z}} S \to A \otimes_{\mathbb{Z}} G \to A \otimes_{\mathbb{Z}} (G/S) \to 0$ is exact for every abelian group A. Corollary 3.72 shows why purity is so important. For example, suppose that $S = \langle s \rangle$ is a subgroup of a finite abelian p-group G, where s has maximal order in G. It is not difficult to prove that S is a pure subgroup of G, and so Proposition 3.71 says that S is a direct summand. In other words, we can prove S is a direct summand by solving equations instead of by constructing a complement. An interesting result of Kulikov (Fuchs, *Infinite Abelian Groups* I, p. 120) is that a pure exact sequence of abelian groups $0 \to S \to G \to G/S \to 0$ in which G/S is a direct sum of cyclic groups must be split. It follows that if G is finitely generated, then a subgroup S of G is pure if and only if it is a direct summand. This is false in general, for Exercise 3.31 shows that the torsion subgroup (which is always pure) need not be a direct summand.

Exercises

3.27 Prove that $\mathbb{I}_2$ is not a flat $\mathbb{Z}$-module.

3.28 Let k be a commutative ring, and let P and Q be flat k-modules. Prove that $P \otimes_k Q$ is a flat k-module.

3.29 Let R be a PID, let $Q = \text{Frac}(R)$, and let M be a torsion-free R-module.

    (i) Prove that M can be imbedded in $Q \otimes_R M$.

    (ii) Prove that $Q \otimes_R M \cong \text{Env}(M)$, the injective envelope of M.

3.30 If R is a commutative ring (not necessarily a domain), define

$$tM = \{ m \in M : rm = 0 \text{ for some nonzero } r \in R \}.$$

    (i) Let $R = \mathbb{I}_6$, and regard R as a module over itself. Prove that $[1] \notin t\mathbb{I}_6$.

    (ii) Prove that $t\mathbb{I}_6$ is not a submodule of $\mathbb{I}_6$.
        Hint. Both $[2], [3] \in t\mathbb{I}_6$, but $[3] - [2] \notin t\mathbb{I}_6$.

*3.31   **(i)** Let P be the set of all primes in $\mathbb{Z}$. Prove that $\bigoplus_{p \in P} \mathbb{I}_p$ is the torsion subgroup of $\prod_{p \in P} \mathbb{I}_p$.

    (ii) Prove that $\left(\prod_{p \in P} \mathbb{I}_p \right) / \left(\bigoplus_{p \in P} \mathbb{I}_p \right)$ is divisible.

    (iii) Prove that $t\left(\prod_{p \in P} \mathbb{I}_p \right)$ is not a direct summand of $\prod_{p \in P} \mathbb{I}_p$.

*3.32 Let $0 \to A \to B \to C \to 0$ be an exact sequence of right R-modules, for some ring R. If both A and C are flat modules, prove that B is a flat module.
    Hint. This result is routine if one uses the derived functor Tor.

*3.33 Let R be a domain, let T be a torsion R-module, and let D be a divisible R-module. Prove that $T \otimes_R D = \{0\}$. (See Proposition 2.73.)

*3.34 Let $B = {}_R B$, so that $\operatorname{Hom}_R(B, R)$ is a right R-module. If C is a left R-module, define $\nu \colon \operatorname{Hom}_R(B, R) \otimes_R C \to \operatorname{Hom}_R(B, C)$ by $\nu \colon f \otimes c \mapsto \widehat{f}$, where $\widehat{f}(b) = f(b)c$ for all $b \in B$ and $c \in C$.

    (i) Prove that ν is natural in B.

    (ii) Prove that ν is an isomorphism if B is finitely generated free.

    (iii) If B is a finitely presented left R-module and C is a flat left R-module, prove that ν is an isomorphism.

3.35 A right R-module B is called *faithfully flat* if

    (i) B is a flat module,

    (ii) for all left R-modules X, if $B \otimes_R X = \{0\}$, then $X = \{0\}$.

Prove that $R[x]$ is a faithfully flat R-module (if R is not commutative, then $R[x]$ is the polynomial ring in which the indeterminate x commutes with each coefficient in R).

3.36 Prove that a right R-module B is faithfully flat if and only if B is flat and $B \otimes_R (R/I) \neq \{0\}$ for all proper left ideals I of R.

3.37   **(i)** Prove that a right R-module B is flat if and only if exactness of any sequence of left R-modules $A' \xrightarrow{i} A \xrightarrow{p} A''$ implies exactness of $B \otimes_R A' \xrightarrow{1 \otimes i} B \otimes_R A \xrightarrow{1 \otimes p} B \otimes_R A''$.

    (ii) Prove that a right R-module B is faithfully flat if and only if it is flat and $B \otimes_R A' \xrightarrow{1 \otimes i} B \otimes_R A \xrightarrow{1 \otimes p} B \otimes_R A''$ exact implies $A' \xrightarrow{i} A \xrightarrow{p} A''$ is exact.

3.38 Prove that if B is a faithfully flat module and C is a flat module, then $B \oplus C$ is faithfully flat.

3.39   **(i)** Prove that $\mathbb{Q}$ is a flat $\mathbb{Z}$-module that is not faithfully flat.

    (ii) Prove that an abelian group G is a faithfully flat $\mathbb{Z}$-module if and only if it is torsion-free and $pG \neq G$ for all primes p.

3.40 Let $0 \to A \to B \to C \to 0$ be an exact sequence of right R-modules, for some ring R. If both A and C are flat modules and if one of them if faithfully flat, prove that B is a faithfully flat module.

3.41 Prove that if $B = {}_R B_S$ is a bimodule that is R-flat, and if $C = C_S$ is S-injective, then $\operatorname{Hom}_S(B, C)$ is an injective left R-module.
Hint. The composite of exact functors is an exact functor.

*3.42 Prove that an exact sequence $0 \to B' \to B \to B'' \to 0$ of left R-modules is pure exact if and only if it remains exact after tensoring by all finitely presented right R-modules A.

Hint. That an element lies in $\ker(A \otimes_R B' \to A \otimes_R B)$ involves only finitely many elements of A.

3.43 **(Kulikov)** If H and K are torsion abelian groups, prove that $H \otimes_{\mathbb{Z}} K$ is a direct sum of cyclic groups.

Hint. Use Kulikov's Theorem: if G is a p-primary abelian group, then there exists a pure exact sequence $0 \to B \to G \to D \to 0$ with B a direct sum of cyclic groups and D divisible. Such a pure subgroup B is called a **basic subgroup** of G. See Rotman, *An Introduction to the Theory of Groups*, p. 327.

3.44 If G is a finite abelian group, prove that a subgroup $S \subseteq G$ is a direct summand of G if and only if S is a pure subgroup of G.

Hint. Proposition 3.71.

*3.45 Let G be an abelian group, and let $S \subseteq G$ be a pure subgroup. If $S \subseteq H \subseteq G$, prove that H is a pure subgroup of G if and only if H/S is a pure subgroup of G/S.

4

Specific Rings

We consider two general problems in this chapter: if conditions are imposed on projective, injective, or flat R-modules, how does this affect R; if conditions are imposed on a ring R, how does this affect these special R-modules? We have already encountered several instances of these questions. A ring R is left noetherian if and only if every direct sum of injective left R-modules is injective [Theorem 3.39]. If R is a PID, then an R-module is injective if and only if it is divisible [Corollary 3.35(ii)], while an R-module is flat if and only if it is torsion-free [Corollary 3.51] (we will soon see that if R is a PID, then an R-module is projective if and only if it is free).

4.1 Semisimple Rings

If k is a field, then k-modules are vector spaces. It follows that all k-modules are projective (even free, for every vector space has a basis). Indeed, every k-module is injective and flat as well. We now describe all rings for which this is true.

Definition. Let R be a ring. A left R-module M is **simple** (or **irreducible**) if $M \neq \{0\}$ and if M has no proper nonzero submodules; we say that M is **semisimple** (or **completely reducible**) if it is a direct sum of (possibly infinitely many) simple modules.

The zero module is not simple, but it is semisimple, for $\{0\} = \bigoplus_{i \in \varnothing} S_i$.

 J.J. Rotman, *An Introduction to Homological Algebra*, Universitext, DOI 10.1007/978-0-387-68324-9_4, © Springer Science+Business Media LLC 2009

Proposition 4.1. *A left R-module M is semisimple if and only if every submodule is a direct summand.*

Proof. If M is semisimple, then $M = \bigoplus_{j \in J} S_j$, where every S_j is simple. Given a subset $I \subseteq J$, define $S_I = \bigoplus_{j \in I} S_j$. If N is a submodule of M, we see, using Zorn's Lemma, that there exists a subset I of J maximal with $S_I \cap N = \{0\}$. We claim that $M = N \oplus S_I$, which will follow if we prove that $S_j \subseteq N + S_I$ for all $j \in J$. This inclusion holds, obviously, if $j \in I$. If $j \notin I$, then the maximality of I gives $(S_j + S_I) \cap N \neq \{0\}$. Thus, $s_j + s_I = n \neq 0$ for some $s_j \in S_j, s_I \in S_I$, and $n \in N$, so that $s_j = n - s_I \in (N + S_I) \cap S_j$. Now $s_j \neq 0$, lest $s_I \in S_I \cap N = \{0\}$. Since S_j is simple, we have $(N + S_I) \cap S_j = S_j$; that is, $S_j \subseteq N + S_I$.

Suppose, conversely, that every submodule of M is a direct summand. We begin by showing that each nonzero submodule N contains a simple submodule. Let $x \in N$ be nonzero; by Zorn's Lemma, there is a submodule $Z \subseteq N$ maximal with $x \notin Z$. Now Z is a direct summand of M, by hypothesis, and so Z is a direct summand of N, by Corollary 2.24; say, $N = Z \oplus Y$. We claim that Y is simple. If Y' is a proper nonzero submodule of Y, then $Y = Y' \oplus Y''$ and $N = Z \oplus Y = Z \oplus Y' \oplus Y''$. Either $Z \oplus Y'$ or $Z \oplus Y''$ does not contain x [lest $x \in (Z \oplus Y') \cap (Z \oplus Y'') = Z$], contradicting the maximality of Z. Next, we show that M is semisimple. By Zorn's Lemma, there is a family $(S_k)_{k \in K}$ of simple submodules of M maximal with the property that they generate their direct sum $D = \bigoplus_{k \in K} S_k$. By hypothesis, $M = D \oplus E$ for some submodule E. If $E = \{0\}$, we are done. Otherwise, $E = S \oplus E'$ for some simple submodule S, by the first part of our argument. But now the family $\{S\} \cup (S_k)_{k \in K}$ violates the maximality of $(S_k)_{k \in K}$, a contradiction. $\bullet$

Corollary 4.2. *Every submodule and every quotient module of a semisimple module M is semisimple.*

Proof. Let N be a submodule of M. Every submodule of N is a direct summand of M, by Proposition 4.1, so that Corollary 2.24 shows that every submodule of N is a direct summand of N; therefore, N is semisimple. A quotient M/N is semisimple, for $M = N \oplus Q$ for some submodule Q of M. But $M/N \cong Q$, and Q is semisimple, as we have just seen. $\bullet$

Lemma 4.3. *If a ring R is a direct sum of left ideals, say, $R = \bigoplus_{i \in I} L_i$, then only finitely many L_i are nonzero.*

Proof. Each element in a direct sum has finite support; in particular, the unit element can be written as $1 = e_1 + \cdots + e_n$, where $e_i \in L_i$. If $a \in L_j$ for some $j \neq 1, \ldots, n$, then

$$a = a1 = ae_1 + \cdots + ae_n \in L_j \cap (L_1 \oplus \cdots \oplus L_n) = \{0\}.$$

Therefore, $L_j = \{0\}$, and $R = L_1 \oplus \cdots \oplus L_n$. $\bullet$

Definition.   A ring R is *left semisimple* if it is semisimple as a left R-module.

When viewing R as a left R-module, its submodules are its left ideals. Now a simple submodule is a *minimal left ideal*, for it is a nonzero ideal containing no proper nonzero left ideals. (Such ideals may not exist; for example, $\mathbb{Z}$ has no minimal left ideals.) By the lemma, a left semisimple ring is a direct sum of finitely many minimal left ideals.

Example 4.4.

(i) The Wedderburn–Artin Theorem (see Rotman, *Advanced Modern Algebra*, pp. 562 and 567) says that every left semisimple ring R is (isomorphic to) a finite direct product of matrix rings:

$$R \cong \mathrm{Mat}_{n_1}(\Delta_1) \times \cdots \times \mathrm{Mat}_{n_t}(\Delta_t),$$

where Δ_i are division rings. Moreover, the division rings Δ_i and the integers $t, n_1, \ldots, n_t$ are a complete set of invariants of R.

(ii) Every left semisimple ring is also right semisimple, and we call such rings *semisimple*, dropping the adjective *left* or *right* (*Advanced Modern Algebra*, p. 563). Moreover, semisimple rings are left and right noetherian.

(iii) Maschke's Theorem (*Advanced Modern Algebra*, p. 556) says that if G is a finite group and k is a field, then the group ring kG is semisimple if and only if the characteristic of k does not divide $|G|$. If k is algebraically closed, then $kG \cong \mathrm{Mat}_{n_1}(k) \times \cdots \times \mathrm{Mat}_{n_t}(k)$ (Molien's Theorem, *Advanced Modern Algebra*, p. 568).

(iv) If k is a field of characteristic 0, then $R = k[x]/(x^n - 1)$ is semisimple, for $R \cong kG$, where G is a cyclic group of order n.

(v) A finite direct product of fields is semisimple; in particular, $R = \mathbb{I}_n$ is semisimple if and only if n is squarefree.   ◄

Here is the reason we have introduced semisimple rings here.

Proposition 4.5.   *The following conditions on a ring R are equivalent.*

(i) *R is semisimple.*

(ii) *Every left (or right) R-module M is a semisimple module.*

(iii) *Every left (or right) R-module M is injective.*

(iv) *Every short exact sequence of left (or right) R-modules splits.*

(v) *Every left (or right) R-module M is projective.*

Proof.

(i) $\Rightarrow$ (ii). Since R is semisimple, it is semisimple as a module over itself; hence, every free left R-module is a semisimple module. Now M is a quotient of a free module, by Theorem 2.35, and so Corollary 4.2 gives M semisimple.

(ii) $\Rightarrow$ (iii). If E is a left R-module, then Proposition 3.40 says that E is injective if every exact sequence $0 \to E \to B \to C \to 0$ splits. By hypothesis, B is a semisimple module, and so Proposition 4.1 implies that the sequence splits; thus, E is injective.

(iii) $\Rightarrow$ (iv). If $0 \to A \to B \to C \to 0$ is an exact sequence, then it must split because, as every module, A is injective (see Corollary 3.27).

(iv) $\Rightarrow$ (v). Given a module M, there is an exact sequence

$$0 \to F' \to F \to M \to 0,$$

where F is free. By hypothesis, this sequence splits and $F \cong M \oplus F'$. Therefore, M is a direct summand of a free module, and hence it is projective, by Theorem 3.5.

(v) $\Rightarrow$ (i). If I is a left ideal of R, then

$$0 \to I \to R \to R/I \to 0$$

is an exact sequence. By hypothesis, R/I is projective, and so this sequence splits, by Proposition 3.3; that is, I is a direct summand of R. By Proposition 4.1, R is a semisimple left R-module. Therefore, R is a left semisimple ring. $\bullet$

Semisimple rings are so nice that there is a notion of *global dimension* of a ring R, defined in Chapter 8, which measures how far R is from being semisimple.

Galois Theory has been generalized from field extensions to extensions of commutative rings, by Chase, Harrison, and Rosenberg, *Galois Theory and Cohomology of Commutative Rings*; see also De Meyer–Ingraham, *Separable Algebras over Commutative Rings*. Here is a connection between projective modules and separable field extensions.

Recall that if L is a commutative k-algebra, then its *enveloping algebra* is $L^e = L \otimes_k L$; multiplication in L^e is given by

$$(a \otimes b)(a' \otimes b') = aa' \otimes bb'.$$

Theorem 4.6. *If L and k are fields and L is a finite separable extension of k, then L is a projective L^e-module, where L^e is the enveloping algebra.*

Proof. Now L is an (L, L)-bimodule, so that L is an L^e-module (Corollary 2.61). It suffices to prove that $L \otimes_k L$ is a direct product of fields, for then it is a semisimple ring and every module is projective.

Since L is a finite separable extension of k, the Theorem of the Primitive Element (see Rotman, *Advanced Modern Algebra*, p. 230) provides an element $\alpha \in L$ with $L = k(\alpha)$. If $f(x) \in k[x]$ is the irreducible polynomial of α, then there is an exact sequence of k-modules

$$0 \to (f) \xrightarrow{i} k[x] \xrightarrow{v} L \to 0,$$

where i is the inclusion, v is a k-algebra map, $v: x \mapsto \alpha$, and (f) is the principal ideal generated by $f(x)$. Since k is a field, the vector space L is a free k-module, and hence it is flat. Thus, the following sequence is exact.

$$0 \to L \otimes_k (f) \xrightarrow{1 \otimes i} L \otimes_k k[x] \xrightarrow{1 \otimes v} L \otimes_k L \to 0.$$

Of course, $L \otimes_k L = L^e$ (for L is commutative), and it is easily checked that $1_L \otimes v$ is a k-algebra map; thus, im $1_L \otimes i$ is an ideal in $k[x] \otimes_k L$. Let $L[y]$ be the polynomial ring in an indeterminate y, and define $\theta: L \otimes_k k[x] \to L[y]$ by $a \otimes g(x) \mapsto ag(y)$; the map θ is an isomorphism, and the following diagram is commutative and has exact rows.

$$
\begin{array}{ccccccccc}
0 & \longrightarrow & L \otimes_k (f) & \xrightarrow{1 \otimes i} & L \otimes_k k[x] & \xrightarrow{1 \otimes v} & L^e & \longrightarrow & 0 \\
& & \downarrow & & \downarrow{\scriptstyle \theta} & & \downarrow & & \\
0 & \longrightarrow & (f) & \longrightarrow & L[y] & \longrightarrow & L[y]/(f) & \longrightarrow & 0.
\end{array}
$$

By Proposition 2.70, there is a k-isomorphism $L^e \to L[y]/(f)$, which is easily seen to be a k-algebra isomorphism.

Now f, though irreducible over k, may factor in $L[y]$, and separability says that there are no repeated factors:

$$f(y) = \prod_i p_i(y),$$

where the $p_i(y)$ are distinct irreducible polynomials in $L[y]$. The ideals (p_i) are thus distinct maximal ideals in $L[y]$, and the Chinese Remainder Theorem gives a k-algebra isomorphism

$$L^e \cong L[y]/(f(y)) \cong \prod_i L[y]/(p_i).$$

Since each $L[y]/(p_i)$ is a field, L^e is a semisimple ring.   ●

The converse of Theorem 4.6 is true (see De Meyer-Ingraham, p. 49), and generalizations of Galois Theory to commutative k-algebras R (where k is a commutative ring) define R to be *separable* over k if R is a projective R^e-module (Chase-Harrison-Rosenberg, *Galois Theory and Cohomology of Commutative Rings*).

4.2   von Neumann Regular Rings

We have just seen that every R-module is projective (or injective) if and only if R is semisimple. What if every R-module is flat?

Definition.   A ring R is *von Neumann regular* if, for each $r \in R$, there is $r' \in R$ with $rr'r = r$.

Informally, one may think of r' as a generalized inverse of r.

Example 4.7.

(i) A ring R is a *Boolean ring* if every element $r \in R$ is *idempotent*; that is, $r^2 = r$. Boolean rings are von Neumann regular: if $r \in R$, define $r' = r$. Boolean rings are commutative.

(ii) Here is a proof that if V is a (possibly infinite-dimensional) vector space over a field k, then $R = \mathrm{End}_k(V)$ is von Neumann regular. Given a linear transformation $\varphi\colon V \to V$, we have $V = \ker \varphi \oplus W$, for every subspace of a vector space is a direct summand. Let X be a basis of $\ker \varphi$ and let Y be a basis of W, so that $X \cup Y$ is a basis of V. Now $\varphi(Y)$ is a linearly independent subset (because $W \cap \ker \varphi = \{0\}$), and so it can be extended to a basis $\varphi(Y) \cup Z$ of V. If we define $\varphi'\colon V \to V$ by $\varphi'(\varphi(y)) = y$ for all $y \in Y$ and $\varphi'(z) = 0$ for all $z \in Z$, then $\varphi\varphi'\varphi = \varphi$. (Example 2.36 shows that von Neumann regular rings may not have IBN; on the other hand, the uniqueness part of the Wedderburn–Artin Theorem shows that semisimple rings do have IBN.) ◄

Lemma 4.8.   *If R is a von Neumann regular ring, then every finitely generated left (or right) ideal is principal, and it is generated by an idempotent.*

Proof.   Denote a principal left ideal by $Ra = \{ra : r \in R\}$. If $a' \in R$ satisfies $a = aa'a$, then $e = aa'$ is idempotent; moreover, $a \in Re$ and $e \in Ra$, so that $Ra = Re$ is generated by an idempotent.

To prove that every finitely generated left ideal is principal, it suffices to prove that $I = Ra + Rb$ is principal. There is an idempotent e with

$Ra = Re$; we claim that $Re + Rb = Re + Rb(1 - e)$: both e and b lie in $Re + Rb(1 - e)$; both e and $b(1 - e)$ lie in $Re + Rb$. There is an idempotent f with $Rb(1 - e) = Rf$, so that $f = rb(1 - e)$ for some $r \in R$. It follows that $fe = rb(1 - e)e = 0$. We do not know whether $ef = 0$, and so we adjust f. Define $g = (1 - e)f$. Now g is idempotent, for

$$g^2 = (1 - e)f(1 - e)f = (1 - e)(f - fe)f = (1 - e)f^2 = (1 - e)f = g.$$

It is easily checked that $ge = 0 = eg$ and that $Rg = Rf$, so that $Ra + Rb = Re + Rg$. We claim that $Re + Rg = R(e + g)$. Clearly, $R(e + g) \subseteq Re + Rg$. For the reverse inclusion, if $u, v \in R$, then $(ue + vg)(e + g) = ue^2 + ueg + vge + vg^2 = ue + vg$; hence, $Re + Rg \subseteq R(e + g)$. A similar argument proves that every finitely generated right ideal is principal.   •

Theorem 4.9 (Harada).   *A ring R is von Neumann regular if and only if every right R-module is flat.*

Proof.   Assume that R is von Neumann regular and that B is a right R-module. If $0 \to K \to F \to B \to 0$ is an exact sequence of right R-modules with F free, then Lemma 3.60 says that B is flat if $KI = K \cap FI$ for every finitely generated left ideal I. By Lemma 4.8, I is principal, say, $I = Ra$. We must show that if $k \in K$ and $k = fa \in Fa$, then $k \in Ka$. But $k = fa = faa'a = ka'a \in Ka$. Therefore, B is flat.

For the converse, take $a \in R$. By hypothesis, the cyclic right R-module R/aR is flat. Since R is free, Lemma 3.60 applies to the exact sequence $0 \to aR \to R \to R/aR \to 0$ to give $(aR)I = aR \cap RI = aR \cap I$ for every left ideal I. In particular, if $I = Ra$, then $aRa = aR \cap Ra$. Thus, there is some $a' \in R$ with $a = aa'a$, and so R is von Neumann regular.   •

Corollary 4.10.   *Every semisimple ring is von Neumann regular.*

Proof.   If a ring is semisimple, then every module is projective and, hence, every module is flat.   •

4.3   Hereditary and Dedekind Rings

We have seen that assuming every R-module is "special" (projective, injective, or flat) constrains R. Moreover, interesting rings are characterized in this way. We now assume that every ideal is special.

Assuming that every left ideal is injective gives nothing new.

Proposition 4.11.   *Every left ideal in a ring R is injective if and only if R is semisimple.*

Proof.   The submodules of R are its left ideals. As each left ideal is injective, it is a direct summand, by Corollary 3.27. Proposition 4.1 now says that R is a semisimple left R-module; that is, R is a (left) semisimple ring. Conversely, if R is semisimple, then every left ideal is injective, by Proposition 4.5.   •

Definition.   A ring R is **left hereditary** if every left ideal is projective; a ring R is **right hereditary** if every right ideal is projective. A **Dedekind ring** is a hereditary domain.

Example 4.12.

(i) Every semisimple ring is both left and right hereditary.

(ii) Small's example of a right noetherian ring that is not left noetherian (see Exercise 3.8 on page 114) is right hereditary but not left hereditary.

(iii) Every PID R is hereditary (for nonzero principal ideals in a domain are isomorphic to R), and so they are Dedekind rings.

(iv) The ring of integers in an algebraic number field is a Dedekind ring (Zariski–Samuel, *Commutative Algebra* I, p. 283). Thus, there are Dedekind rings that are not PIDs. For example, $R = \{a + b\sqrt{-5} : a, b \in \mathbb{Z}\}$ is a Dedekind ring that is not a PID.

(v) If k is a field, then $R = k\langle x, y \rangle$, the ring of polynomials in noncommuting variables, is both left and right hereditary (see Cohn, *Free Rings and Their Relations*, p. 106; every one-sided ideal is a free R-module). This ring is neither right nor left noetherian, so there exist non-noetherian hereditary rings. However, Dedekind rings are always noetherian; in fact, every ideal in a Dedekind ring can be generated by two elements (Rotman, *Advanced Modern Algebra*, p. 959).

(vi) If R is a domain, then certain R-algebras, called R-*orders*, arise in the theory of integral representations of finite groups (see Reiner, *Maximal Orders*). When R is a Dedekind ring, then **maximal R-orders** are hereditary rings.   ◄

The following theorem, well-known for modules over Dedekind rings, was generalized by Kaplansky for left hereditary rings.

Theorem 4.13 (Kaplansky). *If R is left hereditary, then every submodule A of a free left R-module F is isomorphic to a direct sum of left ideals.*

Proof. Let $\{x_k : k \in K\}$ be a basis of F; by the Axiom of Choice, we may assume that the index set K is well-ordered. Define $F_0 = \{0\}$, where 0 is the smallest index in K and, for each $k \in K$, define

$$F_k = \bigoplus_{i < k} Rx_i \quad \text{and} \quad \overline{F}_k = \bigoplus_{i \leq k} Rx_i = F_k \oplus Rx_k.$$

It follows that $\overline{F}_0 = Rx_0$. Each element $a \in A \cap \overline{F}_k$ has a unique expression $a = b + rx_k$, where $b \in F_k$ and $r \in R$, so that $\varphi_k : A \cap \overline{F}_k \to R$, given by $a \mapsto r$, is well-defined. There is an exact sequence

$$0 \to A \cap F_k \to A \cap \overline{F}_k \to \operatorname{im} \varphi_k \to 0.$$

Since $\operatorname{im} \varphi_k$ is a left ideal, it is projective, and so this sequence splits:

$$A \cap \overline{F}_k = (A \cap F_k) \oplus C_k,$$

where $C_k \cong \operatorname{im} \varphi_k$. We claim that $A = \bigoplus_{k \in K} C_k$, which will complete the proof.

(i) $A = \langle \bigcup_{k \in K} C_k \rangle$: Since $F = \bigcup_{k \in K} \overline{F}_k$, each $a \in A$ (as any element of F) lies in some $\overline{F}_k$; let $\mu(a)$ be the smallest index k with $a \in \overline{F}_k$. Define $C = \langle \bigcup_{k \in K} C_k \rangle \subseteq A$. If $C \subsetneq A$, then $J = \{\mu(a) : a \in A - C\} \neq \varnothing$. Let j be the smallest element in J, and let $y \in A - C$ have $\mu(y) = j$. Now $y \in A \cap \overline{F}_j = (A \cap F_j) \oplus C_j$, so that $y = b + c$, where $b \in A \cap F_j$ and $c \in C_j$. Hence, $b = y - c \in A$, $b \notin C$ (lest $y \in C$), and $\mu(b) < j$, a contradiction. Therefore, $A = C = \langle \bigcup_{k \in K} C_k \rangle$.

(ii) Uniqueness of expression: Suppose that $c_1 + \cdots + c_n = 0$, where $c_i \in C_{k_i}$, $k_1 < \cdots < k_n$, and k_n is minimal (among all such equations). Then

$$c_1 + \cdots + c_{n-1} = -c_n \in (A \cap F_{k_n}) \cap C_{k_n} = \{0\}.$$

It follows that $c_n = 0$, contradicting the minimality of k_n. $\bullet$

Corollary 4.14. *If R is a left hereditary ring, then every submodule S of a projective left R-module P is projective.*

Proof. Since P is projective, it is a submodule, even a direct summand, of a free module, by Theorem 3.5. Therefore, S is a submodule of a free module, and so S is a direct sum of ideals, each of which is projective, by Theorem 4.13. Therefore, S is projective, by Corollary 3.6(ii). $\bullet$

Corollary 4.15. *Let R be a* PID.

(i) *If A is a submodule of a free R-module F, then A is a free R-module and* $\text{rank}(A) \leq \text{rank}(F)$.

(ii) *If* $B = \langle b_1, \ldots, b_n \rangle$ *is a finitely generated R-module and* $B' \subseteq B$ *is a submodule, then* B' *is finitely generated and it can be generated by n or fewer elements.*

Proof.

(i) In the notation of Theorem 4.13, if F has a basis $\{x_k : k \in K\}$, then $A = \bigoplus_{k \in K} C_k$, where C_k is isomorphic to an ideal in R. Since R is a PID, every nonzero ideal is isomorphic to R: either $C_k = \{0\}$ or $C_k \cong R$. Therefore, A is free and $\text{rank}(A) \leq |K| = \text{rank}(F)$.

(ii) Let F be a free R-module with basis $\{x_1, \ldots, x_n\}$. Define $\varphi \colon F \to B$ by $x_i \mapsto b_i$ for all i, define $A = \varphi^{-1}(B')$, and note that $\varphi|A \colon A \to B'$ is surjective. By part (i), A is free of rank $m \leq n$, and so B' can be generated by m elements. •

We remark that part (ii) of Corollary 4.15 may be false for more general domains. First, if R is a domain that is not noetherian, then it has an ideal I that is not finitely generated; that is, I is a submodule of a cyclic module that is not finitely generated. Second, if B can be generated by n elements and $B' \subseteq B$ is finitely generated, B' still may require more than n generators. For example, if k is a field and $R = k[x, y]$, then R is not a PID, and so there is some ideal I that is not principal; that is, R is generated by one element and its submodule I cannot be generated by one element.

Corollary 4.16. *If R is a* PID, *then every projective R-module is free.*

Proof. This follows at once from Corollary 4.15(i), for every projective module is a submodule (even a summand) of a free module. •

If R is a Dedekind ring, then we have just shown, in Theorem 4.13, that every finitely generated projective R-module P is (isomorphic to) a direct sum of ideals: $P \cong I_1 \oplus \cdots \oplus I_n$. This decomposition is not unique: $P \cong F \oplus J$, where F is free and J is an ideal (in fact, J is the product ideal $I_1 \cdots I_n$). Steinitz proved that this latter decomposition is unique to isomorphism (see Rotman, *Advanced Modern Algebra*, p. 967).

Let us show that a direct product of projectives need not be projective.

Theorem 4.17 (Baer). *The direct product* $\mathbb{Z}^{\mathbb{N}}$ *of infinitely many copies of* $\mathbb{Z}$ *is not free (and, hence, it is not projective).*

Proof.   Let us write the elements of $\mathbb{Z}^{\mathbb{N}}$ as sequences (m_n), where $m_n \in \mathbb{Z}$. It suffices, by Corollary 4.15, to exhibit a subgroup $S \subseteq \mathbb{Z}^{\mathbb{N}}$ that is not free. Choose a prime p, and define S by

$$S = \left\{ (m_n) \in \mathbb{Z}^{\mathbb{N}} : \text{for each } k \geq 1, \text{ we have } p^k \mid m_n \text{ for almost all } n \right\}^{1}.$$

Thus, p divides almost all m_n, p^2 divides almost all m_n, and so forth. For example, $s = (1, p, p^2, p^3, \ldots) \in S$. It is easy to check that S is a subgroup of $\mathbb{Z}^{\mathbb{N}}$. We claim that if $s = (m_n) \in S$ and $s = ps^*$ for some $s^* \in \mathbb{Z}^{\mathbb{N}}$, then $s^* \in S$. If $s^* = (d_n)$, then $pd_n = m_n$ for all n; since $p^{k+1} \mid m_n$ for almost all n, we have $p^k \mid d_n$ for almost all n.

If $(m_n) \in S$, then so is $(\epsilon_n m_n)$, where $\epsilon = \pm 1$, so that S is uncountable. Were S a free abelian group, then S/pS would be uncountable, for $S = \bigoplus_{j \in J} C_j$ implies $S/pS \cong \bigoplus_{j \in J} (C_j/pC_j)$. We complete the proof by showing that $\dim(S/pS)$ is countable, which gives the contradiction S/pS countable. Let $e_n = (0, \ldots, 0, 1, 0, \ldots)$, where 1 is in the nth spot; note that $e_n \in S$. We claim that the countable family of cosets $\{e_n + pS : n \in \mathbb{N}\}$ spans S/pS. If $s = (m_n) \in S$, then almost all m_n are divisible by p. Hence, there is an integer N so that $s - \sum_{n=0}^{N} m_n e_n = ps^*$, and s^* lies in S. Thus, in S/pS, the coset $s + pS$ is a finite linear combination of cosets of e_n, and so $\dim(S/pS)$ is countable.   •

We have just seen that $\mathbb{Z}^{\mathbb{N}}$, the direct product of countably many copies of $\mathbb{Z}$, is not free abelian, but we saw, in Exercise 3.4 on page 114, that every *countable* subgroup of $\mathbb{Z}^{\mathbb{N}}$ is a free abelian group. A theorem of Specker–Nobeling (see Fuchs, *Infinite Abelian Groups* II, p. 175) shows that the subgroup B of all bounded sequences,

$$B = \{(m_n) \in \mathbb{Z}^{\mathbb{N}} : \text{there exists } N \text{ with } |m_n| \leq N \text{ for all } n\},$$

is a free abelian group (in fact, this is true for $\mathbb{Z}^I$ for any index set I).

We are going to show that Corollary 4.14 characterizes left hereditary rings, but we begin with a lemma.

Lemma 4.18.   *A left R-module P is projective if and only if every diagram with exact row and with Q injective can be completed to a commutative diagram; that is, every map $f : P \to Q''$ can be lifted. The dual is also true.*

---

[1]For readers familiar with the p-adic topology, S consists of null-sequences and it is essentially the p-adic completion of $\mathbb{Z}$.

Proof. If P is projective, then the diagram can always be completed, with no hypothesis on Q.

For the converse, we must find a map $P \to A$ making the following diagram commute.

$$
\begin{array}{ccccccccc}
& & & & & & P & & \\
& & & & & & \downarrow f & & \\
0 & \longrightarrow & A' & \xrightarrow{\ i\ } & A & \xrightarrow{\ \tau\ } & A'' & \longrightarrow & 0.
\end{array}
$$

There are an injective module Q and an imbedding $\sigma : A \to Q$, by Theorem 3.38. Enlarge the diagram to obtain

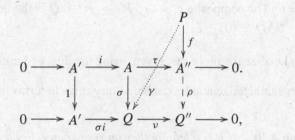

where $Q'' = \operatorname{coker} \sigma i$ and ν is the natural map. By Proposition 2.70, there exists a map $\rho : A'' \to Q''$ making the diagram commute. By hypothesis, the map ρf can be lifted: there exists $\gamma : P \to Q$ with $\nu\gamma = \rho f$. We claim that $\operatorname{im} \gamma \subseteq \operatorname{im} \sigma$, which will complete the proof (because $\operatorname{im} \sigma \cong A$). If $x \in P$, choose $a \in A$ with $\tau a = fx$. Then $\nu\gamma x = \rho f x = \rho \tau a = \nu \sigma a$, so that $\gamma x - \sigma a \in \ker \nu = \operatorname{im} \sigma i$. Hence, there is $a' \in A'$ with $\gamma x - \sigma a = \sigma i a'$, and so $\gamma x = \sigma(a + ia') \in \operatorname{im} \sigma$. $\bullet$

Theorem 4.19 (Cartan–Eilenberg). *The following statements are equivalent for a ring R.*

(i) *R is left hereditary.*

(ii) *Every submodule of a projective module is projective.*

(iii) *Every quotient of an injective module is injective.*

Proof.

(i) $\Rightarrow$ (ii) Corollary 4.14.

(ii) $\Rightarrow$ (i) R is a free R-module, and so it is projective. Therefore, its submodules, the left ideals, are projective, and R is left hereditary.

(iii) $\Rightarrow$ (ii) Consider the diagram with exact rows

$$
\begin{array}{ccccc}
P & \xleftarrow{\ j\ } & P' & \longleftarrow & 0 \\[-2pt]
\scriptstyle k \downarrow \ \scriptstyle h \ \diagdown & \diagup \scriptstyle g & \downarrow \scriptstyle f & & \\[-2pt]
Q & \xrightarrow[\ r\ ]{} & Q'' & \longrightarrow & 0,
\end{array}
$$

where P is projective and Q is injective. By Lemma 4.18, it suffices to find a map $g \colon P' \to Q$ with $rg = f$. Now Q'' is injective, by hypothesis, so that there exists a map $h \colon P \to Q''$ giving commutativity: $hj = f$. Since P is projective, there is a map $k \colon P \to Q$ with $rk = h$. The composite $g = kj \colon P' \to P \to Q$ is the desired map, for $rg = r(kj) = hj = f$.

(ii) $\Rightarrow$ (iii) Dualize the proof just given, using the dual of Lemma 4.18. $\bullet$

We can characterize noetherian hereditary rings in terms of flatness.

Proposition 4.20. *If R is a left noetherian ring, then every left ideal is flat if and only if R is left hereditary.*

Proof. Since R is left noetherian, every left ideal I is finitely presented, and so I flat implies that it is projective, by Corollary 3.57. Hence, R is left hereditary. Conversely, if R is left hereditary, then every left ideal is projective, and so every left ideal is flat, by Proposition 3.46. $\bullet$

Let us now show that our definition of Dedekind ring coincides with more classical definitions.

Definition. Let R be a domain with $Q = \mathrm{Frac}(R)$. An ideal I is ***invertible*** if there are elements $a_1, \ldots, a_n \in I$ and elements $q_1, \ldots, q_n \in Q$ with

(i) $q_i I \subseteq R$ for all $i = 1, \ldots, n$,

(ii) $1 = \sum_{i=1}^{n} q_i a_i$.

For example, every nonzero principal ideal Ra is invertible: define $a_1 = a$ and $q_1 = 1/a$. Note that if I is invertible, then $I \neq (0)$. We show that $I = (a_1, \ldots, a_n)$. Clearly, $(a_1, \ldots, a_n) \subseteq I$. For the reverse inclusion, let $b \in I$. Now $b = b1 = \sum (bq_i)a_i$; since $bq_i \in q_i I \subseteq R$, we have $I \subseteq (a_1, \ldots, a_n)$.

Remark. If R is a domain and $Q = \text{Frac}(R)$, then a *fractional ideal* is a finitely generated nonzero R-submodule of Q. All the fractional ideals in Q form a commutative monoid under the following multiplication: if I, J are fractional ideals, their product is

$$IJ = \left\{\sum_k \alpha_k \gamma_k : \alpha_k \in I \text{ and } \gamma_k \in J\right\}.$$

The unit in this monoid is R. If I is an invertible ideal and I^{-1} is the R-submodule of Q generated by $q_1, \ldots, q_n$, then I^{-1} is a fractional ideal and

$$II^{-1} = R = I^{-1}I$$

[one can show that $I^{-1} \cong \text{Hom}_R(I, R)$]. We will soon see that every nonzero ideal in a Dedekind ring R is invertible, so that the monoid of all fractional ideals is an abelian group (which turns out to be free with basis all nonzero prime ideals). The *class group* of R is defined to be the quotient group of this group by the subgroup of all nonzero principal ideals. ◄

Proposition 4.21. *If R is a domain, then a nonzero ideal I is projective if and only if it is invertible.*

Proof. If I is projective, then Proposition 3.10 says that I has a projective basis: there are $(a_k \in I)_{k \in K}$ and R-maps $(\varphi_k \colon I \to R)_{k \in K}$ such that, (i) for each $b \in I$, almost all $\varphi_k(b) = 0$, (ii) for each $b \in I$, we have $b = \sum_{k \in K}(\varphi_k b)a_k$.

Let $Q = \text{Frac}(R)$. If $b \in I$ and $b \neq 0$, define $q_k \in Q$ by

$$q_k = \varphi_k(b)/b.$$

Note that q_k does not depend on the choice of nonzero b: if $b' \in I$ is nonzero, then $b'\varphi_k(b) = \varphi_k(b'b) = b\varphi_k(b')$, so that $\varphi_k(b')/b' = \varphi_k(b)/b$. It follows that $q_k I \subseteq R$ for all k: if $b \in I$, then $q_k b = [\varphi_k(b)/b]b = \varphi_k(b) \in R$. By condition (i), if $b \in I$, then almost all $\varphi_k(b) = 0$. Since $q_k = \varphi_k(b)/b$ whenever $b \neq 0$, there are only finitely many (nonzero) q_k. Discard all a_k for which $q_k = 0$. Condition (ii) gives, for $b \in I$,

$$b = \sum(\varphi_k b)a_k = \sum(q_k b)a_k = b\left(\sum q_k a_k\right).$$

Cancel b from both sides to obtain $1 = \sum q_k a_k$. Thus, I is invertible.

Conversely, if I is invertible, there are elements $a_i, \ldots, a_n \in I$ and $q_1, \ldots, q_n \in Q$, as in the definition. Define $\varphi_k \colon I \to R$ by $b \mapsto q_k b$ (note that $q_k b \in q_k I \subseteq R$). If $b \in I$, then

$$\sum(\varphi_k b)a_k = \sum q_k b a_k = b \sum q_k a_k = b.$$

Therefore, I has a projective basis and, hence, I is a projective module. ●

Corollary 4.22.   *A domain R is a Dedekind ring if and only if every nonzero ideal in R is invertible.*

Proof.   This follows at once from Proposition 4.21.   ●

Corollary 4.23.   *Every Dedekind ring is noetherian.*

Proof.   Invertible ideals are finitely generated.   ●

We can now generalize Corollary 3.35 from PIDs to Dedekind rings.

Theorem 4.24.   *A domain R is a Dedekind ring if and only if every divisible R-module is injective.*

Proof.   Assume that every divisible R-module is injective. If E is an injective R-module, then E is divisible, by Lemma 3.33. Since every quotient of a divisible module is divisible, every quotient E'' of E is divisible, and so E'' is injective, by hypothesis. Therefore, R is a Dedekind ring, by Theorem 4.19.

Conversely, assume that R is Dedekind and that E is a divisible R-module. By the Baer Criterion, it suffices to complete the diagram

$$
\begin{array}{ccc}
& E & \\
{\scriptstyle f}\uparrow & & \nwarrow \\
0 \longrightarrow & I \xrightarrow[\text{inc}]{} & R,
\end{array}
$$

where I is an ideal and inc is the inclusion. Of course, we may assume that I is nonzero, so that I is invertible: there are elements $a_1, \ldots, a_n \in I$ and $q_1, \ldots, q_n \in \mathrm{Frac}(R)$ with $q_i I \subseteq R$ and $1 = \sum_i q_i a_i$. Since E is divisible, there are elements $e_i \in E$ with $f(a_i) = a_i e_i$. Note, for every $b \in I$, that

$$
f(b) = f\left(\sum_i q_i a_i b\right) = \sum_i (q_i b) f(a_i) = \sum_i (q_i b) a_i e_i = b \sum_i (q_i a_i) e_i.
$$

Hence, if we define $e = \sum_i (q_i a_i) e_i$, then $e \in E$ and $f(b) = be$ for all $b \in I$. Now define $g \colon R \to E$ by $g(r) = re$; since g extends f, the module E is injective.   ●

Lemma 4.25.   *If R is a unique factorization domain, then a nonzero ideal I is projective if and only if it is principal.*

Proof.   Every nonzero principal ideal $I = (b)$ in a domain R is isomorphic to R via $r \mapsto rb$. Thus, I is free and, hence, projective. Conversely, suppose that R is a UFD. If I is a projective ideal, then it is invertible, by Proposition 4.21. There are elements $a_i, \ldots, a_n \in I$ and $q_1, \ldots, q_n \in Q$ with $1 = \sum_i q_i a_i$ and $q_i I \subseteq R$ for all i. Write $q_i = b_i / c_i$ and assume, by unique factorization, that b_i and c_i have no nonunit factors in common. Since $(b_i/c_i)a_j \in R$ for $j = 1, \ldots, n$, we have $c_i \mid a_j$ for all i, j. We claim that $I = (c)$, where $c = \mathrm{lcm}\{c_1, \ldots, c_n\}$. Note that $c \in I$, for $c = c \sum b_i a_i / c_i = \sum (b_i c / c_i) a_i \in I$, for $(b_i c / c_i) \in R$. Hence, $(c) \subseteq I$. For the reverse inclusion, $c_i \mid a_j$ for all i, j implies $c \mid a_j$ for all j, and so $a_j \in (c)$ for all j. Hence, $I \subseteq (c)$.   •

Theorem 4.26.   *A Dedekind ring R is a unique factorization domain if and only if it is a PID.*

Proof.   Every PID is a UFD. Conversely, if R is a Dedekind ring, then every nonzero ideal I is projective. Since R is a UFD, I is principal, by Lemma 4.25, and so R is a PID.   •

4.4   Semihereditary and Prüfer Rings

We now investigate rings in which all finitely generated ideals are special.

Definition.   A ring R is **left semihereditary** if every finitely generated left ideal is projective. A semihereditary domain is called a **Prüfer ring**.

Example 4.27.

  (i) Every left hereditary ring is left semihereditary (of course, these notions coincide for left noetherian rings).

 (ii) Chase gave an example of a left semihereditary ring that is not right semihereditary (see Lam, *Lectures on Modules and Rings*, p. 47). A theorem of Small says that a one-sided noetherian ring is left semihereditary if and only if it is right semihereditary (see Lam, p. 268).

(iii) Every von Neumann regular ring is both left and right semihereditary. By Lemma 4.8, every finitely generated left (or right) ideal I is principal; say, $I = (a)$. If $aa'a = a$, the map $\varphi \colon R \to I$, defined by $\varphi(r) = ra'a$, is a retraction. Therefore, I is a direct summand of R and, hence, I is projective.   ◄

Definition.   A ring R is a **Bézout ring** if it is a domain in which every finitely generated ideal is principal.

It is clear that every Bézout ring is a Prüfer ring; i.e., it is semihereditary.

Example 4.28.

(i) A *valuation ring* is a domain R in which, for all $a, b \in R$, either $a \mid b$ or $b \mid a$. Every valuation ring is a Bézout ring.

(ii) A domain R is a Prüfer ring if and only if, for every maximal ideal $\mathfrak{m}$, the localization $R_{\mathfrak{m}}$ is a valuation ring (see Kaplansky, *Commutative Rings*, p. 39).

(iii) Let X be a noncompact Riemann surface, and let R be the ring of all complex-valued holomorphic functions on X. Helmer [see "Divisibility properties of integral functions," *Duke Math J.* 6 (1940), 345–356] proved that R is a Bézout ring.

(iv) The ring of all algebraic integers (in $\mathbb{C}$) is a Bézout ring (see Kaplansky, *Commutative Rings*, p. 72). ◄

Proposition 4.29. *If R is a left semihereditary ring, then every finitely generated submodule of a free module is a direct sum of a finite number of finitely generated left ideals.*

Proof. Let F be a free left R-module, let $\{x_k : k \in K\}$ be a basis, and let $A = \langle a_1, \ldots, a_m \rangle$ be a finitely generated submodule of F. Each a_i, when expressed as a linear combination of the x_k, has finite support, so that $X = \bigcup_k \text{supp}(a_k)$ is finite and $A \subseteq \langle X \rangle$. Now $\langle X \rangle$ is a free submodule of F, and so we may assume that F is finitely generated with basis $\{x_1, \ldots, x_n\}$.

We prove, by induction on $n \geq 1$, that A is (isomorphic to) a direct sum of finitely generated left ideals. If $n = 1$, then A is isomorphic to a finitely generated left ideal. If $n > 1$, define $B = A \cap (Rx_1 + \cdots + Rx_{n-1})$; by the inductive hypothesis, B is a direct sum of a finite number of finitely generated left ideals. Now each $a \in A$ has a unique expression of the form $a = b + rx_n$, where $b \in B$ and $r \in R$; define $\varphi: A \to R$ by $a \mapsto r$, and note that im φ is a finitely generated left ideal in R. There is an exact sequence $0 \to B \to A \to \text{im } \varphi \to 0$, and this sequence splits because im φ is projective: $A \cong B \oplus \text{im } \varphi$. Therefore, A is a direct sum of finitely many finitely generated left ideals.   •

The reader has probably observed that the proof just given is merely that of Theorem 4.13 stripped of its transfinite apparel. Albrecht [see "On projective modules over a semihereditary ring," *Proc. AMS* 12 (1961), 638–639] proved that if R is left semihereditary, then every (not necessarily finitely generated) projective R-module is a direct sum of finitely generated left ideals.

Proposition 4.30 (Albrecht). *A ring R is left semihereditary if and only if every finitely generated submodule A of a projective left R-module P is projective.*

Proof.   Now P is a submodule, even a summand, of a free left R-module, so that A is a finitely generated submodule of a free module. By Proposition 4.29, A is a direct sum of finitely generated left ideals. As each of these ideals is projective, A is projective.

Conversely, every finitely generated left ideal is a finitely generated submodule of the free R-module R. Hence, such ideals are projective, and R is left semihereditary.   •

Definition.   A right R-module A is *torsionless* if it is isomorphic to a submodule of a direct product R^X for some index set X.

Example 4.31.

(i) Every projective right R-module is torsionless.

(ii) Every right ideal I is torsionless.

(iii) If R is a domain, then every torsionless R-module is torsion-free. The converse is false. For example, $\mathbb{Q}$ is not a submodule of $\mathbb{Z}^X$ for any set X.

(iv) If A is a left R-module, then $\operatorname{Hom}_R(A, R) \subseteq R^A$ is a torsionless right R-module.   ◄

In the midst of proving the next theorem, we are going to use Corollary 8.26: if every left ideal in a ring R is flat, then every submodule of a flat left R-module is flat.

Theorem 4.32 (Chase).   *The following statements are equivalent.*

(i) *R is left semihereditary.*

(ii) *R is left coherent and every submodule of a flat left R-module is flat.*

(iii) *Every torsionless right R-module is flat.*

Proof.   (i) $\Rightarrow$ (ii) If R is left semihereditary, then every finitely generated left ideal I is projective; hence, I is finitely presented, by Proposition 3.11. Therefore, Theorem 3.66 gives R left coherent. Since every finitely generated left ideal is projective, it is flat. It follows from Proposition 3.48 that every left ideal is flat, and so Corollary 8.26 applies to show that every submodule of a flat module is itself flat.

(ii) $\Rightarrow$ (i) If I is a finitely generated left ideal, then I is a submodule of the flat module R, and so I is flat; since R is left coherent, I is also finitely presented. Hence, I is projective, by Theorem 3.56, and so R is left semihereditary.

(ii) $\Rightarrow$ (iii) Since R is left coherent, Theorem 3.66 says that the right R-modules R^X are flat, for any X. By definition, every torsionless right R-module is a submodule of some R^X, and so it is flat.

(iii) $\Rightarrow$ (ii) For every set X, the right R-module R^X is torsionless, and so it is flat, by hypothesis. It follows from Theorem 3.66 that R is left coherent. Every left ideal is torsionless, so it, too, is flat. Thus, Corollary 8.26 says that every submodule of a flat module is flat.   •

Let us now consider Prüfer rings.

Recall that if R is any domain, then the *torsion submodule* tM of an R-module M is $tM = \{m \in M : rm = 0 \text{ for some nonzero } r \in R\}$. Note that tM is a submodule of M and that M/tM is *torsion-free*; that is, its torsion submodule is $\{0\}$.

Lemma 4.33.   *Let R be a domain with fraction field Q.*

(i) *If A is a torsion-free R-module, then there is an exact sequence*

$$0 \to A \to V \to T \to 0,$$

where V is a vector space over Q and T is torsion.

(ii) *If A is finitely generated and torsion-free, then A can be imbedded in a finitely generated free R-module.*

Proof.

(i) Let $V = \text{Env}(A)$, the injective envelope of A. If $v \in V$, then there is $r \in R$ with $rv \neq 0$ and $rv \in A$. It follows that A torsion-free implies V torsion-free, and that V/A is torsion. Finally, Exercise 2.38 on page 97 shows that V is a vector space over Q, for it is torsion-free and divisible.

(ii) Let $A = \langle a_1, \ldots, a_n \rangle$. By part (i), A is imbedded in a vector space V over Q. If X is a basis of V, then each a_i is a linear combination of finitely many basis vectors in X. It follows that A is imbedded in the finite-dimensional vector space with basis $B = \{x_1, \ldots, x_m\}$ consisting of all $x \in X$ involved in expressing any of the a_i. For each a_i, there are $r_{ij}, s_{ij} \in R$ with $a_i = \sum_j (r_{ij}/s_{ij})x_j$. If $s = \prod_{i,j} s_{ij}$, then $s^{-1}B = \{s^{-1}x_1, \ldots, s^{-1}x_m\}$ is a basis of V. In fact, the R-submodule of V generated by $s^{-1}B$ is free with basis $s^{-1}B$, and it contains A.   •

Theorem 4.34.   *A domain R is a Prüfer ring if and only if every finitely generated torsion-free R-module A is projective.*

Proof. By Lemma 4.33(ii), A can be imbedded as a submodule of a free R-module. Since R is a Prüfer ring, Proposition 4.30 says that A is projective.

Conversely, let I be a finitely generated ideal in a domain R. Since R is a torsion-free R-module, the hypothesis says that I is projective. Therefore, R is a Prüfer ring. •

In Corollary 3.51, we saw that if R is a PID, then R-modules are flat if and only if they are torsion-free. We now generalize this to Prüfer rings.

Theorem 4.35. *If R is a Prüfer ring, then an R-module B is flat if and only if B is torsion-free.*

Proof. We proved, in Proposition 3.49, that if R is any domain, then flat R-modules are torsion-free. Conversely, assume that B is a torsion-free R-module. By Proposition 3.48, it suffices to prove that every finitely generated submodule $B' \subseteq B$ is flat. Since R is a Prüfer ring and B' is torsion-free, Theorem 4.34 says that B' is projective. Hence, B' is flat, and so B is flat. •

We now combine the two previous results to give another characterizion of Prüfer rings.

Corollary 4.36. *Let R be a domain. Then R is a Prüfer ring if and only if every torsion-free R-module is flat.*

Proof. If R is a Prüfer ring and B is a torsion-free R-module, then B is flat, by Theorem 4.35. Conversely, we prove that every torsionless R-module is flat. Now R^X is torsion-free, because R is a domain, and so every torsionless R-module, being a submodule of some R^X, is flat, by hypothesis. Therefore, R is a Prüfer ring, by Theorem 4.35. •

4.5 Quasi-Frobenius Rings

We are now going to assume that a ring R is *self-injective*; that is, R is injective as a left R-module. (There is no need to consider *self-projective* or *self-flat*, for the left R-module R is always projective, and hence it is always flat.) Self-injectivity is most interesting when it is coupled with chain conditions.

Definition. A ring R is *quasi-Frobenius* if it is left and right noetherian and R is an injective left R-module.

It can be shown that the apparent asymmetry of the definition is only virtual: if R is quasi-Frobenius, then R is an injective right R-module (see Jans, *Rings and Homology*, p. 78, or Lam, *Lectures on Modules and Rings*, p. 409).

Clearly, semisimple rings are quasi-Frobenius (for they are both left and right noetherian, and every module is injective); in particular, kG is quasi-Frobenius when G is a finite group and k is a field whose characteristic does not divide $|G|$. Although there are other examples, as we shall see, the most important examples of quasi-Frobenius rings are group rings kG for G finite and k a field of any characteristic (see Theorem 4.46). Such rings arise naturally in the theory of modular group representations. For example, if G is a finite solvable group, then a minimal normal subgroup V of G is a vector space over $\mathbb{F}_p$ for some prime p (see Rotman, *An Introduction to the Theory of Groups*, p. 105). Since $V \lhd G$, the group G acts on V by conjugation, and so V is an $\mathbb{F}_p G$-module.

Proposition 4.37.   *If R is a PID and I is a nonzero proper ideal, then R/I is quasi-Frobenius.*

Proof.   It is clear that R/I is noetherian, and so we need show only that R/I is an injective (R/I)-module. By Baer's Criterion, it suffices to extend a map

$$
\begin{array}{ccc}
 & R/I & \\
f \uparrow & \nwarrow & \\
0 \longrightarrow J/I \xrightarrow[\text{inc}]{} R/I, &
\end{array}
$$

where $f: J' \to R/I$ from an ideal J' to a map $R/I \to R/I$. By the Correspondence Theorem, $J' = J/I$, where J is an ideal in R containing I; note that $J = Rb$, because R is a PID. Let $I = Ra$. Since $Ra = I \subseteq J = Rb$, we have $bc = a$ for some $c \in R$. The R/I-module R/I is cyclic with generator $x = 1 + I$, and J/I is cyclic with generator bx.

Now $f(bx) = sx$ for some $s \in R$. Since $bcx = ax = 0$, we have $0 = cf(bx) = csx$, so that $cs \in Ra$ (because $x = 1 + Ra$). Therefore, $cs = ra = rbc$ for some $r \in R$. Canceling c gives $s = rb$, so that $f(bx) = sx = rbx$. Define $g: R/I \to R/I$ to be multiplication by r. Now g extends f, for $g(bx) = rbx = f(bx)$. Therefore, R/I is self-injective, and R/I is quasi-Frobenius.   •

Compare the next result with Examples 4.4(iv) and (v).

Corollary 4.38.   *The rings $\mathbb{I}_n$, where $n > 1$, and the rings $k[x]/I$, where k is a field and I is a nonzero ideal, are quasi-Frobenius rings.*

Proposition 4.39. *If R is left and right noetherian, then R is quasi-Frobenius if and only if every projective left R-module is injective.*

Proof. If R is quasi-Frobenius, then every free left R-module F is a direct sum of injectives (for R is injective). Since R is left noetherian, Proposition 3.31 says that F is injective. If P is projective, then it is a direct summand of a free module; here, P is a direct summand of an injective module and, hence, it is injective.

Conversely, the left R-module $_RR$ is projective, and so it is injective, by hypothesis. Since R is left and right noetherian, it is quasi-Frobenius.   •

One of the standard proofs of the Basis Theorem for finite abelian groups has as its crucial step the observation that a cyclic subgroup of largest order is a direct summand.

Corollary 4.40 (Basis Theorem).   *Every finite abelian group G is a direct sum of cyclic groups.*

Proof. By the primary decomposition theorem, we may assume that G is a p-primary group for some prime p. If p^n is the largest order of elements in G, then $p^n g = 0$ for all $g \in G$, and so G is an $\mathbb{I}_{p^n}$-module. If $x \in G$ has order p^n, then $S = \langle x \rangle \cong \mathbb{I}_{p^n}$. Hence, S is injective, for $\mathbb{I}_{p^n}$ is quasi-Frobenius, by Corollary 4.38. But injective submodules are always direct summands, and so $G = S \oplus T$ for some submodule T. By induction on $|G|$, the complement T is a direct sum of cyclic groups.   •

There is another chain condition that is dual, in the lattice-theoretic sense, to noetherian rings.

Definition.   A left R-module M (over any ring R) has **DCC** (*descending chain condition*) if every descending chain of submodules

$$M = M_0 \supseteq M_1 \supseteq M_2 \supseteq \cdots$$

stops; that is, there is an integer n with $M_n = M_{n+1} = M_{n+2} = \cdots$.

Definition.   A ring R is **left artinian** if it has DCC on left ideals.

Example 4.41.

(i) There exist left artinian rings that are not right artinian (see Lam, *A First Course in Noncommutative Rings*, p. 22).

(ii) The Hopkins–Levitzki Theorem (see Rotman, *Advanced Modern Algebra*, p. 555) says that every left artinian ring is left noetherian.

(iii) If k is a field, then every finite-dimensional k-algebra is left and right artinian. In particular, if G is a finite group, then kG is left and right artinian.

(iv) Every semisimple ring is left and right artinian.

(v) Every finite ring is left and right artinian.

(vi) A left R-module M has **both chain conditions** (DCC and ACC on submodules) if and only if M has a **composition series**; that is, there is a chain of submodules

$$M = M_0 \supseteq M_1 \supseteq M_2 \supseteq \cdots \supseteq M_n = \{0\}$$

in which every factor module M_i/M_{i-1} is a simple module.

(vii) Every quasi-Frobenius ring is left and right artinian (see Lam, *Lectures on Modules and Rings*, p. 409). ◄

Proposition 4.42.

(i) *A ring R is left artinian if and only if R satisfies the **minimum condition**: every nonempty family $\mathcal{F}$ of left ideals in R has a minimal element.*

(ii) *If R is left artinian, then every nonzero left ideal I contains a minimal left ideal.*

Proof. The proof of part (i) is dual to that of Corollary 3.16, and it is left to the reader. To prove (ii), let I be a nonzero left ideal, and define $\mathcal{F}$ to be the family of all the nonzero left ideals J contained in I. The reader may show that a minimal element of $\mathcal{F}$ is a minimal left ideal. •

Corollary 4.43. *Every quotient ring of a left artinian ring R is left artinian.*

Proof. Let I be a two-sided ideal in R, so that R/I is a ring. By the Correspondence Theorem, any descending chain of left ideals in R/I corresponds to a descending chain of left ideals in R (which contain I). This chain in R stops, and so the original chain in R/I stops. •

We are now going to show, for every finite group G and every field k (of any characteristic), that kG is quasi-Frobenius.

Definition. Let R be a finite-dimensional algebra over a field k. Then R is called a **Frobenius algebra** if $R \cong \mathrm{Hom}_k(R_R, k)$ as left R-modules.

Observe that the dual space $\mathrm{Hom}_k(R_R, k)$ is a left R-module, as in Proposition 2.54.

Proposition 4.44.   *Every Frobenius algebra R is quasi-Frobenius.*

Proof.   Every finite-dimensional k-algebra is left and right noetherian. Now $R \cong \operatorname{Hom}_k(R_R, k)$, by hypothesis. On the other hand, Lemma 3.37 shows that $\operatorname{Hom}_k(R_R, k)$ is injective. Therefore, $R \cong \operatorname{Hom}_k(R_R, k)$ is injective, and so R is quasi-Frobenius.   •

Lemma 4.45.   *Let R be a finite-dimensional algebra over a field k. If there is a linear functional $f : R \to k$ whose kernel contains no nonzero left ideals, then R is a Frobenius algebra.*

Proof.   Define $\theta : R \to \operatorname{Hom}_k(R, k)$ by $\theta_r(x) = f(xr)$ for all $x \in R$; it is easy to check that each θ_r is a k-map and that $r'\theta_r = \theta_{r'r}$; that is, θ is an R-map. We claim that θ is injective. If $\theta_r = 0$, then $0 = \theta_r(x) = f(xr)$ for all $x \in R$. But this says that $Rr \subseteq \ker f$; by hypothesis, $r = 0$. Finally, if $\dim_k(R) = n$, then $\dim_k \operatorname{Hom}_k(R, k) = n$ [for $\operatorname{Hom}_k(R, k)$ is just the dual space of $_R R$]. Therefore, θ must be surjective, being an injection between two n-dimensional spaces.   •

Theorem 4.46.   *If G is a finite group and k is any field, then kG is a Frobenius algebra, and hence it is quasi-Frobenius.*

Proof.   By Lemma 4.45, it suffices to give a linear functional $f : kG \to k$ whose kernel contains no nonzero left ideals. Each $r \in kG$ has a unique expression

$$r = \sum_{x \in G} r_x x, \quad \text{where } r_x \in k.$$

Define $f : kG \to k$ by $f : r \mapsto r_1$, the coefficient of 1. Suppose that $\ker f$ contains a left ideal I. If $r = \sum r_x x \in I$, then $0 = f(x^{-1}r) = r_x$. Hence, $r_x = 0$ for all $x \in G$ and $r = 0$. Therefore, $I = \{0\}$.   •

Definition.   A module M is **indecomposable** if $M \neq \{0\}$ and M has no nonzero direct summands.

Every simple module is indecomposable, but the converse is false. For example, if p is a prime, then the abelian group $\mathbb{I}_{p^2}$ is indecomposable, but it is not simple.

Proposition 4.47.   *Let R be a ring. If a left R-module M has either chain condition on submodules, then M is a direct sum of a finite number of indecomposable submodules.*

Proof. Call a module *good* if it is a direct sum of a finite number of inde-composable submodules; call it *bad* otherwise. An indecomposable module is good and, if both A and B are good, then $A \oplus B$ is good. Therefore, if M is a bad module, then $M = U \oplus V$, where U, V are proper submodules at least one of which is bad.

If M is a bad module, define $N_0 = M$. By induction, for every $n \geq 0$, there are bad submodules $N_0, N_1, \ldots, N_n$ with each N_i a proper bad direct summand of N_{i-1}. There is a strictly decreasing sequence of submodules:

$$M = N_0 \supsetneq N_1 \supsetneq N_2 \supsetneq \cdots.$$

If M has DCC, we have reached a contradiction.

Suppose M is a bad module having ACC. Since each N_i is a direct sum-mand of N_{i-1}, there are complements L_i with $N_{i-1} = N_i \oplus L_i$. This gives a strictly ascending sequence of submodules of M,

$$L_1 \subsetneq L_1 \oplus L_2 \subsetneq L_1 \oplus L_2 \oplus L_3 \subsetneq \cdots,$$

another contradiction.   •

Definition. If a ring R is a direct sum of indecomposable modules, say, $R = \bigoplus_i L_i$, then any module M isomorphic to some L_i is called a ***principal indecomposable module***.

By Proposition 4.47, a ring with either chain condition has principal in-decomposable modules. Indeed, every indecomposable direct summand of R is such a module. In particular, quasi-Frobenius rings, being left noetherian, have principal indecomposable modules.

Recall that minimal left ideals are, by definition, nonzero.

Proposition 4.48. *If R is quasi-Frobenius, then there is a bijection between its minimal left ideals and its principal indecomposable modules.*

Proof. Let I be a minimal left ideal in R. Since $R = {}_R R$ is injective, The-orem 3.45(ii) shows that we may assume its injective envelope, $\mathrm{Env}(I)$, is a submodule of R; that is, $\mathrm{Env}(I)$ is a left ideal. We claim that $\mathrm{Env}(I)$ is a prin-cipal indecomposable module. As $\mathrm{Env}(I)$ is injective, it is a direct summand of R. Suppose that $\mathrm{Env}(I)$ is not indecomposable; that is, $\mathrm{Env}(I) = A \oplus B$, where A and B are nonzero. If $I \cap A \neq \{0\}$ and $I \cap B \neq \{0\}$, then min-imality of I gives $I \cap A = I = I \cap B$; that is, $I \subseteq A \cap B = \{0\}$, a contradiction. Hence, either $I \cap A = \{0\}$ or $I \cap B = \{0\}$; but either of these contradicts $\mathrm{Env}(I)$ being an essential extension of I. Thus, the func-tion $\varphi: \{\text{minimal left ideals}\} \to \{\text{principal indecomposable modules}\}$, given by $\varphi: I \to \mathrm{Env}(I)$, is well-defined.

We show that φ is surjective. If E is a principal indecomposable module, then it is injective, for it is a direct summand of R. Since R is left artinian, E (viewed as a left ideal) contains a minimal left ideal I. By Theorem 3.45(ii) (which applies because E is injective), we may assume that $\text{Env}(I)$ is a submodule of E. As $\text{Env}(I)$ is injective, it is a direct summand of E; but E is indecomposable, and so $E = \text{Env}(I) = \varphi(I)$.

We show that $\varphi: I \mapsto \text{Env}(I)$ is injective. If $\text{Env}(I) = \text{Env}(I')$, where I and I' are distinct minimal left ideals, then $\text{Env}(I)$ cannot be an essential extension of I because it contains a nonzero submodule I' with $I \cap I' = \{0\}$. Therefore, φ is a bijection. •

This last result takes on more interest when we observe that every simple module over a quasi-Frobenius ring is isomorphic to a minimal left ideal (see Curtis–Reiner, *Representation Theory of Finite Groups and Associative Algebras*, p. 401). Modular Representation Theory investigates the group ring kG of a finite group G when $|G|$ is divisible by the characteristic of k. This last result suggests that the role of minimal left ideals in semisimple rings is played by principal indecomposable modules in the modular case.

4.6   Semiperfect Rings

There is a notion dual to that of injective envelope, called *projective cover*. In contrast to injective envelopes, which exist for modules over any ring, projective covers exist only for certain rings, called *perfect*. A *semiperfect* ring is one for which every finitely generated module has a projective cover. We shall see that local rings and artinian rings are semiperfect.

We begin with some basic ring theory.

Definition.   If R is a ring, then its *Jacobson radical* $J(R)$ is defined to be the intersection of all the maximal left ideals in R.

Clearly, we can define another Jacobson radical: the intersection of all the maximal *right* ideals. It turns out, however, that both of these coincide (see Rotman, *Advanced Modern Algebra*, p. 547), so that $J(R)$ is a two-sided ideal. Consequently, $R/J(R)$ is a ring.

Example 4.49.

(i) The maximal ideals in $\mathbb{Z}$ are the nonzero prime ideals (p), and so $J(\mathbb{Z}) = \bigcap_{p \text{ prime}} (p) = \{0\}$, for a nonzero integer is divisible by only finitely many primes.

(ii) Let k be a field and let $R = \text{Mat}_n(k)$. For any ℓ between 1 and n, let $\text{COL}(\ell) = \{[a_{ij}] \in R : a_{ij} = 0 \text{ for all } j \neq \ell\}$, and let $\text{COL}^*(\ell) = \sum_{i \neq \ell} \text{COL}(i)$. Now $\text{COL}(\ell)$ is a minimal left ideal, hence, a simple left R-module. Since $R/\text{COL}^*(\ell) \cong \text{COL}(\ell)$, we see that $\text{COL}^*(\ell)$ is a maximal left ideal. Therefore, $J(R) \subseteq \bigcap_\ell \text{COL}^*(\ell) = \{0\}$. ◄

We can characterize the elements in the Jacobson radical.

Proposition 4.50. *If x is an element in a ring R, then $x \in J(R)$ if and only if, for each $a \in R$, the element $1 - ax$ has a left inverse; that is, there is $u \in R$ with $u(1 - ax) = 1$.*

Proof. If $R(1 - ax)$ is a proper left ideal, then Zorn's Lemma shows that there is some maximal left ideal containing it; say, $R(1 - ax) \subseteq M$. By definition, $ax \in J \subseteq M$, so that $1 = (1 - ax) + ax \in M$, contradicting M being a proper ideal. Therefore, $R(1 - ax) = R$, and so there is $u \in R$ with $u(1 - ax) = 1$.

Conversely, if $x \notin J$, then there is a maximal left ideal M with $x \notin M$. Since $M \subsetneq M + Rx$, we have $M + Rx = R$, so that there are $m \in M$ and $a \in R$ with $m + ax = 1$. If $m = 1 - ax$ has a left inverse u, then $1 = um \in M$, contradicting M being a proper left ideal. ●

Proposition 4.51 (Nakayama's Lemma). *If M is a finitely generated left R-module, and if $JM = M$, where $J = J(R)$ is the Jacobson radical, then $M = \{0\}$.*

Proof. Let $m_1, \ldots, m_n$ be a generating set of M that is minimal in the sense that no proper subset generates M. Since $JM = M$, we have $m_1 = \sum_{i=1}^n r_i m_i$, where $r_i \in J$. It follows that

$$(1 - r_1)m_1 = \sum_{i=2}^n r_i m_i.$$

Since $r_1 \in J$, Proposition 4.50 says that $1 - r_1$ has a left inverse, say, u, and so $m_1 = \sum_{i=2}^n u r_i m_i$. This is a contradiction, for now M can be generated by the proper subset $\{m_2, \ldots, m_n\}$. ●

Remark. The hypothesis in Nakayama's lemma that the module M be finitely generated is necessary. For example, it is easy to check that $\mathbb{Z}_{(2)} = \{a/b \in \mathbb{Q} : b \text{ is odd}\}$ has a unique maximal ideal, namely, $P = \mathbb{Z}_{(2)}2$, so that $J(\mathbb{Z}_{(2)}) = P$. But $\mathbb{Q}$ is a $\mathbb{Z}_{(2)}$-module with $P\mathbb{Q} = 2\mathbb{Q} = \mathbb{Q}$. ◄

Proposition 4.52. *Let R be a left artinian ring.*

(i) *$J = J(R)$ is nilpotent; that is, there is $n > 0$ with $J^n = \{0\}$.*

(ii) *If $a \in J$, then $a^n = 0$.*

Proof.

(i) Since R is left artinian, the descending chain $J \supseteq J^2 \supseteq J^3 \supseteq \cdots$ must stop: there is $n > 0$ with $J^n = J^{n+1} = J^{n+2} = \cdots$. We claim that $J^n = \{0\}$. Otherwise, $J^n J = J^{n+1} = J^n \neq \{0\}$, and so $\mathcal{F} = \left\{ I : I \text{ is a left ideal}, J^n I \neq \{0\} \right\} \neq \varnothing$ (for $J \in \mathcal{F}$). Left artinian rings satisfy the minimum condition, so that $\mathcal{F}$ has a minimal element: there is a left ideal I_m minimal such that $J^n I_m \neq \{0\}$. Of course, $I_m \neq \{0\}$; choose $y \in I_m$ with $J^n y \neq \{0\}$, so that $J^n y \in \mathcal{F}$. Now $J^n y \subseteq I_m$, so that minimality gives $J^n y = I_m$. But $J^n y \subseteq Ry \subseteq I_m$, so that $J^n y = Ry$; hence, $J^n y$ is finitely generated (even cyclic).[2] Finally, $J^n Ry = J(J^n y) = J^{n+1} y = J^n y = Ry$, and Nakayama's Lemma gives $Ry = \{0\}$, a contradiction. Therefore, $J^n = \{0\}$.

(ii) Since $J^n = \{0\}$, every product $a_1 \cdots a_n$ having n factors $a_i \in J$ is 0. In particular, if every $a_i = a \in J$, then $a^n = 0$. •

Definition.   An *idempotent* in a ring R is an element e with $e^2 = e$. If I is a two-sided ideal in a ring R, then an idempotent $g + I \in R/I$ can be **lifted mod** I if there is an idempotent $e \in R$ with $e + I = g + I$.

Proposition 4.53.   *If R is left artinian with Jacobson radical $J = J(R)$, then every idempotent can be lifted mod J.*

Proof.   By Proposition 4.52, we may assume that $J^n = \{0\}$. Let $g + J \in R/J$ be an idempotent: $g + J = g^2 + J$. Then $g - g^2 \in J$, and

$$
\begin{aligned}
0 &= (g - g^2)^n \\
&= \sum_{k=0}^{n} \binom{n}{k} g^{n-k} (-g^2)^k \\
&= \sum_{k=0}^{n} (-1)^k \binom{n}{k} g^{n+k} \\
&= g^n - g^{n+1} \left[\sum_{k=1}^{n} (-1)^{k-1} \binom{n}{k} g^k \right].
\end{aligned}
$$

---

[2]The **Hopkins–Levitzki Theorem** states that every left artinian ring is left noetherian. Had we proved this, we could have used it here to show that $J^n y$ is finitely generated.

If we define $h = \sum_{k=1}^{n}(-1)^{k-1}\binom{n}{k}g^k$, then

$$g^n = g^{n+1}h \qquad \text{and} \qquad gh = hg.$$

Define

$$e = g^n h^n.$$

We claim that e is idempotent: $e = g^n h^n = (g^{n+1}h)h^n = g^{n+1}h^{n+1}$; iterating, $e = g^{n+2}h^{n+2} = \cdots = g^{2n}h^{2n} = e^2$. Finally, we show that $e+J = g+J$. The equation $g+J = g^2+J$ gives $g+J = g^n+J = g^{n+1}+J$. Now

$$\begin{aligned} g + J &= g^n + J \\ &= g^{n+1}h + J \\ &= (g^{n+1} + J)(h + J) \\ &= (g + J)(h + J) \\ &= gh + J. \end{aligned}$$

Hence, $g + J = g^n + J = (g + J)^n = (gh + J)^n = g^n h^n + J = e + J$.   •

Definition.   A (not necessarily commutative) ring is *local* if it has a unique maximal left ideal.

Many authors who use the term *local ring* assume that the ring is commutative, and many of these assume further that it is noetherian.

It appears that local rings should be called *left local*, but it can be shown that a ring has a unique left ideal if and only if it has a unique right ideal, in which case they coincide (each is the Jacobson radical). If R is a local ring, then $J(R)$ is its unique maximal left (or right) ideal.

Example 4.54.

(i) Division rings and fields are local with unique maximal left ideal $\{0\}$.

(ii) If p is a prime, then $\mathbb{I}_{p^n} = \mathbb{Z}/p^n\mathbb{Z}$ is local with unique maximal ideal $(p + p^n\mathbb{Z})$.

(iii) If k is a field, then $k[[x]]$ is local with unique maximal ideal (x).

(iv) If p is a prime, then $\mathbb{Z}_{(p)} = \{a/b \in \mathbb{Q} : p \nmid b\}$ is a local ring with unique maximal ideal $\{ap/b \in \mathbb{Q} : p \nmid b\}$.

(v) If E is an indecomposable injective R-module, then $\text{End}_R(E)$ is a local ring with unique maximal left ideal $\{\varphi : E \to E : \ker \varphi \neq \{0\}\}$ (Lam, *Lectures on Modules and Rings*, p. 84).   ◄

Corollary 4.55.   *If R is a local ring with maximal left ideal P, and if M is a finitely generated R-module with PM = M, then M = {0}.*

Proof.   The statement follows at once from Nakayama's Lemma, because $J(R) = P$ when R is a local ring with unique maximal left ideal P.   •

Here are some properties of local rings.

Proposition 4.56.   *Let R be a local ring with maximal left ideal J.*

(i) *If $r \in R$ and $r \notin J$, then r has a left inverse in R.*

(ii) *If R is a local ring with maximal left ideal J, then J is a two-sided ideal and R/J is a division ring.*

(iii) *R has IBN.*

Proof.

(i) If $r \notin J$, then $Rr \subsetneq J$. Now every proper left ideal is contained in some maximal left ideal. Since R has only one maximal left ideal, namely, J, we conclude that Rr is not a proper left ideal: $Rr = R$. Therefore, there is $u \in R$ with $ur = 1$.

(ii) The unique maximal ideal J is the Jacobson radical $J(R)$, which is a two-sided ideal, and so R/J is a ring. If $r + J \neq 0$ in R/J, then $r \notin J$. By part (i), r has a left inverse in R, and so $r + J$ has a left inverse in R/J. It follows that the nonzero elements in R/J form a multiplicative group[3]; that is, R/J is a division ring.

(iii) As we remarked on page 60, the proof that nonzero commutative rings have IBN generalizes to rings R having a two-sided ideal J for which R/J is a division ring.   •

Theorem 4.57.   *Let R be a local ring with maximal left ideal J, Let M be a finitely generated left R-module, and let $B = \{m_1, \ldots, m_n\}$ be a minimal set of generators of M (that is, M cannot be generated by a proper subset of B). If F is a free left R-module with basis $x_1, \ldots, x_n$, and if $\varphi: F \to M$ is given by $\varphi(x_i) = m_i$ for all i, then $\ker \varphi \subseteq JF$.*

Proof.   There is an exact sequence

$$0 \to K \to F \xrightarrow{\varphi} M \to 0, \tag{1}$$

[3]To prove that a monoid is a group, it suffices to assume the existence of left inverses.

where $K = \ker \varphi$. If $K \subsetneq JF$, there is an element $y = \sum_{i=1}^{n} r_i x_i \in K$ that is not in JF; that is, some coefficient, say, $r_1 \notin J$. Thus, r_1 has a left inverse $u \in R$, by Proposition 4.56(i); that is, $ur_1 = 1$. Now $y \in K = \ker \varphi$ gives $\sum r_i m_i = 0$. Hence, $m_1 = -u\left(\sum_{i=2}^{n} r_i m_i\right)$, which implies that $M = \langle m_2, \ldots, m_n \rangle$, contradicting the minimality of the original generating set.   •

Theorem 4.58.   *If R is a local ring, then every finitely generated projective left R-module M is free.*

Proof.   Returning to exact sequence (1), projectivity of M gives $F = K \oplus M'$, where M' is a submodule of F with $M' \cong M$. Hence, $JF = JK \oplus JM'$. Since $JK \subseteq K \subseteq JF$, Corollary 2.24 gives

$$K = JK \oplus (K \cap JM').$$

But $K \cap JM' \subseteq K \cap M' = \{0\}$, so that $K = JK$. The submodule K is finitely generated, being a summand (and hence a homomorphic image) of the finitely generated module F, so that Nakayama's Lemma gives $K = \{0\}$. Therefore, φ is an isomorphism and M is free.   •

After proving Corollary 3.9(ii) [if every countably generated projective left R-module is free (for a ring R), then every projective left R-module is free], Kaplansky proved that every countably generated projective left R-module over a local ring R is free ["Projective modules," *Annals Math.* 68 (1958), 372–377)]. Thus, the finiteness hypothesis in Theorem 4.58 is unnecessary.

We now discuss projective covers.

Definition.   A submodule S of a module M is **superfluous** if, whenever $L \subseteq M$ is a submodule with $L + S = M$, then $L = M$.

One often calls elements of a superfluous submodule *nongenerators*, for if $M = \langle x_1, \ldots, x_n, s_1, \ldots, s_k \rangle$, then $M = \langle x_1, \ldots, x_n \rangle$; discarding them from a generating set of M leaves a generating set of M. It is clear that any submodule of a superfluous submodule is itself superfluous.

Lemma 4.59.

(i) *Let S be superfluous in M. If $M \subseteq N$, then S is superfluous in N.*[4]

(ii) *If S_i is superfluous in M_i for $i = 1, \ldots, n$, then $\bigoplus S_i$ is superfluous in $\bigoplus M_i$.*

---

[4]The converse is false: if $S \subseteq M \subseteq N$ and S is superfluous in N, then S need not be superfluous in M. For example, if S is superfluous in N and $S \neq \{0\}$, take $S = M$.

Proof.

(i) Let $L + S = N$. We claim that $(L \cap M) + S = M$. The inclusion $\subseteq$
   is obvious; for the reverse inclusion, if $m \in M$, then $m = \ell + s$ for
   $\ell \in L$ and $s \in S$. Now $\ell = m - s \in L \cap M$ (because $S \subseteq M$), and so
   $m \in (L \cap M) + S$. Since S is superfluous in M, we have $M = L \cap M$;
   that is, $M \subseteq L$. Hence, $N = S + L = L$ (because $S \subseteq M \subseteq L$).

(ii) By induction, we may assume that $n = 2$. Since S_1 is superfluous in
    M_1, it is superfluous in $M_1 \oplus M_2$, by part (i); similarly, S_2 is superfluous
    in $M_1 \oplus M_2$. Suppose that $L \subseteq M_1 \oplus M_2$ and $L + (S_1 + S_2) = M_1 \oplus M_2$.
    Now $L + (S_1 + S_2) = (L + S_1) + S_2$, so that S_2 superfluous in $M_1 \oplus M_2$
    gives $L + S_1 = M_1 \oplus M_2$; finally, S_1 superfluous in $M_1 \oplus M_2$ gives
    $L = M_1 \oplus M_2$. •

Lemma 4.60.

(i) *Let R be a ring with Jacobson radical J. If M is a finitely generated
    left R-module, then JM is superfluous in M.*

(ii) *Let R be a local ring with maximal left ideal J. If M is a finitely gener-
     ated left R-module, then JM is superfluous in M.*

Proof.

(i) If L is a submodule of M such that $L + JM = M$, then $M/L =
   (L + JM)/L = J(M/L) \subseteq M/L$; hence, $M/L = J(M/L)$. Since
   M/L is finitely generated, Nakayama's Lemma gives $M/L = \{0\}$; that
   is, $L = M$.

(ii) In this case, J is the Jacobson radical. •

The notion of superfluous submodule should be compared to that of essential extension. Using lattice-theoretic notation, a submodule $S \subseteq M$ is
superfluous if $S \vee L = M$ implies $L = M$, while $M \supseteq T$ is essential if
$T \wedge L = 0$ implies $L = 0$. An injection $i : T \to E$ is called *essential* if
$M \supseteq \operatorname{im} i$ is an essential extension; dually (in the categorical sense), a surjection $\varphi : F \to T$ is called *essential* if $\ker \varphi$ is a superfluous submodule of F.

Definition.   A *projective cover* of a module B is an ordered pair (P, φ),
where P is projective and $\varphi : P \to B$ is a surjective map with $\ker \varphi$ a superfluous submodule of P.

Example 4.61.   The $\mathbb{Z}$-module $\mathbb{I}_2$ does not have a projective cover. Let
$\varphi : F \to \mathbb{I}_2$ be a surjection, where F is a free abelian group. If $\varphi(x) = a$,
where $\mathbb{I}_2 = \langle a \rangle$, then $\varphi(3x) = a$. Hence, $F = \ker \varphi + \langle 3x \rangle$. If $\ker \varphi$ is
superfluous in F, then $F = \langle 3x \rangle$, which is not so. ◄

Theorem 4.62. *If R is a local ring with maximal left ideal J, then every finitely generated left R-module B has a projective cover: there is an exact sequence*

$$0 \to K \to F \xrightarrow{\varphi} B \to 0$$

with F a finitely generated free left R-module and K ⊆ JF.

Proof. There exist a free module F and a surjective $\varphi: F \to B$ with $K \subseteq JF$, by Theorem 4.57. But K is superfluous, by Lemma 4.60, and so (F, φ) is a projective cover of B. •

Definition. A ring R is called **left perfect** if every left R-module has a projective cover. A ring R is **semiperfect** if every finitely generated left R-module has a projective cover.

Theorem 4.62 says that local rings are semiperfect.

There are left perfect rings that are not right perfect (Lam, *A First Course in Noncommutative Rings*, p. 356). Lam calls a ring R *semiperfect* if R/J is semisimple and idempotents in R/J can be lifted mod J, a definition not needing any left/right distinction. However, Theorem 24.16 on p. 364 of Lam's book shows that this definition is equivalent to our definition in terms of projective covers. Therefore, the notions of left semiperfect ring and right semiperfect ring coincide, and we will write *semiperfect ring* without the adjectives *left* or *right*.

Theorem (Bass). *The following conditions are equivalent for a ring R.*

(i) *R is left perfect.*

(ii) *R has the DCC on principal right ideals.*

(iii) *Every flat left R-module is projective.*

Proof. See Lam, *A First Course in Noncommutative Rings*, p. 354. •

The following two results can be found in S. U. Chase, "Direct products of modules," *Trans. AMS* 97 (1960), 457–473.

Theorem (Chase). *Every direct product of projective left R-modules is projective if and only if R is left perfect and right coherent.*

Theorem (Chase). *If R is commutative, then every direct product of projective R-modules is projective if and only if R is artinian.*

Let us return to projective covers.

Lemma 4.63. *Let (P, φ) be a projective cover of a module B. If Q is projective and $\psi: Q \to B$ is surjective, then any lifting $\sigma: Q \to P$ is surjective. Moreover, P is (isomorphic to) a direct summand of Q.*

$$
\begin{array}{ccc}
 & & Q \\
 & {}^{\sigma}\nearrow & \downarrow {\scriptstyle \psi} \\
P & \xrightarrow{\ \varphi\ } & B \longrightarrow 0.
\end{array}
$$

Proof. Since Q is projective, there exists a map $\sigma: Q \to P$ with $\varphi\sigma = \psi$. Now $\varphi(\operatorname{im}\sigma) = \operatorname{im}\psi = B$, because ψ is surjective. Therefore, $P = \operatorname{im}\sigma + \ker\varphi$, by Exercise 2.18 on page 67. But $\ker\varphi$ is a superfluous submodule of P, so that $\operatorname{im}\sigma = P$. The second statement follows from the projectivity of P. •

Projective covers, when they exist, are unique.

Proposition 4.64.

(i) *Let P and Q be R-modules having either chain condition on submodules. If each of P, Q is isomorphic to a direct summand of the other, then $P \cong Q$.*

(ii) *Let R have either chain condition on left ideals. If $\varphi: P \to B$ and $\psi: Q \to B$ are projective covers of a finitely generated left R-module B, then there is an isomorphism $\sigma: Q \to P$ with $\varphi\sigma = \psi$.*

Proof.

(i) Let $P = Q_1 \oplus A_1$ and $Q_1 = P_1 \oplus B_1$, where $P_1 \cong P$ and $Q_1 \cong Q$. Hence, $P = Q_1 \oplus A_1 = P_1 \oplus B_1 \oplus A_1$; write $B_1 \oplus A_1 = C_1$. Now repeat: by induction, there are direct summands

$$ P \supseteq P_1 \supseteq P_2 \supseteq P_3 \supseteq \cdots , $$

with $P_n \cong P$ for all $n \geq 1$, and this violates the DCC. If we display the complements, we have a sequence

$$ P = P_1 \oplus C_1 = P_2 \oplus C_2 \oplus C_1 = P_3 \oplus C_3 \oplus C_2 \oplus C_1 = \cdots , $$

where $P_{n-1} = P_n \oplus C_n$ for all $n \geq 1$. The ascending sequence $C_1 \subseteq C_2 \oplus C_1 \subseteq C_3 \oplus C_2 \oplus C_1 \subseteq \cdots$ violates the ACC. Therefore, $A_1 = \{0\} = B_1$, and $P \cong Q$.

(ii) Since R has either chain condtion on left ideals, then P being finitely generated forces P to have either chain condition on submodules. By Lemma 4.63, each of P and Q is a direct summand of the other. By part (i), $P \cong Q$. •

We now prove that left artinian rings are semiperfect.

Lemma 4.65.   *Let R be a left artinian ring with Jacobson radical J. If e' is an idempotent in R/J, then there is a projective cover of $(R/J)e'$.*

Proof.   By Proposition 4.53, the idempotent e' can be lifted to an idempotent $e \in R$. There is an exact sequence

$$0 \to Je \to Re \xrightarrow{\varphi} (R/J)e' \to 0,$$

where $\varphi \colon re \mapsto re'$ (the reader may verify that $\ker \varphi = Je$). Now $P = Re$ is a direct summand of R, because e is idempotent, and so $P = Re$ is projective. By Lemma 4.60, $Je = J\langle e \rangle$ is superfluous in Re, and so we see that (Re, φ) is a projective cover of $(R/J)e'$.   •

Theorem 4.66.   *Every left artinian ring R is semiperfect; that is, every finitely generated left R-module M has a projective cover.*

Proof.   Let M be a finitely generated left R-module. If J is the Jacobson radical of R, then the kernel of the natural map $\pi \colon M \to M/JM$ is JM, which is superfluous in M, by Lemma 4.60. Since R/J is semisimple, the (R/J)-module M/JM is a direct sum of simple modules: $M/JM = \bigoplus S_i$, where each S_i is isomorphic to a minimal left ideal of R/J. But $S_i = (R/J)e_i'$ for some idempotent e_i' (because S_i, as every submodule of R/J, is a direct summand). By Lemma 4.65, there are projective covers $\varphi_i \colon P_i \to S_i$ for all i. Finally, Lemma 4.59 shows that $(\bigoplus P_i, \oplus \varphi_i)$ is a projective cover of $\bigoplus S_i = M$.   •

4.7   Localization

All rings in this section are commutative.

The ring $\mathbb{Z}$ has infinitely many prime ideals, but the ring $\mathbb{Z}_{(2)} = \{a/b \in \mathbb{Q} : b \text{ is odd}\}$ has only one prime ideal, namely, (2) (all other primes in $\mathbb{Z}$ are invertible in $\mathbb{Z}_{(2)}$). Now $\mathbb{Z}_{(2)}$-modules are much simpler than $\mathbb{Z}$-modules. For example, there are only two $\mathbb{Z}_{(2)}$-submodules of $\mathbb{Q}$ (to isomorphism): $\mathbb{Z}_{(2)}$ and $\mathbb{Q}$. On the other hand, there are uncountably many nonisomorphic subgroups of $\mathbb{Q}$. Similar observations lead to a localization-globalization strategy to attack algebraic and number-theoretic problems. The fundamental assumption underlying this strategy is that the local case is simpler than the global. Evidence for this can be seen in the structure of projective R-modules: for arbitrary commutative rings R, projectives can be quite complicated, but Theorem 4.58 says that projective modules over local rings are always free. Given a prime ideal $\mathfrak{p}$ in a commutative ring R, we will construct local rings $R_{\mathfrak{p}}$. Localization looks at problems involving the rings $R_{\mathfrak{p}}$, while globalization uses all such local information to answer questions about R.

Definition. A subset $S \subseteq R$ of a commutative ring R is ***multiplicative*** if S is a monoid not containing 0; that is, $0 \notin S$, $1 \in S$, and S is closed under multiplication: if $s, s' \in S$, then $ss' \in S$.

Example 4.67.

(i) If $\mathfrak{p}$ is a prime ideal in R, then its set-theoretic complement $S = R - \mathfrak{p}$ is multiplicative.

(ii) If R is a domain, then the set $S = R^{\times}$ of all its nonzero elements is multiplicative [this is a special case of part (i), for $\{0\}$ is a prime ideal in a domain].

(iii) If $a \in R$ is not nilpotent, then the set of its powers $S = \{a^n : n \geq 0\}$ is multiplicative. More generally, any submonoid of R not containing 0 is multiplicative. ◀

Definition. If $S \subseteq R$ is multiplicative, consider $\mathcal{C}(S)$, all ordered pairs (A, φ), where A is a commutative R-algebra, $\varphi \colon R \to A$ is an R-algebra map, and $\varphi(s)$ is invertible in A for all $s \in S$. An ordered pair $(S^{-1}R, h)$ in $\mathcal{C}(S)$ is a ***localization*** of R if it is a solution to the following universal mapping problem.

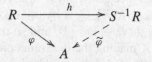

If $(A, \varphi) \in \mathcal{C}(S)$, then there exists a unique R-algebra map $\widetilde{\varphi} \colon S^{-1}R \to A$ with $\widetilde{\varphi}h = \varphi$. The map h is called the ***localization map***.

A localization $S^{-1}R$, as any solution to a universal mapping problem, is unique to isomorphism if it exists, and we call $S^{-1}R$ *the* localization at S.

The reason for excluding 0 from a multiplicative set is now apparent, for 0 is invertible only in the zero ring.

Given a multiplicative subset $S \subseteq R$, most authors construct the localization $S^{-1}R$ by generalizing the (tedious) construction of the fraction field of a domain R. They define a relation on $R \times S$ by $(r, s) \equiv (r', s')$ if there exists $s'' \in S$ with $s''(rs' - r's) = 0$ (this definition reduces to the usual definition involving cross multiplication when R is a domain and $S = R^{\times}$ is the subset of its nonzero elements). After proving that $\equiv$ is an equivalence relation, $S^{-1}R$ is defined to be the set of all equivalence classes, addition and multiplication are defined and proved to be well-defined, all the R-algebra axioms are verified, and the elements of S are shown to be invertible. Our exposition follows that of M. Artin; we develop the existence and first properties of $S^{-1}R$ in a less tedious way, which will show that the equivalence relation generalizing cross multiplication arises naturally.

Theorem 4.68. *If $S \subseteq R$ is multiplicative, then the localization $S^{-1}R$ exists.*

Proof. Let $X = \{x_s : s \in S\}$ be a set with $x_s \mapsto s$ a bijection $X \to S$, and let $R[X]$ be the polynomial ring over R with indeterminates X. Define

$$S^{-1}R = R[X]/J,$$

where J is the ideal generated by $\{sx_s - 1 : s \in S\}$, and define $h \colon R \to S^{-1}R$ by $h \colon r \mapsto r + J$, where r is a constant polynomial. It is clear that $S^{-1}R$ is an R-algebra, that h is an R-algebra map, and that each $h(s)$ is invertible. Assume now that A is an R-algebra and that $\varphi \colon R \to A$ is an R-algebra map with $\varphi(s)$ invertible for all $s \in S$. Consider the diagram in which the top arrow $\iota \colon R \to R[X]$ sends each $r \in R$ to the constant polynomial r and $v \colon R[X] \to R[X]/J = S^{-1}R$ is the natural map.

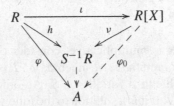

The top triangle commutes because both h and $v\iota$ send $r \in R$ to $r + J$. Define an R-algebra map $\varphi_0 \colon R[X] \to A$ by $\varphi_0(x_s) = \varphi(s)^{-1}$ for all $x_s \in X$. Clearly, $J \subseteq \ker\varphi_0$, for $\varphi_0(sx_s - 1) = 0$, and so there is an R-algebra map $\widetilde{\varphi} \colon S^{-1}R = R[X]/J \to A$ making the diagram commute. The map $\widetilde{\varphi}$ is the unique such map because $S^{-1}R$ is generated by $\operatorname{im} h \cup \{h(s)^{-1} : s \in S\}$ as an R-algebra. ●

We now describe the elements in $S^{-1}R$.

Proposition 4.69. *If $S \subseteq R$ is multiplicative, then each $y \in S^{-1}R$ has a (not necessarily unique) factorization $y = h(r)h(s)^{-1}$, where $h \colon R \to S^{-1}R$ is the localization map, $r \in R$, and $s \in S$.*

Proof. Define $A = \{y \in S^{-1}R : y = h(r)h(s)^{-1}$, for $r \in R$ and $s \in S\}$. It is routine to check that A is an R-subalgebra of $S^{-1}R$ containing $\operatorname{im} h$. Since $\operatorname{im} h \subseteq A$, there is an R-algebra map $h' \colon R \to A$ that is obtained from h by changing its target. Consider the diagram

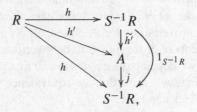

where $j: A \to S^{-1}R$ is the inclusion and $\widetilde{h'}: S^{-1}R \to A$ is given by universality (so the top triangle commutes). The lower triangle commutes, because $h(r) = h'(r)$ for all $r \in R$, and so the large triangle commutes: $(j\widetilde{h'})h = h$. But $1_{S^{-1}R}$ also makes this diagram commute, so that uniqueness gives $j\widetilde{h'} = 1_{S^{-1}R}$. By set theory, j is surjective; that is, $S^{-1}R = A$. •

In light of this proposition, the elements of $S^{-1}R$ can be regarded as "fractions" $h(r)h(s)^{-1}$, where $r \in R$ and $s \in S$.

Notation. Let $h: R \to S^{-1}R$ be the localization map. If $r \in R$ and $s \in S$, define

$$r/s = h(r)h(s)^{-1}.$$

In particular, $r/1 = h(r)$.

Is the localization map $h: r \mapsto r/1$ an injection?

Proposition 4.70. *If $S \subseteq R$ is multiplicative and $h: R \to S^{-1}R$ is the localization map, then*

$$\ker h = \{r \in R : sr = 0 \text{ for some } s \in S\}.$$

Proof. If $sr = 0$, then $0 = h(s)h(r)$ in $S^{-1}R$. Since $h(s)$ is a unit, we have $0 = h(s)^{-1}h(s)h(r) = h(r)$, and so $r \in \ker h$.

Conversely, suppose that $h(r) = 0$ in $S^{-1}R$. Since $S^{-1}R = R[X]/J$, where $J = (sx_s - 1 : s \in S)$, there is an equation $r = \sum_{i=1}^{n} f_i(X)(s_i x_{s_i} - 1)$ in $R[X]$ that involves only finitely many elements $\{s_1, \dots, s_n\} \subseteq S$; let S_0 be the submonoid of S they generate. If $h_0: R \to S_0^{-1}R$ is the localization map, then $r \in \ker h_0$. In fact, if $s = s_1 \cdots s_n$ and $h': R \to \langle s \rangle^{-1}R$ is the localization map (where $\langle s \rangle = \{s^n : n \geq 0\}$), then every $h'(s_i)$ is invertible, for $s_i^{-1} = s^{-1}s_1 \cdots \widehat{s_i} \cdots s_n$. Now $\langle s \rangle^{-1}R = R[x]/(sx - 1)$, so that $r \in \ker h'$ says that there is $f(x) = \sum_{i=0}^{m} a_i x^i \in R[x]$ with

$$r = f(x)(sx - 1) = \left(\sum_{i=0}^{m} a_i x^i \right) (sx - 1) = \sum_{i=0}^{m} (sa_i x^{i+1} - a_i x^i) \text{ in } R[x].$$

Expanding and equating coefficients of like powers of x gives

$$r = -a_0, \quad sa_0 = a_1, \quad \dots, \quad sa_{m-1} = a_m, \quad sa_m = 0.$$

Hence, $sr = -sa_0 = -a_1$, and, by induction, $s^i r = -a_i$ for all i. In particular, $s^m r = -a_m$, and so $s^{m+1}r = -sa_m = 0$, as desired. •

When are two "fractions" r/s and r'/s' equal?

Corollary 4.71. *Let $S \subseteq R$ be multiplicative. If both $r/s, r'/s' \in S^{-1}R$, where $s, s' \in S$, then $r/s = r'/s'$ if and only if there exists $s'' \in S$ with $s''(rs' - r's) = 0$ in R.*

Remark. If S contains no zero-divisors, then $s''(rs' - r's) = 0$ if and only if $rs' - r's = 0$, because s'' is a unit, and so $rs' = r's$. ◄

Proof. If $r/s = r'/s'$, then multiplying by ss' gives $(rs' - r's)/1 = 0$ in $S^{-1}R$. Hence, $rs' - r's \in \ker h$, and Proposition 4.70 gives $s'' \in S$ with $s''(rs' - r's) = 0$ in R.

Conversely, if $s''(rs' - r's) = 0$ in R for some $s'' \in S$. then we have $h(s'')h(rs' - r's) = 0$ in $S^{-1}R$. As $h(s'')$ is a unit, we have $h(r)h(s') = h(r')h(s)$; as $h(s)$ and $h(s')$ are units, $h(r)h(s)^{-1} = h(r')h(s')^{-1}$; that is, $r/s = r'/s'$. •

Corollary 4.72. *Let $S \subseteq R$ be multiplicative.*

(i) *If S contains no zero-divisors, then the localization map $h \colon R \to S^{-1}R$ is an injection.*

(ii) *If R is a domain with $Q = \mathrm{Frac}(R)$, then $S^{-1}R \subseteq Q$. Moreover, if $S = R^{\times}$, then $S^{-1}R = Q$.*

Proof.

(i) This follows easily from Proposition 4.70.

(ii) The localization map $h \colon R \to S^{-1}R$ is an injection, by Proposition 4.70. The result now follows from Proposition 4.69. •

If R is a domain and $S \subseteq R$ is multiplicative, then Corollary 4.72 says that $S^{-1}R$ consists of all elements $a/s \in \mathrm{Frac}(R)$ with $a \in R$ and $s \in S$.

Let us now investigate the ideals in $S^{-1}R$.

Notation. If $S \subseteq R$ is multiplicative and I is an ideal in R, then we denote the ideal in $S^{-1}R$ generated by $h(I)$ by $S^{-1}I$.

Example 4.73.

(i) If $S \subseteq R$ is multiplicative and I is an ideal in R containing an element $s \in S$ (that is, $I \cap S \neq \varnothing$), then $S^{-1}I$ contains $s/s = 1$, and so $S^{-1}I = S^{-1}R$.

(ii) Let S consist of all the odd integers [that is, S is the complement of the prime ideal (2)], let $I = (3)$, and let $I' = (5)$. Then $S^{-1}I = S^{-1}\mathbb{Z} = S^{-1}I'$. Therefore, the function from the ideals in $\mathbb{Z}$ to the ideals in $S^{-1}\mathbb{Z} = \mathbb{Z}_{(2)} = \{a/b \in \mathbb{Q} : b \text{ is odd}\}$, given by $I \mapsto S^{-1}I$, is not injective. ◄

Corollary 4.74. *Let $S \subseteq R$ be multiplicative.*

(i) *Every ideal J in $S^{-1}R$ is of the form $S^{-1}I$ for some ideal I in R. In fact, if R is a domain and $I = J \cap R$, then $J = S^{-1}I$; in the general case, if $I = h^{-1}(h(R) \cap J)$, then $J = S^{-1}I$.*

(ii) *If I is an ideal in R, then $S^{-1}I = S^{-1}R$ if and only if $I \cap S \neq \varnothing$.*

(iii) *If $\mathfrak{q}$ is a prime ideal in R with $\mathfrak{q} \cap S = \varnothing$, then $S^{-1}\mathfrak{q}$ is a prime ideal in $S^{-1}R$.*

(iv) *The function $f : \mathfrak{q} \mapsto S^{-1}\mathfrak{q}$ is a bijection from the family of all prime ideals in R disjoint from S to the family of all prime ideals in $S^{-1}R$.*

(v) *If R is noetherian, then $S^{-1}R$ is also noetherian.*

Proof.

(i) Let $J = (j_\lambda : \lambda \in \Lambda)$. By Proposition 4.69, we have $j_\lambda = h(r_\lambda)h(s_\lambda)^{-1}$, where $r_\lambda \in R$ and $s_\lambda \in S$. Define I to be the ideal in R generated by $\{r_\lambda : \lambda \in \Lambda\}$; that is, $I = h^{-1}(h(R) \cap J)$. It is clear that $S^{-1}I = J$; in fact, since all s_λ are units in $S^{-1}R$, we have $J = (h(r_\lambda) : \lambda \in \Lambda)$.

(ii) If $s \in I \cap S$, then $s/1 \in S^{-1}I$. But $s/1$ is a unit in $S^{-1}R$, and so $S^{-1}I = S^{-1}R$. Conversely, if $S^{-1}I = S^{-1}R$, then $h(a)h(s)^{-1} = 1$ for some $a \in I$ and $s \in S$. Therefore, $s - a \in \ker h$, and so there is $s'' \in S$ with $s''(s - a) = 0$. Therefore, $s''s = s''a \in I$. Since S is multiplicatively closed, $s''s \in I \cap S$.

(iii) Suppose that $\mathfrak{q}$ is a prime ideal in R. First, $S^{-1}\mathfrak{q}$ is a proper ideal, for $\mathfrak{q} \cap S = \varnothing$. If $(a/s)(b/t) = c/u$, where $a, b \in R, c \in \mathfrak{q}$, and $s, t, u \in S$, then there is $s'' \in S$ with $s''(uab - stc) = 0$. Hence, $s''uab \in \mathfrak{q}$. Now $s''u \notin \mathfrak{q}$ (because $s''u \in S$ and $S \cap \mathfrak{q} = \varnothing$); hence, $ab \in \mathfrak{q}$ (because $\mathfrak{q}$ is prime). Thus, either a or b lies in $\mathfrak{q}$, and either a/s or b/t lies in $S^{-1}\mathfrak{q}$. Therefore, $S^{-1}\mathfrak{q}$ is a prime ideal.

(iv) Suppose that $\mathfrak{p}$ and $\mathfrak{q}$ are prime ideals in R with $f(\mathfrak{p}) = S^{-1}\mathfrak{p} = S^{-1}\mathfrak{q} = f(\mathfrak{q})$; we may assume that $\mathfrak{p} \cap S = \varnothing = \mathfrak{q} \cap S$. If $a \in \mathfrak{p}$, then there are $b \in \mathfrak{q}$ and $s \in S$ with $a/1 = b/s$. Hence, $sa - b \in \ker h$, where h is the localization map, and so there is $s' \in S$ with $s'sa = s'b \in \mathfrak{q}$.

But $s's \in S$, so that $s's \notin \mathfrak{q}$. Since $\mathfrak{q}$ is prime, we have $a \in \mathfrak{q}$; that is, $\mathfrak{p} \subseteq \mathfrak{q}$. The reverse inclusion is proved similarly. Thus, f is injective.

Let $\mathfrak{P}$ be a prime ideal in $S^{-1}R$. By part (i), there is some ideal I in R with $\mathfrak{P} = S^{-1}I$. We must show that I can be chosen to be a prime ideal in R. Now $h(R) \cap \mathfrak{P}$ is a prime ideal in $h(R)$, and so $\mathfrak{p} = h^{-1}(h(R) \cap \mathfrak{P})$ is a prime ideal in R. By part (i), $\mathfrak{P} = S^{-1}\mathfrak{p}$, and so f is surjective.

(v) If J is an ideal in $S^{-1}R$, then part (i) shows that $J = S^{-1}I$ for some ideal I in R. Since R is noetherian, we have $I = (r_1, \ldots, r_n)$, and so $J = (r_1/1, \ldots, r_n/1)$. Hence, every ideal in $S^{-1}R$ is finitely generated, and so $S^{-1}R$ is noetherian. •

Notation. If $\mathfrak{p}$ is a prime ideal in a commutative ring R and $S = R - \mathfrak{p}$, then $S^{-1}R$ is denoted by $R_{\mathfrak{p}}$.

The next proposition explains why $S^{-1}R$ is called localization.

Theorem 4.75. *If $\mathfrak{p}$ is a prime ideal in a commutative ring R, then $R_{\mathfrak{p}}$ is a local ring with unique maximal ideal $\mathfrak{p}R_{\mathfrak{p}} = \{r/s : r \in \mathfrak{p} \text{ and } s \notin \mathfrak{p}\}$.*

Proof. If $x \in R_{\mathfrak{p}}$, then $x = r/s$, where $r \in R$ and $s \notin \mathfrak{p}$. If $r \notin \mathfrak{p}$, then r/s is a unit in $R_{\mathfrak{p}}$; that is, all nonunits lie in $\mathfrak{p}R_{\mathfrak{p}}$. Hence, if I is any ideal in $R_{\mathfrak{p}}$ that contains an element r/s with $r \notin \mathfrak{p}$, then $I = R_{\mathfrak{p}}$. It follows that every proper ideal in $R_{\mathfrak{p}}$ is contained in $\mathfrak{p}R_{\mathfrak{p}}$, and so $R_{\mathfrak{p}}$ is a local ring with unique maximal ideal $\mathfrak{p}R_{\mathfrak{p}}$. •

Here is an application of localization.

Definition. A prime ideal $\mathfrak{p}$ in a commutative ring R is a ***minimal prime ideal*** if there is no prime ideal strictly contained in it.

In a domain, (0) is a minimal prime ideal, and it is the unique such.

Proposition 4.76. *Let R be a commutative ring.*

(i) *If $S \subseteq R$ is multiplicative, then any ideal I maximal with $I \cap S = \varnothing$ is a prime ideal.*

(ii) *If $\mathfrak{p}$ is a minimal prime ideal, then every $x \in \mathfrak{p}$ is nilpotent; that is, $x^n = 0$ for some $n = n(x) \geq 1$.*

Proof.

(i) If $ab \in I$ and neither a nor b lies in I, then $I \subsetneq I + Ra$ and $I \subsetneq I + Rb$. By maximality, $(I + Ra) \cap S \neq \varnothing$ and $(I + Rb) \cap S \neq \varnothing$, so there are $r, r' \in R$ and $i, i' \in I$ with $i + ra = s \in S$ and $i' + r'b = s' \in S$. Hence,

$$ss' = (i + ra)(i' + r'b) = ii' + ir'b + rai' + rr'ab \in S \cap I,$$

a contradiction.

(ii) Let $x \in \mathfrak{p}$ be nonzero. By Corollary 4.74(iv), there is only one prime ideal in $R_{\mathfrak{p}}$, namely, $\mathfrak{p}R_{\mathfrak{p}}$, and $x/1$ is a nonzero element in it. Indeed, x is nilpotent if and only if $x/1$ is nilpotent, by Proposition 4.70. Thus, we have normalized the problem; we may now assume that $x \in \mathfrak{p}$ and that $\mathfrak{p}$ is the only prime ideal in R. If x is not nilpotent, then $S = \{1, x, x^2, \ldots\}$ is multiplicative. We can prove, with Zorn's Lemma, that there exists an ideal I in R maximal with $I \cap S = \varnothing$. Now part (i) says that I is a prime ideal; that is, $\mathfrak{p} = I$. But $x \in S \cap \mathfrak{p} = S \cap I = \varnothing$, a contradiction. Therefore, x is nilpotent. $\bullet$

Having localized a commutative ring, we now localize its modules. If M is an R-module and $s \in R$, let $\mu_s : M \to M$ denote the multiplication map $m \mapsto sm$. For a subset $S \subseteq R$, the map μ_s is invertible for every $s \in S$ (that is, every μ_s is an automorphism) if and only if M is an $S^{-1}R$-module.

Definition. Let M be an R-module and let $S \subseteq R$ be multiplicative. A *localization* of M is an ordered pair $(S^{-1}M, h_M)$, where $S^{-1}M$ is an $S^{-1}R$-module and $h_M : M \to S^{-1}M$ is an R-map (called the *localization map*), which is a solution to the following universal mapping problem:

if M' is an $S^{-1}R$-module and $\varphi : M \to M'$ is an R-map, then there exists a unique $S^{-1}R$-map $\widetilde{\varphi} : S^{-1}M \to M'$ with $\widetilde{\varphi}h_M = \varphi$.

The obvious candidate for $(S^{-1}M, h_M)$, namely, $(S^{-1}R \otimes_R M, h \otimes 1_M)$, where $h : R \to S^{-1}R$ is the localization map, actually is the localization.

Proposition 4.77. *Let $S \subseteq R$ be multiplicative and let M be an R-module. Then $S^{-1}R \otimes_R M$ and the R-map $h_M = h \otimes 1_M : M \to S^{-1}R \otimes_R M$, given by $m \mapsto 1 \otimes m$, is a localization of M.*

Proof. Let $\varphi : M \to M'$ be an R-map, where M' is an $S^{-1}R$-module. The function $S^{-1}R \times M \to M'$, defined by $(r/s, m) \mapsto (r/s)\varphi(m)$, where $r \in R$ and $s \in S$, is easily seen to be R-bilinear. Hence, there is a unique R-map $\widetilde{\varphi} : S^{-1}R \otimes_R M \to M'$ with $\widetilde{\varphi}h_M = \varphi$. Now M' is an $S^{-1}R$-module, by Proposition 2.51. We let the reader check that $\widetilde{\varphi}$ is an $S^{-1}R$-map. $\bullet$

One of the most important properties of $S^{-1}R$ is that it is flat as an R-module. To prove this, we first generalize the argument in Proposition 4.70.

Proposition 4.78. *Let $S \subseteq R$ be multiplicative. If M is an R-module and $h_M \colon M \to S^{-1}M$ is the localization map, then*

$$\ker h_M = \{m \in M : sm = 0 \text{ for some } s \in S\}.$$

Proof. Denote $\{m \in M : sm = 0 \text{ for some } s \in S\}$ by K. If $sm = 0$, for $m \in M$ and $s \in S$, then $h_M(m) = (1/s)h_M(sm) = 0$, and so $K \subseteq \ker h_M$. For the reverse inclusion, proceed as in Proposition 4.70: if $m \in K$, there is $s \in S$ with $sm = 0$. Reduce to $S = \langle s \rangle$ for some $s \in S$, where $\langle s \rangle = \{s^n : n \geq 0\}$, so that $S^{-1}R = R[x]/(sx - 1)$. Now $R[x] \otimes_R M \cong \sum_i Rx^i \otimes_R M$, because $R[x]$ is the free R-module with basis $\{1, x, x^2, \ldots\}$. Hence, each element in $R[x] \otimes_R M$ has a unique expression of the form $\sum_i x^i \otimes m_i$, where $m_i \in M$. Hence,

$$\ker h_M = \Big\{ m \in M : 1 \otimes m = (sx - 1) \sum_{i=0}^{n} x^i \otimes m_i \Big\}.$$

The proof now finishes as the proof of Proposition 4.70. Expanding and equating coefficients gives equations

$$1 \otimes m = -1 \otimes m_0, \quad x \otimes sm_0 = x \otimes m_1, \ldots,$$
$$x^n \otimes sm_{n-1} = x^n \otimes m_n, \quad x^{n+1} \otimes sm_n = 0.$$

It follows that

$$m = -m_0, \quad sm_0 = m_1, \quad \ldots, \quad sm_{n-1} = m_n, \quad sm_n = 0.$$

Hence, $sm = -sm_0 = -m_1$, and, by induction, $s^i m = -m_i$ for all i. In particular, $s^n m = -m_n$ and so $s^{n+1} m = -sm_n = 0$ in M. Therefore, $\ker h_M \subseteq K$, as desired.   •

Corollary 4.79. *Let $S \subseteq R$ be multiplicative and let M be an R-module.*

(i) *Every element $u \in S^{-1}M = S^{-1} \otimes_R M$ has the form $u = s^{-1} \otimes m$ for some $s \in S$ and some $m \in M$.*

(ii) *Every $S^{-1}R$-module A is isomorphic to $S^{-1}M$ for some R-module M.*

(iii) *$s_1^{-1} \otimes m_1 = s_2^{-1} \otimes m_2$ in $S^{-1} \otimes_R M$ if and only if $s(s_2 m_1 - s_1 m_2)$ in M for some $s \in S$.*

Proof.

(i) If $u \in S^{-1} \otimes_R M$, then $u = \sum_i (r_i/s_i) \otimes m_i$, where $r_i \in R$, $s_i \in S$, and $m_i \in M$. If we define $s = \prod s_i$ and $\hat{s}_i = \prod_{j \neq i} s_j$, then

$$
\begin{aligned}
u &= \sum (1/s_i) r_i \otimes m_i \\
&= \sum (\hat{s}_i/s) r_i \otimes m_i \\
&= (1/s) \sum \hat{s}_i r_i \otimes m_i \\
&= (1/s) \otimes \sum \hat{s}_i r_i m_i \\
&= (1/s) \otimes m,
\end{aligned}
$$

where $m = \sum \hat{s}_i r_i m_i \in M$.

(ii) The localization map $h \colon R \to S^{-1}R$ allows us to view A as an R-module [define $ra = h(r)a$ for $r \in R$ and $a \in A$]; denote this R-module by ${}_h A$. We claim that $A \cong S^{-1}R \otimes_R {}_h A$ as $S^{-1}R$-modules. Define $f \colon A \to S^{-1}R \otimes_R {}_h A$ by $a \mapsto 1 \otimes a$. Now f is an $S^{-1}R$-map: if $s \in S$ and $a \in A$, then

$$
f(s^{-1}a) = 1 \otimes s^{-1}a = s^{-1}s \otimes s^{-1}a = s^{-1} \otimes a = s^{-1}f(a).
$$

To see that f is an isomorphism, we construct its inverse. Since A is an $S^{-1}R$-module, the function $S^{-1}R \times {}_h A \to A$, defined by $(rs^{-1}, a) \mapsto (rs^{-1})a$, is a well-defined R-bilinear function, and so it induces an R-map $S^{-1}R \otimes_R {}_h A \to A$, which is obviously inverse to f.

(iii) If $s \in S$ with $s(s_2 m_1 - s_1 m_2) = 0$ in M, then $(s/1)(s_2 \otimes m_1 - s_1 \otimes m_2) = 0$ in $S^{-1}R \otimes_R M$. As $s/1$ is a unit, $s_2 \otimes m_1 - s_1 \otimes m_2 = 0$, and so $s_1^{-1} \otimes m_1 = s_2^{-1} \otimes m_2$.

Conversely, if $s_1^{-1} \otimes m_1 = s_2^{-1} \otimes m_2$ in $S^{-1} \otimes_R M$, then we have $(1/s_1 s_2)(s_2 \otimes m_1 - s_1 \otimes m_2) = 0$. Since $1/s_1 s_2$ is a unit, we have $(s_2 \otimes m_1 - s_1 \otimes m_2) = 0$ and $s_2 m_1 - s_1 m_2 \in \ker h_M$. By Proposition 4.78, there exists $s \in S$ with $s(s_2 m_1 - s_1 m_2) = 0$ in M.   •

Theorem 4.80.   *If $S \subseteq R$ is multiplicative, then $S^{-1}R$ is a flat R-module.*

Proof.   We must show that if $0 \to A \xrightarrow{f} B$ is exact, then so is

$$
0 \to S^{-1}R \otimes_R A \xrightarrow{1 \otimes f} S^{-1}R \otimes_R B.
$$

By Corollary 4.79, every $u \in S^{-1}A$ has the form $u = s^{-1} \otimes a$ for some $s \in S$ and $a \in A$. In particular, if $u \in \ker(1 \otimes f)$, then $(1 \otimes f)(u) = s^{-1} \otimes f(a) = 0$.

Multiplying by s gives $1 \otimes f(a) = 0$ in $S^{-1}B$; that is, $f(a) \in \ker h_B$. By Proposition 4.78, there is $t \in S$ with $0 = tf(a) = f(ta)$. Since f is an injection, $ta \in \ker f = \{0\}$. Hence, $0 = 1 \otimes ta = t(1 \otimes a)$. But t is a unit in $S^{-1}R$, so that $1 \otimes a = 0$ in $S^{-1}A$. Therefore, $1 \otimes f$ is an injection, and $S^{-1}R$ is a flat R-module.   •

Corollary 4.81.   *If $S \subseteq R$ is multiplicative, then localization $M \mapsto S^{-1}M = S^{-1}R \otimes_R M$ defines an exact functor $_R\mathbf{Mod} \to {}_{S^{-1}R}\mathbf{Mod}$.*

Proof.   Localization is the functor $S^{-1}R \otimes_R \square$, and it is exact because $S^{-1}R$ is a flat R-module.   •

Since tensor product commutes with direct sums, it is clear that if M is a free (or projective) R-module, then $S^{-1}M$ is a free (or projective) $S^{-1}R$-module.

Example 4.82.   Let R be a Dedekind ring that is not a PID, and let $\mathfrak{p}$ be a nonzero prime ideal in R. Then $R_\mathfrak{p}$ is a local PID (see Rotman, *Advanced Modern Algebra*, p. 950). Hence, if P is a projective R-module, then $P_\mathfrak{p}$ is a projective $R_\mathfrak{p}$-module, and so it is free. In particular, if $\mathfrak{b}$ is a nonprincipal ideal in R, then $\mathfrak{b}$ is not free even though all its localizations are free.   ◄

Proposition 4.83.   *Let $S \subseteq R$ be multiplicative. If B is a flat R-module, then $S^{-1}B$ is a flat $S^{-1}R$-module.*

Proof.   For any $S^{-1}R$-module A, there is an isomorphism

$$S^{-1}B \otimes_{S^{-1}R} A = (S^{-1}R \otimes_R B) \otimes_{S^{-1}R} A \cong S^{-1}R \otimes_R (B \otimes_R A),$$

which can be used to give a natural isomorphism

$$S^{-1}B \otimes_{S^{-1}R} \square \cong (S^{-1}R \otimes_R \square)(B \otimes_R \square).$$

As each factor is an exact functor, the composite $S^{-1}B \otimes_{S^{-1}R} \square$ is also an exact functor; that is, $S^{-1}B$ is flat.   •

We now investigate localization of injective modules, and we begin with some identities.

Proposition 4.84.   *If A and B are R-modules, then there is a natural isomorphism*

$$\varphi: S^{-1}(B \otimes_R A) \to S^{-1}B \otimes_{S^{-1}R} S^{-1}A.$$

Proof. Every element $u \in S^{-1}(B \otimes_R A)$ has the form $u = s^{-1}m$ for $s \in S$ and $m \in B \otimes_R A$, by Corollary 4.79; hence, $u = s^{-1} \sum a_i \otimes b_i$.

$$A \times B \longrightarrow A \otimes_R B \longrightarrow S^{-1}(A \otimes_R B)$$
$$\searrow \qquad \downarrow \qquad \nearrow$$
$$S^{-1}B \otimes_{S^{-1}R} S^{-1}A$$

The idea is to define $\varphi(u) = \sum s^{-1}a_i \otimes b_i$, but, as usual with tensor product, the problem is whether obvious maps are well-defined. We suggest that the reader use the universal property of localization to complete the proof.   •

The analogous isomorphism for Hom,

$$S^{-1}\operatorname{Hom}_R(B, A) \cong \operatorname{Hom}_{S^{-1}R}(S^{-1}B, S^{-1}A),$$

may not hold. However, there is such an isomorphism when B and R are restricted.

Lemma 4.85. *Let A be an R-algebra, and let N be a finitely presented R-module. For every A-module M, there is a natural isomorphism*

$$\theta \colon \operatorname{Hom}_R(N, M) \to \operatorname{Hom}_A(N \otimes_R A, M),$$

given by $\theta \colon f \mapsto \widetilde{f}$, where $\widetilde{f}(n \otimes 1) = f(n)$ for all $n \in N$.

Proof. Assume first that N is a finitely generated free R-module, say, with basis $\{e_1, \ldots, e_n\}$; then $\{e_1 \otimes 1, \ldots, e_n \otimes 1\}$ is a basis of the free A-module $N \otimes_R A$. If $f \colon N \to M$ is an R-map, define an A-map $\widetilde{f} \colon N \otimes_R A \to M$ by $\widetilde{f}(e_i \otimes 1) = f(e_i)$ for all i. It is easy to see that $\theta \colon f \mapsto \widetilde{f}$ is a well-defined natural isomorphism $\operatorname{Hom}_R(N, M) \to \operatorname{Hom}_A(N \otimes_R A, M)$.

Assume now that N is finitely presented, so that there is an exact sequence

$$R^k \to R^n \to N \to 0.$$

This gives rise to the commutative diagram with exact rows.

$$
\begin{array}{ccccccc}
0 & \longrightarrow & \operatorname{Hom}_R(N, M) & \longrightarrow & \operatorname{Hom}_R(R^n, M) & \longrightarrow & \operatorname{Hom}_R(R^k, M) \\
 & & \big\downarrow & & \big\downarrow{\scriptstyle\theta} & & \big\downarrow{\scriptstyle\theta} \\
0 & \longrightarrow & \operatorname{Hom}_A(N \otimes_R A, M) & \longrightarrow & \operatorname{Hom}_A(A^n, M) & \longrightarrow & \operatorname{Hom}_A(A^k, M)
\end{array}
$$

The vertical maps θ are isomorphisms, and so the dashed vertical arrow [which exists by diagram chasing (Proposition 2.71) and which has the desired formula] is also an isomorphism. The reader may prove naturality.   •

Lemma 4.86. *Let B be a flat R-module, and let N be a finitely presented R-module. For every R-module M, there is a natural isomorphism*

$$\psi : B \otimes_R \operatorname{Hom}_R(N, M) \to \operatorname{Hom}_R(N, M \otimes_R B),$$

given by $b \otimes g \mapsto g_b$, where $g_b(n) = g(n) \otimes b$ for all $n \in N$.

Proof. The reader may check that ψ arises from the R-bilinear function with domain $B \times \operatorname{Hom}_R(N, M)$ that sends $(b, g) \mapsto g_b$; this map is natural in N, and it is an isomorphism when N is finitely generated and free. If N is finitely presented, there is an exact sequence

$$R^k \to R^n \to N \to 0.$$

Since B is flat, there is a commutative diagram with exact rows.

$$
\begin{array}{ccccccc}
0 \to & B \otimes \operatorname{Hom}(N, M) & \to & B \otimes \operatorname{Hom}(R^n, M) & \to & B \otimes \operatorname{Hom}(R^k, M) \\
& \downarrow{\psi} & & \downarrow{\psi} & & \downarrow{\psi} \\
0 \to & \operatorname{Hom}(N, M \otimes B) & \to & \operatorname{Hom}(R^n, M \otimes B) & \to & \operatorname{Hom}(R^k, M \otimes B)
\end{array}
$$

By the Five Lemma, the first vertical map is an isomorphism because the second two are.   •

Lemma 4.87. *Let $S \subseteq R$ be multiplicative, and let N be a finitely presented R-module. For every R-module M, there is a natural isomorphism*

$$\varphi : S^{-1} \operatorname{Hom}_R(N, M) \to \operatorname{Hom}_{S^{-1}R}(S^{-1}N, S^{-1}M),$$

given by $g/1 \mapsto \widehat{g}$, where $\widehat{g}(n/1) = g(n) \otimes 1$ for all $n \in N$.

Proof. By definition, $S^{-1} \operatorname{Hom}_R(N, M) = S^{-1}R \otimes_R \operatorname{Hom}_R(N, M)$. Since $S^{-1}R$ is a flat R-module, Lemma 4.86 gives an isomorphism

$$S^{-1} \operatorname{Hom}_R(N, M) \cong \operatorname{Hom}_R(N, S^{-1}M);$$

but $S^{-1}R$ is an R-algebra, and so Lemma 4.85 gives an isomorphism

$$\operatorname{Hom}_R(N, S^{-1}M) \cong \operatorname{Hom}_{S^{-1}R}(S^{-1}N, S^{-1}M).$$

The composite is an isomorphism with the desired formula.   •

Theorem 4.88. *Let R be noetherian and let $S \subseteq R$ be multiplicative. If E is an injective R-module, then $S^{-1}E$ is an injective $S^{-1}R$-module.*

Proof. By Baer's Criterion, it suffices to extend any map $I \to S^{-1}E$ to a map $S^{-1}R \to S^{-1}E$, where I is an ideal in $S^{-1}R$; that is, if $i: I \to S^{-1}R$ is the inclusion, then the induced map

$$i^*: \operatorname{Hom}_{S^{-1}R}(S^{-1}R, S^{-1}E) \to \operatorname{Hom}_{S^{-1}R}(I, S^{-1}E)$$

is a surjection. Now $S^{-1}R$ is noetherian because R is, and so I is finitely generated; say, $I = (r_1/s_1, \ldots, r_n/s_n)$, where $r_i \in R$ and $s_i \in S$. There is an ideal J in R, namely, $J = (r_1, \ldots, r_n)$, with $I = S^{-1}J$. Naturality of the isomorphism in Lemma 4.87 gives a commutative diagram

$$
\begin{array}{ccc}
S^{-1}\operatorname{Hom}_R(R, E) & \longrightarrow & S^{-1}\operatorname{Hom}_R(J, E) \\
\downarrow & & \downarrow \\
\operatorname{Hom}_{S^{-1}R}(S^{-1}R, S^{-1}E) & \longrightarrow & \operatorname{Hom}_{S^{-1}R}(S^{-1}J, S^{-1}E).
\end{array}
$$

Now $\operatorname{Hom}_R(R, E) \to \operatorname{Hom}_R(J, E)$ is a surjection, because E is an injective R-module, and so $S^{-1} = S^{-1}R \otimes_R \square$ being right exact implies that the top arrow is also a surjection. But the vertical maps are isomorphisms, and so the bottom arrow is a surjection; that is, $S^{-1}E$ is an injective $S^{-1}R$-module. •

Remark. Theorem 4.88 may be false if R is not noetherian. E. C. Dade, "Localization of injective modules," *J. Algebra* 69 (1981), 415–425, showed, for every commutative ring k, that if $R = k[X]$, where X is an uncountable set of indeterminates, then there are a multiplicative subset $S \subseteq R$ and an injective R-module E such that $S^{-1}E$ is not an injective $S^{-1}R$-module.

If, however, $R = k[X]$, where k is noetherian and X is countable, then $S^{-1}E$ is an injective $S^{-1}R$-module for every injective R-module E and every multiplicative subset $S \subseteq R$. ◄

Here are some globalization tools.

Notation. In the special case $S = R - \mathfrak{p}$, where $\mathfrak{p}$ is a prime ideal in R, we write
$$S^{-1}M = S^{-1} \otimes_R M = M_{\mathfrak{p}}.$$
If $f: M \to N$ is an R-map, write $f_{\mathfrak{p}}: M_{\mathfrak{p}} \to N_{\mathfrak{p}}$, where $f_{\mathfrak{p}} = 1_{R_{\mathfrak{p}}} \otimes f$.

We restate Corollary 4.74(iv) in this notation. The function $f: \mathfrak{q} \mapsto \mathfrak{q}_{\mathfrak{p}}$ is a bijection from the family of all prime ideals in R that are contained in $\mathfrak{p}$ to the family of prime ideals in $R_{\mathfrak{p}}$.

Proposition 4.89. *Let I and J be ideals in a domain R. If $I_{\mathfrak{m}} = J_{\mathfrak{m}}$ for every maximal ideal $\mathfrak{m}$, then $I = J$.*

Proof. Take $b \in J$, and define

$$(I : b) = \{r \in R : rb \in I\}.$$

Let $\mathfrak{m}$ be a maximal ideal in R. Since $I_\mathfrak{m} = J_\mathfrak{m}$, there are $a \in I$ and $s \notin \mathfrak{m}$ with $b/1 = a/s$. As R is a domain, $sb = a \in I$, so that $s \in (I : b)$; but $s \notin \mathfrak{m}$, so that $(I : b) \subsetneq \mathfrak{m}$. Thus, $(I : b)$ cannot be a proper ideal, for it is not contained in any maximal ideal. Therefore, $(I : b) = R$; hence, $1 \in (I : b)$ and $b = 1b \in I$. We have proved that $J \subseteq I$, and the reverse inclusion is proved similarly.   •

Proposition 4.90.   *Let R be a commutative ring.*

(i) *If M is an R-module with $M_\mathfrak{m} = \{0\}$ for every maximal ideal $\mathfrak{m}$, then $M = \{0\}$.*

(ii) *If $f : M \to N$ is an R-map and $f_\mathfrak{m} : M_\mathfrak{m} \to N_\mathfrak{m}$ is an injection for every maximal ideal $\mathfrak{m}$, then f is an injection.*

(iii) *If $f : M \to N$ is an R-map and $f_\mathfrak{m} : M_\mathfrak{m} \to N_\mathfrak{m}$ is a surjection for every maximal ideal $\mathfrak{m}$, then f is a surjection.*

(iv) *If $f : M \to N$ is an R-map and $f_\mathfrak{m} : M_\mathfrak{m} \to N_\mathfrak{m}$ is an isomorphism for every maximal ideal $\mathfrak{m}$, then f is an isomorphism.*

Proof.

(i) If $M \neq \{0\}$, then there is $m \in M$ with $m \neq 0$. It follows that the annihilator $I = \{r \in R : rm = 0\}$ is a proper ideal in R, for $1 \notin I$, and so there is some maximal ideal $\mathfrak{m}$ containing I. Now $1 \otimes m = 0$ in $M_\mathfrak{m}$, so that $m \in \ker h_M$. Proposition 4.78 gives $s \notin \mathfrak{m}$ with $sm = 0$ in M. Hence, $s \in I \subseteq \mathfrak{m}$, and this is a contradiction. Therefore, $M = \{0\}$.

(ii) There is an exact sequence $0 \to K \to M \xrightarrow{f} N$, where $K = \ker f$. Since localization is an exact functor, there is an exact sequence

$$0 \to K_\mathfrak{m} \to M_\mathfrak{m} \xrightarrow{f_\mathfrak{m}} N_\mathfrak{m}$$

for every maximal ideal $\mathfrak{m}$. By hypothesis, each $f_\mathfrak{m}$ is an injection, so that $K_\mathfrak{m} = \{0\}$ for all maximal ideals $\mathfrak{m}$. Part (i) now shows that $K = \{0\}$, and so f is an injection.

(iii) There is an exact sequence $M \xrightarrow{f} N \to C \to 0$, where $C = \operatorname{coker} f = N/\operatorname{im} f$. Since tensor product is right exact, $C_\mathfrak{m} = \{0\}$ for all maximal ideals $\mathfrak{m}$, and so $C = \{0\}$. But f is surjective if and only if $C = \operatorname{coker} f = \{0\}$.

(iv) This follows at once from parts (ii) and (iii).   •

4.8 Polynomial Rings

In the mid-1950s, Serre proved that if $R = k[x_1, \ldots, x_m]$, where k is a field, then every finitely generated projective R-module P has a finitely generated free complement F; that is, $P \oplus F$ is free. Serre wondered[5] whether every projective $k[x_1, \ldots, x_m]$-module is free (for projective and free modules over R have natural interpetations in algebraic geometry). This problem was the subject of much investigation until 1976, when it was solved in the affirmative by Quillen and Suslin, independently. We refer the reader to T. Y. Lam, *Serre's Problem on Projective Modules*, for a more thorough account. The last chapter of Lam's book describes recent work, after 1976, inspired by and flowing out of the work of Serre, Quillen, and Suslin.

Definition. Let R be a commutative ring and let R^n denote the free R-module of rank n (recall that every commutative ring has IBN). A *unimodular column* is an element $\alpha = (a_1, \ldots, a_n) \in R^n$ for which there exist $b_i \in R$ with $a_1 b_1 + \cdots + a_n b_n = 1$.

A commutative ring R has the *unimodular column property* if, for every n, every unimodular column is the first column of some $n \times n$ invertible matrix over R.

If ε_1 denotes the column vector having first coordinate 1 and all other entries 0, then $\alpha \in R^n$ is the first column of a matrix M over R if and only if

$$\alpha = M\varepsilon_1.$$

The first column $\alpha = [a_{i1}]$ of an invertible matrix $M = [a_{ij}]$ is always unimodular. Since M is invertible, $\det(M) = u$, where u is a unit in R, and Laplace expansion down the first column gives $\det(M) = u = \sum_i a_{i1} d_i$. Hence, $\sum_i a_{i1}(u^{-1} d_i) = 1$, and $\alpha = M\varepsilon_1$ is a unimodular column. The unimodular column property for R asserts the converse.

Proposition 4.91. *Let R be a commutative ring. If every finitely generated projective R-module is free, then R has the unimodular column property.*

Proof. If $\alpha = (a_1, \ldots, a_n) \in R^n$ is a unimodular column, then there exist $b_i \in R$ with $\sum_i a_i b_i = 1$. Define $\varphi \colon R^n \to R$ by $(r_1, \ldots, r_n) \mapsto \sum_i r_i b_i$. Since $\varphi(\alpha) = 1$, there is an exact sequence

$$0 \to K \to R^n \xrightarrow{\varphi} R \to 0,$$

---

[5]Serre wrote, on page 243 of "Faisceaux algébriques cohérents," *Annals Math.* 61 (1955), 197–278, "... on ignore s'il existe des A-modules projectifs de type fini qui ne soient pas libres" (here, $A = k[x_1, \ldots, x_n]$).

where $K = \ker \varphi$. As R is projective, this sequence splits and

$$R^n = K \oplus \langle \alpha \rangle.$$

By hypothesis, K is free (of rank $n - 1$). If $\alpha_2, \ldots, \alpha_n$ is a basis of K, then adjoining α gives a basis of R^n. If $\varepsilon_1, \ldots, \varepsilon_n$ is the **standard basis** of R^n (i.e., ε_i has ith coordinate 1 and all other coordinates 0), then the R-map $T : R^n \to R^n$, with $T\varepsilon_1 = \alpha$ and $T\varepsilon_i = \alpha_i$ for $i \geq 2$, is invertible, and the matrix of T with respect to the standard basis has first column α.   •

In general, the converse of Proposition 4.91 is false. For example, we know that $\mathbb{I}_6$ has nonfree projectives, yet it is not difficult to see that $\mathbb{I}_6$ does have the unimodular column property.

Definition.   A finitely generated R-module P is **stably free** if there exists a finitely generated free R-module F with $P \oplus F$ free.

Example 4.92.

(i) Every finitely generated free module is stably free.

(ii) If P and Q are stably free, then $P \oplus Q$ is stably free.

(iii) Every stably free module is projective, for it is a direct summand of a free module. However, there are (finitely generated) projective R-modules that are not stably free. For example, if $R = \mathbb{I}_6$, then $R = I \oplus J$, where $I \cong \mathbb{I}_2$ and $J \cong \mathbb{I}_3$. An easy counting argument shows that there is no finitely generated free $\mathbb{I}_6$-module F with $I \oplus F$ free.

(iv) A direct summand of a stably free module need not be stably free. After all, every projective module is a direct summand of a free (hence, stably free) module, yet we have seen in (iii) that projectives need not be stably free. However, a complement of a stably free direct summand is stably free: if $K = K' \oplus K''$, where both K and K' are stably free, then it is easy to see that K'' is also stably free.

(v) Kaplansky exhibited a stably free R-module that is not free, where R is the ring of all continuous real-valued functions on the 2-sphere [see R. G. Swan, "Vector bundles and projective modules," *Trans. AMS* 105 (1962), 264–277]. This example has been modified, and there is a completely algebraic proof that if $R = \mathbb{Z}[x, y, z]/(x^2 + y^2 + z^2 - 1)$, then there is a stably free R-module that is not free [see M. Kong, "Euler classes of inner product modules," *J. Algebra* 49 (1977), 276–303].

(vi) Eilenbergs's observation (see Exercise 3.5 on page 114) that every projective module has a free complement (which, of course, need not be
finitely generated) shows why we assume, in the definition of stably
free, that complements are finitely generated.   ◄

Theorem 4.93 (Serre).   *If k is a field, then finitely generated projective
$k[x_1, \ldots, x_m]$-modules are stably free.*

Proof.   See Theorem 8.48.   •

Proposition 4.94.   *If a commutative ring R has the unimodular column property, then every stably free R-module P is free.*

Proof.   By induction on the rank of a free complement, it suffices to prove
that if $P \oplus R = R^n$, then P is free. Let $\varepsilon_1, \ldots, \varepsilon_n$ be the standard basis of
R^n, and let $\pi: R^n \to R$ be an R-map with $\ker \pi = P$. Since π is surjective,
there exists $\alpha = (a_1, \ldots, a_n) \in R^n$ with $1 = \pi(\alpha) = \sum a_i \pi(\varepsilon_i)$; that is,
α is a unimodular column. By hypothesis, there is an invertible matrix M
with first column α, and with other columns, say, $\alpha_2, \ldots, \alpha_n$. Define an R-
map $T: R^n \to R^n$ by $T(\varepsilon_i) = M\varepsilon_i$. If $\pi(\alpha_j) = \lambda_j \in R$ for $j \geq 2$,
then the elementary column operations $\alpha_j \to \alpha'_j$, where $\alpha'_j = \alpha_j - \lambda_j \alpha$,
yield the invertible matrix M' having columns $\alpha, \alpha'_2, \ldots, \alpha'_n$. If $j \geq 2$, then
$\pi(\alpha'_j) = \pi(\alpha_j) - \lambda_j \pi(\alpha) = 0$ [for $\pi(\alpha) = 1$]. Thus, the R-isomorphism
determined by M', namely, $T'(\varepsilon_1) = \alpha$ and $T'(\varepsilon_j) = \alpha'_j$ for $j \geq 2$, satisfies
$\alpha'_j = T'(\varepsilon_j) \in \ker \pi = P$ for all $j \geq 2$.

We claim that the restriction $T^* = T'|\langle \varepsilon_2, \ldots, \varepsilon_n \rangle \colon \langle \varepsilon_2, \ldots, \varepsilon_n \rangle \to P$
is an R-isomorphism. We have just seen that $\operatorname{im} T^* \subseteq P$. Of course, T^*
is injective, for T' is. To see that T^* is surjective, take $\beta \in P$. Now $\beta =
T'(r_1 \varepsilon_1 + \delta)$, where $\delta = \sum_{i=2}^{n} r_i \varepsilon_i$. Since $\beta \in P$, we have $\beta - T'(\delta) =
r_1 T'(\varepsilon_1) = r_1 \alpha \in P \cap \langle \alpha \rangle = \{0\}$. Hence, $\beta \in \operatorname{im} T^*$, T^* is an isomorphism,
and P is free.   •

Corollary 4.95.   *If $k[x_1, \ldots, x_m]$ has the unimodular column property, where
k is a field, then every finitely generated projective $k[x_1, \ldots, x_m]$-module is
free.*

Proof.   This follows at once from Theorem 4.93.   •

We are now going to give Suslin's solution of Serre's problem; afterward,
we will sketch Quillen's solution. Let us begin with a technical result that
is the heart of the classical *Noether Normalization Lemma*. Recall that the
total degree of a monomial $r x_1^{j_1} \cdots x_m^{j_m}$ is $j_1 + \cdots + j_m$; a general polynomial
$a \in k[x_1, \ldots, x_m]$ is a sum of monomials, and its total degree is defined as
the largest total degree of its monomial summands.

Proposition 4.96 (Noether). *Let $R = k[x_1, \ldots, x_m]$, where k is a field, let $a \in R$ have (total) degree δ, and let $b = \delta + 1$. Define*

$$y = x_m,$$

and, for $1 \leq i \leq m - 1$, define

$$y_i = x_i - x_m^{b^{m-i}} = x_i - y^{b^{m-i}}. \qquad (1)$$

Then $a = ra'$, where $r \in k$ and $a' \in \left(k[y_1, \ldots, y_{m-1}]\right)[y]$ is monic.

Proof. Note that $k[y_1, \ldots, y_{m-1}]$ is a polynomial ring, that is, the ys are independent transcendentals, for the defining equations give an automorphism of R (with inverse given by $x_m \mapsto x_m$ and $x_i \mapsto x_i + x_m^{b^{m-i}}$ for $1 \leq i \leq m-1$) that restricts to an isomorphism $k[x_1, \ldots, x_{m-1}] \to k[y_1, \ldots, y_{m-1}]$.

Denote m-tuples $(j_1, \ldots, j_m) \in \mathbb{N}^m$ by (j), and equip $\mathbb{N}^m$ with the usual dot product: $(j) \cdot (j') = j_1 j_1' + \cdots + j_m j_m'$. We denote the specific m-tuple $(b^{m-1}, b^{m-2}, \ldots, b, 1)$ by v. Of course, the exponents of the monomial $x_1^{j_1} \cdots x_m^{j_m}$ give rise to (j). Write the polynomial a in this notation:

$$a = \sum_{(j)} r_{(j)} x_1^{j_1} \cdots x_m^{j_m},$$

where $r_{(j)} \in k$ and $r_{(j)} \neq 0$. Substituting the equations in Eq. (1), the (j)th monomial is

$$r_{(j)}(y_1 + y^{b^{m-1}})^{j_1}(y_2 + y^{b^{m-2}})^{j_2} \cdots (y_1 + y^b)^{j_{m-1}} y^{j_m}.$$

Expand and separate the "pure" power of y from the rest to obtain

$$r_{(j)}\left(y^{(j) \cdot v} + f_{(j)}(y_1, \ldots, y_{m-1}, y)\right),$$

where the polynomial $f_{(j)}$ has at least one positive power of some y_i; the highest exponent of any such y_i is at most the total degree, and so it is strictly less than b. Thus, $0 \leq j_i < b$ for each j_i occurring in any (j). Each $(j) \cdot v = j_1 b^{m-1} + j_2 b^{m-2} + \cdots + j_{m-1} b + j_m$ is the expression of a positive integer in base b; the uniqueness of this b-adic expression says that if $(j) \neq (j')$, then $(j) \cdot v \neq (j') \cdot v$. Thus, there is no cancellation of terms in $\sum_{(j)} r_{(j)} y^{(j) \cdot v}$, the pure part of a. If D is the largest $(j) \cdot v$, then

$$a = r_D y^D + g(y_1, \ldots, y_{m-1}, y),$$

where the largest exponent of y occurring in $g = \sum_{(j) \cdot v < D} f_{(j)}$ is smaller than D. As r_D is a nonzero element of the field k, we have $a = r_D a'$, where $a' = y^D + r_D^{-1} g(y_1, \ldots, y_{m-1}, y)$, a monic polynomial in y. •

Lemma 4.97 (Suslin Lemma). *Let B be a commutative ring, let $s \geq 1$, and consider polynomials in $B[y]$:*

$$f(x) = y^s + a_1 y^{s-1} + \cdots + a_s;$$
$$g(y) = \qquad b_1 y^{s-1} + \cdots + b_s.$$

Then, for each j with $1 \leq j \leq s - 1$, the ideal $(f, g) \subseteq B[y]$ contains a polynomial of degree $\leq s - 1$ having leading coefficient b_j.

Proof. Let I be the set consisting of 0 and all leading coefficients of $h(y) \in (f, g)$ with $\deg(h) \leq s - 1$; it is clear that I is an ideal in B containing b_1. We prove, by induction on $j \geq 1$, that I contains $b_1, \ldots, b_j$ for all $j \leq s$. Define $g'(y) \in (f, g)$ by

$$g'(y) = y g(y) - b_1 f(y) = \sum_{i=1}^{s} (b_{i+1} - b_1 a_i) y^{s-i}.$$

By induction, I contains the first $j - 1$ coefficients of $g'(y)$, the last of which is $b_j - b_1 a_{j-1}$. It follows that $b_j \in I$. •

Observe that performing elementary *row* operations on any $n \times q$ matrix L yields a matrix NL, where $N \in \mathrm{GL}(n, R)$, the group of all invertible $n \times n$ matrices over R. Thus, if $\alpha \in R^n$ is an $n \times 1$ column vector and $N\alpha$ is the first column of some invertible matrix M, then $\alpha = (N^{-1}M)\varepsilon_1$ is also the first column of some invertible matrix.

Proposition 4.98 (Horrocks). *Let $R = B[y]$, where B is a local ring, and let $\alpha = (a_1, \ldots, a_n) \in R^n$ be a unimodular column. If some a_i is monic, then α is the first column of some invertible matrix in $\mathrm{GL}(n, R)$.*

Proof. **(Suslin)** If $n = 1$ or 2, then the conclusion is true for any commutative ring R. For example, if $\alpha = (a_1, a_2)$ and $a_1 b_1 + a_2 b_2 = 1$, then α is the first column of $\begin{bmatrix} a_1 & -b_2 \\ a_2 & b_1 \end{bmatrix}$. Therefore, we may assume $n \geq 3$. We do an induction on s, the degree of the monic polynomial a_1 (there is no loss in generality assuming a_1 is monic). If $s = 0$, then $a_1 = 1$, and our preceding remark about row operations, applied here to the column vector α, yields $N\alpha = \varepsilon_1$, which is the first column of the $n \times n$ identity matrix. Thus, we may assume that $s > 0$; moreover, after applying elementary row operations to the column α, we may assume that $\deg(a_i) \leq s - 1$ for all $i \geq 2$.

Let $\mathfrak{m}$ be the (unique) maximal ideal in B. Now $\mathfrak{m}R$ consists of those polynomials having all coefficients in $\mathfrak{m}$. The column

$$\overline{\alpha} = (a_1 + \mathfrak{m}R, \ldots, a_n + \mathfrak{m}R) \in R^n / \mathfrak{m}R^n$$

is unimodular over $(B/\mathfrak{m})[y]$ (for α is unimodular). If $a_i \in \mathfrak{m}R$ for all $i \geq 2$, then $a_1 + \mathfrak{m}$ would be a unit in $(B/\mathfrak{m})[y]$; but $B/\mathfrak{m}$ is a field, and the non-constant polynomial $a_1(y)$ in the PID $(B/\mathfrak{m})[y]$ cannot be a unit. Thus, we may assume that $a_2(y) \notin \mathfrak{m}R$; say, $a_2(y) = b_1 y^{s-1} + \cdots + b_s$, where some $b_j \notin \mathfrak{m}$. Since B is a local ring, b_j is a unit. By Suslin's Lemma, the ideal $(a_1, a_2) \subseteq R$ contains a monic polynomial of degree $\leq s-1$. Since $n \geq 3$, we may perform an elementary row operation of adding a linear combination of a_1 and a_2 to a_3 to obtain a monic polynomial in the third coordinate of degree $\leq s - 1$. The inductive hypothesis applies to this new version of α, and this completes the proof. $\quad \bullet$

Let $R = B[y]$, where B is a commutative ring. In the next proposition, we will denote a matrix M over R by $M(y)$ to emphasize the fact that its entries are polynomials in y. If $r \in R$, then the notation $M(r)$ means that every occurrence of y in $M(y)$ has been changed to r.

The next proof is ingenious!

Proposition 4.99. *Let B be a domain, let $R = B[y]$, and let $\alpha(y)$ be a unimodular column at least one of whose coordinates is monic, say, $\alpha(y) = (\alpha_1(y), \ldots, \alpha_n(y))$. Then*

$$\alpha(y) = M(y)\beta,$$

where $M(y) \in \mathrm{GL}(n, R)$ and β is a unimodular column over B.

Proof. (**Vaserstein**) Define

$$I = \{b \in B : \mathrm{GL}(n, R)\alpha(u + bv) = \mathrm{GL}(n, R)\alpha(u) \text{ for all } u, v \in R\}.$$

One checks that I is an ideal in B; for example, if $b, b' \in I$, then

$$\mathrm{GL}(n, R)\alpha(u + bv + b'v) = \mathrm{GL}(n, R)\alpha(u + bv) = \mathrm{GL}(n, R)\alpha(u),$$

so that $b + b' \in I$.

If $I = B$, then $1 \in I$; set $u = y$, $b = 1$, and $v = -y$, and obtain

$$\mathrm{GL}(n, R)\alpha(y) = \mathrm{GL}(n, R)\alpha(0).$$

Thus, $\alpha(y) = M(y)\alpha(0)$ for some $M(y) \in \mathrm{GL}(n, R)$. Since $\alpha(0)$ is a unimodular column over B, the proposition is true in this case.

We may now assume that I is a proper ideal in B, and so $I \subseteq J$ for some maximal ideal J in B. Since B is a domain, B is a subring of the localization B_J. As B_J is a local ring and $\alpha(y)$ is a unimodular column over $B_J[y]$ having a monic coordinate, Proposition 4.98 applies:

$$\alpha(y) = M(y)\varepsilon_1,$$

for some $M(y) = [m_{ij}(y)] \in \mathrm{GL}(n, B_J[y])$. Adjoin a new indeterminate z to $B_J[y]$, and define a matrix

$$N(y, z) = M(y)M(y + z)^{-1} \in \mathrm{GL}(n, B_J[y, z]).$$

Note that the matrix $M(y + z)$ is invertible: if $M(y)^{-1} = [h_{ij}(y)]$, then it is easy to see that $M(y + z)^{-1} = [h_{ij}(y + z)]$, for the map $B_J[y] \to B_J[y, z]$, given by $y \mapsto y + z$, is a B_J-algebra map.

Now $N(y, 0) = I_n$, the $n \times n$ identity matrix. Since $\alpha(y) = M(y)\varepsilon_1$, it follows that $\alpha(y + z) = M(y + z)\varepsilon_1$, and

$$N(y, z)\alpha(y + z) = N(y, z)M(y + z)\varepsilon_1 = M(y)\varepsilon_1 = \alpha(y). \qquad (2)$$

Each entry of $N(y, z)$ has the form $f_{ij}(y) + g_{ij}(y, z)$, where each monomial in $g_{ij}(y, z)$ involves a positive power of z; that is, $g_{ij}(y, 0) = 0$. Since $N(y, 0) = I_n$, it follows that each $f_{ij}(y) = 0$ or 1. Hence, the entries of $N(y, z)$ contain no nonzero terms of the form λy^i for $i > 0$ and $\lambda \in B_J$; that is, $h_{ij}(y, z) = r + g_{ij}(y, z)$, where $r = 0$ or $r = 1$. If b is the product of the denominators of all the coefficients of the entries in $N(y, z)$, then $h_{ij}(y, bz)$ has all its coefficients in $B[y, z]$. The definition of the localization B_J shows that $b \notin J$, and so $b \notin I$. Equation (2), with bz playing the role of z, gives

$$\mathrm{GL}(n, B[y, z])\alpha(y + bz) = \mathrm{GL}(n, B[y, z])\alpha(y).$$

For fixed $u, v \in R = B[y]$, define a B-algebra map $\varphi \colon B[y, z] \to B[y]$ by $\varphi(y) = u$ and $\varphi(z) = v$. Applying φ to the last displayed matrix equation gives

$$\mathrm{GL}(n, R)\alpha(u + bv) = \mathrm{GL}(n, R)\alpha(u),$$

and this contradicts $b \notin I$. $\quad\bullet$

Theorem 4.100 (Quillen–Suslin). *If k is a field, then every finitely generated projective $k[x_1, \ldots, x_m]$-module is free.*

Proof. Corollary 4.95 says that it suffices to prove that $R = k[x_1, \ldots, x_m]$ has the unimodular column property, and we prove this by induction on m. The base step is true, for Proposition 4.91 says that $k[x_1]$, as any PID, has the unimodular column property. For the inductive step, let $\alpha = (a_1, \ldots, a_n)$ be a unimodular column over $k[x_1, \ldots, x_m]$. We may assume that $a_1 \neq 0$; hence, by Noether's Proposition, there is $r \in k$ with $a_1 = ra_1'$ for $a_1' \in k[y_1, \ldots, y_{m-1}](y)$ a monic polynomial in y (where y and the y_i are defined in Proposition 4.96). Since k is a field and $r \in k$ is nonzero, r is a unit, and so there is no loss in generality in assuming that $a_1 = a_1'$; that is, a_1 is monic. Thus, Proposition 4.99 applies, and

$$\alpha = M\beta,$$

where $M \in \mathrm{GL}(n, k[x_1, \ldots, x_m])$ and β is a unimodular column over $B = k[y_1, \ldots, y_{m-1}]$. By induction, B has the unimodular column property, so that $\beta = N\varepsilon_1$ for some $N \in \mathrm{GL}(n, B)$. But $MN \in \mathrm{GL}(n, k[x_1, \ldots, x_m])$, because $B \subseteq k[x_1, \ldots, x_m]$, and so

$$\alpha = (MN)\varepsilon_1;$$

that is, α is the first column of an invertible matrix over $k[x_1, \ldots, x_m]$.   •

Suslin's proof generalizes to polynomial rings with coefficient rings other than fields once one relaxes the hypotheses of Noether's Proposition. We now sketch Quillen's solution, which also applies to more general coefficient rings.

In 1958, Seshadri proved that if R is a PID, then finitely generated projective $R[x]$-modules are free; it follows that if k is a field, then finitely generated $k[x, y]$-modules are free. One of the ideas used by Seshadri is that of an *extended* module.

Definition.   If R is a commutative ring and A is an R-algebra, then an A-module P is **extended from** R if there is an R-module P_0 with $P \cong A \otimes_R P_0$.

Example 4.101.

(i) If V is a free R-module, then $A \otimes_R V$ is a free A-module, because tensor product commutes with direct sums. Similarly, since a projective R-module is a direct summand of a free module, any A-module extended from a projective R-module is itself projective.

(ii) Every free A-module F is extended from R: if $B = \{e_i : i \in I\}$ is a basis of F and F_0 is the free R-module with basis B, then $F \cong A \otimes_R F_0$.

(iii) Not every module is extended. For example, if $A = k[x]$, where k is a field, then every k-module V is a vector space over k and, hence, every extended module is free. Thus, any nonfree $k[x]$-module is not extended.   ◄

A projective A-module need not be extended, but Quillen proved the following result.

Theorem (Quillen).   *Let R be a commutative ring, and let P be a finitely generated projective $R[x]$-module. If every $R_{\mathfrak{m}}[x] \otimes_R P$ is extended from the localization $R_{\mathfrak{m}}$, where $\mathfrak{m}$ is a maximal ideal in R, then P is extended from R.*

Proof.   D. Quillen, "Projective modules over polynomial rings," *Invent. Math.* 36 (1976), 167–171.   •

Here is Quillen's main theorem.

Theorem (Quillen).   *Let $\mathfrak{C}$ be a class of commutative rings such that*

(i) *if $R \in \mathfrak{C}$, then $R(x) \in \mathfrak{C}$, where $R(x)$ denotes the ring of all rational functions of the form $f(x)/g(x)$ with $f(x), g(x) \in R[x]$,*

(ii) *if $R \in \mathfrak{C}$ and $\mathfrak{m}$ is a maximal ideal in R, then projective $R_\mathfrak{m}[x]$-modules are free.*

Then finitely generated projective $R[x_1, \ldots, x_m]$-modules are extended from R for all $R \in \mathfrak{C}$ and all $m \geq 1$.

Proof.   See Lam, *Serre's Problem on Projective Modules*, p. 177.   •

Theorem 4.102 (Quillen–Suslin).   *If k is a field, then every finitely generated projective $k[x_1, \ldots, x_m]$-module is free.*

Proof.   The class $\mathfrak{C}$ of all fields satisfies the conditions of Quillen's theorem. If R is a field, then $R(x) = \mathrm{Frac}(R[x]) \in \mathfrak{C}$. To check (ii), note that the only maximal ideal $\mathfrak{m}$ in a field R is $\mathfrak{m} = \{0\}$, so that $R_\mathfrak{m} = R$, $R_\mathfrak{m}[x] = R[x]$ is a PID, and (finitely generated) projective $R[x]$-modules are free. Since R is a field, every R-module is a vector space, hence is free, and so every module extended from R is free.   •

It can be shown that the classes of all PIDs as well as of all Dedekind rings satisfy Quillen's condtions; hence, finitely generated $R[x_1, \ldots, x_m]$-modules are extended from R. If R is a PID, it follows that every such module is free. If R is Dedekind, then finitely generated projective R-modules P_0 need not be free; however, one knows that $P_0 \cong F_0 \oplus J$, where F_0 is a free R-module and J is an ideal in R (see Rotman, *Advanced Modern Algebra*, p. 964). Thus, every finitely generated projective $R[x_1, \ldots, x_m]$-module has the form $F \oplus Q$, where F is free and $Q \cong R[x_1, \ldots, x_m] \otimes_R J$. If R is a Prüfer ring, then every finitely generated projective $R[x_1, \ldots, x_n]$-module is extended from R (see Fontana–Huckaba–Papick, *Prüfer Domains*, p. 211); in particular, if R is a Bézout ring, then every such module is free.

The noncommutative version of Quillen's Theorem is false: Ojanguren and Sridharan, "Cancellation of Azumaya algebras," *J. Algebra* 18 (1971), 501–505, gave an example of a division ring D and a nonfree projective $D[x, y]$-module (where the indeterminates x and y commute with constants in D).

The following problem remains open. Let R be a commutative noetherian ring. We say that R has **Property (E)** if, for all $n \geq 1$, every finitely generated projective $R[x_1, \ldots, x_n]$-module is extended from R; we say that R is *regular* if every finitely generated R-module has a projective resolution of finite length. The **Bass-Quillen Conjecture** asks whether every regular ring of finite

Krull dimension has property (E). See the last chapter of Lam, *Serre's Problem on Projective Modules*, for an account of recent progress on this problem.

We summarize some results we have proved about special R-modules when R is constrained.

(i) R semisimple: every R-module is projective, injective, and flat.

(ii) R von Neumann regular: every R-module is flat.

(iii) R hereditary: every R-submodule of a projective is projective; every quotient of an injective R-module is injective.

(iv) R Dedekind (= hereditary domain): divisible R-modules are injective.

(v) R semihereditary: finitely generated R-submodules of projectives are projective; submodules of flat R-modules are flat.

(vi) R Prüfer (= semihereditary domain): torsion-free R-modules are flat; finitely generated torsion-free R-modules are projective.

(vii) R quasi-Frobenius: projective R-modules are injective.

(viii) R semiperfect: projective covers exist; direct products of projective R-modules are projective.

(ix) R local: projective R-modules are free.

(x) $R = k[X]$ = polynomial ring in several variables over a field k: finitely generated projective R-modules are free.

(xi) R noetherian: direct sums of injective R-modules are injective.

5

Setting the Stage

We plan to use Homological Algebra to prove results about modules, groups, and sheaves. A common context for discussing these topics is that of *abelian categories*; moreover, categories of complexes, the essential ingredient needed to define homology, are also abelian categories. This chapter is devoted to discussing this circle of ideas.

5.1 Categorical Constructions

Imagine a set theory whose primitive terms, instead of *set* and *element*, are *set* and *function*. How could we define bijection, cartesian product, union, and intersection? Category Theory forces us to think in this way, for functors do not recognize elements. One nice aspect of thinking categorically is that we can see unexpected analogies; for example, we shall soon see that disjoint union in **Sets**, direct sum in $_R$**Mod**, and tensor product in **ComRings** are special cases of the same categorical notion. We now set ourselves the task of describing various constructions in **Sets** or in **Ab** in such a way that they make sense in arbitrary categories.

Let us begin by investigating the notion of *disjoint union* of subsets. Two subsets A and B of a set can be forced to be disjoint. Consider the cartesian product $(A \cup B) \times \{1, 2\}$ and its subsets $A' = A \times \{1\}$ and $B' = B \times \{2\}$. It is plain that $A' \cap B' = \varnothing$, for a point in the intersection would have coordinates $(a, 1) = (b, 2)$, which cannot be, for their second coordinates are not equal. We call $A' \cup B'$ the **disjoint union** of A and B, and we note that it comes equipped with two functions, namely, $\alpha \colon A \to A' \cup B'$ and $\beta \colon B \to A' \cup B'$, defined by $\alpha \colon a \mapsto (a, 1)$ and $\beta \colon b \mapsto (b, 2)$. Denote $A' \cup B'$ by $A \sqcup B$.

J.J. Rotman, *An Introduction to Homological Algebra*, Universitext, 213
DOI 10.1007/978-0-387-68324-9_5, © Springer Science+Business Media LLC 2009

Given functions $f: A \to X$ and $g: B \to X$, for some set X, there is a unique function $\theta: A \sqcup B \to X$ that extends both f and g; namely,

$$\theta(u) = \begin{cases} f(a) & \text{if } u = (a, 1) \in A', \\ g(b) & \text{if } u = (b, 2) \in B'. \end{cases}$$

The function θ is well-defined because $A' \cap B' = \varnothing$. We have described disjoint union *categorically* (i.e., with diagrams).

In Category Theory, we often view objects, not in isolation, but together with morphisms relating them to other objects; for example, objects may arise as solutions to *universal mapping problems*.

Definition. If A and B are objects in a category $\mathcal{C}$, then their **coproduct** is a triple $(A \sqcup B, \alpha, \beta)$, where $A \sqcup B$ is an object in $\mathcal{C}$ and $\alpha: A \to A \sqcup B$, $\beta: B \to A \sqcup B$ are morphisms, called **injections**,[1] such that, for every object X in $\mathcal{C}$ and every pair of morphisms $f: A \to X$ and $g: B \to X$, there exists a unique morphism $\theta: A \sqcup B \to X$ making the diagram commute: $\theta\alpha = f$ and $\theta\beta = g$.

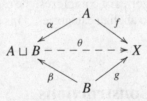

Here is a proof that the disjoint union $A \sqcup B = A' \cup B' \subseteq (A \cup B) \times \{1, 2\}$ is a coproduct in **Sets**. If X is any set and $f: A \to X$ and $g: B \to X$ are functions, we have already seen that the function $\theta: A \sqcup B \to X$, given by $\theta(a, 1) = f(a)$ and $\theta(b, 2) = g(b)$, makes the diagram commute. Let us prove uniqueness of θ. If $\psi: A \sqcup B \to X$ satisfies $\psi\alpha = f$ and $\psi\beta = g$, then $\psi(a, 1) = f(a) = \theta(a, 1)$ and $\psi(b, 2) = g(b) = \theta(b, 2)$. Therefore, ψ agrees with θ on $A' \cup B' = A \sqcup B$, and so $\psi = \theta$. We have proved that a coproduct of two sets exists in **Sets**, and that it is the disjoint union.

It is not true that coproducts always exist; in fact, it is easy to construct examples of categories in which a pair of objects does not have a coproduct (see Exercise 5.6 on page 227).

Proposition 5.1. *If A and B are left R-modules, then their coproduct in $_R$**Mod** exists and is the direct sum $A \oplus B$.*

---

[1] We will introduce the notion of *monomorphism* later. Exercise 5.57 on page 321 says that the injections α, β are, in fact, monomorphisms.

Proof. The statement of the proposition is not complete, for a coproduct requires morphisms α and β. The underlying set of $C = A \oplus B$ is the cartesian product $A \times B$, and we define $\alpha: A \to C$ by $\alpha: a \mapsto (a, 0)$ and $\beta: B \to C$ by $\beta: b \mapsto (0, b)$. Of course, α and β are R-maps.

Now let X be a module, and let $f: A \to X$ and $g: B \to X$ be R-maps. Define $\theta: C \to X$ by $\theta: (a, b) \mapsto f(a) + g(b)$. First, the diagram commutes: if $a \in A$, then $\theta\alpha(a) = \theta(a, 0) = f(a)$ and, similarly, if $b \in B$, then $\theta\beta(b) = \theta(0, b) = g(b)$. Second, θ is unique. If $\psi: C \to X$ makes the diagram commute, then $\psi(a, 0) = f(a)$ for all $a \in A$ and $\psi(0, b) = g(b)$ for all $b \in B$. Since ψ is an R-map, we have

$$\psi(a, b) = \psi[(a, 0) + (0, b)] = \psi(a, 0) + \psi(0, b) = f(a) + g(b).$$

Therefore, $\psi = \theta$. •

Proposition 5.2. *If k is a commutative ring and A and B are commutative k-algebras, then $A \otimes_k B$ is the coproduct in the category of commutative k-algebras.*

Proof. Define $\alpha: A \to A \otimes_k B$ by $\alpha: a \mapsto a \otimes 1$, and define $\beta: B \to A \otimes_k B$ by $\beta: b \mapsto 1 \otimes b$; note that both α and β are k-algebra maps. Let X be a commutative k-algebra, and consider the diagram

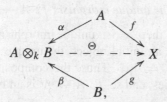

where f and g are k-algebra maps. The function $\varphi: A \times B \to X$, given by $(a, b) \mapsto f(a)g(b)$, is easily seen to be k-bilinear, and so there is a unique map of k-modules $\Theta: A \otimes_k B \to X$ with $\Theta(a \otimes b) = f(a)g(b)$. To prove that Θ is a k-algebra map, it suffices to prove that $\Theta\big((a \otimes b)(a' \otimes b')\big) = \Theta(a \otimes b)\Theta(a' \otimes b')$. Now

$$\Theta\big((a \otimes b)(a' \otimes b')\big) = \Theta(aa' \otimes bb') = f(a)f(a')g(b)g(b').$$

On the other hand, $\Theta(a \otimes b)\Theta(a' \otimes b') = f(a)g(b)f(a')g(b')$. Since X is commutative, $f(a')g(b) = g(b)f(a')$, and so Θ does preserve multiplication.

To prove uniqueness of Θ, let $\Phi: A \otimes_k B \to X$ be a k-algebra map making the diagram commute. In $A \otimes_k B$, we have $a \otimes b = (a \otimes 1)(1 \otimes b) = \alpha(a)\beta(b)$, where $a \in A$ and $b \in B$. Thus,

$$\Phi(a \otimes b) = \Phi(\alpha(a)\beta(b)) = \Phi(\alpha(a))\Phi(\beta(b)) = f(a)g(b) = \Theta(a \otimes b).$$

Since $A \otimes_k B$ is generated as a k-module by all $a \otimes b$, we have $\Psi = \Theta$. •

We know that if $\langle x \rangle$ and $\langle y \rangle$ are infinite cyclic groups, then each is a free abelian group and $\langle x \rangle \oplus \langle y \rangle$ is a free abelian group with basis $\{x, y\}$. If k is a commutative ring, then the polynomial ring $k[x]$ is a free k-algebra with basis $\{x\}$ (given any k-algebra A and any $a \in A$, there exists a unique k-algebra map $\varphi \colon k[x] \to A$ with $\varphi(x) = a$). In light of Proposition 5.2, we expect that $k[x] \otimes_k k[y]$ is a free commutative k-algebra with basis $\{x, y\}$. Exercise 5.5 on page 226 says that this is, in fact, true; moreover, $k[x] \otimes_k k[y] \cong k[x, y]$.

In Chapter 2, we used the structure of the proof of Proposition 2.44 as a strategy for proving uniqueness to isomorphism of solutions to universal mapping problems. We are now going to describe a second strategy.

Definition.   An object A in a category $\mathcal{C}$ is called an ***initial object*** if, for every object X in $\mathcal{C}$, there exists a unique morphism $A \to X$.

The empty set $\varnothing$ is an initial object in **Sets**, the zero module $\{0\}$ is an initial object in $_R$**Mod**, and 0 is the initial object of the natural numbers $\mathbb{N}$ viewed as a partially ordered set. On the other hand, a category may not have an initial object; for example, $\mathbb{Z}$, viewed as a partially ordered set, has no initial object.

Lemma 5.3.   *Any two initial objects A, A' in a category $\mathcal{C}$, should they exist, are isomorphic. In fact, the unique morphism $f \colon A \to A'$ is an isomorphism.*

Proof.   By hypothesis, there exist unique morphisms $f \colon A \to A'$ and $g \colon A' \to A$. Since A is an initial object, the unique morphism $h \colon A \to A$ must be the identity: $h = 1_A$. Thus, the composites $gf \colon A \to A$ and $fg \colon A' \to A'$ are identities, and so $f \colon A \to A'$ is an isomorphism.   •

Proposition 5.4.   *If $\mathcal{C}$ is a category and if A and B are objects in $\mathcal{C}$, then any two coproducts of A and B, should they exist, are isomorphic.*

Proof.   If C is a coproduct of A and B, then there are morphisms $\alpha \colon A \to C$ and $\beta \colon B \to C$. Define a new category $\mathcal{D}$ whose objects are diagrams

$$A \xrightarrow{\gamma} X \xleftarrow{\delta} B,$$

where X is in obj$(\mathcal{C})$ and $\gamma \colon A \to X$ and $\delta \colon B \to X$ are morphisms. Define a morphism in $\mathcal{D}$ to be a triple $(1_A, \theta, 1_C)$, where θ is a morphism in $\mathcal{C}$ making the following diagram commute:

$$
\begin{array}{ccccc}
A & \xrightarrow{\gamma} & X & \xleftarrow{\delta} & B \\
{\scriptstyle 1_A}\downarrow & & \downarrow{\scriptstyle \theta} & & \downarrow{\scriptstyle 1_B} \\
A & \xrightarrow[\gamma']{} & X' & \xleftarrow[\delta']{} & B.
\end{array}
$$

Define composition in $\mathcal{D}$ by $(1_A, \psi, 1_C)(1_A, \theta, 1_C) = (1_A, \psi\theta, 1_C)$. It is easy to check that $\mathcal{D}$ is a category and that a coproduct in $\mathcal{C}$ is an initial object in $\mathcal{D}$. By Lemma 5.3, coproducts in a category are unique to (unique) isomorphism if they exist. •

Here is an illustration of this uniqueness. If A, B are submodules of a module M with $A + B = M$ and $A \cap B = \{0\}$, then their internal direct sum $A \oplus B$ is isomorphic to their external direct sum via $a + b \mapsto (a, b)$.

Remark.   Informally, an ordered pair $(S, (\varphi_i)_{i \in I})$ occurring in a commutative diagram,, where S is an object and the φ_i are morphisms having domain or target S, is a *solution* to a *universal mapping problem* if every other such ordered pair factors uniquely through S. There is a "metatheorem" which states that solutions, if they exist, are unique to isomorphism; indeed, such an isomorphism is itself unique. The proof just given is the prototype for proving the metatheorem (if we wax categorical, then the statement of the metatheorem can be made precise, and we can then prove it; see Mac Lane, *Categories for the Working Mathematician*, Chapter III, for appropriate definitions, statement, and proof). ◄

Here is the dual definition (obtained by reversing all arrows).

Definition.   If A and B are objects in a category $\mathcal{C}$, then their *product* is a triple $(A \sqcap B, p, q)$, where $A \sqcap B$ is an object in $\mathcal{C}$ and $p: A \sqcap B \to A$, $q: A \sqcap B \to B$ are morphisms, called *projections*,[2] such that, for every object $X \in \mathcal{C}$ and every pair of morphisms $f: X \to A$ and $g: X \to B$, there exists a unique morphism $\theta: X \to A \sqcap B$ making the diagram commute.

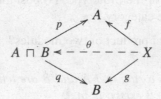

Example 5.5.   We claim that the (categorical) product of two sets A and B in **Sets** is their cartesian product $A \times B$. Define projections $p: A \times B \to A$ by $p: (a, b) \mapsto a$ and $q: A \times B \to B$ by $q: (a, b) \mapsto b$. If X is a set and $f: X \to A$ and $g: X \to B$ are functions, then the reader may show that $\theta: X \to A \times B$, defined by $\theta: x \mapsto (f(x), g(x)) \in A \times B$, is the unique function making the diagram commute. ◄

---

[2]We will introduce the notion of *epimorphism* later. Exercise 5.57 on page 321 says that the projections p, q are, in fact, epimorphisms.

Definition.   An object Ω in a category $\mathcal{C}$ is called a ***terminal object*** if, for every object C in $\mathcal{C}$, there exists a unique morphism $X \to \Omega$.

Every one-point set is a terminal object in **Sets** [the empty set $\varnothing$ is not a terminal object in **Sets** because $\mathrm{Hom}_{\mathbf{Sets}}(X, \varnothing) = \varnothing$ for every nonempty set X]. The zero module $\{0\}$ is a terminal object in $_R\mathbf{Mod}$, but $\mathbb{N}$, viewed as a partially ordered set, has no terminal object.

The solution to a universal mapping problem is an object (with morphisms) defined in terms of a diagram $\mathcal{D}$; the ***dual object*** is defined as the solution to the universal mapping problem posed by the dual diagram; that is, the diagram obtained from $\mathcal{D}$ by reversing all its arrows. For example, initial and terminal objects are dual, as are coproducts and products. There are examples of categories in which an object and its dual object both exist, there are categories in which neither exists, and there are categories in which an object exists while its dual does not exist. For example, $\mathbb{N}$ has an initial object but no terminal object; $-\mathbb{N} = \{-n : n \in \mathbb{N}\}$ has a terminal object but no initial object; $\mathbb{Z}$ has neither initial objects nor terminal objects.

Lemma 5.6.   *Any two terminal objects Ω, Ω' in a category $\mathcal{C}$, should they exist, are isomorphic. In fact, the unique morphism $f : \Omega' \to \Omega$ is an isomorphism.*

Proof.   Just reverse all the arrows in the proof of Lemma 5.3; that is, apply Lemma 5.3 to the opposite category $\mathcal{C}^{op}$.   •

Proposition 5.7.   *If A and B are objects in a category $\mathcal{C}$, then any two products of A and B, should they exist, are isomorphic.*

Proof.   Adapt the proof of the prototype, Proposition 5.4; products are terminal objects in a suitable category.   •

What is the categorical product of two modules?

Proposition 5.8.   *If R is a ring and A and B are left R-modules, then their (categorical) product $A \sqcap B$ exists; in fact,*

$$A \sqcap B \cong A \oplus B.$$

Remark.   Thus, the product and coproduct of two objects, though distinct in **Sets**, coincide in $_R\mathbf{Mod}$.   ◄

Proof.   In Proposition 2.20(iii), we characterized $M \cong A \oplus B$ by the existence of projections and injections

$$A \underset{p}{\overset{i}{\rightleftarrows}} M \underset{j}{\overset{q}{\rightleftarrows}} B$$

satisfying the equations

$$pi = 1_A, \; qj = 1_B, \; pj = 0, \; qi = 0, \quad \text{and} \quad ip + jq = 1_M. \qquad (1)$$

If X is a module and $f \colon X \to A$ and $g \colon X \to B$ are homomorphisms, define $\theta \colon X \to A \oplus B$ by $\theta = if + jg$. The product diagram commutes:

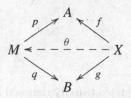

for all $x \in X$, $p\theta(x) = pif(x) + pjg(x) = pif(x) = f(x)$ [using Eq. (1)], and, similarly, $q\theta(x) = g(x)$. To prove the uniqueness of θ, note that $1_M = ip + jq$, so that $\psi = 1_M\psi = ip\psi + jq\psi = if + jg = \theta$. •

Coproducts exist in **Groups**; if G, H are groups, then their coproduct is called a *free product*, and it is denoted by $G * H$; free groups turn out to be free products of infinite cyclic groups (analogous to free abelian groups being direct sums of infinite cyclic groups). If G and H are groups, then Exercise 5.9 on page 227 shows that the direct product $G \times H$ is the categorical product $G \sqcap H$ in **Groups**. Thus, coproduct and product are distinct in **Groups**.

There is a practical value in recognizing coproducts in categories. For example, Exercise 5.7 on page 227 shows that wedge, $(X \vee Y, z_0)$, is the coproduct in **Top**$_*$, the category of pointed spaces. The fundamental group is a functor $\pi_1 \colon$ **Top**$_* \to$ **Groups**. Since free product is the coproduct in **Groups**, a reasonable guess (which is often correct) is that $\pi_1(X \vee Y, z_0) \cong \pi_1(X, x_0) * \pi_1(Y, y_0)$. With a mild hypothesis on $X \vee Y$, this is a special case of *van Kampen's theorem*.

If $F \colon \mathcal{C} \to \mathcal{D}$ is a contravariant functor and $a \sqcup b \in \mathrm{obj}(\mathcal{C})$, then it is reasonable to guess (often true) that $F(a \sqcup b) = a \sqcap b$. For example, Exercise 5.6 on page 227 says that if a partially ordered set X is viewed as a category and if $a, b \in X$, then their coproduct is their least upper bound $a \vee b$ and their product is their greatest lower bound $a \wedge b$. Let E/k be a finite Galois extension, let $\mathcal{L}$ be the family of all intermediate fields B; that is, $k \subseteq B \subseteq E$, and let $\mathcal{S}$ be the family of all the subgroups of the Galois group $\mathrm{Gal}(E/k)$; of course, both $\mathcal{L}$ and $\mathcal{S}$ are partially ordered sets. Part of the statement of the Fundamental Theorem of Galois Theory is that the function $g \colon \mathcal{L} \to \mathcal{S}$, given by $B \mapsto \mathrm{Gal}(E/B)$, is an order-reversing function and that $B \vee C \mapsto \mathrm{Gal}(E/B) \cap \mathrm{Gal}(E/C)$. Example 1.12 shows that $g \colon \mathcal{L} \to \mathcal{S}$ is a contravariant functor, and we now see that g converts coproducts to products.

We now extend the definitions of coproduct and product of two objects to (possibly infinite) families of objects.

Definition. Let C be a category, and let $(A_i)_{i \in I}$ be a family of objects in C indexed by a set I. A *coproduct* is an ordered pair $(C, (\alpha_i : A_i \to C)_{i \in I})$, consisting of an object C and a family $(\alpha_i : A_i \to C)_{i \in I}$ of morphisms, called *injections*, that is a solution to the following universal mapping problem: for every object X equipped with morphisms $(f_i : A_i \to X)_{i \in I}$, there exists a unique morphism $\theta : C \to X$ making the diagram commute for each i.

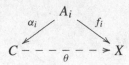

Should it exist, a coproduct is usually denoted by $\bigsqcup_{i \in I} A_i$ (the injections are not mentioned). A coproduct is unique to isomorphism, for it is an initial object in a suitable category.

We sketch the existence of the disjoint union of sets $(A_i)_{i \in I}$. Form the set $U = (\bigcup_{i \in I} A_i) \times I$, and define $A_i' = \{(a_i, i) \in U : a_i \in A_i\}$. The *disjoint union* is $((\bigsqcup_i A_i, (\alpha_i : A_i \to \bigsqcup_i A_i)_{i \in I})$, where $\bigsqcup_{i \in I} A_i = \bigcup_{i \in I} A_i'$ and $\alpha_i : a_i \mapsto (a_i, i) \in \bigsqcup_i A_i$ (of course, the disjoint union of two sets is a special case of this construction). The reader may show that this is a coproduct in **Sets**; that is, it is a solution to the universal mapping problem.

Proposition 5.9. *If $(A_i)_{i \in I}$ is a family of left R-modules, then the direct sum $\bigoplus_{i \in I} A_i$ is their coproduct in $_R$**Mod***.

Proof. The statement of the proposition is incomplete, for a coproduct requires injections α_i. Write $C = \bigoplus_{i \in I} A_i$, and define $\alpha_i : A_i \to C$ to be the function that assigns to each $a_i \in A_i$ the I-tuple whose ith coordinate is a_i and whose other coordinates are zero. Note that each α_i is an R-map.

Let X be a module and, for each $i \in I$, let $f_i : A_i \to X$ be an R-map. If $(a_i) \in C = \bigoplus_i A_i$, then only finitely many a_i are nonzero, and $(a_i) = \sum_i \alpha_i a_i$. Define $\theta : C \to X$ by $\theta : (a_i) \mapsto \sum_i f_i a_i$. The coproduct diagram does commute: if $a_i \in A_i$, then $\theta \alpha_i a_i = f_i a_i$. We now prove that θ is unique. If $\psi : C \to X$ makes the coproduct diagram commute, then

$$\psi\big((a_i)\big) = \psi\Big(\sum_i \alpha_i a_i\Big) = \sum_i \psi \alpha_i a_i = \sum_i f_i(a_i) = \theta\big((a_i)\big).$$

Therefore, $\psi = \theta$. •

Here is the dual notion.

Definition. Let C be a category, and let $(A_i)_{i \in I}$ be a family of objects in C indexed by a set I. A *product* is an ordered pair $(C, (p_i : C \to A_i)_{i \in I})$, consisting of an object C and a family $(p_i : C \to A_i)_{i \in I}$ of *projections*, that is a solution to the following universal mapping problem: for every object X equipped with morphisms $f_i \colon X \to A_i$, there exists a unique morphism $\theta \colon X \to C$ making the diagram commute for each i.

$$
\begin{array}{ccc}
 & A_i & \\
{}^{p_i}\nearrow & & \nwarrow{}^{f_i} \\
C \overset{\theta}{\longleftarrow}\!\!\dashleftarrow\!\!-\!-\!-\!-\! & & X
\end{array}
$$

Should it exist, a product is denoted by $\prod_{i \in I} A_i$, and it is unique to isomorphism, for it is a terminal object in a suitable category.

We let the reader prove that cartesian product is the product in **Sets**.

Proposition 5.10.   *If $(A_i)_{i \in I}$ is a family of left R-modules, then the direct product $C = \prod_{i \in I} A_i$ is their product in $_R$**Mod**.*

Proof.   The statement of the proposition is incomplete, for a product requires projections. For each $j \in I$, define $p_j \colon C \to A_j$ by $p_j \colon (a_i) \mapsto a_j \in A_j$. Note that each p_j is an R-map.

Now let X be a module and, for each $i \in I$, let $f_i \colon X \to A_i$ be an R-map. Define $\theta \colon X \to C$ by $\theta \colon x \mapsto (f_i(x))$. First, the diagram commutes: if $x \in X$, then $p_i \theta(x) = f_i(x)$. Second, θ is unique. If $\psi \colon X \to C$ makes the diagram commute, then $p_i \psi(x) = f_i(x)$ for all i; that is, for each i, the ith coordinate of $\psi(x)$ is $f_i(x)$, which is also the ith coordinate of $\theta(x)$. Therefore, $\psi(x) = \theta(x)$ for all $x \in X$, and so $\psi = \theta$. •

We now present another pair of dual constructions.

Definition.   Given two morphisms $f \colon B \to A$ and $g \colon C \to A$ in a category $\mathcal{C}$, a ***pullback*** (or ***fibered product***) is a triple (D, α, β) with $g\alpha = f\beta$ that is a solution to the universal mapping problem: for every (X, α', β') with $g\alpha' = f\beta'$, there exists a unique morphism $\theta \colon X \to D$ making the diagram commute. The pullback is often denoted by $B \sqcap_A C$.

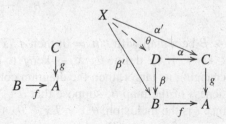

Pullbacks, when they exist, are unique to isomorphism, for they are terminal objects in a suitable category. We now prove the existence of pullbacks in $_R$**Mod**.

Proposition 5.11. *The pullback of two maps $f : B \to A$ and $g : C \to A$ in $_R\mathbf{Mod}$ exists.*

Proof. Define

$$D = \{(b, c) \in B \oplus C : f(b) = g(c)\},$$

define $\alpha : D \to C$ to be the restriction of the projection $(b, c) \mapsto c$, and define $\beta : D \to B$ to be the restriction of the projection $(b, c) \mapsto b$. It is easy to see that (D, α, β) satisfies $g\alpha = f\beta$.

If (X, α', β') satisfies $g\alpha' = f\beta'$, define a map $\theta : X \to D$ by $\theta : x \mapsto (\beta'(x), \alpha'(x))$. The values of θ do lie in D, for $f\beta'(x) = g\alpha'(x)$. We let the reader prove that the diagram commutes and that θ is unique. Thus, (D, α, β) is a solution to the universal mapping problem. •

Example 5.12.

(i) That B and C are submodules of a module A can be restated as saying that there are inclusion maps $i : B \to A$ and $j : C \to A$. The reader will enjoy proving that the pullback D exists in $_R\mathbf{Mod}$ and that $D = B \cap C$.

(ii) Pullbacks exist in **Groups**: they are subgroups of a direct product defined as in the proof of Proposition 5.11.

(iii) We show that kernel is a pullback. More precisely, if $f : B \to A$ is a homomorphism in $_R\mathbf{Mod}$, then the pullback of the first diagram below is $(\ker f, i)$, where $i : \ker f \to B$ is the inclusion.

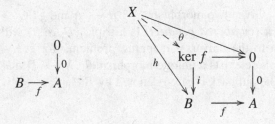

Let $h : X \to B$ be a map with $fh = 0$; then $fhx = 0$ for all $x \in X$, and so $hx \in \ker f$. If we define $\theta : X \to \ker f$ to be the map obtained from h by changing its target, then the diagram commutes: $i\theta = h$. To prove uniqueness of the map θ, suppose that $\theta' : X \to \ker f$ satisfies $i\theta' = h$. Since i is the inclusion, $\theta'x = hx = \theta x$ for all $x \in X$, and so $\theta' = \theta$. Thus, $(\ker f, i)$ is a pullback. ◀

Here is the dual construction; we have already seen it in Lemma 3.41 when we were discussing injective modules.

Definition.   Given two morphisms $f: A \to B$ and $g: A \to C$ in a category $\mathcal{C}$, a *pushout* (or *fibered sum*) is a triple (D, α, β) with $\beta g = \alpha f$ that is a solution to the universal mapping problem: for every triple (Y, α', β') with $\beta' g = \alpha' g$, there exists a unique morphism $\theta: D \to Y$ making the diagram commute. The pushout is often denoted by $B \cup_A C$.

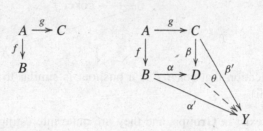

Pushouts are unique to isomorphism when they exist, for they are initial objects in a suitable category.

Proposition 5.13.   *The pushout of two maps $f: A \to B$ and $g: A \to C$ in $_R\mathbf{Mod}$ exists.*

Proof.   It is easy to see that

$$S = \big\{ (f(a), -g(a)) \in B \oplus C : a \in A \big\}$$

is a submodule of $B \oplus C$. Define $D = (B \oplus C)/S$, define $\alpha: B \to D$ by $b \mapsto (b, 0) + S$, define $\beta: C \to D$ by $c \mapsto (0, c) + S$; it is easy to see that $\beta g = \alpha f$, for if $a \in A$, then $\alpha f a - \beta g a = (fa, -ga) + S = S$. Given another triple (X, α', β') with $\beta' g = \alpha' f$, define

$$\theta: D \to X \quad \text{by} \quad \theta: (b, c) + S \mapsto \alpha'(b) + \beta'(c).$$

We let the reader prove commutativity of the diagram and uniqueness of θ.   •

Example 5.14.

(i) If B and C are submodules of a left R-module U, there are inclusions $f: B \cap C \to B$ and $g: B \cap C \to C$. The reader will enjoy proving that the pushout D exists in $_R\mathbf{Mod}$ and that D is $B + C$.

(ii) If B and C are subsets of a set U, there are inclusions $f: B \cap C \to B$ and $g: B \cap C \to C$. The pushout in **Sets** is the union $B \cup C$.

(iii) If $f: A \to B$ is a homomorphism in $_R\mathbf{Mod}$, then coker f is the pushout of the first diagram below.

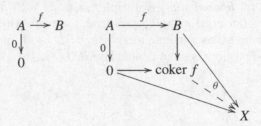

The verification that coker f is a pushout is similar to that in Example 5.12(iii).

(iv) Pushouts exist in **Groups**, and they are quite interesting; for example, the pushout of two injective homomorphisms is called a *free product with amalgamation*. If K_1 and K_2 are subcomplexes of a connected simplicial complex K with $K_1 \cup K_2 = K$ and $K_1 \cap K_2$ connected, then *van Kampen's Theorem* says that $\pi_1(K)$ is the free product of $\pi_1(K_1)$ and $\pi_1(K_2)$ with $\pi_1(K_1 \cap K_2)$ amalgamated (see Spanier, *Algebraic Topology*, p. 151). ◄

Here is another dual pair of useful constructions.

Definition.   Given two morphisms $f, g : B \to C$, then their *coequalizer* is an ordered pair (Z, e) with $ef = eg$ that is universal with this property: if $p : C \to X$ satisfies $pf = pg$, then there exists a unique $p' : Z \to X$ with $p'e = p$.

$$B \underset{g}{\overset{f}{\rightrightarrows}} C \overset{e}{\longrightarrow} Z$$
$$p \searrow \quad \downarrow p'$$
$$X.$$

More generally, if $(f_i : B \to C)_{i \in I}$ is a family of morphisms, then the coequalizer is an ordered pair (Z, e) with $ef_i = ef_j$ for all $i, j \in I$ that is universal with this property.

We can prove the existence of coequalizers in **Sets** using the notion of *orbit space*.

Definition.   Let $\sim$ be an equivalence relation on a set X. The *orbit space* $X/\sim$ is the set of all equivalence classes:

$$X/\sim = \{[x] : x \in X\},$$

where $[x]$ is the equivalence class containing x. The function $\nu : X \to X/\sim$, defined by $\nu(x) = [x]$, is called the *natural map*.

If $f, g \colon B \to C$ in **Sets** and $\sim$ is the equivalence relation on C generated by $\{(fb, gb) : b \in B\}$, then the coequalizer is the ordered pair $(C/\sim, \nu)$. In $_R$**Mod**, the coequalizer is $(\operatorname{coker}(f - g), \nu)$, where $\nu \colon C \to \operatorname{coker}(f - g)$ is the natural map. Hence, $\operatorname{coker} f$ is the coequalizer of f and 0.

Definition. Given two morphisms $f, g \colon B \to C$, then their *equalizer* is an ordered pair (A, e) with $fe = ge$ that is universal with this property: if $q \colon X \to B$ satisfies $fq = gq$, then there exists a unique $q' \colon X \to B$ with $eq' = q$.

$$A \xrightarrow{\ e\ } B \underset{g}{\overset{f}{\rightrightarrows}} C$$

$$q' \Big\uparrow \quad \nearrow q$$

$$X.$$

More generally, if $(f_i \colon B \to C)_{i \in I}$ is a family of morphisms, then the equalizer is an ordered pair (A, e) with $f_i e = f_j e$ for all $i, j \in I$ that is universal with this property.

In **Sets**, the equalizer is (E, e), where $E = \{b \in B : fb = gb\} \subseteq B$ and $e \colon E \to C$ is the inclusion. In $_R$**Mod**, the equalizer of f and g is $(\ker(f - g), e)$, where $e \colon \ker(f - g) \to B$ is the inclusion. Hence, $\ker f$ is the equalizer of f and 0.

Example 5.15. If X is a topological space and $(U_i)_{i \in I}$ is a family of open subsets of X, write $U = \bigcup_{i \in I} U_i$ and $U_{ij} = U_i \cap U_j$ for $i, j \in I$.

(i) If $f, g \colon U \to \mathbb{R}$ are continuous functions such that $f|U_i = g|U_i$ for all $i \in I$, then $f = g$.

If $x \in U$, then $x \in U_i$ for some i, and $f(x) = (f|U_i)x = (g|U_i)x = g(x)$. Hence, $f = g$.

(ii) If $(f_i \colon U_i)_{i \in I}$ is a family of continuous real-valued functions such that $f_i|U_{ij} = f_j|U_{ij}$ for all i, j, then there exists a unique continuous $f \colon U \to \mathbb{R}$ with $f|U_i = f_i$ for all $i \in I$.

If $x \in U$, then $x \in U_i$ for some i; define $f \colon U \to Y$ by $f(x) = f_i(x)$. The condition on overlaps $U_{(i,j)}$ shows that f is a well-defined function; it is obviously the unique function $U \to Y$ satisfying $f|U_i = f_i$ for all $i \in I$. We prove the continuity of f. If V is an open subset of Y, then $f^{-1}(V) = U \cap f^{-1}(V) = (\bigcup_i U_i) \cap f^{-1}(V) = \bigcup_i (U_i \cap f^{-1}(V)) = \bigcup_i f_i^{-1}(V)$. Continuity of f_i says that $f_i^{-1}(V)$ is open in U_i for all i, hence is open in U; thus, $f^{-1}(V)$ is open in U, and f is continuous.

For every open $V \subseteq X$, define $\Gamma(V) = \{\text{continuous } f: V \to \mathbb{R}\}$ and, if $V \subseteq W$, where $W \subseteq X$ is another open subset, define $\Gamma(W) \to \Gamma(V)$ to be the restriction map $f \mapsto f|V$. Then properties (i) and (ii) say that $\Gamma(U)$ is the equalizer of the family of maps $\Gamma(U_i) \to \Gamma(U_{ij})$. ◄

Exercises

*5.1   **(i)** Prove that $\varnothing$ is an initial object in **Sets**.

   (ii) Prove that any one-point set $\Omega = \{x_0\}$ is a terminal object in **Sets**. In particular, what is the function $\varnothing \to \Omega$?

*5.2 A **zero object** in a category $\mathcal{C}$ is an object that is both an initial object and a terminal object.

   (i) Prove the uniqueness to isomorphism of initial, terminal, and zero objects, if they exist.

   (ii) Prove that $\{0\}$ is a zero object in $_R\mathbf{Mod}$ and that $\{1\}$ is a zero object in **Groups**.

   (iii) Prove that neither **Sets** nor **Top** has a zero object.

   (iv) Prove that if $A = \{a\}$ is a set with one element, then (A, a) is a zero object in **Sets**$_*$, the category of pointed sets. If A is given the discrete topology, prove that (A, a) is a zero object in **Top**$_*$, the category of pointed topological spaces.

5.3   **(i)** Prove that the zero ring is not an initial object in **ComRings**.

   (ii) If k is a commutative ring, prove that k is an initial object in **ComAlg**$_k$, the category of all commutative k-algebras.

   (iii) In **ComRings**, prove that $\mathbb{Z}$ is an initial object and that the zero ring $\{0\}$ is a terminal object.

5.4 For every commutative ring k, prove that the direct product $R \times S$ is the categorical product in **ComAlg**$_k$ (in particular, direct product is the categorical product in **ComAlg**$_\mathbb{Z}$ = **ComRings**).

*5.5 Let k be a commutative ring.

   (i) Prove that $k[x, y]$ is a free commutative k-algebra with basis $\{x, y\}$.

   Hint. If A is any commutative k-algebra, and if $a, b \in A$, there exists a unique k-algebra map $\varphi: k[x, y] \to A$ with $\varphi(x) = a$ and $\varphi(y) = b$.

   (ii) Use Proposition 5.2 to prove that $k[x] \otimes_k k[y]$ is a free k-algebra with basis $\{x, y\}$.

   (iii) Use Proposition 5.4 to prove that $k[x] \otimes_k k[y] \cong k[x, y]$ as k-algebras.

*5.6   **(i)**   Let Y be a set, and let $\mathcal{P}(Y)$ denote its **power set**; that is, $\mathcal{P}(Y)$ is the partially ordered set of all the subsets of Y. As in Example 1.3(iii), view $\mathcal{P}(Y)$ as a category. If $A, B \in \mathcal{P}(Y)$, prove that the coproduct $A \sqcup B = A \cup B$ and that the product $A \sqcap B = A \cap B$.

    (ii)   Generalize part (i) as follows. If X is a partially ordered set viewed as a category, and $a, b \in X$, prove that the coproduct $a \sqcup b$ is the least upper bound of a and b, and that the product $a \sqcap b$ is the greatest lower bound.

    (iii)   Give an example of a category in which there are two objects whose coproduct does not exist.

       Hint. Let Ω be a set with at least two elements, and let $\mathcal{C}$ be the category whose objects are its proper subsets, partially ordered by inclusion. If A is a nonempty subset of Ω, then the coproduct of A and its complement does not exist in $\mathcal{C}$.

*5.7   Define the **wedge** of pointed spaces $(X, x_0), (Y, y_0) \in \textbf{Top}_*$ to be $(X \vee Y, z_0)$, where $X \vee Y$ is the quotient space of the disjoint union $X \sqcup Y$ in which the basepoints are identified to z_0. Prove that wedge is the coproduct in $\textbf{Top}_*$.

5.8   Give an example of a covariant functor that does not preserve coproducts.

*5.9   If A and B are (not necessarily abelian) groups, prove that $A \sqcap B = A \times B$ (direct product) in **Groups**. For readers familiar with group theory, prove that $A \sqcup B = A * B$ (free product) in **Groups**.

*5.10   **(i)**   Given a pushout diagram in $_R\textbf{Mod}$:

    prove that g injective implies α injective and that g surjective implies α surjective. Thus, parallel arrows have the same properties.

    (ii)   Given a pullback diagram in $_R\textbf{Mod}$:

$$D \xrightarrow{\alpha} C$$
$$\beta \downarrow \qquad \downarrow g$$
$$B \xrightarrow{f} A$$

    prove that f injective implies α injective and that f surjective implies α surjective. Thus, parallel arrows have the same properties.

5.11   **(i)**   Assuming that coproducts exist, prove commutativity:

$$A \sqcup B \cong B \sqcup A.$$

(ii)   Assuming that coproducts exist, prove associativity:

$$A \sqcup (B \sqcup C) \cong (A \sqcup B) \sqcup C.$$

5.12   **(i)**   Assuming that products exist, prove commutativity:

$$A \sqcap B \cong B \sqcap A.$$

(ii)   Assuming that products exist, prove associativity:

$$A \sqcap (B \sqcap C) \cong (A \sqcap B) \sqcap C.$$

*5.13   **(i)**   If Ω is a terminal object in a category $\mathcal{C}$, prove, for any $G \in \mathrm{obj}(\mathcal{C})$, that the projections $\lambda \colon G \sqcap \Omega \to G$ and $\rho \colon \Omega \sqcap G \to G$ are isomorphisms.

(ii)   If A is an initial object in a category $\mathcal{C}$, prove, for any $G \in \mathrm{obj}(\mathcal{C})$, that the injections $\lambda \colon G \to G \sqcup \Omega$ and $\rho \colon G \to \Omega \sqcup G$ are isomorphisms.

*5.14   Let C_1, C_2, D_1, D_2 be objects in a category $\mathcal{C}$.

(i)   If there are morphisms $f_i \colon C_i \to D_i$, for $i = 1, 2$, and if $C_1 \sqcap C_2$ and $D_1 \sqcap D_2$ exist, prove that there exists a unique morphism $f_1 \sqcap f_2$ making the following diagram commute for $i = 1, 2$:

$$
\begin{array}{ccc}
C_1 \sqcap C_2 & \xrightarrow{\ f_1 \sqcap f_2\ } & D_1 \sqcap D_2 \\
{\scriptstyle p_i}\downarrow & & \downarrow{\scriptstyle q_i} \\
C_i & \xrightarrow[\ f_i\ ]{} & D_i,
\end{array}
$$

where p_i and q_i are projections.

(ii)   If there are morphisms $g_i \colon X \to C_i$, where X is an object in $\mathcal{C}$ and $i = 1, 2$, prove that there is a unique morphism (g_1, g_2) making the following diagram commute:

$$
\begin{array}{ccccc}
 & & X & & \\
 & {\scriptstyle g_1}\swarrow & \downarrow{\scriptstyle (g_1, g_2)} & {\scriptstyle g_2}\searrow & \\
C_1 & \xleftarrow[\ p_1\ ]{} & C_1 \sqcap C_2 & \xrightarrow[\ p_2\ ]{} & C_2,
\end{array}
$$

where the p_i are projections.

Hint. Define an analog of the diagonal $\Delta_X \colon X \to X \times X$ in **Sets**, given by $x \mapsto (x, x)$, and then define $(g_1, g_2) = (g_1 \sqcap g_2)\Delta_X$.

5.15 Let C be a category having finite products and a terminal object Ω. A **group object** in C is a quadruple (G, μ, η, ϵ), where G is an object in C, $\mu: G \bigsqcap G \to G$, $\eta: G \to G$, and $\epsilon: \Omega \to G$ are morphisms, so that the following diagrams commute:

Associativity:

$$
\begin{array}{ccc}
G \sqcap G \sqcap G & \xrightarrow{1 \sqcap \mu} & G \sqcap G \\
{\scriptstyle \mu \sqcap 1}\downarrow & & \downarrow{\scriptstyle \mu} \\
G \sqcap G & \xrightarrow{\ \mu\ } & G,
\end{array}
$$

Identity:

$$
\begin{array}{ccccc}
G \sqcap \Omega & \xrightarrow{1 \sqcap \epsilon} & G \sqcap G & \xleftarrow{\epsilon \sqcap 1} & \Omega \sqcap G \\
& {\scriptstyle \lambda}\searrow & \downarrow{\scriptstyle \mu} & \swarrow{\scriptstyle \rho} & \\
& & G, & &
\end{array}
$$

where λ and ρ are the isomorphisms in Exercise 5.13.

Inverse:

$$
\begin{array}{ccccc}
G & \xrightarrow{(1,\eta)} & G \sqcap G & \xleftarrow{(\eta,1)} & G \\
{\scriptstyle \omega}\downarrow & & \downarrow{\scriptstyle \mu} & & \downarrow{\scriptstyle \omega} \\
\Omega & \xrightarrow{\ \epsilon\ } & G & \xleftarrow{\ \epsilon\ } & \Omega,
\end{array}
$$

where $\omega: G \to \Omega$ is the unique morphism to the terminal object.

(i)   Prove that a group object in **Sets** is a group.

(ii) Prove that a group object in **Groups** is an abelian group.

(iii) Define a morphism between group objects in a category C, and prove that all the group objects form a subcategory of C.

(iv) Define the dual notion **cogroup object**, and prove the dual of (iii).

*5.16 Prove that every left exact covariant functor $T: {}_R\mathbf{Mod} \to \mathbf{Ab}$ preserves pullbacks. Conclude that if B and C are submodules of a module A, then for every module M, we have

$$
\operatorname{Hom}_R(M, B \cap C) = \operatorname{Hom}_R(M, B) \cap \operatorname{Hom}_R(M, C).
$$

5.2   Limits

We now discuss inverse limit, a construction generalizing products, pullbacks, kernels, equalizers, and intersections, and direct limit, which generalizes coproducts, pushouts, cokernels, coequalizers, and unions.

Definition.   Given a partially ordered set I and a category $\mathcal{C}$, an ***inverse system in*** $\mathcal{C}$ is an ordered pair $\left((M_i)_{i \in I}, (\psi_i^j)_{j \succeq i}\right)$, abbreviated $\{M_i, \psi_i^j\}$, where $(M_i)_{i \in I}$ is an indexed family of objects in $\mathcal{C}$ and $(\psi_i^j : M_j \to M_i)_{j \succeq i}$ is an indexed family of morphisms for which $\psi_i^i = 1_{M_i}$ for all i, and such that the following diagram commutes whenever $k \succeq j \succeq i$.

$$
\begin{array}{ccc}
M_k & \xrightarrow{\;\;\psi_i^k\;\;} & M_i \\
& {\scriptstyle \psi_j^k} \searrow \quad \nearrow {\scriptstyle \psi_i^j} & \\
& M_j &
\end{array}
$$

A partially ordered set I, when viewed as a category, has as its objects the elements of I and as its morphisms exactly one morphism $\kappa_j^i : i \to j$ whenever $i \preceq j$. It is easy to see that inverse systems in $\mathcal{C}$ over I are merely contravariant functors $M : I \to \mathcal{C}$; in our original notation, $M(i) = M_i$ and $M(\kappa_j^i) = \psi_i^j$.

Example 5.16.

(i) If $I = \{1, 2, 3\}$ is the partially ordered set in which $1 \preceq 2$ and $1 \preceq 3$, then an inverse system over I is a diagram of the form

$$
\begin{array}{ccc}
& & A \\
& & \downarrow{\scriptstyle g} \\
B & \xrightarrow{\;\;f\;\;} & C.
\end{array}
$$

(ii) If $\mathcal{M}$ is a family of submodules of a module A, then it can be partially ordered under *reverse inclusion*; that is, $M \preceq M'$ in case $M \supseteq M'$. For $M \preceq M'$, the inclusion map $M' \to M$ is defined, and it is easy to see that the family of all $M \in \mathcal{M}$ with inclusion maps is an inverse system.

(iii) If I is equipped with the ***discrete*** partial order, that is, $i \preceq j$ if and only if $i = j$, then an inverse system over I is just an indexed family of modules.

(iv) If $\mathbb{N}$ is the natural numbers with the usual partial order, then an inverse system over $\mathbb{N}$ is a diagram

$$ M_0 \leftarrow M_1 \leftarrow M_2 \leftarrow \cdots $$

(we have hidden identities $M_n \to M_n$ and composites $M_{n+k} \to M_n$).

(v) If J is an ideal in a commutative ring R, then its nth power is defined by

$$ J^n = \left\{ \sum a_1 \cdots a_n : a_i \in J \right\}. $$

Each J^n is an ideal, and there is a descending sequence

$$R \supseteq J \supseteq J^2 \supseteq J^3 \supseteq \cdots .$$

If A is an R-module, there is a descending sequence of submodules

$$A \supseteq JA \supseteq J^2 A \supseteq J^3 A \supseteq \cdots .$$

If $m \geq n$, define $\psi_n^m : A/J^m A \to A/J^n A$ by

$$\psi_n^m : a + J^m A \mapsto a + J^n A$$

(these maps are well-defined, for $m \geq n$ implies $J^m A \subseteq J^n A$). It is easy to see that $\{A/J^n A, \psi_n^m\}$ is an inverse system over $\mathbb{N}$.

(vi) Let G be a group and let $\mathcal{N}$ be the family of all the normal subgroups N of G having finite index. Partially order $\mathcal{N}$ by reverse inclusion: if $N \preceq N'$ in $\mathcal{N}$, then $N' \subseteq N$, and define $\psi_N^{N'} : G/N' \to G/N$ by $gN' \mapsto gN$. It is easy to see that the family of all such quotients together with the maps $\psi_N^{N'}$ form an inverse system over $\mathcal{N}$. ◄

Definition. Let I be a partially ordered set, let $\mathcal{C}$ be a category, and let $\{M_i, \psi_i^j\}$ be an inverse system in $\mathcal{C}$ over I. The **inverse limit** (also called **projective limit** or **limit**) is an object $\varprojlim M_i$ and a family of **projections** $(\alpha_i : \varprojlim M_i \to M_i)_{i \in I}$ such that

(i) $\psi_i^j \alpha_j = \alpha_i$ whenever $i \preceq j$,

(ii) for every $X \in \mathrm{obj}(\mathcal{C})$ and all morphisms $f_i : X \to M_i$ satisfying $\psi_i^j f_j = f_i$ for all $i \preceq j$, there exists a unique morphism $\theta : X \to \varprojlim M_i$ making the diagram commute.

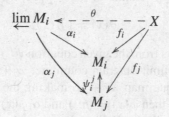

The notation $\varprojlim M_i$ for an inverse limit is deficient in that it does not display the morphisms of the corresponding inverse system (and $\varprojlim M_i$ does depend on them; see Exercise 5.17 on page 254). However, this is standard practice.

As with any object defined as a solution to a universal mapping problem, the inverse limit of an inverse system is unique (to unique isomorphism) if it exists; it is a terminal object in a suitable category.

Here is a fancy rephrasing of inverse limit, using the notion of *constant functor*.

Definition.   Let $\mathcal{C}$ be a category and let $A \in \mathrm{obj}(\mathcal{C})$. The ***constant functor*** at A is the functor $F : \mathcal{C} \to \mathcal{C}$ with $F(C) = A$ for every $C \in \mathrm{obj}(\mathcal{C})$ and with $F(f) = 1_A$ for every morphism f in $\mathcal{C}$.

Remark.   View the partially ordered index set I as a category, so that an inverse system $\{M_i, \varphi_i^j\}$ is a contravariant functor $M : I \to \mathcal{C}$, where $M(i) = M_i$ for all $i \in I$. If $L = \varprojlim M_i$, then its projections $\alpha_i : L \to M_i$ give commutative diagrams

$$\begin{array}{ccc} L & \xrightarrow{\ \alpha_i\ } & M_i \\ {\scriptstyle 1_L}\uparrow & & \uparrow{\scriptstyle \varphi_i^j} \\ L & \xrightarrow[\ \alpha_j\ ]{} & M_j. \end{array}$$

More concisely, the projections constitute a natural transformation $\alpha : |L| \to M$, where $|L| : I \to \mathcal{C}$ is the constant functor at L. Thus, the inverse limit is the ordered pair $(L, \alpha) \in \mathcal{C} \times (\mathcal{C}^{\mathrm{op}})^I$.   ◀

Proposition 5.17.   *The inverse limit of any inverse system $\{M_i, \psi_i^j\}$ of left R-modules over a partially ordered index set I exists.*

Proof.   Define a ***thread*** to be an element $(m_i) \in \prod M_i$ such that $m_i = \psi_i^j(m_j)$ whenever $i \preceq j$, and define L to be the set of all threads. It is easy to check that L is a submodule of $\prod_i M_i$. If p_i is the projection of the direct product to M_i, define $\alpha_i : L \to M_i$ to be the restriction $p_i | L$. It is clear that $\psi_i^j \alpha_j = \alpha_i$ when $i \preceq j$.

Let X be a module, and let there be given R-maps $f_i : X \to M_i$ satisfying $\psi_i^j f_j = f_i$ for all $i \preceq j$. Define $\theta : X \to \prod M_i$ by

$$\theta(x) = (f_i(x)).$$

That $\mathrm{im}\,\theta \subseteq L$ follows from the given equation $\psi_i^j f_j = f_i$ for all $i \preceq j$. Also, θ makes the inverse limit diagram commute: $\alpha_i \theta : x \mapsto (f_i(x)) \mapsto f_i(x)$. Finally, θ is the unique map $X \to L$ making the diagram commute for all $i \preceq j$. If $\varphi : X \to L$, then $\varphi(x) = (m_i)$ and $\alpha_i \varphi(x) = m_i$. Thus, if φ satisfies $\alpha_i \varphi(x) = f_i(x)$ for all i and all x, then $m_i = f_i(x)$, and so $\varphi = \theta$. We conclude that $L \cong \varprojlim M_i$.   •

Inverse limits in categories other than module categories may exist; for example, inverse limits of commutative rings always exist, as do inverse limits of groups or of topological spaces.

Example 5.18.   The reader should supply verifications of the following assertions in which we describe the inverse limit of inverse systems in Example 5.16.

(i) If I is the partially ordered set $\{1, 2, 3\}$ with $1 \succeq 3$ and $2 \succeq 3$, then an inverse system is a diagram

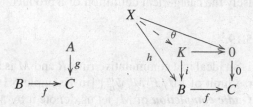

and the inverse limit is the pullback. In $_R\mathbf{Mod}$, if $A = \{0\}$, then $\ker f$ is a pullback (and is an equalizer), and so $K = \ker f$ may be regarded as an inverse limit. In more detail, the kernel of f is an ordered pair (K, i), where $i : K \to B$ satisfies $fi = 0$ and, given any map $h : X \to B$ with $fh = 0$, there exists a unique $\theta : X \to K$ with $i\theta = h$.

(ii) We have seen that the intersection of two submodules of a module is a special case of pullback. If $\mathcal{M}$ is a family of submodules M of a module, then $\mathcal{M}$ and inclusion maps form an inverse system if $\mathcal{M}$ is partially ordered by reverse inclusion [see Example 5.16(ii)].

Let us first consider the special case when $\mathcal{M}$ is *closed under finite intersections*; that is, if $M, N \in \mathcal{M}$, then $M \cap N \in \mathcal{M}$. For example, a *nested family* $\mathcal{M}$ (if $M, N \in \mathcal{M}$, then either $M \subseteq N$ or $N \subseteq M$) is closed under finite intersections.

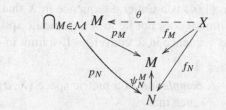

Let us show that $\bigcap_{M \in \mathcal{M}} M$ is the inverse limit of $\{M, \psi_N^M\}$ (where $\psi_N^M : M \to N$ is the inclusion when $M \preceq N$; that is, when $N \subseteq M$). For each $M \in \mathcal{M}$, define $p_M : \bigcap_{M \in \mathcal{M}} M \to M$ to be the inclusion map. If $x \in X$ and $M \in \mathcal{M}$, then $f_M(x) \in M$. We claim that $f_M(x) = f_N(x)$ for all $M, N \in \mathcal{M}$. If $D \in \mathcal{M}$ and $D \subseteq M$, then $f_D = \psi_M^D f_M$ and $f_D(x) = f_M(x)$. Since $\mathcal{M}$ is closed under finite intersections, $M, N \in \mathcal{M}$ implies $D = M \cap N \in \mathcal{M}$, and so $f_M(x) = f_{M \cap N}(x) = f_N(x)$. Define $\theta : X \to \bigcap_{M \in \mathcal{M}} M$ by $\theta(x) = f_M(x)$. We have just shown that $\theta(x)$ is independent of the choice of M; moreover, $\theta(x) = f_M(x) \in M$ for all $M \in \mathcal{M}$, and so $\theta(x) \in \bigcap_{M \in \mathcal{M}} M$.

In general, define $\mathcal{M}' = \{M_{i_1} \cap \cdots \cap M_{i_n} : M_{i_j} \in \mathcal{M}, n \geq 1\}$; then $\mathcal{M}'$ is closed under finite intersections and $\bigcap_{M \in \mathcal{M}'} M = \bigcap_{M \in \mathcal{M}} M$.

(iii) If I is a discrete index set, then the inverse system $\{M_i : i \in I\}$ in $_R\mathbf{Mod}$ has the direct product $\prod_i M_i$ as its inverse limit. Indeed, this is precisely the categorical definition of a product. ◀

Example 5.19.

(i) If J is an ideal in a commutative ring R and M is an R-module, then the inverse limit of $\{M/J^n M, \psi_n^m\}$ [in Example 5.16(v)] is usually called the *J-adic completion* of M; let us denote it by $\widehat{M}$.

Recall that a sequence (x_n) in a metric space X with metric d *converges* to a *limit* $y \in X$ if, for every $\epsilon > 0$, there is an integer N so that $d(x_n, y) < \epsilon$ whenever $n \geq N$; we denote (x_n) converging to y by

$$x_n \to y.$$

A difficulty with this definition is that we cannot tell if a sequence is convergent without knowing what its limit is. A sequence (x_n) is a *Cauchy sequence* if, for every $\epsilon > 0$, there is N so that $d(x_m, x_n) < \epsilon$ whenever $m, n \geq N$. The virtue of this condition on a sequence is that it involves only the terms of the sequence and not its limit. If $X = \mathbb{R}$, then a sequence is convergent if and only if it is a Cauchy sequence. In general metric spaces, however, we can prove that convergent sequences are Cauchy sequences, but the converse may be false. For example, if X is the set of all strictly positive real numbers with the usual metric $|x - y|$, then $(1/n)$ is a Cauchy sequence in X that does not converge (because its limit 0 does not lie in X). A metric space X is *complete* if every Cauchy sequence in X converges to a limit in X.

Definition. A *completion* of a metric space (X, d) is an ordered pair $(\widehat{X}, \varphi \colon X \to \widehat{X})$ such that

(a) $(\widehat{X}, \widehat{d})$ is a *complete metric space*,

(b) φ is an *isometry*; that is, $\widehat{d}(\varphi(x), \varphi(y)) = d(x, y)$ for all $x, y \in X$,

(c) $\varphi(X)$ is a *dense* subspace of $\widehat{X}$; that is, for every $\widehat{x} \in \widehat{X}$, there is a sequence (x_n) in X with $\varphi(x_n) \to \widehat{x}$.

It can be proved that completions exist (Kaplansky, *Set Theory and Metric Spaces*, p. 92) and that any two completions of a metric space X are *isometric*: if $(\widehat{X}, \varphi)$ and (Y, ψ) are completions of X, then there exists a unique bijective isometry $\theta \colon \widehat{X} \to Y$ with $\psi = \theta\varphi$. Indeed, a completion of X is just a solution to the obvious universal mapping problem (density of im φ gives the required uniqueness of θ). One way to prove existence of a completion is to define its elements as equivalence classes of Cauchy sequences (x_n) in X, where we define $(x_n) \equiv (y_n)$ if $d(x_n, y_n) \to 0$.

Let us return to the inverse system $\{M/J^n M, \psi_n^m\}$. A thread

$$(a_1 + JM, a_2 + J^2 M, a_3 + J^3 M, \ldots) \in \varprojlim(M/J^n M)$$

satisfies the condition $\psi_n^m(a_m + J^m M) = a_m + J^n M$ for all $m \geq n$, so that

$$a_m - a_n \in J^n M \quad \text{whenever } m \geq n.$$

This suggests the following metric on M in the (most important) special case when $\bigcap_{n=1}^{\infty} J^n M = \{0\}$. If $x \in M$ and $x \neq 0$, then there is i with $x \in J^i M$ and $x \notin J^{i+1} M$; define $\|x\| = 2^{-i}$; define $\|0\| = 0$. It is a routine calculation to see that $d(x, y) = \|x - y\|$ is a metric on M (without the intersection condition, $\|x\|$ would not be defined for a nonzero $x \in \bigcap_{n=1}^{\infty} J^n M$). Moreover, if a sequence (a_n) in M is a Cauchy sequence, then it is easy to construct an element $(b_n + JM) \in \varprojlim M/J^n M$ that is a limit of $(\varphi(a_n))$. In particular, when $M = \mathbb{Z}$ and $J = (p)$, where p is a prime, then the completion $\mathbb{Z}_p$ is called the *ring of p-adic integers*. It turns out that $\mathbb{Z}_p$ is a domain, and $\mathbb{Q}_p = \text{Frac}(\mathbb{Z}_p)$ is called the *field of p-adic numbers*.

(ii) We have seen, in Example 5.16(vi), that the family $\mathcal{N}$ of all normal subgroups of finite index in a group G forms an inverse system; the inverse limit $\varprojlim G/N$, denoted by $\widehat{G}$, is called the *profinite completion* of G. There is a map $G \to \widehat{G}$, namely, $g \mapsto (gN)$, and it is an injection if and only if G is *residually finite*; that is, $\bigcap_{N \in \mathcal{N}} N = \{1\}$. It is known, for example, that every free group is residually finite.

There are some lovely results obtained making use of profinite completions. If r is a positive integer, a group G is said to have *rank r* if every subgroup of G can be generated by r or fewer elements. If G is a p-group (every element in G has order a power of p) of rank r that is residually finite, then G is isomorphic to a subgroup of $\text{GL}(n, \mathbb{Z}_p)$ for some n (not every residually finite group admits such a linear imbedding). See Dixon–du Sautoy–Mann–Segal, *Analytic Pro-p Groups*, p. 98. ◄

Example 5.20 (Griffith). Injective envelope is not an additive functor; that is, there is no additive functor $T \colon \mathbf{Ab} \to \mathbf{Ab}$ with $T(G) = \text{Env}(G)$ for all $G \in \text{obj}(\mathbf{Ab})$. If such a functor exists, then Exercise 2.21 on page 68 says that the function $T_* \colon \text{End}(G) \to \text{End}(\text{Env}(G))$, given by $f \mapsto Tf$, is a nonzero ring homomorphism. Now if $G = \mathbb{I}_p$, then $\text{Env}(\mathbb{I}_p) \cong \mathbb{Z}(p^\infty)$ (the Prüfer group), $\text{End}(\mathbb{I}_p) \cong \mathbb{F}_p$, and $\text{End}(\mathbb{Z}(p^\infty)) \cong \mathbb{Z}_p$ (by Exercise 5.20 on page 254). But the additive group of $\mathbb{F}_p$ is finite, while the additive group of

$\mathbb{Z}_p$ is torsion-free (by Exercise 5.20). Hence, there can be no nonzero additive map $\operatorname{End}(\mathbb{I}_p) \to \operatorname{End}(\operatorname{Env}(\mathbb{I}_p))$, and so Env is not an additive functor.   ◄

We now prove that covariant Hom functors preserve inverse limits; Theorem 2.30 is the special case involving a discrete index set. We will give another proof of this after we introduce adjoint functors.

Proposition 5.21.   *If $\{M_i, \psi_i^j\}$ is an inverse system of left R-modules, then there is a natural isomorphism*

$$\operatorname{Hom}_R(A, \varprojlim M_i) \cong \varprojlim \operatorname{Hom}_R(A, M_i)$$

for every left R-module A.

Proof.   This statement follows from inverse limit solving a universal mapping problem. In more detail, $\operatorname{Hom}_R(A, \square)$ carries the inverse system $\{M_i, \psi_i^j\}$ into the inverse system $\{\operatorname{Hom}_R(A, M_i), \psi_{i*}^j\}$. Consider the diagram

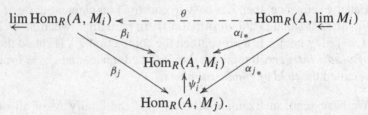

We may assume that $\varprojlim \operatorname{Hom}_R(A, M_i)$ is constructed as in Proposition 5.17, so that its elements are threads $(g_i) \in \prod_i \operatorname{Hom}_R(A, M_i)$ and $\beta_i : (g_i) \mapsto g_i$. The maps $\alpha_i : \varprojlim M_i \to M_i$ are the projections $(m_i) \mapsto m_i$, and α_{i*} are the induced maps.

Define $\theta : \operatorname{Hom}_R(A, \varprojlim M_i) \to \varprojlim \operatorname{Hom}_R(A, M_i)$ by $f \mapsto (\alpha_i f)$; it is easy to check that $\theta(f)$ is a thread and that θ is a homomorphism. The diagram commutes, for if $f \in \operatorname{Hom}_R(A, \varprojlim M_i)$, then

$$\beta_i \theta(f) = \beta_i\big((\alpha_i f)\big) = \alpha_i f = \alpha_{i*}(f).$$

To see that θ is injective, let $\theta(f) = 0$, where $f \in \operatorname{Hom}_R(A, \varprojlim M_i)$. Then $0 = \theta(f) = (\alpha_i f)$, so that $\alpha_i f = 0$ for all i. If $a \in A$, then $f(a) = (m_i)$, say. Hence, $\alpha_i f(a) = m_i = 0$ and $f = 0$.

To see that θ is surjective, take $g = (g_i) \in \varprojlim \operatorname{Hom}_R(A, M_i)$. Since (g_i) is a thread, the right-hand triangle in the following diagram commutes.

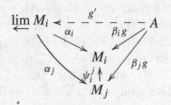

But this says that $g = (g_i) = (\alpha_i g') = \theta(g')$; that is, θ is surjective.

Naturality means that if $\varphi \colon A \to B$, then the following diagram commutes.

$$
\begin{array}{ccc}
\operatorname{Hom}_R(B, \varprojlim M_i) & \xrightarrow{\ \theta\ } & \varprojlim \operatorname{Hom}_R(B, M_i) \\
\varphi^* \uparrow & & \uparrow \varphi^* \\
\operatorname{Hom}_R(A, \varprojlim M_i) & \xrightarrow{\ \theta\ } & \varprojlim \operatorname{Hom}_R(A, M_i)
\end{array}
$$

The straightforward proof is left to the reader.   •

Remark.   Once we define a morphism of inverse systems, then it will be clear that the isomorphism in Proposition 5.21 is also natural in the second variable.   ◄

We now consider the dual construction.

Definition.   Given a partially ordered set I and a category $\mathcal{C}$, a ***direct system in*** $\mathcal{C}$ is an ordered pair $\big((M_i)_{i \in I}, (\varphi_j^i)_{i \preceq j}\big)$, abbreviated $\{M_i, \varphi_j^i\}$, where $(M_i)_{i \in I}$ is an indexed family of objects in $\mathcal{C}$ and $(\varphi_j^i \colon M_j \to M_i)_{i \preceq j}$ is an indexed family of morphisms for which $\varphi_i^i = 1_{M_i}$ for all i, and such that the following diagram commutes whenever $i \preceq j \preceq k$.

$$
\begin{array}{ccc}
M_i & \xrightarrow{\ \varphi_k^i\ } & M_k \\
& \searrow_{\varphi_j^i} \quad \nearrow_{\varphi_k^j} & \\
& M_j &
\end{array}
$$

A partially ordered set I, when viewed as a category, has as its objects the elements of I and as its morphisms exactly one morphism κ_j^i when $i \preceq j$. It is easy to see that direct systems in $\mathcal{C}$ over I are merely covariant functors $M \colon I \to \mathcal{C}$; in our original notation, $M(i) = M_i$ and $M(\kappa_j^i) = \varphi_j^i$.

Example 5.22.

(i) If $I = \{1, 2, 3\}$ is the partially ordered set in which $1 \preceq 2$ and $1 \preceq 3$, then a direct system over I is a diagram of the form

$$
\begin{array}{ccc}
A & \xrightarrow{\ f\ } & B \\
g \downarrow & & \\
C. & &
\end{array}
$$

(ii) If $\mathcal{I}$ is a family of submodules of a module A, then it can be partially ordered under inclusion; that is, $M \preceq M'$ in case $M \subseteq M'$. For $M \preceq M'$, the inclusion map $M \to M'$ is defined, and it is easy to see that the family of all $M \in \mathcal{I}$ with inclusion maps is a direct system.

(iii) If I is equipped with the discrete partial order, then a direct system over I is just a family of modules indexed by I. ◄

Definition.   Let I be a partially ordered set, let $\mathcal{C}$ be a category, and let $\{M_i, \varphi_j^i\}$ be a direct system in $\mathcal{C}$ over I. The *direct limit* (also called *inductive limit* or *colimit*) is an object $\varinjlim M_i$ and *insertion morphisms* $(\alpha_i : M_i \to \varinjlim M_i)_{i \in I}$ such that

(i) $\alpha_j \varphi_j^i = \alpha_i$ whenever $i \preceq j$,

(ii) Let $X \in \mathrm{obj}(\mathcal{C})$, and let there be given morphisms $f_i : M_i \to X$ satisfying $f_j \varphi_j^i = f_i$ for all $i \preceq j$. There exists a unique morphism $\theta : \varinjlim M_i \to X$ making the diagram commute.

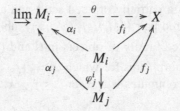

The notation $\varinjlim M_i$ for a direct limit is deficient in that it does not display the morphisms of the corresponding direct system (and $\varinjlim M_i$ does depend on them; see Exercise 5.17 on page 254). However, this is standard practice.

As with any object defined as a solution to a universal mapping problem, the direct limit of a direct system is unique (to unique isomorphism) if it exists; it is an initial object in a suitable category.

Here is a fancy rephrasing of direct limit similar to the remark on page 232. View the partially ordered index set I as a category, so that a direct system $\{M_i, \varphi_j^i\}$ is a (covariant) functor $M : I \to \mathcal{C}$, where $M(i) = M_i$ for all $i \in I$. If $L = \varinjlim M_i$, then its insertion morphisms $\alpha_i : M_i \to L$ give commutative diagrams:

$$
\begin{array}{ccc}
L & \xleftarrow{\ \alpha_i\ } & M_i \\
{\scriptstyle 1_L}\big\downarrow & & \big\downarrow{\scriptstyle \varphi_j^i} \\
L & \xleftarrow{\ \alpha_j\ } & M_j.
\end{array}
$$

More concisely, the insertion morphisms constitute a natural transformation $\alpha : M \to |L|$, where $|L| : I \to \mathcal{C}$ is the constant functor at L (constant functors are both covariant and contravariant). Thus, the direct limit is the ordered pair $(L, \alpha) \in \mathcal{C} \times \mathcal{C}^I$.

Proposition 5.23. *The direct limit of any direct system* $\{M_i, \varphi_j^i\}$ *of left R-modules over a partially ordered index set I exists.*

Proof. For each $i \in I$, let λ_i be the morphism of M_i into the direct sum $\bigoplus_i M_i$. Define

$$D = \left(\bigoplus_i M_i \right) / S,$$

where S is the submodule of $\bigoplus M_i$ generated by all elements $\lambda_j \varphi_j^i m_i - \lambda_i m_i$ with $m_i \in M_i$ and $i \preceq j$. Now define insertion morphisms $\alpha_i : M_i \to D$ by

$$\alpha_i : m_i \mapsto \lambda_i(m_i) + S.$$

It is routine to check that D solves the universal mapping problem, and so $D \cong \varinjlim M_i$. •

Thus, each element of $\varinjlim M_i$ has a representative of the form $\sum \lambda_i m_i + S$.

The argument in Proposition 5.23 can be modified to prove that direct limits in other categories exist; for example, direct limits of commutative rings, of groups, or of topological spaces always exist.

Example 5.24. The reader should supply verifications of the following assertions, in which we describe the direct limit of some direct systems in Example 5.22.

(i) If I is the partially ordered set $\{1, 2, 3\}$ with $1 \preceq 2$ and $1 \preceq 3$, then the diagram

$$A \xrightarrow{f} B$$
$$\left. g \right\downarrow$$
$$C$$

is a direct system and its direct limit is the pushout. In particular, if $g = 0$, then coker f is a pushout, and so cokernel may be regarded as a direct limit.

(ii) If I is a discrete index set, then the direct system is just the indexed family $\{M_i : i \in I\}$, and the direct limit is the direct sum: $\varinjlim M_i \cong \bigoplus_i M_i$, for the submodule S in the construction of $\varinjlim M_i$ is $\{0\}$. Alternatively, this is just the categorical definition of a coproduct. ◄

Definition. A covariant functor $F : \mathcal{A} \to \mathcal{C}$ *preserves direct limits* if, whenever $(\varinjlim A_i, (\alpha_i : A_i \to \varinjlim A_i))$ is a direct limit of a direct system $\{A_i, \varphi_j^i\}$ in $\mathcal{A}$, then $(F(\varinjlim A_i), (F\alpha_i : FA_i \to F(\varinjlim A_i)))$ is a direct limit of the direct system $\{FA_i, F\varphi_j^i\}$ in $\mathcal{C}$.

Dually, a covariant functor $F: \mathcal{A} \to \mathcal{C}$ *preserves inverse limits* if, whenever $(\varprojlim A_i, (\alpha_i: \varprojlim A_i \to A_i))$ is an inverse limit of an inverse system $\{A_i, \psi_i^j\}$ in $\mathcal{A}$, then $(F(\varprojlim A_i), (F\alpha_i: F(\varprojlim A_i) \to FA_i))$ is an inverse limit of the inverse system $\{FA_i, F\psi_i^j\}$ in $\mathcal{C}$.

A contravariant functor $F: \mathcal{A} \to \mathcal{C}$ *converts direct limits to inverse limits* if, whenever $\{A_i, \varphi_j^i\}$ is a direct system and $\varinjlim A_i$ has insertion morphisms $\alpha_i: A_i \to \varinjlim A_i$, then $F(\varinjlim A_i) \cong \varprojlim FA_i$ and its projections are $F\alpha_i: F(\varinjlim A_i) \to FA_i$. There is a similar definition of a contravariant functor converting inverse limits to direct limits.

Let us illustrate these definitions.

Proposition 5.25.   *Let $F: {}_R\mathbf{Mod} \to \mathbf{Ab}$ be a covariant functor. Then F preserves kernels if and only if F is left exact, and F preserves cokernels if and only if F is right exact.*

Proof.   Let $0 \to A' \xrightarrow{i} A \xrightarrow{p} A''$ be exact. If F preserves kernels, then (FA', Fi) is a kernel of Fp, and Example 5.12(iii) shows that Fi is an injection; that is, F is left exact. Conversely, if F is left exact, then $(\ker Fp, Fi)$ is a kernel of Fp, and so F preserves kernels. The proof that right exactness and preserving cokernels are equivalent is dual.   •

The reader may wonder why we have mentioned kernel and cokernel but not image. If $f: A \to B$ in ${}_R\mathbf{Mod}$, then coker $f = B/\operatorname{im} f$, and so

$$\operatorname{im} f = \ker(B \to \operatorname{coker} f).$$

When the index set is discrete, it makes sense to say that a functor *preserves direct products* or *converts direct sums*. In Proposition 5.21, we proved that covariant Hom functors preserve inverse limits. We now generalize Theorem 2.31, which says that $\operatorname{Hom}_R(\bigoplus_i A_i, B) \cong \prod_i \operatorname{Hom}_R(A_i, B)$, by showing that $\operatorname{Hom}_R(\square, B)$ converts direct limits to inverse limits.

Proposition 5.26.   *If $\{M_i, \varphi_j^i\}$ is a direct system of left R-modules, then there is an isomorphism*

$$\theta: \operatorname{Hom}_R(\varinjlim M_i, B) \to \varprojlim \operatorname{Hom}_R(M_i, B)$$

for every left R-module B.

Proof.   Since $\operatorname{Hom}_R(\square, B)$ is a contravariant functor, $\{\operatorname{Hom}_R(M_i, B), \varphi_j^{i*}\}$ is an inverse system. The isomorphism θ is defined by $f \mapsto (f\alpha_i)$, where α_i are the insertion morphisms of the direct limit $\varinjlim M_i$; the proof, modeled on that of Proposition 5.21, is left to the reader.   •

Remark. Once we define a morphism of direct systems, then it will be clear that the isomorphism in Proposition 5.26 is natural. ◄

We now prove that $A \otimes_R \square$ preserves direct limits. This also follows from Theorem 5.43, a result about adjoint functors, but the proof here is based on the construction of direct limits.

Theorem 5.27. *If A is a right R-module, then $A \otimes_R \square$ preserves direct limits. Thus, if $\{B_i, \varphi^i_j\}$ is a direct system of left R-modules over a partially ordered index set I, then there is a natural isomorphism*

$$A \otimes_R \varinjlim B_i \cong \varinjlim (A \otimes_R B_i).$$

Proof. Note that Exercise 5.18 on page 254 shows that $\{A \otimes_R B_i, 1 \otimes \varphi^i_j\}$ is a direct system, so that $\varinjlim (A \otimes_R B_i)$ makes sense.

We begin by constructing $\varinjlim B_i$ as the cokernel of a certain map between direct sums. For each pair $i, j \in I$ with $i \preceq j$ in the partially ordered index set I, define $B_{ij} = B_i \times \{j\}$, and denote its elements (b_i, j) by b_{ij}. View B_{ij} as a module isomorphic to B_i via the map $b_i \mapsto b_{ij}$, where $b_i \in B_i$, and define $\sigma : \bigoplus_{ij} B_{ij} \to \bigoplus_i B_i$ by

$$\sigma : b_{ij} \mapsto \lambda_j \varphi^i_j b_i - \lambda_i b_i,$$

where λ_i is the injection of B_i into the direct sum. Note that $\operatorname{im} \sigma = S$, the submodule arising in the construction of $\varinjlim B_i$ in Proposition 5.23. Thus, $\operatorname{coker} \sigma = (\bigoplus B_i)/S \cong \varinjlim B_i$, and there is an exact sequence[3]

$$\bigoplus B_{ij} \xrightarrow{\sigma} \bigoplus B_i \to \varinjlim B_i \to 0.$$

Right exactness of $A \otimes_R \square$ gives exactness of

$$A \otimes_R \left(\bigoplus B_{ij} \right) \xrightarrow{1 \otimes \sigma} A \otimes_R \left(\bigoplus B_i \right) \to A \otimes_R (\varinjlim B_i) \to 0.$$

By Theorem 2.65, the map $\tau : A \otimes_R (\bigoplus_i B_i) \to \bigoplus_i (A \otimes_R B_i)$, given by

$$\tau : a \otimes (b_i) \mapsto (a \otimes b_i),$$

is a natural isomorphism, and so there is a commutative diagram

$$
\begin{array}{ccccccc}
A \otimes \bigoplus B_{ij} & \xrightarrow{1 \otimes \sigma} & A \otimes \bigoplus B_i & \longrightarrow & A \otimes \varinjlim B_i & \longrightarrow & 0 \\
\downarrow{\scriptstyle \tau} & & \downarrow{\scriptstyle \tau'} & & \downarrow & & \\
\bigoplus (A \otimes B_{ij}) & \xrightarrow{\tilde{\sigma}} & \bigoplus (A \otimes B_i) & \longrightarrow & \varinjlim (A \otimes B_i) & \longrightarrow & 0,
\end{array}
$$

---

[3] The astute reader will recognize $\varinjlim B_i$ as a coequalizer.

where τ' is another instance of the isomorphism of Theorem 2.65, and

$$\widetilde{\sigma} : a \otimes b_{ij} \mapsto (1 \otimes \lambda_j)(a \otimes \varphi_j^i b_i) - (1 \otimes \lambda_i)(a' \otimes b_i).$$

By the Five Lemma, there is an isomorphism $A \otimes_R \varinjlim B_i \to \operatorname{coker} \widetilde{\sigma} \cong \varinjlim(A \otimes_R B_i)$, the direct limit of the direct system $\{A \otimes_R B_i, 1 \otimes \varphi_j^i\}$.

The proof of naturality is left to the reader.   •

The reader may have observed that the hypothesis of Theorem 5.27 is too strong. We really proved that any right exact functor that preserves direct sums must preserve all direct limits (for these are the only properties of $A \otimes_R \Box$ that we used). But this generalization is only virtual, for we will soon prove, in Theorem 5.45, that such functors must be tensor products. The dual of Theorem 2.65 also holds, and it has a similar proof; every left exact functor that preserves products must preserve all inverse limits (see Exercise 5.28 on page 256).

There is a special kind of partially ordered index set that is useful for direct limits.

Definition.   A *directed set* is a partially ordered set I such that, for every $i, j \in I$, there is $k \in I$ with $i \preceq k$ and $j \preceq k$.

Example 5.28.   If I is the partially ordered set $\{1, 2, 3\}$ with $1 \preceq 2$ and $1 \preceq 3$, then I is *not* a directed set.   ◄

Example 5.29.

  (i) If $\mathcal{I}$ is a simply ordered family of submodules of a module A (that is, if $M, M' \in \mathcal{I}$, then either $M \subseteq M'$ or $M' \subseteq M$), then $\mathcal{I}$ is a directed set.

 (ii) If $\mathcal{I}$ is a family of submodules of a left R-module M, then it can be partially ordered by inclusion; that is, $S \preceq S'$ if and only if $S \subseteq S'$. If $S \preceq S'$, then the inclusion map $S \to S'$ is defined. If $S, S' \in \mathcal{I}$, then $S + S' \in \mathcal{I}$, and so the family of all $S \in \mathcal{I}$ is a directed set.

(iii) If $(M_i)_{i \in I}$ is some family of modules, and if I is a discrete partially ordered index set, then I is not directed. However, if we consider the family $\mathcal{F}$ of all *finite partial sums*

$$M_{i_1} \oplus \cdots \oplus M_{i_n},$$

then $\mathcal{F}$ is a directed set under inclusion.

(iv) If A is a left R-module, then the family $\operatorname{Fin}(A)$ of all the finitely generated submodules of A is partially ordered by inclusion, as in part (ii), and is a directed set.

(v) If R is a domain and $Q = \text{Frac}(R)$, then the family of all cyclic R-submodules Q of the form $\langle 1/r \rangle$, where $r \in R$ and $r \neq 0$, is a partially ordered set, as in part (ii); here, it is a directed set under inclusion, for given $\langle 1/r \rangle$ and $\langle 1/s \rangle$, then each is contained in $\langle 1/rs \rangle$.

(vi) Let X be a topological space. If $x \in X$, let $\Phi(x)$ be the family of all those open sets containing x. Partially order $\Phi(x)$ by reverse inclusion:

$$U \preceq V \quad \text{if} \quad V \subseteq U.$$

Notice that $\Phi(x)$ is directed: given $U, V \in \Phi(x)$, then $U \cap V \in \mathcal{U}_0$ and $U \preceq U \cap V$ and $V \preceq U \cap V$.

(vii) Abstract simplicial complexes have *geometric realizations* that are topological spaces. Finite simplicial complexes are homeomorphic to certain subspaces of Euclidean space, while an infinite simplicial complex X is, by definition, a **CW-complex**; that is, it is a Hausdorff space that is **closure finite** with the **weak topology**. Direct limits exist in **Top**$_2$, the category of all Hausdorff spaces (they are quotients of coproducts), and $X \approx \varinjlim_{\mathcal{K}} K$, where $\mathcal{K}$ is the family of all finite simplicial complexes in X. ◄

There are two reasons to consider direct systems with directed index sets. The first is that a simpler description of the elements in the direct limit can be given; the second is that $\varinjlim$ preserves short exact sequences.

Lemma 5.30. *Let $\{M_i, \varphi_j^i\}$ be a direct system of left R-modules over a directed index set I, and let $\lambda_i : M_i \to \bigoplus M_i$ be the ith injection, so that $\varinjlim M_i = (\bigoplus M_i)/S$, where $S = \langle \lambda_j \varphi_j^i m_i - \lambda_i m_i : m_i \in M_i \text{ and } i \preceq j \rangle$.*

(i) *Each element of $\varinjlim M_i$ has a representative of the form $\lambda_i m_i + S$ (instead of $\sum_i \lambda_i m_i + S$).*

(ii) *$\lambda_i m_i + S = 0$ if and only if $\varphi_t^i(m_i) = 0$ for some $t \succeq i$.*

Proof.

(i) As in the proof Proposition 5.23, the existence of direct limits, $\varinjlim M_i = (\bigoplus M_i)/S$, and so a typical element $x \in \varinjlim M_i$ has the form $x = \sum \lambda_i m_i + S$. Since I is directed, there is an index j with $j \succeq i$ for all i occurring in the sum for x. For each such i, define $b^i = \varphi_j^i m_i \in M_j$, so that the element b, defined by $b = \sum_i b^i$, lies in M_j. It follows that

$$\sum \lambda_i m_i - \lambda_j b = \sum (\lambda_i m_i - \lambda_j b^i)$$
$$= \sum (\lambda_i m_i - \lambda_j \varphi_j^i m_i) \in S.$$

Therefore, $x = \sum \lambda_i m_i + S = \lambda_j b + S$, as desired.

(ii) If $\varphi_t^i m_i = 0$ for some $t \succeq i$, then

$$\lambda_i m_i + S = \lambda_i m_i + (\lambda_t \varphi_t^i m_i - \lambda_i m_i) + S = S.$$

Conversely, if $\lambda_i m_i + S = 0$, then $\lambda_i m_i \in S$, and there is an expression

$$\lambda_i m_i = \sum_j a_j (\lambda_k \varphi_k^j m_j - \lambda_j m_j) \in S,$$

where $a_j \in R$. We are going to normalize this expression; first, we introduce the following notation for relators: if $j \preceq k$, define

$$r(j, k, m_j) = \lambda_k \varphi_k^j m_j - \lambda_j m_j.$$

Since $a_j r(j, k, m_j) = r(j, k, a_j m_j)$, we may assume that the notation has been adjusted so that

$$\lambda_i m_i = \sum_j r(j, k, m_j).$$

As I is directed, we may choose an index $t \in I$ larger than any of the indices i, j, k occurring in the last equation. Now

$$\lambda_t \varphi_t^i m_i = (\lambda_t \varphi_t^i m_i - \lambda_i m_i) + \lambda_i m_i$$
$$= r(i, t, m_i) + \lambda_i m_i$$
$$= r(i, t, m_i) + \sum_j r(j, k, m_j).$$

Next,

$$r(j, k, m_j) = \lambda_k \varphi_k^j m_j - \lambda_j m_j$$
$$= (\lambda_t \varphi_t^j m_j - \lambda_j m_j) + [\lambda_t \varphi_t^k (-\varphi_k^j m_j) - \lambda_k(-\varphi_k^j m_j)]$$
$$= r(j, t, m_j) + r(k, t, -\varphi_k^j m_j),$$

because $\varphi_t^k \varphi_k^i = \varphi_t^i$, by the definition of direct system. Hence,

$$\lambda_t \varphi_t^i m_i = \sum_\ell r(\ell, t, x_\ell),$$

where $x_\ell \in M_\ell$. But it is easily checked, for $\ell \preceq t$, that

$$r(\ell, t, m_\ell) + r(\ell, t, m'_\ell) = r(\ell, t, m_\ell + m'_\ell).$$

Therefore, we may amalgamate all relators with the same smaller index ℓ and write

$$\lambda_t \varphi_t^i m_i = \sum_\ell r(\ell, t, x_\ell)$$
$$= \sum_\ell \lambda_t \varphi_t^\ell x_\ell - \lambda_\ell x_\ell$$
$$= \lambda_t \Big(\sum_\ell \varphi_t^\ell x_\ell \Big) - \sum_\ell \lambda_\ell x_\ell,$$

where $x_\ell \in M_\ell$ and all the indices ℓ are distinct. The unique expression of an element in a direct sum allows us to conclude, if $\ell \neq t$, that $\lambda_\ell x_\ell = 0$; it follows that $x_\ell = 0$, for λ_ℓ is an injection. The right side simplifies to $\lambda_t \varphi_t^t m_t - \lambda_t m_t = 0$, because φ_t^t is the identity. Thus, the right side is 0 and $\lambda_t \varphi_t^i m_i = 0$. Since λ_t is an injection, we have $\varphi_t^i m_i = 0$, as desired.   ●

When the index set is directed, there is a simpler description of direct limits in terms of orbit spaces.

Corollary 5.31.

(i) *Let $\{M_i, \varphi_j^i\}$ be a direct system of left R-modules over a directed index set I, and let $\bigsqcup_i M_i$ be their disjoint union. For $m_i \in M_i, m_j \in M_j$, define $m_i \sim m_j$ if they have a **common successor**; that is, there exists an index k with $k \succeq i, j$ such that $\varphi_k^i m_i = \varphi_k^j m_j$. Then $\sim$ is an equivalence relation on $\bigsqcup_i M_i$.*

(ii) *The orbit space $L = (\bigsqcup_i M_i)/\sim$ is a left R-module.*

(iii) *$L \cong \varinjlim M_i$; hence, elements of $\varinjlim M_i$ are equivalence classes $[m_i]$, where $m_i \in M_i$, and $[m_i] + [m_j'] = [\varphi_k^i m_i + \varphi_k^j m_j']$, where $k \succeq i, j$.*

Proof.

(i) Reflexivity and symmetry are obvious. For transitivity, assume that $\varphi_p^i m_i = \varphi_p^j m_j$ for some $p \succeq i, j$ and $\varphi_q^j m_j = \varphi_q^k m_k$ for some $q \succeq j, k$. Since I is directed, there is an index $r \succeq p, q$. Using the commutativity relation between the maps of the direct system, we have $\varphi_r^i m_i = \varphi_r^k m_k$.

(ii) Denote the equivalence class of m_i by $[m_i]$. It is routine to check that the operations

$$r[m_i] = [rm_i] \text{ if } r \in R,$$
$$[m_i] + [m_j'] = [\varphi_k^i m_i + \varphi_k^j m_j'], \text{ where } k \succeq i, j,$$

are well-defined and that they give L the structure of a left R-module.

(iii) As in the proof of Proposition 5.23, let $\varprojlim M_i = \left(\bigoplus_i M_i \right)/S$, where S is generated by $\{ \lambda_j \varphi^i_j m_i - \lambda_i m_i \; : \; m_i \in M_i \text{ and } i \leq j \}$. Define $f : L \to \varinjlim M_i$ by $f : [m_i] \mapsto m_i + S$. It is routine to check that f is a well-defined R-map. Since I is directed, Lemma 5.30(i) shows that f is surjective. If $0 = f([m_i]) = m_i + S$, then Lemma 5.30(ii) says that $\varphi^i_t m_i = 0$ for some $t \geq i$; that is, $[m_i] = [0]$. Therefore, f is an isomorphism. •

Example 5.32.   We now compute the direct limits of some of the direct systems in Example 5.29.

 (i) Let $\mathcal{I}$ be a simply ordered family of submodules of a module A; that is, if $M, M' \in \mathcal{I}$, then either $M \subseteq M'$ or $M' \subseteq M$. Then $\mathcal{I}$ is a directed set, and $\varinjlim M_i \cong \bigcup_i M_i$.

 (ii) If $(M_i)_{i \in I}$ is some family of modules, then $\mathcal{F}$, all finite partial sums, is a directed set under inclusion, and $\varinjlim M_i \cong \bigoplus_i M_i$.

(iii) If A is a module, then the family $\mathrm{Fin}(A)$ of all the finitely generated submodules of A is a directed set and $\varinjlim M_i \cong A$.

(iv) If R is a domain and $Q = \mathrm{Frac}(R)$, then the family of all cyclic R-submodules $M_r \subseteq Q$ of the form $\langle 1/r \rangle$, where $r \in R$ and $r \neq 0$, forms a directed set under inclusion, and $\varinjlim M_r \cong Q$; that is, Q is a direct limit of modules $M_r \cong R$.

 (v) In Example 1.14, we considered the presheaf $\mathcal{P}$ over a topological space X, defined on an open $U \subseteq X$ by $\mathcal{P}(U) = \{ \text{continuous } f : U \to \mathbb{R} \}$. For a point $p \in X$, let $\mathcal{U}$ be the family of all open neighborhoods U of p partially ordered by reverse inclusion. As in Example 5.29(vi), $\mathcal{U}$ is a directed set. If f, g are continuous functions $U, U' \to \mathbb{R}$, where $U, U' \in \mathcal{U}$, define $f \sim g$ in case there is some neighborhood W of p with $W \subseteq U \cap U'$ such that $f|W = g|W$. Now $\sim$ is an equivalence relation (transitivity uses the hypothesis that $\mathcal{U}$ is directed), and the equivalence class $[f, p]$ of f is called the **germ** of f at p. By Corollary 5.31, a germ $[f, p]$ is just an element of $\varinjlim \mathcal{P}(W)$, and we may view $[f, p]$ as a typical element of this direct limit. ◄

Recall that a direct system $\{ A_i, \alpha^i_j \}$ in a category $\mathcal{C}$ over a partially ordered index set I can be construed as a covariant functor $A : I \to \mathcal{C}$, where $A(i) = A_i$ and $A(\kappa^i_j) = \alpha^i_j$.

Definition.   Let $A = \{ A_i, \alpha^i_j \}$ and $B = \{ B_i, \beta^i_j \}$ be direct systems over the same (not necessarily directed) index set I. A **morphism of direct systems** is a natural transformation $r : A \to B$.

In more detail, r is an indexed family of homomorphisms

$$r = (r_i : A_i \to B_i)_{i \in I}$$

making the following diagrams commute for all $i \preceq j$:

$$
\begin{array}{ccc}
A_i & \xrightarrow{\;r_i\;} & B_i \\
\alpha^i_j \downarrow & & \downarrow \beta^i_j \\
A_j & \xrightarrow{\;r_j\;} & B_j.
\end{array}
$$

A morphism of direct systems $r : (A_i, \alpha^i_j)\} \to \{B_i, \beta^i_j\}$ determines a homomorphism

$$\overrightarrow{r} : \varinjlim A_i \to \varinjlim B_i$$

by

$$\overrightarrow{r} : \sum \lambda_i a_i + S \mapsto \sum \mu_i r_i a_i + T,$$

where $S \subseteq \bigoplus A_i$ and $T \subseteq \bigoplus B_i$ are the relation submodules in the construction of $\varinjlim A_i$ and $\varinjlim B_i$, respectively, and λ_i and μ_i are the injections of A_i and B_i, respectively, into the direct sums. The reader should check that r being a morphism of direct systems implies that $\overrightarrow{r}$ is independent of the choice of coset representative, and hence $\overrightarrow{r}$ is a well-defined function. One can, in a similar way, define a morphism of inverse systems, and such a morphism induces a homomorphism between the inverse limits. With these definitions, the reader may state and prove the naturality assertions for Theorems 5.21, 5.26, and 5.27.

Proposition 5.33. *Let I be a directed set, and let $\{A_i, \alpha^i_j\}$, $\{B_i, \beta^i_j\}$, and $\{C_i, \gamma^i_j\}$ be direct systems of left R-modules over I. If $r : \{A_i, \alpha^i_j\} \to \{B_i, \beta^i_j\}$ and $s : \{B_i, \beta^i_j\} \to \{C_i, \gamma^i_j\}$ are morphisms of direct systems, and if*

$$0 \to A_i \xrightarrow{\;r_i\;} B_i \xrightarrow{\;s_i\;} C_i \to 0$$

is exact for each $i \in I$, then there is an exact sequence

$$0 \to \varinjlim A_i \xrightarrow{\;\overrightarrow{r}\;} \varinjlim B_i \xrightarrow{\;\overrightarrow{s}\;} \varinjlim C_i \to 0.$$

Proof. We prove only that $\overrightarrow{r}$ is an injection, for the proof of exactness of the rest is routine; moreover, the hypothesis that I be directed enters the proof only in showing that $\overrightarrow{r}$ is an injection.

Suppose that $\overrightarrow{r}(x) = 0$, where $x \in \varinjlim A_i$. Since the index set I is directed, Lemma 5.30(i) allows us to write $x = \lambda_i a_i + S$ (where $S \subseteq \bigoplus A_i$ is the relation submodule and λ_i is the injection of A_i into the direct sum). By definition, $\overrightarrow{r}(x+S) = \mu_i r_i a_i + T$ (where $T \subseteq \bigoplus B_i$ is the relation submodule and μ_i is the injection of B_i into the direct sum). Now Lemma 5.30(ii) shows that $\mu_i r_i a_i + T = 0$ in $\varinjlim B_i$ implies that there is an index $k \succeq i$ with $\beta_k^i r_i a_i = 0$. Since r is a morphism of direct systems, we have

$$0 = \beta_k^i r_i a_i = r_k \alpha_k^i a_i.$$

Finally, since r_k is an injection, we have $\alpha_k^i a_i = 0$ and, hence, that $x = \lambda_i a_i + S = 0$. Therefore, $\overrightarrow{r}$ is an injection.   •

The next result generalizes Proposition 3.48.

Proposition 5.34.   *If $\{F_i, \varphi_j^i\}$ is a direct system of flat right R-modules over a directed index set I, then $\varinjlim F_i$ is also flat.*

Proof.   Let $0 \to A \xrightarrow{k} B$ be an exact sequence of left R-modules. Since each F_i is flat, the sequence

$$0 \to F_i \otimes_R A \xrightarrow{1_i \otimes k} F_i \otimes_R B$$

is exact for every i, where 1_i abbreviates 1_{F_i}. Consider the commutative diagram

$$
\begin{array}{ccccc}
0 & \longrightarrow & \varinjlim(F_i \otimes A) & \xrightarrow{\vec{k}} & \varinjlim(F_i \otimes B) \\
 & & \varphi \downarrow & & \downarrow \psi \\
0 & \longrightarrow & (\varinjlim F_i) \otimes A & \xrightarrow{1 \otimes k} & (\varinjlim F_i) \otimes B,
\end{array}
$$

where the vertical maps φ and ψ are the isomorphisms of Theorem 5.27, the map $\vec{k}$ is induced from the morphism of direct systems $\{1_i \otimes k\}$, and 1 is the identity map on $\varinjlim F_i$. Since each F_i is flat, the maps $1_i \otimes k$ are injections; since the index set I is directed, the top row is exact, by Proposition 5.33. Therefore, $1 \otimes k \colon (\varinjlim F_i) \otimes A \to (\varinjlim F_i) \otimes B$ is an injection, for it is the composite of injections $\psi \vec{k} \varphi^{-1}$. Therefore, $\varinjlim F_i$ is flat.   •

Corollary 5.35.

(i) *If R is a domain with $Q = \mathrm{Frac}(R)$, then Q is a flat R-module.*

(ii) *If every finitely generated submodule of a right R-module M is flat, then M is flat.*

Proof.

(i) In Example 5.29(v), we saw that Q is a direct limit, over a directed index set, of cyclic submodules, each of which is isomorphic to R. Since R is projective, hence flat, the result follows from Proposition 5.34.

(ii) In Example 5.29(iii), we saw that M is a direct limit, over a directed index set, of its finitely generated submodules. Since every finitely generated submodule is flat, by hypothesis, the result follows from Proposition 5.34. We have given another proof of Proposition 3.48.   •

Example 5.36.   The generalization of Corollary 5.35(ii): a right R-module M is flat if every finitely presented submodule of M is flat, may not be true. Let k be a field, let $R = k[X]$, where X is an infinite set of indeterminates, and let m be the ideal generated by X. Now m is a maximal ideal because $R/\mathfrak{m} \cong k$. Hence, the module $M = R/\mathfrak{m}$ is a simple module; that is, it has no submodules other than $\{0\}$ and M. As in Example 3.14, M is not finitely presented, and so the only finitely presented submodule of M is $\{0\}$. Thus, every finitely presented submodule of M is flat. However, we claim that M is not flat. By Proposition 3.49, every flat module over a domain is torsion-free. But $R = k[X]$ is a domain and $M = R/\mathfrak{m}$ is not torsion-free, for if $x \in X$, then $x(1 + \mathfrak{m}) = 0$. Therefore, M is not flat.   ◄

We are going to prove the surprising result that every flat module is a direct limit of free modules. We often think of direct limits as generalized unions, but this can be misleading. After all, $\mathbb{I}_6 = \mathbb{I}_2 \oplus \mathbb{I}_3$, so that $\mathbb{I}_2$ is a projective $\mathbb{I}_6$-module and, hence, it is flat; but $\mathbb{I}_2$ is surely not a union of free modules, each of which has at least six elements. Our exposition follows that in Osborne, *Basic Homological Algebra*.

We begin with a technical definition.

Definition.   Let D be a submodule of a module C, let $\mathcal{A}$ be a set of submodules of C partially ordered by inclusion, and let $\mathcal{B}$ be a set of submodules of D partially ordered by inclusion. Then $(C, D, \mathcal{A}, \mathcal{B})$ is a (C, D)-*system* if

(i) $\mathcal{A}$ and $\mathcal{B}$ are directed sets,

(ii) $C = \bigcup_{A \in \mathcal{A}} A$, and $D = \bigcup_{B \in \mathcal{B}} B$,

(iii) $\mathcal{A}$ *dominates* $\mathcal{B}$; that is, for each $B \in \mathcal{B}$, there exists $A \in \mathcal{A}$ with $A \supseteq B$.

Every (C, D)-system determines a directed set. Define

$$I = I(C, D, \mathcal{A}, \mathcal{B}) = \{(A, B) \in \mathcal{A} \times \mathcal{B} : A \supseteq B\}.$$

Partially order I by

$$(A, B) \leq (A', B') \quad \text{if } A \subseteq A' \text{ and } B \subseteq B'.$$

Proposition 5.37.   *Let $(C, D, \mathcal{A}, \mathcal{B})$ be a (C, D)-system. Then the set $I = I(C, D, \mathcal{A}, \mathcal{B})$ is a directed set and*

$$\varinjlim_{(A,B)\in I} (A/B) \cong C/D.$$

Proof.   To see that I is directed, let $(A, B), (A', B') \in I$. Since $\mathcal{B}$ is directed, there is $B'' \in \mathcal{B}$ with $B, B' \subseteq B''$. But $\mathcal{A}$ dominates $\mathcal{B}$, so there is $A'' \in \mathcal{A}$ with $B'' \subseteq A''$. Finally, since $\mathcal{A}$ is directed, there is $A^* \in \mathcal{A}$ with $A, A', A'' \subseteq A^*$. Then $(A, B), (A', B') \le (A^*, B'')$.

The indexed family $(A/B)_{(A,B)\in I}$ is a direct system if we define $\varphi^{(A,B)}_{(A',B')} : A/B \to A'/B'$, whenever $(A, B) \le (A', B')$, as the composite $A/B \to A'/B \to A'/B'$, where the first arrow is inclusion and the second is enlargement of coset.

Consider the diagram in which $(A, B) \le (A', B')$:

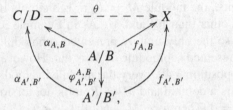

where $\alpha_{A,B} : A/B \to C/D$ is the composite $A/B \to C/B \to C/D$ (inclusion followed by enlargement of coset); that is, $\alpha_{A,B} : a + B \to a + D$.

Uniqueness of a map $\theta : C/D \to X$ is easy. Let $c + D \in C/D$. By (ii), there is $A \in \mathcal{A}$ with $c \in A$; if $B \in \mathcal{B}$, there is $A' \in \mathcal{A}$ with $A' \supseteq B$, by (iii); finally, there is $A'' \in \mathcal{A}$ with $A'' \supseteq A, A'$, by (i). Thus, $(A'', B) \in I$ and $c \in A''$. Commutativity of the completed diagram would give

$$\theta(c + D) = \theta\alpha_{A'',B}(c + D) = f_{A'',B}(c + D).$$

It is straightforward to prove that this formula gives a well-defined homomorphism; commutativity of the triangles shows that it is independent of the choice of index (A'', B) and, if $c \in D$, that $\theta(c + D) = 0$.   •

Corollary 5.38.   *Let A be a left R-module, let J be a directed set, and let $(A_j)_{j\in J}$ be a family of submodules of A. Then*

$$\varinjlim_J A_j = \bigcup_J A_j \quad and \quad \varinjlim_J A/A_j = A/\bigcup A_j.$$

Proof.   For the first result, apply the proposition when $C = \bigcup_J A_j$, $D = \{0\}$, $\mathcal{A} = (A_j)_{j\in J}$, and $\mathcal{B} = \{D\}$. For the second result, apply the proposition when $\mathcal{A} = \{A\}$ and $\mathcal{B} = (A_j)_{j\in J}$.   •

We will need the following technical result.

Lemma 5.39 (Lazard).   *Let R be a ring and let M be a left R-module.*

(i) *Then $M \cong \varinjlim_I G_i$, where every G_i is finitely presented and I is a (C, D)-system (hence is directed) with $M \cong C/D$.*

(ii) *The (C, D)-system I is universal with respect to all homomorphisms from finitely presented modules to M: given $(A, B) \in I$, a finitely presented X, and maps $\rho: X \to M$ and $\sigma: A/B \to X$ with $\alpha_{A,B} = \rho\sigma$, there exist $(A', B') \in I$ with $(A', B') \geq (A, B)$ and an isomorphism $\tau: X \to A'/B'$ making the following diagram commute.*

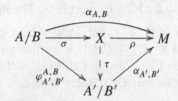

Proof.

(i) Let C be the free left R-module with basis[4] $M \times \mathbb{N}$, let $\pi: C \to M$ be defined by $(m, n) \mapsto m$, and let $D = \ker \pi$; since π is surjective, we have $C/D \cong M$. Define $\mathcal{A}$ to be all those submodules of C generated by a finite subset of $M \times \mathbb{N}$ (so that each $A \in \mathcal{A}$ is finitely generated free), and define $\mathcal{B}$ to be all the finitely generated submodules of D. It is easy to check that $(C, D, \mathcal{A}, \mathcal{B})$ is a (C, D)-system, so that $I = I(C, D, \mathcal{A}, \mathcal{B})$ is a directed set. If $(A, B) \in I$, then A/B is finitely presented (for both A and B are, by definition, finitely generated). Therefore, $\varinjlim_I (A/B) \cong C/D \cong M$, by Proposition 5.37.

(ii) Given the top triangle, our task is to find $(A', B') \in I$ with $(A', B') \geq (A, B)$ and an isomorphism $\tau: X \to A'/B'$ making the augmented diagram commute. We first construct the following auxiliary diagram for some $(A^*, B^*) \in I$.

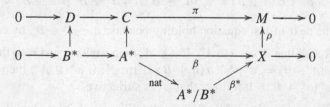

Since $A \in \mathcal{A}$, it is a free module with basis $\{(m_1, n_1), \ldots, (m_\ell, n_\ell)\}$; choose $N > n_i$ for $i = 1, \ldots, \ell$. Now X is finitely presented; let $X =$

---

[4]This part of the lemma can be proved if C were defined as the free module with basis M; the reason for the larger basis $M \times \mathbb{N}$ will appear in part (ii).

$\langle x_1, \dots, x_p \rangle$. For each $j \leq p$, let $\rho(x_j) = \pi(y_j)$, where $y_j \in C$; define A^* to be the free submodule of C with basis $\{(\pi y_j, N + j)$ for $j \leq p\}$; note that $A^* \in \mathcal{A}$. If we define $\beta \colon A^* \to X$ by $\beta \colon (\pi y_j, N + j) \mapsto x_j$, then the right-hand square commutes:

$$\rho\beta \colon (\pi y_j, N + j) \mapsto p(x_j) = \pi(y_j).$$

The map β is surjective, and its kernel B^* is finitely generated, by Corollary 3.13; thus, $B^* \in \mathcal{B}$ and $(A^*, B^*) \in I$. The map induced by β, namely, $\beta^* \colon a^* + B^* \mapsto \beta a^*$ for all $a^* \in A^*$, is an isomorphism $A^*/B^* \to X$, by the First Isomorphism Theorem.

Now $(A, B) \not\leq (A^*, B^*)$, but we use it to construct (A', B') and τ. Define $A' = A \oplus A^* \subseteq C$ (A' is an internal direct sum because the bases of A and A^* are disjoint subsets of the basis of C). Since the isomorphism β^* is surjective, there exist $z_i \in A^*$, for each $i \leq \ell$, with

$$\sigma\big((m_i, n_i) + B\big) = \beta^*(z_i + B^*).$$

Define $B' = \big\langle B, B^*, (m_i, n_i) - z_i$ for $i \leq \ell \big\rangle$. Obviously, B' is finitely generated. We claim that if $a \in A$, $a^* \in A^*$, and $\sigma(a + B) = \beta(a^*)$ in X, then the element $a - a^* \in A \oplus A^* \subseteq C$ lies in $\ker \pi$.

$$a + D = \alpha(a + B) = \rho\sigma(a + B) = \rho\beta(a^*)$$
$$= \rho\beta^*(a^* + B^*) = \alpha(a^* + B^*) = a^* + D.$$

Hence, $B' \subseteq D = \ker \pi$, $(A', B') \in I$, and $(A, B) \leq (A', B')$.

Write $\varphi = \varphi^{A,B}_{A',B'} \colon A/B \to A'/B'$ and $\varphi^* = \varphi^{A^*,B^*}_{A',B'} \colon A^*/B^* \to A'/B'$; thus, if $a \in A$ and $a^* \in A^*$, then $\varphi \colon a + B \mapsto a + B'$ and $\varphi^* \colon a^* + B^* \mapsto a^* + B'$. Let us see that $\varphi^* \beta^{*-1} \sigma = \varphi$ and that φ^* is an isomorphism. We saw above that if $a = (m_i, n_i) \in A$ and $a^* = z_i \in A^*$, then $\sigma(a + B) = \beta^*(a^* + B^*)$. Hence,

$$\varphi^* \beta^{*-1} \sigma(a + B) = \varphi^*(a^* + B^*) = a^* + B' = a + B' = \varphi(a + B),$$

the next-to-last equation holding because $a - a^* \in B'$, by construction.

Recall that $A' = A \oplus A^*$. If $a \in A$, then we have just seen that $a + B' = \varphi(a + B) = \varphi^* \beta^{*-1} \sigma(a + B) \in \operatorname{im} \varphi^*$; if $a^* \in A^*$, then $a^* + B' = \varphi^*(a^* + B^*)$. It follows that φ^* is surjective.

Suppose that $\varphi^*(a^* + B^*) = a^* + B' = 0$; that is, $a^* \in B' \cap A^*$. Now $a^* \in B'$ says that $a^* = b + b^* + \sum_i r_i\big((m_i, n_i) - z_i\big)$; that is,

$$a^* = \Big(b + \sum_i r_i(m_i, n_i)\Big) + \Big(b^* - \sum_i r_i z_i\Big), \qquad (1)$$

where $b \in B$, $b^* \in B^*$, and $r_i \in R$. Since $a^* \in A^*$, unique expression in $A' = A \oplus A^*$ gives $b + \sum_i r_i(m_i, n_i) = 0$, so that $\sum_i r_i(m_i, n_i) \in B$. Now

$$\sum_i r_i z_i + B^* = \sum_i r_i \beta^{*-1} \sigma \big((m_i, n_i) + B\big)$$

$$= \beta^{*-1} \sigma \Big(\sum_i r_i(m_i, n_i) + B\Big) = 0 + B^*,$$

so that $\sum_i r_i z_i \in B^*$. Equation (1) says $a^* = b^* - \sum_i r_i z_i \in B^*$; that is, $a^* \in B^*$. Therefore, φ^* is an isomorphism.

Assemble the maps into a commutative diagram:

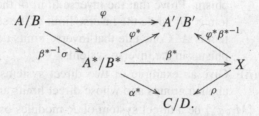

Define $\tau: X \to A'/B'$ by $\tau = \varphi^* \beta^{*-1}$. Now τ is an isomorphism, for both β^* and φ^* are isomorphisms. Consider the diagram in the statement of this proposition. The left triangle commutes, for $\tau\sigma = \varphi^* \beta^{*-1} \sigma = \varphi$. Finally, the right-hand triangle commutes: $\alpha'\tau = \alpha'\varphi^*\beta^{*-1} = \alpha'\varphi^*\alpha^{*-1}\rho = (\alpha'\varphi^*)\alpha^{*-1}\rho = \rho$, the last equation holding because $\alpha'\varphi^* = \alpha^*$ in the direct system.   •

Theorem 5.40 (Lazard).   *For any ring R, a left R-module M is flat if and only if it is a direct limit (over a directed index set) of finitely generated free left R-modules.*

Proof.   Sufficiency is Proposition 5.34. For the converse, let M be flat, and let $I = I(C, D, \mathcal{A}, \mathcal{B})$ be the (C, D)-system in Lemma 5.39(i) with $M \cong C/D$. Define

$$J = \{(A, B) \in I : A/B \text{ is free}\}.$$

The result will follow from Exercise 5.22 on page 255 if we show that J is a cofinal subset of I.

Since C/D is flat, the map $\operatorname{Hom}_R(X, R) \otimes_R C/D \to \operatorname{Hom}_R(X, C/D)$, given by $f \otimes (c + D) \mapsto [\varphi : x \mapsto f(x)(c+D)]$, is an isomorphism for every finitely presented left R-module X, by Exercise 3.34 on page 152. Given $(A, B) \in I$, the insertion $\alpha: A/B \to C/D$, taking $a + B \mapsto a + D$, lies in $\operatorname{Hom}_R(A/B, C/D) \cong \operatorname{Hom}_R(A/B, R) \otimes_R C/D$. If we write $\bar{c} = c + D$ for $c \in C$ and $\bar{a} = a + B$ for $a \in A$, then there exist $\sigma_1, \ldots, \sigma_n \in \operatorname{Hom}_R(A/B, R)$

and $\overline{y_1}, \ldots, \overline{y_n} \in C/D$ with $\alpha(\overline{a}) = \sum_i \sigma_i(\overline{a}) \overline{y_i}$. Define $\sigma : A/B \to R^n$ by $\sigma(\overline{a}) = (\sigma_i(\overline{a}), \ldots, \sigma_n(\overline{a}))$, and $\rho : R^n \to C/D$ by $\rho(r_1, \ldots, r_n) = \sum_i r_i \overline{y_i}$. Now $\alpha = \rho\sigma$. By Lemma 5.39(ii), there are $(A', B') \in I$ with $(A, B) \leq (A', B')$ and an isomorphism $A'/B' \cong R^n$. Therefore, $(A', B') \in J$, and J is cofinal in I, as desired.   •

Exercises

*5.17   (i)   Let $(A_n)_{n \in \mathbb{N}}$ be a family of isomorphic abelian groups; say, $A_n \cong A$ for all n. Consider inverse systems $\{A_n, f_n^m\}$ and $\{A_n, g_n^m\}$, where each $f_n^m = 0$ and each g_n^m is an isomorphism. Prove that the inverse limit of the first inverse system is $\{0\}$ while the inverse limit of the second inverse system is A. Conclude that inverse limits depend on the morphisms in the inverse systems.

(ii)   Give an example of two direct systems having the same abelian groups and whose direct limits are not isomorphic.

*5.18   Let $\{M_i, \varphi_j^i\}$ be a direct system of R-modules over an index set I, and let $F : {}_R\mathbf{Mod} \to \mathcal{C}$ be a functor to some category $\mathcal{C}$. Prove that $\{FM_i, F\varphi_j^i\}$ is a direct system in $\mathcal{C}$ if F is covariant, while it is an inverse system if F is contravariant.

Hint. If we regard the direct system as a functor $D : I \to {}_R\mathbf{Mod}$, then the composite FD is a functor $I \to \mathcal{C}$.

5.19   Give an example of a direct system of modules, $\{A_i, \alpha_j^i\}$, over some directed index set I, for which $A_i \neq \{0\}$ for all i and $\varinjlim A_i = \{0\}$.

*5.20   (i)   Prove that $\mathrm{End}(\mathbb{Z}(p^\infty)) \cong \mathbb{Z}_p$ as rings, where $\mathbb{Z}_p$ is the ring of p-adic integers.

Hint. A presentation for $\mathbb{Z}(p^\infty)$ is

$$(a_0, a_1, a_2 \ldots, \mid pa_0 = 0, pa_n = a_{n-1} \text{ for } n \geq 1).$$

(ii)   Prove that the additive group of $\mathbb{Z}_p$ is torsion-free.

Hint. View $\mathbb{Z}_p$ as a subgroup of $\prod_n \mathbb{I}_{p^n}$.

*5.21   Let $0 \to U \to V \to V/U \to 0$ be an exact sequence of left R-modules.

(i)   Let $\{U_i, \alpha_j^i\}$ be a direct system of submodules of U, where $(\alpha_j^i : U_i \to U_j)_{i \leq j}$ are inclusions. Prove that $\{V/U_i, e_j^i\}$ is a direct system, where each $e_j^i : V/U_i \to V/U_j$ is enlargement of coset.

(ii)   If $\varinjlim U_i = U$, prove that $\varinjlim(V/U_i) \cong V/U$.

*5.22   (i)   Let K be a *cofinal subset* of a directed index set I (that is, for each $i \in I$, there is $k \in K$ with $i \preceq k$). Let $\{M_i, \varphi^i_j\}$ be a direct system over I, and let $\{M_i, \varphi^i_j\}$ be the subdirect system whose indices lie in K. Prove that the direct limit over I is isomorphic to the direct limit over K.

(ii)   Let K be a cofinal subset of a directed index set I, let $\{M_i, \varphi^i_j\}$ be an inverse system over I, and let $\{M_i, \varphi^i_j\}$ be the subinverse system whose indices lie in K. Prove that the inverse limit over I is isomorphic to the inverse limit over K.

(iii)   A partially ordered set I has a *top element* if there exists $\infty \in I$ with $i \preceq \infty$ for all $i \in I$. If $\{M_i, \varphi^i_j\}$ is a direct system over I, prove that

$$\varinjlim M_i \cong M_\infty.$$

(iv)   Show that part (i) may not be true if the index set is not directed.

Hint. Pushout.

*5.23 Prove that a ring R is left noetherian if and only if every direct limit (with directed index set) of injective left R-modules is itself injective.
Hint. See Theorem 3.39.

*5.24 Let

$$\begin{array}{ccc} D & \xrightarrow{\alpha} & C \\ \beta \downarrow & & \downarrow g \\ B & \xrightarrow{f} & A \end{array}$$

be a pullback diagram in **Ab**. If there are $c \in C$ and $b \in B$ with $gc = fb$, prove that there exists $d \in D$ with $c\alpha(d)$ and $b = \beta(d)$.
Hint. Define $p: \mathbb{Z} \to C$ by $p(n) = nc$, and define $q: \mathbb{Z} \to B$ by $q(n) = nb$. There is a map $\theta: \mathbb{Z} \to D$ making the diagram commute; define $d = \theta(1)$.

5.25 Consider the ideal $J = (x)$ in $k[x]$, where k is a commutative ring. Prove that the completion $\varprojlim(k[x]/J^n)$ of the polynomial ring $k[x]$ is $k[[x]]$, the ring of formal power series.

5.26 In $_R\mathbf{Mod}$, let $r: \{A_i, \alpha^i_j\} \to \{B_i, \beta^i_j\}$ and $s: \{B_i, \beta^i_j\} \to \{C_i, \gamma^i_j\}$ be morphisms of inverse systems over any (not necessarily directed) index set I. If

$$0 \to A_i \xrightarrow{r_i} B_i \xrightarrow{s_i} C_i$$

is exact for each $i \in I$, prove that there are homomorphisms $\overleftarrow{r}$, $\overleftarrow{s}$ given by the universal property of inverse limits, and an exact sequence

$$0 \to \varprojlim A_i \xrightarrow{\overleftarrow{r}} \varprojlim B_i \xrightarrow{\overleftarrow{s}} \varprojlim C_i.$$

5.27 Definition. A category $\mathcal{C}$ is *complete* if $\varprojlim A_i$ exists in $\mathcal{C}$ for every inverse system $\{A_i, \psi_i^j\}$ in $\mathcal{C}$; a category $\mathcal{C}$ is *cocomplete* if $\varinjlim A_i$ exists in $\mathcal{C}$ for every direct system $\{A_i, \varphi_j^i\}$ in $\mathcal{C}$.

Prove that a category is complete if and only if it has equalizers and products (over any index set). Dually, prove that a category is cocomplete if and only if it has coequalizers and coproducts (over any index set).

*5.28 Prove that if $T : {}_R\mathbf{Mod} \to \mathbf{Ab}$ is an additive left exact functor preserving direct products, then T preserves inverse limits.
Hint. Consider an inverse limit as the kernel of a map between direct products.

5.3   Adjoint Functor Theorem for Modules

Recall the adjoint isomorphism, Theorem 2.75: given modules A_R, ${}_RB_S$, and C_S, there is a natural isomorphism

$$\tau_{A,B,C} : \operatorname{Hom}_S(A \otimes_R B, C) \to \operatorname{Hom}_R(A, \operatorname{Hom}_S(B, C)).$$

Write $F = \square \otimes_R B$ and $G = \operatorname{Hom}_S(B, \square)$, so that the isomorphism reads

$$\operatorname{Hom}_S(FA, C) \cong \operatorname{Hom}_R(A, GC).$$

If we pretend that $\operatorname{Hom}(\square, \square)$ is an inner product, then this reminds us of the definition of adjoint pairs in Linear Algebra: if $T : V \to W$ is a linear transformation between vector spaces equipped with inner products, then its *adjoint* is the linear transformation $T^* : W \to V$ such that

$$(Tv, w) = (v, T^*w)$$

for all $v \in V$ and $w \in W$. This analogy explains why the isomorphism τ is called the *adjoint isomorphism*.

Definition. Let $F\colon \mathcal{C} \to \mathcal{D}$ and $G\colon \mathcal{D} \to \mathcal{C}$ be covariant functors. The ordered pair (F, G) is an **adjoint pair** if, for each $C \in \mathrm{obj}(\mathcal{C})$ and $D \in \mathrm{obj}(\mathcal{D})$, there are bijections

$$\tau_{C,D}\colon \mathrm{Hom}_{\mathcal{D}}(FC, D) \to \mathrm{Hom}_{\mathcal{C}}(C, GD)$$

that are natural transformations in C and in D.

In more detail, naturality says that the following two diagrams commute for all $f\colon C' \to C$ in $\mathcal{C}$ and $g\colon D \to D'$ in $\mathcal{D}$:

$$
\begin{array}{ccc}
\mathrm{Hom}_{\mathcal{D}}(FC, D) & \xrightarrow{(Ff)^*} & \mathrm{Hom}_{\mathcal{D}}(FC', D) \\
{\scriptstyle \tau_{C,D}}\downarrow & & \downarrow{\scriptstyle \tau_{C',D}} \\
\mathrm{Hom}_{\mathcal{C}}(C, GD) & \xrightarrow{f^*} & \mathrm{Hom}_{\mathcal{C}}(C', GD);
\end{array}
$$

$$
\begin{array}{ccc}
\mathrm{Hom}_{\mathcal{D}}(FC, D) & \xrightarrow{g_*} & \mathrm{Hom}_{\mathcal{D}}(FC, D') \\
{\scriptstyle \tau_{C,D}}\downarrow & & \downarrow{\scriptstyle \tau_{C,D'}} \\
\mathrm{Hom}_{\mathcal{C}}(C, GD) & \xrightarrow{(Gg)_*} & \mathrm{Hom}_{\mathcal{C}}(C, GD').
\end{array}
$$

Example 5.41.

(i) If $B = {}_RB_S$ is a bimodule, then $\big(\square \otimes_R B, \mathrm{Hom}_S(B, \square)\big)$ is an adjoint pair, by Theorem 2.75. Similarly, if $B = {}_SB_R$ is a bimodule, then $\big(B \otimes_R \square, \mathrm{Hom}_S(B, \square)\big)$ is an adjoint pair, by Theorem 2.76.

(ii) Let $U\colon$ **Groups** $\to$ **Sets** be the *forgetful functor* which assigns to each group G its underlying set and views each homomorphism as a mere function. Let $F\colon$ **Sets** $\to$ **Groups** be the *free functor* defined in Exercise 1.6 on page 33, which assigns to each set X the free group FX having basis X. The function

$$\tau_{X,H}\colon \mathrm{Hom}_{\mathbf{Groups}}(FX, H) \to \mathrm{Hom}_{\mathbf{Sets}}(X, UH),$$

given by $f \mapsto f|X$, is a bijection (its inverse is $\varphi \mapsto \widetilde{\varphi}$, where X, being a basis of FX, says that every function $\varphi\colon X \to H$ corresponds to a unique homomorphism $\widetilde{\varphi}\colon FX \to H$). Indeed, $\tau_{X,H}$ is a natural bijection, showing that (F, U) is an adjoint pair of functors. This example can be generalized by replacing **Groups** by other categories having free objects; e.g., ${}_R\mathbf{Mod}$ or $\mathbf{Mod}_R$.

(iii) If $U\colon$ **ComRings** $\to$ **Sets** is the forgetful functor, then (F, U) is an adjoint pair where, for any set X, we have $F(X) = \mathbb{Z}[X]$, the ring of all polynomials in commuting variables X. More generally, if k is a commutative ring and $\mathbf{ComAlg}_k$ is the category of all commutative k-algebras, then $F(X) = k[X]$, the polynomial ring over k. This is essentially the same example as in part (ii), for $k[X]$ is the free k-algebra on X. ◄

For many examples of adjoint pairs of functors, see Herrlich–Strecker, *Category Theory*, p. 197, and Mac Lane, *Categories for the Working Mathematician*, Chapter 4, especially pp. 85–86.

Example 5.42.   Adjointness is a property of an *ordered pair* of functors; if (F, G) is an adjoint pair of functors, it does not follow that (G, F) is also an adjoint pair. For example, if $F = \square \otimes B$ and $G = \mathrm{Hom}(B, \square)$, then the adjoint isomorphism says that $\mathrm{Hom}(A, B \otimes C) \cong \mathrm{Hom}(A, \mathrm{Hom}(B, C))$ for all A and C; that is, $\mathrm{Hom}(FA, C) \cong \mathrm{Hom}(A, GC)$. It does not say that there an isomorphism (natural or not) $\mathrm{Hom}(\mathrm{Hom}(B, A), C) \cong \mathrm{Hom}(A, B \otimes C)$. Indeed, if $A = \mathbb{Q}$, $B = \mathbb{Q}/\mathbb{Z}$, and $C = \mathbb{Z}$, then $\mathrm{Hom}(G\mathbb{Q}, \mathbb{Z}) \not\cong \mathrm{Hom}(\mathbb{Q}, F\mathbb{Z})$; that is,

$$\mathrm{Hom}\big(\mathrm{Hom}(\mathbb{Q}/\mathbb{Z}, \mathbb{Q}), \mathbb{Z}\big) \not\cong \mathrm{Hom}\big(\mathbb{Q}, (\mathbb{Q}/\mathbb{Z}) \otimes \mathbb{Z}\big),$$

for the left side is $\{0\}$, while the right side is isomorphic to $\mathrm{Hom}(\mathbb{Q}, \mathbb{Q}/\mathbb{Z})$, which contains the natural map $\mathbb{Q} \to \mathbb{Q}/\mathbb{Z}$. ◄

Definition.   Let $\mathcal{C} \underset{G}{\overset{F}{\rightleftarrows}} \mathcal{D}$ be functors. If (F, G) is an adjoint pair, then we say that F has a **right adjoint** and that G has a **left adjoint**.

Let (F, G) be an adjoint pair, where $F \colon \mathcal{C} \to \mathcal{D}$ and $G \colon \mathcal{D} \to \mathcal{C}$. If $C \in \mathrm{obj}(\mathcal{C})$, then setting $D = FC$ gives a bijection $\tau \colon \mathrm{Hom}_{\mathcal{D}}(FC, FC) \to \mathrm{Hom}_{\mathcal{C}}(C, GFC)$, so that η_C, defined by

$$\eta_C = \tau(1_{FC}),$$

is a morphism $C \to GFC$. Exercise 5.30 on page 271 says that $\eta \colon 1_{\mathcal{C}} \to GF$ is a natural transformation; it is called the **unit** of the adjoint pair.

Theorem 5.43.   *Let (F, G) be an adjoint pair of functors, where $F \colon \mathcal{C} \to \mathcal{D}$ and $G \colon \mathcal{D} \to \mathcal{C}$. Then F preserves direct limits and G preserves inverse limits.*

Remark.   There is no restriction on the index sets of the limits; in particular, they need not be directed. ◄

Proof. Let I be a partially ordered set, and let $\{C_i, \varphi_j^i\}$ be a direct system in C over I. By Exercise 5.18 on page 254, $\{FC_i, F\varphi_j^i\}$ is a direct system in $\mathcal{D}$ over I. Consider the following diagram in $\mathcal{D}$:

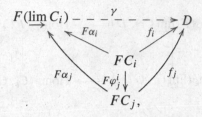

where $\alpha_i: C_i \to \varinjlim C_i$ are the maps in the definition of direct limit. We must show that there exists a unique morphism $\gamma: F(\varinjlim C_i) \to D$ making the diagram commute. The idea is to apply G to this diagram, and use the unit $\eta: 1_C \to GF$ to replace $GF(\varinjlim C_i)$ and GFC_i by $\varinjlim C_i$ and C_i, respectively. In more detail, Exercise 5.30 on page 271 gives morphisms η and η_i making the following diagram commute:

$$\begin{array}{ccc} \varinjlim C_i & \xrightarrow{\ \eta\ } & GF(\varinjlim C_i) \\ {\scriptstyle \alpha_i}\uparrow & & \uparrow{\scriptstyle GF\alpha_i} \\ C_i & \xrightarrow{\ \eta_i\ } & GFC_i. \end{array}$$

Apply G to the original diagram and adjoin this diagram to its left:

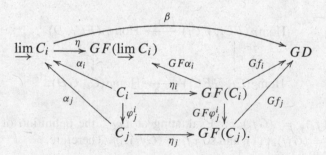

This diagram commutes: we know that $(GF\varphi_j^i)\eta_i = \eta_j\varphi_j^i$, since η is natural, and $Gf_i = Gf_j(GF\varphi_j^i)$, since G is a functor; therefore, $Gf_i\eta_i = Gf_j(GF\varphi_j^i)\eta_i = Gf_j\eta_j\varphi_j^i$. By the definition of direct limit, there exists a unique $\beta: \varinjlim C_i \to GD$ [that is, $\beta \in \text{Hom}_C(\varinjlim C_i, GD)$] making the diagram commute. Since (F, G) is an adjoint pair, there exists a natural bijection

$$\tau: \text{Hom}_{\mathcal{D}}(F(\varinjlim C_i), D) \to \text{Hom}_C(\varinjlim C_i, GD).$$

Define

$$\gamma = \tau^{-1}(\beta) \in \text{Hom}_{\mathcal{D}}(F(\varinjlim C_i), D).$$

We claim that $\gamma : F(\varinjlim C_i) \to D$ makes the first diagram commute. The first commutative square in the definition of adjointness, which involves the morphism $\alpha_i : C_i \to \varinjlim C_i$, gives commutativity of

$$
\begin{array}{ccc}
\operatorname{Hom}_{\mathcal{D}}(F(\varinjlim C_i), D) & \xrightarrow{(F\alpha_i)^*} & \operatorname{Hom}_{\mathcal{D}}(FC_i, D) \\
\tau \downarrow & & \downarrow \tau \\
\operatorname{Hom}_{\mathcal{C}}(\varinjlim C_i, GD) & \xrightarrow[\alpha_i^*]{} & \operatorname{Hom}_{\mathcal{C}}(C_i, GD).
\end{array}
$$

Thus, $\tau(F\alpha_i)^* = \alpha_i^* \tau$, and so $\tau^{-1}\alpha_i^* = (F\alpha_i)^* \tau^{-1}$. Evaluating on β, we have

$$(F\alpha_i)^* \tau^{-1}(\beta) = (F\alpha_i)^* \gamma = \gamma F\alpha_i.$$

On the other hand, since $\beta\alpha_i \doteq (Gf_i)\eta_i$, we have

$$\tau^{-1}\alpha_i^*(\beta) = \tau^{-1}(\beta\alpha_i) = \tau^{-1}((Gf_i)\eta_i).$$

Therefore,

$$\gamma F\alpha_i = \tau^{-1}((Gf_i)\eta_i).$$

The second commutative square in the definition of adjointness, for the morphism $f_i : FC_i \to D$, gives commutativity of

$$
\begin{array}{ccc}
\operatorname{Hom}_{\mathcal{D}}(FC_i, FC_i) & \xrightarrow{(f_i)_*} & \operatorname{Hom}_{\mathcal{D}}(FC_i, D) \\
\tau \downarrow & & \downarrow \tau \\
\operatorname{Hom}_{\mathcal{C}}(C_i, GFC_i) & \xrightarrow[(Gf_i)_*]{} & \operatorname{Hom}_{\mathcal{C}}(C_i, GD);
\end{array}
$$

that is, $\tau(f_i)_* = (Gf_i)_*\tau$. Evaluating at 1_{FC_i}, the definition of η_i gives $\tau(f_i)_*(1) = (Gf_i)_*\tau(1)$, and so $\tau f_i = (Gf_i)_*\eta_i$. Therefore,

$$\gamma F\alpha_i = \tau^{-1}((Gf_i)\eta_i) = \tau^{-1}\tau f_i = f_i,$$

so that γ makes the original diagram commute. We leave the proof of the uniqueness of γ as an exercise for the reader, with the hint to use the uniqueness of β.

The dual proof shows that G preserves inverse limits. $\bullet$

We are now going to characterize the Hom and tensor functors on module categories, yielding a necessary and sufficient condition for a functor on such categories to be half of an adjoint pair (Theorems 5.51 and 5.52).

Lemma 5.44.

(i) *If M is a right R-module and $m \in M$, then $\varphi_m \colon R \to M$, defined by $r \mapsto mr$, is a map of right R-modules. In particular, if $M = R$ and $u \in R$, then $\varphi_u \colon R \to R$ is a map of right R-modules.*

(ii) *If M is a right R-module, $m \in M$, and $u \in R$, then*

$$\varphi_{mu} = \varphi_m \varphi_u.$$

(iii) *Let $f \colon M \to N$ be an R-map between right R-modules. If $m \in M$, then*

$$\varphi_{fm} = f \varphi_m.$$

Proof.

(i) φ_m is additive because $m(r + s) = mr + ms$; φ_m preserves scalar multiplication on the right because $\varphi_m(rs) = m(rs) = (mr)s = \varphi_m(r)s$.

(ii) Now $\varphi_{mr} \colon u \mapsto (mr)u$, while $\varphi_m \varphi_r \colon u \mapsto \varphi_m(ru) = m(ru)$. These values agree because M is a right R-module.

(iii) Now $\varphi_{fm} \colon u \mapsto (fm)u$, while $f \varphi_m \colon u \mapsto f(mu)$. These values agree because f is an R-map.   •

Theorem 5.45 (Watts).   *If $F \colon \mathbf{Mod}_R \to \mathbf{Ab}$ is a right exact additive functor that preserves direct sums, then F is naturally isomorphic to $\square \otimes_R B$, where B is $F(R)$ made into a left R-module.*

Proof.   We begin by making the abelian group FR [our abbreviation for $F(R)$] into a left R-module. If M is a right R-module and $m \in M$, then $\varphi_m \colon R \to M$, defined by $r \mapsto mr$, is an R-map, by Lemma 5.44(i), and so the $\mathbb{Z}$-map $F\varphi_m \colon FR \to FM$ is defined. In particular, if $M = R$ and $u \in R$, then $\varphi_u \colon R \to R$ and, for all $x \in FR$, we define ux by

$$ux = (F\varphi_u)x.$$

Let us show that this scalar multiplication makes FR into a left R-module. If $M = R$ and $u, v \in R$, then $F\varphi_u, F\varphi_v \colon FR \to FR$, and Lemma 5.44(ii) gives $\varphi_{uv} = \varphi_u \varphi_v$. Hence,

$$(uv)x = (F\varphi_{uv})x = F(\varphi_u \varphi_v)x = (F\varphi_u)(F\varphi_v)x = u(vx).$$

Denote the left R-module FR by B, so that $\square \otimes_R B \colon \mathbf{Mod}_R \to \mathbf{Ab}$. We claim that $\tau_M \colon M \times FR \to FM$, defined by $(m, x) \mapsto (F\varphi_m)x$, is R-biadditive; that is, $\tau_M(mu, x) = \tau_M(m, ux)$ for all $u \in R$. Now

$$\tau_M(mu, x) = (F\varphi_{mu})x = F(\varphi_m \varphi_u)x,$$

by Lemma 5.44(ii). On the other hand,

$$\tau_M(m, ux) = (F\varphi_m)ux = (F\varphi_m)(F\varphi_u)x = (F\varphi_{mu})x.$$

Thus, τ_M induces a homomorphism $\sigma_M \colon M \otimes_R B \to FM$. We claim that $\sigma \colon \square \otimes_R B \to F$ is a natural transformation; that is, the following diagram commutes for R-maps $f \colon M \to N$.

$$
\begin{array}{ccc}
M \otimes_R B & \xrightarrow{\ \sigma_M\ } & FM \\
{\scriptstyle f\otimes 1}\downarrow & & \downarrow{\scriptstyle Ff} \\
N \otimes_R B & \xrightarrow[\ \sigma_N\ ]{} & FN
\end{array}
$$

Going clockwise, $m \otimes x \mapsto (F\varphi_m)x \mapsto (Ff)(F\varphi_m)x$; going counterclockwise, $m \otimes x \mapsto f(m) \otimes x \mapsto (F\varphi_{fm})x = F(f\varphi_m)x = (Ff)(F\varphi_m)x$, by Lemma 5.44(iii).

Now $\sigma_R \colon R \otimes_R B \to FR$ is an isomorphism (because $B = FR$); moreover, since both $\square \otimes_R B$ and F preserve direct sums, $\sigma_A \colon A \otimes_R B \to FA$ is an isomorphism for every free right R-module A. Let M be any right R-module. There are a free right R-module A and a short exact sequence

$$0 \to K \xrightarrow{\ i\ } A \to M \to 0;$$

there is also a surjection $f \colon C \to K$ for some free right R-module C. Splicing these together, there is an exact sequence

$$C \xrightarrow{\ if\ } A \to M \to 0.$$

Now the following commutative diagram has exact rows, for both $\square \otimes_R B$ and F are right exact.

$$
\begin{array}{ccccccc}
C \otimes_R B & \longrightarrow & A \otimes_R B & \longrightarrow & M \otimes_R B & \longrightarrow & 0 \\
{\scriptstyle \sigma_C}\downarrow & & \downarrow{\scriptstyle \sigma_A} & & \downarrow{\scriptstyle \sigma_M} & & \\
FC & \longrightarrow & FA & \longrightarrow & FM & \longrightarrow & 0
\end{array}
$$

Since σ_C and σ_A are isomorphisms, the Five Lemma shows that σ_M is an isomorphism. Therefore, σ is a natural isomorphism. $\bullet$

Remark.   If, in Theorem 5.45, F takes values in $\mathbf{Mod}_S$ instead of in $\mathbf{Ab}$, then the first paragraph of the proof can be modified to prove that the right S-module FR may be construed as an (R, S)-bimodule; thus, the theorem remains true if $\mathbf{Ab}$ is replaced by $\mathbf{Mod}_S$.   ◄

Example 5.46. If R is a commutative ring and $r \in R$, then there is a functor $F: {}_R\mathbf{Mod} \to {}_R\mathbf{Mod}$ that takes an R-module M to M/rM [if $\varphi: M \to N$ is an R-map, define $F\varphi: M/rM \to N/rN$ by $m + rM \mapsto \varphi(m) + rN$]. The reader may check that F is a right exact functor preserving direct sums, and so it follows from Watts' Theorem that F is naturally isomorphic to $\square \otimes_R (R/rR)$, for $FR = R/rR$. This generalizes Proposition 2.68.   ◄

Corollary 5.47. *Let R be a right noetherian ring, and let $\mathcal{F}_R$ be the category of all finitely generated right R-modules. If $F: \mathcal{F}_R \to \mathbf{Mod}_S$ is a right exact additive functor, then F is naturally isomorphic to $\square \otimes_R B$, where B is $F(R)$ made into a left R-module.*

Proof. The proof is almost the same as that of Theorem 5.45 coupled with the remark after it. Given a finitely generated right R-module M, we can choose a finitely generated free right R-module A mapping onto M. Moreover, since R is right noetherian, Proposition 3.18 shows that the kernel K of the surjection $A \to M$ is also finitely generated (if K were not finitely generated, then there would be no free right R-module in the category $\mathcal{F}_R$ mapping onto K). Finally, we need not assume that F preserves finite direct sums, for Corollary 2.21 shows that this follows from the additivity of F.   •

We now characterize contravariant Hom functors.

Theorem 5.48 (Watts). *If $H: {}_R\mathbf{Mod} \to \mathbf{Ab}$ is a contravariant left exact additive functor that converts direct sums to direct products, then H is naturally isomorphic to $\mathrm{Hom}_R(\square, B)$, where B is $H(R)$ made into a right R-module.*

Proof. We begin by making the abelian group HR into a right R-module. As in the beginning of the proof of Theorem 5.45, if M is a right R-module and $m \in M$, then the function $\varphi_m: R \to M$, defined by $r \mapsto mr$, is an R-map. In particular, if $M = R$ and $u \in R$, then $H\varphi_u: HR \to HR$, and Lemma 5.44(ii) gives $\varphi_{uv} = \varphi_u \varphi_v$ for all $u, v \in R$. If $x \in HR$, define

$$ux = (H\varphi_u)x.$$

Here, HR is a *right* R-module, for the contravariance of H gives

$$(uv)x = (H\varphi_{uv})x = H(\varphi_u \varphi_v)x = (H\varphi_v)(H\varphi_u)x = v(ux).$$

Define $\sigma_M: HM \to \mathrm{Hom}_R(M, B)$ by $\sigma_M(x): m \mapsto (H\varphi_m)x$, where $x \in HM$. It is easy to check that $\sigma: H \to \mathrm{Hom}_R(\square, B)$ is a natural transformation and that σ_R is an isomorphism. The remainder of the proof proceeds, mutatis mutandis, as that of Theorem 5.45.   •

We can characterize covariant Hom functors, but the proof is a bit more complicated.

Definition.   A left R-module C is called a *cogenerator* of $_R\mathbf{Mod}$ if, for every left R-module M and every nonzero $m \in M$, there exists an R-map $g: M \to C$ with $g(m) \neq 0$.

Exercise 3.19(i) on page 130 can be restated to say that $\mathbb{Q}/\mathbb{Z}$ is an injective cogenerator of **Ab**.

Lemma 5.49.   *There exists an injective cogenerator of $_R\mathbf{Mod}$.*

Proof.   Define C to be an injective left R-module containing $\bigoplus_I R/I$, where I varies over all the left ideals in R (the module C exists, by Theorem 3.38). If M is a left R-module and $m \in M$ is nonzero, then $\langle m \rangle \cong R/J$ for some left ideal J. Consider the diagram

$$
\begin{array}{ccc}
 & C & \\
{\scriptstyle f}\Big\uparrow & \nwarrow{\scriptstyle g} & \\
0 \longrightarrow \langle m \rangle & \xrightarrow{\;i\;} & M,
\end{array}
$$

where i is the inclusion and f is an isomorphism of $\langle m \rangle$ to some submodule of C isomorphic to R/J. Since C is injective, there is an R-map $g: M \to C$ extending f, and so $g(m) \neq 0$.   •

An analysis of the proof of Proposition 5.21 shows that it can be generalized by replacing $\mathrm{Hom}(A, \square)$ by any left exact functor that preserves direct products. However, this added generality is only illusory, in light of the following theorem of Watts characterizing representable functors on module categories.

Theorem 5.50 (Watts).   *If $G: {}_R\mathbf{Mod} \to \mathbf{Ab}$ is a covariant additive functor preserving inverse limits, then G is naturally isomorphic to $\mathrm{Hom}_R(B, \square)$ for some left R-module B.*

Proof.   For a module M and a set X, let M^X denote the direct product of copies of M indexed by X; more precisely, M^X is the set of all functions $X \to M$. In particular, $1_M \in M^M$, and we write $e = 1_M \in M^M$. If $m \in M$ and $\pi_m: M^M \to M$ is the mth projection, then the mth coordinate of e is $\pi_m(e) = m$.

Choose an injective cogenerator C of $_R\mathbf{Mod}$. Let $\Pi = C^{GC}$, and let its projection maps be $p_x: \Pi \to C$ for all $x \in GC$. Since G preserves inverse limits, it preserves direct products, and so $G\Pi$ is a direct product with projection maps Gp_x. More precisely, if $\pi_x: (GC)^{GC} \to GC$ are the projection maps, then there is a unique isomorphism θ making the following diagrams commute for all $x \in GC$:

$$G\Pi \xleftarrow{\quad \theta \quad} (GC)^{GC}$$

with maps Gp_x and π_x going to GC.

Thus, $(Gp_x)\theta = \pi_x$ for all $x \in GC$. Write

$$e = 1_{GC} \in (GC)^{GC}.$$

Define $\tau \colon \operatorname{Hom}_R(\Pi, C) \to GC$ by

$$\tau \colon f \mapsto (Gf)(\theta e).$$

If $f \colon \Pi \to C$, then $Gf \colon G\Pi \to GC$; since $\theta e \in G\Pi$, $\tau(f) = (Gf)(\theta e)$ makes sense.

The map τ is surjective, for if $x \in GC$, then $\tau(p_x) = (Gp_x)(\theta e) = \pi_x(e) = x$. We now describe $\ker \tau$. If $S \subseteq \Pi$, denote the inclusion $S \to \Pi$ by i_S. Define

$$B = \bigcap_{S \in \mathcal{S}} S, \quad \text{where } \mathcal{S} = \{\text{submodules } S \subseteq \Pi : \theta e \in \operatorname{im} G(i_S)\}.$$

We show that $\mathcal{S}$ is closed under finite intersections. All the maps in the first diagram below are inclusions, so that $i_S \lambda = i_{S \cap T}$. Since G preserves inverse limits, it preserves pullbacks; since the first diagram is a pullback, the second diagram is also a pullback.

$$\begin{array}{ccc} S \cap T \xrightarrow{\ \lambda\ } S & \qquad & G(S \cap T) \xrightarrow{\ G\lambda\ } GS \\ \mu \downarrow \quad\ \downarrow i_S & & G\mu \downarrow \qquad\quad \downarrow G(i_S) \\ T \xrightarrow{\ i_T\ } \Pi & & GT \xrightarrow{\ G(i_T)\ } G\Pi \end{array}$$

By the definition of $\mathcal{S}$, there are $u \in GS$ with $(Gi_S)u = \theta e$ and $v \in GT$ with $(Gi_T)v = \theta e$. By Exercise 5.24 on page 255, there is $d \in G(S \cap T)$ with $(Gi_S)(G\lambda)d = \theta e$. But $(Gi_S)(G\lambda) = Gi_{S \cap T}$, so that $\theta e \in \operatorname{im} G(i_{S \cap T})$ and $S \cap T \in \mathcal{S}$. It now follows from Example 5.18(ii) that $B = \bigcap S \cong \varprojlim S$, so that $B \in \mathcal{S}$.

Now G is left exact, so that exactness of $0 \to \ker f \xrightarrow{\ v\ } \Pi \xrightarrow{\ f\ } C$ gives exactness of $0 \to G(\ker f) \xrightarrow{\ Gv\ } G\Pi \xrightarrow{\ Gf\ } GC$. Thus, $\operatorname{im} Gv = \ker(Gf)$. If

$$j \colon B \to \Pi$$

is the inclusion, then $\ker \tau = \ker j^*$, where $j^* \colon f \mapsto fj$ is the induced map $j^* \colon \operatorname{Hom}_R(\Pi, C) \to \operatorname{Hom}_R(B, C)$: if $f \in \ker \tau$, then $(Gf)\theta e = 0$, and

$\theta e \in \ker Gf = \operatorname{im} Gv$; thus, $\ker f \in \mathcal{S}$. Hence, $B \subseteq \ker f$, $fj = 0$, $f \in \ker j^*$, and $\ker \tau \subseteq \ker j^*$. For the reverse inclusion, assume that $f \in \ker j^*$, so that $B \subseteq \ker f$. Then $\operatorname{im} Gj \subseteq \operatorname{im} Gv = \ker Gf$. But $\theta e \in \ker Gf$; that is, $(Gf)\theta e = 0$, and $f \in \ker \tau$. Therefore, $\ker j^* = \ker \tau$.

In the diagram

$$
\begin{array}{ccccccccc}
0 & \longrightarrow & \operatorname{Hom}_R(\Pi/B, C) & \longrightarrow & \operatorname{Hom}_R(\Pi, C) & \overset{j^*}{\longrightarrow} & \operatorname{Hom}_R(B, C) & \longrightarrow & 0 \\
& & \Big\| & & \Big\| & & \Big\downarrow{\sigma_C} & & \\
0 & \longrightarrow & \operatorname{Hom}_R(\Pi/B, C) & \longrightarrow & \operatorname{Hom}_R(\Pi, C) & \underset{\tau}{\longrightarrow} & GC & \longrightarrow & 0,
\end{array}
$$

the first two vertical arrows are identities, so that the diagram commutes. Exactness of $0 \to B \overset{j}{\longrightarrow} \Pi \to \Pi/B \to 0$ and injectivity of C give exactness of the top row, while the bottom row is exact because τ is surjective and $\ker \tau = \ker j^*$. It follows that the two cokernels are isomorphic: there is an isomorphism

$$\sigma_C \colon \operatorname{Hom}_R(B, C) \to GC,$$

given by $\sigma_C \colon f \mapsto (Gf)\theta e$ (for the fussy reader, this is Proposition 2.70).

For any module M, there is a map $M \to C^{\operatorname{Hom}_R(M,C)}$ given by $m \mapsto (fm)$, that "vector" whose fth coordinate is fm; this map is an injection because C is a cogenerator. Similarly, if $N = \operatorname{coker}(M \to C^{\operatorname{Hom}_R(M,C)})$, there is an injection $N \to C^Y$ for some set Y; splicing these together gives an exact sequence

$$
\begin{array}{ccccc}
0 \longrightarrow M \longrightarrow & C^{\operatorname{Hom}_R(M,C)} & \text{-- -} \!\!\!\! & \to & C^Y \\
& & \searrow & & \uparrow \\
& & & N. &
\end{array}
$$

Since both G and $\operatorname{Hom}_R(B, \square)$ are left exact, there is a commutative diagram with exact rows

$$
\begin{array}{ccccccc}
0 & \longrightarrow & \operatorname{Hom}_R(B, M) & \longrightarrow & \operatorname{Hom}_R(B, C^{\operatorname{Hom}_R(M,C)}) & \longrightarrow & \operatorname{Hom}_R(B, C^Y) \\
& & \Big\downarrow{\sigma_M} & & \Big\downarrow{\sigma_C{}^{\operatorname{Hom}_R(M,C)}} & & \Big\downarrow{\sigma_C{}^Y} \\
0 & \longrightarrow & GM & \longrightarrow & GC^{\operatorname{Hom}_R(M,C)} & \longrightarrow & GC^Y.
\end{array}
$$

The vertical maps $\sigma_C{}^{\operatorname{Hom}_R(M,C)}$ and $\sigma_C{}^Y$ are isomorphisms, so that Proposition 2.71 gives a unique isomorphism $\sigma_M \colon \operatorname{Hom}_R(B, M) \to GM$. It remains to prove that the isomorphisms σ_M constitute a natural transformation. Recall, for any set X, that $\operatorname{Hom}_R(B, C^X) \cong \operatorname{Hom}_R(B, C)^X$ via $f \mapsto (p_x f)$, where p_x is the xth projection. The map $\sigma_{C^X} \colon \operatorname{Hom}_R(B, C^X) \to GC^X$ is given by $f \mapsto ((Gp_x f)\theta e) = ((Gp_x f))\theta e = (Gf)\theta e$. Therefore, $\sigma_M \colon \operatorname{Hom}_R(B, M) \to GM$ is given by $f \mapsto (Gf)\theta e$, and Yoneda's Lemma, Theorem 1.17, shows that σ is a natural isomorphism.   ●

Remark. No easy description of the module B is known. However, we know that B is not $G(R)$. For example, if $G = \text{Hom}_{\mathbb{Z}}(\mathbb{Q}, \square)$, then Watts' Theorem applies to give $\text{Hom}_{\mathbb{Z}}(B, \square) \cong \text{Hom}_{\mathbb{Z}}(\mathbb{Q}, \square)$. Now Corollary 1.18(iii) says that $B \cong \mathbb{Q}$, but $B \not\cong G(\mathbb{Z}) = \text{Hom}_{\mathbb{Z}}(\mathbb{Q}, \mathbb{Z}) = \{0\}$. ◄

Theorem 5.51. *If* $F: \textbf{Mod}_R \to \textbf{Ab}$ *is an additive functor, then the following statements are equivalent.*

(i) *F preserves direct limits.*

(ii) *F is right exact and preserves direct sums.*

(iii) *$F \cong \square \otimes_R B$ for some left R-module B.*

(iv) *F has a right adjoint; there is a functor $G: \textbf{Ab} \to \textbf{Mod}_R$ so that (F, G) is an adjoint pair.*

Proof.

(i) $\Rightarrow$ (ii) Cokernels and direct sums are direct limits.

(ii) $\Rightarrow$ (iii) Theorem 5.45.

(iii) $\Rightarrow$ (iv) Take $G = \text{Hom}_R(B, \square)$ in the adjoint isomorphism theorem.

(iv) $\Rightarrow$ (i) Theorem 5.43. ●

Theorem 5.52. *If* $G: {}_R\textbf{Mod} \to \textbf{Ab}$ *is an additive functor, then the following statements are equivalent.*

(i) *G preserves inverse limits.*

(ii) *G is left exact and preserves direct products.*

(iii) *G is representable; i.e., $G \cong \text{Hom}_R(B, \square)$ for some left R-module B.*

(iv) *G has a left adjoint; there is a functor $F: \textbf{Ab} \to {}_R\textbf{Mod}$ so that (F, G) is an adjoint pair.*

Proof.

(i) $\Rightarrow$ (ii) Kernels and direct products are inverse limits.

(ii) $\Rightarrow$ (iii) Theorem 5.50.

(iii) $\Rightarrow$ (iv) Take $F = \square \otimes_R B$ in the adjoint isomorphism theorem.

(iv) $\Rightarrow$ (i) Exercise 5.28 on page 256. ●

The ***Adjoint Functor Theorem*** says that a functor G on an arbitrary category has a left adjoint [that is, there exists a functor F so that (F, G) is an adjoint pair] if and only if G preserves inverse limits and G satisfies a "solution set condition" [Mac Lane, *Categories for the Working Mathematician*, pp. 116–127 and 230]. One consequence is a proof of the existence of free objects when a forgetful functor has a left adjoint; see M. Barr, "The existence of free groups," *Amer. Math. Monthly*, 79 (1972), 364–367. The Adjoint Functor Theorem also says that F has a right adjoint if and only if F preserves all direct limits and satisfies a solution set condition. Theorems 5.51 and 5.52 are special cases of the Adjoint Functor Theorem.

It can be proved that adjoints are unique if they exist: if (F, G) and (F, G') are adjoint pairs, where $F: \mathcal{A} \to \mathcal{B}$ and $G, G': \mathcal{B} \to \mathcal{A}$, then $G \cong G'$; similarly, if (F, G) and (F', G) are adjoint pairs, then $F \cong F'$ (Mac Lane, *Categories for the Working Mathematician*, p. 83, or May, *Simplicial Objects in Algebraic Topology*, p. 61). Here is the special case for module categories.

Proposition 5.53.   *Let* $F: {}_R\mathbf{Mod} \to \mathbf{Ab}$ *and* $G, G': \mathbf{Ab} \to {}_R\mathbf{Mod}$ *be functors. If* (F, G) *and* (F, G') *are adjoint pairs, then* $G \cong G'$.

Proof.   For every left R-module C, there are natural isomorphisms

$$\operatorname{Hom}_R(C, G\square) \cong \operatorname{Hom}_{\mathbb{Z}}(FC, \square) \cong \operatorname{Hom}_R(C, G'\square).$$

Thus, $\operatorname{Hom}_R(C, \square) \circ G \cong \operatorname{Hom}_R(C, \square) \circ G'$ for every left R-module C. In particular, if $C = R$, then $\operatorname{Hom}_R(R, \square) \cong 1$, the identity functor on ${}_R\mathbf{Mod}$, and so $G \cong G'$. •

Remark.   In Functional Analysis, one works with topological vector spaces; moreover, there are many different topologies imposed on vector spaces, depending on the sort of problem being considered. We know that if A, B, C are modules, then the Adjoint Isomorphism, Theorem 2.75, gives a natural isomorphism

$$\operatorname{Hom}(A \otimes B, C) \cong \operatorname{Hom}(A, \operatorname{Hom}(B, C)).$$

Thus, $\square \otimes B$ is the left adjoint of $\operatorname{Hom}(B, \square)$. In the category of topological vector spaces, Grothendieck defined ***topological tensor products*** as left adjoints of $\operatorname{Hom}(B, \square)$. Since the Hom sets consist of *continuous* linear transformations, they depend on the topology, and so topological tensor products also depend on the topology. ◄

The Wedderburn–Artin theorems can be better understood in the context of determining those abstract categories that are isomorphic to module categories.

Definition.   A module P is *small* if the covariant Hom functor $\text{Hom}(P, \Box)$ preserves (possibly infinite) direct sums.

In more detail, if P is small and $B = \bigoplus_{i \in I} B_i$ has injections $\lambda_i : B_i \to B$, then $\text{Hom}\left(P, \bigoplus_{i \in I} B_i\right) = \bigoplus_{i \in I} \text{Hom}(P, B_i)$ has as injections the induced maps $(\lambda_i)_* : \text{Hom}(P, B_i) \to \text{Hom}(P, B)$.

Example 5.54.

 (i) Any finite direct sum of small modules is small, and any direct summand of a small module is small.

 (ii) Since every ring R is a small R-module, by Exercise 2.13 on page 66, it follows from part (i) that every finitely generated projective R-module is small.   ◀

Definition.   A right R-module P is a *generator* of $\textbf{Mod}_R$ if every right R-module M is a quotient of a direct sum of copies of P.

It is clear that R is a generator of $\textbf{Mod}_R$, as is any free right R-module. However, a projective right R-module may not be a generator. For example, if $R = \mathbb{I}_6$, then $R = P \oplus Q$, where $P \cong \mathbb{I}_3$. The projective module P is not a generator, for $Q \cong \mathbb{I}_2$ is not a quotient of a direct sum of copies of P.

Recall that a functor $F : \mathcal{C} \to \mathcal{D}$ is an *isomorphism* if there is a functor $G : \mathcal{D} \to \mathcal{C}$ such that the composites GF and FG are naturally isomorphic to the identity functors $1_\mathcal{C}$ and $1_\mathcal{D}$, respectively.

Theorem 5.55 (Morita).   *Let R be a ring and let P be a small projective generator of* $\textbf{Mod}_R$. *If $S = \text{End}_R(P)$, then there is an isomorphism*

$$F : \textbf{Mod}_S \to \textbf{Mod}_R$$

given by $M \mapsto M \otimes_S P$.

Proof.   Notice that P is a left S-module, for if $x \in P$ and $f, g \in S = \text{End}_R(P)$, then $(g \circ f)x = g(fx)$. In fact, P is an (S, R)-bimodule, for associativity $f(xr) = (fx)r$, where $r \in R$, is just the statement that f is an R-map. It now follows from Corollary 2.53 that the functor $F : \textbf{Mod}_S \to \textbf{Ab}$, defined by $F = \Box \otimes_S P$, actually takes values in $\textbf{Mod}_R$. Proposition 2.4 shows that the functor $G : \text{Hom}_R(P, \Box) : \textbf{Mod}_R \to \textbf{Ab}$ actually takes values in $\textbf{Mod}_S$. As (F, G) is an adjoint pair, Exercise 5.30 on page 271 gives natural transformations $FG \to 1_R$ and $1_S \to GF$, where 1_R and 1_S denote identity functors on the categories $\textbf{Mod}_R$ and $\textbf{Mod}_S$, respectively. It suffices to prove that each of these natural transformations is a natural isomorphism.

Since P is a projective right R-module, the functor $G = \operatorname{Hom}_R(P, \square)$ is exact; since P is small, G preserves direct sums. Now $F = \square \otimes_S P$, as any tensor product functor, is right exact and preserve sums. Therefore, both composites GF and FG preserve direct sums and are right exact.

Note that

$$FG(P) = F(\operatorname{Hom}_R(P, P)) = F(S) = S \otimes_S P \cong P.$$

Since P is a generator of $\mathbf{Mod}_R$, every right R-module M is a quotient of some direct sum of copies of P. There is an exact sequence $K \to \bigoplus P \xrightarrow{f} M \to 0$, where $K = \ker f$. There is also some direct sum of copies of P mapping onto K, and so there is an exact sequence

$$\bigoplus P \to \bigoplus P \to M \to 0.$$

Hence, there is a commutative diagram (by naturality of the upward maps) with exact rows

$$
\begin{array}{ccccccc}
\bigoplus P & \longrightarrow & \bigoplus P & \longrightarrow & M & \longrightarrow & 0 \\
\uparrow & & \uparrow & & \uparrow & & \\
\bigoplus FG(P) & \longrightarrow & \bigoplus FG(P) & \longrightarrow & FG(M) & \longrightarrow & 0.
\end{array}
$$

We know that the first two vertical maps are isomorphisms, and so the Five Lemma gives the other vertical map an isomorphism (just extend both rows to the right by adjoining $\to 0$, and insert two vertical arrows $0 \to 0$). Thus, $FG(M) \cong M$, and so $1_R \cong FG$.

For the other composite, note that

$$GF(S) = G(S \otimes_S P) \cong G(P) = \operatorname{Hom}_R(P, P) = S.$$

If N is any left S-module, there is an exact sequence of the form

$$\bigoplus S \to \bigoplus S \to N \to 0,$$

because every module is a quotient of a free module. The argument now concludes as that just done.   •

Corollary 5.56.

(i) *If R is a ring and $n \geq 1$, there is an isomorphism of categories*

$$\mathbf{Mod}_R \cong \mathbf{Mod}_{\operatorname{Mat}_n(R)}.$$

(ii) *If R is a semisimple ring and $n \geq 1$, then $\operatorname{Mat}_n(R)$ is semisimple.*

Proof. For any integer $n \geq 1$, the free module $P = \bigoplus_{i=1}^{n} R_i$, where $R_i \cong R$, is a small projective generator of $\mathbf{Mod}_R$, and $S = \mathrm{End}_R(P) \cong \mathrm{Mat}_n(R)$. The isomorphism $F \colon \mathbf{Mod}_R \to \mathbf{Mod}_{\mathrm{Mat}_n(R)}$ in Morita's Theorem carries $M \mapsto M \otimes_S P \cong \bigoplus_i M_i$, where $M_i \cong M$ for all i. Hence, if M is a projective right R-module, then $F(M)$ is also projective. But every module in $\mathbf{Mod}_{\mathrm{Mat}_n(R)}$ is projective, by Proposition 4.5 (a ring R is semisimple if and only if every R-module is projective). Therefore, $\mathrm{Mat}_n(R)$ is semisimple, •

There is a lovely part of ring theory, **Morita theory** (after K. Morita), developing these ideas. A category C is isomorphic to a module category if and only if it is an *abelian category* (see Section 5.5) containing a small projective generator P, and which is closed under infinite coproducts (see Mitchell, *Theory of Categories*, p. 104, or Pareigis, *Categories and Functors*, p. 211). Given this hypothesis, then $C \cong \mathbf{Mod}_S$, where $S = \mathrm{End}(P)$ (the proof is essentially that given for Theorem 5.55). Two rings R and S are called **Morita equivalent** if $\mathbf{Mod}_R \cong \mathbf{Mod}_S$. If R and S are Morita equivalent, then $Z(R) \cong Z(S)$; that is, they have isomorphic centers (the proof actually identifies all the possible isomorphisms between the categories). In particular, two *commutative* rings are Morita equivalent if and only if they are isomorphic. See Jacobson, *Basic Algebra* II, pp. 177–184, Lam, *Lectures on Modules and Rings*, Chapters 18 and 19, or Reiner, *Maximal Orders*, Chapter 4.

Exercises

5.29 Give an example of an additive functor $H \colon \mathbf{Ab} \to \mathbf{Ab}$ that has neither a left nor a right adjoint.

*5.30 Let (F, G) be an adjoint pair, where $F \colon C \to D$ and $G \colon D \to C$, and let $\tau_{C,D} \colon \mathrm{Hom}(FC, D) \to \mathrm{Hom}(C, GC)$ be the natural bijection.

    (i) If $D = FC$, there is a natural bijection

$$\tau_{C,FC} \colon \mathrm{Hom}(FC, FC) \to \mathrm{Hom}(C, GFC)$$

    with $\tau(1_{FC}) = \eta_C \colon C \to GFC$. Prove that $\eta \colon 1_C \to GF$ is a natural transformation.

    (ii) If $C = GD$, there is a natural bijection

$$\tau_{GD,D}^{-1} \colon \mathrm{Hom}(GD, GD) \to \mathrm{Hom}(FGD, D)$$

    with $\tau^{-1}(1_D) = \varepsilon_D \colon FGD \to D$. Prove that $\varepsilon \colon FG \to 1_D$ is a natural transformation. (We call ε the **counit** of the adjoint pair.)

5.31 Let (F, G) be an adjoint pair of functors between module categories. Prove that if G is exact, then F preserves projectives; that is, if P is a projective module, then FP is projective. Dually, prove that if F is exact, then G preserves injectives.

5.32 (i) Let $F: \mathbf{Groups} \to \mathbf{Ab}$ be the functor with $F(G) = G/G'$, where G' is the commutator subgroup of a group G, and let $U: \mathbf{Ab} \to \mathbf{Groups}$ be the functor taking every abelian group A into itself (that is, UA regards A as a not necessarily abelian group). Prove that (F, U) is an adjoint pair of functors.

(ii) Prove that the unit of the adjoint pair (F, U) is the natural map $G \to G/G'$.

*5.33 (i) If I is a partially ordered set, let $\mathbf{Dir}(I, {}_R\mathbf{Mod})$ denote all direct systems of left R-modules over I. Prove that $\mathbf{Dir}(I, {}_R\mathbf{Mod})$ is a category and that $\varinjlim: \mathbf{Dir}(I, {}_R\mathbf{Mod}) \to {}_R\mathbf{Mod}$ is a functor.

(ii) In Example 1.19(ii), we saw that constant functors define a functor $|\square|: \mathcal{C} \to \mathcal{C}^{\mathcal{D}}$; to each object C in $\mathcal{C}$ assign the constant functor $|C|$, and to each morphism $\varphi: C \to C'$ in $\mathcal{C}$, assign the natural transformation $|\varphi|: |C| \to |C'|$ defined by $|\varphi|_D = \varphi$. If $\mathcal{C}$ is cocomplete, prove that $(\varinjlim, |\square|)$ is an adjoint pair, and conclude that $\varinjlim$ preserves direct limits.

(iii) Let I be a partially ordered set and let $\mathbf{Inv}(I, {}_R\mathbf{Mod})$ denote the class of all inverse systems, together with their morphisms, of left R-modules over I. Prove that $\mathbf{Inv}(I, {}_R\mathbf{Mod})$ is a category and that $\varprojlim: \mathbf{Inv}(I, {}_R\mathbf{Mod}) \to {}_R\mathbf{Mod}$ is a functor.

(iv) Prove that if $\mathcal{C}$ is complete, then $(|\square|, \varprojlim)$ is an adjoint pair and $\varprojlim$ preserves inverse limits.

5.34 (i) If $A_1 \subseteq A_2 \subseteq A_3 \subseteq \cdots$ is an ascending sequence of submodules of a module A, prove that $A/\bigcup A_i \cong \bigcup A/A_i$; that is, $\mathrm{coker}(\varinjlim A_i \subseteq A) \cong \varinjlim \mathrm{coker}(A_i \to A)$.

(ii) Generalize part (i): prove that any two direct limits (perhaps with distinct index sets) commute.

(iii) Prove that any two inverse limits (perhaps with distinct index sets) commute.

(iv) Give an example in which direct limit and inverse limit do not commute.

5.35 (i) Define ACC in ${}_R\mathbf{Mod}$, and prove that if ${}_S\mathbf{Mod} \cong {}_R\mathbf{Mod}$, then ${}_S\mathbf{Mod}$ has ACC. Conclude that if R is left noetherian, then S is left noetherian.

     (ii) Give an example showing that $_R\textbf{Mod}$ and $\textbf{Mod}_R$ are not isomorphic.

5.36   **(i)** Recall that a *cogenerator* of a category C is an object C such that $\text{Hom}(\square, C)\colon C \to \textbf{Sets}$ is a faithful functor; that is, if $f, g\colon A \to B$ are distinct morphisms in C, then there exists a morphism $h\colon B \to C$ with $hf \neq hg$. Prove, when $C = _R\textbf{Mod}$, that this definition coincides with the definition of cogenerator on page 264.

     (ii) A *generator* of a category C is an object G such that $\text{Hom}(G, \square)\colon C \to \textbf{Sets}$ is a faithful functor; that is, if $f, g\colon A \to B$ are distinct morphisms in C, then there exists a morphism $h\colon G \to A$ with $fh \neq gh$. Prove, when $C = _R\textbf{Mod}$, that this definition coincides with the definition of cogenerator on page 269.

5.37 We call a functor $F\colon \mathcal{A} \to \mathcal{B}$ a *strong isomorphism* if there exists a functor $G\colon \mathcal{B} \to \mathcal{A}$ with $GF = 1_{\mathcal{A}}$ and $FG = 1_{\mathcal{B}}$. If R is a ring, show that $\text{Hom}_R(R, \square)\colon {}_R\textbf{Mod} \to {}_R\textbf{Mod}$ (which is naturally isomorphic to $1_{R\textbf{Mod}}$, by Exercise 2.13 on page 66) is not a strong isomorphism. Conclude that strong isomorphism is not an interesting idea.

5.4 Sheaves

At the beginning of his book, *The Theory of Sheaves*, Swan asks, "What are sheaves good for? The obvious answer is that sheaves are very useful in proving theorems." He then lists interesting applications of sheaves to Topology, Complex Variables, and Algebraic Geometry, and he concludes, "the importance of the theory of sheaves is simply that it gives relations (quite strong relations, in fact) between the local and global properties of a space." We proceed to the definition of sheaves.

Definition. A continuous map $p\colon E \to X$ between topological spaces E and X is called a *local homeomorphism* if, for each $e \in E$, there is an open neighborhood S of e, called a *sheet*, with $p(S)$ open in X and $p|S\colon S \to p(S)$ a homeomorphism. The triple (E, p, X) is called a *protosheaf*[5] if the local homeomorphism p is surjective.

---

[5] This term, with a slightly different meaning, was used by Swan in *The Theory of Sheaves*. Since *protosheaf* has not been widely adopted (Swan's book appeared in 1964), our usage should not cause any confusion.

Each of the ingredients of a protosheaf has a name. The space E is called the *sheaf space*, p is the *projection*, and X is the *base space*. For each $x \in X$, the fiber $p^{-1}(x)$ is denoted by E_x and is called the *stalk* over x.

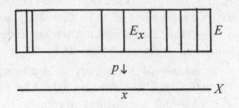

Fig. 5.1   Protosheaf.

Here are several examples of protosheaves.

Example 5.57.

(i) If X is a topological space and Y is a discrete space, define $E = X \times Y$, and define $p \colon E \to X$ by $p \colon (x, y) \mapsto x$. If $e = (x, y) \in E$ and V is an open neighborhood of x, then $S = \{(v, y) : v \in V\}$ is an open neighborhood of e (because $\{y\}$ is open in Y) and $p|S \colon S \to V = p(S)$ is a homeomorphism. The triple (E, p, X) is called a *constant protosheaf*.

(ii) The triple $(\mathbb{R}, p, S^1)$ is a protosheaf, where $p(x) = e^{2\pi i x}$.

(iii) A *covering space* is a triple (E, p, X) in which each $x \in X$ has an open neighborhood V such that $p^{-1}(V) = \bigcup_i S_i$, a disjoint union of open subsets of E with $p|S_i \colon S_i \to V$ a homeomorphism for each i. Every covering space is a protosheaf.

(iv) If G is a topological group and H is a discrete normal subgroup of G, then $(G, p, G/H)$ is a covering space, where p is the natural map.

(v) The protosheaf in part (ii) (which is actually a covering space) gives rise to an example showing that the converse is false. Let $E = (0, 3) \subseteq \mathbb{R}$ and let $p' = p|E$, where $p \colon \mathbb{R} \to S^1$ is the projection in part (ii). The map p' is a local homeomorphism, being a restriction of such, but (E, p', S^1) is not a covering space because there is no open neighborhood V of $(1, 0) \in S^1$ with $p^{-1}(V)$ a disjoint union $\bigcup_i S_i$. ◄

Proposition 5.58. *Let (E, p, X) be a protosheaf.*

(i) *The sheets form a base of open sets for E.*

(ii) *p is an open map.*

(iii) *Each stalk E_x is discrete.*

(iv) *Let $(U_i)_{i \in I}$ be a family of open subsets of X, and let $U = \bigcup_{i \in I} U_i$. If $f, g \colon U \to Y$ for some space Y and $f|U_i = g|U_i$ for all $i \in I$, then $f = g$.*

(v) *Let $(U_i)_{i \in I}$ be a family of open subsets of X and, for $(i, j) \in I \times I$, define $U_{(i,j)} = U_i \cap U_j$. If $(f_i \colon U_i \to Y)_{i \in I}$ are continuous maps satisfying $f_i|U_{(i,j)} = f_j|U_{(i,j)}$ for all $(i, j) \in I \times I$, then there exists a unique continuous $f \colon U \to Y$ with $f|U_i = f_i$ for all $i \in I$.*

Proof.

(i) Since, for each $e \in E$, there is a sheet S containing e, the sheaf space E is the union of all the sheets: $E = \bigcup_S S$. If $U \subseteq E$ is open, then $U \cap S$ is open for every sheet S, and so $U = \bigcup_S (U \cap S)$. But every open subset of a sheet is also a sheet, and so U is a union of sheets; that is, the sheets comprise a base for the topology of E.

(ii) If $U \subseteq E$ is open, then $p(U) = \bigcup_S p(U \cap S)$. But $p(U \cap S)$ is open in X, because p is a local homeomorphism; thus, $p(U)$ is open, for it is a union of open sets.

(iii) Let $e \in E_x$, and let S be a sheet containing e. If $e' \in E_x$ and $e' \neq e$, then $e' \notin S$, for $p|S$ is injective and $p(e') = x = p(e)$. Therefore, $S \cap E_x = \{e\}$, and so E_x is discrete.

(iv) If $x \in U$, then $x \in U_i$ for some i, and $f(x) = (f|U_i)x = (g|U_i)x = g(x)$. Hence, $f = g$.

(v) This is proved in Example 5.15. •

There are two equivalent versions of *sheaf*: the first, defined as a special kind of protosheaf, we call an *etale-sheaf* (the French term for *sheaf* space is *espace étalé*); the second, defined as a special kind of presheaf, we call a *sheaf* (recall Example 1.14: if $\mathcal{C}$ is a category and $\mathcal{U}$ is the topology of a topological space X viewed as a category, then a *presheaf on X* is a contravariant functor $\mathcal{P} \colon \mathcal{U} \to \mathcal{C}$).

Definition.   If $p: E \to X$ is continuous, where X and E are topological spaces, then $S = (E, p, X)$ is an ***etale-sheaf of abelian groups*** if

(i) (E, p, X) is a protosheaf,

(ii) the stalk E_x is an abelian group for each $x \in X$,

(iii) inversion and addition are continuous.

The meaning of continuity of inversion $e \mapsto -e$ is clear, but we elaborate on the definition of continuity of addition. Define

$$E + E = \bigcup_{x \in X} (E_x \times E_x) = \{(e, e') \in E \times E : p(e) = p(e')\}.$$

Addition $\alpha: E + E \to E$ is given by $\alpha: (e, e') \mapsto e + e'$, and continuity means, of course, that for every open neighborhood V of $e + e'$ in E, there exists an open neighborhood U of (e, e') in $E + E$ with $\alpha(U) \subseteq V$. Since $E \times E$ has the product topology and $E + E \subseteq E \times E$, there are open neighborhoods $H \subseteq E$ of e and $K \subseteq E$ of e', with $\alpha\big((H \times K) \cap (E + E)\big) \subseteq V$. If we define $H + K = \{(h, k) \in H \times K : p(h) = p(k)\}$, then $\alpha(H + K) \subseteq V$.

The definition of etale-sheaf can be modified so that its stalks lie in algebraic categories other than **Ab**, such as $_R\text{Mod}$ or **ComRings**. For example, the *structure sheaf* of a commutative ring R has base space $\text{Spec}(R)$ with the Zariski topology, sheaf space $E = \bigcup_{\mathfrak{p} \in \text{Spec}(R)} R_\mathfrak{p}$ suitably topologized, and projection $p: E \to \text{Spec}(R)$ defined by $p(e) = \mathfrak{p}$ for all $e \in R_\mathfrak{p}$ (see Example 5.95). Of course, axiom (iii) is modified so that all the algebraic operations are continuous. Even though the results in this section hold in more generality, we assume throughout that stalks are merely abelian groups.

Definition.   Let $S = (E, p, X)$ and $S' = (E', p', X)$ be etale-sheaves over a space X. An ***etale-map*** $\varphi: S \to S'$ is a continuous map $\varphi: E \to E'$ such that $p'\varphi = p$ (so that $\varphi|E_x: E_x \to E'_x$ for all $x \in X$), and each $\varphi|E_x$ is a homomorphism. We write $\text{Hom}_{et}(S, S')$ for the set of all etale-maps.

It is easy to check that all etale-sheaves of abelian groups over a topological space X form a category, which we denote by

$$\mathbf{Sh}_{et}(X, \mathbf{Ab}).$$

Proposition 5.59.   *Let $S = (E, p, X)$ and $S' = (E', p', X)$ be etale-sheaves over a topological space X.*

(i) $\text{Hom}_{et}(S, S')$ *is an additive abelian group.*

(ii) *The distributive laws hold: given etale-maps*

$$\mathcal{X} \xrightarrow{\alpha} \mathcal{S} \underset{\psi}{\overset{\varphi}{\rightrightarrows}} \mathcal{S}' \xrightarrow{\beta} \mathcal{Y},$$

where $\mathcal{X}$ *and* $\mathcal{Y}$ *are etale-sheaves over* X, *then*

$$\beta(\varphi + \psi) = \beta\varphi + \beta\psi \quad and \quad (\varphi + \psi)\alpha = \varphi\alpha + \psi\alpha.$$

(iii) *Every etale-map* $\varphi \colon \mathcal{S} \to \mathcal{S}'$ *is an open map* $E \to E'$.

Proof. Define $\varphi + \psi \colon E \to E'$ by $\varphi + \psi \colon e \mapsto \varphi(e) + \psi(e)$. Verification of the first two statements is routine. The third follows from Proposition 5.58(i), which says that the sheets form a base of open sets for E. •

It will be simpler to give examples of etale-sheaves once we see the (equivalent) definition of sheaf in terms of presheaves, and so we merely describe an example without verifying that all the particulars in the definition of etale-sheaf actually hold.

Example 5.60.

(i) Let X be a topological space, and let A be an abelian group equipped with the discrete topology. Define the ***constant etale-sheaf at*** A to be $(X \times A, p, X)$, where $X \times A$ has the product topology and $p \colon (x, a) \mapsto x$ is the projection. In particular, if $A = \{0\}$, then the constant sheaf at A is called the ***zero sheaf***.

(ii) The protosheaf $(\mathbb{R}, p, S^1)$, where $p \colon \mathbb{R} \to S^1$ is the local homeomorphism given by $x \mapsto e^{2\pi i x}$, is *not* an etale-sheaf of abelian groups (for its stalks are not abelian groups). ◄

Definition. Let $\mathcal{S}' = (E', p', X)$ and $\mathcal{S} = (E, p, X)$ be etale-sheaves. Then $\mathcal{S}'$ is an ***subetale-sheaf*** of $\mathcal{S}$ if $E' \subseteq E$ and the inclusion $\iota \colon E' \to E$ is an etale-map.

By Proposition 5.59, if (E', p', X) is a subetale-sheaf of (E, p, X), then E' is an open subset of E. The reader may prove the converse: if E' is an open subset of E and $p' = p|E'$, then (E', p', X) is a subetale-sheaf of (E, p, X).

Even though the next proposition is obvious, we state it explicitly.

Proposition 5.61. *Two subetale-sheaves* (E, p, X) *and* (E', p', X) *of an etale-sheaf* $\mathcal{S}$ *are equal if and only if they have the same stalks; that is,* $E_x = E'_x$ *for all* $x \in X$.

Proof. This is true because $E = \bigcup_{x \in X} E_x$. •

We now introduce the *sections* of an etale-sheaf.

Definition. If $S = (E, p, X)$ is an etale-sheaf of abelian groups and $U \subseteq X$ is an open set, then a *section over* U is a continuous map $\sigma : U \to E$ such that $p\sigma = 1_U$; call σ a *global section* if $U = X$. Define $\Gamma(\varnothing, S) = \{0\}$ and, if $U \neq \varnothing$, define

$$\Gamma(U, S) = \{\text{sections } \sigma : U \to E\}.$$

Sections $\Gamma(U, S)$ may be viewed as describing local properties of a base space X, while $\Gamma(X, S)$ describes the corresponding global properties.

Proposition 5.62. *Let $S = (E, p, X)$ be an etale-sheaf of abelian groups, and let $\mathcal{F} = \Gamma(\square, S)$.*

(i) *$\mathcal{F}(U)$ is an abelian group for each open $U \subseteq X$.*

(ii) *$\mathcal{F} = \Gamma(\square, S)$ is a presheaf of abelian groups on X, called the **sheaf of sections** of S.*

(iii) *The function $z : X \to E$, defined by $z(x) = 0_x \in E_x$ (called the **zero section**), is a global section.*

Proof.

(i) Let us show that $\mathcal{F}(U) \neq \varnothing$ for every open set $U \subseteq X$. If $U = \varnothing$, then $\mathcal{F}(U) = \{0\}$, by definition. If $x \in U$, choose $e \in E_x$ and a sheet S containing e. Since p is an open map, $p(S) \cap U$ is an open neighborhood of x. Now $(p|S)^{-1} : p(S) \to S \subseteq E$ is a section; define σ_S to be its restriction to $p(S) \cap U$. The family of all such $p(S) \cap U$ is an open cover of U; since the maps σ_S agree on overlaps, Proposition 5.58(v) shows that they may be glued together to give a section in $\mathcal{F}(U)$.

If $\sigma, \tau \in \mathcal{F}(U)$, then $(\sigma, \tau) : x \mapsto (\sigma x, \tau x)$ is a continuous map $U \to E + E$; composing with the continuous map $(\sigma x, \tau x) \mapsto \sigma x + \tau x$ shows that $\sigma + \tau : x \mapsto \sigma x + \tau x \in \mathcal{F}(U)$. That $\mathcal{F}(U)$ is an abelian group now follows from inversion $E \to E$ being continuous, for $\sigma \in \mathcal{F}(U)$ implies $-\sigma \in \mathcal{F}(U)$.

(ii) If $U \subseteq V$ are open sets, then the restriction $\sigma \to \sigma|U$ is the required group homomorphism $\mathcal{F}(V) \to \mathcal{F}(U)$.

(iii) If $U = X$, then z is the identity element of the group $\mathcal{F}(X)$. •

If S is an etale-sheaf, then the presheaf $\mathcal{F} = \Gamma(\square, S)$ satisfies a special property not shared by arbitrary presheaves.

Proposition 5.63. *Let $\mathcal{S} = (E, p, X)$ be an etale-sheaf with sheaf of sections $\Gamma(\square, \mathcal{S})$, let U be an open set and let $(U_i)_{i \in I}$ be an open cover of it: $U = \bigcup_{i \in I} U_i$.*

(i) *If $\sigma, \tau \in \Gamma(U, \mathcal{S})$ and $\sigma|U_i = \tau|U_i$ for all $i \in I$, then $\sigma = \tau$.*

(ii) *If $(\sigma_i \in \Gamma(U_i, \mathcal{S}))_{i \in I}$ satisfies $\sigma_i|(U_i \cap U_j) = \sigma_j|(U_i \cap U_j)$ for all $(i, j) \in I \times I$, then there exists a unique $\sigma \in \Gamma(U, \mathcal{S})$ with $\sigma|U_i = \sigma_i$ for all $i \in I$.*

Proof. Propositions 5.58(iv) and 5.58(v). ●

Definition. A presheaf $\{\mathcal{F}, \rho_U^V\}$ of abelian groups on a space X satisfies the *equalizer conditon* if

(i) (**Uniqueness**) for every open set U and open cover $U = \bigcup_{i \in I} U_i$, if $\sigma, \tau \in \mathcal{F}(U)$ satisfy $\sigma|U_i = \tau|U_i$ for all $i \in I$, then $\sigma = \tau$ [we have written $\sigma|U_i$ instead of $\rho_{U_i}^U(\sigma)$],

(ii) (**Gluing**) for every open set U and open cover $U = \bigcup_{i \in I} U_i$, if $\sigma_i \in \mathcal{F}(U_i)$ satisfy $\sigma_i|(U_i \cap U_j) = \sigma_j|(U_i \cap U_j)$ for all i, j, then there exists a unique $\sigma \in \mathcal{F}(U)$ with $\sigma|U_i = \sigma_i$ for all $i \in I$.

Proposition 5.63 shows that the sheaf of sections of an etale-sheaf satisfies the equalizer condition, but there are presheaves that do not satisfy it.

Example 5.64. Let $X = \mathbb{R}^2$ and, for each open $U \subseteq \mathbb{R}^2$, define

$$\mathcal{P}(U) = \{f : U \to \mathbb{R} \mid f \text{ is constant}\};$$

if $U \subseteq V$, define $\rho_U^V : \mathcal{P}(V) \to \mathcal{P}(U)$ to be the restriction map $\sigma \mapsto \sigma|U$. It is easy to check that $\mathcal{P}$ is a presheaf of abelian groups over $\mathbb{R}^2$, but $\mathcal{P}$ does not satisfy the equalizer condition. For example, let $U = U_1 \cup U_2$, where U_1, U_2 are disjoint nonempty open sets. Define $\sigma_1 \in \mathcal{P}(U_1)$ by $\sigma_1(u_1) = 0$ for all $u_1 \in U_1$, and define $\sigma_2 \in \mathcal{P}(U_2)$ by $\sigma_2(u_2) = 5$ for all $u_2 \in U_2$. The overlap condition here is vacuous, because $U_1 \cap U_2 = \varnothing$, but there is no *constant* function $\sigma \in \mathcal{P}(U)$ with $\sigma|U_i = \sigma_i$ for $i = 1, 2$. ◄

The equalizer condition can be restated in a more categorical way; see Example 5.15.

Corollary 5.65. *Let $\mathcal{S}$ be an etale-sheaf with sheaf of sections $\mathcal{F} = \Gamma(\square, \mathcal{S})$. Given a family $(U_i)_{i \in I}$ of open subsets of X, write $U = \bigcup_{i \in I} U_i$ and $U_{(i,j)} = U_i \cap U_j$ for $i, j \in I$. Then there is an exact sequence*

$$0 \to \mathcal{F}(U) \xrightarrow{\alpha} \prod_{i \in I} \mathcal{F}(U_i) \xrightarrow{\beta} \prod_{(i,j) \in I \times I} \mathcal{F}(U_{(i,j)});$$

if $\sigma \in \mathcal{F}(U)$, *the ith coordinate of* $\alpha(\sigma)$ *is* $\sigma|U_i$; *if* $(\sigma_i) \in \prod_{i \in I} \mathcal{F}(U_i)$, *the* (i, j)*th coordinate of* $\beta((\sigma_i))$ *is* $\sigma_i|U_{(i,j)} - \sigma_j|U_{(i,j)}$.

Proof.   Proposition 5.63(i) shows that α is an injection. Now $\operatorname{im}\alpha \subseteq \ker\beta$, for $\beta\alpha(\sigma)$ has (i, j) coordinate $\sigma|(U_{(i,j)} - \sigma|(i, j) = 0$. The reverse inclusion follows from Proposition 5.63(ii).   •

It follows that $\mathcal{P}(U)$ is an equalizer of

$$\mathcal{P}(U) \xrightarrow{\alpha} \prod_{i \in I} \mathcal{P}(U_i) \underset{\beta''}{\overset{\beta'}{\rightrightarrows}} \prod_{(i,j) \in I \times I} \mathcal{P}(U_{(i,j)}),$$

where $\alpha \colon \sigma \mapsto (\sigma|U_i)$, $\beta' \colon (\sigma_i) \mapsto (\sigma_i|U_{(i,j)})$, and $\beta'' \colon (\sigma_i) \mapsto (\sigma_i|U_{(j,i)})$.

We now adopt another point of view, one that is preferred by every serious user of sheaves.

Definition.   A *sheaf of abelian groups* over a space X is a presheaf[6] $\mathcal{F}$ on X that satisfies the equalizer condition. We shall always assume that $\mathcal{F}(\varnothing) = \{0\}$.

As with etale-sheaves, we may define sheaves with values in categories other than **Ab**.

Corollary 5.66.   *If $\mathcal{F}$ is a sheaf of abelian groups over a space X, then every family* $\sigma = (\sigma_i) \in \prod_{i \in I} \mathcal{F}(U_i)$ *for which* $\sigma_i|(U_i \cap U_j) = \sigma_j|(U_i \cap U_j)$ *corresponds to a unique global section in* $\mathcal{F}(X)$.

Proof.   Such an element σ lies in $\ker\beta$, where $\beta = \beta' - \beta''$, where β, β' are the maps in the equalizer diagram above.   •

Sheaves arise naturally when encoding local information (sheaf cohomology, discussed in the next chapter, is the way to globalize this data), as we shall see in the subsection on manifolds.

Definition.   If $\{\mathcal{F}, \rho_U^V\}$, $\{\mathcal{G}, \tau_U^V\}$ are sheaves over X, a *sheaf map* $\varphi \colon \mathcal{F} \to \mathcal{G}$ is a natural transformation; that is, φ is a one-parameter family of homomorphisms $\varphi_U \colon \mathcal{F}(U) \to \mathcal{G}(U)$, indexed by the open sets U in X, such that there is a commutative diagram whenever $U \subseteq V$:

$$
\begin{array}{ccc}
\mathcal{F}(V) & \xrightarrow{\ \varphi_V\ } & \mathcal{G}(V) \\
{\scriptstyle \rho_U^V}\big\downarrow & & \big\downarrow{\scriptstyle \tau_U^V} \\
\mathcal{F}(U) & \xrightarrow[\ \varphi_U\ ]{} & \mathcal{G}(U).
\end{array}
$$

---

[6]We denote a sheaf by $\mathcal{F}$ because F is the initial letter of the French term *faisceau*.

If $\mathcal{U}$ is the topology on X, then $\mathcal{U}$ is a set, and so $\mathcal{U}$ is a small category; as in Example 1.19(iv), all presheaves form a category $\mathbf{pSh}(X, \mathbf{Ab})$, with morphisms $\mathrm{Hom}(\mathcal{P}, \mathcal{Q}) = \mathrm{Nat}(\mathcal{P}, \mathcal{Q})$. We call morphisms $\mathcal{P} \to \mathcal{Q}$ *presheaf maps*. It follows that if $\mathcal{F}$ and $\mathcal{G}$ are sheaves, then every presheaf map $\mathcal{F} \to \mathcal{G}$ is a sheaf map.

Notation. Define $\mathbf{Sh}(X, \mathbf{Ab})$ to be the full subcategory of $\mathbf{pSh}(X, \mathbf{Ab})$ generated by all sheaves over a space X. We denote the Hom sets by

$$\mathrm{Hom}_{\mathrm{sh}}(\mathcal{F}, \mathcal{F}') = \mathrm{Nat}(\mathcal{F}, \mathcal{F}').$$

Example 5.67. For each open set U of a topological space X, define

$$\mathcal{F}(U) = \{\text{continuous } f \colon U \to \mathbb{R}\}.$$

It is routine to see that $\mathcal{F}(U)$ is an abelian group under pointwise addition: $f + g \colon x \mapsto f(x) + g(x)$, and that $\mathcal{F}$ is a presheaf over X. For each $x \in X$, define an equivalence relation on $\bigcup_{U \ni x} \mathcal{F}(U)$ by $f \sim g$ if there is some open set W containing x with $f|W = g|W$. The equivalence class of f, denoted by $[x, f]$, is called a *germ* at x. Define E_x to be the family of all germs at x, define $E = \bigcup_{x \in X} E_x$, and define $p \colon E \to X$ by $p \colon [x, f] \mapsto x$. In our coming discussion of *associated etale-sheaves*, we will see how to topologize E so that (E, p, X) is an etale-sheaf (called the ***sheaf of germs of continuous functions over*** X). The stalks E_x of this etale-sheaf can be viewed as direct limits: the family of all open sets U containing x is a directed partially ordered set and, by Corollary 5.31, a germ $[x, f]$ is just an element of the direct limit $\varinjlim_{U \ni x} \mathcal{F}(U)$. Variations of this construction are the sheaves of ***germs of differentiable functions*** and of ***germs of holomorphic functions***. ◀

Example 5.67 generalizes; we shall see, in Theorem 5.68, that the stalks of every etale-sheaf are direct limits.

We now construct an etale-sheaf from any presheaf $\mathcal{P}$ (we do not assume that $\mathcal{P}$ is the sheaf of sections of an etale-sheaf). The next result shows that there is no essential difference between sheaves and etale-sheaves.

Theorem 5.68.

(i) *The sheaf of sections defines a functor* $\Gamma \colon \mathbf{Sh}_{\mathrm{et}}(X, \mathbf{Ab}) \to \mathbf{pSh}(X, \mathbf{Ab})$, *and* $\mathrm{im}\, \Gamma \subseteq \mathbf{Sh}(X, \mathbf{Ab})$.

(ii) *There are a functor* $\Phi \colon \mathbf{pSh}(X, \mathbf{Ab}) \to \mathbf{Sh}_{\mathrm{et}}(X, \mathbf{Ab})$ *(which is injective on objects) and a natural transformation* $\nu \colon 1_{\mathbf{pSh}(X, \mathbf{Ab})} \to \Gamma\Phi$ *such that* $\nu_{\mathcal{F}} \colon \mathcal{F} \to \Gamma\Phi(\mathcal{F})$ *is an isomorphism whenever* $\mathcal{F}$ *is a sheaf.*

(iii) *The restriction* $\Phi | \mathbf{Sh}(X, \mathbf{Ab})$ *is an isomorphism of categories*:

$$\mathbf{Sh}(X, \mathbf{Ab}) \cong \mathbf{Sh}_{\mathrm{et}}(X, \mathbf{Ab}).$$

Proof.

(i) If $\varphi: \mathcal{S} \to \mathcal{S}'$, define $\Gamma(\varphi): \Gamma(U, \mathcal{S}) \to \Gamma(U, \mathcal{S}')$ by $\sigma \mapsto \varphi\sigma$. The reader may check that Γ is a functor. Proposition 5.63 says that $\Gamma(\square, \mathcal{S})$ is a sheaf.

(ii) Given a presheaf $\mathcal{P}$ of abelian groups over a space X, we first construct its **associated etale-sheaf** $\mathcal{P}^{\mathrm{et}} = (E^{\mathrm{et}}, p^{\mathrm{et}}, X)$. For each $x \in X$, the index set consisting of all open neighborhoods $U \ni x$, partially ordered by reverse inclusion, is a directed set. Define $E_x^{\mathrm{et}} = \varinjlim_{U \ni x} \mathcal{P}(U)$ (generalizing the stalks of the sheaf of germs in Example 5.67).

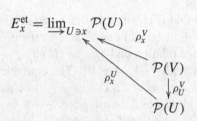

Since the index set is directed, Corollary 5.31(iii) says that the elements of $E_x^{\mathrm{et}} = \varinjlim \mathcal{P}(U)$ are equivalence classes $[\rho_x^U(\sigma)]$, where $U \ni x$, $\sigma \in \mathcal{P}(U)$, and $\rho_x^U: \mathcal{P}(U) \to E_x^{\mathrm{et}}$ is an insertion morphism of the direct limit; moreover, $[\rho_x^U(\sigma)] + [\rho_x^{U'}(\sigma')] = [\rho_x^W \rho_W^U(\sigma) + \rho_x^W \rho_W^{U'}(\sigma')]$, where $W \subseteq U \cap U'$ (thus, $[\rho_x^U(\sigma)]$ generalizes $[x, f]$ in Example 5.67). Define $E^{\mathrm{et}} = \bigcup_{x \in X} E_x^{\mathrm{et}}$, and define a surjection $p^{\mathrm{et}}: E^{\mathrm{et}} \to X$ by $[\rho_x^U(\sigma)] \mapsto x$.

If $U \subseteq X$ is a nonempty open set and $\sigma \in \mathcal{P}(U)$, define

$$\langle \sigma, U \rangle = \{[\rho_x^U(\sigma)] : x \in U\}.$$

We claim that $\langle \sigma, U \rangle \cap \langle \sigma', U' \rangle$ either is empty or contains a subset of the same form. If $e \in \langle \sigma, U \rangle \cap \langle \sigma', U' \rangle$, then $e = [\rho_x^U(\sigma)] = [\rho_y^{U'}(\sigma')]$, where $x \in U$, $\sigma \in \mathcal{P}(U)$, and $y \in U'$, $\sigma' \in \mathcal{P}(U')$. But $x = p^{\mathrm{et}}[\rho_x^U(\sigma)] = p^{\mathrm{et}}[\rho_y^{U'}(\sigma')] = y$, so that $x \in U \cap U'$. By Lemma 5.30(ii), there is an open $W \subseteq U \cap U'$ with $W \ni x$ and $[\rho_W^U \rho_x^W(\sigma)] = [\rho_W^{U'} \rho_x^W(\sigma')]$; call this element $[\tau]$; note that $\langle \tau, W \rangle \subseteq \langle \sigma, U \rangle \cap \langle \sigma', U' \rangle$, as desired. Equip E^{et} with the topology[7] generated

---

[7]This is the coarsest topology on E that makes all sections continuous.

by all $\langle \sigma, U \rangle$; it follows that these sets form a base for the topology; that is, every open set is a union of $\langle \sigma, U \rangle$s.

To see that $(E^{\text{et}}, p^{\text{et}}, X)$ is a protosheaf, we must show that the surjection p^{et} is a local homeomorphism. If $e \in E^{\text{et}}$, then $e = [\rho_x^U(\sigma)]$ for some $x \in X$, where U is an open neighborhood of x and $\sigma \in \mathcal{P}(U)$. If $S = \langle \sigma, U \rangle$, then S is an open neighborhood of e, and it is routine to see that $p^{\text{et}}|S \colon S \to U$ is a homeomorphism.

Now each stalk E_x^{et} is an abelian group. To see that addition is continuous, take $(e, e') \in E^{\text{et}} + E^{\text{et}}$; that is, $e = [\rho_x^U(\sigma)]$ and $e' = [\rho_x^{U'}(\sigma')]$. We may assume the representatives have been chosen so that $\sigma, \sigma' \in \mathcal{P}(U)$ for some U, so that $e + e' = [\rho_x^U(\sigma + \sigma')]$. Let $V^{\text{et}} = \langle \sigma + \sigma', V \rangle$ be a basic open neighborhood of $e + e'$. If $\alpha \colon E^{\text{et}} + E^{\text{et}} \to E^{\text{et}}$ is addition, then it is easy to see that if $U^{\text{et}} = [\langle \tau, W \rangle \times \langle \tau', W \rangle] \cap (E^{\text{et}} + E^{\text{et}})$, then $\alpha(U^{\text{et}}) \subseteq V^{\text{et}}$. Thus, α is continuous. As inversion $E^{\text{et}} \to E^{\text{et}}$ is also continuous, $\mathcal{P}^{\text{et}} = (E^{\text{et}}, p^{\text{et}}, X)$ is an etale-sheaf.

Define $\Phi \colon \mathbf{pSh}(X, \mathbf{Ab}) \to \mathbf{Sh}_{\text{et}}(X, \mathbf{Ab})$ on objects by $\Phi(\mathcal{P}) = \mathcal{P}^{\text{et}} = (E^{\text{et}}, p^{\text{et}}, X)$. Note that Φ is injective on objects, for if $\mathcal{P} \neq \mathcal{P}'$, then $\{\varinjlim_{U \ni x} \mathcal{P}(U)\} \neq \{\varinjlim_{U \ni x} \mathcal{P}'(U)\}$, and so their direct limits are distinct (of course, they may be isomorphic). Hence, $\mathcal{P}^{\text{et}} \neq \mathcal{P}'^{\text{et}}$ and $\Phi \mathcal{P} \neq \Phi \mathcal{P}'$. To define Φ on morphisms, let $\varphi \colon \mathcal{P}_1 \to \mathcal{P}_2$ be a presheaf map, and let $\mathcal{P}_i^{\text{et}} = (E_i^{\text{et}}, p_i^{\text{et}}, X)$ for $i = 1, 2$. For each $x \in X$, φ induces a morphism of direct systems $\{\mathcal{P}_1(U) : U \ni x\} \to \{\mathcal{P}_2(U) : U \ni x\}$ and, hence, a homomorphism $\varphi_x \colon \varinjlim_{U \ni x} \mathcal{P}_1(U) \to \varinjlim_{U \ni x} \mathcal{P}_1(U)$; that is, $\varphi_x \colon (E_1^{\text{et}})_x \to (E_2^{\text{et}})_x$. Finally, define $\Phi(\varphi) \colon E_1^{\text{et}} \to E_2^{\text{et}}$ by $e_x \mapsto \varphi_x(e_x)$ for all $e_x \in (E_1^{\text{et}})_x$. We let the reader prove that $\Phi(\varphi)$ is an etale-map and that Φ is a functor.

Given a presheaf $\{\mathcal{P}, \rho_U^V\}$ and an open subset $U \subseteq X$ (that is, $U \in \mathcal{U}$), a base for the topology of E^{et} consists of all $\langle \sigma, U \rangle = \{[\rho_x^U(\sigma)] : x \in U\}$. Define $\sigma^{\text{et}} \colon U \to E^{\text{et}}$ by $\sigma^{\text{et}}(x) = [\rho_x^U(\sigma)]$; Exercise 5.39(i) on page 301 now says that $\sigma^{\text{et}} \in \Gamma(U, \mathcal{P}^{\text{et}})$. Define $\nu_U \colon \mathcal{P}(U) \to \Gamma(U, \mathcal{P}^{\text{et}})$ by $\sigma \mapsto \sigma^{\text{et}}$. If V is an open set containing U, then it easy to see that $\nu_V = \nu_U \rho_U^V$, so that the family $\{\nu_U : U \in \mathcal{U}\}$ gives a presheaf map $\nu_{\mathcal{P}} \colon \mathcal{P} \to \Gamma(\square, \mathcal{P}^{\text{et}})$. We let the reader check that $\nu = (\nu_U)$ is a natural transformation $1_{\mathbf{pSh}(X, \mathbf{Ab})} \to \Gamma \Phi$.

If $\mathcal{F}$ is a sheaf, we show that $\nu_{\mathcal{F}} \colon \mathcal{F} \to \Gamma(\square, \mathcal{F}^{\text{et}})$ is an isomorphism using Exercise 5.41 on page 301. It suffices to prove, for each open U, that $\nu_U \colon \mathcal{F}(U) \to \Gamma(U, \mathcal{F}^{\text{et}})$, given by $\sigma \to \sigma^{\text{et}}$, is a bijection. To see that ν_U is injective, suppose that $\sigma, \tau \in \mathcal{F}(U)$ and $\sigma^{\text{et}} = \tau^{\text{et}}$. For each $x \in U$, we have $\rho_x^U(\sigma) = \rho_x^U(\tau)$; that is, there is an open neighborhood W_x of x with $\sigma|W_x = \tau|W_x$. The family of all such W_x is an open cover of U, and so Proposition 5.58(iv) gives $\sigma = \tau$. To see that ν_U is surjective, let $\beta \in \Gamma(U, \mathcal{F}^{\text{et}})$. For each $x \in U$, there is a basic open set $\langle U, \sigma_x \rangle$ containing $\beta(x)$, where $\sigma_x \in \mathcal{F}(U_x)$. The gluing condition, Proposition 5.58(v), shows that there is $\sigma \in \mathcal{F}(U)$ with $\sigma|U_x = \sigma_x$ for all $x \in U$, and another application of Proposition 5.58(iv) gives $\beta = \sigma^{\text{et}}$. Thus, ν_U is a bijection.

(iii) This follows easily from parts (i) and (ii).   •

The stalks of the etale-sheaf of germs in Example 5.67 are direct limits, as are the stalks of $\mathcal{P}^{\text{et}}$; we now define the stalks of an arbitrary presheaf.

Definition.   If $\mathcal{P}$ is a presheaf on a space X, then the *stalk* at $x \in X$ is

$$\mathcal{P}_x = \varinjlim_{U \ni x} \mathcal{P}(U).$$

For each $x \in X$, the presheaf map $\varphi \colon \mathcal{P} \to \mathcal{Q}$ induces a morphism of direct systems $\{\mathcal{P}(U) : U \ni x\} \to \{\mathcal{Q}(U) : U \ni x\}$, which, in turn, gives the homomorphism $\varphi_x \colon \varinjlim_{U \ni x} \mathcal{P}(U) \to \varinjlim_{U \ni x} \mathcal{Q}(U)$ defined by $\varphi_x \colon [\sigma] \mapsto [\varphi\sigma]$, where $\sigma \in \mathcal{P}(U)$ and $x \in U$. Exercise 5.33 on page 272 shows that $\varinjlim$ is a functor $\mathbf{Dir}(I, \mathbf{Ab}) \to \mathbf{Ab}$, where $\mathbf{Dir}(I, \mathbf{Ab})$ is the category of direct systems of abelian groups over $I = \{U \ni x\}$. Hence, if $\mathcal{P} \xrightarrow{\varphi} \mathcal{Q} \xrightarrow{\psi} \mathcal{R}$ are presheaf maps, then $(\psi\varphi)_x = \psi_x \varphi_x$. See Exercise 5.45 on page 302 for a description of ν_x, where $\nu \colon \mathcal{P} \to \Gamma(\square, \mathcal{P}^{\text{et}})$ is the natural map in Theorem 5.68.

Lemma 5.69.   *Let $\varphi, \psi \colon \mathcal{P} \to \mathcal{F}$ be presheaf maps, where $\mathcal{P}$ is a presheaf and $\mathcal{F}$ is a sheaf. If φ, ψ agree on stalks, that is, $\varphi_x = \psi_x$ for all $x \in X$, then $\varphi = \psi$.*

Proof.   We must show that $\varphi_U = \psi_U$ for all open U. Given U, choose $x \in U$ and $e_x = [\sigma_x] \in \mathcal{P}_x$, where $\sigma_x \in \mathcal{P}(U_x)$ for some open $U_x \ni x$ with $U_x \subseteq U$. By hypothesis,

$$[\varphi\sigma_x] = \varphi_x([\sigma_x]) = \psi_x([\sigma_x]) = [\psi\sigma_x] \text{ in } \varinjlim_{U \ni x} \mathcal{F}(U).$$

By the definition of equality in direct limits, there are open neighborhoods W_x of x with $\varphi\sigma_x|W_x = \psi\sigma_x|W_x$, and $(W_x)_{x \in U}$ is an open cover of U. Since the equalizer condition holds for the sheaf $\mathcal{F}$, the restrictions determine a unique section; that is, $\varphi\sigma_x = \psi\sigma_x$. Hence, $\varphi_U = \psi_U$ and $\varphi = \psi$.   •

Theorem 5.70.   *Let $\mathcal{P} = \{\mathcal{P}(U), \rho_U^V\}$ be a presheaf of abelian groups over a space X, let $\mathcal{P}^{\text{et}} = (E^{\text{et}}, p^{\text{et}}, X)$ be its associated etale-sheaf, and let*

$\mathcal{P}^* = \Gamma(\square, \mathcal{P}^{\text{et}})$ *be the sheaf of sections of* $\mathcal{P}^{\text{et}}$. *There exists a presheaf map* $v \colon \mathcal{P} \to \mathcal{P}^*$ *that solves the following universal mapping problem:*

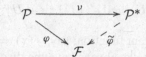

for every presheaf map $\varphi \colon \mathcal{P} \to \mathcal{F}$, *where* $\mathcal{F}$ *is a sheaf over* X, *there exists a unique sheaf map* $\widetilde{\varphi} \colon \mathcal{P}^* \to \mathcal{F}$ *with* $\widetilde{\varphi} v = \varphi$.

Proof.   Applying Φ gives an etale-map $\Phi(\varphi) \colon \Phi(\mathcal{P}) \to \Phi(\mathcal{F})$, and applying Γ gives a sheaf map $\Gamma\Phi(\mathcal{P}) \to \Gamma\Phi(\mathcal{F})$. But $\mathcal{P}^* = \Gamma\Phi(\mathcal{P})$, by definition, while $v_{\mathcal{F}} \colon \mathcal{F} \to \Gamma\Phi(\mathcal{F})$ is a natural isomorphism, because $\mathcal{F}$ is a sheaf. Therefore, the diagram commutes if we define $\widetilde{\varphi} = v_{\mathcal{F}}^{-1}\Gamma\Phi(\varphi)$.

By Lemma 5.69, it suffices to see that $\widetilde{\varphi}_x = \psi_x$ for all $x \in X$. But $\mathcal{P}_x^* = \varinjlim_{U \ni x} \mathcal{P}(U)$,

$$\mathcal{P}(U) \xrightarrow{\;\;v_x\;\;} \mathcal{P}_x^* = \varinjlim \mathcal{P}(U)$$
$$\varphi_x \searrow \qquad \nearrow \widetilde{\varphi}_x$$
$$\mathcal{F}_x = \varinjlim \mathcal{F}(U),$$

and the universal property of direct limit gives a unique map making the diagram commute.   ●

Definition.   If $\mathcal{P}$ is a presheaf of abelian groups, then its *sheafification* is the sheaf $\mathcal{P}^* = \Gamma(\square, \mathcal{P}^{\text{et}})$, where Γ is the sheaf of sections of $\mathcal{P}^{\text{et}}$, the associated etale-sheaf of $\mathcal{P}$.

The construction of the associated etale-sheaf in Theorem 5.68 shows that a presheaf $\mathcal{P}$ and its sheafification $\mathcal{P}^*$ have the same stalks.

Example 5.71.   Let A be an abelian group and let X be a topological space. The *constant presheaf at* A over X is defined on a nonempty open set $U \subseteq X$ by

$$\mathcal{P}(U) = \{f \colon U \to A \mid f \text{ is constant}\};$$

define $\mathcal{P}(\varnothing) = \{0\}$ and, if $U \subseteq V$, define $\rho_U^V \colon \mathcal{P}(V) \to \mathcal{P}(U)$ by $f \mapsto f|U$ [an equivalent description of $\mathcal{P}$ has $\mathcal{P}(U) = A$ for every nonempty open U and $\rho_U^V = 1_A$]. As in Example 5.64, $\mathcal{P}$ is not a sheaf, for we may not be able to glue sections defined on disjoint open sets.

Let the protosheaf of $\mathcal{P}^{\text{et}}$ be (E, p, X). Now the stalk $E_x = A$, and so the underlying set of E is $X \times A$; what is the topology on $X \times A$ making $\mathcal{P}^{\text{et}}$ an etale-sheaf? As in the proof of Theorem 5.68, a base for open sets consists of all the subsets of the form

$$\langle f, U \rangle = \{[\rho_x^U(f)] : x \in U\};$$

here, $f: U \to A$ is a constant function; say, $f(x) = a$ for all $x \in U$. Thus, $\langle f, U \rangle = U \times \{a\}$. Since stalks always have the discrete topology, it follows that $X \times A$ has the product topology. Since there may be nonconstant functions $f: U \to X \times A$ with $pf = 1_U$ (see Example 5.64), the constant presheaf $\mathcal{P}$ is not a sheaf.

Define the ***constant sheaf at*** A to be the sheafification $\mathcal{P}^*$ of the constant presheaf $\mathcal{P}$. Recall that a function $f: X \to Y$ between spaces is ***locally constant*** if each $x \in X$ has an open neighborhood U_x with $f|U_x$ constant. If Y is discrete, then every continuous $f: X \to Y$ is locally constant; since $\{f(x)\}$ is open, we may take $U_x = f^{-1}(\{f(x)\})$. The reader may check that the constant sheaf $\mathcal{P}^*$ has sections

$$\mathcal{P}(U) = \{f: U \to A \mid f \text{ is locally constant}\}.$$

It follows that $\mathcal{P} \neq \mathcal{P}^*$. ◄

Here is an example showing that both protosheaf and presheaf views of a sheaf are useful.

Example 5.72.

(i) Let A be an abelian group, X a topological space, and $x \in X$. Define a presheaf by

$$x_* A(U) = \begin{cases} A & \text{if } x \in U, \\ \{0\} & \text{otherwise.} \end{cases}$$

If $U \subseteq V$, then the restriction map ρ_U^V is either 1_A or 0. It is easy to check that $x_* A$ is a sheaf; it is called a ***skyscraper sheaf***. Its name arises because all the stalks of $x_* A$ are $\{0\}$ except $(x_* A)_x$, which is A.

(ii) Let X be the unit circle, which we view as $\{z \in \mathbb{C} : |z| = 1\}$, and let $p: X \to X$ be defined by $p: z \mapsto z^2$. If we set $E = X$, then we have defined a protosheaf $\mathcal{S} = (E, p, X)$. Now $\mathcal{S}$ is an etale-sheaf with all stalks isomorphic to $\mathbb{I}_2$, which we call the ***double cover***. An interesting feature of the sheaf of sections $\Gamma(\square, \mathcal{S})$ is that it has the same stalks as the constant sheaf at $\mathbb{I}_2$, yet the two sheaves are not isomorphic. The nonisomorphism merely reflects the obvious fact that different spaces can be the same locally. ◄

Fig. 5.2 Double cover.

Definition.   Let $\mathcal{P}'$ and $\mathcal{P}$ be presheaves of abelian groups on a topological space X such that $\mathcal{P}'(U) \subseteq \mathcal{P}(U)$ for every open set U in X; that is, there are inclusions $\iota_U \colon \mathcal{P}'(U) \to \mathcal{P}(U)$. Then $\mathcal{P}'$ is a *subpresheaf* of $\mathcal{P}$ if the *inclusion* $\iota \colon \mathcal{P}' \to \mathcal{P}$ is a presheaf map..

If $\mathcal{F}$ is a sheaf, then $\mathcal{P}'$ is a *subsheaf* of $\mathcal{F}$ if $\mathcal{P}'$ is a subpresheaf that is also a sheaf.

Example 5.73.

(i) The zero sheaf [see Example 5.60(i)] is a subsheaf of every sheaf.

(ii) Let $\mathcal{F}$ be the sheaf of germs of continuous functions on an n-manifold X [see Example 5.67], let $\mathcal{F}'$ be the sheaf of germs of differentiable functions on X, and let $\mathcal{F}''$ be the sheaf of germs of holomorphic functions on X. Then $\mathcal{F}''$ is a subsheaf of $\mathcal{F}'$, and $\mathcal{F}'$ is a subsheaf of $\mathcal{F}$.

(iii) Let $\mathcal{F}$ be the sheaf of germs of continuous functions on a space X. Define $\mathcal{G}$ by setting $\mathcal{G}(U) = \mathcal{F}(U)$ for all open sets U and by setting restrictions ψ_U^V to be identically 0. Then $\mathcal{G}$ is a presheaf, but $\mathcal{G}$ is not a subpresheaf of $\mathcal{F}$ (for the inclusion is not a presheaf map). ◄

It is clear that subpresheaves $\mathcal{F}$ and $\mathcal{F}'$ of a presheaf $\mathcal{G}$ are equal if and only if $\mathcal{F}(U) = \mathcal{F}'(U)$ for all open U. This simplifies for sheaves.

Proposition 5.74.

(i) *If* $\Phi \colon \mathbf{pSh}(X, \mathbf{Ab}) \to \mathbf{Sh}_{\text{et}}(X, \mathbf{Ab})$ *is the functor in* Theorem 5.68, *and if* $\mathcal{F}$ *is a subsheaf of a sheaf* $\mathcal{G}$, *then* $\Phi\mathcal{F}$ *is a subetale-sheaf of* $\Phi\mathcal{G}$.

(ii) *If all the stalks of a sheaf* $\mathcal{F}$ *are* $\{0\}$, *then* $\mathcal{F}$ *is the zero sheaf.*

(iii) *If* $\mathcal{F}$ *and* $\mathcal{F}'$ *are subsheaves of a sheaf* $\mathcal{G}$, *then* $\mathcal{F} = \mathcal{F}'$ *if and only if they have the same stalks; that is,* $\mathcal{F}_x = \mathcal{F}'_x$ *for all* $x \in X$.

Remark. Example 5.81 shows that part (ii) is false for presheaves, and so the hypothesis in part (iii) that $\mathcal{F}$ and $\mathcal{F}'$ both be subsheaves of a sheaf $\mathcal{G}$ is necessary. ◄

Proof.

(i) If $\iota\colon \mathcal{F} \to \mathcal{G}$ is the inclusion, then $\Phi(\iota)\colon \Phi\mathcal{F} \to \Phi\mathcal{F}$ is an etale-map. Since each $\iota_U\colon \mathcal{F}(U) \to \mathcal{G}(U)$ is an injection, the induced map on stalks $\iota_x\colon \mathcal{F}_x \to \mathcal{G}_x$ is an injection (Proposition 5.33 applies because {open $W \subseteq U : W \ni x$} is a directed set).

(ii) The construction in Theorem 5.68 shows, for any presheaf $\mathcal{P}$, that the stalks of $\mathcal{P}$ and $\Gamma(\square, \mathcal{P}^{\text{et}})$ are the same. It follows that if all the stalks of a sheaf $\mathcal{F}$ are {0}, then $\Gamma(\square, \mathcal{F}^{\text{et}})$ are {0}. But the restriction of the functor $\Phi\colon \mathbf{pSh}(X, \mathbf{Ab}) \to \mathbf{Sh}_{\text{et}}(X, \mathbf{Ab})$ to $\mathbf{Sh}(X, \mathbf{Ab})$ is an isomorphism, and so $\Phi(\mathcal{F}) = \Phi(0)$ is the zero sheaf.

(iii) By Proposition 5.61, if $\mathcal{F}$ and $\mathcal{F}'$ have the same stalks, then $\Phi\mathcal{F} = \Phi\mathcal{F}'$. But Φ is injective on objects, and so $\mathcal{F} = \mathcal{F}'$. •

5.4.1 Manifolds

In his book *Differential Geometry: Cartan's Generalization of Klein's Erlangen Program* (Springer-Verlag, New York, 1997, p. 52), R. W. Sharpe writes,

> Let us begin with a rough and ready description of p-forms for $p \leq 2$. The 0-forms (with values in a finite-dimensional vector space V) on a manifold M are just the V-valued functions on M. The 1-forms generalize the derivatives of a function on M. The 2-forms are used as a way of formalizing the necessary conditions on a 1-form for it to be the derivative of a function.

When we first learn Calculus, it is natural for us to regard differential forms dx as being very small and, hence, to regard $(dx)^2 = dxdx$ as being neglible. Taking this observation seriously leads to *Grassmann algebras*.

Definition. Let V_n be an n-dimensional vector space over $\mathbb{R}$, and label a basis of V_n as $dx_1, \ldots, dx_n$. The ***Grassmann algebra***[8] $G(V_n)$ is the (associative) $\mathbb{R}$-algebra with generators $dx_1, \ldots, dx_n$ and relations $v^2 = 0$ for all $v \in V_n$.

---

[8]This is a special case of an *exterior algebra*. The product of two elements is usually denoted by $dx \wedge dy$ instead of by $dxdy$, and the maps $d^p\colon G^p(V_n) \to G^{p+1}(V_n)$ defined below are special cases of *exterior derivatives*.

Of course, $(dx_i)^2 = 0$ in $G(V_n)$; note that $(dx_i + dx_j)^2 = 0$ gives $0 = (dx_i)^2 + dx_i dx_j + dx_j dx_i + (dx_j)^2 = dx_i dx_j + dx_j dx_i$, so that

$$dx_i dx_j = -dx_j dx_i.$$

In particular, products of dx_is can be rewritten in the form $\pm dx_{i_1} dx_{i_2} \cdots dx_{i_p}$, where $1 \le i_1 < i_2 < \cdots < i_p \le n$. For example, $dx_3 dx_1 dx_2 = -dx_1 dx_3 dx_2 = dx_1 dx_2 dx_3$. If $I = (i_1, \ldots, i_p)$, write

$$dx_I = dx_{i_1} dx_{i_2} \cdots dx_{i_p}.$$

We can prove that $G(V_n)$ is a graded algebra:

$$G(V_n) = \bigoplus_{p \ge 0} G^p(V_n),$$

where $G^0(V_n) = \mathbb{R}$ and $G^p(V_n)$, for $p \ge 1$, is the vector space over $\mathbb{R}$ generated by all $dx_{i_1} dx_{i_2} \cdots dx_{i_p}$. It follows that $G^p(V_n) = \{0\}$ if $p > n$, because any product of dx_is having more than n factors must have some dx_j repeated. In fact, $\dim(G^p(V_n)) = \binom{n}{p}$, with basis all dx_I with $I = (i_1, \ldots, i_p)$ and $1 \le i_1 < i_2 < \cdots < i_p \le n$ (Rotman, *Advanced Modern Algebra*, p. 749).

Definition. A *euclidean m-chart* is an ordered pair (U, φ), where U is a topological space, called a *coordinate neighborhood*, and $\varphi \colon U \to \mathbb{R}^m$ is a homeomorphism. A function $f \colon U \to \mathbb{R}$ is *smooth* if $f\varphi^{-1} \colon \varphi(U) \to \mathbb{R}$ is smooth (if $V \subseteq \mathbb{R}^m$ is open, then a function $f \colon V \to \mathbb{R}$ is *smooth* if all its mixed partials exist). All these smooth functions form a commutative ring,

$$C(U, \varphi).$$

Definition. An *m-manifold* is a Hausdorff space X such that every $x \in X$ has an open neighborhood homeomorphic to $\mathbb{R}^m$.

If U_x is an open neighborhood of X and $\varphi_x \colon U_x \to \mathbb{R}^m$ is a homeomorphism, then (U_x, φ_x) is an m-chart.

Definition. Given a euclidean m-chart (U, φ), define

$$\Omega^p(U, \varphi) = C(U, \varphi) \otimes_{\mathbb{R}} G^p(V_m).$$

Now $\Omega^0(U, \varphi) = C(U, \varphi)$ and, when $p \ge 1$, $\Omega^p(U, \varphi)$ is the free $C(U, \varphi)$-module with basis all dx_I with $I = (i_1, \ldots, i_p)$ and $1 \le i_1 < i_2 < \cdots < i_p \le m$. The elements $\omega \in \Omega^p(U, \varphi)$ are called *real p-forms*; each has a unique expression $\omega = \sum_I f_I \, dx_I$, where $f_I \in C(U, \varphi)$ (we write dx_I instead of $1 \otimes dx_I$ and, more generally, $f_I \, dx_I$ instead of $f_I \otimes dx_I$).

Definition.   If (U, φ) is a euclidean chart, then the *de Rham complex* $\Omega^\bullet(U, \varphi)$ is the sequence

$$0 \to \Omega^0(U, \varphi) \xrightarrow{d^0} \Omega^1(U, \varphi) \xrightarrow{d^1} \Omega^2(U, \varphi) \to \cdots ,$$

where $d^p: \Omega^p(U, \varphi) \to \Omega^{p+1}(U, \varphi)$ is defined[9] as follows:

$$\text{if } f \in \Omega^0(U, \varphi) = C, \text{ then } d^0 f = \sum_i \frac{\partial f}{\partial x_i} \, dx_i;$$

$$\text{if } \omega = \sum_I f_I \, dx_I \in \Omega^p(U, \varphi), \text{ then } d^p(\omega) = \sum_I (d^0 f_I) \, dx_I.$$

The de Rham complex is a complex; that is, $d^{p+1} d^p = 0$; the proof depends on the fact that the mixed partials of a smooth function are equal (see Bott–Tu, *Differential Forms in Algebraic Topology*, p. 15).

Example 5.75.   If $U \subseteq \mathbb{R}^3$, then the de Rham complex is just the familiar one of Advanced Calculus:[10]

$$0 \to \Omega^0(U) \xrightarrow{\text{grad}} \Omega^1(U) \xrightarrow{\text{curl}} \Omega^2(U) \xrightarrow{\text{div}} \Omega^3(U) \to 0. \quad \blacktriangleleft$$

As in Advanced Calculus, we say that a p-form ω is *closed* if $d^p \omega = 0$, and it is *exact* if $\omega = d^{p-1} \zeta$ for some $(p - 1)$-form ζ. Let $Z^p(U, \varphi)$ denote the subspace of $\Omega^p(U, \varphi)$ comprised of all closed p-forms, and let $B^p(U, \varphi)$ denote the subspace of $\Omega^p(U, \varphi)$ comprised of all exact p-forms. Since $d^p d^{p-1} = 0$, every exact p-form is closed; that is, $B^p(U, \varphi) \subseteq Z^p(U, \varphi)$, and we define

$$H^p(\Omega^\bullet(U, \varphi)) = Z^p(U, \varphi)/B^p(U, \varphi).$$

The cohomology groups $H^p(\Omega^\bullet(U, \varphi))$ of the de Rham complex are isomorphic to the singular cohomology groups $H^p(U, \mathbb{R})$.

We shall see, in Example 5.77, that sheaves will allow us to generalize this discussion from charts to manifolds.

The following general construction is useful.

---

[9]Strictly speaking, we should write $\partial(f\varphi^{-1})/\partial x_i$ instead of $\partial f/\partial x_i$.

[10]Recall: $\text{grad} f = \frac{\partial f}{\partial x} \, dx + \frac{\partial f}{\partial y} \, dy + \frac{\partial f}{\partial z} \, dz$; $\text{curl}(f_1 \, dx + f_2 \, dy + f_3 \, dz) = \left(\frac{\partial f_3}{\partial y} - \frac{\partial f_2}{\partial z}\right) dydz - \left(\frac{\partial f_1}{\partial z} - \frac{\partial f_3}{\partial x}\right) dxdz + \left(\frac{\partial f_2}{\partial x} - \frac{\partial f_1}{\partial y}\right) dxdy$; $\text{div}(f_1 \, dydz - f_2 \, dxdz + f_3 \, dxdy) = \left(\frac{\partial f_1}{\partial x} + \frac{\partial f_2}{\partial y} + \frac{\partial f_3}{\partial z}\right) dxdydz$.

Definition. Let $\mathcal{F}$ be a sheaf of abelian groups over X, and let U be an open subset of X. If $V \subseteq U$ is an open set, define a presheaf $\mathcal{F}|U : V \mapsto \mathcal{F}(V)$. It is easy to see that $\mathcal{F}|U$ is a sheaf; it is called the **restriction sheaf**.

If $\mathcal{F}$ is a sheaf over a space X and $U \subseteq X$ is open, we may say "$\mathcal{F}$ over U" instead of "the restriction sheaf $\mathcal{F}|U$."

Here is the geometric picture of the restriction $\mathcal{F}|U$. If $\mathcal{F}^{\text{et}} = (E, p, X)$ is the etale-sheaf of $\mathcal{F}$, then the etale-sheaf of $\mathcal{F}|U$ is $(E', p|E', U)$, where $E' = p^{-1}(U)$.

Given a euclidean m-chart (U, φ), we have seen how to define the the de Rham complex $\Omega^p(U, \varphi)$. We can generalize this construction from m-charts to *smooth manifolds*, which are the most interesting manifolds for Geometry and Analysis.

Definition. Let X be an m-manifold. An **atlas** is a family of m-charts $((U_i, \varphi_i))_{i \in I}$ with $\mathcal{U} = (U_i)_{i \in I}$ an open cover of X.

Let $((U_i, \varphi_i))_{i \in I}$ be an atlas of an m-manifold. If $p \in U_i$, then φ_i equips p with coordinates, namely, $\varphi_i(p)$. Write $U_{ij} = U_i \cap U_j$. If $U_{ij} \neq \varnothing$, then every $p \in U_{ij}$ has two sets of coordinates: $\varphi_i(p)$ and $\varphi_j(p)$.

Definition. If $((U_i, \varphi_i))_{i \in I}$ is an atlas, then its **transition functions** are

$$h_{ij} = \varphi_i \varphi_j^{-1} : \varphi_j(U_{ij}) \to \varphi_i(U_{ij}).$$

Transition functions compare the two sets of coordinates of $p \in U_{ij}$. If $(y_1, \ldots, y_m) = \varphi_j(p)$ and $(x_1, \ldots, x_m) = \varphi_i(p)$, then $h_{ij} : (y_1, \ldots, y_m) \mapsto (x_1(y_1, \ldots, y_m), \ldots, x_m(y_1, \ldots, y_m))$. If V and W are open subsets of $\mathbb{R}^m$ and $h : V \to W$, then

$$h : (y_1, \ldots, y_m) \mapsto (x_1(y_1, \ldots, y_m), \ldots, x_m(y_1, \ldots, y_m)).$$

Call h **smooth** if all its mixed partials exist; that is, for each $1 \leq q \leq m$, the coordinate function $(y_1, \ldots, y_m) \mapsto x_q(y_1, \ldots, y_m)$ is smooth.

Definition. A *smooth m-manifold* is an m-manifold having an atlas whose transition functions are smooth.

The next proposition will allow us to define (global) smooth functions on X.

Proposition 5.76 (Gluing Lemma). *Let $\mathcal{U} = (U_i)_{i \in I}$ be an open cover of a space X. For each $i \in I$, let $\mathcal{F}_i$ be a sheaf of abelian groups over U_i and, for each $i, j \in I$, let there be sheaf isomorphisms $\theta_{ij} : \mathcal{F}_j|U_{ij} \to \mathcal{F}_i|U_{ij}$ such that*

(i) $\theta_{ii} = 1_{\mathcal{F}_i|U_{ij}}$,

(ii) *for all* $i, j, k \in I$, *the restrictions to* $U_i \cap U_j \cap U_k$ *satisfy*

$$\theta_{ik} = \theta_{ij}\theta_{jk}.$$

Then there exist a unique sheaf $\mathcal{F}$ *over* X *and isomorphisms* $\eta_i : \mathcal{F}|U_i \to \mathcal{F}_i$ *such that* $\eta_i \eta_j^{-1} = \theta_{ij}$ *over* U_{ij} *for all* i, j.

Proof. Uniqueness of $\mathcal{F}$ (to isomorphism) is left to the reader. For existence, note that if $V \subseteq X$ is open, then

$$V = V \cap X = V \cap \bigcup_{i \in I} U_i = \bigcup_{i \in I} (V \cap U_i).$$

Define $\mathcal{F}(V) = \varinjlim \mathcal{F}_i(V \cap U_i)$. It is routine to check that the presheaf $\mathcal{F}$ is a sheaf that satisfies the stated properties. •

Example 5.77. Let $((U_i, \varphi_i))_{i \in I}$ be an atlas of an m-manifold X. For each $i \in I$, define a commutative ring over U_i by setting, for each open $W_i \subseteq U_i$,

$$\mathcal{F}_i(W_i) = \{f|W_i : f \in C(U_i, \varphi_i)\},$$

where $C(U_i, \varphi_i)$ is the commutative ring of all smooth functions on (U_i, φ_i). If $W_i' \subseteq W_i$ are open, define $\mathcal{F}_i(W_i) \to \mathcal{F}_i(W_i')$ to be "honest" restriction: $f|W_i \mapsto f|W_i'$. It is easy to see that $\mathcal{F}_i$ is a sheaf of commutative rings over U_i. The reader can define sheaf maps θ_{ij} that satisfy the hypotheses of the Gluing Lemma, Proposition 5.76, yielding a sheaf $\mathcal{C}$ over X. In light of Corollary 5.66, define a smooth function on X to be a global section; that is, define $\mathcal{C}(X) = \Gamma(X, \mathcal{C})$. With this definition, the smooth functions on X form a commutative ring. Let (E, p, X) be the etale-sheaf corresponding to $\mathcal{C}$. Locally, smooth functions have values in $\mathbb{R}^m$; however, smooth functions on X correspond to global sections; that is, they take values in E.

A similar construction allows us to define (global) p-forms on X, and the de Rham complex can be defined for manifolds (see Bott–Tu, *Differential Forms in Algebraic Topology*, for more details). ◄

Here are two important constructions, called ***change of base*** (but which we will not be using in the text); the first generalizes the construction of the restriction sheaf just given.

(i) Given a continuous $f : Y \to X$, there is a change of base construction, called ***inverse image***, that constructs a sheaf over Y from a sheaf over X. It is simplest to define inverse image in terms of etale-sheaves. If

$\mathcal{P}$ is a presheaf, then $\mathcal{P}^{\text{et}} = (E, p, X)$ is an etale-sheaf. Construct the pullback (in **Top**)

$$
\begin{array}{ccc}
E' & \dashrightarrow & E \\
p' \downarrow & & \downarrow p \\
Y & \xrightarrow{\ f\ } & X.
\end{array}
$$

Then $\mathcal{S}' = (E', p', Y)$ is an etale-sheaf over Y, and define $f^*\mathcal{P} = \Gamma(\square, \mathcal{S}')$. There is a functor $f^* \colon \mathbf{pSh}(X, \mathbf{Ab}) \to \mathbf{pSh}(Y, \mathbf{Ab})$, called *inverse image*, with $f^* \colon \mathcal{P} \mapsto \Gamma(\square, \mathcal{P}^{\text{et}})$. For example, if $f \colon U \to X$ is the inclusion of an open subset, then $f^*\mathcal{F}$ is the restriction sheaf $\mathcal{F}|U$.

(ii) Given a continuous $f \colon Y \to X$, the second change of base construction constructs a presheaf over X from a presheaf over Y. If $\mathcal{P}$ is a presheaf over Y, define the *direct image* $f_*\mathcal{P}$ by $f_*\mathcal{P}(V) = \mathcal{P}(f^{-1}V)$ for every open subset $V \subseteq X$ and, if $W \subseteq V$, then $\rho_V^W = \rho_{f^{-1}V}^{f^{-1}W}$. Then $f_*\mathcal{P}$ is a presheaf over X; if $\mathcal{P}$ is also a sheaf, then $f_*\mathcal{P}$ is a sheaf as well. Moreover, $f_* \colon \mathbf{pSh}(Y; \mathbf{Ab}) \to \mathbf{pSh}(X, \mathbf{Ab})$ is a functor.

As an example, let $i \colon Y \to X$ be the inclusion of a subspace, and let $\mathcal{F}$ be a sheaf of abelian groups over Y. If Y is *closed*, then we can prove (see Tennison, *Sheaf Theory*, p. 64) that the stalks of $i_*\mathcal{F}$ are

$$
(i_*\mathcal{F})_x = \begin{cases} \mathcal{F}_x & \text{if } x \in Y, \\ \{0\} & \text{if } x \notin Y. \end{cases}
$$

If Y is an open subset, then the stalks of $i_*\mathcal{F}$ are $\{0\}$ outside $\overline{Y}$, the closure of Y.

Theorem. *If $f \colon X \to Y$ is continuous and $\mathcal{P}$ is a presheaf over Y, then $f^*\mathcal{P}$ is a sheaf over X. Moreover, there is a presheaf map $v \colon \mathcal{P} \to f_*f^*\mathcal{P}$ solving the universal mapping problem (where $\mathcal{F}$ is a sheaf)*

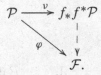

$$
\begin{array}{ccc}
\mathcal{P} & \xrightarrow{\ v\ } & f_*f^*\mathcal{P} \\
& \varphi \searrow & \downarrow \\
& & \mathcal{F}.
\end{array}
$$

Proof. Tennison, *Sheaf Theory*, p. 60. ●

This generalizes Theorem 5.70, for the sheafification of a presheaf $\mathcal{P}$ (which we constructed using etale-sheaves) turns out to be $\mathcal{P}^* = f_*f^*\mathcal{P}$, where $f = 1_X \colon X \to X$.

Corollary. *If $f: X \to Y$ is continuous, then (f^*, f_*) is an adjoint pair of functors* $\mathbf{pSh}(Y, \mathbf{Ab}) \to \mathbf{pSh}(X, \mathbf{Ab})$.

Proof.   Tennison, *Sheaf Theory*, p. 61.   •

The reader should note that $\mathcal{P} \mapsto f_* f^* \mathcal{P}$ is the unit of this adjoint pair of functors.

5.4.2 Sheaf Constructions

We now show that many constructions made for abelian groups can be generalized to presheaves and to sheaves. It turns out that finite products and finite coproducts exist in the categories $\mathbf{pSh}(X, \mathbf{Ab})$ and $\mathbf{Sh}(X, \mathbf{Ab})$ and, as in $_R\mathbf{Mod}$, they coincide.

Proposition 5.78. *Let $\mathcal{P}$ and $\mathcal{Q}$ be presheaves of abelian groups on a space X.*

(i) *If, for every open $U \subseteq X$, we define*

$$(\mathcal{P} \oplus \mathcal{Q})(U) = \mathcal{P}(U) \oplus \mathcal{Q}(U),$$

then $\mathcal{P} \oplus \mathcal{Q}$ is both a product and a coproduct in $\mathbf{pSh}(X, \mathbf{Ab})$.

(ii) *If both $\mathcal{P}$ and $\mathcal{Q}$ are sheaves, then $\mathcal{P} \oplus \mathcal{Q}$ is a sheaf, and it is both a product and a coproduct in* $\mathbf{Sh}(X, \mathbf{Ab})$.

Proof.

(i) It is easy to generalize Proposition 2.20(iii) from modules to presheaves; $\mathcal{P} \oplus \mathcal{Q}$ is a coproduct if and only if there are projection and injection presheaf maps $\mathcal{P} \overset{i}{\underset{p}{\rightleftarrows}} \mathcal{P} \oplus \mathcal{Q} \overset{q}{\underset{j}{\rightleftarrows}} \mathcal{Q}$ satisfying the equations $pi = 1_\mathcal{P}, qj = 1_\mathcal{Q}, pj = 0, qi = 0$, and $ip + jq = 1_{\mathcal{P} \oplus \mathcal{Q}}$. If $\mathcal{X}$ is a presheaf and $\alpha: \mathcal{P} \to \mathcal{X}$ and $\beta: \mathcal{Q} \to \mathcal{X}$ are presheaf maps, define $\theta: \mathcal{P} \oplus \mathcal{Q} \to \mathcal{X}$ by $\theta = \alpha i + \beta j$. We conclude that $\mathcal{P} \oplus \mathcal{Q}$ is a coproduct.

Similarly, $\mathcal{P} \oplus \mathcal{Q}$ is a product, for if $s: \mathcal{X} \to \mathcal{P}$ and $t: \mathcal{X} \to \mathcal{Q}$ are presheaf maps, define $\theta': \mathcal{X} \to \mathcal{P} \oplus \mathcal{Q}$ by $\theta' = is + tj$.

(ii) If $\mathcal{P}$ and $\mathcal{Q}$ are sheaves, then the equalizer condition holds for each, from which it follows that there is an exact sequence of abelian groups

$$0 \to (\mathcal{P} \oplus \mathcal{Q})(U) \xrightarrow{\alpha} \prod_{i \in I} (\mathcal{P} \oplus \mathcal{Q})(U_i) \underset{\beta''}{\overset{\beta'}{\rightrightarrows}} \prod_{(i,j) \in I \times I} (\mathcal{P} \oplus \mathcal{Q})(U_{(i,j)}).$$

Thus, $\mathcal{P} \oplus \mathcal{Q}$ is a sheaf. That $\mathcal{P} \oplus \mathcal{Q}$ is both a product and a coproduct in **pSh** implies that this also holds in the subcategory **Sh** (for there are more diagrams to complete in the large category).[11] •

Remark. If $\mathcal{P}$ and $\mathcal{Q}$ are presheaves (or sheaves), we call $\mathcal{P} \oplus \mathcal{Q}$ their **direct sum**. Note that the stalks of $\mathcal{P} \oplus \mathcal{Q}$ are $(\mathcal{P} \oplus \mathcal{Q})_x = \mathcal{P}_x \oplus \mathcal{Q}_x$. ◄

Definition. If $(\mathcal{F}_i)_{i \in I}$ is a family of presheaves of abelian groups over a space X with restriction maps ρ_i, define the **direct product** presheaf $\mathcal{P} = \prod_{i \in I} \mathcal{F}_i$ as follows. For every open $U \subseteq X$, define

$$\mathcal{P}(U) = \prod_{i \in I} \mathcal{F}_i(U);$$

if $U \subseteq V$ are open, define the restriction $\mathcal{P}(V) \to \mathcal{P}(U)$ coordinatewise:

$$(s_i) \mapsto (\rho_i(s_i)).$$

Proposition 5.79. *If $(\mathcal{F}_i)_{i \in I}$ is a family of sheaves of abelian groups over a space X, then $\prod_{i \in I} \mathcal{F}_i$ is a sheaf, and it is a categorical direct product in* $\mathbf{Sh}(X, \mathbf{Ab})$.

Proof. The straightforward checking is left to the reader. •

We want to define exact sequences of sheaves; defining the kernel of a presheaf map is straightforward.

Definition. The **kernel** of a presheaf map $\varphi \colon \mathcal{P} \to \mathcal{Q}$ is defined by

$$(\ker \varphi)(U) = \ker(\varphi_U).$$

It is easy to check that $\ker \varphi$ is a subpresheaf of $\mathcal{P}$. Note that the inclusions $\iota_U \colon \ker(\varphi_U) \to \mathcal{P}$ constitute a presheaf map $\iota \colon \ker \varphi \to \mathcal{P}$; we call ι the **inclusion**.

---

[11]However, a product in a subcategory need not be a product in a larger category. For example, let $\mathcal{T}$ be the category of all torsion abelian groups. If $(G_n)_{n \geq 1}$ is a family of torsion groups, then $t(\prod_{n \geq 1} G_n)$ is the categorical direct product in $\mathcal{T}$, while $\prod_{n \geq 1} G_n$ is the categorical direct product in **Ab**.

Proposition 5.80. *Let $\{\mathcal{P}, \rho_U^V\}$ and $\{\mathcal{Q}, \psi_U^V\}$ be presheaves over a space X, and let $\varphi\colon \mathcal{P} \to \mathcal{Q}$ be a presheaf map.*

(i) *The inclusion $\iota\colon \ker\varphi \to \mathcal{P}$ is a categorical kernel of φ in $\mathbf{pSh}(X, \mathbf{Ab})$; that is, ι solves the universal mapping problem*

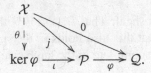

(ii) *If $\mathcal{P}$ and $\mathcal{Q}$ are sheaves, then $\ker\varphi$ is a sheaf, and the sheaf map $\iota\colon \ker\varphi \to \mathcal{P}$ is a categorical kernel of φ in $\mathbf{Sh}(X, \mathbf{Ab})$.*

(iii) *If $\mathcal{P}$ and $\mathcal{Q}$ are sheaves, then $(\ker\varphi)_x = \ker(\varphi_x)$.*

Proof.

(i) For each U, there is a unique map $\theta_U\colon \mathcal{X}(U) \to (\ker\varphi)(U)$ for each U, because $(\ker\varphi)(U) = \ker(\varphi_U)$ solves the universal problem in $\mathbf{Ab}$, as in Example 5.12(iii).

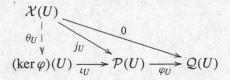

The reader may check that $\theta\colon \mathcal{X} \to \ker\varphi$ is a presheaf map.

(ii) It suffices to prove that $\ker\varphi$ is a sheaf. Assume that $(U_i)_{i \in I}$ is an open cover of an open $U \subseteq X$. If $\sigma, \sigma' \in \mathcal{F}(U)$ agree on overlaps [i.e., $\rho_{U_i \cap U_j}^{U_i}(\sigma) = \rho_{U_i \cap U_j}^{U_i}(\sigma')$ for all i, j], then $\sigma = \sigma'$, because $\mathcal{P}$ is a sheaf. For the gluing axiom, suppose that $\sigma_i \in \ker\varphi_{U_i}$ for all i satisfy the compatibility condition $\rho_{U_i \cap U_j}^{U_i}(\sigma_i) = \rho_{U_i \cap U_j}^{U_j}(\sigma_j)$ for all i, j. Since $\mathcal{P}$ is a sheaf, there is $\sigma \in \mathcal{P}(U)$ with $\rho_{U_i}^{U}(\sigma) = \sigma_i$ for all i. Now

$$\psi_{U_i}^{U}\varphi_U(\sigma) = \varphi_{U_i}\rho_{U_i}^{U}(\sigma) = \varphi_{U_i}(\sigma_i) = 0,$$

because $\sigma_i \in \ker\varphi_{U_i}$. Since $\mathcal{Q}$ is a sheaf, $\varphi_U(\sigma) = 0$; that is, $\sigma \in \ker\varphi_U$. Therefore, $\sigma \in (\ker\varphi)(U)$, and so $\ker\varphi$ is a sheaf.

(iii) Since $(\ker\varphi)(U) = \ker(\varphi_U) \subseteq \mathcal{P}(U)$ for all U, we have $(\ker\varphi)_x = \varinjlim \ker(\varphi_U) \subseteq \varinjlim \mathcal{P}(U)$, because $\{U : U \ni x\}$ is directed. Now $\varphi_x\colon \varinjlim_{U \ni x} \mathcal{P}(U) \to \varinjlim_{U \ni x} \mathcal{Q}(U)$ is defined by $\varphi_x\colon [\sigma_U] \mapsto [\varphi_U\sigma_U]$, and so $\ker(\varphi_x) \subseteq \varinjlim_{U \ni x} \mathcal{P}(U)$. Since both $(\ker \varphi)_x$ and $\ker(\varphi_x)$ are subsets of $\varinjlim_{U \ni x} \mathcal{P}(U)$, it makes sense to assert their equality. If $[\sigma_U] \in \ker(\varphi_U)_x = \varinjlim \ker(\varphi_U)$, then $\varphi_U \sigma_U = 0$ for all U; hence, $\varinjlim \ker(\varphi_U) \subseteq \ker(\varphi_x)$. For the reverse inclusion, if $[\sigma_U] \in \ker(\varphi_x)$, then $[\varphi_U \sigma_U] = 0$ in $\varinjlim \mathcal{Q}(U)$, and so there is some $W \subseteq U$ with $\varphi_W(\rho_W^U \sigma_U) = 0$, by Lemma 5.30(ii). Hence, $\varphi_V \sigma_V = 0$ for all $V \subseteq W$, and so $[\sigma_U] \in \varinjlim \ker(\varphi_U)$.   •

Defining the image of a presheaf map and, hence, defining exactness, is straightforward for presheaves. If $\varphi \colon \mathcal{P} \to \mathcal{Q}$ is a presheaf map, define $\operatorname{im} \varphi$ by $(\operatorname{im} \varphi)(U) = \operatorname{im} \varphi_U$; note that $\operatorname{im} \varphi$ is a subpresheaf of $\mathcal{Q}$.

Definition.   A sequence of presheaves $\mathcal{P}' \xrightarrow{\varphi} \mathcal{P} \xrightarrow{\psi} \mathcal{P}''$ is *exact* in **pSh**$(X, \mathbf{Ab})$ if

$$\operatorname{im} \varphi = \ker \psi.$$

It is easy to see that $\mathcal{P}' \xrightarrow{\varphi} \mathcal{P} \xrightarrow{\psi} \mathcal{P}''$ is an exact sequence of presheaves if and only if $\mathcal{P}'(U) \xrightarrow{\varphi_U} \mathcal{P}(U) \xrightarrow{\psi_U} \mathcal{P}''(U)$ is an exact sequence of abelian groups for every open set U.

If $\mathcal{P}'$ is a subpresheaf of a presheaf $\mathcal{P}$, define the **quotient presheaf** by $(\mathcal{P}/\mathcal{P}')(U) = \mathcal{P}(U)/\mathcal{P}'(U)$. It is easy to see that $\mathcal{P}/\mathcal{P}'$ is a presheaf and that the **natural map** $\pi \colon \mathcal{P} \to \mathcal{P}/\mathcal{P}'$ [with $\pi_U \colon \mathcal{P}(U) \to \mathcal{P}(U)/\mathcal{P}'(U)$] is a presheaf map. If $\varphi \colon \mathcal{P} \to \mathcal{Q}$ is a presheaf map, then $\mathcal{Q}/\operatorname{im} \varphi$ is a *cokernel*. The **First Isomorphism Theorem**, $\mathcal{P}/\ker \varphi \cong \operatorname{im} \varphi$, holds as well.

Example 5.81.   If $\mathcal{F}$ is a sheaf over a space X with every stalk $\{0\}$, then $\mathcal{F}$ is the zero sheaf. However, this is not true for presheaves. Let $\mathcal{P}$ be a presheaf that is not a sheaf. There is an exact sequence of presheaves

$$\mathcal{P} \xrightarrow{\nu} \mathcal{P}^* \to \mathcal{P}^*/\mathcal{P} \to 0,$$

where $\mathcal{P}^*$ is the sheafification of $\mathcal{P}$. For each $x \in X$, the index set $\{U : U \ni x\}$ is a directed set, and so there is an exact sequence

$$\mathcal{P}_x \xrightarrow{\nu_x} \mathcal{P}_x^* \to (\mathcal{P}^*/\mathcal{P})_x \to 0,$$

where ν_x is the identity on $\mathcal{P}_x$. Thus, $(\mathcal{P}^*/\mathcal{P})_x = \{0\}$ for all $x \in X$.   ◀

Alas, the image of a sheaf map need not be a sheaf.

Example 5.82.   Let $\mathcal{O}$ be the sheaf of germs of complex holomorphic functions on the punctured plane $X = \mathbb{C} - \{0\}$; thus,

$$\mathcal{O}(U) = \{\text{holomorphic } f \colon U \to \mathbb{C}\}.$$

Let $\mathcal{O}^\times$ be the sheaf on X defined by $\mathcal{O}^\times(U) = \{\text{holomorphic } f : U \to \mathbb{C}^\times\}$; that is, $f(z) \neq 0$ for all $z \in U$. If $\varphi : \mathcal{O} \to \mathcal{O}^\times$ is the sheaf map defined by $\varphi_U : f \mapsto e^{2\pi i f}$, then $\ker \varphi \cong \mathbb{Z}$, the constant sheaf at $\mathbb{Z}$ on X, and, for each U, there is an exact sequence of presheaves

$$0 \to \mathbb{Z} \to \mathcal{O} \xrightarrow{\varphi} \mathcal{O}^\times \to 0.$$

We claim that $\operatorname{im} \varphi$ is not a sheaf. Let $(U_i)_{i \in I}$ be an open cover of X by disks. Define $f_i \in \mathcal{O}^\times(U_i)$ by $f_i(z) = z$ for all $z \in U_i$. Of course, this family agrees on overlaps, and the unique global section they determine is $f = 1_X$. Now each $f_i \in \varphi(U_i)$, for there is a logarithm $\ell_i(z)$ defined on U_i with $e^{\ell_i(z)} = z$ (because the disk U_i is simply connected). However, it is well-known that there is no complex holomorphic logarithm defined on all of X, and so 1_X is not a global section of $\operatorname{im} \varphi$. Therefore, $\operatorname{im} \varphi$ is not a sheaf. ◀

Example 5.82 shows that we must be more careful with sheaves than with presheaves, for the image of a sheaf map $\mathcal{P} \to \mathcal{Q}$ need not be a sheaf, even when both $\mathcal{P}$ and $\mathcal{Q}$ are sheaves. There is also a problem with quotients: if $\mathcal{F}'$ is a subsheaf of a sheaf $\mathcal{F}$, then the quotient $\mathcal{F}/\mathcal{F}'$ is a presheaf that need not be a sheaf. If $\varphi : \mathcal{F} \to \mathcal{G}$ is a sheaf map between sheaves, then the First Isomorphism Theorem for presheaves gives an isomorphism $\mathcal{F}/\ker \varphi \cong \operatorname{im} \varphi$. Hence, if $\operatorname{im} \varphi$ is not a sheaf, then the quotient presheaf $\mathcal{F}/\ker \varphi$ is not a sheaf either. The definition of quotient sheaf is given in Exercise 5.49 on page 302.

How should an image sheaf be defined? In **Ab**, the cokernel of a map $f : A \to B$ is $B/\operatorname{im} f$; that is, then $\operatorname{im} f = \ker \pi$, where $\pi : B \to B/\operatorname{im} f$ is the natural map. Thus,

$$\operatorname{im}(A \xrightarrow{f} B) = \ker(\operatorname{coker}(A \xrightarrow{f} B)).$$

This remark is interesting because $\operatorname{coker} f$ can be defined as a solution to a universal mapping problem in **Ab** that does not mention $\operatorname{im} f$.

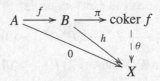

Note that the natural map $\pi : B \to \operatorname{coker} f$ is needed to pose the universal problem. In fact, it is more convenient to think of π as the cokernel rather than $B/\operatorname{im} f$. We could call π the **categorical cokernel** to avoid confusion.

We now define the cokernel of a sheaf map. Recall that $\operatorname{Hom}(\mathcal{F}, \mathcal{X})$ is an abelian group whose identity element is the sheaf map $0 : \mathcal{F} \to \mathcal{X}$.

Definition. A *cokernel* of a sheaf map $\varphi \colon \mathcal{F} \to \mathcal{G}$ is a sheaf map $\pi \colon \mathcal{G} \to \mathcal{C}$ with $\pi \varphi = 0$, where $\mathcal{C} \in \mathrm{obj}(\mathbf{Sh})$, that solves the universal mapping problem: for every sheaf map $\eta \colon \mathcal{G} \to \mathcal{X}$ with $\eta \varphi = 0$, there exists a unique sheaf map $\theta \colon \mathcal{C} \to \mathcal{X}$ with $\theta \pi = \eta$.

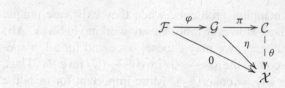

Proposition 5.83. *If $\mathcal{F}, \mathcal{G}$ are sheaves and $\varphi \colon \mathcal{F} \to \mathcal{G}$ is a sheaf map, then* coker φ *exists in* $\mathbf{Sh}(X, \mathbf{Ab})$; *it is the composite* $\mathcal{G} \to \mathcal{G}/\mathrm{im}\,\varphi \to (\mathcal{G}/\mathrm{im}\,\varphi)^*$, *where $(\mathcal{G}/\mathrm{im}\,\varphi)^*$ is the sheafification of $\mathcal{G}/\mathrm{im}\,\varphi$.*

Proof. Consider the diagram of presheaves:

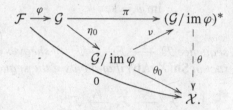

Since $\eta \varphi = 0$, there are presheaf maps $\theta_0 \colon \mathcal{G}/\mathrm{im}\,\varphi \to \mathcal{X}$ and $\eta_0 \colon \mathcal{G} \to \mathcal{G}/\mathrm{im}\,\varphi$ with $\eta = \theta_0 \eta_0$ (because $\mathcal{G}/\mathrm{im}\,\varphi$ is a cokernel in $\mathbf{pSh}$). There is a presheaf map $\nu \colon \mathcal{G}/\mathrm{im}\,\varphi \to (\mathcal{G}/\mathrm{im}\,\varphi)^*$, and we define $\pi \colon \mathcal{G} \to (\mathcal{G}/\mathrm{im}\,\varphi)^*$ by $\pi = \nu \eta_0$. By Theorem 5.70, there is a sheaf map $\theta \colon (\mathcal{G}/\mathrm{im}\,\varphi)^* \to \mathcal{X}$ making the diagram commute. Uniqueness of the sheaf map θ now follows from Lemma 5.69. •

Proposition 5.84. *If $\mathcal{F}, \mathcal{G}$ are sheaves and $\varphi \colon \mathcal{F} \to \mathcal{G}$ is a sheaf map, then* $\ker \varphi = 0$ *if and only if $\varphi_x = 0$ for all $x \in X$, and* coker $\varphi = 0$ *if and only if* $\mathrm{im}\,\varphi_x = \mathcal{G}_x$ *for all $x \in X$.*

Proof. If $\ker \varphi = 0$, then $(\ker \varphi)_x = \{0\}$ for all x, for Proposition 5.74(ii) says that sheaves are zero if and only if all their stalks are zero. But $(\ker \varphi)_x = \ker(\varphi_x)$; that is, $\varphi_x = 0$ for all x. Conversely, if $\varphi_x = 0$ for all $x \in X$, then $(\ker \varphi)_x = 0$ for all x, and so $\ker \varphi = 0$, by Proposition 5.74(ii). This argument can be repeated for coker φ, for $(\mathrm{coker}\,\varphi)_x = \mathrm{coker}(\varphi_x)$. •

Definition. If $\varphi \colon \mathcal{F} \to \mathcal{G}$ is a sheaf map, then

$$\mathrm{im}\,\varphi = \ker(\mathcal{G} \xrightarrow{\ \pi\ } \mathrm{coker}\,\varphi).$$

We are abusing notation: we have defined a sheaf, namely, $\operatorname{coker}\varphi$, that is equal to $(\mathcal{G}/\operatorname{im}\varphi)^*$. By definition, the cokernel of φ is a morphism (not a sheaf!), namely, the sheaf map $\pi: \mathcal{G} \to (\mathcal{G}/\operatorname{im}\varphi)^*$. Thus, the definition says that $\operatorname{im}\varphi = \ker\pi$.

Solutions to universal mapping problems, when they exist, are unique only to (unique!) isomorphism. However, since we are working in $\mathbf{Sh}(X, \mathbf{Ab})$ and not in some abstract category, we may choose, once and for all, a specific solution, namely, the composite $\mathcal{G} \to \mathcal{G}/\operatorname{im}\varphi \to (\mathcal{G}/\operatorname{im}\varphi)^*$. Thus, we may assume that $(\operatorname{coker}\varphi)_x = \operatorname{coker}(\varphi_x)$. More important for us is the consequence:

$$(\operatorname{im}\varphi)_x = (\ker\pi)_x = \ker(\pi_x) = \operatorname{im}(\varphi_x).$$

Definition.   A sequence $\mathcal{F}' \xrightarrow{\varphi} \mathcal{F} \xrightarrow{\psi} \mathcal{F}''$ of sheaves of abelian groups over a space X is **exact** if

$$\operatorname{im}\varphi = \ker\psi.$$

Theorem 5.85.   *A sequence $\mathcal{F}' \xrightarrow{\varphi} \mathcal{F} \xrightarrow{\psi} \mathcal{F}''$ of sheaves of abelian groups over a space X is exact in $\mathbf{Sh}(X, \mathbf{Ab})$ if and only if the sequence of stalks*

$$\mathcal{F}'_x \xrightarrow{\varphi_x} \mathcal{F}_x \xrightarrow{\psi_x} \mathcal{F}''_x$$

is exact in $\mathbf{Ab}$ for all $x \in X$.

Proof.   If the sequence of sheaves is exact, then $\ker\varphi = \operatorname{im}\psi$ and $(\ker\varphi)_x = (\operatorname{im}\psi)_x$ for all $x \in X$. But $(\ker\varphi)_x = \ker(\varphi_x)$ and $(\operatorname{im}\psi)_x = \operatorname{im}(\psi_x)$, and so the sequence of stalks is exact.

If the sequences of stalks are exact (for each $x \in X$), then Proposition 5.74 gives $\operatorname{im}\varphi = \ker\psi$, for both $\operatorname{im}\varphi$ and $\ker\psi$ are subsheaves of $\mathcal{F}$, and so the sequence of sheaves is exact.   •

Corollary 5.86.   *If $\mathcal{F}, \mathcal{G}$ are sheaves and $\varphi: \mathcal{F} \to \mathcal{G}$ is a sheaf map, then there is an exact sequence*

$$0 \to \mathcal{K} \xrightarrow{\iota} \mathcal{F} \xrightarrow{\varphi} \mathcal{G} \xrightarrow{\nu} \mathcal{K}' \to 0$$

with $\iota = \ker\varphi$ and $\nu = \operatorname{coker}\varphi$.

Proof.   This follows at once from Theorem 5.85.   •

Thus, exactness of a sequence of sheaves means exactness at each stalk. In contrast, exactness of a sequence of presheaves means exactness at each open set. If a sequence of *sheaves* $\mathcal{F}' \to \mathcal{F} \to \mathcal{F}''$ is exact in $\mathbf{pSh}(X, \mathbf{Ab})$, then it is also exact in $\mathbf{Sh}(X, \mathbf{Ab})$, for Proposition 5.33 says that exactness of $\mathcal{F}'(U) \to$

$\mathcal{F}(U) \to \mathcal{F}''(U)$ for all open U implies exactness of $\varinjlim_{U \ni x} \mathcal{F}'(U) \to \varinjlim_{U \ni x} \mathcal{F}(U) \to \varinjlim_{U \ni x} \mathcal{F}''(U)$ for all $x \in X$; that is, $\mathcal{F}'_x \to \mathcal{F}_x \to \mathcal{F}''_x$ is exact for all $x \in X$. We have seen that the converse is false.

Exercises

*5.38   **(i)**   Prove that the zero sheaf is a zero object in $\mathbf{Sh}(X, \mathbf{Ab})$ and in $\mathbf{pSh}(X, \mathbf{Ab})$.

(ii)   Prove that $\operatorname{Hom}(\mathcal{P}, \mathcal{P}')$ is an additive abelian group when $\mathcal{P}, \mathcal{P}'$ are presheaves or when $\mathcal{P}, \mathcal{P}'$ are sheaves.

(iii)   The distributive laws hold: given presheaf maps

$$\mathcal{X} \xrightarrow{\alpha} \mathcal{P} \underset{\psi}{\overset{\varphi}{\rightrightarrows}} \mathcal{Q} \xrightarrow{\beta} \mathcal{Y},$$

where $\mathcal{X}$ and $\mathcal{Y}$ are presheaves over a space X, prove that

$$\beta(\varphi + \psi) = \beta\varphi + \beta\psi \quad \text{and} \quad (\varphi + \psi)\alpha = \varphi\alpha + \psi\alpha.$$

*5.39   Let (E, p, X) be an etale-sheaf, and let $\mathcal{F}$ be its sheaf of sections.

(i)   Prove that a *subset* $G \subseteq E$ is a sheet if and only if $G = \sigma(U)$ for some open $U \subseteq X$ and $\sigma \in \mathcal{F}(U)$.

(ii)   Prove that $G \subseteq E$ is a sheet if and only if G is an open subset of E and $p|G$ is a homeomorphism.

(iii)   If $G = \sigma(U)$ and $H = \tau(V)$ are sheets, where $\sigma \in \mathcal{F}(U)$ and $\tau \in \mathcal{F}(V)$, prove that $G \cap H$ is a sheet.

(iv)   If $\sigma \in \mathcal{F}(U)$, prove that

$$\operatorname{supp}(\sigma) = \{x \in X : \sigma(x) \neq 0_x \in E_x\}$$

is a closed subset of X.

Hint.   Consider $\sigma(U) \cap z(U)$, where $z \in \mathcal{F}(U)$ is the zero section.

5.40   Prove that an etale-map $\varphi \colon (E, p, X) \to (E', p', X)$ is an isomorphism in $\mathbf{Sh}_{\mathrm{et}}(X, \mathbf{Ab})$ if and only if $\varphi \colon E \to E'$ is a homeomorphism.

*5.41   Let $\varphi \colon \mathcal{P} \to \mathcal{P}'$ be a presheaf map. Prove that the following statements are equivalent:

(i) φ is an isomorphism;

(ii) $\varphi|\mathcal{P}(U) \colon \mathcal{P}(U) \to \mathcal{P}'(U)$ is an isomorphism for every open set U;

(iii) $\varphi|\mathcal{P}(U)\colon \mathcal{P}(U) \to \mathcal{P}'(U)$ is a bijection for every open set U.

5.42 Prove that every presheaf of abelian groups $\mathcal{P}$ on a discrete space X is a sheaf.

5.43 If $X = \{x\}$ is a space with only one point, prove that

$$\mathbf{pSh}(X, \mathbf{Ab}) = \mathbf{Sh}(X, \mathbf{Ab}) \cong \mathbf{Ab}.$$

*5.44 Let x_*A be a skyscraper sheaf, as in Example 5.72.

(i) Prove, for every sheaf $\mathcal{G}$, that there is an isomorphism

$$\mathrm{Hom}_{\mathbb{Z}}(\mathcal{G}_x, A) \cong \mathrm{Hom}_{\mathbf{Sh}}(\mathcal{G}, x_*A)$$

that is natural in $\mathcal{G}$.

(ii) Every sheaf map $\varphi\colon \mathcal{F} \to \mathcal{G}$ induces homomorphisms of stalks $\varphi_y\colon \mathcal{F}_y \to \mathcal{G}_y$ for all $y \in X$. Choose $x \in X$. If $\mathcal{F}$ is a sheaf over X with stalk $\mathcal{F}_x = A$, prove that there is a sheaf map $\varphi\colon \mathcal{F} \to x_*A$ with $\varphi_x = 1_A$.

*5.45 Let $\nu\colon \mathcal{P} \to \Gamma(\square, \mathcal{P}^{\mathrm{et}})$ be the natural map in Theorem 5.68: in the notation of this proposition, if U is an open set in X, then $\nu_U\colon \mathcal{P}(U) \to \Gamma(U, \mathcal{P}^{\mathrm{et}})$ is given by $\sigma \mapsto \sigma^{\mathrm{et}}$. If $x \in X$, prove that $\nu_x\colon \sigma(x) \mapsto \sigma^{\mathrm{et}}(x) = \sigma(x)$.

*5.46 Let X be a topological space and let $\mathcal{B}$ be a base for the topology $\mathcal{U}$ on X. Viewing $\mathcal{B}$ as a partially ordered set, we may define a presheaf on $\mathcal{B}$ to be a contravariant functor $\mathcal{Q}\colon \mathcal{B} \to \mathbf{Ab}$. Prove that $\mathcal{Q}$ can be extended to a presheaf $\widetilde{\mathcal{Q}}\colon \mathcal{U} \to \mathbf{Ab}$ by defining

$$\widetilde{\mathcal{Q}}(U) = \varprojlim_{\substack{V \in \mathcal{B} \\ V \subseteq U}} \mathcal{Q}(V).$$

If $U \in \mathcal{B}$, prove that $\widetilde{\mathcal{Q}}(U)$ is canonically isomorphic to $\mathcal{Q}(U)$.

5.47   (i) If $f\colon A \to B$ is a homomorphism in $\mathbf{Ab}$ and $K = \ker f$ is the usual kernel (which is a subgroup!), prove that the inclusion $i\colon K \to A$ is a categorical kernel of f.

(ii) If $f\colon A \to B$ is a homomorphism in $\mathbf{Ab}$ and $C = B/\mathrm{im}\, f$ is the usual cokernel (which is a quotient group), then the natural map $p\colon B \to C$ is a categorical cokernel of f. Note that $\mathrm{im}\, f = \ker p$.

5.48 Let $\mathcal{S} = (E, p, X)$ be an etale-sheaf and let $\mathcal{G} = (G, p|G, X)$, where $G \subseteq E$. Prove that $\Gamma(\square, \mathcal{G})$ is a subsheaf of $\Gamma(\square, \mathcal{S})$ if and only if G is open in E and $G_x = G \cap E_x$ is a subgroup for all $x \in X$.

*5.49 If $\mathcal{F}$ is a subsheaf of a sheaf $\mathcal{G}$, define the **quotient sheaf** as $(\mathcal{G}/\mathcal{F})^*$, the sheafification of the presheaf $\mathcal{G}/\mathcal{F}$. Define the **natural map** to be the composite $\pi\colon \mathcal{G} \to \mathcal{G}/\mathcal{F} \to (\mathcal{G}/\mathcal{F})^*$. Prove that if $\iota\colon \mathcal{F} \to \mathcal{G}$ is the inclusion, then the natural map is $\mathrm{coker}\,\iota$.

5.50 Denote the sheafification functor $\mathbf{pSh}(X, \mathbf{Ab}) \to \mathbf{Sh}(X, \mathbf{Ab})$ by $\mathcal{P} \mapsto \mathcal{P}^*$. Prove that $*$ is left adjoint to the inclusion functor $\mathbf{Sh}(X, \mathbf{Ab}) \to \mathbf{pSh}(X, \mathbf{Ab})$. [Either prove this directly or use the fact that $f_* f^*$ is the unit of the adjoint pair (f_*, f^*).]

5.5 Abelian Categories

The most interesting categories for Homological Algebra are *abelian categories*, so called because of their resemblance to **Ab**. Our discussion will apply to categories of modules, sheaves, and chain complexes.

Definition. A category $\mathcal{C}$ is ***additive*** if

(i) $\mathrm{Hom}(A, B)$ is an (additive) abelian group for every $A, B \in \mathrm{obj}(\mathcal{C})$,

(ii) the distributive laws hold: given morphisms

$$X \xrightarrow{a} A \underset{g}{\overset{f}{\rightrightarrows}} B \xrightarrow{b} Y,$$

where X and $Y \in \mathrm{obj}(\mathcal{C})$, then

$$b(f + g) = bf + bg \quad \text{and} \quad (f + g)a = fa + ga,$$

(iii) $\mathcal{C}$ has a zero object (recall that a *zero object* is an object that is both initial and terminal),

(iv) $\mathcal{C}$ has finite products and finite coproducts: for all objects A, B in $\mathcal{C}$, both $A \sqcap B$ and $A \sqcup B$ exist in $\mathrm{obj}(\mathcal{C})$.

If $\mathcal{C}$ and $\mathcal{D}$ are additive categories, a functor $T : \mathcal{C} \to \mathcal{D}$ (of either variance) is ***additive*** if, for all A, B and all $f, g \in \mathrm{Hom}(A, B)$, we have

$$T(f + g) = Tf + Tg;$$

that is, the function $\mathrm{Hom}_{\mathcal{C}}(A, B) \to \mathrm{Hom}_{\mathcal{D}}(TA, TB)$, given by $f \mapsto Tf$, is a homomorphism of abelian groups.

Of course, if T is an additive functor, then $T(0) = 0$, where 0 is either a zero object or a zero morphism.

Lemma 2.3 shows that $_R\mathbf{Mod}$ and $\mathbf{Mod}_R$ are additive categories, while Exercise 5.38 on page 301 and Proposition 5.78 show that both $\mathbf{pSh}(X, \mathbf{Ab})$ and $\mathbf{Sh}(X, \mathbf{Ab})$ are additive categories. Of course, $\mathbf{Sh}_{\mathrm{et}}(X, \mathbf{Ab}) \cong \mathbf{Sh}(X, \mathbf{Ab})$ is also additive. On the other hand, neither **Groups** nor **ComRings** is an additive category.

Proposition 2.4 shows that the Hom functors $_R\mathbf{Mod} \to \mathbf{Ab}$ are additive functors, while Theorem 2.48 shows that the tensor product functors are additive.

That finite coproducts and products coincide for modules and for sheaves is a special case of a more general fact: finite products and finite coproducts coincide in all additive categories.

Lemma 5.87. *Let C be an additive category, and let $M, A, B \in \text{obj}(C)$. Then $M \cong A \sqcap B$ if and only if there are morphisms $i: A \to M$, $j: B \to M$, $p: M \to A$, and $q: M \to B$ such that*

$$pi = 1_A, \quad qj = 1_B, \quad pj = 0, \quad qi = 0, \quad and \quad ip + jq = 1_M.$$

Moreover, $A \sqcap B$ is also a coproduct with injections i and j, and so

$$A \sqcap B \cong A \sqcup B.$$

Proof. The proof of the first statement, left to the reader, is a variation of the proof of Proposition 2.20. The proof of the second statement is a variation of the proof of Proposition 5.8, and it, too, is left to the reader. The last statement holds because two coproducts, here $A \sqcup B$ and $A \sqcap B$, must be isomorphic. •

If A and B are objects in an additive category, then $A \sqcap B \cong A \sqcup B$; their common value, denoted by $A \oplus B$, is called their ***direct sum*** (or ***biproduct***).

Corollary 5.88. *If C and D are additive categories and $T: C \to D$ is an additive functor of either variance, then $T(A \oplus B) \cong T(A) \oplus T(B)$ for all $A, B \in \text{obj}(C)$.*

Proof. Modify the proof of Corollary 2.21 using Lemma 5.87. •

We have been reluctant to discuss injections and surjections in categories; after all, morphisms in a category need not be functions. On the other hand, it is often convenient to have them.

Definition. A morphism $u: B \to C$ in a category C is a ***monomorphism***[12] (or is ***monic***) if u can be canceled from the left; that is, for all objects A and all morphisms $f, g: A \to B$, we have that $uf = ug$ implies $f = g$.

$$A \underset{g}{\overset{f}{\rightrightarrows}} B \xrightarrow{u} C.$$

It is clear that $u: B \to C$ is monic if and only if, for all A, the induced map $u_*: \text{Hom}(A, B) \to \text{Hom}(A, C)$ is an injection. In an additive category,

---

[12]A useful notation for a monomorphism $f: A \to B$ is $A \rightarrowtail B$, while a notation for an epimorphism $g: B \to C$ is $B \twoheadrightarrow C$.

$\mathrm{Hom}(A, B)$ is an abelian group, and so u is monic if and only if $ug = 0$ implies $g = 0$. Exercise 5.55 on page 321 shows that monomorphisms and injections coincide in **Sets** and in $_R$**Mod**. Every injective homomorphism in **Groups** is monic, but we must be clever to show this (see Exercise 5.59). The underlying function of a monomorphism in a concrete category is usually (but not always) an injection; Exercise 5.56 gives an example of a concrete category in which these two notions are distinct.

Here is the dual definition.

Definition.   A morphism $v: B \to C$ in a category $\mathcal{C}$ is an *epimorphism* (or is *epic*) if v can be canceled from the right; that is, for all objects D and all morphisms $h, k: C \to D$, we have that $hv = kv$ implies $h = k$.

$$B \xrightarrow{v} C \underset{k}{\overset{h}{\rightrightarrows}} D.$$

It is clear that $v: B \to C$ is epic if and only if, for all D, the induced map $v^*: \mathrm{Hom}(C, D) \to \mathrm{Hom}(B, D)$ is an injection. In an additive category, $\mathrm{Hom}(A, B)$ is an abelian group, and so v is monic if and only if $gv = 0$ implies $g = 0$. The relation between an epimorphism in a concrete category and the surjectivity of its underlying function is not clear. Exercise 5.55 on page 321 shows that epimorphisms and surjections coincide in **Sets** and in $_R$**Mod**. On the other hand, if R is a domain, then the ring homomorphism $\varphi: R \to \mathrm{Frac}(R)$, given by $r \mapsto r/1$, is an epimorphism in **ComRings**; if A is a commutative ring and $h, k: \mathrm{Frac}(R) \to A$ are ring homomorphisms agreeing on R, then $h = k$. However, φ is a surjective function only when R is a field. Another example is provided by **Top**$_2$, the category of Hausdorff spaces. A continuous $f: X \to Y$ with $\mathrm{im} f$ a dense subspace of Y is an epimorphism, for any two continuous functions agreeing on a dense subspace must be equal. Recognizing presheaf monomorphisms and epimorphisms will follow from the upcoming discussion of abelian categories.

Definition.   If $u: A \to B$ is a morphism in an additive category $\mathcal{A}$, then its *kernel* $\ker u$ is a morphism $i: K \to A$ that satisfies the following universal mapping property: $ui = 0$ and, for every $g: X \to A$ with $ug = 0$, there exists a unique $\theta: X \to K$ with $i\theta = g$.

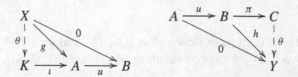

There is a dual definition for *cokernel* (the morphism π in the diagram).

Proposition 5.89. *Let* $u\colon A \to B$ *be a morphism in an additive category* $\mathcal{A}$.

(i) *If* ker u *exists, then* u *is monic if and only if* ker $u = 0$.

(ii) *Dually, if* coker u *exists, then* u *is epic if and only if* coker $u = 0$.

Proof.   We refer to the diagrams in the definitions of kernel and cokernel. Let ker u be $\iota\colon K \to A$, and assume that $\iota = 0$. If $g\colon X \to A$ satisfies $ug = 0$, then the universal property of kernel provides a morphism $\theta\colon X \to K$ with $g = \iota\theta = 0$ (because $\iota = 0$). Hence, u is monic. Conversely, if u is monic, consider

$$K \underset{0}{\overset{\iota}{\rightrightarrows}} A \overset{u}{\longrightarrow} B.$$

Since $u\iota = 0 = u0$, we have $\iota = 0$. The proof for epimorphisms and cokers is dual.   ●

Categorical kernels and cokernels are equivalence classeses of morphisms even though, in our heart of hearts, we think of them as subobjects. However, we saw, in Example 1.3(xi), that objects in a category may not have subobjects in a naive sense, for objects in an arbitrary category are not comprised of elements—there are only other objects and morphisms. Let us try, nevertheless, to define a *subobject* of an object B. It must, obviously, involve an object, say, A, but this is not enough; we need a morphism $i\colon A \to B$ (indeed, a monomorphism i) to relate A to B. Defining a subobject A of B to be an ordered pair (A, i), where $i\colon A \to B$ is monic, is inadequate. In **Ab**, for example, let $B = \mathbb{Q}$ and $A = \mathbb{Z}$. The homomorphisms $i, j\colon A \to B$, defined by $i(1) = 1$ and $j(1) = -1$, are both monic. The ordered pairs (A, i) and (A, j) are distinct, but, intuitively, we want them to be the same; the subgroups $\langle 1 \rangle$ and $\langle -1 \rangle$ are equal, after all.

Definition.   If B is an object in an additive category $\mathcal{A}$, consider all ordered pairs (A, f), where $f\colon A \to B$ is a monomorphism. Call two such pairs (A, f) and (A', f') *equivalent* if there exists an isomorphism $g\colon A' \to A$ with $f' = fg$.

$$
\begin{array}{ccc}
A & \overset{f}{\longrightarrow} & B \\
{\scriptstyle g}\big\uparrow & \nearrow_{f'} & \\
A' & &
\end{array}
$$

A *subgadget of* B is an equivalence class $[(A, f)]$, and we call A a *subobject of* B. Note that if (A', f') is equivalent to (A, f), then $A' \cong A$.

Even though kernels are morphisms, we may regard them as subobjects—just choose a pair (A, f) from the equivalence class. In a general category $\mathcal{C}$, there may not exist any monomorphisms $X \to B$, where $X \in \mathrm{obj}(\mathcal{C})$, except 1_B, and so B may have no subobjects other than $[1_B]$ (see Exercise 5.63 on page 321). Here is the dual notion of *quotient*.

Definition. If B is an object in an additive category $\mathcal{A}$, consider all ordered pairs (f, C), where $f: B \to C$ is an epimorphism. Call two such pairs (f, C) and (f', C') **equivalent** if there exists an isomorphism $g: C \to C'$ with $f' = gf$. A quotient of B is an equivalence class $[(f, C)]$, and we call C a **quotient object** of B. Note that if (f', C') is equivalent to (f, C), then $C' \cong C$.

We may now regard cokernels as quotient objects—just choose a pair (f, C) from the equivalence class.

Abelian categories are additive categories in which a reasonable notion of exactness can be defined. In Proposition 5.89, we saw that if $u: A \to B$ is a morphism in an additive category and $\ker u$ exists, then u is monic if and only if $\ker u = 0$; dually, if $\mathrm{coker}\, u$ exists, then u is epic if and only if $\mathrm{coker}\, u = 0$.

Definition. A category $\mathcal{C}$ is an **abelian category** if it is an additive category such that

  (i) every morphism has a kernel and a cokernel,

  (ii) every monomorphism is a kernel and every epimorphism is a cokernel.[13]

One consequence of the existence of finite direct sums, kernels, and cokernels in an abelian category is the existence of finite direct and inverse limits (see Exercise 5.60 on page 321).

Remark. Abelian categories are **self-dual** in the sense that the dual of every axiom in its definition is itself an axiom; it follows that if $\mathcal{A}$ is an abelian category, then so is its opposite $\mathcal{A}^{\mathrm{op}}$. A theorem using only the axioms in its proof is true in every abelian category; moreover, its dual is also a theorem in every abelian category, and its proof is dual to the original proof. The categories $_R\mathbf{Mod}$ and $\mathbf{Mod}_R$ are abelian categories having extra properties [a category is isomorphic to $_R\mathbf{Mod}$ for some ring R if and only if it is a cocomplete abelian category having a small projective generator P; in this case, $R \cong \mathrm{End}(P)$ (see Pareigis, *Categories and Functors*, p. 241)]. Module categories are not self-dual, because they have these additional properties. This explains why a theorem and its dual that are true in every module category may have very

---

[13]Exercise 5.53 on page 320 says, in any category having a zero object, that kernels are monic and cokernels are epic. The converse is true in abelian categories.

different proofs. For example, the statements "every module is a quotient of a projective module" and "every module can be imbedded in an injective module" are dual and are always true. The proofs are not dual, because these statements are not true in every abelian category. Exercise 5.64 on page 322 shows that the abelian category of all torsion abelian groups has no nonzero projectives, and Exercise 5.65 shows that the abelian category of all finitely generated abelian groups has no nonzero injectives. ◄

Example 5.90.

(i) For every ring R, both $_R\mathbf{Mod}$ and $\mathbf{Mod}_R$ are abelian categories. In particular, $_{\mathbb{Z}}\mathbf{Mod} = \mathbf{Ab}$ is abelian.

(ii) The full subcategory $\mathcal{G}$ of $\mathbf{Ab}$ of all finitely generated abelian groups is an abelian category, as is the full subcategory $\mathcal{T}$ of all torsion abelian groups.

(iii) The full subcategory of $\mathbf{Ab}$ of all torsion-free abelian groups is not an abelian category, for there are morphisms having no cokernel; for example, the inclusion $2\mathbb{Z} \to \mathbb{Z}$ has cokernel $\mathbb{I}_2$, which is not torsion-free.

(iv) Quillen introduced a more general notion that is adequate for Algebraic K-Theory.

> **Definition.** A category $\mathcal{P}$ is an ***exact category*** if $\mathcal{P}$ is a full subcategory of some abelian category $\mathcal{A}$ and if $\mathcal{P}$ is *closed under extensions*; that is, if $0 \to P' \to A \to P'' \to 0$ is an exact sequence in $\mathcal{A}$, and if $P', P'' \in \mathrm{obj}(\mathcal{P})$, then $A \in \mathrm{obj}(\mathcal{P})$.

> Every abelian category is an exact category. The full subcategory of $\mathbf{Ab}$ consisting of all torsion-free abelian groups is an exact category, but it is not an abelian category.

(v) The category **Groups** is not abelian (it is not even additive). If $S \subseteq G$ is a nonnormal subgroup of a group G, then the inclusion $i \colon S \to G$ has no cokernel. However, if K is a normal subgroup of G with inclusion $j \colon K \to G$, then $\mathrm{coker}\, j$ does exist. Thus, axiom (ii) essentially says that every subobject in an abelian category is normal. ◄

Definition. Let $f \colon A \to B$ be a morphism in an abelian category, and let $\mathrm{coker}\, f$ be $\tau \colon B \to C$ for some object C. Then its ***image*** is

$$\mathrm{im}\, f = \ker(\mathrm{coker}\, f) = \ker \tau.$$

In more suggestive notation,

$$\mathrm{im}(A \xrightarrow{f} B) = \ker(\mathrm{coker}\, A \xrightarrow{f} B) = \ker \tau.$$

A sequence $A \xrightarrow{f} B \xrightarrow{g} C$ in $\mathcal{A}$ is ***exact*** if there is equality of subobjects

$$\ker g = \mathrm{im}\, f.$$

Remark. Here is another way to view exactness. If $f \colon A \to B$ is a morphism in an abelian category, then $f = me$, where $m = \ker(\mathrm{coker}\, f)$ is monic and $e = \mathrm{coker}(\ker f)$ is epic. Moreover, this factorization is unique in the following sense. If $f = m'e'$, where m' is monic and e' is epic, then there is equality of subobjects $[m] = [m']$ and $[e] = [e']$ (see Mac Lane, *Categories for the Working Mathematician*, Chapter VIII, Sections 1 and 3). In light of this, we may redefine exactness of a sequence in an abelian category. If $f = me$ and $g = m'e'$, where m, m' are monic and e, e' are epic, then $A \xrightarrow{f} B \xrightarrow{g} C$ is exact if and only if $[e] = [m']$. ◀

Recall that if X is a topological space with topology $\mathcal{U}$, then a *presheaf* of abelian groups over X is a contravariant functor $\mathcal{U} \to \mathbf{Ab}$, and a *sheaf* is a presheaf that satisfies the equalizer condition. We now generalize this notion by replacing $\mathbf{Ab}$ by an abelian category.

Definition. If $\mathcal{A}$ is an abelian category, then a ***presheaf*** on X with values in $\mathcal{A}$ is a contravariant functor $\mathcal{P} \colon \mathcal{U} \to \mathcal{A}$; we shall always assume that $\mathcal{P}(\varnothing) = 0$. A ***sheaf*** is a presheaf that satisfies the equalizer condition. A ***(pre)sheaf map*** is a natural transformation, and all presheaves form the category $\mathbf{pSh}(X, \mathcal{A})$. All sheaves form the full subcategory

$$\mathbf{Sh}(X, \mathcal{A}).$$

Theorem 5.91. *If $\mathcal{A}$ is an abelian category, then $\mathbf{Sh}(X, \mathcal{A})$ is an abelian category.*

Proof. The theorems in the previous section for $\mathbf{Sh}(X, \mathbf{Ab})$ generalize to $\mathbf{Sh}(X, \mathcal{A})$, for the only properties of $\mathbf{Ab}$ that were used hold in every abelian category. Now $\mathbf{Sh}(X, \mathcal{A})$ is an additive category, by Exercise 5.38 on page 301 and Proposition 5.78, and it has kernels and cokernels, by Propositions 5.80 and 5.83. It remains to show that monomorphisms φ are kernels and epimorphisms ψ are cokernels. Given a sheaf map $\varphi \colon \mathcal{F} \to \mathcal{G}$, then $\mathrm{coker}\, \varphi$ equals $\psi \colon \mathcal{G} \to (\mathcal{G}/\mathrm{im}\, \varphi)^*$.

By Corollary 5.86, there is an exact sequence of sheaves

$$0 \to \mathcal{K} \xrightarrow{\iota} \mathcal{F} \xrightarrow{\varphi} \mathcal{G} \xrightarrow{\nu} \mathcal{K}' \to 0.$$

If φ is monic, then ker $\varphi = 0$, by Proposition 5.89, for $\mathbf{Sh}(X, \mathcal{A})$ is an additive category, and so $0 \to \mathcal{F} \xrightarrow{\varphi} \mathcal{G} \xrightarrow{\nu} \mathcal{K}'$ is an exact sequence in $\mathbf{Sh}(X, \mathcal{A})$. Hence, there is an exact sequence $0 \to \mathcal{F}_x \xrightarrow{\varphi_x} \mathcal{G}_x \xrightarrow{\nu_x} \mathcal{K}'_x$ in $\mathcal{A}$; that is, im $\varphi_x = \ker \nu_x$ for all x. Therefore, $\varphi = \ker \nu$, by Lemma 5.69. A dual argument shows that epimorphisms are cokernels. •

We remark that minor changes in the proof of Theorem 5.91 show that $\mathbf{pSh}(X, \mathcal{A})$ is an abelian category. However, we will give another proof of this fact in Corollary 5.94.

The next two propositions construct new abelian categories from old ones.

Proposition 5.92. *Let $\mathcal{S}$ be a full subcategory of an abelian category $\mathcal{A}$. If, for all $A, B \in \text{obj}(\mathcal{S})$ and all $f: A \to B$,*

(i) *a zero object in $\mathcal{A}$ lies in $\mathcal{S}$,*

(ii) *the direct sum $A \oplus B$ in $\mathcal{A}$ lies in $\mathcal{S}$,*

(iii) *both ker f and coker f lie in $\mathcal{S}$,*

then $\mathcal{S}$ is an abelian category. Moreover, if $A \xrightarrow{f} B \xrightarrow{g} C$ is an exact sequence in $\mathcal{S}$, then it is an exact sequence in $\mathcal{A}$.

Proof. The hypothesis gives $\mathcal{S}$ additive, by Exercise 5.54 on page 320, so that $\mathcal{S}$ is abelian if axiom (ii) in the definition of abelian category holds. If $f: A \to B$ is a monomorphism in $\mathcal{S}$, then ker $f = 0$, by Proposition 5.89. But ker f is the same in $\mathcal{A}$ as in $\mathcal{S}$, by hypothesis, so that f is monic in $\mathcal{A}$. By hypothesis, coker f is a morphism in $\mathcal{S}$. As $\mathcal{A}$ is abelian, there is a morphism $g: B \to C$ with $f = \ker g$. But g is a morphism in $\mathcal{S}$, because $\mathcal{S}$ contains cokernels, and so $f = \ker g$ in $\mathcal{S}$. The dual argument shows that epimorphisms in $\mathcal{S}$ are cokernels.

Finally, since kernels and cokernels are the same in $\mathcal{S}$ as in $\mathcal{A}$, images are also the same, and so exactness in $\mathcal{S}$ implies exactness in $\mathcal{A}$. •

Proposition 5.93. *If $\mathcal{A}$ is an abelian category and $\mathcal{C}$ is a small category, then the functor category $\mathcal{A}^{\mathcal{C}}$ is an abelian category.*

Proof. We assume that $\mathcal{C}$ is small to guarantee that the Hom sets $\text{Hom}(F, G)$, where $F, G: \mathcal{C} \to \mathcal{A}$, are sets, not proper classes (see the discussion on page 18). The zero object in $\mathcal{A}^{\mathcal{C}}$ is the constant functor with value 0, where 0 is a zero object in $\mathcal{A}$. If $\tau, \sigma \in \text{Hom}(F, G) = \text{Nat}(F, G)$, where $F, G: \mathcal{C} \to \mathcal{A}$ are functors, define $\tau + \sigma: F \to G$ by $(\tau + \sigma)_C = \tau_C + \sigma_C: FC \to GC$ for all $C \in \text{obj}(\mathcal{C})$. Finally, define $F \oplus G$ by $(F \oplus G)C = FC \oplus GC$. It is straightforward to check that these definitions make $\mathcal{A}^{\mathcal{C}}$ an additive category.

If $\tau : F \to G$, define K by

$$KC = \ker(\tau_C).$$

In the following commutative diagram with exact rows, where $f : C \to C'$ in $\mathcal{C}$, there is a unique $Kf : KC \to KC'$ making the augmented diagram commute.

$$
\begin{array}{ccccccc}
0 & \longrightarrow & KC & \xrightarrow{\ \iota_C\ } & FC & \longrightarrow & GC \\
& & {\scriptstyle Kf}\big\downarrow & & {\scriptstyle Ff}\big\downarrow & & {\scriptstyle Gf}\big\downarrow \\
0 & \longrightarrow & KC' & \xrightarrow[\ \iota_{C'}\ ]{} & FC' & \longrightarrow & GC'
\end{array}
$$

The reader may check that K is a functor, $\iota : K \to F$ is a natural transformation, and $\iota = \ker \tau$; dually, cokernels exist in $\mathcal{A}^{\mathcal{C}}$. Verification of the axioms is routine. •

Combining these propositions gives the following examples.

Corollary 5.94.   *Let $\mathcal{A}$ be an abelian category.*

(i) *The category $\mathbf{pSh}(X, \mathcal{A})$ of presheaves over a space X with values in $\mathcal{A}$ is abelian.*

(ii) *The categories of direct systems $\mathbf{Dir}(I, \mathcal{A})$ and inverse systems $\mathbf{Inv}(I, \mathcal{A})$ are abelian.*

Proof.

(i) $\mathbf{pSh}(X, \mathcal{A})$ is the functor category $\mathcal{A}^{\mathcal{U}^{\mathrm{op}}}$, where $\mathcal{U}$ is the topology on X. The contravariance of a presheaf is encoded by the "exponent" being the small opposite category $\mathcal{U}^{\mathrm{op}}$.

(ii) $\mathbf{Dir}(I, \mathcal{A})$ is the functor category $\mathcal{A}^{I}$, where the partially ordered set I is viewed as a (small) category; similarly, $\mathbf{Inv}(I, \mathcal{A})$ is the functor category $\mathcal{A}^{I^{\mathrm{op}}}$. •

Theorem 5.91, which says that categories of sheaves are abelian, does not follow from Corollary 5.94 and Proposition 5.92, for the category of presheaves and its subcategory of sheaves do not satisfy the conditions of Proposition 5.92; sheaf cokernels may be different than presheaf cokernels.

Example 5.95.   If R is a commutative ring, then $X = \mathrm{Spec}(R)$ is the set of all of its prime ideals. The *Zariski topology* has as *closed sets* those subsets of the form

$$V(S) = \{\mathfrak{p} \in \mathrm{Spec}(R) : S \subseteq \mathfrak{p}\},$$

where S is any subset of R. Of course, open sets are complements of closed sets. A base[14] of the Zariski topology turns out to be all $D(s) = X - V(\{s\})$, where $s \in R$ is nonzero. Thus,

$$D(s) = \{\mathfrak{p} \in \mathrm{Spec}(R) : s \notin \mathfrak{p}\}.$$

Exercise 5.46 on page 302 shows that we can define a presheaf on a space X by giving its values on basic open sets. If $D(t) \subseteq D(s)$, then $t \in \sqrt{Rs}$, by Hilbert's Nullstellensatz, and so $t^n = rs$ for some $r \in R$ and $n \geq 0$. The **structure sheaf of R** is the presheaf $\mathcal{O}$ over $X = \mathrm{Spec}(R)$ of commutative rings having sections $\mathcal{O}(D(s)) = s^{-1}R$ and restriction maps $\rho_{D(t)}^{D(s)} : s^{-1}R \to t^{-1}R$ defined by $u/s^m \mapsto ur^m/t^{nm}$ (recall that $t^n = rs$). The structure sheaf $\mathcal{O}$ is a sheaf of commutative rings, and the stalk $\mathcal{O}_{\mathfrak{p}}$ is the localization $R_{\mathfrak{p}}$. (See Hartshorne, *Algebraic Geometry*, p. 71.) ◄

Example 5.96.   Serre ["Faisceaux algébriques cohérents," *Annals Math* 61 (1955), pp. 197–278] developed the theory of sheaves over spaces X that need not be Hausdorff, enabling him to apply sheaves in Algebraic Geometry. For example, the structure sheaf $\mathcal{O}$ of a commutative ring R is a sheaf of commutative rings over $X = \mathrm{Spec}(R)$, and $\mathrm{Spec}(R)$ is rarely Hausdorff. Because of the importance of Serre's paper, it has acquired a nickname; it is usually referred to as FAC.

Definition.   An $\mathcal{O}$-**Module** (note the capital M), where $\mathcal{O}$ is a sheaf of commutative rings over a space X, is a sheaf $\mathcal{F}$ of abelian groups over X such that

(i) $\mathcal{F}(U)$ is an $\mathcal{O}(U)$-Module for every open $U \subseteq X$,

(ii) if $U \subseteq V$, then $\mathcal{F}(U)$ is also an $\mathcal{O}(V)$-Module, and the restriction $\rho_U^V : \mathcal{F}(V) \to \mathcal{F}(U)$ is an $\mathcal{O}(V)$-Module homomorphism.

If $\mathcal{F}$ and $\mathcal{G}$ are $\mathcal{O}$-Modules, then an $\mathcal{O}$-**morphism** $\tau : \mathcal{F} \to \mathcal{G}$ is a sheaf map such that $\tau_U : \mathcal{F}(U) \to \mathcal{G}(U)$ is an $\mathcal{O}(U)$-map for every open set U.

For example, if $\mathcal{O}$ is the structure sheaf of a commutative ring R, then every R-module M gives rise to an $\mathcal{O}$-Module $\widetilde{M}$ over $\mathrm{Spec}(R)$ whose stalk over $\mathfrak{p} \in \mathrm{Spec}(R)$ is $M_{\mathfrak{p}} = R_{\mathfrak{p}} \otimes_R M$.

All $\mathcal{O}$-Modules and $\mathcal{O}$-morphisms form an abelian category $_{\mathcal{O}}\mathbf{Mod}$ which has a version of tensor product. If $\mathcal{F}$ and $\mathcal{G}$ are $\mathcal{O}$-Modules, then $U \mapsto \mathcal{F}(U) \otimes_{\mathcal{O}(U)} \mathcal{G}(U)$ is a presheaf, and the **tensor product** $\mathcal{F} \otimes_{\mathcal{O}} \mathcal{G}$ is defined to be its sheafification. There is a faithful exact functor $_R\mathbf{Mod} \to {}_{\mathcal{O}}\mathbf{Mod}$

---

[14]Recall that a **base** of a topology is a family of open subsets $\mathcal{B}$ such that every open set is a union of sets in $\mathcal{B}$.

with $M \mapsto \widetilde{M}$, and $_R\mathbf{Mod}$ is isomorphic to the full subcategory of $_O\mathbf{Mod}$ generated by all $\widetilde{M}$ (see Hartshorne, *Algebraic Geometry*, II §5).

Definition. If $\mathcal{O}$ is a sheaf of commutative rings over a space X, then an $\mathcal{O}$-Module $\mathcal{F}$ is ***coherent***[15] if there is an exact sequence

$$\mathcal{O}^s \rightarrow \mathcal{O}^r \rightarrow \mathcal{F} \rightarrow 0,$$

where r, s are natural numbers and $\mathcal{O}^r$ is the direct sum of r copies of $\mathcal{O}$. (We remark that $\mathcal{O}^r$ is *not* a projective object in the category of $\mathcal{O}$-Modules.)

If $\mathcal{F}$ is an $\mathcal{O}$-Module over X, then an ***r-chart*** is an ordered pair (U, φ), where $\varphi \colon \mathcal{F}|U \rightarrow \mathcal{O}^r|U$ is an $\mathcal{O}$-isomorphism of $\mathcal{O}$-Modules; we call U the ***coordinate neighborhood*** of the chart. An $\mathcal{O}$-Module $\mathcal{F}$ is ***locally free of rank*** r if there is a family $(U_i, \varphi_i)_{i \in I}$ of r-charts, called an ***atlas***, whose coordinate neighborhoods form an open cover of X. An ***invertible sheaf***[16] is a locally free $\mathcal{O}$-Module of rank 1.

Let $\mathcal{F}$ be a locally free $\mathcal{O}$-Module over a space X, and let $(U_i, \varphi_i)_{i \in I}$ be an atlas. Whenever an intersection $U_{ij} = U_i \cap U_j$ is nonempty, we can define $\mathcal{O}|U_{ij}$-isomorphisms $\varphi_i \colon (\mathcal{F}|U_i)|U_{ij} \rightarrow \mathcal{O}^r|U_{ij}$ and $\varphi_j \colon (\mathcal{F}|U_j)|U_{ij} \rightarrow \mathcal{O}^r|U_{ij}$ (these isomorphisms are really restrictions of φ_i and φ_j). Now define $\mathcal{O}|U_{ij}$-automorphisms of $\mathcal{O}^r|U_{ij}$

$$g_{ij} = \varphi_i \varphi_j^{-1},$$

called ***transition functions***. Transition functions satisfy the ***cocycle conditions***:

(i) $g_{ij} g_{jk} g_{ki} = 1_{\mathcal{O}^r|U_{ij}}$ for all $i, j \in I$ $(\varphi_i \varphi_j^{-1} \varphi_j \varphi_k^{-1} \varphi_k \varphi_i^{-1} = 1)$;

(ii) $g_{ii} = 1_{\mathcal{O}^r|U_i}$ for all $i \in I$.

Of course, transition functions depend on the choice of atlas $(U_i, \varphi_i)_{i \in I}$. Consider new transition functions arising from a new atlas $(U_i, \widetilde{\varphi}_i)_{i \in I}$ in which we vary only the $\mathcal{O}|U_i$-isomorphisms, keeping the same coordinate neighborhoods. If we define h_i by $\widetilde{\varphi}_i = h_i \varphi_i^{-1}$, then the new transition functions are

$$\widetilde{g}_{ij} = \widetilde{\varphi}_i \widetilde{\varphi}_j^{-1} = h_i \varphi_i \varphi_j^{-1} h_j^{-1} = h_i g_{ij} h_j^{-1}.$$

Let (g_{ij}), $(\widetilde{g}_{ij})$, where $g_{ij}, \widetilde{g}_{ij} \in \mathrm{Aut}(\mathcal{O}^r|U_{ij})$, be two families that may not have arisen as transition functions of a locally free $\mathcal{O}$-Module of rank r. Call (g_{ij}), $(\widetilde{g}_{ij})$ ***equivalent*** if there are $\mathcal{O}(U_{ij})$-isomorphisms h_i such that

$$\widetilde{g}_{ij} = h_i g_{ij} h_j^{-1}.$$

---

[15]A coherent $\mathcal{F}$-Module is an analog of a finitely presented module, and coherent rings are so called because their finitely generated modules are analogous to coherent $\mathcal{O}$-Modules.

[16]If $\mathcal{F}$ is an invertible sheaf, then there exists an (invertible) sheaf $\mathcal{G}$ with $\mathcal{F} \otimes_{\mathcal{O}} \mathcal{G} \cong \mathcal{O}$.

Definition.   Locally free $\mathcal{O}$-Modules $\mathcal{F}$ and $\mathcal{G}$ of rank r are **isomorphic** if their transition functions (g_{ij}) and $(\tilde{g}_{ij})$ are equivalent.

Given a family (g_{ij}), where $g_{ij} \in \mathrm{Aut}(\mathcal{O}^r | U_{ij})$, that satisfies the cocycle conditions, it is not hard to see that there is a unique (to isomorphism) locally free $\mathcal{O}$-Module $\mathcal{F}$ whose transition functions are the given family. In particular, if $\mathcal{F}$ is the constant sheaf with $\mathcal{F}(U)$ of rank r, then there is an open cover $\mathcal{U}$ giving transition functions $g_{ij} = h_i h_j^{-1}$.

A locally free $\mathcal{O}$-Module of rank r is almost classified by an equivalence class of cocycles (g_{ij}); we must still investigate transition functions that arise from an atlas having different families of coordinate neighborhoods. It turns out that transition functions are elements of a certain cohomology *set* of a sheaf with coefficients in the general linear group $\mathrm{GL}(r, k)$ (cohomology need not be a group when coefficients lie in a nonabelian group).   ◄

Projectives and injectives can be defined in any category. However, recognizing epimorphisms and monomorphisms in general categories is too difficult, and we usually restrict attention to abelian categories.

Definition.   An object P in an abelian category $\mathcal{A}$ is **projective** if, for every epic $g: B \to C$ and every $f: P \to C$, there exists $h: P \to B$ with $f = gh$.

$$
\begin{array}{ccc}
 & P & \\
{}^{h}\nearrow & \downarrow f & \\
B \xrightarrow[g]{} & C &
\end{array}
\qquad
\begin{array}{ccc}
 & E & \\
{}^{f}\uparrow & \nwarrow^{h} & \\
A \xrightarrow[g]{} & & B
\end{array}
$$

An object E in an abelian category $\mathcal{A}$ is **injective** if, for every monic $g: A \to B$ and every $f: A \to E$, there exists $h: B \to E$ with $f = hg$.

Definition.   An abelian category $\mathcal{A}$ has **enough injectives** if, for every $A \in \mathrm{obj}(\mathcal{A})$, there exist an injective E and a monic $A \to E$. Dually, $\mathcal{A}$ has **enough projectives** if, for every $A \in \mathrm{obj}(\mathcal{A})$, there exist a projective P and an epic $P \to A$.

We saw in Theorem 2.35 that $_R\mathbf{Mod}$ has enough projectives, and we saw in Theorem 3.38 that $_R\mathbf{Mod}$ has enough injectives.

Proposition 5.97.   *If $\mathcal{A}$ is an abelian category that is closed under products and that has enough injectives, then $\mathbf{Sh}(X, \mathcal{A})$ has enough injectives.*[17]

---

[17] In *The Theory of Sheaves*, Swan writes "... if the base space X is not discrete, I know of no examples of projective sheaves except the zero sheaf." In Bredon, *Sheaf Theory*, McGraw-Hill, New York, 1967, Exercise 4 on p. 20 reads: show that on a locally connected Hausdorff space without isolated points, the only projective sheaf is 0.

Proof. In Example 5.72(i), we defined a *skyscraper sheaf* x_*A, where A is an abelian group and $x \in X$, by

$$(x_*A)(U) = \begin{cases} A & \text{if } x \in U, \\ \{0\} & \text{otherwise.} \end{cases}$$

Of course, we may generalize to $A \in \text{obj}(\mathcal{A})$. By Exercise 5.44 on page 302, there is an isomorphism $\text{Hom}_{\mathcal{A}}(\mathcal{G}_x, A) \cong \text{Hom}_{\textbf{Sh}}(\mathcal{G}, x_*A)$ that is natural in $\mathcal{G}$. In particular, if A is injective, then $\text{Hom}_{\mathcal{A}}(\square, A)$ is an exact functor. It follows that $\text{Hom}_{\textbf{Sh}}(\square, x_*A)$ is also an exact functor; that is, x_*A is an injective sheaf. The proof of Proposition 3.28 shows that any product of injectives is injective; therefore, if A_x is injective, then so is the sheaf $\prod_{x \in X}(x_*A_x)$.

Let $\mathcal{F}$ be a sheaf. By hypothesis, for every $x \in X$, there are an injective $A_x \in \text{obj}(\mathcal{A})$ and a monic $\lambda_x : \mathcal{F}_x \to A_x$. Assemble these morphisms into a sheaf map $\lambda : \prod_{x \in X}(x_*\mathcal{F}_x) \to \prod_{x \in X}(x_*A_x)$. By Exercise 5.44 on page 302, there is a sheaf map $\varphi : \mathcal{F} \to x_*\mathcal{F}_x$ with $\varphi_x : \mathcal{F}_x \to \mathcal{F}_x$ the identity. By the universal property of product, there is a sheaf map

$$\prod_{x \in X}(x_*\mathcal{F}_x) \xleftarrow{\quad \theta \quad} \mathcal{F}$$

with maps p_x and φ_x to $x_*\mathcal{F}_x$.

The composite $\lambda\theta : \mathcal{F} \to \prod_{x \in X}(x_*A_x)$ is a monic sheaf map: $\ker \lambda\theta = 0$ because $(\lambda\theta)_x = \lambda_x\theta_x$, each of whose factors is monic in $\mathcal{A}$. •

The thrust of the next theorem is that it allows us to do diagram chasing in abelian categories.

Definition. A functor $F : \mathcal{C} \to \mathcal{D}$, where $\mathcal{C}, \mathcal{D}$ are categories, is *faithful* if, for each $A, B \in \text{obj}(\mathcal{C})$, the function $\text{Hom}_{\mathcal{C}}(A, B) \to \text{Hom}_{\mathcal{D}}(FA, FB)$, given by $f \mapsto Ff$, is an injection; F is *full* if the function $\text{Hom}_{\mathcal{C}}(A, B) \to \text{Hom}_{\mathcal{D}}(FA, FB)$ is surjective.

If $\mathcal{A}$ is an abelian category, then a functor $F : \mathcal{A} \to \textbf{Ab}$ is *exact* if $A' \to A \to A''$ exact in $\mathcal{A}$ implies $FA' \to FA \to FA''$ exact in $\textbf{Ab}$.

Theorem 5.98 (Freyd–Heron[18]–Lubkin).   *If $\mathcal{A}$ is a small abelian category, then there is a covariant faithful exact functor $F : \mathcal{A} \to \textbf{Ab}$.*

Sketch of Proof. Recall the Yoneda Imbedding, Corollary 1.20: if $\mathcal{A}$ is a small category, then the functor $Y : \mathcal{A}^{\text{op}} \to \textbf{Sets}^{\mathcal{A}}$, which sends each $A \in \text{obj}(\mathcal{A})$ to the representable functor $\text{Hom}(A, \square)$, imbeds $\mathcal{A}$ as a subcategory

---

[18]I have been unable to find any data about Heron other than that he was a student at Oxford around 1960.

of **Sets**$^{\mathcal{A}}$. In the functor category, $\mathrm{Hom}(F, G) = \mathrm{Nat}(F, G)$. The Yoneda Lemma, Theorem 1.17, says that

$$\mathrm{Hom}(Y(A), Y(B)) = \mathrm{Nat}(\mathrm{Hom}(A, \Box), \mathrm{Hom}(B, \Box)) = \mathrm{Hom}(B, A),$$

from which it follows that if $\mathcal{A}$ is an additive category, then im $Y \subseteq \mathbf{Ab}^{\mathcal{A}}$ and Y is an additive functor.

One observes next that the functor category $\mathbf{Ab}^{\mathcal{A}}$ has a generator, namely, $U = \bigoplus_{A \in \mathrm{obj}(\mathcal{A})} \mathrm{Hom}(A, \Box)$, and this is used to prove that every functor $F \in \mathbf{Ab}^{\mathcal{A}}$ has an injective envelope, $\mathrm{Env}(F)$ (however, Example 5.20 shows that Env is not, in general, a functor). In particular, $\mathrm{Env}(U)\colon \mathcal{A} \to \mathbf{Ab}$ turns out to be an exact faithful functor. For details, see Mitchell, *Theory of Categories*, p. 101. •

The imbedding theorem can be improved so that its image is a *full* subcategory of **Ab**.

Theorem 5.99 (Mitchell).   *If $\mathcal{A}$ is a small abelian category, then there is a covariant full faithful exact functor $F\colon \mathcal{A} \to \mathbf{Ab}$.*

Proof.   Mitchell, *Theory of Categories*, p. 151. •

In his *Theory of Categories*, p. 94, Mitchell writes,

> Let us say that a statement about a diagram in an abelian category is **categorical** if it states that certain parts of the diagram are or are not commutative, that certain sequences in the diagram are or are not exact, and that certain parts of the diagram are or are not (inverse) limits or (direct) limits. Then we have the following metatheorem.

Metatheorem.   *Let $\mathcal{A}$ be an abelian category.*

(i) *If a statement is of the form "p implies q," where p and q are categorical statements about a diagram in $\mathcal{A}$, and if the statement is true in **Ab**, then the statement is true in $\mathcal{A}$.*

(ii) *Consider a statement of the form "p implies q," where p is a categorical statement concerning a diagram in $\mathcal{A}$, and q states that additional morphisms exist between certain objects in the diagram and that some categorical statement is true of the extended diagram. If the statement can be proved in **Ab** by constructing the additional morphisms through diagram chasing, then the statement is true in $\mathcal{A}$.*

Proof.   See Mitchell, *Theory of Categories*, p. 97. The category $\mathcal{A}$ need not be a small category, for the metatheorem follows from the Imbedding Theorems with $\mathcal{A}$ replaced by its full subcategory having objects occurring in a diagram. •

Part (i) follows from the Freyd–Heron–Lubkin Imbedding Theorem. To illustrate, the Five Lemma is true in **Ab**, and so it is true in every abelian category. Another example is the 3×3 Lemma (see Exercise 2.32 on page 96), which we now see holds in every abelian category.

Part (ii) follows from Mitchell's Full Imbedding Theorem. To illustrate, recall Proposition 2.70: given a commutative diagram of abelian groups with exact rows,

$$
\begin{array}{ccccccc}
A' & \xrightarrow{\;i\;} & A & \xrightarrow{\;p\;} & A'' & \longrightarrow & 0 \\
{\scriptstyle f}\downarrow & & {\scriptstyle g}\downarrow & & \downarrow{\scriptstyle h} & & \\
B' & \xrightarrow[\;j\;]{} & B & \xrightarrow[\;q\;]{} & B'' & \longrightarrow & 0,
\end{array}
$$

there exists a unique map $h \colon A'' \to B''$ making the augmented diagram commute. Suppose now that the diagram lies in an abelian category $\mathcal{A}$. Applying the imbedding functor $F \colon \mathcal{A} \to$ **Ab** of the Full Imbedding Theorem, we have a diagram in **Ab** as above, and so there is a homomorphism in **Ab**, say, $h \colon F(A'') \to F(B'')$, making the diagram commute: $F(q)F(g) = hF(p)$. Since F is a full imbedding, there exists $\eta \in \mathrm{Hom}_{\mathcal{A}}(A'', B'')$ with $h = F(\eta)$; hence, $F(qg) = F(q)F(g) = hF(p) = F(\eta)F(p) = F(\eta p)$. But F is faithful, so that $qg = \eta p$. Other examples are given in the next chapter: the Snake Lemma, which constructs the *connecting homomorphism* in homology; the Comparison Theorem; the Horseshoe Lemma.

5.5.1 Complexes

The singular homology groups $H_n(X)$ of a topological space X, for $n \geq 0$, are constructed in two steps: first, construct the singular complex

$$
\mathbf{S}_\bullet(X) = \to C_{n+1}(X) \xrightarrow{\partial_{n+1}} C_n(X) \xrightarrow{\partial_n} C_{n-1}(X) \to ;
$$

second, define $H_n(X) = \ker \partial_n / \mathrm{im}\, \partial_{n+1}$. The first step is geometric; the second is algebraic, and it is this second step that is the *raison d'etre* of Homological Algebra. For any abelian category $\mathcal{A}$, we are now going to construct another abelian category **Comp**$(\mathcal{A})$, the category of *complexes over* $\mathcal{A}$; the assignment $X \mapsto \mathbf{S}_\bullet(X)$ will then be a functor **Top** $\to$ **Comp**(**Ab**). In the next chapter, we will construct homology functors $H_n \colon$ **Comp**$(\mathcal{A}) \to \mathcal{A}$, for all n, and singular homology is the composite of these two functors.

Definition. A *complex* (abbreviating *chain complex*) in an abelian category $\mathcal{A}$ is a sequence of objects and morphisms in $\mathcal{A}$ (called *differentials*),

$$
(\mathbf{C}_\bullet, d_\bullet) = \to A_{n+1} \xrightarrow{d_{n+1}} A_n \xrightarrow{d_n} A_{n-1} \to ,
$$

such that the composite of adjacent morphisms is 0:

$$d_n d_{n+1} = 0 \qquad \text{for all } n \in \mathbb{Z}.$$

We usually simplify the notation, writing $\mathbf{C_\bullet}$ or even $\mathbf{C}$ instead of $(\mathbf{C_\bullet}, d_\bullet)$.

It is convenient to consider the category of all complexes, and so we introduce its morphisms.

Definition. If $(\mathbf{C_\bullet}, d_\bullet)$ and $(\mathbf{C'_\bullet}, d'_\bullet)$ are complexes, then a **chain map**

$$f = f_\bullet : (\mathbf{C_\bullet}, d_\bullet) \to (\mathbf{C'_\bullet}, d'_\bullet)$$

is a sequence of morphisms $f_n : C_n \to C'_n$ for all $n \in \mathbb{Z}$ making the following diagram commute:

$$
\begin{array}{ccccccc}
\cdots \longrightarrow & C_{n+1} & \xrightarrow{d_{n+1}} & C_n & \xrightarrow{d_n} & C_{n-1} & \longrightarrow \cdots \\
& \downarrow{f_{n+1}} & & \downarrow{f_n} & & \downarrow{f_{n-1}} & \\
\cdots \longrightarrow & C'_{n+1} & \xrightarrow{d'_{n+1}} & C'_n & \xrightarrow{d'_n} & C'_{n-1} & \longrightarrow \cdots .
\end{array}
$$

It is easy to check that the composite gf of two chain maps

$$f_\bullet : (\mathbf{C_\bullet}, d_\bullet) \to (\mathbf{C'_\bullet}, d'_\bullet) \quad \text{and} \quad g_\bullet : (\mathbf{C'_\bullet}, d'_\bullet) \to (\mathbf{C''_\bullet}, d''_\bullet)$$

is itself a chain map, where $(gf)_n = g_n f_n$. The identity chain map $1_{\mathbf{C_\bullet}}$ on $(\mathbf{C_\bullet}, d_\bullet)$ is the sequence of identity morphisms $1_{C_n} : C_n \to C_n$.

The singular complex $\mathbf{S_\bullet}(X)$ of a topological space X is an example of a complex of abelian groups.

Definition. If $\mathcal{A}$ is an abelian category, then the category of all complexes in $\mathcal{A}$ is denoted by $\mathbf{Comp}(\mathcal{A})$. If R is a ring, then $\mathbf{Comp}(_R\mathbf{Mod})$ is denoted by $_R\mathbf{Comp}$ and $\mathbf{Comp}(\mathbf{Mod}_R)$ is denoted by $\mathbf{Comp}_R$. If the category $\mathcal{A}$ (or the ring R) is understood, we may simply write $\mathbf{Comp}$.

The most important example of $\mathbf{Comp}(\mathcal{A})$ is $\mathbf{Comp} = \mathbf{Comp}(\mathbf{Ab})$, but it is also interesting when $\mathcal{A} = \mathbf{Sh}(X, \mathbf{Ab})$, which arises when one defines cohomology of a topological space with sheaf coefficients. Although everything we say in this subsection holds for general abelian categories, we assume here that complexes are complexes of abelian groups, leaving the reader to generalize using the Metatheorem on page 316.

A complex $(\mathbf{A_\bullet}, \delta_\bullet)$ is defined to be a **subcomplex** of a complex $(\mathbf{C_\bullet}, d_\bullet)$ if there is a chain map $i : \mathbf{A_\bullet} \to \mathbf{C_\bullet}$ with each i_n monic. In $_R\mathbf{Comp}$, we have that $(\mathbf{A_\bullet}, \delta_\bullet)$ is a subcomplex of $(\mathbf{C_\bullet}, \delta_\bullet)$ if A_n is a submodule of C_n and $\delta_n = d_n | A_n$ for every $n \in \mathbb{Z}$.

Proposition 5.100.   *If $\mathcal{A}$ is an abelian category, then $\mathbf{Comp}(\mathcal{A})$ is an abelian category.*

Proof.   As we have just said, we only prove this when $\mathcal{A} = \mathbf{Ab}$. View $\mathbb{Z}$ first as a partially ordered set under reverse inequality and then as a small category (with morphisms $m \to n$ if $m \geq n$). By Proposition 5.93, the functor category $\mathbf{Ab}^{\mathbb{Z}}$ is an abelian category. Proposition 5.92 says that $\mathcal{S}$ is abelian if $\mathcal{S}$ is a full subcategory of $\mathbf{Ab}^{\mathbb{Z}}$ containing a zero object, the direct sum $A \oplus B$ of $A, B \in \mathrm{obj}(\mathcal{S})$, and both $\ker f$ and $\mathrm{coker}\, f$, where f is a morphism in $\mathcal{S}$. All the steps are routine. Note that $\mathbf{Comp}$ is, by definition, a full subcategory of $\mathbf{Ab}^{\mathbb{Z}}$. The *zero complex* is the complex each of whose terms is 0, while $(\mathbf{C_\bullet}, d_\bullet) \oplus (\mathbf{C'_\bullet}, d'_\bullet)$ is, by definition, the complex whose nth term is $C_n \oplus C'_n$ and whose nth differential is $d_n \oplus d'_n$. If $f_\bullet \colon (\mathbf{C_\bullet}, d_\bullet) \to (\mathbf{C'_\bullet}, d'_\bullet)$ is a chain map, define

$$\mathbf{ker}\, f = \to \ker f_{n+1} \xrightarrow{\delta_{n+1}} \ker f_n \xrightarrow{\delta_n} \ker f_{n-1} \to,$$

where $\delta_n = d_n | \ker f_n$, and

$$\mathbf{im}\, f = \to \mathrm{im}\, f_{n+1} \xrightarrow{\Delta_{n+1}} \mathrm{im}\, f_n \xrightarrow{\Delta_n} \mathrm{im}\, f_{n-1} \to,$$

where $\Delta_n = d'_n | \mathrm{im}\, f_n$. Then $\mathbf{ker}\, f$ is a subcomplex of $\mathbf{C_\bullet}$, and $\mathbf{im}\, f$ is a subcomplex of $\mathbf{C'_\bullet}$. If $\mathbf{A_\bullet}$ is a subcomplex of $\mathbf{C_\bullet}$, define the *quotient complex* to be

$$\mathbf{C_\bullet}/\mathbf{A_\bullet} = \to C_n/A_n \xrightarrow{\overline{d}_n} C_{n-1}/A_{n-1} \to,$$

where $\overline{d}_n \colon c_n + A_n \mapsto d_n c_n + A_{n-1}$ (it must be shown that $\overline{d}_n$ is well-defined: if $c_n + A_n = b_n + A_n$, then $d_n c_n + A_{n-1} = d_n b_n + A_{n-1}$). If $p_n \colon C_n \to C_n/A_n$ is the natural map, then $p \colon \mathbf{C_\bullet} \to \mathbf{C_\bullet}/\mathbf{A_\bullet}$ is a chain map. Finally, define

$$\mathbf{coker}\, f = \to C_{n+1}/\mathrm{im}\, \partial_{n+2} \xrightarrow{\overline{\partial}_{n+1}} C_n/\mathrm{im}\, \partial_{n+1} \xrightarrow{\overline{\partial}_n} C_{n-1}/\mathrm{im}\, \partial_n \xrightarrow{\overline{\partial}_{n-1}} \to.$$

The reader must verify that the definitions just given agree with the categorical definitions of ker and coker in $\mathbf{Comp}$.   •

Let us make some other important items in $\mathbf{Comp}$ explicit.

(i) An *isomorphism* in $\mathbf{Comp}$ is a chain map $f \colon \mathbf{C_\bullet} \to \mathbf{C'_\bullet}$ for which $f_n \colon C_n \to C'_n$ is an isomorphism in $\mathcal{A}$ for all $n \in \mathbb{Z}$ (note that the sequence of inverses f_n^{-1} is a chain map; that is, the appropriate diagram commutes).

(ii) If $((\mathbf{C_\bullet^i}, d_\bullet^i))_{i \in I}$ is a family of complexes, then their *direct sum* is the complex

$$\bigoplus_i \mathbf{C_\bullet^i} = \to \bigoplus_i C_{n+1}^i \xrightarrow{\oplus_i d_n^i} \bigoplus_i C_n^i \xrightarrow{\oplus_i d_{n-1}^i} \bigoplus_i C_{n-1}^i \to,$$

where $\bigoplus_i d_n^i$ acts coordinatewise; that is, $\bigoplus_i d_n^i \colon (c_n^i) \mapsto (d_n^i c_n^i)$.

(iii) It is easy to see that ***direct limits*** and ***inverse limits*** of complexes exist in **Comp**($\mathcal{A}$) if they exist in $\mathcal{A}$.

(iv) A sequence of complexes and chain maps

$$\cdots \to C^{m+1}_{\bullet} \xrightarrow{f^{m+1}} C^m_{\bullet} \xrightarrow{f^m} C^{m-1}_{\bullet} \to \cdots$$

is *exact* if $\operatorname{im} f^{m+1} = \ker f^m$ for all $m \in \mathbb{Z}$.

The reader should realize that this notation is very compact. For example, if we write a complex as a column, then a short exact sequence of complexes is really the infinite commutative diagram having three columns and exact rows:

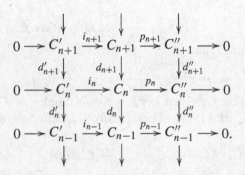

A sequence of complexes $\to C^{n+1}_{\bullet} \xrightarrow{f^{n+1}} C^n_{\bullet} \xrightarrow{f^n} C^{n-1}_{\bullet} \to$ is exact if and only if each row $\to C^{n+1}_m \to C^n_m \to C^{n-1}_m \to$ is an exact sequence of modules.

Exercises

5.51 If $\mathcal{C}$ is an additive category with zero object 0, prove that the unique morphism $A \to 0$ [where $A \in \operatorname{obj}(\mathcal{C})$] and the unique morphism $0 \to A$ are the identity elements of the abelian groups $\operatorname{Hom}_{\mathcal{C}}(A, 0)$ and $\operatorname{Hom}_{\mathcal{C}}(0, A)$.

5.52 If $\mathcal{C}$ is an additive category and $C \in \operatorname{obj}(\mathcal{C})$, prove that $\operatorname{Hom}(C, C)$ is a ring with composition as product.

*5.53 In any category having a zero object, prove that every kernel is a monomorphism and, dually, every cokernel is an epimorphism.

*5.54 Let $\mathcal{C}$ be an additive category and let $\mathcal{S}$ be a subcategory. Prove that $\mathcal{S}$ is an additive category if $\mathcal{S}$ is full, contains a zero object of $\mathcal{C}$, and contains the direct sum $A \oplus B$ (in $\mathcal{C}$) of all $A, B \in \operatorname{obj}(\mathcal{S})$.

*5.55   **(i)**   Prove that a function is epic in **Sets** if and only if it is surjective and that a function is monic in **Sets** if and only if it is injective.

(ii)   Prove that an R-map is epic in $_R$**Mod** if and only if it is surjective and that an R-map is monic in $_R$**Mod** if and only if it is injective.

*5.56 Let $\mathcal{C}$ be the category of all divisible abelian groups.

(i)   Prove that the natural map $\mathbb{Q} \to \mathbb{Q}/\mathbb{Z}$ is monic in $\mathcal{C}$.

(ii)   Conclude that $\mathcal{C}$ is a concrete category in which monomorphisms and injections do not coincide.

*5.57 Prove, in every category, that the injections of a coproduct are monic and the projections of a product are epic.

*5.58   **(i)**   Prove that every isomorphism in an additive category is both monic and epic.

(ii)   Prove that a morphism in an abelian category is an isomorphism if and only if it is both monic and epic.

(iii)   Prove, in **ComRings**, that $\varphi : R \to \mathrm{Frac}(R)$ is both monic and epic, but that φ is not an isomorphism.

*5.59 (***Eilenberg–Moore***) Let G be a (possibly nonabelian) group.

(i)   If H is a proper subgroup of a group G, prove that there exist a group L and distinct homomorphisms $f, g : G \to L$ with $f|H = g|H$.

Hint. Define $L = S_X$, where X denotes the family of all the left cosets of H in G together with an additional element, denoted ∞. If $a \in G$, define $f(a) = f_a \in S_X$ by $f_a(\infty) = \infty$ and $f_a(bH) = abH$. Define $g : G \to S_X$ by $g = \gamma f$, where $\gamma \in S_X$ is conjugation by the transposition (H, ∞).

(ii)   Prove that a homomorphism $\varphi : A \to G$, where A and G are groups, is surjective if and only if it is an epimorphism in **Groups**.

*5.60 We call $\varinjlim_I F$ or $\varprojlim_I F$ *finite* if the index set I is finite. Prove that if $\mathcal{A}$ is an additive category having kernels and cokernels, then $\mathcal{A}$ has all finite inverse limits and direct limits. Conclude that $\mathcal{A}$ has pullbacks, pushouts, equalizers, and coequalizers.

5.61 State and prove the ***First Isomorphism Theorem*** in an abelian category $\mathcal{A}$.

5.62 Prove that every object in **Sets** is projective and injective.

*5.63   **(i)**   Let X be a set and, for each subset $Y \subseteq X$, let $i_Y : Y \to X$ be the inclusion. If 2^X is the family of all subsets of X, prove that the function $2^X \to \{[i_y] : Y \subseteq X\}$, given by $Y \mapsto [i_Y]$, is a bijection, where $[i_Y]$ is the categorical subobject of X.

   (ii) Prove that the analog of (i) is true for **Groups**, **Rings**, and $_R$**Mod**, but it is false for **Top**.

*5.64   (i) Prove that the category $\mathcal{T}$ of all torsion abelian groups is an abelian category having no nonzero projective objects.

   (ii) Prove, for every index set I, that $\mathcal{T}$ has a product $\prod_{i \in I} G_i$.

*5.65   (i) Prove that the full subcategory $\mathcal{T}$ of **Ab** consisting of all torsion abelian groups is an abelian category that is closed under (infinite) coproducts.

   (ii) Prove that $\mathcal{T}$ has enough injectives.

   (iii) Prove that $\mathcal{T}$ has no nonzero projective objects. Conclude that $\mathcal{T}$ is not isomorphic to a category of modules.

*5.66 If $\mathcal{A}$ is an abelian category, prove that a morphism $f = (f_n)$ in $\mathrm{Comp}(\mathcal{A})$ [i.e., a chain map] is monic (or epic) if and only if each f_n is monic (or epic) in $\mathcal{A}$.

5.67 Let $\mathcal{A}$ be an abelian category with enough projectives, and let $\mathfrak{C} \subseteq \mathrm{obj}(\mathcal{A})$ satisfy

   (i) for every object A in $\mathcal{A}$, there exists $C \in \mathfrak{C}$ and an epimorphism $C \to A$;

   (ii) if $C \in \mathfrak{C}$, then every direct summand of C also lies in $\mathfrak{C}$.

Prove that every projective lies in $\mathfrak{C}$. The dual result also holds.

6

Homology

At the end of Chapter 1, we saw that the construction of homology groups of topological spaces has a geometric half and an algebraic half. More precisely, for each $n \geq 0$, the nth singular homology functor $H_n \colon \textbf{Top} \to \textbf{Ab}$ is a composite $\textbf{Top} \to \textbf{Comp(Ab)} \to \textbf{Ab}$, where $\textbf{Comp(Ab)}$ is the category of all complexes of abelian groups. We now focus on the algebraic portion of this construction.

The theorems in this chapter are true for abelian categories $\mathcal{A}$ and additive functors between them (the most interesting categories for us are categories of modules, but the extra generality allows us to apply results to sheaves and to complexes). Even though some of these results hold for arbitrary abelian categories, we will usually assume that $\mathcal{A}$ has enough projectives or injectives. In light of the Metatheorem on page 316, however, it suffices to prove these theorems for the special case $\mathcal{A} = \textbf{Ab}$.

6.1 Homology Functors

Recall that a *complex* in an abelian category $\mathcal{A}$ is a sequence of morphisms (called *differentials*),

$$(\textbf{C}_\bullet, d_\bullet) = \to C_{n+1} \xrightarrow{d_{n+1}} C_n \xrightarrow{d_n} C_{n-1} \to,$$

with the composite of adjacent morphisms being 0:

$$d_n d_{n+1} = 0 \qquad \text{for all } n \in \mathbb{Z}.$$

We usually simplify notation and write $\textbf{C}_\bullet$ or $\textbf{C}$ instead of $(\textbf{C}_\bullet, d_\bullet)$.

J.J. Rotman, *An Introduction to Homological Algebra*, Universitext,
DOI 10.1007/978-0-387-68324-9_6, © Springer Science+Business Media LLC 2009

In $_R\mathbf{Mod}$, the condition $d_n d_{n+1} = 0$ is equivalent to $\operatorname{im} d_{n+1} \subseteq \ker d_n$.

The category $\mathbf{Comp}(\mathcal{A})$ has as *objects* complexes whose terms and differentials are in $\mathcal{A}$, as *morphisms* chain maps $f = (f_n) : (\mathbf{C}_\bullet, d_\bullet) \to (\mathbf{C}'_\bullet, d'_\bullet)$ making the following diagram commute:

$$\begin{array}{ccccccc}
\longrightarrow & C_{n+1} & \xrightarrow{d_{n+1}} & C_n & \xrightarrow{d_n} & C_{n-1} & \xrightarrow{d_{n-1}} \\
 & \downarrow f_{n+1} & & \downarrow f_n & & \downarrow f_{n-1} & \\
\longrightarrow & C'_{n+1} & \xrightarrow[d'_{n+1}]{} & C'_n & \xrightarrow[d'_n]{} & C'_{n-1} & \xrightarrow[d'_{n-1}]{} ,
\end{array}$$

and as *composition* $(g_n)(f_n) = (g_n f_n)$ (i.e., coordinatewise composition). We proved in Chapter 5 that $\mathbf{Comp}(\mathcal{A})$ is an abelian category when $\mathcal{A}$ is.

Example 6.1.

(i) Every exact sequence is a complex, for the equalities $\operatorname{im} d_{n+1} = \ker d_n$ imply $d_n d_{n+1} = 0$.

(ii) If X is a topological space, then its singular chain groups and boundary maps form a complex of abelian groups (called the *singular complex*)

$$\mathbf{S}_\bullet(X) = \to S_{n+1}(X) \xrightarrow{\partial_{n+1}} S_n(X) \xrightarrow{\partial_n} S_{n-1}(X) \to .$$

However, the definition of singular complex is incomplete, for $S_n(X)$ was defined only for $n \geq 0$. To complete the definition, set $S_n(X) = \{0\}$ for all $n \leq -1$. The differentials $S_n \to S_{n-1}$ for negative n are necessarily 0, and so the lengthened sequence is, indeed, a complex. This device of adding 0s is always available.

Similarly, if K is a simplicial complex, then $\mathbf{C}_\bullet(K)$ is a complex, where $C_n(K)$ is the group of all simplicial n-chains.

(iii) If $A \in \operatorname{obj}(\mathcal{A})$ and $k \in \mathbb{Z}$ is a fixed integer, then the sequence $\varrho^k(A)$ whose kth term is A, whose other terms are 0, and whose differentials are zero maps is a complex, called A *concentrated in degree* k.

(iv) Every morphism $f : A \to B$ is a differential; in more detail, form a complex $\Sigma^k(f)$ whose kth term is A, whose $(k-1)$st term is B, whose other terms are 0, and whose kth diffentiation is f:

$$\Sigma^k(f) = \to 0 \to 0 \to A \xrightarrow{f} B \to 0 \to 0 \to .$$

Call $\Sigma^k(f)$ the complex having f *concentrated in degrees* $(k, k-1)$.

(v) A short exact sequence can be made into a complex by adding 0s to the left and right:

$$\to 0 \to 0 \to A \xrightarrow{i} B \xrightarrow{p} C \to 0 \to 0 \to .$$

Usually, one assumes that A is term 2, B is term 1, and C is term 0. This is a complex because $pi = 0$.

(vi) Every sequence of objects (M_n) occurs in a complex, namely, $(\mathbf{M}_\bullet, d_\bullet)$, in which all the differentials d_n are 0. ◀

Let us begin by seeing that the idea of describing a module by generators and relations gives rise to complexes. Recall our discussion in How to Read This Book on pages xi and xii: if A is an R-module, then a **presentation** of A is (X, Y), where F is a free R-module mapping onto A, X is a basis of F, and Y generates $K = \ker(F \to A)$. A presentation allows us to treat equations in A as if they were equations in the free module F. Computations in F, especially those involved in whether elements of F lie in K, become much simpler when K is also free and Y is a basis. We know, however, that submodules of free modules need not be free. It is natural to iterate taking generators and relations: map a free module F_1 onto K, and let (X_1, Y_1) be a presentation of K; that is, X_1 is a basis of F_1 and Y_1 generates $K_1 = \ker(F_1 \to K)$. If K_1 is free, we stop; if K_1 is not free, we continue with a presentation of it. Thus, a *free resolution* of a module A is a generalized presentation; it is our way of treating equations in A by a sequence of equations in free modules. We can now begin to appreciate Theorem 8.37, Hilbert's Theorem on Syzygies, which says that if $R = k[x_1, \ldots, x_n]$ (where k is a field) and A is an R-module, then the kernel K_n after n iterations must be free.

Definition.   A *projective resolution* of $A \in \mathrm{obj}(\mathcal{A})$, where $\mathcal{A}$ is an abelian category, is an exact sequence

$$\mathbf{P} = \to P_2 \xrightarrow{d_2} P_1 \xrightarrow{d_1} P_0 \xrightarrow{\varepsilon} A \to 0$$

in which each P_n is projective. If $\mathcal{A}$ is $_R\mathbf{Mod}$ or $\mathbf{Mod}_R$, then a *free resolution* of a module A is a projective resolution in which each P_n is free; a *flat resolution* is an exact sequence in which each P_n is flat.

If $\mathbf{P}$ is a projective resolution of A, then its *deleted projective resolution* is the complex

$$\mathbf{P}_A = \to P_2 \xrightarrow{d_2} P_1 \xrightarrow{d_1} P_0 \to 0.$$

A projective (or free or flat) resolution is a complex if we assume it has been lengthened by adding 0s to the right. Of course, a deleted resolution is no longer exact if $A \neq 0$, for $\mathrm{im}\, d_1 = \ker \varepsilon \neq \ker(P_0 \to 0) = P_0$.

Deleting A loses no information: $A \cong \operatorname{coker} d_1$; the inverse operation, restoring A to $\mathbf{P}_A$, is called **augmenting**. Deleted resolutions should be regarded as glorified presentations.

Proposition 6.2. *Every (left or right) R-module A has a free resolution (which is necessarily a projective resolution and a flat resolution).*

Proof. There are a free module F_0 and an exact sequence

$$0 \to K_1 \xrightarrow{i_1} F_0 \xrightarrow{\varepsilon} A \to 0.$$

Similarly, there are a free module F_1, a surjection $\varepsilon_1 \colon F_1 \to K_1$, and an exact sequence

$$0 \to K_2 \xrightarrow{i_2} F_1 \xrightarrow{\varepsilon_1} K_1 \to 0.$$

Splice these together: define $d_1 \colon F_1 \to F_0$ to be the composite $i_1\varepsilon_1$. It is plain that $\operatorname{im} d_1 = K_1 = \ker \varepsilon$ and $\ker d_1 = K_2$, yielding the exact row

This construction can be iterated for all $n \geq 0$, and the ultimate exact sequence is infinitely long.

The parenthetical statement follows because free $\Rightarrow$ projective $\Rightarrow$ flat.   •

We have actually proved more.

Corollary 6.3. *If $\mathcal{A}$ is an abelian category with enough projectives, then every $A \in \operatorname{obj}(\mathcal{A})$ has a projective resolution.*

Proof. Apply the proof of Proposition 6.2, mutatis muntandis.   •

Definition. An **injective resolution** of $A \in \operatorname{obj}(\mathcal{A})$, where $\mathcal{A}$ is an abelian category, is an exact sequence

$$\mathbf{E} = 0 \to A \xrightarrow{\eta} E^0 \xrightarrow{d^0} E^1 \xrightarrow{d^1} E^2 \to$$

in which each E^n is injective.

If $\mathbf{E}$ is an injective resolution of A, then its **deleted injective resolution** is the complex

$$\mathbf{E}^A = 0 \to E^0 \xrightarrow{d^0} E^1 \xrightarrow{d^1} E^2 \to .$$

Deleting A loses no information, for $A \cong \ker d^0$.

Proposition 6.4. *Every (left or right) R-module A has an injective resolution.*

Proof. By Theorem 3.38, every module can be imbedded as a submodule of an injective module. Thus, there are an injective module E^0, an injection $\eta \colon A \to E^0$, and an exact sequence

$$0 \to A \xrightarrow{\eta} E^0 \xrightarrow{\pi} V^0 \to 0,$$

where $V^0 = \operatorname{coker} \eta$ and π is the natural map. Repeat: there are an injective module E^1 and an imbedding $\eta^1 \colon V^0 \to E^1$, yielding the exact row

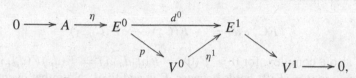

where d^0 is the composite $d^0 = \eta^1 p$. This construction can be iterated. ●

Deleted injective resolutions should be regarded as duals of presentations.

Corollary 6.5. *If $\mathcal{A}$ is an abelian category with enough injectives, then every $A \in \operatorname{obj}(\mathcal{A})$ has an injective resolution.*

In particular, every sheaf with values in $\mathcal{A}$ has an injective resolution.

Proof. By Theorem 5.91, the category $\mathbf{Sh}(X, \mathcal{A})$ of sheaves is an abelian category; by Proposition 5.97, it has enough injectives. ●

Most categories of sheaves do not have enough projectives.

We may lengthen an injective resolution by adding 0s to the left, but this does not yet make it a complex, for the definition of complex says that the indices must decrease if we go the right. The simplest way to satisfy the definition is to use negative indices: define $C_{-n} = E^n$, and

$$0 \to A \to C_0 \to C_{-1} \to C_{-2} \to$$

is a complex.

Definition. Given a projective resolution in an abelian category $\mathcal{A}$,

$$\mathbf{P} = \to P_n \xrightarrow{d_n} P_{n-1} \to \cdots \to P_1 \xrightarrow{d_1} P_0 \xrightarrow{\varepsilon} A \to 0,$$

define $K_0 = \ker \varepsilon$ and $K_n = \ker d_n$, for $n \geq 1$. We call K_n the ***nth syzygy*** of $\mathbf{P}$. Given an injective resolution

$$\mathbf{E} = 0 \to A \xrightarrow{\eta} E^0 \xrightarrow{d^0} E^1 \to \cdots \to E^n \xrightarrow{d^n} E^{n+1} \to,$$

define $V_0 = \operatorname{coker} \eta$ and $V^n = \operatorname{coker} d^{n-1}$, for $n \geq 1$. We call V^n the ***nth cosyzygy*** of $\mathbf{E}$.

Injective resolutions are not the only way in which complexes with indices going up can occur.

Example 6.6.   Let $\mathcal{A}$ be an abelian category.

(i) Let $F : \mathcal{A} \to \mathbf{Ab}$ be a covariant additive functor, and let

$$\mathbf{C} = \to C_n \xrightarrow{d_n} C_{n-1} \to$$

be a complex. Then

$$(F\mathbf{C}, Fd) = \to F(C_n) \xrightarrow{Fd_n} F(C_{n-1}) \to$$

is also a complex, for $0 = F(0) = F(d_n d_{n+1}) = F(d_n) F(d_{n+1})$ [the equation $0 = F(0)$ holds because F is additive]. Note that even if the original complex is exact, the functored complex $F\mathbf{C}$ may not be exact.

(ii) If F is a contravariant additive functor, it is also true that $F\mathbf{C}$ is a complex, but we have to arrange notation so that differentials lower indices. In more detail, after applying F, we have

$$F\mathbf{C} = \leftarrow F(C_n) \xleftarrow{Fd_n} F(C_{n-1}) \leftarrow;$$

the differentials Fd_n increase indices by 1. Introducing negative indices almost solves the problem. Define

$$X_{-n} = F(C_n),$$

so that the sequence becomes

$$\leftarrow X_{-n} \xleftarrow{Fd_n} X_{-n+1} \leftarrow,$$

or $\to X_{-n+1} \xrightarrow{Fd_n} X_{-n} \to$. The index on the map should be $-n + 1$, and not n. Define

$$\delta_{-n+1} = Fd_n.$$

The relabeled sequence now reads properly:

$$F\mathbf{C} = \to X_{-n+1} \xrightarrow{\delta_{-n+1}} X_{-n} \to . \quad \blacktriangleleft$$

Definition.   A complex $\mathbf{C}$ is a *positive complex* if $C_n = 0$ for all $n < 0$. Thus, a positive complex looks like

$$\to C_n \to C_{n-1} \to \cdots \to C_1 \to C_0 \to 0.$$

All positive complexes form the full subcategory $\mathbf{Comp}_{\geq 0}(\mathcal{A})$ of $\mathbf{Comp}(\mathcal{A})$. A complex $\mathbf{C}$ is a ***negative complex*** (or ***cochain complex***) if $C_n = 0$ for all $n > 0$. A negative complex looks like

$$0 \to C_0 \to C_{-1} \to \cdots \to C_{-n} \to C_{-n-1} \to .$$

All negative complexes form the full subcategory $\mathbf{Comp}^{\leq 0}(\mathcal{A})$ of $\mathbf{Comp}(\mathcal{A})$. As in Example 6.6(ii), we usually raise indices and change sign in this case:

$$0 \to C^0 \to C^1 \to \cdots \to C^n \to C^{n+1} \to .$$

Projective resolutions are positive complexes, and injective resolutions are negative complexes.

Since $\mathbf{Comp}(\mathcal{A})$ is an abelian category when $\mathcal{A}$ is, its Hom sets are abelian groups; addition is given by

$$f + g = (f_n + g_n), \text{ where } f = (f_n) \text{ and } g = (g_n).$$

The following definitions imitate the construction of homology groups of topological spaces, which we described in Section 1.3.

Definition.   If $(\mathbf{C}, d)$ is a complex in $\mathbf{Comp}(\mathcal{A})$, where $\mathcal{A}$ is an abelian category, define

$$n\text{-}\textit{chains} = C_n,$$
$$n\text{-}\textit{cycles} = Z_n(\mathbf{C}) = \ker d_n,$$
$$n\text{-}\textit{boundaries} = B_n(\mathbf{C}) = \operatorname{im} d_{n+1}.$$

Notice that C_n, Z_n, and B_n all lie in $\mathcal{A}$.

In $_R\mathbf{Mod}$, the equation $d_n d_{n+1} = 0$ in a complex is equivalent to the condition $\operatorname{im} d_{n+1} \subseteq \ker d_n$; hence, $B_n(\mathbf{C}) \subseteq Z_n(\mathbf{C})$ for every complex $\mathbf{C}$. This is also true in an abelian category:

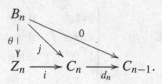

Definition.   If $\mathbf{C}$ is a complex in $\mathbf{Comp}(\mathcal{A})$, where $\mathcal{A}$ is an abelian category, and $n \in \mathbb{Z}$, its nth ***homology*** is

$$H_n(\mathbf{C}) = Z_n(\mathbf{C})/B_n(\mathbf{C}).$$

Now $H_n(\mathbf{C})$ lies in $\operatorname{obj}(\mathcal{A})$ if quotients are viewed as objects, as on page 307. However, if we recognize $\mathcal{A}$ as a full subcategory of $\mathbf{Ab}$, then an element of $H_n(\mathbf{C})$ is a coset $z + B_n(\mathbf{C})$; we call this element a ***homology class***, and often denote it by $\operatorname{cls}(z)$.

Example 6.7.

(i) A complex $\mathbf{C}$ is an exact sequence if and only if $H_n(\mathbf{C}) = 0$ for all n. Thus, homology measures the deviation of a complex from being an exact sequence. An exact sequence is often called an *acyclic complex*; *acyclic* means "no cycles"; that is, no cycles that are not boundaries.

(ii) There are two *fundamental exact sequences* arising from a complex $(\mathbf{C}, d)$: for each $n \in \mathbb{Z}$,

$$0 \to B_n \xrightarrow{i_n} Z_n \to H_n(\mathbf{C}) \to 0$$

and

$$0 \to Z_n \xrightarrow{j_n} C_n \xrightarrow{d_n'} B_{n-1} \to 0,$$

where i_n, j_n are inclusions and $j_{n-1}i_{n-1}d_n' = d_n$; that is, d_n' is just d_n with its target changed from C_{n-1} to $\operatorname{im} d_n = B_{n-1}$.

(iii) If $(\mathbf{C}, d)$ is a complex with all $d_n = 0$, then $H_n(\mathbf{C}) = C_n$ for all $n \in \mathbb{Z}$, for

$$H_n(\mathbf{C}) = \ker d_n / \operatorname{im} d_{n+1} = \ker d_n = C_n.$$

In particular, the subcomplexes $\mathbf{Z}$ of cycles and $\mathbf{B}$ of boundaries have all differentials 0, and so $H_n(\mathbf{Z}) = Z_n$ and $H_n(\mathbf{B}) = B_n$.

(iv) In Example 6.1(iv), we saw that every morphism $f: A \to B$ can be viewed as the complex $\Sigma^1(f)$ concentrated in degrees 1, 0:

$$\Sigma^1(f) = \to 0 \to 0 \xrightarrow{d_2} A \xrightarrow{d_1} B \xrightarrow{d_0} 0 \to 0,$$

with A term 1 and B term 0. Now $d_2 = 0$ implies $\operatorname{im} d_2 = 0$, and $d_0 = 0$ implies $\ker d_0 = B$; it follows that

$$H_n(\Sigma^1(f)) = \begin{cases} \ker f & \text{if } n = 1, \\ \operatorname{coker} f & \text{if } n = 0, \\ 0 & \text{otherwise.} \end{cases} \quad \blacktriangleleft$$

Proposition 6.8. *If $\mathcal{A}$ is an abelian category, then $H_n: \mathbf{Comp}(\mathcal{A}) \to \mathcal{A}$ is an additive functor for each $n \in \mathbb{Z}$.*

Proof. In light of the Metatheorem on page 316, a consequence of Theorem 5.99, Mitchell's Full Imbedding Theorem, it suffices to prove this proposition when $\mathcal{A} = \mathbf{Ab}$. We have just defined H_n on objects; it remains to define H_n on morphisms. If $f: (\mathbf{C}, d) \to (\mathbf{C}', d')$ is a chain map, define $H_n(f): H_n(\mathbf{C}) \to H_n(\mathbf{C}')$ by

$$H_n(f): \operatorname{cls}(z_n) \mapsto \operatorname{cls}(f_n z_n).$$

We must show that $f_n z_n$ is a cycle and that $H_n(f)$ is independent of the choice of cycle z_n; both of these follow from f being a chain map; that is, from commutativity of the following diagram:

$$
\begin{array}{ccccc}
C_{n+1} & \xrightarrow{d_{n+1}} & C_n & \xrightarrow{d_n} & C_{n-1} \\
{\scriptstyle f_{n+1}}\downarrow & & {\scriptstyle f_n}\downarrow & & \downarrow{\scriptstyle f_{n-1}} \\
C'_{n+1} & \xrightarrow[d'_{n+1}]{} & C'_n & \xrightarrow[d'_n]{} & C'_{n-1}.
\end{array}
$$

First, let z be an n-cycle in $Z_n(\mathbf{C})$, so that $d_n z = 0$. Then commutativity of the diagram gives $d'_n f_n z = f_{n-1} d_n z = 0$, so that $f_n z$ is an n-cycle.

Next, assume that $z + B_n(\mathbf{C}) = y + B_n(\mathbf{C})$; hence, $z - y \in B_n(\mathbf{C})$;

$$z - y = d_{n+1} c$$

for some $c \in C_{n+1}$. Applying f_n gives

$$f_n z - f_n y = f_n d_{n+1} c = d'_{n+1} f_{n+1} c \in B_n(\mathbf{C}').$$

Thus, $\operatorname{cls}(f_n z) = \operatorname{cls}(f_n y)$, and $H_n(f)$ is well-defined.

Let us now see that H_n is a functor. It is obvious that $H_n(1_{\mathbf{C}})$ is the identity. If f and g are chain maps whose composite gf is defined, then for every n-cycle z, we have (with obvious abbreviations)

$$
\begin{aligned}
H_n(gf) \colon \operatorname{cls}(z) &\mapsto (gf)_n \operatorname{cls}(z) \\
&= g_n f_n(\operatorname{cls}(z)) \\
&= H_n(g)(\operatorname{cls}(f_n z)) \\
&= H_n(g) H_n(f)(\operatorname{cls}(z)).
\end{aligned}
$$

Finally, H_n is additive: if $f, g \colon (\mathbf{C}, d) \to (\mathbf{C}', d')$ are chain maps, then

$$
\begin{aligned}
H_n(f + g) \colon \operatorname{cls}(z) &\mapsto (f_n + g_n) \operatorname{cls}(z) \\
&= \operatorname{cls}(f_n z + g_n z) \\
&= \big(H_n(f) + H_n(g)\big) \operatorname{cls}(z). \quad \bullet
\end{aligned}
$$

Proposition 6.8 says that if $\mathbf{C}$ is a complex in an abelian category $\mathcal{A}$, then $H_n(\mathbf{C}) \in \operatorname{obj}(\mathcal{A})$ for all n; in particular, if $\mathcal{A}$ is the category of all sheaves of abelian groups over a space X, then $H_n(\mathbf{C})$ is a sheaf. In this case, one often denotes $H_n(\mathbf{C})$ by $\mathcal{H}_n(\mathbf{C})$.

Definition. We call $H_n(f)$ the **induced map**, and we usually denote it by f_{n*}, or even by f_*.

The following elementary construction is fundamental; it gives a relation between different homologies. The proof is a series of diagram chases, which is legitimate because of the metatheorem on page 316. Ordinarily, we would just say that the proof is routine, but, because of the importance of the result, we present (perhaps too many) details; as a sign that the proof is routine, we drop many subscripts.

Proposition 6.9.   *Let $\mathcal{A}$ be an abelian category. If*

$$0 \to \mathbf{C}' \xrightarrow{i} \mathbf{C} \xrightarrow{p} \mathbf{C}'' \to 0$$

is an exact sequence in **Comp**$(\mathcal{A})$, *then, for each $n \in \mathbb{Z}$, there is a morphism in $\mathcal{A}$*

$$\partial_n : H_n(\mathbf{C}'') \to H_{n-1}(\mathbf{C}')$$

defined by

$$\partial_n : \mathrm{cls}(z_n'') \mapsto \mathrm{cls}(i_{n-1}^{-1} d_n p_n^{-1} z_n'').$$

Proof.   We will make many notational abbreviations in this proof. Consider the commutative diagram having exact rows:

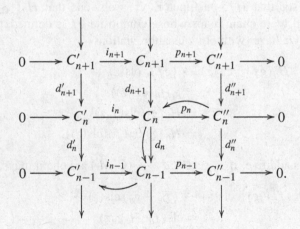

Let $z'' \in C_n''$ and $d'' z'' = 0$. Since p_n is surjective, there is $c \in C_n$ with $pc = z''$. Now push c down to $dc \in C_{n-1}$. By commutativity, $p_{n-1} dc = d'' p_n c = d'' z'' = 0$, so that $dc \in \ker p_{n-1} = \operatorname{im} i_{n-1}$. Therefore, there is a unique $c' \in C_{n-1}'$ with $i_{n-1} c' = dc$, for i_{n-1} is an injection. Thus, $i_{n-1}^{-1} d p_n^{-1} z''$ makes sense; that is, the claim is that

$$\partial_n(\mathrm{cls}(z'')) = \mathrm{cls}(c')$$

is a well-defined homomorphism.

First, let us show independence of the choice of lifting. Suppose that $p_n \check{c} = z''$, where $\check{c} \in C_n$. Then $c - \check{c} \in \ker p_n = \operatorname{im} i_n$, so that there is $u' \in C'_n$ with $i_n u' = c - \check{c}$. By commutativity of the first square, we have

$$i_{n-1} d' u' = d i_n u' = dc - d\check{c}.$$

Hence, $i^{-1} dc - i^{-1} d\check{c} = d'u' \in B'_{n-1}$; that is, $\operatorname{cls}(i^{-1} dc) = \operatorname{cls}(i^{-1} d\check{c})$. Thus, the formula gives a well-defined function

$$Z''_n \to C'_{n-1}/B'_{n-1}.$$

Second, the function $Z''_n \to C'_{n-1}/B'_{n-1}$ is a homomorphism. If z'', $z''_1 \in Z''_n$, let $pc = z''$ and $pc_1 = z''_1$. Since the definition of ∂ is independent of the choice of lifting, choose $c + c_1$ as a lifting of $z'' + z''_1$. This step may now be completed in a routine way.

Third, we show that if $i_{n-1} c' = dc$, then c' is a cycle: $0 = ddc = dic' = idc'$, and so $d'c' = 0$ because i is an injection. Hence, the formula gives a homomorphism

$$Z'' \to Z'/B' = H_{n-1}.$$

Finally, the subgroup B''_n goes into B'_{n-1}. Suppose that $z'' = d''c''$, where $c'' \in C''_{n+1}$, and let $pu = c''$, where $u \in C_{n+1}$. Commutativity gives $pdu = d'' pu = d'' c'' = z''$. Since $\partial(z'')$ is independent of the choice of lifting, we choose du with $pdu = z''$, and so $\partial(\operatorname{cls}(z'')) = \operatorname{cls}(i^{-1} d(du)) = \operatorname{cls}(0)$. Thus, the formula gives a homomorphism $\partial_n : H_n(\mathbf{C}'') \to H_{n-1}(\mathbf{C}')$. •

Definition. The morphisms $\partial_n : H_n(\mathbf{C}'') \to H_{n-1}(\mathbf{C}')$ are called *connecting homomorphisms*.

The first question we ask is what homology functors do to a short exact sequence of complexes. The next theorem is also proved by diagram chasing and, again, we give too many details because of the importance of the result. The reader should try to prove the theorem before looking at the proof.

Theorem 6.10 (Long Exact Sequence). *Let $\mathcal{A}$ be an abelian category. If*

$$0 \to \mathbf{C}' \xrightarrow{i} \mathbf{C} \xrightarrow{p} \mathbf{C}'' \to 0$$

is an exact sequence in $\mathbf{Comp}(\mathcal{A})$, *then there is an exact sequence in* $\mathcal{A}$

$$\to H_{n+1}(\mathbf{C}'') \xrightarrow{\partial_{n+1}} H_n(\mathbf{C}') \xrightarrow{i_*} H_n(\mathbf{C}) \xrightarrow{p_*} H_n(\mathbf{C}'') \xrightarrow{\partial_n} H_{n-1}(\mathbf{C}') \to .$$

Proof. This proof is also routine and, again, it suffices to prove it when $\mathcal{A} = \mathbf{Ab}$. Our notation is abbreviated, and there are six inclusions to verify.

(i) $\operatorname{im} i_* \subseteq \ker p_*$ because $p_* i_* = (pi)_* = 0_* = 0$.

(ii) $\ker p_* \subseteq \operatorname{im} i_*$: If $p_* \operatorname{cls}(z) = \operatorname{cls}(pz) = \operatorname{cls}(0)$, then $pz = d''c''$ for some $c'' \in C''_{n+1}$. But p surjective gives $c'' = pc$ for some $c \in C_{n+1}$, so that $pz = d''pc = pdc$, because p is a chain map, and so $p(z-dc) = 0$. By exactness, there is $c' \in C'_n$ with $ic' = z - dc$. Now c' is a cycle, for $id'c' = dic' = dz - ddc = 0$, because z is a cycle; since i is injective, $d'c' = 0$. Therefore, $i_* \operatorname{cls}(c') = \operatorname{cls}(ic') = \operatorname{cls}(z - dc) = \operatorname{cls}(z)$.

(iii) $\operatorname{im} p_* \subseteq \ker \partial$: If $p_* \operatorname{cls}(c) = \operatorname{cls}(pc) \in \operatorname{im} p_*$, then $\partial \operatorname{cls}(pz) = \operatorname{cls}(z')$, where $iz' = dp^{-1}pz$. Since this formula is independent of the choice of lifing of pz, let us choose $p^{-1}pz = z$. Now $dp^{-1}pz = dz = 0$, because z is a cycle. Thus, $iz' = 0$, and hence $z' = 0$, because i is injective.

(iv) $\ker \partial \subseteq \operatorname{im} p_*$: If $\partial \operatorname{cls}(z'') = \operatorname{cls}(0)$, then $z' = i^{-1}dp^{-1}z'' \in B'$; that is, $z' = d'c'$ for some $c' \in C'$. But $iz' = id'c' = dic' = dp^{-1}z''$, so that $d(p^{-1}z'' - ic') = 0$; that is, $p^{-1}z'' - ic'$ is a cycle. Exactness of the original sequence gives $pi = 0$, so that $p_* \operatorname{cls}(p^{-1}z'' - ic') = \operatorname{cls}(pp^{-1}z'' - pic') = \operatorname{cls}(z'')$.

(v) $\operatorname{im} \partial \subseteq \ker i_*$: We have $i_* \partial \operatorname{cls}(z'') = \operatorname{cls}(iz')$. But $iz' = dp^{-1}z'' \in B$; that is, $i_* \partial = 0$.

(vi) $\ker i_* \subseteq \operatorname{im} \partial$: If $i_* \operatorname{cls}(z') = \operatorname{cls}(iz') = \operatorname{cls}(0)$, then $iz' = dc$ for some $c \in C$. Since p is a chain map, $d''pc = pdc = piz' = 0$, by exactness of the original sequence, and so pc is a cycle. But $\partial \operatorname{cls}(pc) = \operatorname{cls}(i^{-1}dp^{-1}pc) = \operatorname{cls}(i^{-1}dc) = \operatorname{cls}(i^{-1}iz') = \operatorname{cls}(z')$.   •

Theorem 6.10 is often called the **exact triangle** because of the diagram

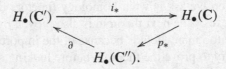

$$H_\bullet(\mathbf{C'}) \xrightarrow{\ i_*\ } H_\bullet(\mathbf{C})$$
$$\partial \qquad\qquad p_*$$
$$H_\bullet(\mathbf{C''}).$$

Example 6.11. Let X be a topological space and let $\mathbf{S}_\bullet(X)$ be its singular complex. If G is an abelian group, we define $H_\bullet(X, G)$, **homology** of X with **coefficients** G, to be the homology groups of the complex $\mathbf{S}_\bullet(X) \otimes_{\mathbb{Z}} G$. Given a short exact sequence $0 \to G' \to G \to G'' \to 0$ of abelian groups, there is a short exact sequence of complexes $0 \to \mathbf{S}_\bullet(X, G') \to \mathbf{S}_\bullet(X, G) \to \mathbf{S}_\bullet(X, G'') \to 0$ [each $S_n(X)$ is free, hence flat], and a long exact sequence

$$\to H_q(X, G') \to H_q(X, G) \to H_q(X, G'') \to H_{q-1}(X, G') \to \cdot$$

In this case, the connecting homomorphism $H_q(X, G'') \to H_{q-1}(X, G')$ is called the **Bockstein homomorphism**. ◄

Corollary 6.12 (Snake Lemma[1]).   *Let $\mathcal{A}$ be an abelian category. Given a commutative diagram in* **Comp**$(\mathcal{A})$ *with exact rows,*

$$
\begin{array}{ccccccccc}
0 & \longrightarrow & A' & \longrightarrow & A & \longrightarrow & A'' & \longrightarrow & 0 \\
 & & \downarrow{\scriptstyle f} & & \downarrow{\scriptstyle g} & & \downarrow{\scriptstyle h} & & \\
0 & \longrightarrow & B' & \longrightarrow & B & \longrightarrow & B'' & \longrightarrow & 0,
\end{array}
$$

there is an exact sequence in $\mathcal{A}$

$$0 \to \ker f \to \ker g \to \ker h \to \operatorname{coker} f \to \operatorname{coker} g \to \operatorname{coker} h \to 0.$$

Proof. If we view each of the vertical maps f, g, and h as a complex concentrated in degrees 1, 0 [as in Example 6.1(iv)], then the given commutative diagram can be viewed as a short exact sequence of complexes. The homology of each of these complexes has only two nonzero terms: for example, Example 6.7(iv) shows that the homology of the first column is $H_1 = \ker f$, $H_0 = \operatorname{coker} f$, and all other $H_n = 0$. The lemma now follows at once from the long exact sequence. •

We have just proved that the Long Exact Sequence implies the Snake Lemma; the converse is contained in Exercise 6.5 on page 338.

Theorem 6.13 (Naturality of ∂).   *Let $\mathcal{A}$ be an abelian category. Given a commutative diagram in* **Comp**$(\mathcal{A})$ *with exact rows,*

$$
\begin{array}{ccccccccc}
0 & \longrightarrow & \mathbf{C}' & \overset{i}{\longrightarrow} & \mathbf{C} & \overset{p}{\longrightarrow} & \mathbf{C}'' & \longrightarrow & 0 \\
 & & \downarrow{\scriptstyle f} & & \downarrow{\scriptstyle g} & & \downarrow{\scriptstyle h} & & \\
0 & \longrightarrow & \mathbf{A}' & \underset{j}{\longrightarrow} & \mathbf{A} & \underset{q}{\longrightarrow} & \mathbf{A}'' & \longrightarrow & 0,
\end{array}
$$

there is a commutative diagram in $\mathcal{A}$ with exact rows,

$$
\begin{array}{ccccccccc}
\longrightarrow & H_n(\mathbf{C}') & \overset{i_*}{\longrightarrow} & H_n(\mathbf{C}) & \overset{p_*}{\longrightarrow} & H_n(\mathbf{C}'') & \overset{\partial}{\longrightarrow} & H_{n-1}(\mathbf{C}') & \longrightarrow \\
 & \downarrow{\scriptstyle f_*} & & \downarrow{\scriptstyle g_*} & & \downarrow{\scriptstyle h_*} & & \downarrow{\scriptstyle f_*} & \\
\longrightarrow & H_n(\mathbf{A}') & \underset{j_*}{\longrightarrow} & H_n(\mathbf{A}) & \underset{q_*}{\longrightarrow} & H_n(\mathbf{A}'') & \underset{\partial'}{\longrightarrow} & H_{n-1}(\mathbf{A}') & \longrightarrow \cdot
\end{array}
$$

Proof. Exactness of the rows is Theorem 6.10, while commutativity of the first two squares follows from H_n being a functor. To prove commutativity of the square involving the connecting homomorphism, let us first display the

---

[1]The Snake Lemma is also called the Serpent Lemma.

chain maps and differentials in one (three-dimensional!) diagram:

$$
\begin{array}{ccccccccc}
0 & \longrightarrow & C'_n & \xrightarrow{\ i\ } & C_n & \xrightarrow{\ p\ } & C''_n & \longrightarrow & 0 \\
 & {}^{d'}\!\!\swarrow & {}^{i}\downarrow f_* & & {}^{p}\!\!\swarrow \; \downarrow g_* & & {}^{d''}\!\!\swarrow \;\downarrow h_* & & \\
0 & \longrightarrow & C'_{n-1} & \xrightarrow{\ i\ } & C_{n-1} & \xrightarrow{\ p\ } & C''_{n-1} & \longrightarrow & 0 \\
 & {}^{f_*}\!\downarrow & & {}^{g_*}\!\downarrow & & {}^{h_*}\!\downarrow & & & \\
0 & \longrightarrow & A'_n & \xrightarrow{\ j\ } & A_n & \xrightarrow{\ q\ } & A''_n & \longrightarrow & 0 \\
 & {}^{\delta'}\!\!\swarrow \downarrow & & {}^{\delta}\!\!\swarrow \downarrow & & {}^{\delta''}\!\!\swarrow \downarrow & & & \\
0 & \longrightarrow & A'_{n-1} & \xrightarrow{\ j\ } & A_{n-1} & \xrightarrow{\ q\ } & A''_{n-1} & \longrightarrow & 0.
\end{array}
$$

If $\mathrm{cls}(z'') \in H_n(\mathbf{C''})$, we must show that $f_* \partial \, \mathrm{cls}(z'') = \partial' h_* \, \mathrm{cls}(z'')$. Let $c \in C_n$ be a lifting of z''; that is, $pc = z''$. Now $\partial \, \mathrm{cls}(z'') = \mathrm{cls}(z')$, where $iz' = dc$. Hence, $f_* \partial \, \mathrm{cls}(z'') = \mathrm{cls}\,\mathrm{cls}(fz')$. On the other hand, since h is a chain map, we have $qgc = hpc = hz''$. In computing $\partial' \, \mathrm{cls}(hz'')$, we choose gc as the lifting of hz''. Hence, $\partial' \, \mathrm{cls}(hz'') = \mathrm{cls}(u')$, where $ju' = \delta gc$. But $jfz' = giz' = gdc = \delta gc = ju'$, and so $fz' = u'$, because j is injective.   •

Remark.   One advantage of having worked in an abelian category is that we can now give a conceptual proof of Theorem 6.13. (Homology in abelian categories will also be very useful when we discuss sheaf cohomology.)

Let $\mathcal{A}$ be an abelian category, and let $\mathcal{D}$ be the category having exactly two objects, $*$ and $\bullet$, and only one nonidentity morphism, $* \to \bullet$. Since $\mathcal{D}$ is a small category, the functor category $\mathcal{A}^{\mathcal{D}}$ is also abelian, by Proposition 5.93 ($\mathcal{A}^{\mathcal{D}}$ is often called an ***arrow category***). Of course, objects in $\mathcal{A}^{\mathcal{D}}$ are morphisms in $\mathcal{A}$, while a morphism $f \to g$ in $\mathcal{A}^{\mathcal{D}}$ is an ordered pair (α, β) of morphisms in $\mathcal{A}$ making the following diagram commute.

$$
\begin{array}{ccc}
A & \xrightarrow{\ \alpha\ } & A' \\
{\scriptstyle f}\downarrow & & \downarrow{\scriptstyle g} \\
B & \xrightarrow[\ \beta\ ]{} & B'
\end{array}
$$

Naturality of the connecting homomorphism may be restated. A 2×3 commutative diagram in $\mathbf{Comp}\,\mathcal{A}$ with exact rows can be viewed as a short exact sequence in $\mathbf{Comp}\,\mathcal{A}^{\mathcal{D}}$, and the corresponding Long Exact Sequence in $\mathcal{A}^{\mathcal{D}}$, when viewed in $\mathcal{A}$, is the usual "ladder" diagram in homology. The proof of Theorem 6.13 shows that the connecting homomorphism in $\mathcal{A}^{\mathcal{D}}$ is just the ordered pair (∂, ∂') of connecting homomorphisms in $\mathcal{A}$.   ◄

There are interesting maps of complexes that are not chain maps.

Definition.   Let $\mathbf{C}$ and $\mathbf{D}$ be complexes, and let $p \in \mathbb{Z}$. A ***map of degree*** p, denoted by $s \colon \mathbf{C} \to \mathbf{D}$, is a sequence $s = (s_n)$ with $s_n \colon C_n \to D_{n+p}$ for all n.

For example, a chain map is a map of degree 0, while the differentials of $(\mathbf{C}, d)$ form a map $d \colon \mathbf{C} \to \mathbf{C}$ of degree -1.

We now introduce a notion that arises in topology.

Definition. Chain maps $f, g \colon (\mathbf{C}, d) \to (\mathbf{C}', d')$ are **homotopic**,[2] denoted by $f \simeq g$, if, for all n, there is a map $s = (s_n) \colon \mathbf{C} \to \mathbf{C}'$ of degree $+1$ with

$$f_n - g_n = d'_{n+1} s_n + s_{n-1} d_n.$$

A chain map $f \colon (\mathbf{C}, d) \to (\mathbf{C}', d')$ is **null-homotopic** if $f \simeq 0$, where 0 is the zero chain map.

Theorem 6.14. *Homotopic chain maps induce the same morphism in homology: if $f, g \colon (\mathbf{C}, d) \to (\mathbf{C}', d')$ are chain maps and $f \simeq g$, then for all n,*

$$f_{*n} = g_{*n} \colon H_n(\mathbf{C}) \to H_n(\mathbf{C}').$$

Proof. If z is an n-cycle, then $d_n z = 0$ and

$$f_n z - g_n z = d'_{n+1} s_n z + s_{n-1} d_n z = d'_{n+1} s_n z.$$

Therefore, $f_n z - g_n z \in B_n(\mathbf{C}')$, and so $f_{*n} = g_{*n}$.   •

Definition. A complex $(\mathbf{C}, d)$ has a **contracting homotopy** if its identity $1_{\mathbf{C}}$ is null-homotopic. A complex $\mathbf{C}$ is **contractible**[3] if its identity $1 = 1_{\mathbf{C}}$ is null-homotopic; that is, there is $s \colon \mathbf{C} \to \mathbf{C}$ of degree $+1$ with $1 = sd + ds$.

Proposition 6.15. *A complex $\mathbf{C}$ having a contracting homotopy is acyclic; that is, it is an exact sequence.*

Proof. We use Example 6.1(i). Now $1_{\mathbf{C}} \colon H_n(\mathbf{C}) \to H_n(\mathbf{C})$ is the identity map, while $0_* \colon H_n(\mathbf{C}) \to H_n(\mathbf{C})$ is the zero map. Since $1_{\mathbf{C}} \simeq 0$, however, these maps are the same. It follows that $H_n(\mathbf{C}) = \{0\}$ for all n; that is, $\ker d_n = \operatorname{im} d_{n+1}$ for all n, and this is the definition of exactness.   •

---

[2]Recall that two continuous functions $f, g \colon X \to Y$ are called *homotopic* if there exists a continuous $F \colon X \times \mathbf{I} \to Y$, where $\mathbf{I} = [0, 1]$ is the closed unit interval, with $F(x, 0) = f(x)$ and $F(x, 1) = g(x)$ for all $x \in X$. If f and g are homotopic, then their induced maps are equal: $f_* = g_* \colon H_n(X) \to H_n(Y)$. The algebraic definition of homotopy given here has been distilled from the proof of this topological theorem.

[3]A topological space is called **contractible** if its identity map is homotopic to a constant map. A contractible space has the same homotopy type as a point.

Exercises

*6.1 If **C** is a complex with $C_n = \{0\}$ for some n, prove that $H_n(\mathbf{C}) = \{0\}$.

6.2 Prove that isomorphic complexes have the same homology: if **C** and **D** are isomorphic, then $H_n(\mathbf{C}) \cong H_n(\mathbf{D})$ for all $n \in \mathbb{Z}$.

6.3 If $f = (f_n) \colon \mathbf{C} \to \mathbf{D}$ is a chain map, prove, for all $n \in \mathbb{Z}$, that

$$f_{n*}Z_n(\mathbf{C}) \subseteq Z_n(\mathbf{D}) \quad \text{and} \quad f_{n*}B_n(\mathbf{C}) \subseteq B_n(\mathbf{D}).$$

*6.4   **(i)** If **P** and **P**$'$ are projective resolutions of a module A with syzygies K_n and K_n' for all $n \geq 0$, prove that there are projective modules Q_n, Q_n' with $K_n \oplus Q_n' \cong K_n' \oplus Q_n$.

          Hint. Schanuel's Lemma.

    (ii) If one projective resolution of a module A has a projective nth syzygy, prove that the nth syzygy of every projective resolution of A is projective.

*6.5 This exercise shows that the Snake Lemma implies Theorem 6.10 (so this theorem should *not* be used in solving this problem).

    Consider the commutative diagram with exact rows (note that two zeros are "missing" from this diagram):

$$
\begin{array}{ccccccc}
A & \longrightarrow & B & \overset{p}{\longrightarrow} & C & \longrightarrow & 0 \\
\downarrow{\scriptstyle\alpha} & & \downarrow{\scriptstyle\beta} & & \downarrow{\scriptstyle\gamma} & & \\
0 & \longrightarrow & A' & \underset{i}{\longrightarrow} & B' & \longrightarrow & C'.
\end{array}
$$

    (i) Prove that $\Delta \colon \ker \gamma \to \operatorname{coker} \alpha$, defined by

$$\Delta \colon z \mapsto i^{-1}\beta p^{-1}z + \operatorname{im}\alpha,$$

       is a well-defined homomorphism.

    (ii) Prove that there is an exact sequence

$$\ker \alpha \to \ker \beta \to \ker \gamma \overset{\Delta}{\to} \operatorname{coker} \alpha \overset{i'}{\to} \operatorname{coker} \beta \to \operatorname{coker} \gamma,$$

       where $i' \colon a' + \operatorname{im}\alpha \mapsto ia' + \operatorname{im}\beta$ for $a' \in A'$.

    (iii) Given a commutative diagram with exact rows,

$$
\begin{array}{ccccccccc}
0 & \longrightarrow & A_n' & \longrightarrow & A_n & \longrightarrow & A_n'' & \longrightarrow & 0 \\
& & \downarrow{\scriptstyle d_n'} & & \downarrow{\scriptstyle d_n} & & \downarrow{\scriptstyle d_n''} & & \\
0 & \longrightarrow & A_{n-1}' & \longrightarrow & A_{n-1} & \longrightarrow & A_{n-1}'' & \longrightarrow & 0,
\end{array}
$$

prove that the following diagram is commutative and has exact rows:

$$A_n'/\operatorname{im} d_{n+1}' \twoheadrightarrow A_n/\operatorname{im} d_{n+1} \twoheadrightarrow A_n''/\operatorname{im} d_{n+1}'' \twoheadrightarrow 0$$

$$d' \downarrow \qquad\qquad \downarrow d \qquad\qquad \downarrow d''$$

$$0 \longrightarrow \ker d_{n-1}' \longrightarrow \ker d_{n-1} \longrightarrow \ker d_{n-1}''.$$

(iv) Use part (ii) and this last diagram to give another proof of Theorem 6.10, the Long Exact Sequence.

6.6 Let $f, g: \mathbf{C} \to \mathbf{C}'$ be chain maps, and let $F: \mathcal{C} \to \mathcal{C}'$ be an additive functor. If $f \simeq g$, prove that $Ff \simeq Fg$; that is, if f and g are homotopic, then Ff and Fg are homotopic.

*6.7 Let $0 \to \mathbf{C}' \xrightarrow{i} \mathbf{C} \xrightarrow{p} \mathbf{C}'' \to 0$ be an exact sequence of complexes in which $\mathbf{C}'$ and $\mathbf{C}''$ are acyclic; prove that $\mathbf{C}$ is also acyclic.

6.8 Let R and A be rings, and let $T: {}_R\mathbf{Mod} \to {}_A\mathbf{Mod}$ be an exact additive functor. Prove that T commutes with homology; that is, for every complex $(\mathbf{C}, d) \in {}_R\mathbf{Comp}$ and for every $n \in \mathbb{Z}$, there is an isomorphism

$$H_n(T\mathbf{C}, Td) \cong TH_n(\mathbf{C}, d).$$

*6.9 **(i)** Prove that homology commutes with direct sums: for all n, there are natural isomorphisms

$$H_n\left(\bigoplus_\alpha \mathbf{C}^\alpha\right) \cong \bigoplus_\alpha H_n(\mathbf{C}^\alpha).$$

(ii) Define a direct system of complexes $(\mathbf{C}^i)_{i\in I}$, $(\varphi_j^i)_{i\leq j}$, and prove that $\varinjlim \mathbf{C}^i$ exists.

(iii) If $(\mathbf{C}^i)_{i\in I}$, $(\varphi_j^i)_{i\leq j}$ is a direct system of complexes over a directed index set, prove, for all $n \geq 0$, that

$$H_n(\varinjlim \mathbf{C}^i) \cong \varinjlim H_n(\mathbf{C}^i).$$

*6.10 Assume that a complex $(\mathbf{C}, d)$ of R-modules has a contracting homotopy in which the maps $s_n: C_n \to C_{n+1}$ satisfying

$$1_{C_n} = d_{n+1}s_n + s_{n-1}d_n$$

are only $\mathbb{Z}$-maps. Prove that $(\mathbf{C}, d)$ is an exact sequence.

*6.11 (*Barratt–Whitehead*). Consider the commutative diagram with exact rows:

$$\to A_n \xrightarrow{i_n} B_n \xrightarrow{p_n} C_n \xrightarrow{\partial_n} A_{n-1} \to B_{n-1} \to C_{n-1} \to$$
$$f_n\downarrow \quad g_n\downarrow \quad h_n\downarrow \quad f_{n-1}\downarrow \quad g_{n-1}\downarrow \quad h_{n-1}\downarrow$$
$$\to A_n' \xrightarrow{j_n} B_n' \xrightarrow{q_n} C_n' \to A_{n-1}' \to B_{n-1}' \to C_{n-1}' \to.$$

If each h_n is an isomorphism, prove that there is an exact sequence

$$\to A_n \xrightarrow{(f_n, i_n)} A'_n \oplus B_n \xrightarrow{j_n - g_n} B'_n \xrightarrow{\partial_n h_n^{-1} q_n} A_{n-1}$$
$$\to A'_{n-1} \oplus B_{n-1} \to B'_{n-1} \to,$$

where

$$(f_n, i_n): a_n \mapsto (f_n a_n, i_n a_n) \text{ and } j_n - g_n: (a'_n, b_n) \mapsto j_n a'_n - g_n b_n.$$

*6.12 (*Mayer–Vietoris*). Given a commutative diagram of complexes with exact rows,

$$
\begin{array}{ccccccccc}
0 & \to & \mathbf{C}' & \xrightarrow{i} & \mathbf{C} & \xrightarrow{p} & \mathbf{C}'' & \to & 0 \\
 & & {\scriptstyle f}\downarrow & & \downarrow{\scriptstyle g} & & \downarrow{\scriptstyle h} & & \\
0 & \to & \mathbf{A}' & \xrightarrow[j]{} & \mathbf{A} & \xrightarrow[q]{} & \mathbf{A}'' & \to & 0,
\end{array}
$$

if every third vertical map h_* in the diagram

$$
\begin{array}{ccccccccc}
\to & H_n(\mathbf{C}') & \xrightarrow{i_*} & H_n(\mathbf{C}) & \xrightarrow{p_*} & H_n(\mathbf{C}'') & \xrightarrow{\partial} & H_{n-1}(\mathbf{C}') & \to \\
 & {\scriptstyle f_*}\downarrow & & \downarrow{\scriptstyle g_*} & & \downarrow{\scriptstyle h_*} & & \downarrow{\scriptstyle f_*} & \\
\to & H_n(\mathbf{A}') & \xrightarrow[j_*]{} & H_n(\mathbf{A}) & \xrightarrow[q_*]{} & H_n(\mathbf{A}'') & \xrightarrow[\partial']{} & H_{n-1}(\mathbf{A}') & \to
\end{array}
$$

is an isomorphism, prove that there is an exact sequence

$$\to H_n(\mathbf{C}') \to H_n(\mathbf{A}') \oplus H_n(\mathbf{C}) \to H_n(\mathbf{A}) \to H_{n-1}(\mathbf{C}') \to \cdot$$

6.2 Derived Functors

In order to apply the general results in the previous section, we need a source of short exact sequences of complexes. The idea is to replace every module by a deleted resolution of it; given a short exact sequence of modules, we shall see that this replacement gives a short exact sequence of complexes. We then apply either Hom or $\otimes$, and the resulting homology modules are called Ext or Tor.

We know that a module has many presentations; since resolutions are generalized presentations, the next result is fundamental.

Theorem 6.16 (Comparison Theorem). *Let $\mathcal{A}$ be an abelian category. Given a morphism $f : A \to A'$ in $\mathcal{A}$, consider the diagram*

$$
\begin{array}{ccccccccc}
\longrightarrow & P_2 & \xrightarrow{d_2} & P_1 & \xrightarrow{d_1} & P_0 & \xrightarrow{\varepsilon} & A & \longrightarrow 0 \\
 & {\scriptstyle \check{f}_2}\downarrow & & {\scriptstyle \check{f}_1}\downarrow & & {\scriptstyle \check{f}_0}\downarrow & & \downarrow{\scriptstyle f} & \\
\longrightarrow & P'_2 & \xrightarrow[d'_2]{} & P'_1 & \xrightarrow[d'_1]{} & P'_0 & \xrightarrow[\varepsilon']{} & A' & \longrightarrow 0,
\end{array}
$$

where the rows are complexes. If each P_n in the top row is projective, and if the bottom row is exact, then there exists a chain map $\check{f} : \mathbf{P}_A \to \mathbf{P}'_{A'}$ making the completed diagram commute. Moreover, any two such chain maps are homotopic.

Remark. The dual of the comparison theorem is also true. Given a morphism $g : A' \to A$, consider the diagram of negative complexes

$$
\begin{array}{ccccccccc}
0 & \longrightarrow & A & \longrightarrow & E^0 & \longrightarrow & E^1 & \longrightarrow & E^2 & \longrightarrow \\
& & \big\uparrow{\scriptstyle g} & & \big\uparrow & & \big\uparrow & & \big\uparrow & \\
0 & \longrightarrow & A' & \longrightarrow & X^0 & \longrightarrow & X^1 & \longrightarrow & X^2 & \longrightarrow \cdot
\end{array}
$$

If the bottom row is exact and each E^n in the top row is injective, then there exists a chain map $\mathbf{X}^{A'} \to \mathbf{E}^A$ making the completed diagram commute. ◄

Proof. Again, it suffices to prove the result when $\mathcal{A} = \mathbf{Ab}$.

(i) We prove the existence of $\check{f}_n$ by induction on $n \geq 0$. For the base step $n = 0$, consider the diagram

$$
\begin{array}{ccc}
 & & P_0 \\
 & {\scriptstyle \check{f}_0} \nearrow & \big\downarrow{\scriptstyle f\varepsilon} \\
P'_0 & \xrightarrow{\;\varepsilon'\;} & A' \longrightarrow 0.
\end{array}
$$

Since ε' is surjective and P_0 is projective, there is a map $\check{f}_0 : P_0 \to P'_0$ with $\varepsilon' \check{f}_0 = f\varepsilon$. For the inductive step, consider the diagram

$$
\begin{array}{ccccc}
P_{n+1} & \xrightarrow{\;d_{n+1}\;} & P_n & \xrightarrow{\;d_n\;} & P_{n-1} \\
 & & {\scriptstyle \check{f}_n}\big\downarrow & & \big\downarrow{\scriptstyle \check{f}_{n-1}} \\
P'_{n+1} & \xrightarrow[\;d'_{n+1}\;]{} & P'_n & \xrightarrow[\;d'_n\;]{} & P'_{n-1}.
\end{array}
$$

If $\operatorname{im} \check{f}_n d_{n+1} \subseteq \operatorname{im} d'_{n+1}$, then we have the diagram

$$
\begin{array}{ccc}
 & & P_{n+1} \\
 & {\scriptstyle \check{f}_{n+1}} \nearrow & \big\downarrow{\scriptstyle \check{f}_n d_{n+1}} \\
P'_{n+1} & \xrightarrow[\;d'_{n+1}\;]{} & \operatorname{im} d'_{n+1} \longrightarrow 0,
\end{array}
$$

and projectivity of P_{n+1} gives $\check{f}_{n+1} : P_{n+1} \to P'_{n+1}$ with $d'_{n+1}\check{f}_{n+1} = \check{f}_n d_{n+1}$. To check that the inclusion holds, note that exactness at P'_n of the bottom row of the original diagram gives $\operatorname{im} d'_{n+1} = \ker d'_n$, and so it suffices to prove that $d'_n \check{f}_n d_{n+1} = 0$. But $d'_n \check{f}_n d_{n+1} = \check{f}_{n-1} d_n d_{n+1} = 0$.

(ii) We prove uniqueness of $\check{f}$ to homotopy. If $h\colon \mathbf{P}_A \to \mathbf{P}'_{A'}$ is another chain map with $\varepsilon' h_0 = f\varepsilon$, we construct the terms $s_n\colon P_n \to P'_{n+1}$ of a homotopy s by induction on $n \geq -1$; that is, we will show that

$$h_n - \check{f}_n = d'_{n+1}s_n + s_{n-1}d_n.$$

For the base step, first view A and $\{0\}$ as being terms -1 and -2 in the top complex, and define $d_0 = \varepsilon$ and $d_{-1} = 0$. Also, view A' and $\{0\}$ as being terms -1 and -2 in the bottom complex, and define $d'_0 = \varepsilon'$ and $d'_{-1} = 0$. Finally, define $\check{f}_{-1} = f = h_{-1}$ and $s_{-2} = 0$.

$$
\begin{array}{ccccccccc}
\longrightarrow & P_1 & \xrightarrow{d_1} & P_0 & \xrightarrow{d_0 = \varepsilon} & A & \xrightarrow{d_{-1}} & 0 \\
 & & & & & & & \\
\longrightarrow & P'_1 & \xrightarrow{d'_1} & P'_0 & \xrightarrow{d'_0 = \varepsilon'} & A' & \xrightarrow{d'_{-1}} & 0
\end{array}
$$

With this notation, defining $s_{-1} = 0$ gives $h_{-1} - \check{f}_{-1} = f - f = 0 = d'_0 s_{-1} + s_{-2}d_{-1}$.

For the inductive step, it suffices to prove, for all $n \geq -1$, that

$$\operatorname{im}(h_{n+1} - \check{f}_{n+1} - s_n d_{n+1}) \subseteq \operatorname{im} d'_{n+2},$$

for then we have a diagram with exact row

$$
\begin{array}{ccc}
 & P_{n+1} & \\
 {\scriptstyle s_{n+1}} \swarrow & \downarrow {\scriptstyle h_{n+1} - \check{f}_{n+1} - s_n d_{n+1}} & \\
P'_{n+2} \xrightarrow{d'_{n+2}} & \operatorname{im} d'_{n+2} \longrightarrow & 0,
\end{array}
$$

and projectivity of P_{n+1} gives a map $s_{n+1}\colon P_{n+1} \to P'_{n+2}$ satisfying the desired equation. As in the proof of part (i), exactness of the bottom row of the original diagram gives $\operatorname{im} d'_{n+2} = \ker d'_{n+1}$, and so it suffices to prove $d'_{n+1}(h_{n+1} - \check{f}_{n+1} - s_n d_{n+1}) = 0$. But

$$
\begin{aligned}
d'_{n+1}(h_{n+1} - \check{f}_{n+1} - s_n d_{n+1}) &= d'_{n+1}(h_{n+1} - \check{f}_{n+1}) - d'_{n+1}s_n d_{n+1} \\
&= d'_{n+1}(h_{n+1} - \check{f}_{n+1}) - (h_n - \check{f}_n - s_{n-1}d_n)d_{n+1} \\
&= d'_{n+1}(h_{n+1} - \check{f}_{n+1}) - (h_n - \check{f}_n)d_{n+1},
\end{aligned}
$$

and the last term is 0 because h and $\check{f}$ are chain maps. $\bullet$

We introduce a term to describe the chain map $\check{f}$ just constructed.

Definition. If $f: A \to A'$ is a morphism and $\mathbf{P}_A$ and $\mathbf{P}'_{A'}$ are deleted projective resolutions of A and A', respectively, then a chain map $\check{f}: \mathbf{P}_A \to \mathbf{P}'_{A'}$ is said to be **over** f if $f\varepsilon = \varepsilon' \check{f}_0$.

$$
\begin{array}{ccccccccc}
\longrightarrow & P_2 & \xrightarrow{d_2} & P_1 & \xrightarrow{d_1} & P_0 & \xrightarrow{\varepsilon} & A & \longrightarrow 0 \\
& \downarrow{\check{f}_2} & & \downarrow{\check{f}_1} & & \downarrow{\check{f}_0} & & \downarrow{f} & \\
\longrightarrow & P'_2 & \xrightarrow[d'_2]{} & P'_1 & \xrightarrow[d'_1]{} & P'_0 & \xrightarrow[\varepsilon']{} & A' & \longrightarrow 0
\end{array}
$$

Given a morphism $f: A \to A'$, the comparison theorem implies that a chain map over f always exists between deleted projective resolutions of A and A'; moreover, such a chain map is unique to homotopy.

6.2.1 Left Derived Functors

In Algebraic Topology, we apply the functor $\square \otimes_{\mathbb{Z}} G$, for an abelian group G, to the singular complex $\mathbf{S}_\bullet(X)$ of a topological space X to get the complex

$$
\to S_{n+1}(X) \otimes G \xrightarrow{\partial_{n+1} \otimes 1_G} S_n(X) \otimes G \xrightarrow{\partial_n \otimes 1_G} S_{n-1}(X) \otimes G \to .
$$

The homology groups $H_n(X, G) = H_n(\mathbf{S}_\bullet(X) \otimes_{\mathbb{Z}} G)$ are called the *homology groups of X with coefficients in G*, as we have seen in Example 6.11. Similarly, applying the contravariant functor $\mathrm{Hom}(\square, G)$ gives the complex

$$
\leftarrow \mathrm{Hom}(S_{n+1}(X), G) \xleftarrow{\partial_n^*} \mathrm{Hom}(S_n(X), G) \xleftarrow{\partial_{n-1}^*} \mathrm{Hom}(S_{n-1}(X), G) \leftarrow;
$$

its homology groups $H^n(X, G)$ are called the **cohomology groups of X with coefficients in G**. This last terminology generalizes. If a contravariant functor T is applied to a complex $\mathbf{C}$, then many of the usual terms involving the complex $T\mathbf{C}$ acquire the prefix "co" and all indices are raised. For example, one has *n-cochains* $C^n = T(C_n)$, *n-cocycles* $Z^n(T\mathbf{C})$, *n-coboundaries* $B^n(T\mathbf{C})$, *n*th **cohomology** $H^n(T\mathbf{C})$, and **induced maps** f^*. Originally, the left derived functors $L_n T$ were called *homology* when T is a covariant functor, and the right derived functors $R^n T$ were called *cohomology* when T is contravariant.[4] Unfortunately, this clear distinction is blurred because the Hom functor is contravariant in one variable but covariant in the other. As a result, derived functors of any variance which involve Hom are often called cohomology.

Given an additive covariant functor $T: \mathcal{A} \to \mathcal{C}$ between abelian categories, where $\mathcal{A}$ has enough projectives, we now construct its *left derived*

---

[4]In their book *Homology Theory*, Hilton and Wylie tried to replace *cohomology* by *contrahomology*, but their suggestion was not adopted.

functors $L_n T: \mathcal{A} \to \mathcal{C}$, for all $n \in \mathbb{Z}$ (we will construct *right derived functors* afterwards). The definition will be in two parts: first on objects; then on morphisms.

Choose, once and for all, one projective resolution[5]

$$\mathbf{P} = \to P_2 \xrightarrow{d_2} P_1 \xrightarrow{d_1} P_0 \xrightarrow{\varepsilon} A \to 0$$

for every object A (thus, the chosen projective resolution of A is the analog of the singular complex of a space X, but it is more fruitful to regard it as a presentation of A). Form the deleted resolution $\mathbf{P}_A$, then form the complex $T\mathbf{P}_A$ (as in Example 6.6), take homology, and define

$$(L_n T)A = H_n(T\mathbf{P}_A).$$

Let $f: A \to A'$ be a morphism. By the comparison theorem, there is a chain map $\check{f}: \mathbf{P}_A \to \mathbf{P}'_{A'}$ over f. Then $T\check{f}: T\mathbf{P}_A \to T\mathbf{P}'_{A'}$ is also a chain map, and we define $(L_n T)f: (L_n T)A \to (L_n T)A'$ by

$$(L_n T)f = H_n(T\check{f}) = (T\check{f})_{n*}.$$

In more detail, if $z \in \ker T d_n$, then

$$(L_n T)f: z + \operatorname{im} T d_{n+1} \mapsto (T\check{f}_n)z + \operatorname{im} T d'_{n+1};$$

that is,

$$(L_n T)f: \operatorname{cls}(z) \mapsto \operatorname{cls}(T\check{f}_n z).$$

In pictures, look at the chosen projective resolutions:

$$\begin{array}{ccccccccc}
\to & P_2 & \to & P_1 & \to & P_0 & \to & A & \to 0 \\
& \downarrow & & \downarrow & & \downarrow & & \downarrow f & \\
\to & P'_2 & \to & P'_1 & \to & P'_0 & \to & A' & \to 0.
\end{array}$$

Fill in a chain map $\check{f}$ over f, delete A and A', apply T to this diagram, and then take the map induced by $T\check{f}$ in homology.

Theorem 6.17. *If $T: \mathcal{A} \to \mathcal{C}$ is an additive covariant functor between abelian categories, where $\mathcal{A}$ has enough projectives, then $L_n T: \mathcal{A} \to \mathcal{C}$ is an additive covariant functor for every $n \in \mathbb{Z}$.*

[5]We will see, in Proposition 6.20, that the definition does not depend on the choice of projective resolution.

Proof. We will prove that $L_n T$ is well-defined on morphisms; it is then routine to check that it is an additive covariant functor [remember that H_n is an additive covariant functor $\mathbf{Comp}(\mathcal{A}) \to \mathcal{A}$].

If $h \colon \mathbf{P}_A \to \mathbf{P}'_{A'}$ is another chain map over f, then the comparison theorem says that $h \simeq \check{f}$; therefore, $T h \simeq T \check{f}$, by Exercise 6.6 on page 339, and so $H_n(T h) = H_n(T \check{f})$, by Theorem 6.14. $\bullet$

Here is a useful computation of an induced map when $\mathcal{A} = {}_R\mathbf{Mod}$. Recall that if $r \in Z(R)$ and A is a left R-module, then multiplication by r, denoted by $\mu_r \colon A \to A$, is an R-map. We say that a functor $T \colon {}_R\mathbf{Mod} \to {}_R\mathbf{Mod}$, of either variance, *preserves multiplications* if, for all $r \in Z(R)$, $T(\mu_r) \colon T A \to T A$ is also multiplication by r. For example, tensor product and Hom preserve multiplications.

Proposition 6.18. *If $T \colon {}_R\mathbf{Mod} \to {}_R\mathbf{Mod}$ is an additive functor that preserves multiplications, then $L_n T \colon {}_R\mathbf{Mod} \to {}_R\mathbf{Mod}$ also preserves multiplications.*

Proof. Given a projective resolution $\to P_1 \xrightarrow{d_1} P_0 \xrightarrow{\varepsilon} A \to 0$, it is easy to see that $\check{\mu}$ is a chain map over μ_r, where every $\check{\mu}_n \colon P_n \to P_n$ is

$$
\begin{array}{ccccccccc}
\longrightarrow & P_2 & \xrightarrow{d_2} & P_1 & \xrightarrow{d_1} & P_0 & \xrightarrow{\varepsilon} & A & \longrightarrow 0 \\
& \downarrow{\check{\mu}_2} & & \downarrow{\check{\mu}_1} & & \downarrow{\check{\mu}_0} & & \downarrow{\mu_r} & \\
\longrightarrow & P_2 & \xrightarrow{d_2} & P_1 & \xrightarrow{d_1} & P_0 & \xrightarrow{\varepsilon} & A & \longrightarrow 0,
\end{array}
$$

multiplication by r. Since T preserves multiplications, the terms $T \check{\mu}_n$ of the chain map $T \check{\mu}$ are also multiplication by r, and so the induced maps in homology are multiplication by r:

$$(T \check{\mu})_* \colon \operatorname{cls}(z_n) \mapsto \operatorname{cls}((T \check{\mu}_n) z_n) = \operatorname{cls}(r z_n) = r \operatorname{cls}(z_n),$$

where $z_n \in \ker T d_n$. $\bullet$

Definition. Given an additive covariant functor $T \colon \mathcal{A} \to \mathcal{C}$ between abelian categories, where $\mathcal{A}$ has enough projectives, the functors $L_n T$ are called the *left derived functors* of T.

Proposition 6.19. *If $T \colon \mathcal{A} \to \mathcal{C}$ is an additive covariant function between abelian categories, then $(L_n T) A = 0$ for all negative n and for all A.*

Proof. By Exercise 6.1 on page 338, we have $(L_n T) A = 0$ because the nth term of $\mathbf{P}_A$ is 0 when n is negative. $\bullet$

The functors $L_n T$ are called *left* derived functors because of the last proposition. Since $L_n T = 0$ on the right, that is, for all negative n, these functors are of interest only on the left; that is, for $n \geq 0$.

Definition. If B is a left R-module and $T = \square \otimes_R B$, define

$$\operatorname{Tor}_n^R(\square, B) = L_n T.$$

Thus, if $\mathbf{P} = \to P_2 \xrightarrow{d_2} P_1 \xrightarrow{d_1} P_0 \xrightarrow{\varepsilon} A \to 0$ is the chosen projective resolution of a right R-module A, then

$$\operatorname{Tor}_n^R(A, B) = H_n(\mathbf{P}_A \otimes_R B) = \frac{\ker(d_n \otimes 1_B)}{\operatorname{im}(d_{n+1} \otimes 1_B)}.$$

The domain of $\operatorname{Tor}_n^R(\square, B)$ is $\mathbf{Mod}_R$, the category of right R-modules, and its target is $\mathbf{Ab}$ [if B is an (R, S)-bimodule, then the target is $\mathbf{Mod}_S$]. In particular, if R is commutative, then $A \otimes_R B$ is an R-module, and so the values of $\operatorname{Tor}_n^R(\square, B)$ lie in $_R\mathbf{Mod}$.

We can also form the left derived functors of $A \otimes_R \square$, obtaining functors $_R\mathbf{Mod} \to \mathbf{Ab}$.

Definition. If A is a right R-module and $T = A \otimes_R \square$, define

$$\operatorname{tor}_n^R(A, \square) = L_n T.$$

Thus, if $\mathbf{Q} = \to Q_2 \xrightarrow{d_2} Q_1 \xrightarrow{d_1} Q_0 \xrightarrow{\eta} B \to 0$ is the chosen projective resolution of a left R-module B, then

$$\operatorname{tor}_n^R(A, B) = H_n(A \otimes_R \mathbf{Q}_B) = \frac{\ker(1_A \otimes d_n)}{\operatorname{im}(1_A \otimes d_{n+1})}.$$

One nice result of Homological Algebra is Theorem 6.32 on page 355: for all left R-modules A, all right R-modules B, and all $n \geq 0$,

$$\operatorname{Tor}_n^R(A, B) \cong \operatorname{tor}_n^R(A, B).$$

Thus, the notation $\operatorname{tor}_n^R(A, B)$ is only temporary.

The definition of $L_n T$ assumes that a choice of projective resolution of every module has been made. Does $L_n T$ depend on this choice?

Proposition 6.20. *Let $\mathcal{A}$ be an abelian category with enough projectives. Assume that new choices $\widetilde{\mathbf{P}}_A$ of deleted projective resolutions have been made, and denote the left derived functors arising from these new choices by $\widetilde{L}_n T$.*

If $T: \mathcal{A} \to \mathcal{C}$ is an additive covariant functor, where $\mathcal{C}$ is an abelian category, then the functors $L_n T$ and $\widetilde{L}_n T$, for each $n \geq 0$, are naturally isomorphic. In particular, for all A, the objects

$$(L_n T)A \cong (\widetilde{L}_n T)A$$

are independent of the choice of projective resolution of A.

Proof. Consider the diagram

$$
\begin{array}{ccccccccc}
\longrightarrow & P_2 & \longrightarrow & P_1 & \longrightarrow & P_0 & \longrightarrow & A & \longrightarrow 0 \\
& & & & & & & \downarrow 1_A & \\
\longrightarrow & \widetilde{P}_2 & \longrightarrow & \widetilde{P}_1 & \longrightarrow & \widetilde{P}_0 & \longrightarrow & A & \longrightarrow 0,
\end{array}
$$

where the top row is the chosen projective resolution of A used to define L_nT and the bottom is that used to define $\widetilde{L}_nT$. By the comparison theorem, there is a chain map $\iota \colon \mathbf{P}_A \to \widetilde{\mathbf{P}}_A$ over 1_A. Applying T gives a chain map $T\iota \colon T\mathbf{P}_A \to T\widetilde{\mathbf{P}}_A$ over $T1_A = 1_{TA}$. This last chain map induces morphisms, one for each n,

$$
\tau_A = (T\iota)_* \colon (L_nT)A \to (\widetilde{L}_nT)A.
$$

We now prove that each τ_A is an isomorphism (thereby proving the last statement in the theorem) by constructing its inverse. Turn the preceding diagram upside down, so that the chosen projective resolution $\mathbf{P}_A \to A \to 0$ is now the bottom row. Again, the comparison theorem gives a chain map, say, $\kappa \colon \widetilde{\mathbf{P}}_A \to \mathbf{P}_A$. Now the composite $\kappa\iota$ is a chain map from $\mathbf{P}_A$ to itself over 1_A. By the uniqueness statement in the comparison theorem, $\kappa\iota \simeq 1_{\mathbf{P}_A}$; similarly, $\iota\kappa \simeq 1_{\widetilde{\mathbf{P}}_A}$. It follows that $T(\iota\kappa) \simeq 1_{T\widetilde{\mathbf{P}}_A}$ and $T(\kappa\iota) \simeq 1_{T\mathbf{P}_A}$. Hence, $1_{(\widetilde{L}_nT)A} = (T\iota\kappa)_* = (T\iota)_*(T\kappa)_*$ and $1_{(L_nT)A} = (T\kappa\iota)_* = (T\kappa)_*(T\iota)_*$. Therefore, $\tau_A = (T\iota)_*$ is an isomorphism.

We now prove that the isomorphisms τ_A constitute a natural isomorphism; that is, if $f \colon A \to B$ is a morphism, then the following diagram commutes.

$$
\begin{array}{ccc}
(L_nT)A & \xrightarrow{\ \tau_A\ } & (\widetilde{L}_nT)A \\
{\scriptstyle (L_nT)f}\downarrow & & \downarrow{\scriptstyle (\widetilde{L}_nT)f} \\
(L_nT)B & \xrightarrow[\ \tau_B\ ]{} & (\widetilde{L}_nT)B
\end{array}
$$

To evaluate in the clockwise direction, consider

$$
\begin{array}{ccccccccc}
\longrightarrow & P_1 & \longrightarrow & P_0 & \longrightarrow & A & \longrightarrow 0 \\
& & & & & & \downarrow 1_A \\
\longrightarrow & \widetilde{P}_1 & \longrightarrow & \widetilde{P}_0 & \longrightarrow & A & \longrightarrow 0 \\
& & & & & & \downarrow f \\
\longrightarrow & \widetilde{Q}_1 & \longrightarrow & \widetilde{Q}_0 & \longrightarrow & B & \longrightarrow 0,
\end{array}
$$

where the bottom row is the new chosen projective resolution of B. The comparison theorem gives a chain map $\mathbf{P}_A \to \widetilde{\mathbf{Q}}_B$ over $f1_A = f$. Going counterclockwise, the picture will now have the original chosen projective resolution of B as its middle row, and we get a chain map $\mathbf{P}_A \to \widetilde{\mathbf{Q}}_B$ over $1_B f = f$. The uniqueness statement in the comparison theorem tells us that these two chain maps are homotopic, and so they induce the same morphism in homology. Thus, the appropriate diagram commutes, showing that $\tau : L_n T \to \tilde{L}_n T$ is a natural isomorphism.   •

Corollary 6.21.   *The modules* $\operatorname{Tor}_n^R(A, B)$ *are independent of the choice of projective resolution of A, and the modules* $\operatorname{tor}_n^R(A, B)$ *are independent of the choice of projective resolution of B.*

Proof.   Both $\operatorname{Tor}_n^R(\square, B)$ and $\operatorname{tor}_n^R(A, \square)$ are left derived functors, and so Proposition 6.20 applies to each of them.   •

Corollary 6.22.   *Let* $T : {}_R\mathbf{Mod} \to {}_S\mathbf{Mod}$ *be an additive covariant functor. If P is a projective module, then* $(L_n T)P = \{0\}$ *for all* $n \geq 1$*. In particular, if A and P are right R-modules with P projective, and if B and Q are left R-modules with Q projective, then for all* $n \geq 1$,

$$\operatorname{Tor}_n^R(P, B) = \{0\} \quad and \quad \operatorname{tor}_n^R(A, Q) = \{0\}.$$

Proof.   Since P is projective, a projective resolution is $\mathbf{P}$, the complex with 1_P concentrated in degrees $0, -1$. The corresponding deleted projective resolution $\mathbf{P}_P$ is $\varrho^0(P)$, the complex with P concentrated in degree 0. Hence, $T\mathbf{P}_P$ has nth term $\{0\}$ for all $n \geq 1$, and so $(L_n T)P = H_n(T\mathbf{P}_P) = \{0\}$ for all $n \geq 1$, by Exercise 6.1 on page 338.   •

Corollary 6.23.   *Let* $\mathcal{A}$ *be an abelian category with enough projectives. Let*

$$\mathbf{P} = \to P_2 \xrightarrow{d_2} P_1 \xrightarrow{d_1} P_0 \xrightarrow{\varepsilon} A \to 0$$

be a projective resolution of $A \in \operatorname{obj} \mathcal{A}$*. Define* $K_0 = \ker \varepsilon$*, and define* $K_n = \ker d_n$ *for all* $n \geq 1$*. Then*

$$(L_{n+1} T)A \cong (L_n T)K_0 \cong (L_{n-1}T)K_1 \cong \cdots \cong (L_1 T)K_{n-1}.$$

In particular, if $\mathcal{A} = \mathbf{Mod}_R$ *and B is a left R-module,*

$$\operatorname{Tor}_{n+1}^R(A, B) \cong \operatorname{Tor}_n^R(K_0, B) \cong \cdots \cong \operatorname{Tor}_1^R(K_{n-1}, B).$$

Similarly, if A is a right R-module, let $\mathbf{P}' = \to P_1' \xrightarrow{d_1'} P_0' \xrightarrow{\varepsilon'} B \to 0$
be a projective resolution of a left R-module B, and define $V_0 = \ker \varepsilon'$ *and* $V_n = \ker d_n'$ *for all* $n \geq 1$*. Then*

$$\operatorname{tor}_{n+1}^R(A, B) \cong \operatorname{tor}_n^R(A, V_0) \cong \cdots \cong \operatorname{tor}_1^R(A, V_{n-1}).$$

Proof. By exactness of **P**, we have $K_0 = \ker \varepsilon = \operatorname{im} d_1$, and so

$$\mathbf{Q} = \to P_2 \xrightarrow{d_2} P_1 \xrightarrow{d_1} K_0 \to 0$$

is a projective resolution of K_0 if we relabel the indices; replace each n by $n - 1$, and define $Q_n = P_{n+1}$ and $\delta_n = d_{n+1}$ for all $n \geq 0$. Since the value of $L_n T$ on a module is independent of the choice of projective resolution, we have

$$(L_n T) K_0 \cong H_n(T \mathbf{Q}_{K_0}) = \frac{\ker T \delta_n}{\operatorname{im} T \delta_{n+1}}$$

$$= \frac{\ker T d_{n+1}}{\operatorname{im} T d_{n+2}} = H_{n+1}(T \mathbf{P}_A) \cong (L_{n+1} T) A.$$

The remaining isomorphisms are obtained by iteration. •

We are now going to show that there is a long exact sequence of left derived functors. We begin with a useful lemma; it says that if we are given a short exact sequence $0 \to A' \to A \to A'' \to 0$ as well as projective resolutions of A' and A'', then we can "fill in the horseshoe"; that is, there is a projective resolution of A that fits in the middle.

Proposition 6.24 (Horseshoe Lemma). *Given a diagram in an abelian category* $\mathcal{A}$ *with enough projectives,*

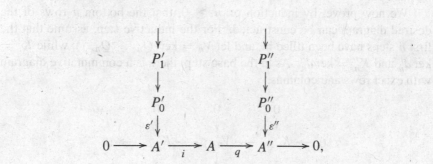

where the columns are projective resolutions and the row is exact, then there exist a projective resolution of A and chain maps so that the three columns form an exact sequence of complexes.

Remark. The dual theorem, in which projective resolutions are replaced by injective resolutions, is also true. ◄

Proof.   We show first that there are a projective Q_0 and a commutative 3×3 diagram with exact columns and rows:

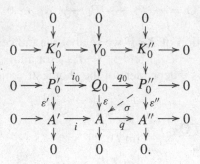

Define $Q_0 = P_0' \oplus P_0''$; it is projective because both P_0' and P_0'' are projective. Define $i_0 \colon P_0' \to P_0' \oplus P_0''$ by $x' \mapsto (x', 0)$, and define $q_0 \colon P_0' \oplus P_0'' \to P_0''$ by $(x', x'') \mapsto x''$. It is clear that

$$0 \to P_0' \xrightarrow{i_0} Q_0 \xrightarrow{q_0} P_0'' \to 0$$

is exact. Since P_0'' is projective, there exists a map $\sigma \colon P_0'' \to A$ with $q\sigma = \varepsilon''$. Now define $\varepsilon \colon Q_0 \to A$ by $\varepsilon \colon (x', x'') \mapsto i\varepsilon'x' + \sigma x''$ (the map σ makes the square with base $A \xrightarrow{q} A''$ commute). Surjectivity of ε follows from the Five Lemma. It is a routine exercise that if $V_0 = \ker \varepsilon$, then there are maps $K_0' \to K_0$ and $K_0 \to K_0''$ (where $K_0' = \ker \varepsilon'$ and $K_0'' = \ker \varepsilon''$), so that the resulting 3×3 diagram commutes. Exactness of the top row is Exercise 2.32 on page 96.

We now prove, by induction on $n \geq 0$, that the bottom n rows of the desired diagram can be constructed. For the inductive step, assume that the first n steps have been filled in, and let $V_n = \ker(Q_n \to Q_{n-1})$, while $K_n' = \ker d_n'$ and $K_n'' = \ker d_n''$. As in the base step, there is a commutative diagram with exact rows and columns.

$$
\begin{array}{ccccccccc}
& & 0 & & 0 & & 0 & & \\
& & \downarrow & & \downarrow & & \downarrow & & \\
0 & \to & K_{n+1}' & \to & V_{n+1} & \to & K_{n+1}'' & \to & 0 \\
& & \downarrow & & \downarrow & & \downarrow & & \\
0 & \to & P_{n+1}' & \xrightarrow{i_{n+1}} & Q_{n+1} & \xrightarrow{q_{n+1}} & P_{n+1}'' & \to & 0 \\
& & d_{n+1}' \downarrow & & \downarrow \delta_{n+1} & & \downarrow d_{n+1}'' & & \\
0 & \to & K_n' & \to & V_n & \to & K_n'' & \to & 0 \\
& & \downarrow & & \downarrow & & \downarrow & & \\
& & 0 & & 0 & & 0 & &
\end{array}
$$

Now splice this diagram to the nth diagram by defining $\delta_{n+1} \colon Q_{n+1} \to Q_n$ as the composite $Q_{n+1} \to V_n \to Q_n$.   •

Corollary 6.25. *Let* $0 \to A' \to A \to A'' \to 0$ *be an exact sequence of left R-modules. If both A' and A'' are finitely presented, then A is finitely presented.*

Proof. There are exact sequences $0 \to K_0' \to P_0' \to A' \to 0$ and $0 \to K_0'' \to P_0'' \to A'' \to 0$, where $P_0', P_0'', K_0'', P_0''$ are finitely generated and P_0', P_0'' are projective. As in the beginning of the proof of Proposition 6.24, there is a 3×3 diagram, with Q_0 projective, whose rows and columns are exact.

$$
\begin{array}{ccccccccc}
& & 0 & & 0 & & 0 & & \\
& & \downarrow & & \downarrow & & \downarrow & & \\
0 & \to & K_0' & \to & V_0 & \to & K_0'' & \to & 0 \\
& & \downarrow & & \downarrow & & \downarrow & & \\
0 & \to & P_0' & \to & Q_0 & \to & P_0'' & \to & 0 \\
& & \downarrow & & \downarrow & & \downarrow & & \\
0 & \to & A' & \to & A & \to & A'' & \to & 0 \\
& & \downarrow & & \downarrow & & \downarrow & & \\
& & 0 & & 0 & & 0 & &
\end{array}
$$

Both Q_0 and V_0 are finitely generated, being extensions of finitely generated modules, and so A is finitely presented. •

Theorem 6.26. *Given a commutative diagram of right R-modules having exact rows,*

$$
\begin{array}{ccccccccc}
0 & \to & A' & \xrightarrow{i} & A & \xrightarrow{p} & A'' & \to & 0 \\
& & f\downarrow & & g\downarrow & & h\downarrow & & \\
0 & \to & C' & \xrightarrow{j} & C & \xrightarrow{q} & C'' & \to & 0,
\end{array}
$$

there is a commutative diagram with exact rows for every left R-module B,

$$
\begin{array}{ccccccc}
\mathrm{Tor}_n^R(A',B) & \xrightarrow{i_*} & \mathrm{Tor}_n^R(A,B) & \xrightarrow{p_*} & \mathrm{Tor}_n^R(A'',B) & \xrightarrow{\partial_n} & \mathrm{Tor}_{n-1}^R(A',B) \\
f_*\downarrow & & g_*\downarrow & & h_*\downarrow & & f_*\downarrow \\
\mathrm{Tor}_n^R(C',B) & \xrightarrow{j_*} & \mathrm{Tor}_n^R(C,B) & \xrightarrow{q_*} & \mathrm{Tor}_n^R(C'',B) & \xrightarrow{\partial_n'} & \mathrm{Tor}_{n-1}^R(C',B).
\end{array}
$$

The similar statement for $\mathrm{tor}_n^R(A, \square)$ *is also true.*

Proof. Exactness of $0 \to A' \to A \to A'' \to 0$ gives exactness of the sequence of deleted complexes $0 \to \mathbf{P}_{A'} \to \mathbf{P}_A \to \mathbf{P}_{A''} \to 0$. If $T = \square \otimes_R B$, then $0 \to T\mathbf{P}_{A'} \to T\mathbf{P}_A \to T\mathbf{P}_{A''} \to 0$ is still exact, for every row splits because each term of $\mathbf{P}_{A''}$ is projective. Therefore, the naturality of the connecting homomorphism, Theorem 6.13, applies at once. •

We now show that a short exact sequence gives a long exact sequence of left derived functors.

Theorem 6.27. *Let A be an abelian category with enough projectives. If $0 \to A' \xrightarrow{i} A \xrightarrow{p} A'' \to 0$ is an exact sequence in A and $T : A \to C$ is an additive covariant functor, where C is an abelian category, then there is a long exact sequence in C*

$$\to (L_nT)A' \xrightarrow{(L_nT)i} (L_nT)A \xrightarrow{(L_nT)p} (L_nT)A'' \xrightarrow{\partial_n}$$

$$(L_{n-1}T)A' \xrightarrow{(L_{n-1}T)i} (L_{n-1}T)A \xrightarrow{(L_{n-1}T)p} (L_{n-1}T)A'' \xrightarrow{\partial_{n-1}}$$

which ends with

$$\to (L_0T)A' \to (L_0T)A \to (L_0T)A'' \to 0.$$

Proof. Let $\mathbf{P}'$ and $\mathbf{P}''$ be the chosen projective resolutions of A' and A'', respectively. By the Horseshoe Lemma, there is a projective resolution $\widetilde{\mathbf{P}}$ of A with

$$0 \to \mathbf{P}'_{A'} \xrightarrow{j} \widetilde{\mathbf{P}}_A \xrightarrow{q} \mathbf{P}''_{A''} \to 0.$$

Here, j is a chain map over i and q is a chain map over p. Applying T gives the sequence of complexes

$$0 \to T\mathbf{P}'_{A'} \xrightarrow{Tj} T\widetilde{\mathbf{P}}_A \xrightarrow{Tq} T\mathbf{P}''_{A''} \to 0.$$

This sequence is exact, for each row $0 \to P'_n \xrightarrow{j_n} \widetilde{P}_n \xrightarrow{q_n} P''_n \to 0$ is a split exact sequence (because P''_n is projective), and additive functors preserve split short exact sequences.[6] There is thus a long exact sequence

$$\to H_n(T\mathbf{P}'_{A'}) \xrightarrow{(Tj)_*} H_n(T\widetilde{\mathbf{P}}_A) \xrightarrow{(Tq)_*} H_n(T\mathbf{P}''_{A''}) \xrightarrow{\partial_n} H_{n-1}(T\mathbf{P}'_{A'}) \to;$$

that is, there is an exact sequence

$$\to (L_nT)A' \xrightarrow{(Tj)_*} (\widetilde{L}_nT)A \xrightarrow{(Tq)_*} (L_nT)A'' \xrightarrow{\partial_n} (L_{n-1}T)A' \to .$$

The sequence does terminate with 0, for $L_{-1}T$ is zero for all negative n, by Proposition 6.19.

We do not know that $\widetilde{\mathbf{P}}_A$ arises from the projective resolution of A originally chosen, and so we must change it into the sequence we seek. There are chain maps $\kappa : \mathbf{P}_A \to \widetilde{\mathbf{P}}_A$ and $\lambda : \widetilde{\mathbf{P}}_A \to \mathbf{P}_A$, where both κ, λ are chain maps over 1_A in opposite directions. Indeed, as in the proof of Proposition 6.20, $T\kappa T\lambda$ and $T\lambda T\kappa$ are chain maps over 1_{TA} in opposite directions, whose induced maps in homology are isomorphisms; in fact, $(T\lambda)_* : \widetilde{L}_nT \to L_nT$ is

---

[6]The exact sequence of complexes may *not* split, because the sequence of splitting maps need not constitute a chain map $\mathbf{P}''_{A''} \to \widetilde{\mathbf{P}}_A$.

the inverse of $(T\kappa)_*$. Now $\check{i}$ is a chain map over i, and $\check{p}$ is a chain map over p, while κ, λ are chain maps over 1_A.

The diagram displaying these chain maps is *not* commutative!

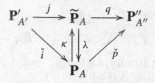

Consider this diagram after applying T and taking homology.

$$\longrightarrow H_n(T\mathbf{P}'_{A'}) \xrightarrow{(Tj)_*} H_n(T\widetilde{\mathbf{P}}_A) \xrightarrow{(Tq)_*} H_n(T\mathbf{P}''_{A''}) \longrightarrow$$
$$\underset{(T\check{i})_*}{\searrow} \quad \underset{(T\kappa)_*}{\big\uparrow} \Big\Vert (T\lambda)_* \quad \underset{(T\check{p})_*}{\nearrow}$$
$$H_n(T\mathbf{P}_A)$$

The noncommutative diagram remains noncommutative after applying T, but the last diagram is commutative. Now $T\lambda Tj \simeq T\check{i}$, because both are chain maps $T\mathbf{P}'_{A'} \to T\mathbf{P}_A$ over Ti; hence, $(T\lambda Tj)_* = (T\check{i})_*$, because homotopic chain maps induce the same homomorphism in homology. But $(T\lambda Tj)_* = (T\lambda)_*(Tj)_*$, and so

$$(T\lambda)_*(Tj)_* = (T\check{i})_* = (L_n T)i.$$

Similarly, $(Tq)_*(T\kappa)_* = (T\check{p})_* = (L_n T)p$. The proof that

$$(L_n T)A' \xrightarrow{(L_n T)i} (L_n T)A \xrightarrow{(L_n T)p} (L_n T)A''$$

is exact can be completed using Exercise 6.14 on page 376. ●

Corollary 6.28. *If $T: {}_R\mathbf{Mod} \to {}_S\mathbf{Mod}$ is an additive covariant functor, then the functor $L_0 T$ is right exact.*

Proof. If $A \to B \to C \to 0$ is exact, then $(L_0 T)A \to (L_0 T)B \to (L_0 T)C \to 0$ is exact. ●

Theorem 6.29.

(i) *If an additive covariant functor $T: \mathcal{A} \to \mathcal{B}$ is right exact, where $\mathcal{A}, \mathcal{B}$ are abelian categories and $\mathcal{A}$ has enough projectives, then T is naturally isomorphic to $L_0 T$.*

(ii) *The functor $\square \otimes_R B$ is naturally isomorphic to $\mathrm{Tor}_0^R(\square, B)$, and the functor $A \otimes_R \square$ is naturally isomorphic to $\mathrm{tor}_0^R(A, \square)$. Hence, for all right R-modules A and left R-modules B, there are isomorphisms*

$$\mathrm{Tor}_0^R(A, B) \cong A \otimes_R B \cong \mathrm{tor}_0^R(A, B).$$

Proof.

(i) Let $\mathbf{P} = \to P_1 \xrightarrow{d_1} P_0 \xrightarrow{\varepsilon} A \to 0$ be the chosen projective resolution of A. By definition, $(L_0 T)A = \operatorname{coker} T d_1$. But right exactness of T gives an exact sequence

$$T P_1 \xrightarrow{T d_1} T P_0 \xrightarrow{T\varepsilon} T A \to 0.$$

Now $T\varepsilon$ induces an isomorphism $\sigma_A \colon \operatorname{coker} T d_1 \to T A$, by the First Isomorphism Theorem; that is,

$$\operatorname{coker} T d_1 = T P_0 / \operatorname{im} T d_1 = T P_0 / \ker T\varepsilon \xrightarrow{\sigma_A} \operatorname{im} T\varepsilon = T A.$$

It is easy to prove that $\sigma = (\sigma_A)_{A \in \mathrm{obj}(\mathbf{Mod}_R)} \colon L_0 T \to T$ is a natural isomorphism.

(ii) Immediate from part (i), for both $\square \otimes_R B$ and $A \otimes_R \square$ are additive covariant right exact functors.   •

Corollary 6.30.   *If $0 \to A' \to A \to A'' \to 0$ is a short exact sequence of right R-modules, then there is a long exact sequence for every left R-module B,*

$$\to \operatorname{Tor}_2^R(A', B) \to \operatorname{Tor}_2^R(A, B) \to \operatorname{Tor}_2^R(A'', B)$$
$$\to \operatorname{Tor}_1^R(A', B) \to \operatorname{Tor}_1^R(A, B) \to \operatorname{Tor}_1^R(A'', B)$$
$$\to A' \otimes_R B \to A \otimes_R B \to A'' \otimes_R B \to 0.$$

The similar statement for $\operatorname{tor}_n^R(A, \square)$ is also true.

Thus, the Tor sequence repairs the loss of exactness after tensoring a short exact sequence.

We now prove that Tor and tor are the same, and we begin with a variation of the Snake Lemma.

Lemma 6.31.   *Given the commutative diagram with exact rows and columns in an abelian category $\mathcal{A}$,*

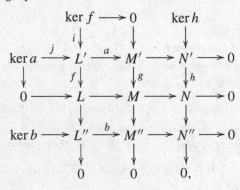

then $\ker f \cong \ker a$ *and* $\ker h \cong \ker b$.

Proof.   Apply the version of the Snake Lemma in Exercise 6.5 on page 338 (with two "missing" zeros) to the maps f, g, h, obtaining exactness of

$$\ker g \to \ker h \to \operatorname{coker} f \to \operatorname{coker} g.$$

Now $\ker g = \{0\}$, $\operatorname{coker} f = L''$, $\operatorname{coker} g = M''$, and we may assume $\operatorname{coker} f \to \operatorname{coker} g$ is b (as in Exercise 6.5). Thus, $0 \to \ker h \to L'' \xrightarrow{b} M''$ is exact, and we conclude that $\ker h \cong \ker b$.

We may assume that i and j are inclusions. Commutativity of the square with corner $\ker a$ gives $fj = 0$; that is, $\ker a = \operatorname{im} j \subseteq \ker f = \operatorname{im} i$; commutativity of the square with corner $\ker f$ gives $ai = 0$; that is, $\ker f = \operatorname{im} i \subseteq \ker a = \operatorname{im} j$. Therefore, $\operatorname{im} i = \operatorname{im} j$ and $\ker f = \ker a$.   •

Theorem 6.32.   *Let A be a right R-module, let B be a left R-module, and let*

$$\mathbf{P} = \to P_1 \xrightarrow{d_1} P_0 \xrightarrow{\varepsilon} A \to 0 \quad and \quad \mathbf{Q} = \to Q_1 \xrightarrow{d_1'} Q_0 \xrightarrow{\varepsilon'} B \to 0$$

be projective resolutions. Then $H_n(\mathbf{P}_A \otimes_R B) \cong H_n(A \otimes_R \mathbf{Q}_B)$ for all $n \geq 0$; that is,

$$\operatorname{Tor}_n^R(A, B) \cong \operatorname{tor}_n^R(A, B).$$

Proof.   (**A. Zaks**) The proof is by induction on $n \geq 0$. The base step $n = 0$ is true, by Theorem 6.29(ii). Let us display the syzygies of $\mathbf{P}$ by "factoring" it into short exact sequences:

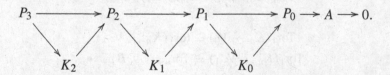

There are exact sequences $0 \to K_i \to P_i \to K_{i-1} \to 0$ for all $i \geq 0$ if we write $A = K_{-1}$ (so that $0 \to K_0 \to P_0 \to A \to 0$ has the same notation as the others). Similarly, we display the syzygies of $\mathbf{Q}$ by factoring it into short exact sequences $0 \to V_j \to Q_j \to V_{j-1} \to 0$ for all $j \geq 0$. Since tensor is a functor of two variables (see Exercise 2.35 on page 96), the following diagram commutes for each $i, j \geq 0$.

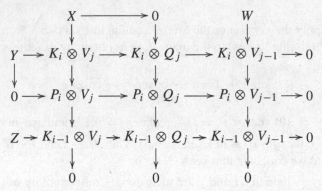

The rows and columns are exact because tensor is right exact; the modules W, X, Y, Z are, by definition, kernels of obvious arrows. Zeros flank the middle row and column because P_i and Q_j are flat (they are even projective). Now $W = \text{Tor}_1(K_{i-1}, V_{j-1})$, $X = \text{Tor}_1(K_{i-1}, V_j)$, $Y = \text{tor}_1(K_i, V_{j-1})$, and $Z = \text{tor}_1(K_{i-1}, V_{j-1})$. By Lemma 6.31, we conclude, for all $i, j \geq -1$,

$$\text{Tor}_1(K_{i-1}, V_{j-1}) \cong \text{tor}_1(K_{i-1}, V_{j-1}).$$

If $i = 0 = j$, then $\text{Tor}_1(A, B) \cong \text{tor}_1(A, B)$ because $K_{-1} = A$ and $V_{-1} = B$. The theorem has been proved for $n = 1$.

We now prove the inductive step. Corollary 6.23 gives

$$\text{tor}_{n+1}(A, B) \cong \text{tor}_1(A, V_{n-1}) = \text{tor}_1(K_{-1}, V_{n-1}),$$
$$\text{Tor}_{n+1}(A, B) \cong \text{Tor}_1(K_{n-1}, B) = \text{Tor}_1(K_{n-1}, V_{-1}).$$

Use these isomorphisms and the isomorphism $X \cong Y$; i.e.,

$$\text{Tor}_1(K_{i-1}, V_j) \cong \text{tor}_1(K_i, V_{j-1}).$$

To go from any equation to the one below it, use the theorem for $n = 1$:

$$\text{tor}_{n+1}(A, B) \cong \text{tor}_1(K_{-1}, V_{n-1});$$
$$\text{Tor}_1(K_{-1}, V_{n-1}) \cong \text{tor}_1(K_0, V_{n-2});$$
$$\text{Tor}_1(K_0, V_{n-2}) \cong \text{tor}_1(K_1, V_{n-3});$$

$$\cdots$$

$$\text{Tor}_1(K_{n-2}, V_0) \cong \text{tor}_1(K_{n-1}, V_{-1});$$
$$\text{Tor}_1(K_{n-1}, V_{-1}) \cong \text{Tor}_{n+1}(A, B). \quad \bullet$$

This last proof is ingenious; we will give a straightforward proof of the theorem once we have spectral sequences.

Remark. The fact that the proof of Theorem 6.32 uses only the flatness of the terms in a projective resolution suggests that Tor can be defined using flat resolutions. See Theorem 7.5 for a proof of this. ◄

6.2.2 Axioms

Here is a set of axioms characterizing the sequence of functors $\mathrm{Tor}_n^R(\square, M)$.

Theorem 6.33 (Axioms for Tor).   *Let* $(T_n\colon \mathbf{Mod}_R \to \mathbf{Ab})_{n\geq 0}$ *be a sequence of additive covariant functors. If,*

(i) *for every short exact sequence* $0 \to A \to B \to C \to 0$ *of right R-modules, there is a long exact sequence with natural connecting homomorphisms*

$$\to T_{n+1}(C) \xrightarrow{\Delta_{n+1}} T_n(A) \to T_n(B) \to T_n(C) \xrightarrow{\Delta_n} T_{n-1}(A) \to,$$

(ii) T_0 *is naturally isomorphic to* $\square \otimes_R M$ *for some left R-module M,*

(iii) $T_n(P) = \{0\}$ *for all projective right R-modules P and all* $n \geq 1$,

then T_n *is naturally isomorphic to* $\mathrm{Tor}_n^R(\square, M)$ *for all* $n \geq 0$.

Proof.   We proceed by induction on $n \geq 0$. The step $n = 0$ is axiom (ii). For the step $n = 1$, given a right R-module A, there is an exact sequence

$$0 \to K \to P \to A \to 0,$$

where P is projective. By axiom (i), there is a diagram with exact rows:

$$
\begin{array}{ccccccc}
\longrightarrow & T_1(P) & \longrightarrow & T_1(A) & \xrightarrow{\Delta_1} & T_0(K) & \longrightarrow & T_0(P) \\
& & & \Big\downarrow{\scriptstyle \tau_{1A}} & & \Big\downarrow{\scriptstyle \tau_{0K}} & & \Big\downarrow{\scriptstyle \tau_{0P}} \\
\longrightarrow & \mathrm{Tor}_1^R(P, M) & \longrightarrow & \mathrm{Tor}_1^R(A, M) & \xrightarrow{\delta_1} & \mathrm{Tor}_0^R(K, M) & \longrightarrow & \mathrm{Tor}_0^R(P, M),
\end{array}
$$

where the maps τ_{0K} and τ_{0P} are the natural isomorphisms given by axiom (ii). Of course, naturality gives commutativity of the square on the right. Axiom (iii) gives $T_1(P) = \{0\} = \mathrm{Tor}_1^R(P, M)$, so that the maps Δ_1 and δ_1 are injective. Diagram chasing, Proposition 2.71, gives an isomorphism τ_{1A} making the augmented diagram commute.

We now prove the inductive step, and we may assume that $n \geq 1$. Look further out in the long exact sequence. By axiom (i), there is a commutative diagram with exact rows

$$
\begin{array}{ccccccc}
T_{n+1}(P) & \longrightarrow & T_{n+1}(A) & \xrightarrow{\Delta_{n+1}} & T_n(K) & \longrightarrow & T_n(P) \\
& & \Big\downarrow{\scriptstyle \tau_{n+1,A}} & & \Big\downarrow{\scriptstyle \tau_{nK}} & & \Big\downarrow \\
\mathrm{Tor}_{n+1}^R(P, M) & \longrightarrow & \mathrm{Tor}_{n+1}^R(A, M) & \xrightarrow{\delta_{n+1}} & \mathrm{Tor}_n^R(K, M) & \longrightarrow & \mathrm{Tor}_n^R(P, M),
\end{array}
$$

where $\tau_{nK} : T_n(K) \to \text{Tor}_n^R(K, M)$ is an isomorphism given by the inductive hypothesis. Since $n \geq 1$, all four terms involving the projective P are $\{0\}$. It follows from exactness of the rows that both Δ_n and ∂_n are isomorphisms. Therefore, the composite, $\tau_{n+1,A} = \delta_{n+1}^{-1} \tau_{nK} \Delta_{n+1} : T_{n+1}(A) \to F_{n+1}(A)$ is an isomorphism.

That the isomorphisms $\tau_{n+1,A}$ constitute a natural isomorphism $T_{n+1} \to \text{Tor}_{n+1}^R(\square, M)$ is left to the reader with the remark that the proof uses the assumed naturality of the connecting homomorphisms Δ and δ.   •

The strategy of the proof of Theorem 6.33 occurs frequently, and it is called *dimension shifting*. Choose a short exact sequence $0 \to A \to X \to C \to 0$ whose middle term X forces higher homology groups to vanish; for example, X might be projective or injective. The proof proceeds as a slow starting induction, proving results for $n = 0$ and $n = 1$ before proving the inductive step.

The theorem can be generalized.

Corollary 6.34. *Let $(T_n)_{n \geq 0}, (T_n')_{n \geq 0}$ be sequences of additive covariant functors $\mathcal{A} \to \mathcal{B}$, where $\mathcal{A}, \mathcal{B}$ are abelian categories and $\mathcal{A}$ has enough projectives. If,*

   (i) *for every short exact sequence $0 \to A \to B \to C \to 0$ in $\mathcal{A}$, there are long exact sequences with natural connecting homomorphisms,*

   (ii) *T_0 is naturally isomorphic to T_0',*

   (iii) *$T_n(P) = 0 = T_n'(P)$ for all projectives P and all $n \geq 1$,*

then T_n is naturally isomorphic to T_n' for all $n \geq 0$.

Remark. Notice that this corollary does *not* assume that the sequences $(T_n)_{n \geq 0}, (T_n')_{n \geq 0}$ are derived functors.   ◀

Proof. A harmless rewriting of the proof of Theorem 6.33.   •

This corollary can itself be generalized.

Definition. Let $(T_n : \mathcal{A} \to \mathcal{B})_{n \geq 0}$ be a sequence of additive functors, where $\mathcal{A}$ and $\mathcal{B}$ are abelian categories. If $\mathcal{X}$ is a class of objects in $\mathcal{A}$, then we say that $\mathcal{A}$ has *enough $\mathcal{X}$-objects* if every object in $\mathcal{A}$ is a quotient of an object in $\mathcal{X}$. We call $(T_n)_{n \geq 0}$ *$\mathcal{X}$-effaceable* if $T_n(X) = 0$ for all $X \in \mathcal{X}$ and $n \geq 1$.

We could call $\mathcal{X}$-objects *acyclic* or *relatively projective*.

Corollary 6.35. *Let* $(T_n \colon \mathcal{A} \to \mathcal{B})_{n \geq 0}$, $(T_n' \colon \mathcal{A} \to \mathcal{B})_{n \geq 0}$ *be sequences of additive covariant functors, where* $\mathcal{A}, \mathcal{B}$ *are abelian categories,* $\mathcal{X}$ *is a class of objects in* $\mathcal{A}$, *and* $\mathcal{A}$ *has enough* $\mathcal{X}$-*objects. If,*

(i) *for every short exact sequence* $0 \to A \to B \to C \to 0$ *in* $\mathcal{A}$, *there are long exact sequences with natural connecting homomorphisms,*

(ii) T_0 *is naturally isomorphic to* T_0',

(iii) *both* $(T_n)_{n \geq 0}$, $(T_n')_{n \geq 0}$ *are* $\mathcal{X}$-*effaceable,*

then T_n *is naturally isomorphic to* T_n' *for all* $n \geq 0$.

Grothendieck made this last corollary into a definition, for it displays the fundamental properties of homology [see "Sur quelques points d'algèbre homologique," *Tohoku Math J.*, 1957, pp. 119–183, pp. 185–221].

Definition. If $\mathcal{A}, \mathcal{B}$ are abelian categories, then a sequence of additive functors $(T_n \colon \mathcal{A} \to \mathcal{B})_{n \geq 0}$ is a **homological** ∂-**functor** if, for every short exact sequence $0 \to A \to B \to C \to 0$ in $\mathcal{A}$, there is a long exact sequence

$$\to T_n(A) \to T_n(B) \to T_n(C) \xrightarrow{\partial_n} T_{n-1}(A) \to$$

ending $\to T_0(A) \to T_0(B) \to T_0(C)$ [7] and having natural connecting homomorphisms $\partial_n \colon T_n(C) \to T_{n-1}(A)$; that is, if $0 \to A' \to B' \to C' \to 0$ is exact in $\mathcal{A}$, then the following diagram commutes.

$$\begin{array}{ccc} T_n(C) & \xrightarrow{\partial'} & T_{n-1}(A) \\ \downarrow & & \downarrow \\ T_n(C') & \xrightarrow{\partial} & T_{n-1}(A') \end{array}$$

A **morphism** $\tau \colon (T_n)_{n \geq 0} \to (H_n)_{n \geq 0}$ of homological ∂-functors is a sequence of natural transformations $\tau_n \colon T_n \to H_n$, for $n \geq 0$, such that the following diagram commutes:

$$\begin{array}{ccc} T_n(C) & \xrightarrow{\partial} & T_{n-1}(A) \\ {\scriptstyle \tau_{n,C}} \downarrow & & \downarrow {\scriptstyle \tau_{n-1,A}} \\ H_n(C) & \xrightarrow{\partial} & H_{n-1}(A) \end{array}$$

for every short exact sequence $0 \to A \to B \to C \to 0$ in $\mathcal{A}$.

We can now give a useful variation of Corollary 6.35.

---

[7] Most authors assume further, as part of the definition, that the long exact sequence ends with $\to T_0(A) \to T_0(B) \to T_0(C) \to 0$; that is, that T_0 is right exact.

Definition. If $F: \mathcal{A} \to \mathcal{B}$ is an additive functor, then a homological ∂-functor $(T_n: \mathcal{A} \to \mathcal{B})_{n \geq 0}$ is a **homological extension** of F if there is a natural isomorphism $\tau: F \to T_0$.

For example, $(\operatorname{Tor}_n^R(A, \square))_{n \geq 0}$ is a homological extension of $A \otimes_R \square$.

Theorem 6.36. *Let $\mathcal{A}$ be an abelian category with enough projectives.*

(i) *If (T_n) and (H_n) are homological ∂-functors $\mathcal{A} \to \mathbf{Ab}$ with $H_n(P) = \{0\}$ for all projective P and $n \geq 1$, and if $\tau_0: T_0 \to H_0$ is a natural transformation, then there exists a unique morphism $\tau: (T_n) \to (H_n)$. Moreover, if τ_0 is a natural isomorphism, then τ_n is a natural isomorphism for all $n \geq 0$.*

(ii) *If $F: \mathcal{A} \to \mathbf{Ab}$ is a right exact additive covariant functor, then there exists a unique homological extension $(H_n)_{n \geq 0}$ of F with $H_n(P) = \{0\}$ for all projective P and all $n \geq 1$.*

Proof.

(i) If $0 \to A \to B \to C \to 0$ is exact, we construct $\tau_n: T_n \to H_n$ by induction on $n \geq 0$. We are assuming the existence of τ_0, and so we may assume that $n > 0$ and that there is a natural $\tau_{n-1}: T_{n-1} \to H_{n-1}$. Since there are enough projectives, there is an exact sequence

$$0 \to K \to P \to C \to 0 \tag{1}$$

with P projective. In the commutative diagram

$$
\begin{array}{ccccc}
T_n(C) & \longrightarrow & T_{n-1}(K) & \longrightarrow & T_{n-1}(P) \\
\Big\downarrow & & \Big\downarrow & & \Big\downarrow \\
H_n(P) \longrightarrow H_n(C) & \longrightarrow & H_{n-1}(K) & \longrightarrow & H_{n-1}(P),
\end{array}
$$

the map $H_n(C) \to H_{n-1}(K)$ is an isomorphism [for $H_q(P) = \{0\}$ for $n \geq 1$], and there exists a unique homomorphism $\tau_{n,C}: T_n(C) \to H_n(C)$ (the clockwise composite) making the first square commute. We claim that $\tau_n = (\tau_{n,C}: T_n \to H_n)_{C \in \operatorname{obj}(\mathcal{A})}$ is natural, that it is well-defined [it does not depend on the choice of sequence (1)], and that it commutes with the connecting homomorphisms of (T_n) and (H_n).

We prove these assertions with the aid of the following fact. We claim, given the diagram

$$
\begin{array}{c}
C \\
\Big\downarrow f \\
0 \longrightarrow A' \longrightarrow A \longrightarrow A'' \longrightarrow 0
\end{array}
$$

with exact row, that there is a commutative diagram

$$
\begin{array}{ccc}
T_n(C) \xrightarrow{T_n f} T_n(A'') \xrightarrow{\ \partial\ } T_{n-1}(A') \\
\Big\downarrow{\scriptstyle \tau_{n,C}} \qquad\qquad\qquad \Big\downarrow{\scriptstyle \tau_{n-1,A'}} \\
H_n(C) \xrightarrow[H_n f]{} H_n(A'') \xrightarrow[\partial]{} H_{n-1}(A').
\end{array}
$$

Since P is projective, there is a commutative diagram

$$
\begin{array}{ccccccccc}
0 & \longrightarrow & K & \longrightarrow & P & \longrightarrow & C & \longrightarrow & 0 \\
& & \Big\downarrow{\scriptstyle g} & & \Big\downarrow & & \Big\downarrow{\scriptstyle f} & & \\
0 & \longrightarrow & A' & \longrightarrow & A & \longrightarrow & A'' & \longrightarrow & 0.
\end{array}
$$

Consider the following diagram (which can also be drawn as a cube).

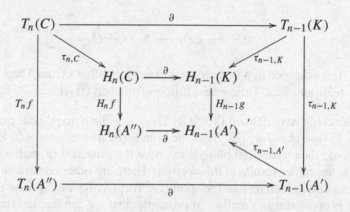

The upper trapezoid commutes, by construction of $\tau_{n,C}$; the inner square commutes because (H_n) is a ∂-functor; the right trapezoid commutes, by naturality of τ_{n-1}; finally, the outer square commutes because (T_n) is a ∂-functor. It follows that the rectangle of the claim commutes.

We now show that τ_n is natural. Let $X_1, X_2 \in \mathrm{obj}(\mathcal{A})$, and choose exact sequences $0 \to K_i \to P_i \to X_i \to 0$ with P_i projective, for $i = 1, 2$. Given $f: X_1 \to X_2$, define $\tau_{n,X_i}: T_n(X_i) \to H_n(X_i)$, as in the first paragraph of this proof. Apply the fact just proved to

$$
\begin{array}{c}
X_1 \\
\Big\downarrow{\scriptstyle f} \\
0 \longrightarrow K_2 \longrightarrow P_2 \longrightarrow X_2 \longrightarrow 0,
\end{array}
$$

and obtain the diagram in which the perimeter commutes:

$$
\begin{array}{ccc}
T_n(X_1) & \xrightarrow{\;T_n f\;} T_n(X_2) \xrightarrow{\;\partial\;} & T_{n-1}(X_2) \\
\Big\downarrow{\scriptstyle \tau_{n,X_1}} & \Big\downarrow{\scriptstyle \tau_{n,X_2}} & \Big\downarrow{\scriptstyle \tau_{n-1,K_2}} \\
H_n(X_1) & \xrightarrow[\;H_n f\;]{} H_n(X_2) \xrightarrow[\;\partial\;]{} & H_{n-1}(K_2).
\end{array}
$$

The right square commutes, by construction of τ_{n,X_2}, while the connecting homomorphism ∂ in the bottom row is injective because $H_n(P_2) = \{0\}$. It now follows that the left square commutes; that is, $\tau_n = (\tau_{n,X})$ is natural. To see that $\tau_{n,X}$ does not depend on the choice of sequence (1), apply this argument with $X_1 = X_2$ and $f = 1_{X_i}$. Finally, to see that τ_n commutes with connecting homomorphisms, take any exact sequence $0 \to A \to B \to C \to 0$ and apply the fact to the diagram

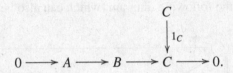

$$
\begin{array}{ccccccccc}
& & & & & & C & & \\
& & & & & & \Big\downarrow{\scriptstyle 1_C} & & \\
0 & \longrightarrow & A & \longrightarrow & B & \longrightarrow & C & \longrightarrow & 0.
\end{array}
$$

(ii) The existence of a homological extension follows from Theorems 6.10, 6.26, and 6.29. Uniqueness follows from part (i) with τ_0 the identity.   •

We could have defined $(\mathrm{Tor}_n^R(A, \square))_{n \geq 0}$ as the homological extension of $A \otimes_R \square$ [and also $\mathrm{Tor}_n^R(\square, B)$ as the homological extension of $\square \otimes_R B$], but we would then have been obliged to prove the existence of such a sequence; that is, the earlier results in this section. There are other constructions of Tor (for example, generators and relations for $\mathrm{Tor}_1^{\mathbb{Z}}(A, B)$ are given on page 411), but it is comforting to realize, in principle, that we can use the functors Tor without being constantly aware of the details of their construction as left derived functors.

Remark. In the 1930s, there were many constructions of homology groups and cohomology groups associated to topological spaces (e.g., simplicial homology, singular homology, cubical homology, Čech cohomology), each invented for a specific purpose, and it was natural to ask whether these groups coincided with other homology groups. The first axiomatic characterization of homology was due to Eilenberg and Steenrod, "Axiomatic approach to homology theory," *Proc. Nat. Acad. Sci. U.S.A.*, 31 (1945), 117–120.

Definition. Let $\mathbf{Top}^2$ be the category having as objects all pairs (X, A) of topological spaces with A a subspace of X, as morphisms $(X, A) \to (X', A')$

all continuous functions $f: X \to X'$ for which $f(A) \subseteq A'$, and usual composition. The **Eilenberg–Steenrod Axioms** for a sequence of covariant functors $(G_n: \textbf{Top}^2 \to \textbf{Ab})$ and natural maps $(\partial_n: G_n(X, A) \to G_{n-1}(A, \varnothing))$ are

(i) **Homotopy Axiom**. If $f_0, f_1: (X, A) \to (X', A')$ are homotopic, then, for all $n \geq 0$,

$$G_n(f_0) = G_n(f_1): G_n(X, A) \to G_n(X', A');$$

(ii) **Exactness Axiom**. Write $G_n(X, \varnothing) = G_n(X)$. Given a pair (X, A) and inclusion maps $i: (A, \varnothing) \to (X, \varnothing)$ and $j: (X, \varnothing) \to (X, A)$, there is a long exact sequence

$$\xrightarrow{G_{n+1}(j)} G_{n+1}(X, A) \xrightarrow{\partial_{n+1}} G_n(A) \xrightarrow{G_n(i)} G_n(X) \xrightarrow{G_n(j)} G_n(X, A) \xrightarrow{\partial_n};$$

(iii) **Excision Axiom**. Given a pair (X, A) and an open $U \subseteq X$ such that $\overline{U} \subseteq \text{interior}(A)$, then the inclusion $(X - U, A - U) \to (X, A)$ induces isomorphisms $G_n(X - U, A - U) \to G_n(X, A)$ for all $n \geq 0$;

(iv) **Dimension Axiom**. For every one-point space P, we have

$$G_n(P) = \begin{cases} \{0\} & \text{if } n > 0, \\ \mathbb{Z} & \text{if } n = 0. \end{cases}$$

In more detail, on the full subcategory of all one-point subspaces, there is a natural isomorphism $\tau: G_0 \to \mathbb{Z}$ (where $\mathbb{Z}$ denotes the constant functor at $\mathbb{Z}$). Thus, if P and Q are one-point spaces and $f: P \to Q$, then there is a commutative diagram

$$\begin{array}{ccc} G_0(P) & \xrightarrow{\tau_P} & \mathbb{Z} \\ G_0(f) \downarrow & & \downarrow 1_{\mathbb{Z}} \\ G_0(Q) & \xrightarrow{\tau_Q} & \mathbb{Z}. \end{array}$$

Theorem (Eilenberg–Steenrod). *If two sequences of covariant functors* $(G_n, H_n: \textbf{Top}^2 \to \textbf{Ab})_{n \geq 0}$ *and natural maps satisfy the Eilenberg–Steenrod Axioms, then*

$$G_n \cong H_n \text{ for all } n \geq 0. \quad \blacktriangleleft$$

It follows that the singular and simplicial homology of a simplicial complex K are the same.

6.2.3 Covariant Right Derived Functors

Left derived functors $L_n T$ satisfy Corollary 6.28: if T is right exact, then $L_0 T$ is naturally isomorphic to T. For this reason, it is natural to take left derived functors of tensor. We are now going to define *right derived functors* $R^n T$, where $T : \mathcal{A} \to \mathcal{C}$ is an additive covariant functor between abelian categories. We will prove the analog of Corollary 6.28, which says that $R^0 T \cong T$ when T is left exact. Thus, this construction is appropriate for Hom functors.

Choose, once for all, an injective resolution

$$\mathbf{E} = 0 \to B \xrightarrow{\eta} E^0 \xrightarrow{d^0} E^1 \xrightarrow{d^1} E^2 \xrightarrow{d^2} E^3 \to$$

of every object B, form the complex $T\mathbf{E}^B$, where $\mathbf{E}^B$ is the deleted injective resolution, and take homology:

$$(R^n T)B = H^n(T\mathbf{E}^B) = \frac{\ker T d^n}{\operatorname{im} T d^{n-1}}.$$

The reader should reread Example 6.6(ii); if we relabel E^n as E_{-n} and d^n as d_{-n}, then the definition is

$$(R^n T)B = H_{-n}(T\mathbf{E}^B) = \frac{\ker T d_{-n}}{\operatorname{im} T d_{-n+1}}.$$

Notice that the indices on homology are now superscripts; we write H^n instead of H_{-n}.

The definition of $(R^n T)f$, where $f : B \to B'$ is a homomorphism, is similar to that for left derived functors. By the dual of the comparison theorem, there is a chain map $\check{f} : \mathbf{E}^B \to \mathbf{E}'^{B'}$ over f, unique to homotopy, and so there is a well-defined map $(R^n T)f : H^n(T\mathbf{E}^B) \to H^n(T\mathbf{E}^{B'})$ induced in homology, namely, $(T\check{f})_{n*}$.

In pictures, look at the chosen injective resolutions:

$$
\begin{array}{ccccccc}
0 & \longrightarrow & B' & \longrightarrow & E'^0 & \longrightarrow & E'^1 \longrightarrow \\
 & & {\scriptstyle f}\uparrow & & & & \\
0 & \longrightarrow & B & \longrightarrow & E^0 & \longrightarrow & E^1 \longrightarrow \cdot
\end{array}
$$

Fill in a chain map $\check{f}$ over f, then apply T to this diagram, and then take the map induced by $T\check{f}$ in homology.

The proofs of the following propositions about right derived functors are essentially duals of the proofs we have given for left derived functors, and so they will be omitted.

Theorem 6.37. *If $T : \mathcal{A} \to \mathcal{C}$ is an additive covariant functor between abelian categories, where $\mathcal{A}$ has enough injectives, then $R^n T : \mathcal{A} \to \mathcal{C}$ is an additive covariant functor for every $n \in \mathbb{Z}$.*

Definition. If $T: \mathcal{A} \to \mathcal{C}$ is an additive covariant functor between abelian categories, where $\mathcal{A}$ has enough injectives, the functors $R^n T$ are called the *right derived functors* of T.

Proposition 6.38. *If $T: {}_R\mathbf{Mod} \to {}_R\mathbf{Mod}$ is an additive covariant functor that preserves multiplications, then $R^n T: {}_R\mathbf{Mod} \to {}_R\mathbf{Mod}$ also preserves multiplications.*

The next proposition shows that $R^n T$ is of interest only for $n \geq 0$.

Proposition 6.39. *If $T: \mathcal{A} \to \mathcal{C}$ is an additive covariant functor between abelian categories, where $\mathcal{A}$ has enough injectives, then $(R^n T)B = 0$ for all negative n and for all B.*

Definition. If $T = \mathrm{Hom}_R(A, \square)$, define $\mathrm{Ext}^n_R(A, \square) = R^n T$. If the chosen injective resolution of B is $\mathbf{E} = 0 \to B \xrightarrow{\eta} E^0 \xrightarrow{d^0} E^1 \xrightarrow{d^1} E^2 \to$, then

$$\mathrm{Ext}^n_R(A, B) = H^n(\mathrm{Hom}_R(A, \mathbf{E}^B)) = \frac{\ker d^n_*}{\mathrm{im}\, d^{n-1}_*},$$

where $d^n_*: \mathrm{Hom}_R(A, E^n) \to \mathrm{Hom}_R(A, E^{n+1})$ is defined, as usual, by

$$d^n_*: f \mapsto d^n f.$$

The domain of $\mathrm{Ext}^n_R(A, \square)$ is ${}_R\mathbf{Mod}$, the category of left R-modules, and its target is $\mathbf{Ab}$ (there are also Ext^n functors defined on $\mathbf{Mod}_R$ if the Hom functor T acts on right modules). If R is commutative, then $\mathrm{Hom}_R(A, B)$ is an R-module, and so the values of $\mathrm{Ext}^n_R(A, \square)$ lie in ${}_R\mathbf{Mod}$.

Assume that new choices $\widetilde{\mathbf{E}}$ of injective resolutions have been made; denote the right derived functors arising from these new choices by $\widetilde{R}^n T$.

Proposition 6.40. *If $T: \mathcal{A} \to \mathcal{C}$ is an additive covariant functor between abelian categories, where $\mathcal{A}$ has enough injectives, then the functors $R^n T$ and $\widetilde{R}^n T$ are naturally isomorphic for each n. In particular, for all $A \in \mathrm{obj}(\mathcal{A})$,*

$$(R^n T)B \cong (\widetilde{R}^n T)B,$$

and so these objects are independent of the choice of injective resolution of B. In particular, if $T: {}_R\mathbf{Mod} \to {}_S\mathbf{Mod}$, where R and S are rings, then the module $\mathrm{Ext}^n_R(A, B)$ is independent of the choice of injective resolution of B.

Corollary 6.41. *Let $T: \mathcal{A} \to \mathcal{C}$ be an additive covariant functor between abelian categories, where $\mathcal{A}$ has enough injectives. If E is injective, then $(R^n T)E = \{0\}$ for all $n \geq 1$. In particular, if E is an injective left R-module, then $\mathrm{Ext}^n_R(A, E) = \{0\}$ for all $n \geq 1$ and all left R-modules A.*

Corollary 6.42. *Let $\mathcal{A}$ be an abelian category with enough injectives, let $B \in \mathrm{obj}(\mathcal{A})$, and let $\mathbf{E} = 0 \to B \xrightarrow{\eta} E^0 \xrightarrow{d^0} E^1 \xrightarrow{d^1} E^2 \xrightarrow{d^2} E^3 \to be$ an injective resolution of B. Define $V_0 = \mathrm{im}\,\eta$ and define $V_n = \mathrm{im}\,d^{n-1}$ for all $n \geq 1$. Then*

$$(R^{n+1}T)B \cong (R^nT)V_0 \cong (R^{n-1}T)V_1 \cong \cdots \cong (R^1T)V_{n-1}.$$

In particular, for any left R-modules A and B,

$$\mathrm{Ext}_R^{n+1}(A, B) \cong \mathrm{Ext}_R^n(A, V_0) \cong \cdots \cong \mathrm{Ext}_R^1(A, V_{n-1}).$$

Theorem 6.43. *If $0 \to B' \xrightarrow{i} B \xrightarrow{p} B'' \to 0$ is an exact sequence in an abelian category $\mathcal{A}$ with enough injectives, and if $T : \mathcal{A} \to \mathcal{C}$ is an additive covariant functor, where $\mathcal{C}$ is an abelian category, then there is a long exact sequence*

$$\to (R^nT)B' \xrightarrow{(R^nT)i} (R^nT)B \xrightarrow{(R^nT)p} (R^nT)B'' \xrightarrow{\partial^n}$$

$$(R^{n+1}T)B' \xrightarrow{(R^{n+1}T)i} (R^{n+1}T)B \xrightarrow{(R^{n+1}T)p} (R^{n+1}T)B'' \xrightarrow{\partial^{n+1}}$$

that begins with

$$0 \to (R^0T)B' \to (R^0T)B \to (R^0T)B'' \to .$$

Corollary 6.44. *If $T : \mathcal{A} \to \mathcal{C}$ is an additive covariant functor between abelian categories, where $\mathcal{A}$ has enough injectives, then the functor R^0T is left exact.*

Theorem 6.45.

(i) *If an additive covariant functor $T : \mathcal{A} \to \mathcal{C}$ is left exact, where $\mathcal{A}$ and $\mathcal{C}$ are abelian categories and $\mathcal{A}$ has enough injectives, then T is naturally isomorphic to R^0T.*

(ii) *If A is a left R-module, then the functor $\mathrm{Hom}_R(A, \square)$ is naturally isomorphic to $\mathrm{Ext}_R^0(A, \square)$. Hence, for all left R-modules B, there is an isomorphism*

$$\mathrm{Hom}_R(A, B) \cong \mathrm{Ext}_R^0(A, B).$$

Corollary 6.46. *If $0 \to B' \to B \to B'' \to 0$ is a short exact sequence of left R-modules, then for every left R-module A, there is a long exact sequence* *of abelian groups,*

$$0 \to \operatorname{Hom}_R(A, B') \to \operatorname{Hom}_R(A, B) \to \operatorname{Hom}_R(A, B'')$$
$$\to \operatorname{Ext}^1_R(A, B') \to \operatorname{Ext}^1_R(A, B) \to \operatorname{Ext}^1_R(A, B'')$$
$$\to \operatorname{Ext}^2_R(A, B') \to \operatorname{Ext}^2_R(A, B) \to \operatorname{Ext}^2_R(A, B'') \to .$$

Thus, Ext repairs the loss of exactness after applying $\operatorname{Hom}_R(A, \square)$ to a short exact sequence.

Theorem 6.47. *Given a commutative diagram of left R-modules having exact rows,*

$$
\begin{array}{ccccccccc}
0 & \longrightarrow & B' & \xrightarrow{\ i\ } & B & \xrightarrow{\ p\ } & B'' & \longrightarrow & 0 \\
& & \downarrow{\scriptstyle f} & & \downarrow{\scriptstyle g} & & \downarrow{\scriptstyle h} & & \\
0 & \longrightarrow & C' & \xrightarrow[\ j\ ]{} & C & \xrightarrow[\ q\ ]{} & C'' & \longrightarrow & 0,
\end{array}
$$

there is a commutative diagram of abelian groups with exact rows,

$$
\begin{array}{ccccccc}
\operatorname{Ext}^n_R(A, B') & \xrightarrow{i_*} & \operatorname{Ext}^n_R(A, B) & \xrightarrow{p_*} & \operatorname{Ext}^n_R(A, B'') & \xrightarrow{\partial^n} & \operatorname{Ext}^{n+1}_R(A, B') \\
\downarrow{\scriptstyle f_*} & & \downarrow{\scriptstyle g_*} & & \downarrow{\scriptstyle h_*} & & \downarrow{\scriptstyle f_*} \\
\operatorname{Ext}^n_R(A, C') & \xrightarrow[j_*]{} & \operatorname{Ext}^n_R(A, C) & \xrightarrow[q_*]{} & \operatorname{Ext}^n_R(A, C'') & \xrightarrow[\partial'^n]{} & \operatorname{Ext}^{n+1}_R(A, C').
\end{array}
$$

Theorem 6.48 (Axioms for Covariant Ext). *Let $(F^n \colon {}_R\mathbf{Mod} \to \mathbf{Ab})_{n \geq 0}$ be a sequence of additive covariant functors. If,*

(i) *for every short exact sequence $0 \to A \to B \to C \to 0$, there is a long exact sequence with natural connecting homomorphisms*

$$\to F^{n-1}(C) \xrightarrow{\Delta_{n-1}} F^n(A) \to F^n(B) \to F^n(C) \xrightarrow{\Delta_n} F^{n+1}(A) \to,$$

(ii) *there is a left R-module M such that F^0 and $\operatorname{Hom}_R(M, \square)$ are naturally isomorphic,*

(iii) *$F^n(E) = \{0\}$ for all injective left R-modules E and all $n \geq 1$,*

then F^n is naturally isomorphic to $\operatorname{Ext}^n_R(M, \square)$ for all $n \geq 0$.

Proof. See the proof of Theorem 6.33. ●

Corollary 6.49. *Let* $(F^n \colon \mathcal{A} \to \mathcal{B})_{n \geq 0}$, $(F'^n)_{n \geq 0}$ *be sequences of additive covariant functors, where* $\mathcal{A}, \mathcal{B}$ *are abelian categories and* $\mathcal{A}$ *has enough injectives. If,*

(i) *for every short exact sequence* $0 \to A \to B \to C \to 0$, *there are long exact sequences with natural connecting homomorphisms,*

(ii) F^0 *is naturally isomorphic to* F'^0,

(iii) $F^n(E) = 0 = F'^n(E)$ *for all injective objects* E *and all* $n \geq 1$,

then F^n *is naturally isomorphic to* F'^n *for all* $n \geq 0$.

Definition. Let $(F^n \colon \mathcal{A} \to \mathcal{B})_{n \geq 0}$ be a sequence of additive functors, where $\mathcal{A}$ and $\mathcal{B}$ are abelian categories. If $\mathcal{Y}$ is a class of objects in $\mathcal{A}$, then we say that $\mathcal{A}$ has **enough co-$\mathcal{Y}$-objects** if every object can be imbedded in a $\mathcal{Y}$-object. We call $(F^n)_{n \geq 0}$ $\mathcal{Y}$-**coeffaceable** if $F^n(Y) = 0$ for all $Y \in \mathcal{Y}$ and all $n \geq 1$.

We could call objects in $\mathcal{Y}$ *acyclic* or *relatively injective*.

Corollary 6.50. *Let* $(F^n \colon \mathcal{A} \to \mathcal{B})_{n \geq 0}$, $(F'^n \colon \mathcal{A} \to \mathcal{B})_{n \geq 0}$ *be sequences of additive covariant functors, where* $\mathcal{A}, \mathcal{B}$ *are abelian categories,* $\mathcal{Y}$ *is a class of objects in* $\mathcal{A}$, *and* $\mathcal{A}$ *has enough co-$\mathcal{Y}$-objects. If,*

(i) *for every short exact sequence* $0 \to A \to B \to C \to 0$ *in* $\mathcal{A}$, *there are long exact sequences with natural connecting homomorphisms,*

(ii) F^0 *is naturally isomorphic to* F'^0,

(iii) *both* $(F^n)_{n \geq 0}$, $(F'^n)_{n \geq 0}$ *are* $\mathcal{Y}$-*coeffaceable,*

then F^n *is naturally isomorphic to* F'^n *for all* $n \geq 0$.

Definition. If $\mathcal{A}, \mathcal{B}$ are abelian categories, then a sequence of additive functors $(T^n \colon \mathcal{A} \to \mathcal{B})_{n \geq 0}$ is a **cohomological ∂-functor** if, for every short exact sequence $0 \to A \to B \to C \to 0$ in $\mathcal{A}$, there is a long exact sequence

$$\to T^n(A) \to T^n(B) \to T^n(C) \xrightarrow{\partial^n} T^{n+1}(A) \to$$

beginning with $T^0(A) \to T^0(B) \to T^0(C)$[8] having natural connecting homomorphisms $\partial^n \colon T^n(C) \to T^{n+1}(A)$; that is, if $0 \to A' \to B' \to C' \to 0$

---

[8]Most authors assume further, as part of the definition, that the long exact sequence starts with $0 \to T_0(A) \to T_0(B) \to T_0(C)$; that is, that T_0 is left exact.

is exact in $\mathcal{A}$, then the following diagram commutes.

$$
\begin{array}{ccc}
T^n(C) & \xrightarrow{\ \partial'\ } & T^{n+1}(A) \\
\downarrow & & \downarrow \\
T^n(C') & \xrightarrow[\ \partial\ ]{} & T^{n+1}(A')
\end{array}
$$

A *morphism* $\tau\colon (H^n)_{n\geq 0} \to (T^n)_{n\geq 0}$ of cohomological ∂-functors is a sequence of natural transformations $\tau^n\colon H^n \to T^n$, for $n \geq 0$, such that the following diagram commutes.

$$
\begin{array}{ccc}
H^n(C) & \xrightarrow{\ \partial\ } & H^{n+1}(A) \\
\tau^n_C \downarrow & & \downarrow \tau^{n+1}_A \\
T^n(C) & \xrightarrow[\ \partial\ ]{} & T^{n+1}(A)
\end{array}
$$

for every short exact sequence $0 \to A \to B \to C \to 0$ in $\mathcal{A}$.

Definition. If $F\colon \mathcal{A} \to \mathcal{B}$ is an additive functor, then a cohomological ∂-functor $(T^n\colon \mathcal{A} \to \mathcal{B})_{n\geq 0}$ is a *cohomological extension* of F if there is a natural isomorphism $\tau\colon F \to T^0$.

$(\operatorname{Ext}^n_R(A, \square))_{n\geq 0}$ is a cohomological extension of $\operatorname{Hom}_R(A, \square)$.

Theorem 6.51. *Let $\mathcal{A}$ be an abelian category with enough injectives.*

(i) *If (H^n) and (T^n) are cohomological ∂-functors with $H^n(E) = \{0\}$ for all injective E and $n \geq 1$, and if $\tau^0\colon H^0 \to T^0$ is a natural transformation, then there exists a unique morphism $\tau\colon (H^n) \to (T^n)$. Moreover, if τ^0 is a natural isomorphism, then τ^n is a natural isomorphism for all $n \geq 0$.*

(ii) *If F is a left exact covariant additive functor, then there exists a unique cohomological extension $(H^n)_{n\geq 0}$ of F with $H^n(E) = \{0\}$ for all injective E and all $n \geq 1$.*

Proof. Dual to the proof of Theorem 6.36. $\quad\bullet$

6.2.4 Contravariant Right Derived Functors

We now discuss right derived functors $R^n T$ of an additive contravariant functor T. Given a resolution $\mathbf{C}$, we want $T\mathbf{C}$ to have only negative indices. Thus, we start with a projective resolution **P** of A, for the contravariance of T puts $T\mathbf{P}_A$ on the right.[9]

Given an additive contravariant functor $T: \mathcal{A} \to \mathcal{C}$ between abelian categories, we are now going to construct, for all $n \in \mathbb{Z}$, its *right derived functors* $R^n T: \mathcal{A} \to \mathcal{C}$.

Choose, once and for all, a projective resolution $\mathbf{P} = \to P_1 \xrightarrow{d_1} P_0 \xrightarrow{\varepsilon} A \to 0$ for every object A, form the complex $T\mathbf{P}_A$, and take homology:

$$(R^n T)A = H^n(T\mathbf{P}_A) = \frac{\ker T d_{n+1}}{\operatorname{im} T d_n}.$$

If $f: A \to A'$, define $(R^n T)f: (R^n T)A' \to (R^n T)A$ as we did for left derived functors. There is a chain map $\check{f}: \mathbf{P}_A \to \mathbf{P}'_{A'}$ over f, unique to homotopy, that induces a map $(R^n T)f: H^n(T\mathbf{P}'_{A'}) \to H^n(T\mathbf{P}_A)$ in homology, and we define $(R^n T)f = (T\check{f})_{n*}$.

Theorem 6.52. *If $T: \mathcal{A} \to \mathcal{C}$ is an additive contravariant functor between abelian categories, where $\mathcal{A}$ has enough projectives, then $R^n T: \mathcal{A} \to \mathcal{C}$ is an additive contravariant functor for every $n \in \mathbb{Z}$.*

Definition. If $T: \mathcal{A} \to \mathcal{C}$ is an additive contravariant functor between abelian categories, where $\mathcal{A}$ has enough projectives, the functors $R^n T$ are called the **right derived functors** of T.

Proposition 6.53. *If $T: {}_R\mathbf{Mod} \to {}_R\mathbf{Mod}$ is an additive functor that preserves multiplications, then $R^n T: {}_R\mathbf{Mod} \to {}_R\mathbf{Mod}$ also preserves multiplications.*

Definition. If $T = \operatorname{Hom}_R(\square, B)$, define $\operatorname{ext}_R^n(\square, B) = R^n T$. If the chosen projective resolution of A is $\mathbf{P} = \to P_2 \xrightarrow{d_2} P_1 \xrightarrow{d_1} P_0 \xrightarrow{\varepsilon} A \to 0$, then

$$\operatorname{ext}_R^n(A, B) = H^n(\operatorname{Hom}_R(\mathbf{P}_A, B)) = \frac{\ker d^{n*}}{\operatorname{im} d^{n-1,*}},$$

where $d^{n*}: \operatorname{Hom}_R(P_{n-1}, B) \to \operatorname{Hom}_R(P_n, B)$ is defined, as usual, by

$$d^{n*}: f \mapsto f d^n.$$

---

[9]If we cared about left derived functors of a contravariant T (we do not, because there are few interesting examples, but see Exercise 7.13 on page 435), then we would use injective resolutions **E**, for the contravariance of T would put all the nonzero terms of $T\mathbf{E}$ on the left.

Proposition 6.54. *If $T : \mathcal{A} \to \mathcal{C}$ is an additive contravariant functor, where $\mathcal{A}, \mathcal{C}$ are abelian categories and $\mathcal{A}$ has enough projectives, then $(R^n T)A = 0$ for all negative n and for all A.*

Corollary 6.55. *Let $\mathcal{A}$ be an abelian category with enough projectives, and let $\mathbf{P} = \to P_2 \xrightarrow{d_2} P_1 \xrightarrow{d_1} P_0 \xrightarrow{\varepsilon} A \to 0$ be a projective resolution of A. Define $K_0 = \ker \varepsilon$ and $K_n = \ker d_n$ for all $n \geq 1$. Then*

$$(R^{n+1}T)A \cong (R^n T)K_0 \cong (R^{n-1}T)K_1 \cong \cdots \cong (R^1 T)K_{n-1}.$$

In particular, for any left R-module B,

$$\operatorname{ext}_R^{n+1}(A, B) \cong \operatorname{ext}_R^n(K_0, B) \cong \cdots \cong \operatorname{ext}_R^1(K_{n-1}, B).$$

Assume that new choices $\widetilde{\mathbf{P}}_A$ of deleted projective resolutions have been made, and denote the right derived functors arising from these new choices by $\widetilde{R}^n T$.

Proposition 6.56. *If $T : \mathcal{A} \to \mathcal{C}$ is an additive contravariant functor between abelian categories, where $\mathcal{A}$ has enough projectives, then for each $n \in \mathbb{Z}$, the functors $R^n T$ and $\widetilde{R}^n T$ are naturally isomorphic. In particular, for all A,*

$$(R^n T)A \cong (\widetilde{R}^n T)A,$$

and so these objects are independent of the choice of projective resolution of A.

Corollary 6.57. *The module $\operatorname{ext}_R^n(A, B)$ is independent of the choice of projective resolution of A.*

Corollary 6.58. *Let $T : \mathcal{A} \to \mathcal{C}$ be an additive contravariant functor, where $\mathcal{A}$ and $\mathcal{C}$ are abelian categories and $\mathcal{A}$ has enough projectives. If P is projective, then $(R^n T)P = 0$ for all $n \geq 1$.*

In particular, if $T : {}_R\mathbf{Mod} \to {}_S\mathbf{Mod}$ and P is a projective left R-module, then $\operatorname{ext}_R^n(P, B) = \{0\}$ for all $n \geq 1$ and all left R-modules B.

Theorem 6.59. *Let $\mathcal{A}$ be an abelian category with enough projectives. If $0 \to A' \xrightarrow{i} A \xrightarrow{p} A'' \to 0$ is an exact sequence and $T : \mathcal{A} \to \mathcal{C}$ is an additive contravariant functor, where $\mathcal{C}$ is an abelian category, then there is a long exact sequence in $\mathcal{C}$,*

$$\to (R^n T)A'' \xrightarrow{(R^n T)p} (R^n T)A \xrightarrow{(R^n T)i} (R^n T)A' \xrightarrow{\partial^n}$$

$$(R^{n+1}T)A'' \xrightarrow{(R^{n+1}T)p} (R^{n+1}T)A \xrightarrow{(R^{n+1}T)i} (R^{n+1}T)A' \xrightarrow{\partial^{n+1}},$$

that begins with

$$0 \to (R^0 T)A'' \to (R^0 T)A \to (R^0 T)A' \to .$$

Corollary 6.60.   *If* $T \colon \mathcal{A} \to \mathcal{C}$ *is an additive contravariant functor, where* $\mathcal{A}, \mathcal{C}$ *are abelian categories and* $\mathcal{A}$ *has enough projectives, then the functor* $R^0 T$ *is left exact.*

Theorem 6.61.

(i) *If an additive contravariant functor* $T \colon \mathcal{A} \to \mathcal{C}$ *is left exact, where* $\mathcal{A}, \mathcal{C}$ *are abelian categories and* $\mathcal{A}$ *has enough projectives, then* T *is naturally isomorphic to* $R^0 T$.

(ii) *If* B *is a left* R-*module, the functor* $\operatorname{Hom}_R(\square, B)$ *is naturally isomorphic to* $\operatorname{ext}_R^0(\square, B)$. *Hence, for all left* R-*modules* A, *there is an isomorphism*

$$\operatorname{Hom}_R(A, B) \cong \operatorname{ext}_R^0(A, B).$$

Corollary 6.62.   *If* $0 \to A' \to A \to A'' \to 0$ *is a short exact sequence of left* R-*modules, then for every left* R-*module* B, *there is a long exact sequence of abelian groups*

$$0 \to \operatorname{Hom}_R(A'', B) \to \operatorname{Hom}_R(A, B) \to \operatorname{Hom}_R(A', B)$$
$$\to \operatorname{ext}_R^1(A'', B) \to \operatorname{ext}_R^1(A, B) \to \operatorname{ext}_R^1(A', B)$$
$$\to \operatorname{ext}_R^2(A'', B) \to \operatorname{ext}_R^2(A, B) \to \operatorname{ext}_R^2(A', B) \to .$$

Proposition 6.63.   *Given a commutative diagram of left* R-*modules having exact rows,*

$$
\begin{array}{ccccccccc}
0 & \longrightarrow & A' & \overset{i}{\longrightarrow} & A & \overset{p}{\longrightarrow} & A'' & \longrightarrow & 0 \\
& & {\scriptstyle f}\downarrow & & {\scriptstyle g}\downarrow & & {\scriptstyle h}\downarrow & & \\
0 & \longrightarrow & C' & \underset{j}{\longrightarrow} & C & \underset{q}{\longrightarrow} & C'' & \longrightarrow & 0,
\end{array}
$$

then for every left R-*module* B, *there is a commutative diagram of abelian groups with exact rows*

$$
\begin{array}{ccccccc}
\operatorname{ext}_R^n(A'', B) & \overset{p^*}{\longrightarrow} & \operatorname{ext}_R^n(A, B) & \overset{i^*}{\longrightarrow} & \operatorname{ext}_R^n(A', B) & \overset{\partial^n}{\longrightarrow} & \operatorname{ext}_R^{n+1}(A'', B) \\
{\scriptstyle h^*}\uparrow & & {\scriptstyle g^*}\uparrow & & {\scriptstyle f^*}\uparrow & & {\scriptstyle h^*}\uparrow \\
\operatorname{ext}_R^n(C'', B) & \underset{q^*}{\longrightarrow} & \operatorname{ext}_R^n(C, B) & \underset{j^*}{\longrightarrow} & \operatorname{ext}_R^n(C', B) & \underset{\partial'^n}{\longrightarrow} & \operatorname{ext}_R^{n+1}(C'', B).
\end{array}
$$

Theorem 6.64 (Axioms for Contravariant Ext). *Let* $(G^n : {}_R\mathbf{Mod} \to \mathbf{Ab})$ *be a sequence of additive contravariant functors. If,*

(i) *for every short exact sequence* $0 \to A \to B \to C \to 0$, *there is a long exact sequence with natural connecting homomorphisms,*

$$\to G^{n-1}(A) \xrightarrow{\Delta^{n-1}} G^n(C) \to G^n(B) \to G^n(A) \xrightarrow{\Delta^n} G^{n+1}(C) \to,$$

(ii) *there exists a left R-module M with* G^0 *and* $\operatorname{Hom}_R(\square, M)$ *naturally isomorphic,*

(iii) $G^n(P) = \{0\}$ *for all projective left R-modules P and all* $n \geq 1$,

then G^n *is naturally isomorphic to* $\operatorname{Ext}^n_R(\square, M)$ *for all* $n \geq 0$.

Proof.   See the proof of Theorem 6.33.   ●

Remark.   It is easy to see that Theorem 6.64 is true if we replace *projective* in part (iii) by *free*.   ◄

Corollary 6.65.   *Let* $(G^n : \mathcal{A} \to \mathcal{B})_{n \geq 0}$, $(G'^n : \mathcal{A} \to \mathcal{B})_{n \geq 0}$ *be sequences of additive contravariant functors, where* $\mathcal{A}, \mathcal{B}$ *are abelian categories and* $\mathcal{A}$ *has enough projectives. If,*

(i) *for every short exact sequence* $0 \to A \to B \to C \to 0$, *there are long exact sequences with natural connecting homomorphisms,*

(ii) G^0 *is naturally isomorphic to* G'^0,

(iii) $G^n(E) = 0 = G'^n(P)$ *for all projective objects P and all* $n \geq 1$,

then G^n *is naturally isomorphic to* G'^n *for all* $n \geq 0$.

Corollary 6.66.   *Let* $(G^n : \mathcal{A} \to \mathcal{B})_{n \geq 0}$, $(G'^n : \mathcal{A} \to \mathcal{B})_{n \geq 0}$ *be sequences of additive covariant functors, where* $\mathcal{A}, \mathcal{B}$ *are abelian categories,* $\mathcal{X}$ *is a class of objects in* $\mathcal{A}$, *and* $\mathcal{A}$ *has enough* $\mathcal{X}$-*objects. If,*

(i) *for every short exact sequence* $0 \to A \to B \to C \to 0$ *in* $\mathcal{A}$, *there are long exact sequences with natural connecting homomorphisms,*

(ii) G^0 *is naturally isomorphic to* G'^0,

(iii) *both* $(G^n(_{n \geq 0}, (G'^n)_{n \geq 0}$ *are* $\mathcal{X}$-*effaceable,*

then G^n *is naturally isomorphic to* G'^n *for all* $n \geq 0$.

We let the reader give the obvious definition of cohomological extension of a contravariant additive functor, and then prove the analog of Theorem 6.51.

The next theorem shows that Ext and ext are the same.

Theorem 6.67.   *Let A and B be left R-modules, let $\mathbf{E} = 0 \to B \xrightarrow{\eta}$ $E^0 \xrightarrow{d^0} E^1 \xrightarrow{d^1} E^2 \xrightarrow{d^2} E^3 \to$ be an injective resolution of B, and let $\mathbf{P} = \to P_1 \xrightarrow{d_1} P_0 \xrightarrow{\varepsilon} A \to 0$ be a projective resolution of A. Then $H^n(\operatorname{Hom}_R(\mathbf{P}_A, B)) \cong H^n(\operatorname{Hom}_R(A, \mathbf{E}^B))$ for all $n \geq 0$, and so*

$$\operatorname{Ext}_R^n(A, B) \cong \operatorname{ext}_R^n(A, B).$$

Proof.   Use Theorem 6.32.   •

Singular cohomology groups $H^q(X, G)$ of a space X with coefficients in an abelian group G, defined as $H^q(\operatorname{Hom}_{\mathbb{Z}}(\mathbf{S}_\bullet(X), G))$, arise for several reasons.   One application arises in *obstruction theory* (see Spanier, *Algebraic Topology*, §8.4).   Recall that the nth homotopy group $\pi_n(X, x_0)$ of X (with basepoint $x_0 \in X$) essentially consists of homotopy classes of continuous maps $S^n \to X$, where S^n is the n-sphere.   If X is a CW complex, we often construct a continuous map $X \to Y$ by induction on the n-skeletons $X^{(n)}$ of X.   In particular, having defined a map $X^{(n-1)} \to Y$, extending it to an n-cell of X naturally leads to $H^{n+1}(X, \pi_n(Y))$, where $H^n(X, \pi) = H^n(\operatorname{Hom}_{\mathbb{Z}}(\mathbf{S}_\bullet(X), \pi))$.   There is a *Universal Coefficient Theorem* for cohomology, Theorem 7.59, that computes such groups in terms of homology groups.

The most important uses of cohomology involve *products* (see Mac Lane, *Homology*, Chapter VIII).   Assume that the coefficients form a commutative ring R.   Although cup product is defined on singular complexes of spaces X, we will define it only for simplicial complexes K.   If f is a p-cochain and g is a q-cochain, then their ***cup product*** is the $(p + q)$-cochain $f \cup g$ defined on a $(p + q)$-simplex $[v_0, \ldots, v_{p+q}]$ by

$$(f \cup g)[v_0, \ldots, v_{p+q}] = f[v_0, \ldots, v_p]g[v_p, \ldots, v_{p+q}],$$

where fg is the product in the the the ring R [see Munkres, *Elements of Algebraic Topology*, §48].   Now $\delta(f \cup g) = (\delta(f) \cup g) + (-1)^p(f \cup \delta g)$, so that cup product induces a bilinear function $H^p(X, R) \times H^q(X, \mathbb{Z}) \to H^{p+q}(X, R)$. The direct sum $\bigoplus_{n=0}^{\infty} H^n(X, R)$ is now equipped with a multiplication that makes it a graded ring, called the ***cohomology ring*** of X with coefficients in R.   Although there are more important uses of this ring, it can be used to show that even though the torus $X = S^1 \times S^1$ and the wedge of spheres $Y = S^2 \vee S^1 \vee S^1$ have the same cohomology groups (with $\mathbb{Z}$ coefficients), they do not have the same homotopy type, because their cohomology rings are not isomorphic (see Rotman, *An Introduction to Algebraic Topology*, p. 404).

There is also a ***cap product***:

$$H^p(X, R) \otimes H_{p+q}(X, R) \to H_q(X, R).$$

If X is a compact orientable n-manifold without boundary, then it has an *orientation class* $\Gamma \in H_n(X, \mathbb{Z})$, and the cap products with Γ are isomorphisms for all p with $0 \leq p \leq n$, called *Poincaré Duality* (see Munkres, §66, §67):

$$H^p(X, G) \cong H_{n-p}(X, G).$$

Here is another source of complexes giving rise to homology.

Definition. A *simplicial object* in a category $\mathcal{C}$ is a sequence of objects $X_0, X_1, X_2, \ldots$ and two doubly indexed families of morphisms: *face operators*, which are morphisms $d_n^i \colon X_n \to X_{n-1}$ for all $0 \leq i \leq n$ and $1 \leq n$; *degeneracy operators*, which are morphisms $s_n^i \colon X_n \to X_{n+1}$ for all $0 \leq i \leq n$ and $0 \leq n$. These morphisms satisfy the following identities:

$$d_n^i d_{n+1}^j = d_n^{j-1} d_{n+1}^i \qquad \text{if } 0 \leq i < j \leq n+1,$$

$$s_n^j s_{n-1}^i = s_n^i d_{n-1}^{j-1} \qquad \text{if } 0 \leq i < j \leq n,$$

$$d_{n+1}^i s_n^j = \begin{cases} s_{n-1}^{j-1} d_n^i & \text{if } 0 \leq i < j \leq n, \\ 1 & \text{if } 0 \leq i = j \leq n \text{ or } 0 \leq i - 1 = j < n, \\ s_{n-1}^j d_n^{i-1} & \text{if } 0 < j < i - 1 \leq n. \end{cases}$$

We picture face operators as

$$\cdots X_3 \underset{\to}{\overset{\to}{\rightrightarrows}} X_2 \rightrightarrows X_1 \rightrightarrows X_0,$$

and degeneracy operators as

$$\cdots X_3 \underset{\leftarrow}{\overset{\leftarrow}{\leftleftarrows}} X_2 \leftleftarrows X_1 \leftarrow X_0.$$

These operators arise naturally when one constructs the boundary operator for simplicial complexes. Recall that the standard n-simplex Δ^n consists of all *convex combinations* $(t_0, \ldots, t_n)$ in $\mathbb{R}^{n+1}$; that is, $t_i \geq 0$ for all i and $\sum_{i=0}^n t_i = 1$. The face operators are precisely the *face maps* $\epsilon_i^n \colon \Delta^{n-1} \to \Delta^n$ defined in Chapter 1:

$$\epsilon_i^n \colon (t_0, \ldots, t_{n-1}) \mapsto \begin{cases} (0, t_0, \ldots, t_{n-1}) \text{ if } i = 0, \\ (t_0, \ldots, t_{i-1}, 0, t_i, \ldots, t_{n-1}) \text{ if } i > 0. \end{cases}$$

We did not define degeneracies in Chapter 1; they are given by

$$s_i(t_0, \ldots, t_n) = (t_0, \ldots, t_{i-1}, t_i + t_{i+1}, t_{i+2}, \ldots, t_n).$$

Given a simplicial object in an abelian category $\mathcal{A}$, define its *associated complex*

$$\to X_n \xrightarrow{\partial_n} X_{n-1} \to \cdots X_1 \xrightarrow{\partial_1} X_0 \to 0,$$

where $\partial_n = \sum_{i=0}^{n} (-1)^i d_n^i$. The given identities for d_n^i imply $\partial \partial = 0$. Thus, simplicial objects have homology. The degeneracies allow one to construct an abstract version of homotopy groups as well (see Gelfand–Manin, *Methods of Homological Algebra*, May, *Simplicial Objects in Algebraic Topology*, and Weibel, *An Introduction to Homological Algebra*).

Exercises

6.13 If $\tau \colon F \to G$ is a natural transformation between additive functors, prove that τ gives chain maps $\tau_{\mathbf{C}} \colon F\mathbf{C} \to G\mathbf{C}$ for every complex $\mathbf{C}$. If τ is a natural isomorphism, prove that $F\mathbf{C} \cong G\mathbf{C}$.

*6.14 Consider the commutative diagram with exact row

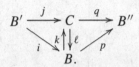

If k is an isomorphism with inverse ℓ, prove exactness of

$$B' \xrightarrow{\ i\ } B \xrightarrow{\ p\ } B''.$$

6.15 Let $T \colon \mathcal{A} \to \mathcal{C}$ be an exact additive functor between abelian categories, and suppose that P projective implies TP projective. If $B \in \mathrm{obj}(\mathcal{A})$ and $\mathbf{P}_B$ is a deleted projective resolution of B, prove that $T\mathbf{P}_{TB}$ is a deleted projective resolution of TB.

6.16 Let R be a k-algebra, where k is a commutative ring, which is flat as a k-module. Prove that if B is an R-module (and hence a k-module), then

$$R \otimes_k \mathrm{Tor}_n^k(B, C) \cong \mathrm{Tor}_n^R(B, R \otimes_k C)$$

for all k-modules C and all $n \geq 0$.

6.17 Let R be a semisimple ring.

    (i) Prove, for all $n \geq 1$, that $\mathrm{Tor}_n^R(A, B) = \{0\}$ for all right R-modules A and all left R-modules B.

    (ii) Prove, for all $n \geq 1$, that $\mathrm{Ext}_R^n(A, B) = \{0\}$ for all left R-modules A and B.

*6.18 If R is a PID, prove, for all $n \geq 2$, that $\mathrm{Tor}_n^R(A, B) = \{0\} = \mathrm{Ext}_R^n(A, B)$ for all R-modules A and B.
    Hint. Use Corollary 4.15.

*6.19 Let R be a domain with fraction field Q, and let A, C be R-modules. If either C or A is a vector space over Q, prove that $\mathrm{Tor}_n^R(C, A)$ and $\mathrm{Ext}_R^n(C, A)$ are also vector spaces over Q.
    Hint. Use Exercise 2.38 on page 97.

*6.20 Let R be a domain and let $Q = \text{Frac}(R)$.

    (i)   If $r \in R$ is nonzero and A is an R-module for which $rA = \{0\}$, that is, $ra = 0$ for all $a \in A$, prove that $\text{Ext}_R^n(Q, A) = \{0\} = \text{Tor}_n^R(Q, A)$ for all $n \geq 0$.

    (ii) Prove that $\text{Ext}_R^n(V, A) = \{0\} = \text{Tor}_n^R(V, A)$ for all $n \geq 0$ whenever V is a vector space over Q and A is an R-module for which $rA = \{0\}$ for some nonzero $r \in R$.

6.21 Let A and B be R-modules, and let A' be a submodule of A. Define the **obstruction** of a map $f : A' \to B$ to be $\partial(f)$, where ∂ is the connecting homomorphism $\text{Hom}_R(A', B) \to \text{Ext}_R^1(A/A', B)$. Prove that f can be extended to a homomorphism $\widetilde{f} : A \to B$ if and only if its obstruction is 0.

6.22 Give an example of an R-module B for which $L_0 \text{Hom}_R(B, \square)$ is not naturally isomorphic to $\text{Hom}_R(B, \square)$, where L_0 is the 0th left derived functor.

6.3   Sheaf Cohomology

Even though there were earlier accounts of abelian categories (for example, Buchsbaum's appendix on *exact categories* in Cartan–Eilenberg, *Homological Algebra*), it was Grothendieck's Tohoku papers that have been most influential. Grothendieck began:

> Ce travail a son origine dans une tentative d'exploiter l'analogie formelle entre la théorie de la cohomologie d'un espace à coéfficients dans un faisceau et la théorie des foncteurs dérivés de foncteurs de modules, pour trouver un cadre commun permettant d'englober ces théories et d'autres.

In a word, sheaf cohomology arises as the right derived functors of global sections. We restrict our discussion to sheaves of abelian groups, but the reader should have no problem extending it to sheaves having values in other abelian categories.

If X is a topological space, the group of global sections defines functors $\Gamma : \mathbf{pSh}(X) \to \mathbf{Ab}$ and $\Gamma : \mathbf{Sh}(X) \to \mathbf{Ab}$. In each case, the functor is defined on objects X by

$$\Gamma : X \mapsto \Gamma(X, \mathcal{F}) = \mathcal{F}(X)$$

and on (pre)sheaf maps $\varphi = (\varphi_U)_{U\text{ open}} : \mathcal{F} \to \mathcal{G}$ by $\Gamma : \varphi s \mapsto \varphi_X(s)$, where $s \in \Gamma(X, \mathcal{F})$. It is clear that each Γ is a (covariant) additive functor.

Lemma 6.68.   *The functors* $\Gamma \colon \mathbf{pSh}(X) \to \mathbf{Ab}$ *and* $\Gamma \colon \mathbf{Sh}(X) \to \mathbf{Ab}$ *are left exact.*

Proof.   Exactness of presheaves $0 \to \mathcal{P}' \xrightarrow{\varphi} \mathcal{P} \xrightarrow{\psi} \mathcal{P}'' \to 0$ is defined as exactness of the abelian groups $0 \to \mathcal{P}'(U) \xrightarrow{\varphi_U} \mathcal{P}(U) \xrightarrow{\psi_U} \mathcal{P}''(U) \to 0$ for every open $U \subseteq X$. In particular, the sequence is exact when $U = X$, and so Γ is even an exact functor on presheaves.

Exactness of sheaves means exactness of stalks, which is usually different from exactness of presheaves. However, if $0 \to \mathcal{F}' \to \mathcal{F} \xrightarrow{\psi} \mathcal{F}''$ is an exact sequence of sheaves, then ψ is a presheaf map, and Proposition 5.80(ii) says that $\ker \psi$ computed in $\mathbf{Sh}(X)$ is the same as $\ker \psi$ computed in $\mathbf{pSh}(X)$. Hence, $0 \to \mathcal{F}' \to \mathcal{F} \to \mathcal{F}''$ is exact in $\mathbf{pSh}(X)$, and the proof in the first paragraph now applies.   •

The next example shows that $\Gamma \colon \mathbf{Sh}(X) \to \mathbf{Ab}$ need not be an exact functor.

Example 6.69.   In Example 5.82, we saw that there is an exact sequence of sheaves over the punctured plane $X = \mathbb{C} - \{0\}$,

$$0 \to \mathbb{Z} \to \mathcal{O} \xrightarrow{\varphi} \mathcal{O}^\times \to 0,$$

where $\mathbb{Z}$ is the constant sheaf, $\mathcal{O}$ is the sheaf of germs of holomorphic functions, $\mathcal{O}^\times$ is the sheaf of nonzero holomorphic functions, and $\varphi_U \colon \mathcal{O}(U) \to \mathcal{O}^*(U)$ is given by $f \mapsto e^{2\pi i f}$. For every open set U, we have $\mathcal{O}(U)$ the additive group of all holomorphic $f \colon U \to \mathbb{C}$ and $\mathcal{O}^\times(U)$ the multiplicative group of all never-zero holomorphic $f \colon U \to \mathbb{C}^\times$. If the function $s(z) = z$ in $\Gamma(X, \mathcal{C}^\times)$ is in $\operatorname{im} \varphi^*$ [where $\varphi^* \colon \Gamma(\mathcal{O}) \to \Gamma(\mathcal{O}^\times)$ is the induced map], then $z = e^{2\pi i f(z)}$; that is, $f(z) = \frac{1}{2\pi i} \log(z)$. This is a contradiction, for no branch of $\log(z)$ on the punctured plane is single-valued. Therefore, Γ is not an exact functor.   ◄

We now define sheaf cohomology as right derived functors of global sections Γ; this is possible because $\mathbf{Sh}(X)$ has enough injectives, by Proposition 5.97. Note that taking derived functors of $\Gamma \colon \mathbf{pSh}(X) \to \mathbf{Ab}$ is uninteresting, for the higher derived functors of an exact functor are trivial.

Definition.   If X is a topological space, then **sheaf cohomology** is defined, for every sheaf $\mathcal{F}$ over X, by

$$H^q(\mathcal{F}) = (R^q \Gamma)(\mathcal{F}).$$

In short, take an injective resolution $\mathbf{E}$ of $\mathcal{F}$, delete $\mathcal{F}$ to obtain $\mathbf{E}^{\mathcal{F}}$, apply Γ, and take homology:

$$H^q(\mathcal{F}) = H^q(\Gamma \mathbf{E}^{\mathcal{F}}).$$

As usual, $H^0(\mathcal{F})$ can be computed.

Proposition 6.70. *If X is a topological space, then*

$$H^0(\mathcal{F}) \cong \Gamma(\mathcal{F})$$

for every sheaf $\mathcal{F}$ over X.

Proof. Since Γ is a left exact functor, the result follows at once from Theorem 6.45. •

Thus, $H^1(\mathcal{F})$ repairs the loss of exactness arising from $\Gamma\colon \mathbf{Sh}(X) \to \mathbf{Ab}$ not being exact; in other words, we may interpret H^1 as obstructions.

Remark. The global section functor $\Gamma = \Gamma(X, \square)$ is often modified.

Definition. A *family of supports* Φ is a family of closed subsets of X such that

(i) whenever $A \in \Phi$ and $B \subseteq A$ is closed, then $B \in \Phi$,

(ii) whenever $A, A' \in \Phi$, then $A \cup A' \in \Phi$.

Define $\Gamma_\Phi(\mathcal{F}) = \{s \in \Gamma(X, \mathcal{F}) : \{x \in X : s(x) \neq 0_x \in E_x\} \in \Phi\}$, where $\mathcal{F}$ has etale-sheaf (E, p, X) and Φ is a family of supports. It is easy to see that $\Gamma_\Phi\colon \mathbf{Sh}(X) \to \mathbf{Ab}$ is a covariant left exact additive functor. One defines *sheaf cohomology H_Φ^q with supports* Φ as the right derived functors of Γ_Φ. The family Φ of all closed subsets is a family of supports, so that $\Gamma_\Phi = \Gamma$ and $H_\Phi^q = H^q$ in this case. ◄

Definition. A sheaf $\mathcal{L}$ over a space X is *acyclic* if $H^q(\mathcal{L}) = \{0\}$ for all $q \geq 1$.

We know that injective sheaves are acyclic, by Corollary 6.41, but there are other examples. Acyclic sheaves become especially interesting when there are enough of them; that is, when every sheaf $\mathcal{F}$ can be imbedded in an acyclic sheaf $\mathcal{L}$. In this case, the short exact sequence $0 \to \mathcal{F} \to \mathcal{L} \to \mathcal{L}/\mathcal{F} \to 0$ can be used in dimension shifting arguments. The most popular acyclic sheaves are *flabby sheaves*.

Definition. A sheaf $\mathcal{L}$ over a space X is *flabby* (or *flasque*) if, for each open $U \subseteq X$, every section $s \in \mathcal{L}(U)$ can be extended to a global section.

A *flabby resolution* of a sheaf $\mathcal{F}$ is an exact sequence

$$0 \to \mathcal{F} \to \mathcal{L}^0 \to \mathcal{L}^1 \to \cdots$$

in which $\mathcal{L}^q$ is flabby for all $q \geq 0$.

A sheaf $\mathcal{L}$ is flabby if and only if the restriction maps $\Gamma(X, \mathcal{L}) \to \Gamma(U, \mathcal{L})$ are all epic; it follows that the restriction maps $\rho_U^V : \Gamma(V, \mathcal{L}) \to \Gamma(U, \mathcal{L})$ are epic for all open sets $U \subseteq V$, because $\rho_V^X \rho_U^V = \rho_U^X$. Hence, if $U \subseteq X$ is open, then $\mathcal{L}$ flabby implies $\mathcal{L}|U$ is also flabby.

Example 6.71.   Every skyscraper sheaf $\mathcal{S} = x_* A$ is flabby. Recall that $\mathcal{S}(U) = A$ if $x \in U$, while $\mathcal{S}(U) = \{0\}$ otherwise, and its restrictions are either 1_A or zero. Hence, if $x \in U \subseteq V$, then $\rho_U^V = 1_A$ is surjective.   ◄

Definition.   If $\mathcal{F}$ is a sheaf of abelian groups over a space X, then its **Godement sheaf** $\mathcal{G}^0 \mathcal{F}$ is defined by

$$\mathcal{G}^0 \mathcal{F}(U) = \prod_{x \in U} \mathcal{F}_x,$$

and $\rho_U^V : \mathcal{G}^0 \mathcal{F}(V) \to \mathcal{G}^0 \mathcal{F}(U)$, for $U \subseteq V$, is given by $s \mapsto s|U$.

It is routine to check that $\mathcal{G}^0 \mathcal{F}$ is a sheaf. In fact, $\mathcal{G}^0$ defines a covariant exact functor $\mathbf{Sh}(X) \to \mathbf{Sh}(X)$: if $0 \to \mathcal{F}' \to \mathcal{F} \to \mathcal{F}'' \to 0$ is an exact sequence of sheaves, then it is clear, for every open $U \subseteq X$, that

$$0 \to \prod_{x \in U} \mathcal{F}'(U) \to \prod_{x \in U} \mathcal{F}(U) \to \prod_{x \in U} \mathcal{F}''(U) \to 0$$

is an exact sequence of abelian groups; that is, $0 \to \mathcal{G}^0 \mathcal{F}'(U) \to \mathcal{G}^0 \mathcal{F}(U) \to \mathcal{G}^0 \mathcal{F}''(U) \to 0$ is exact. Taking the direct limit gives exactness of stalks: $0 \to \mathcal{G}^0 \mathcal{F}'_x \to \mathcal{G}^0 \mathcal{F}_x \to \mathcal{G}^0 \mathcal{F}''_x \to 0$; that is, $0 \to \mathcal{G}^0 \mathcal{F}' \to \mathcal{G}^0 \mathcal{F} \to \mathcal{G}^0 \mathcal{F}'' \to 0$ is an exact sequence of sheaves.

Proposition 6.72.   *The Godement sheaf $\mathcal{G}^0 \mathcal{F}$ of a sheaf $\mathcal{F}$ is flabby.*

Proof.   Since global sections here are merely (not necessarily continuous) functions $X \to \prod_{x \in X} \mathcal{F}_x$, every section s over U extends to a global section s'; for example, define $s'|U = s$ and, if $x \notin U$, define $s'(x) = 0$.   •

Proposition 6.73 (Godement).   *Let $\mathcal{F}$ be a sheaf over a space X.*

(i) *There is a natural imbedding $0 \to \mathcal{F} \to \mathcal{G}^0 \mathcal{F}$.*

(ii) *There is a flabby resolution*

$$\mathcal{G}^\bullet \mathcal{F} = 0 \to \mathcal{F} \to \mathcal{G}^0 \mathcal{F} \to \mathcal{G}^1 \mathcal{F} \to \cdots.$$

Proof.

(i) If $U \subseteq X$ is open, define $\mathcal{F}(U) \to (\mathcal{G}^0\mathcal{F})(U)$ by

$$s \mapsto (s(x)) \in \prod_{x \in U} \mathcal{F}_x = (\mathcal{G}^0\mathcal{F})(U).$$

It is routine to check that this is a natural sheaf monomorphism.

(ii) We prove, by induction on q, that there are flabby sheaves $\mathcal{G}^i\mathcal{F}$ for all $i \leq q$ and sheaf maps $d^i : \mathcal{G}^i\mathcal{F} \to \mathcal{G}^{i+1}\mathcal{F}$ for $i \leq q-1$ such that

$$0 \to \mathcal{F} \to \mathcal{G}^0\mathcal{F} \xrightarrow{d^0} \mathcal{G}^1\mathcal{F} \to \cdots \to \mathcal{G}^{q-1}\mathcal{F} \xrightarrow{d^{q-1}} \mathcal{G}^q\mathcal{F}$$

is exact. We have already defined $\mathcal{G}^0\mathcal{F}$. Define

$$\mathcal{G}^{q+1}\mathcal{F} = \mathcal{G}^0(\operatorname{coker} d^{q-1}),$$

and define $d^q : \mathcal{G}^q\mathcal{F} \to \mathcal{G}^{q+1}\mathcal{F}$ as the composite

$$\mathcal{G}^q\mathcal{F} \to \operatorname{coker} d^{q-1} \to \mathcal{G}^0(\operatorname{coker} d^{q-1}) = \mathcal{G}^{q+1}\mathcal{F}.$$

Now $\mathcal{G}^{q+1}\mathcal{F}$ is flabby because it is $\mathcal{G}^0$ of some sheaf, and the sequence is exact because $\operatorname{coker} d^{q-1} \to \mathcal{G}^{q+1}\mathcal{F}$ is monic. •

Corollary 6.74. *Every injective sheaf $\mathcal{E}$ over a space X is flabby.*

Proof. It is easy to see that every direct summand of a flabby sheaf is flabby. By Proposition 6.73(i), there is an exact sequence $0 \to \mathcal{E} \to \mathcal{G}^0\mathcal{E} \to \mathcal{G}^0\mathcal{E}/\mathcal{E} \to 0$, and $\mathcal{G}^0\mathcal{E}$ is flabby. But this sequence splits, because $\mathcal{E}$ is injective; thus, $\mathcal{E}$ is a direct summand of $\mathcal{G}^0\mathcal{E}$ and, hence, it is flabby. •

Flabby sheaves give another construction of sheaf cohomology.

Definition. The flabby resolution $\mathcal{G}^\bullet\mathcal{F}$ in Proposition 6.73(ii) is called the *Godement resolution* of $\mathcal{F}$.

Proposition 6.75. *Let $\mathcal{F}$ be a sheaf over a space X.*

(i) *If $0 \to \mathcal{F}' \xrightarrow{\iota} \mathcal{F} \xrightarrow{\varphi} \mathcal{F}'' \to 0$ is an exact sequence of sheaves with $\mathcal{F}'$ flabby, then $0 \to \Gamma(\mathcal{F}') \to \Gamma(\mathcal{F}) \to \Gamma(\mathcal{F}'') \to 0$ is an exact sequence of abelian groups.*

(ii) *Let $0 \to \mathcal{L}' \to \mathcal{L} \to \mathcal{Q} \to 0$ be an exact sequences of sheaves. If $\mathcal{L}'$ and $\mathcal{L}$ are flabby, then $\mathcal{Q}$ is flabby.*

(iii) *Flabby sheaves $\mathcal{L}$ are acyclic.*

(iv) $H^q(\Gamma(\mathcal{G}^\bullet \mathcal{F})^{\mathcal{F}}) \cong H^q(\mathcal{F})$ *for all* $q \geq 0$, *where* $(\mathcal{G}^\bullet \mathcal{F})^{\mathcal{F}}$ *is the deleted Godement resolution of* $\mathcal{F}$.

Proof.

(i) It suffices to prove that $\varphi_X : \Gamma(\mathcal{F}) \to \Gamma(\mathcal{F}'')$, given by $\varphi_X : s \mapsto \varphi s$, is epic. Let $s'' \in \mathcal{F}''(X) = \Gamma(\mathcal{F}'')$. Define

$$\mathcal{X} = \{(U, s) : U \subseteq X \text{ is open}, s \in \mathcal{F}(U), \varphi s = s''|U\}.$$

Partially order $\mathcal{X}$ by $(U, s) \preceq (U_1, s_1)$ if $U \subseteq U_1$ and $s_1|U = s$. It is routine to see that chains in $\mathcal{X}$ have upper bounds, and so Zorn's Lemma provides a maximal element (U_0, s_0). If $U_0 = X$, then s_0 is a global section and φ_X is epic. Otherwise, choose $x \in X$ with $x \notin U_0$. Since $\varphi : \mathcal{F} \to \mathcal{F}''$ is an epic sheaf map, it is epic on stalks, and so there are an open $V \subseteq X$ with $V \ni x$ and a section $t \in \mathcal{F}(V)$ with $\varphi t = s''|V$. Now $s - t \in \mathcal{F}'(U \cap V)$ (we regard $\iota : \mathcal{F}' \to \mathcal{F}$ as the inclusion), so that $\mathcal{F}'$ flabby provides $r \in \mathcal{F}'(X)$ extending $s - t$. Hence, $s = t + r|(U \cap V)$ in $\mathcal{F}(U \cap V)$. Therefore, these sections may be glued: there is $\tilde{s} \in \mathcal{F}(U \cup V)$ with $\tilde{s}|U = s$ and $\tilde{s}|V = t + r|(U \cap V)$. But $\varphi(\tilde{s}) = s''$, and this contradicts the maximality of (U_0, s_0).

(ii) Let $U \subseteq X$ be open, and consider the commutative diagram

$$
\begin{array}{ccc}
\mathcal{F}(X) & \xrightarrow{\varphi_X} & \mathcal{F}''(X) \\
\downarrow{\scriptstyle \rho} & & \downarrow{\scriptstyle \rho''} \\
\mathcal{F}(U) & \xrightarrow{\varphi_U} & \mathcal{F}''(U),
\end{array}
$$

where ρ, ρ'' are restriction maps. Since $\mathcal{F}$ is flabby, ρ is epic. We have exactness of $0 \to \mathcal{F}'|U \to \mathcal{F}|U \to \mathcal{F}''|U \to 0$, for exactness of sheaves is stalkwise. As mentioned earlier, $\mathcal{F}'$ flabby implies $\mathcal{F}'|U$ flabby, so that part (i) gives φ_U epic. Therefore, the composite $\varphi_U \rho = \rho'' \varphi_X$ is epic, and hence ρ'' is epic; that is, $\mathcal{F}''$ is flabby.

(iii) Let $\mathcal{L}$ be flabby. Since there are enough injective sheaves, there is an exact sequence $0 \to \mathcal{L} \to \mathcal{E} \to \mathcal{Q} \to 0$ with $\mathcal{E}$ injective. Now $\mathcal{E}$ is flabby, by Corollary 6.74, and so $\mathcal{Q}$ is flabby, by part (ii). We prove that $H^q(\mathcal{L}) = \{0\}$ by induction on $q \geq 1$. If $q = 1$, the long exact cohomology sequence contains the fragment

$$H^0(\mathcal{E}) \to H^0(\mathcal{Q}) \to H^1(\mathcal{L}) \to H^1(\mathcal{E}).$$

Since $H^1(\mathcal{E}) = \{0\}$, we have $H^1(\mathcal{L}) = \text{coker}(\Gamma(\mathcal{E}) \to \Gamma(\mathcal{Q}))$. But this cokernel is 0, by part (i), and so $H^1(\mathcal{L}) = \{0\}$. For the inductive step, consider the fragment

$$H^q(\mathcal{Q}) \to H^{q+1}(\mathcal{L}) \to H^{q+1}(\mathcal{E}).$$

Now $H^{q+1}(\mathcal{E}) = \{0\}$, because $\mathcal{E}$ is injective, while $H^q(\mathcal{Q}) = \{0\}$, by the inductive hypothesis (which applies because $\mathcal{Q}$ is flabby). Therefore, exactness gives $H^{q+1}(\mathcal{L}) = \{0\}$.

(iv) Since the homology functors defined from flabby resolutions are effaceable, by part (iii), the result follows from uniqueness, Corollary 6.66.   ●

Corollary 6.76. *If $\mathcal{S} = x_* A$ is a skyscraper sheaf over a space X, where $x \in X$ and A is an abelian group, then $H^q(X, \mathcal{S}) = \{0\}$ for all $q \geq 1$.*

Proof. Skyscraper sheaves are flabby.   ●

There are other kinds of sheaves that are convenient when the base space X is *paracompact*.

Definition. A topological space X is *paracompact* if it is Hausdorff and every open cover $\mathcal{U}$ of X has a *locally finite* refinement. An open cover $\mathcal{V}$ is *locally finite* if each $x \in X$ has an open neighborhood N that meets only finitely many $V \in \mathcal{V}$; that is, $N \cap V \neq \varnothing$ for only finitely many $V \in \mathcal{V}$.

Of course, compact Hausdorff spaces are paracompact, and a theorem of A. H. Stone ("Paracompactness and product spaces," *Bull. AMS* 54 (1948), 977–982) says that every metric space is paracompact.

Definition. A sheaf $\mathcal{F}$ over a paracompact space X is *fine* if, for every locally finite open cover $\mathcal{U} = (U_i)_{i \in I}$ of X, there exists a family of sheaf morphisms $(\eta_i \colon \mathcal{F} \to \mathcal{F})_{i \in I}$, called a *partition of unity subordinate to* $\mathcal{U}$, such that

(i) for each $i \in I$, there is an open neighborhood V_i of the complement of U_i on which η_i is trivial; that is, $\eta_i \mathcal{F}(W) = \{0\}$ for all open $W \subseteq V_i$,

(ii) $\sum_i \eta_i = 1_{\mathcal{F}}$.

For example, sheaves of differentials on a paracompact manifold are fine; they comprise the de Rham complex, which, by the *Poincaré Lemma*, is a fine resolution of the constant sheaf $\mathbb{R}$ (see Bott–Tu, *Differential Forms in Algebraic Topology*, p. 35). Fine sheaves are acyclic (Gunning, *Lectures on Riemann Surfaces*, p. 36; Wells, *Differential Analysis on Complex Manifolds*, Chapter II §3). Moreover, every sheaf over a paracompact space can be imbedded in a fine sheaf, and so sheaf cohomology can also be computed in terms of fine resolutions.

6.3.1 Čech Cohomology

There is another construction of cohomology of sheaves, called *Čech cohomology*. Although its definition seems complicated, Čech cohomology is more amenable to computation than is sheaf cohomology.

Recall Example 1.3(x): if $\mathcal{U} = (U_i)_{i \in I}$ is an open cover of a topological space X, then the *nerve* $N(\mathcal{U})$ is the abstract simplicial complex with vertices $\text{Vert}(N(\mathcal{U})) = \mathcal{U}$ and q-simplexes all $(q+1)$-tuples σ of distinct open sets, $\sigma = [U_{i_0}, \ldots, U_{i_q}]$, with $\bigcap_{j=0}^{q} U_{i_j} \neq \varnothing$.

Example 6.77.

(i) Let K be an abstract simplicial complex. Recall the complex

$$\mathbf{C}_{\bullet}(K) = \to C_q(K) \xrightarrow{\partial_q} C_{q-1}(K) \to,$$

that we constructed in Chapter 1: the term $C_q(K)$ is the free abelian group with basis all q-simplexes $\sigma = [v_{i_0}, \ldots, v_{i_q}]$, and the differential $\partial_q : C_q(X) \to C_{q-1}(X)$ is

$$\partial_q(\sigma) = \partial_q[v_0, \ldots, v_q] = \sum_{i=0}^{q} (-1)^i [v_0, \ldots, \widehat{v}_i, \ldots, v_q].$$

If G is an abelian group, then $C^q(K, G) = \text{Hom}_{\mathbb{Z}}(C_q(K), G)$ is called the ***simplicial q-cochains*** with coefficients in G. Since $C_q(K)$ is free abelian, a q-cochain $f : C_q(K) \to G$ is determined by its values on the basis $\Sigma_q(K)$, the family of all q-simplexes in K. Thus, we may view f as a function $\Sigma_q(K) \to G$. The differential $\delta^q : C^q(K, G) \to C^{q+1}(K, G)$ is the induced map $f \mapsto f \partial_q$: if f is a q-cochain, then $\delta^q f \in C^{q+1}(K, G)$ is defined on a $(q+1)$-simplex $\tau = [v_{i_0}, \ldots, v_{i_{q+1}}]$ by

$$
\begin{aligned}
(\delta^q f)(\tau) &= f \partial_q(\tau) \\
&= f \partial_q[v_{i_0}, \ldots, v_{i_{q+1}}] \\
&= f \left(\sum_{j=0}^{q+1} (-1)^j [v_{i_0}, \ldots, \widehat{v}_{i_j} \ldots, v_{i_{q+1}}] \right) \\
&= \sum_{j=0}^{q+1} (-1)^j f[v_{i_0}, \ldots, \widehat{v}_{i_j} \ldots, v_{i_{q+1}}].
\end{aligned}
$$

The homology groups of the complex $\text{Hom}_{\mathbb{Z}}(\mathbf{C}_{\bullet}(K), G)$ are called the ***simplicial cohomology groups*** of K with ***coefficients*** in G.

(ii) Recall that the singular complex of a topological space X is $(\mathbf{S}_\bullet(X), \partial)$, whose qth term $S_q(X)$ is the free abelian group with basis Σ_q, the family of all q-simplexes $\sigma : \Delta^q \to X$, where Δ^q is the standard q-simplex. The differential $\partial_q : S_q(X) \to S_{q-1}(X)$ has a formula similar to that in part (i). If G is an abelian group, we define the *singular cohomology groups* of X with *coefficients* in G to be the homology groups of the complex $\mathrm{Hom}_{\mathbb{Z}}(\mathbf{S}_\bullet(X), G)$. ◄

Since $N(\mathcal{U})$, the nerve of an open cover $\mathcal{U}$ of a space X, is an abstract simplicial complex [see Example 1.3(x)], the last example shows how to define cohomology groups of $N(\mathcal{U})$ with coefficients in an abelian group G. If Σ_q is the set of all the q-simplexes in $N(\mathcal{U})$, then a q-*cochain* is a $\mathbb{Z}$-linear combination of functions $f : \Sigma_q \to G$. If $\tau = [U_{i_0}, \ldots, U_{i_{q+1}}]$ is a $(q+1)$-simplex, define

$$(\delta^q f)(\tau) = \sum_{j=0}^{q+1} (-1)^j f[U_{i_0}, \ldots, \widehat{U}_{i_j} \ldots U_{i_{q+1}}].$$

We obtain a complex of abelian groups $C^\bullet(N(\mathcal{U}), G)$.

Definition. The homology groups of the complex $C^\bullet(N(\mathcal{U}), G)$ are called the *cohomology groups of the open cover* $\mathcal{U}$ with *coefficients* G, and they are denoted by

$$H^q(\mathcal{U}, G).$$

We now modify this construction by replacing an abelian group G by a sheaf of abelian groups $\mathcal{F}$ over a space X. Given an open cover $\mathcal{U}$ of X, define the group $C^q(\mathcal{U}, \mathcal{F})$ of q-*cochains* by

$$C^q(\mathcal{U}, \mathcal{F}) = \prod_{[U_{i_0}, \ldots, U_{i_q}]} \mathcal{F}(U_{i_0} \cap \cdots \cap U_{i_q}),$$

where the product is over the set Σ_q of all q-simplexes in $N(\mathcal{U})$. Let us rephrase this. A q-cochain with coefficients in $\mathcal{F}$ is a function

$$f : \Sigma_q \to \bigcup_{\sigma \in \Sigma_q} \mathcal{F}(U_\sigma),$$

where $\sigma = [U_{i_0}, \ldots, U_{i_q}] \in \Sigma_q$ and $U_\sigma = U_{i_0} \cap \cdots \cap U_{i_q}$. Define the differential $\delta^q : C^q(\mathcal{U}, \mathcal{F}) \to C^{q+1}(\mathcal{U}, \mathcal{F})$ by

$$(\delta^q f)([U_{i_0}, \ldots, U_{i_{q+1}}]) = \sum_{j=0}^{q+1} (-1)^j f([U_{i_0}, \ldots, \widehat{U}_{i_j} \ldots, U_{i_{q+1}}]).$$

We obtain a complex of abelian groups $C^\bullet(N(\mathcal{U}), \mathcal{F})$.

Definition. The homology groups of the complex $C^\bullet(N(\mathcal{U}), \mathcal{F})$ are called the ***cohomology groups of the open cover*** $\mathcal{U}$ with ***sheaf coefficients*** $\mathcal{F}$, and they are denoted by

$$\check{H}^q(\mathcal{U}, \mathcal{F}).$$

We would like to use Corollary 6.49 to show that $\check{H}^q(\mathcal{U}, \mathcal{F})$ coincides with sheaf cohomology $H^q(\mathcal{F})$ but, alas, it may not apply, as part (v) of the next example shows: a short exact sequence of sheaves $0 \to \mathcal{F}' \to \mathcal{F} \to \mathcal{F}'' \to 0$ need not give a long exact cohomology sequence

$$\to \check{H}^q(\mathcal{U}, \mathcal{F}') \to \check{H}^q(\mathcal{U}, \mathcal{F}) \to \check{H}^q(\mathcal{U}, \mathcal{F}'') \to \check{H}^{q+1}(\mathcal{U}, \mathcal{F}') \to .$$

This does not say that $\check{H}^q(\mathcal{U}, \square)$ is irrelevant, but it does say that we may have to add some hypotheses to guarantee that the two cohomologies agree.

Example 6.78.

(i) For any any sheaf $\mathcal{F}$ over X and any open cover $\mathcal{U}$, we claim that

$$\check{H}^0(\mathcal{U}, \mathcal{F}) = \Gamma(\mathcal{F}) = \mathcal{F}(X).$$

To see this, it is clearest to describe $\delta^q : C^q(\mathcal{U}, \mathcal{F}) \to C^{q+1}(\mathcal{U}, \mathcal{F})$ more precisely. Formally, a q-cochain f is a function

$$\{q\text{-simplexes } [U_0, \dots, U_q] \in N(\mathcal{U})\} \to \bigcup_{[U_0, \dots, U_q]} \mathcal{F}(U_0 \cap \dots \cap U_q)$$

with $f([U_0, \dots, U_q]) \in \mathcal{F}(U_0 \cap \dots \cap U_q)$. Thus, f lies in the direct product $\prod_{[U_0, \dots, U_q]} \mathcal{F}(U_0 \cap \dots \cap U_q)$, and it can be written as a tuple

$$f = (s_{[U_0, \dots, U_q]}),$$

where $s_{[U_0, \dots, U_q]} \in \mathcal{F}(U_0 \cap \dots \cap U_q)$. Therefore, a 0-cochain is a tuple of sections (s_U), where $s_U \in \mathcal{F}(U)$, while a 1-cochain is a tuple of sections $(t_{[U,V]})$ indexed by all 1-simplexes $[U, V] \in N(\mathcal{U})$, where $t_{[U,V]} \in \mathcal{F}(U \cap V)$. The differential $\delta^0 : C^0(\mathcal{U}, \mathcal{F}) \to C^1(\mathcal{U}, \mathcal{F})$ sends $(s_U) \mapsto (\rho^V_{U \cap V} s_V - \rho^U_{U \cap V} s_U)$, where $\rho^V_{U \cap V}$ is the restriction map $\mathcal{F}(V) \to \mathcal{F}(U \cap V)$. Thus, $(s_U) \in \ker \delta^0$ if the family of sections satisfies the equalizer condition. Since $\mathcal{F}$ is a sheaf, there is a unique global section of $\mathcal{F}$ obtained by gluing these local sections. We conclude that $\check{H}^0(\mathcal{U}, \mathcal{F}) = \ker \delta^0 = \Gamma(\mathcal{F})$.

(ii) If K is a simplicial complex K and G is an abelian group, then $\mathbf{C}_\bullet(K)$ is the complex of simplicial chains and $H^q(K, G)$ is the homology of

$\text{Hom}_{\mathbb{Z}}(\mathbf{C}_{\bullet}(K), G)$. But $C_q(K)$ is the free abelian group with basis all q-simplexes in K, and so a map $f \in \text{Hom}(C_q(K), G)$ is just a function with $f(\sigma) \in G$ for every q-simplex σ in K.

Let $\mathcal{G}$ be the constant sheaf at G over a space X. If $\mathcal{U}$ is an open cover of X, we claim that $\check{H}^q(\mathcal{U}, \mathcal{G})$ is the simplicial cohomology $H^q(N(\mathcal{U}), G)$: after all, a q-cochain $f \in C^q(\mathcal{U}, \mathcal{G})$ is just a function that satisfies $f([U_0, \ldots, U_q]) \in \mathcal{G}(U_0 \cap \cdots \cap U_q) = G$ for every q-simplex $[U_0, \ldots, U_q]$. Thus, $\text{Hom}(C_q(N(\mathcal{U})), G) = C^q(\mathcal{U}, \mathcal{G})$, and we have $H^q(N(\mathcal{U}), G) \cong \check{H}^q(\mathcal{U}, \mathcal{G})$.

(iii) If K is a simplicial complex, then $\dim(K) \leq n$ if K has no $(n+1)$-simplexes. In this case, $H^q(K) = \{0\}$ for all $q > n$. Now $\dim(N(\mathcal{U})) \leq n$ for some open cover $\mathcal{U}$ if $U_0 \cap \cdots \cap U_{n+1} = \varnothing$ whenever all $U_i \in \mathcal{U}$ are distinct. In this case, $\check{H}^q(\mathcal{U}, \mathcal{F}) = \{0\}$ for every sheaf $\mathcal{F}$ and all $q > n$.

(iv) If $\mathcal{U} = \{X\}$, the open cover consisting of X itself, then $\dim(N(\mathcal{U})) \leq 0$, and so part (iii) gives $\check{H}^q(\mathcal{U}, \mathcal{F}) = \{0\}$ for every sheaf $\mathcal{F}$ and all $q \geq 1$.

(v) An exact sequence of sheaves $0 \to \mathcal{F}' \to \mathcal{F} \to \mathcal{F}'' \to 0$ may not give a long exact sequence. For example, $0 \to \mathbb{Z} \to \mathcal{O} \xrightarrow{\exp} \mathcal{O}^{\times} \to 0$ is a short exact sequence of sheaves over the punctured plane $X = \mathbb{C} - \{0\}$ (see Example 6.69). If $\mathcal{U} = \{X\}$ is the open cover consisting of X itself, consider the sequence

$$0 \to \check{H}^0(\mathcal{U}, \mathbb{Z}) \to \check{H}^0(\mathcal{U}, \mathcal{O}) \to \check{H}^0(\mathcal{U}, \mathcal{O}^{\times}) \to \check{H}^1(\mathcal{U}, \mathbb{Z}). \quad (1)$$

By part (i), $\check{H}^0(\mathcal{U}, \mathcal{F}) \cong \Gamma(\mathcal{F})$, while part (iv) gives $\check{H}^1(\mathcal{U}, \mathbb{Z}) = \{0\}$. Example 6.69 shows that $\Gamma(\mathcal{O}) \to \Gamma(\mathcal{O}^{\times})$ is not surjective and, hence, sequence (1) is not exact.

(vi) The cohomology groups $\check{H}^q(\mathcal{U}, \mathcal{F})$ may depend on the open cover. There is an open cover $\mathcal{V} = \{V_1, V_2, V_3\}$ of the punctured plane $X = \mathbb{C} - \{0\}$ with $V_i \cap V_j \neq \varnothing$ for all i, j but with $V_1 \cap V_2 \cap V_3 = \varnothing$. Thus, $N(\mathcal{V})$ is a triangle; that is, $N(\mathcal{V}) \approx S^1$. But if $\mathbb{Z}$ is the constant sheaf $\mathbb{Z}$, then part (ii) gives $\check{H}^1(\mathcal{V}, \mathbb{Z}) \cong H^1(S^1, \mathbb{Z}) \cong \mathbb{Z}$. In contrast, if $\mathcal{U} = \{X\}$, then part (iv) gives $\check{H}^1(\mathcal{U}, \mathbb{Z}) = \{0\}$. ◄

Here is a sketch of a way to compare $\check{H}^q(\mathcal{U}, \mathcal{F})$ and $H^q(\mathcal{F})$.

Lemma 6.79. *Let $\mathcal{U}$ be an open cover of a space X, and let $\mathcal{F}$ be a sheaf of abelian groups over X.*

(i) *There is an exact sequence of sheaves*

$$0 \to \mathcal{F} \to \mathfrak{C}^0(\mathcal{U}, \mathcal{F}) \to \mathfrak{C}^1(\mathcal{U}, \mathcal{F}) \to \mathfrak{C}^2(\mathcal{U}, \mathcal{F}) \to \qquad (2)$$

with $\Gamma(\mathfrak{C}^q(\mathcal{U}, \mathcal{F})) = C^q(\mathcal{U}, \mathcal{F})$ *for all* $q \geq 0$.

(ii) *If* $\mathcal{F}$ *is flabby, then* $\mathfrak{C}^q(\mathcal{U}, \mathcal{F})$ *is flabby for all* $q \geq 0$, *and so* (2) *is a flabby resolution.*

Proof.   Godement, *Topologie Algébrique et Théorie des Faisceaux*, pp. 206–207; see our Section §10.8 for more details.   •

We can now construct a map relating open covers and sheaf cohomology; unfortunately, this construction may not give an isomorphism without additional hypotheses.

Proposition 6.80.   *If* $\mathcal{U} = (U_i)_{i \in I}$ *is an open cover of a space* X *and* $\mathcal{F}$ *is a sheaf of abelian groups over* X, *then for each* $q \geq 0$, *there is a natural map* $\varphi^q : \check{H}^q(\mathcal{U}, \mathcal{F}) \to H^q(\mathcal{F})$.

Proof.   Consider the following diagram of sheaves:

$$
\begin{array}{ccccccccc}
0 & \longrightarrow & \mathcal{F} & \longrightarrow & \mathfrak{C}^0(\mathcal{U}, \mathcal{F}) & \longrightarrow & \mathfrak{C}^1(\mathcal{U}, \mathcal{F}) & \longrightarrow & \mathfrak{C}^2(\mathcal{U}, \mathcal{F})) & \longrightarrow \\
& & \downarrow{\scriptstyle 1_\mathcal{F}} & & \vdots & & \vdots & & \vdots & \\
0 & \longrightarrow & \mathcal{F} & \longrightarrow & \mathcal{E}^0 & \longrightarrow & \mathcal{E}^1 & \longrightarrow & \mathcal{E}^2 & \longrightarrow ,
\end{array}
$$

where the bottom row **E** is an injective resolution of $\mathcal{F}$ in $\mathbf{Sh}(X)$. Let $f = (f^q)$ be a chain map $\mathfrak{C}^\bullet(\mathcal{U}, \mathcal{F}) \to \mathbf{E}$ of sheaves over $1_\mathcal{F}$ arising from the Comparison Theorem. Applying the global section functor Γ gives a chain map $\Gamma f : \Gamma \mathfrak{C}^\bullet(\mathcal{U}, \mathcal{F}) \to \Gamma \mathbf{E}^\mathcal{F}$ of complexes of abelian groups. Note that $\Gamma \mathfrak{C}^\bullet(\mathcal{U}, \mathcal{F}) = C^\bullet(\mathcal{U}, \mathcal{F})$, so that

$$H^\bullet(\Gamma \mathfrak{C}^\bullet(\mathcal{U}, \mathcal{F})) = H^\bullet(C^\bullet(\mathcal{U}, \mathcal{F})) = \check{H}^\bullet(\mathcal{U}, \mathcal{F}).$$

On the other hand, $H^\bullet(\Gamma \mathbf{E}^\mathcal{F}) = (R^\bullet \Gamma)\mathcal{F} = H^\bullet(\mathcal{F})$. Therefore, $\varphi = H^\bullet(\Gamma f)$ maps $\check{H}^\bullet(\mathcal{U}, \mathcal{F})) \to H^\bullet(\mathcal{F})$.   •

Čech cohomology $\check{H}^q(X, \mathcal{F})$ will be defined as a direct limit of $\check{H}^q(\mathcal{U}, \mathcal{F})$ over all open covers $\mathcal{U}$, so that it will be independent of the choice of $\mathcal{U}$. Let us begin by trying to partially order the open covers of X.

Definition.   An open cover $\mathcal{V}$ of X is a **refinement** of an open cover $\mathcal{U}$, denoted by $\mathcal{V} \succeq \mathcal{U}$, if, for each $V \in \mathcal{V}$, there exists $U \in \mathcal{U}$ with $V \subseteq U$. For each $V \in \mathcal{V}$, a choice of $U \in \mathcal{U}$ with $V \subseteq U$ defines a function $r : \mathcal{V} \to \mathcal{U}$ with $r(V) = U$, which we call a **refining map** (there are many refining maps for each pair $\mathcal{V} \succeq \mathcal{U}$).

Of course, every subcover of an open cover $\mathcal{U}$ is a refinement of $\mathcal{U}$.

We want the family of all open covers of X to be a partially ordered set under refinement, but there are two difficulties. The first problem is whether refinement is a partial order. It is easy to see that $\mathcal{V} \succeq \mathcal{U}$ is reflexive and transitive, but it may not be antisymmetric. Recall that an open cover $\mathcal{U}$ is an indexed family $\mathcal{U} = (U_i)_{i \in I}$ [indices are needed to define the simplexes in the nerve $N(\mathcal{U})$, and the ordering of the vertices in a simplex is needed to define differentials]. Suppose that $X = U_1 \cup U_2$, where U_1, U_2 are open sets. The open covers $\{U_1, U_2\}$ and $\{U_2, U_1\}$ are distinct, yet each refines the other. To surmount this difficulty, we will partially order the homology groups $\check{H}^q(\mathcal{U}, \mathcal{F})$ instead of the open covers.

Recall that a *simplicial map* $f \colon K \to L$, for simplicial complexes K and L, is a function $f \colon \mathrm{Vert}(K) \to \mathrm{Vert}(L)$ such that $[fv_0, \ldots, fv_q]$ is a simplex in L for every simplex $[v_0, \ldots, v_q]$ in K (we do not insist that the vertices $fv_0, \ldots, fv_q$ be distinct).

Definition. Simplicial maps $f, g \colon K \to L$ are **contiguous** if, for every simplex $[v_0, \ldots, v_q]$ in K, we have $[fv_0, \ldots, fv_q, gv_0, \ldots, gv_q]$ a simplex in L.

Every refining map r induces a simplicial map $r_\# \colon N(\mathcal{V}) \to N(\mathcal{U})$: if $[V_0, \ldots, V_q]$ is a simplex, define $r_\#([V_0, \ldots, V_q]) = [rV_0, \ldots, rV_q]$. Note that $[rV_0, \ldots, rV_q]$ is a simplex in $N(\mathcal{U})$, for $\bigcap_i rV_i \supseteq \bigcap_i V_i \neq \varnothing$. Indeed, if $r, s \colon \mathcal{V} \to \mathcal{U}$ are refining maps, then $r_\#$ and $s_\#$ are contiguous, for $\bigcap_i rV_i \supseteq \bigcap_i V_i$ and $\bigcap_i sV_i \supseteq \bigcap_i V_i$; therefore, $[rV_0, \ldots, rV_q, sV_0, \ldots, sV_q]$ is a simplex, for $\bigcap_i rV_i \cap \bigcap_i sV_i \supseteq \bigcap_i V_i \neq \varnothing$.

Lemma 6.81.

(i) *If K and L are simplicial complexes and $f, g \colon K \to L$ are contiguous simplicial maps, then $f^* = g^* \colon H^q(K) \to H^q(L)$; that is, their induced maps in cohomology are equal.*

(ii) *If $r \colon \mathcal{U} \to \mathcal{U}$ is a refining map of an open cover of itself, then the induced map $r^* \colon \check{H}^q(\mathcal{U}, \mathcal{F}) \to \check{H}^q(\mathcal{U}, \mathcal{F})$ is the identity.*

Proof.

(i) It is proved in Munkres, *Elements of Algebraic Topology*, p. 67, that the chain maps $f_\#, g_\# \colon \mathbf{C}_\bullet(K) \to \mathbf{C}_\bullet(L)$ induced by f, g are homotopic, and so the maps they induce in cohomology are equal.

(ii) Both $r \colon \mathcal{U} \to \mathcal{U}$ and $1_\mathcal{U}$ are refining maps, and so both induce the same map in cohomology. •

Definition. If $\mathcal{F}$ is a sheaf over a space X, define $\check{H}^q(\mathcal{U}, \mathcal{F}) \preceq \check{H}^q(\mathcal{V}, \mathcal{F})$ if there exists a refining map $r : \mathcal{V} \to \mathcal{U}$.

Lemma 6.82. $\check{H}^q(\mathcal{U}, \mathcal{F}) \preceq \check{H}^q(\mathcal{V}, \mathcal{F})$ *is a partial order.*

Proof. Lemma 6.81 implies that there is at most one map $\check{H}^q(\mathcal{V}, \mathcal{F}) \to \check{H}^q(\mathcal{U}, \mathcal{F})$ induced by a refining map. The class of all $\check{H}^q(\mathcal{U}, \mathcal{F})$ and maps r^* induced by refining maps r is a category, and so Example 1.3(iii) shows that $\{\check{H}^q(\mathcal{U}, \mathcal{F}) : \mathcal{U} \text{ is an open cover of } X\}$ is partially ordered.   •

The second difficulty in dealing with all the open covers of a space is set-theoretical. Given open covers $\mathcal{U}$ and $\mathcal{V}$, there is an open cover $\mathcal{W}$ that refines each: define $\mathcal{W} = (U \cap V)_{U \in \mathcal{U} \text{ and } V \in \mathcal{V}}$. It is possible that $\mathcal{W}$ has many repetitions; for example, the empty set $\varnothing$ can occur many times. Here is the formal definition of an open cover (we have already explained why open covers are indexed sets).

Definition. An *open cover* $\mathcal{U} = (U_i)_{i \in I}$ of a topological space X is an indexed family of open subsets whose union is X; thus, $\mathcal{U}$ is a function $I \to \mathcal{T}$, where $\mathcal{T}$ is the family of all open subsets of X.

Since open covers may have repeated terms, any set is allowed to be an index set. Thus, the number of terms in an open cover can be arbitrarily large, and the class of all open covers of a space X is a proper class! Were it not for this inconvenient fact, the class $\mathcal{K}$ of all $\check{H}^q(\mathcal{U}, \mathcal{F})$ would be a directed set. Here is a way to deal with this. Informally, we say that a class $\mathcal{K}$ is a *directed class* if it is a directed set whose underlying set may be a proper class.

Definition. A class $\mathcal{K}$ is a *directed class* if there is a relation $k \preceq k'$ defined on $\mathcal{K}$ that is reflexive, antisymmetric, and transitive, and, for each $k, k' \in \mathcal{K}$, there is $k^* \in \mathcal{K}$ with $k \preceq k^*$ and $k' \preceq k^*$. We say that a subclass $\mathcal{L} \subseteq \mathcal{K}$ is *cofinal* in $\mathcal{K}$ if, for each $k \in \mathcal{K}$, there exists $\ell \in \mathcal{L}$ with $k \preceq \ell$. We can also define a *direct system* $\{A_i, \varphi_j^i\}$ with indices lying in a directed class I.

Example 6.83. Let $\mathcal{F}$ be a sheaf over a space X. Lemma 6.82 and the paragraph following it show that the class $\mathcal{K}$ of all groups $\check{H}^q(\mathcal{U}, \mathcal{F})$, where $\mathcal{U}$ varies over all open covers $\mathcal{U}$ of a space X, is a directed class. If $\mathcal{U} = (U_i)_{i \in I}$ is an open cover, let $\mathcal{V}$ be obtained from $\mathcal{U}$ by throwing away repetitions; for example, if $\mathcal{U} = \{U_0, U_1, U_1\}$, then $\mathcal{V} = \{U_0, U_1\}$. It is clear that $\mathcal{V}$ is a refinement of $\mathcal{U}$, and so

$$\mathcal{H} = \{\check{H}^q(\mathcal{U}, \mathcal{F}) : \mathcal{U} \text{ is an open cover having no repeated terms}\}$$

is a cofinal subclass of $\mathcal{K}$. Indeed, $\mathcal{H}$ is a set, for if $\check{H}^q(\mathcal{U}, \mathcal{F}) \in \mathcal{H}$, then $\mathcal{U} = (U_i)_{i \in I}$ is an injective function $I \to \mathcal{T}$, and so $|I| \leq |\mathcal{T}| \leq 2^{|X|}$

(because the topology $\mathcal{T}$ is a family of subsets of X). Thus, $\mathcal{H}$ is a directed set for any sheaf $\mathcal{F}$ over X. ◄

The next proposition says, under certain circumstances, that it is possible to form direct limits over directed classes.

Proposition 6.84. *Let $\mathcal{K}$ be a directed class, let $\mathcal{C}$ be a cocomplete category, and let $\{A_k, \varphi_j^k\}$ be a direct system in $\mathcal{C}$ over $\mathcal{K}$. If $\mathcal{L}$ and $\mathcal{M}$ are cofinal in $\mathcal{K}$ and both $\mathcal{L}$ and $\mathcal{M}$ are sets, then $\varinjlim_{\mathcal{L}} A_k \cong \varinjlim_{\mathcal{M}} A_k$.*

Proof. Let $\mathcal{L} \cup \mathcal{M}$ be the partially ordered subset of $\mathcal{K}$ generated by the subsets $\mathcal{L}$ and $\mathcal{M}$. Note that each of $\mathcal{L}$ and $\mathcal{M}$ is cofinal in $\mathcal{L} \cup \mathcal{M}$ because each is cofinal in $\mathcal{K}$; it follows that $\mathcal{L} \cup \mathcal{M}$ is directed. Since $\mathcal{L} \cup \mathcal{M}$ is a set, the direct limit $D = \varinjlim_{\mathcal{L} \cup \mathcal{M}} A_k$ is defined, and Exercise 5.22 on page 255 gives $\varinjlim_{\mathcal{L}} A_k \cong D \cong \varinjlim_{\mathcal{M}} A_k$. •

Definition. *Čech cohomology* of a space X with coefficients in a sheaf $\mathcal{F}$ over X is defined by

$$\check{H}^q(\mathcal{F}) = \varinjlim_{\mathcal{H}} \check{H}^q(\mathcal{U}, \mathcal{F}),$$

where $\mathcal{H}$ is the directed set of all cohomology groups $\check{H}^q(\mathcal{U}, \mathcal{F})$ with $\mathcal{U}$ an open cover of X having no repeated terms.

It follows easily from Example 6.78(i) that $\check{H}^0(\mathcal{F}) = \Gamma(\mathcal{F})$ for every sheaf $\mathcal{F}$, and so $\check{H}^0(\mathcal{F}) \cong H^0(\mathcal{F})$; that is, Čech cohomology and sheaf cohomology agree in degree 0. It is true that they also agree in degree 1: $\check{H}^1(\mathcal{F}) \cong H^1(\mathcal{F})$ (Tennison, *Sheaf Theory*, p. 147), but they can disagree otherwise.

Lemma 6.85. *If $\mathcal{F}$ is an injective sheaf over a space X, then $\check{H}^q(\mathcal{F}) = \{0\}$ for all $q \geq 1$.*

Proof. Tennison, *Sheaf Theory*, p. 145. •

Theorem 6.86 (Serre). *If $0 \to \mathcal{F}' \to \mathcal{F} \to \mathcal{F}'' \to 0$ is a short exact sequence of sheaves over a topological space X, then there is a six term exact sequence in Čech cohomology:*

$$0 \to \check{H}^0(\mathcal{F}') \to \check{H}^0(\mathcal{F}) \to \check{H}^0(\mathcal{F}'')$$
$$\to \check{H}^1(\mathcal{F}') \to \check{H}^1(\mathcal{F}) \to \check{H}^1(\mathcal{F}).$$

Proof. Serre, FAC, p. 217. •

Although Serre's paper FAC is concerned with sheaves over arbitrary spaces, it also contains results about sheaves over Hausdorff spaces.

Given a sheaf $\mathcal{F}$ over a paracompact space X, it is easy to see that the class of all $\check{H}^q(\mathcal{V}, \mathcal{F})$, where $\mathcal{V}$ varies over all locally finite open covers of X with no repeated terms, is cofinal in the directed set $\mathcal{H}$.

Theorem 6.87 (Serre).   *If $0 \to \mathcal{F}' \to \mathcal{F} \to \mathcal{F}'' \to 0$ is a short exact sequence of sheaves over a paracompact space X, then there is an exact sequence in Čech cohomology:*

$$0 \to \check{H}^0(\mathcal{F}') \to \check{H}^0(\mathcal{F}) \to \check{H}^0(\mathcal{F}'')$$
$$\to \check{H}^1(\mathcal{F}') \to \check{H}^1(\mathcal{F}) \to \check{H}^1(\mathcal{F}) \to \cdots$$
$$\to \check{H}^q(\mathcal{F}') \to \check{H}^q(\mathcal{F}) \to \check{H}^q(\mathcal{F}'') \to \check{H}^{q+1}(\mathcal{F}') \to .$$

Proof.   Serre, FAC, p. 218.   •

Theorem 6.88.   *If $\mathcal{F}$ is a sheaf over a paracompact space X, then Čech cohomology agrees with sheaf cohomology: for all $q \geq 0$,*

$$\check{H}^q(\mathcal{F}) \cong H^q(\mathcal{F}).$$

Proof.   Using Lemma 6.85, we see that the hypotheses of Corollary 6.49 hold for Čech cohomology over a paracompact space.   •

The next corollary illustrates how Čech cohomology can be used.

Theorem 6.89.   *Let $\mathcal{F}$ be a sheaf over a paracompact space X. If X has an open cover $\mathcal{U}$ with $\dim(N(\mathcal{U})) \leq n$, then $\check{H}^q(\mathcal{F}) = \{0\}$ for all $q \geq n + 1$.*

Proof.   Swan, *The Theory of Sheaves*, p. 109.   •

Corollary 6.90.   *If X is a compact Hausdorff space, then $H^q(\mathcal{F}) = \{0\}$ for large q.*

Proof.   Since X is compact, every open cover $\mathcal{U}$ of X has a finite subcover $\mathcal{V}$. But $N(\mathcal{V})$ is a finite simplicial complex, and hence it is finite-dimensional. Theorem 6.89 now gives $\check{H}^q(\mathcal{F}) = \{0\}$ for all $q > \dim(N(\mathcal{U}))$, and Theorem 6.88 gives $H^q(\mathcal{F}) = \{0\}$ for all $q > \dim(N(\mathcal{U}))$.   •

6.3.2 Riemann–Roch Theorem

We end this chapter by describing the Riemann–Roch Theorem, first for the **Riemann sphere** $\widehat{\mathbb{C}} = \mathbb{C} \cup \{\infty\}$ and then, more generally, for compact Riemann surfaces. We shall see that the statement of this theorem involves a formula whose ingredients can be better understood in terms of sheaf cohomology. Although there are proofs of this special case of Riemann–Roch without sheaves (see Fulton, *Algebraic Topology*, Chapter 21, or Kendig, *Elementary Algebraic Geometry*, Chapter V.7), both the statement and the proof of its generalizations for higher-dimensional manifolds or for varieties defined over fields of characteristic $p > 0$, involve sheaves in an essential way. This discussion will illustrate how a sheaf, constructed using local data, yields global information.

Recall some definitions from Complex Analysis.

Definition.   Let $U \subseteq \mathbb{C}$ be open. A complex-valued function f is **meromorphic on** U if f is defined on $U - D$, where D is discrete and, for each $p \in U$,

$$f(z) = \sum_{n \geq m} a_n (z - p)^n$$

for all z in some deleted neighborhood of p, where $m \in \mathbb{Z}$, $a_n \in \mathbb{C}$, and $a_m \neq 0$. We write

$$\mathrm{ord}_p(f) = m.$$

If $m > 0$, then p is called a **zero** of f of **order** m, and if $m < 0$, then p is called a **pole** of f of **order** $|m|$. Call f **holomorphic** (or **analytic**) if it is meromorphic and has no poles.

This definition can be extended to the Riemann sphere $\widehat{\mathbb{C}}$: if $p = \infty$, replace z by $1/z$ (basic open neighborhoods of ∞ in $\widehat{\mathbb{C}}$ have the form $U_\infty = \{\infty\} \cup \{z \in \mathbb{C} : |z| > N\}$ for some number N). Thus, there is a pole of order m at ∞ if $f(1/z)$ has a pole of order m at 0.

Later, we will discuss generalizations of these terms for *complex manifolds* instead of $\widehat{\mathbb{C}}$ and, in particular, for *Riemann surfaces*.

The following query was posted on the newsgroup sci.math.

> I know that the Riemann–Roch Theorem is a very famous theorem in Algebraic Geometry. I'm an undergraduate student. I don't know the terms of Algebraic Geometry, but I want to grasp the meaning of the theorem. Can you explain it in an elementary way?

Keith Ramsay posted the following excellent reply.

> There are various forms of the Riemann–Roch Theorem of varying generality. The basic problem is to determine the functions on a space that have prescribed poles. The *space* is typically an algebraic variety, but you might find it easier to learn the version of the theorem which is concerned with compact complex

manifolds. In fact, the special case of compact Riemann surfaces (i.e., compact complex manifolds of complex dimension 1) are of enough interest to start out with.

A compact Riemann surface S has an invariant g, called its *genus*. Roughly speaking, the genus is the number of holes in the surface; a sphere has genus 0, a torus (a doughnut-shaped surface) has genus 1, and so on. Given a list of points $p_1, \ldots, p_n$ on S and a list of integers $m_1, \ldots, m_n$, we'd like to have some information about the meromorphic functions f on S that are holomorphic except at the points $p_1, \ldots, p_n$, where the orders of the poles are prescribed; that is, $\mathrm{ord}_{p_i}(f) \geq -m_i$ for all i. The set of all such functions f is a complex vector space: if f satisfies the given conditions and c is a complex number, then cf also does; if g satisfies the conditions, then so does $f + g$ (the order of the pole at a point can't be any greater than the order of the poles of each of f and g). Write $m_1 p_1 + \cdots + m_n p_n$ as just an abstract notation describing the points and associated orders; this is called a ***divisor***. Note that every nonconstant meromorphic function f determines a divisor: there are only finitely many points $p_1, \ldots, p_n$ at which f has either a pole or a zero, and we define

$$\mathrm{Div}(f) = \mathrm{ord}_{p_1}(f) p_1 + \cdots + \mathrm{ord}_{p_n}(f) p_n.$$

Define the ***degree*** of a divisor $D = m_1 p_1 + \cdots + m_n p_n$ to be $\sum_i m_i$; hence,

$$\deg(\mathrm{Div}(f)) = \mathrm{ord}_{p_1}(f) + \cdots + \mathrm{ord}_{p_n}(f).$$

A theorem of Abel (see Fulton, *Algebraic Topology*, p. 267) says that if f is a nonconstant meromorphic function, then $\mathrm{Div}(f)$ has degree 0. If $D = m_1 p_1 + \cdots + m_n p_n$, write

$$L(D)$$

for the vector space of functions that satisfy the bounds on poles at the points $p_1, \ldots, p_n$. Now $L(D)$ is finite-dimensional, and we define

$$\ell(D) = \dim(L(D)).$$

For example, let S be the Riemann sphere. The meromorphic functions are just the rational functions (any meromorphic function on the complex plane that isn't rational has an essential singularity at ∞). The functions having no poles except at ∞ are the polynomials, by Liouville's Theorem. The space $L(m\infty)$ (here, we are taking only one point, namely, $p_1 = \infty$) is the set of functions that are holomorphic except at ∞, where they're allowed to have a zero of order at most m. That's just the set of polynomials of degree $\leq m$. So $\ell(m\infty) = m + 1$, for the functions $1, z, z^2, \ldots, z^m$ form a basis. Here is a second example. The space $L(1\infty - 1(1 + i))$ (so $p_1 = \infty$, $m_1 = 1$, and $p_2 = 1 + i$, $m_2 = -1$) consists of those meromorphic functions with a zero at $1 + i$ and, at worst, a pole of order 1 at ∞. This is a subspace of $L(1\infty)$, and so it consists of just those functions of the form $c(z - 1 - i)$; hence, $\ell(1\infty - 1(1 + i)) = 1$. Similarly, $L(2\infty - 1(1 + i))$ is the space of all those functions of the form $(z - 1 - i)(az + b)$, so the dimension $\ell(2\infty - 1(1 + i)) = 2$. Generally, $L(m_1 p_1 + \cdots + m_n p_n)$ is the space of meromorphic functions of the form

$$(z - p_1)^{-m_1}(z - p_2)^{-m_2} \cdots (z - p_n)^{-m_n}(c_0 z^d + \cdots + c_d),$$

where $d = m_1 + \cdots + m_n$ if $d \geq 0$. (If $d < 0$, the only function satisfying the conditions is the function identically 0.) So, on the Riemann sphere we always get $\ell(m_1 p_1 + \cdots + m_n p_n) = m_1 + \cdots + m_n + 1 = d + 1$ unless that's negative, in which case we get $\ell(m_1 p_1 + \cdots + m_n p_n) = 0$.

That's a special case of Riemann–Roch for a Riemann surface of genus $g = 0$. In this case, Riemann–Roch is enough to tell us exactly what this dimension ℓ is. If D is a divisor of degree $d < 0$, then there are no functions other than 0 satisfying the conditions, and so $\ell = 0$.

Theorem (Riemann–Roch).   *A compact Riemann surface of genus g has a **canonical divisor** $K = k_1 p_1 + \cdots + k_n p_n$ of $\deg(K) = 2g - 2$ such that*

$$\ell(D) - \ell(K - D) = \deg(D) + 1 - g$$

for every divisor $D = m_1 p_1 + \cdots + m_n p_n$.

Notice that the formula displays a connection between topology on the one hand (the genus) and analysis on the other (meromorphic functions). The equality is, of course, two inequalities. Riemann proved that $\ell(D) \geq d + 1 - g$; a few years later, Roch proved that $\ell(D) = d + 1 - g + \ell(K - D)$. Both mathematicians died in 1866, and both died young; Riemann was 40 and Roch was 26.

Riemann–Roch isn't enough by itself to determine $\ell = \ell(D)$ in every case but, in some special cases, it is enough. In particular, if $d > 2g - 2$, then the dual divisor has degree < 0, which simplifies the formula to

$$\ell = d + 1 - g.$$

So, for example, if $g = 0$ and $d < 0$, we know $\ell = 0$; when $g = 0$ and $d \geq 0$, we have $d > 2g - 2 = -2$ so that $\ell = d + 1 - g = d + 1$, which is just what we figured out for the special case of the Riemann sphere.

If $g = 1$ and $d < 0$, we get $\ell = 0$; when $g = 1$ and $d > 0 = 2g - 2$, we get $\ell = d + 1 - 1 = d$. But if $d = 0$, Riemann–Roch isn't enough to tell us whether $\ell = 0$ or $\ell = 1$, i.e., whether the set of functions is $\{0\}$ or $\{cf : c \in \mathbb{C}\}$ for some nonzero function f. Both cases occur. A curve of genus 1 with a specified basepoint p is called an *elliptic curve* (different from an ellipse in Calculus). The set of functions on an elliptic curve that have no poles is one-dimensional; it consists of constant functions. So $\ell(0p_1) = 1$. But if p_2 is some point other than p_1, the only function on the elliptic curve having no poles except at p_1 (where it has a pole of order at most 1) and a zero at p_2 is the zero function, so that $\ell(1p_1 - 1p_2) = 0$.

This is the start of an interesting analysis of elliptic curves. Since $\ell(1p) = 1$, $L(1p)$ consists just of constant functions c; since $\ell(2p) = 2$, we can see that $L(2p)$ is a set of functions of the form $c_0 + c_1 f$ for some f that has a pole of order exactly 2 at p. Likewise $L(3p)$ is three-dimensional, so there's some independent function g in it that has a pole of order 3 at p; $L(3p) = \{c_0 + c_1 f + c_2 g\}$. Now $L(4p)$ is four-dimensional, but f^2 is in it, and is independent of 1, f, g, so $L(4p) = \{c_0 + c_1 f + c_2 g + c_3 f^2\}$. Then $L(5p) = \{c_0 + c_1 f + c_2 g + c_3 f^2 + c_4 fg\}$. Where it gets interesting is when we get to $L(6p)$, the set of functions having no poles except at p, and having a pole of order at most 6 at p. All of 1, f, g, f^2, fg, f^3, g^2 are in this set. But since the set is only six-dimensional and not seven-dimensional, there is some linear dependence among them:

$$g^2 = c_0 f^3 + c_1 fg + c_2 f^2 + c_3 g + c_4 f + c_5.$$

So, if we plot the values of f and g on the plane, they fall within this algebraic curve given by a cubic equation. In fact, the original Riemann surface is essentially given by the cubic equation, and it turns out to be an algebraic plane curve after all.

We introduce complex manifolds in order to discuss Riemann surfaces. The definitions parallel those for (real) manifolds in Section §5.4.1.

Definition. A *complex n-chart* is an ordered pair (U, φ) with U a topological space, called a *coordinate neighborhood*, and $\varphi \colon U \to \mathbb{C}^n$ a homeomorphism.

We restrict our discussion to complex 1-charts, for we are interested in spaces (Riemann surfaces) that only involve such charts; this simplification allows us to avoid subtleties arising in the passage from functions of one complex variable to several complex variables.

It is clear how to generalize the definition on page 393 of f being *meromorphic* from functions f defined on open subsets of the Riemann sphere to coordinate neighborhoods of complex 1-charts.

Definition. Let (U, φ) be a complex 1-chart. A complex-valued function f is *meromorphic on U* if f is defined on $U - D$, where D is discrete and, for each $p \in U$, $f\varphi^{-1} \colon \operatorname{im} \varphi \to \mathbb{C}$ is meromorphic; that is,

$$f\varphi^{-1}(z) = \sum_{n \geq m} a_n (z - \varphi(p))^n$$

for all z in some deleted neighborhood of $\varphi(p)$, where $m \in \mathbb{Z}$, $a_n \in \mathbb{C}$, and $a_m \neq 0$. We write

$$\operatorname{ord}_p(f) = m.$$

If $m > 0$, then p is called a *zero* of f of *order m*, and if $m < 0$, then p is called a *pole* of f of *order $|m|$*. Call f *holomorphic* if it is meromorphic and has no poles.

All meromorphic functions on (U, φ) form a commutative $\mathbb{R}$-algebra

$$\mathcal{M}(U, \varphi)$$

under pointwise operations, and all holomorphic functions on U form an $\mathbb{R}$-subalgebra:

$$\mathcal{O}(U, \varphi) \subseteq \mathcal{M}(U, \varphi).$$

It follows that $\mathcal{O}(U, \varphi)$ is a domain and that $\operatorname{Frac}(\mathcal{O}(U, \varphi)) \subseteq \mathcal{M}(U, \varphi)$; that is, if f, g are holomorphic and $g \neq 0$, then f/g is meromorphic.

We now pass from complex charts to more interesting topological spaces: complex manifolds.

Definition. A *complex atlas* of a 2-manifold X is a family of complex 1-charts $((U_i, \varphi_i))_{i \in I}$ with $(U_i)_{i \in I}$ an open cover of X.

Let $((U_i, \varphi_i))_{i \in I}$ be a complex atlas of a 2-manifold. If $p \in U_i$, then φ_i equips p with complex coordinates, namely, $\varphi_i(p)$. Write $U_{ij} = U_i \cap U_j$. If $U_{ij} \neq \varnothing$, then every $p \in U_{ij}$ has two coordinates: $\varphi_i(p)$ and $\varphi_j(p)$.

Definition. If $((U_i, \varphi_i))_{i \in I}$ is a complex atlas, then its **transition functions** are the homeomorphisms

$$h_{ij} = \varphi_i \varphi_j^{-1} : \varphi_j(U_{ij}) \to \varphi_i(U_{ij}).$$

Transition functions compare the two coordinates of $p \in U_{ij}$: if $y = \varphi_j(p)$ and $x = \varphi_i(p)$, then $h_{ij} : y \mapsto x(y)$.

Definition. A **complex 1-manifold** is a 2-manifold having holomorphic transition functions, and a **Riemann surface** is a connected complex 1-manifold.

Viewing $\widehat{\mathbb{C}}$ as a Riemann surface, we need not treat ∞ differently from other points (as we did on page 393), for we can use the complex 1-chart $(U_\infty, \varphi_\infty)$, where U_∞ is an open neighborhood of ∞ and $\varphi_\infty : U_\infty \to \mathbb{C}$ is the homeomorphism given by $\infty \mapsto 0$ and $p \mapsto 1/p$ if $p \neq \infty$.

We now generalize the definitions of *meromorphic, holomorphic,* and *1-forms* from charts to Riemann surfaces.

Definition. If $((U_i, \varphi_i))_{i \in I}$ is a complex atlas of a Riemann surface X with transition functions $h_{ij} = \varphi_j \varphi_i^{-1}$, then a family $f = (f_i : U_i \to \mathbb{C})_{i \in I}$ is **meromorphic** (or **holomorphic**) if each f_i is meromorphic (or holomorphic) [i.e., $f \varphi_i^{-1}$ is meromorphic (or holomorphic)] and the f_i are **compatible**; that is, $f_i h_{ij} = f_j$ on $\varphi_i(U_i \cap U_j)$ for all i, j. At any point $p \in U_i$, there is a Laurent expansion $f_i(z) = \sum_{n \geq m} a_n (z - \varphi_i(p))^n$. If f_i is not identically zero, then $\mathrm{ord}_p(f_i)$ does not depend on the choice of the chart containing p, and so, if $f = (f_i)$ and $p \in U_i$, we can define

$$\mathrm{ord}_p(f) = \mathrm{ord}_p(f_i).$$

If $m > 0$, then f has a **zero** of order m at p; if $m < 0$, then f has a **pole** at p of order $|m|$. We say that f is **meromorphic on X** if it is defined and meromorphic on the complement of a discrete subset of X, and f is **holomorphic** if it is meromorphic on X and has no poles.

As in Example 5.77, we may define a sheaf $\mathcal{M}$ of fields over X [construct sheaves $\mathcal{M}_i$ over U_i having global sections $\mathcal{M}(U_i, \varphi_i)$, and glue them together using Proposition 5.76]. Meromorphic functions on a Riemann surface X are the global sections of this sheaf; that is, they are compatible families of locally defined meromorphic functions. Define Ω^0, the **structure sheaf of X**, to be the subsheaf of $\mathcal{M}$ with $\Omega^0(U_i) = \mathcal{O}(U_i, \varphi_i)$ for all $i \in I$. The **ring of holomorphic functions on X** is

$$\Omega^0(X) = \Gamma(X, \Omega^0),$$

the global sections of Ω^0. Now $\Omega^0(X)$ is a subring of the field $\mathcal{M}(X)$ and so it is a domain.

One can define the de Rham complex of a Riemann surface and, indeed, of higher-dimensional complex manifolds (see Bott–Tu, *Differential Forms in Algebraic Topology*). However, as our aim is more modest, we will define complex 1-forms in an ad hoc way. Define

$$\Omega^1(U_i, \varphi_i)$$

to be the free $\mathcal{O}(U_i, \varphi_i)$-module of rank 1 with basis element denoted by dz_i. If $((U_i, \varphi_i))_{i \in I}$ is a complex atlas for X, define Ω^1 to be the sheaf with $\Omega^1(U_i) = \Omega^1(U_i, \varphi_i)$ that is obtained by gluing compatible sheaves over U_i (see Example 5.77). Thus, a **complex 1-form** ω on X is a global section; that is, ω is a compatible family of complex 1-forms $\omega = (f_i \, dz_i)_{i \in I}$ with f_i holomorphic. In a similar way, we may define **meromorphic 1-forms** that, locally, look like $g_i \, dz_i$ with g_i meromorphic. More precisely, define $\mathcal{M}^1 = \mathcal{M} \otimes_{\Omega^0} \Omega^1$.

Define

$$d \colon \Omega^0 \to \Omega^1$$

to be the sheaf map that is defined locally by

$$d \colon f_i \mapsto f_i' \, dz_i,$$

where, if $f = (f_i)$ and $f_i(z) = \sum_{n \geq 0} a_n z^n$, then $f_i'(z) = \sum_{n \geq 1} n a_n z^{n-1}$.

We are now going to discuss the Riemann–Roch Theorem for Riemann surfaces (generalizing the special case for the Riemann sphere $\widehat{\mathbb{C}}$). Given a divisor D, we will see how the number $\ell(D)$ and the canonical divisor K are related to sheaf cohomology. Our account follows that in Serre, *Algebraic Groups and Class Fields*, Chapter II.

If X is a Riemann surface, let $D(X)$ be the free abelian group with basis the points in X. A **divisor** D is an element of $D(X)$:

$$D = \sum_{p \in X} n_p p,$$

where $n_p \in \mathbb{Z}$ and almost all $n_p = 0$. The coefficients n_p of D will also be denoted by $\nu_p(D)$, so that

$$D = \sum_{p \in X} n_p p = \sum_{p \in X} \nu_p(D) p.$$

The **degree** of D is

$$\deg(D) = \sum_{p \in X} \nu_p(D).$$

If $((U_i, \varphi_i))_{i \in I}$ is a complex atlas for a **compact** Riemann surface X, then the open cover $(U_i)_{i \in I}$ has a finite subcover. It follows that a nonzero meromorphic function $f \in \mathcal{M}(X)$ has only finitely many poles and zeros; define

$\text{Div}(f)$, the *divisor of* f, by

$$\text{Div}(f) = \sum_{p \in X} \text{ord}_p(f)p.$$

A divisor D is *positive* (or *effective*) if $v_p(D) \geq 0$ for all $p \in X$. In particular, if f is meromorphic, then $\text{Div}(f)$ is positive if and only if f is holomorphic. Define

$$P(X) = \{D \in D(X) : D = \text{Div}(f) \text{ for some } f \in \mathcal{M}(X)\}$$

[divisors of the form $\text{Div}(f)$ are called *principal divisors*]. Since $\text{Div}(fg) = \text{Div}(f) + \text{Div}(g)$, the subset $P(X)$ is a subgroup of $D(X)$; the quotient group $D(X)/P(X)$ is called the group of *divisor classes*. The subgroup $P(X)$ defines an order relation on $D(X)$. Define

$$D_1 \leq D_2 \quad \text{if} \quad D_2 - D_1 \in P(X).$$

We say that divisors $D_1, D_2 \in D(X)$ are *linearly equivalent* if their cosets in $D(X)/P(X)$ are equal; that is, $D_2 = D_1 + \text{Div}(f)$ for some meromorphic f.

Proposition A. *If* $D \in P(X)$, *then* $\deg(D) = 0$.

Proof.   See Gunning, *Lectures on Riemann Surfaces*, or Wells, *Differential Analysis on Complex Manifolds*.   •

It follows that we may define the *degree* of a divisor class: if D_1, D_2 are linearly equivalent, then $\deg(D_1) = \deg(D_2)$.

Given a divisor D, consider all those positive divisors D' that are linearly equivalent to D; that is, $D' \geq 0$ (where 0 is the divisor identically zero) and $D' = D + \text{Div}(f)$ for some $f \in \mathcal{M}(X)$. Thus, $\text{Div}(f) \geq -D$. Define

$$L(D) = \{0\} \cup \{\text{Div}(f) : f \in \mathcal{M}(X) \text{ and } \text{Div}(f) \geq -D\};$$

that is,

$$L(D) = \{0\} \cup \{\text{Div}(f) : f \in \mathcal{M}(X) \text{ and } \text{ord}_p(f) \geq -v_p(D) \text{ for all } p \in X\}.$$

Recall that the constant sheaf $\mathcal{F}$ at $\mathcal{M}(X)$ has etale-sheaf (E, π, X), where $E_p = \mathcal{M}(X)$, and the sections over an open set U are locally constant functions $s \colon U \to \bigcup_{p \in X} E_p$ with $\pi s = 1_U$. If U is an open neighborhood of a point $p \in X$ and $s \in \mathcal{F}(U)$, then $s(p) \in \mathcal{M}(X)$ has an order: define $\mathcal{L}(D)_p$ to be all locally constant functions $s \colon X \to E$ satisfying $v_p(s(p)) \geq -v_p(D)$. Finally, define $\mathcal{L}(D)$ to be the subsheaf of $\mathcal{F}$ whose stalk over p is $\mathcal{L}(D)_p$.

Proposition B. *The vector spaces* $H^0(X, \mathcal{L}(D))$ *and* $H^1(X, \mathcal{L}(D))$ *are finite-dimensional.*

Proof.   See Gunning or Wells.   •

As in Ramsay's exposition, define $\ell(D) = \dim(L(D))$. Now

$$H^0(X, \mathcal{L}(D)) = \Gamma(X, \mathcal{L}(D)) = L(D),$$

and so

$$\ell(D) = \dim(L(D)) = \dim(H^0(X, \mathcal{L}(D))).$$

Define $i(D) = \dim(H^1(X, \mathcal{L}(D)))$. In particular, if $D = 0$, then $\mathcal{L}(D) = \mathcal{O}$, and so $i(0) = \dim(H^1(X, \mathcal{O}))$.

The following analog of Liouville's Theorem holds for compact Riemann surfaces.

Proposition C. *If X is a compact Riemann surface, then $L(0) \cong \mathbb{C}$.*

Theorem 6.91. *For every divisor D of a compact Riemann surface X, we have*

$$\ell(D) - i(D) = \deg(D) + 1 - i(0),$$

where $i(D) = \dim(H^1(X, \mathcal{L}(D)))$.

Proof. The formula is true when $D = 0$: by Proposition C, we have $\ell(0) = 1$, while $\deg(0) = 0$. Thus, the formula reads

$$1 - i(0) = 0 + 1 - i(0).$$

Since any divisor D can be obtained from 0 in a finite number of steps, each adding or subtracting a point, it suffices to show that if the formula holds for a divisor D, then it also holds for the divisors $D + p$ and $D - p$. Write

$$\chi(D) = \ell(D) - i(D) \quad \text{and} \quad \chi'(D) = \deg(D) + 1 - i(0).$$

Now $\chi'(D+p) = \chi'(D)+1$, so that we must show that $\chi(D+p) = \chi(D)+1$. There is an exact sequence of sheaves

$$0 \to \mathcal{L}(D) \to \mathcal{L}(D + p) \to \mathcal{Q} \to 0,$$

where $\mathcal{Q}$ is the quotient sheaf. But $\mathcal{Q}$ is a skyscraper sheaf, with $\mathcal{Q}_q = \{0\}$ for $q \neq p$ while $\mathcal{Q}_p \cong \mathbb{C}$. The corresponding long exact sequence begins

$$0 \to L(D) \to L(D + p) \to H^0(X, \mathcal{Q})$$
$$\to H^1(X, \mathcal{L}(D)) \to H^1(X, \mathcal{L}(D + p)) \to H^1(X, \mathcal{Q}).$$

But $H^0(X, \mathcal{Q})$ is one-dimensional, and $H^1(X, \mathcal{Q}) = \{0\}$, by Corollary 6.76, for $\mathcal{Q}$ is a skyscraper sheaf. Hence, the alternating sum of the dimensions is 0:

$$\ell(D) - \ell(D + p) + \dim(L(\mathcal{Q})) - i(D) + i(D + p) = 0;$$

that is,

$$\chi(D + p) = \chi(D) + 1.$$

The same argument works for $D - p$. ●

Proposition D. *If X is a compact Riemann surface of genus g, then* $i(0) = g$. *Hence, for every divisor D of X, we have*

$$\ell(D) - i(D) = \deg(D) + 1 - g.$$

Proof. Serre, *Algebraic Groups and Class Fields*. p. 17, proves this assuming that one can recognize the genus as the dimension of a certain space of differential forms. Fulton, *Algebraic Topology: A First Course*, Chapter 21, gives a more detailed discussion. •

The Riemann–Roch Theorem follows from this result once we show that $i(D) = \ell(K - D)$, where K is a canonical divisor (we will define K in a moment).

Now $\mathcal{M}^1(X)$ is a vector space over the field $\mathcal{M}(X)$; we claim that it is one-dimensional. It can be proved that there always exists a nonzero meromorphic 1-form in $\mathcal{M}^1(X)$; choose, once and for all, one such, say, $\omega_0 = (\omega_{0i})_{i \in I}$. It follows that $\mathcal{M}^1(X)$ is a one-dimensional vector space over the field $\mathcal{M}(X)$. Indeed, if $\omega = (\omega_i)_{i \in I} \in \mathcal{M}^1(X)$, then we construct a meromorphic h with $\omega = h\omega_0$ as follows. Locally, $\omega_{0i} = f_i dz_i$ and $\omega_i = g_i dz_i$, where f_i is not identically zero. Then $h_i = g_i/f_i$ in $\mathcal{M}(U_i, \varphi_i)$, and it is straightforward to check that the family $(h_i)_{i \in I}$ can be glued to define a global meromorphic function h with $\omega = h\omega_0$. Note that $h \in \mathcal{M}(X)$ is unique because $\mathcal{M}^1(X)$ is a vector space over $\mathcal{M}(X)$.

Definition.   If ω is a nonzero meromorphic 1-form, then $\omega = h\omega_0$; define

$$\operatorname{ord}_p(\omega) = \operatorname{ord}_p(h).$$

Since X is compact, the set of zeros and poles of a meromorphic function h is finite, and so $\operatorname{ord}_p(h)$ is nonzero for only finitely many points p. Therefore, if $\omega = h\omega_0$ is a nonzero meromorphic 1-form, then $\operatorname{ord}_p(\omega) = \operatorname{ord}_p(h)$ is nonzero for only finitely many p.

Definition.   If $\omega \in \mathcal{M}^1(X)$ is a nonzero meromorphic 1-form, define

$$\operatorname{Div}(\omega) = \sum_{p \in X} \operatorname{ord}_p(\omega) p.$$

If ω is a nonzero meromorphic 1-form, then $\omega = h\omega_0$ and $\operatorname{Div}(\omega) = \operatorname{Div}(h) + \operatorname{Div}(\omega_0)$. Thus, all $\operatorname{Div}(\omega)$ are linearly equivalent, and they form a single divisor class. Call

$$\operatorname{Div}(\omega_0) + P(X)$$

the ***canonical class***; any divisor K in this class is called a ***canonical divisor***.

Definition.   If D is a divisor, define

$$\Omega(D) = \{0\} \cup \{\omega \in \mathcal{M}^1(X) : \mathrm{Div}(\omega) \geq D\}.$$

Now if ω is a nonzero meromorphic 1-form, then $\omega = f\omega_0$ for some meromorphic f. Given a divisor D, the following statements are equivalent for $\omega = f\omega_0$:

$$\omega \in \Omega(D);$$
$$\mathrm{Div}(\omega) \geq D;$$
$$\mathrm{Div}(f) + \mathrm{Div}(\omega_0) \geq D;$$
$$\mathrm{Div}(f) \geq \mathrm{Div}(D) - \mathrm{Div}(\omega_0);$$
$$f \in L(K - D).$$

Thus, $\Omega(D) \cong L(K - D)$.

Serre proved $H^1(X, \mathcal{L}(D)) \cong \Omega(D)$ using **Serre Duality** (see Gunning, *Lectures on Riemann Surfaces*, §5 and §6); actually, Serre Duality is the difficult part of the proof of the Riemann–Roch Theorem. It follows that

$$i(D) = \dim(\Omega(D)) = \ell(K - D).$$

The statement of the Riemann–Roch Theorem for compact Riemann surfaces is the same as for the Riemann sphere.

Theorem (Riemann–Roch). *For every divisor D on a compact Riemann surface of genus g, we have*

$$\ell(D) - \ell(K - D) = \deg(D) + 1 - g.$$

All canonical divisors have the same degree, for they lie in the same divisor class. If $D = 0$, then $\ell(0) = 1$, and the Riemann–Roch Theorem gives $\ell(K) = g$. If $D = K$, then the Riemann–Roch Theorem gives $\deg(K) = 2g - 2$.

There are fancier, more general, versions of the Riemann–Roch Theorem that are needed to cope with complications arising from replacing compact Riemann surfaces by compact complex manifolds of higher dimension or by varieties in Algebraic Geometry defined over fields of positive characteristic. One such version is due to Hirzebruch (Hartshorne, *Algebraic Geometry*, p. 431), and an even more general version is due to Grothendieck, (Ibid., p. 436).

7

Tor and Ext

7.1 Tor

We now examine Tor more closely. As we said in the last chapter, all properties of $\text{Tor}_n^R(A, \Box)$ must follow from Theorem 6.33, the axioms characterizing it. In particular, its construction via derived functors need not be used (now that existence of such functors has been proved). However, it is possible that a proof of some property of Tor using derived functors may be simpler than a proof from the axioms. For example, it was very easy to prove Proposition 6.18: $\text{Tor}_n^R(A, \Box)$ preserves multiplications, but it is not obvious how to give a new proof of this fact from the axioms.

Theorem 7.1.

(i) *If R is a ring, A is a right R-module, and B is a left R-module, then*

$$\text{Tor}_n^R(A, B) \cong \text{Tor}_n^{R^{\text{op}}}(B, A)$$

for all $n \geq 0$, where R^{op} is the opposite ring of R.

(ii) *If R is a commutative ring and A and B are R-modules, then for all $n \geq 0$,*

$$\text{Tor}_n^R(A, B) \cong \text{Tor}_n^R(B, A).$$

J.J. Rotman, *An Introduction to Homological Algebra*, Universitext,
DOI 10.1007/978-0-387-68324-9_7, © Springer Science+Business Media LLC 2009

Proof.

(i) Recall Exercise 1.11 on page 35: every left R-module is a right R^{op}-module, and every right R-module is a left R^{op}-module. Choose a deleted projective resolution $\mathbf{P}_A$ of A. Now $t \colon \mathbf{P}_A \otimes_R B \to B \otimes_{R^{op}} \mathbf{P}_A$ is a chain map of $\mathbb{Z}$-complexes, by Proposition 2.56, where

$$t_n \colon P_n \otimes_R B \to B \otimes_{R^{op}} P_n$$

is given by $t_n \colon x_n \otimes b \mapsto b \otimes x_n$. Since each t_n is an isomorphism of abelian groups (its inverse is $b \otimes x_n \mapsto x_n \otimes b$), the chain map t is an isomorphism of complexes. Since isomorphic complexes have the same homology (because each H_n is a functor),

$$\operatorname{Tor}_n^R(A, B) = H_n(\mathbf{P}_A \otimes_R B) \cong H_n(B \otimes_{R^{op}} \mathbf{P}_A)$$

for all $n \geq 0$. But $\mathbf{P}_A$, viewed as a complex of left R^{op}-modules, is a deleted projective resolution of A qua left R^{op}-module, and so $H_n(B \otimes_{R^{op}} \mathbf{P}_A) \cong \operatorname{Tor}_n^{R^{op}}(B, A)$.

(ii) This is obvious from part (i). •

In light of this result, theorems about $\operatorname{Tor}_n(A, \square)$ will yield results about $\operatorname{Tor}_n(\square, B)$; we will not have to say "similarly in the other variable."

We know that Tor_n vanishes on projectives for all $n \geq 1$; we now show that they vanish on flat modules.

Theorem 7.2. *If a right R-module F is flat, then $\operatorname{Tor}_n^R(F, M) = \{0\}$ for all $n \geq 1$ and every left R-module M. Conversely, if $\operatorname{Tor}_1^R(F, M) = \{0\}$ for every left R-module M, then F is flat.*

Proof. Let $\mathbf{P}$ be a projective resolution of M. Since F is flat, the functor $F \otimes_R \square$ is exact, and so the complex

$$F \otimes_R \mathbf{P}_M = \to F \otimes_R P_2 \to F \otimes_R P_1 \to F \otimes_R P_0 \to 0$$

is exact for all $n \geq 1$. Therefore, $\operatorname{Tor}_n(F, M) = \{0\}$ for all $n \geq 1$.

For the converse, $0 \to A \xrightarrow{i} B$ exact implies exactness of

$$0 = \operatorname{Tor}_1^R(F, B/A) \to F \otimes A \xrightarrow{1 \otimes i} F \otimes B.$$

Hence, $1 \otimes i$ is an injection, and so F is flat. •

Here is another proof of Proposition 3.67. Recall that an exact sequence $0 \to B' \to B \to B'' \to 0$ of left R-modules is *pure exact* if the sequence of abelian groups $0 \to A \otimes_R B' \to A \otimes_R B \to A \otimes_R B'' \to 0$ is exact for every right R-module A.

Corollary 7.3.   *A left R-module B'' is flat if and only if every exact sequence $0 \to B' \to B \to B'' \to 0$ of left R-modules is pure exact.*

Proof.   There is an exact sequence

$$\operatorname{Tor}_1^R(A, B'') \to A \otimes_R B' \to A \otimes_R B \to A \otimes_R B'' \to 0.$$

Since B'' is flat, $\operatorname{Tor}_1^R(A, B'') = \{0\}$, by Theorem 7.2, and so the sequence of Bs is pure exact.

Conversely, choose an exact sequence $0 \to B' \xrightarrow{i} B \to B'' \to 0$ with B free. For every right R-module A, there is an exact sequence

$$\operatorname{Tor}_1^R(A, B) \to \operatorname{Tor}_1^R(A, B'') \to A \otimes_R B' \xrightarrow{1 \otimes i} A \otimes_R B.$$

But $\operatorname{Tor}_1^R(A, B'') = \ker 1 \otimes i$, for B free implies $\operatorname{Tor}_1^R(A, B) = \{0\}$. By purity, $1 \otimes i$ is an injection; hence, $\operatorname{Tor}_1^R(A, B'') = \{0\}$ for all A, and B'' is flat.   •

The next corollary generalizes Exercise 3.32 on page 151.

Corollary 7.4.   *Let $0 \to A \to B \to C \to 0$ be an exact sequence of right R-modules for some ring R. If C is flat, then A is flat if and only if B is flat.*

Proof.   For any left R-module X, there is an exact sequence

$$\operatorname{Tor}_2^R(C, X) \to \operatorname{Tor}_1^R(A, X) \to \operatorname{Tor}_1^R(B, X) \to \operatorname{Tor}_1^R(C, X).$$

Since C is flat, the flanking terms are $\{0\}$, so that $\operatorname{Tor}_1^R(A, X) \cong \operatorname{Tor}_1^R(B, X)$. Therefore, if one of these terms is $\{0\}$, i.e., if one of them is flat, then so is the other.   •

Note that A, B flat does not imply that C is flat: For example, $0 \to \mathbb{Z} \to \mathbb{Z} \to \mathbb{I}_2 \to 0$ is an exact sequence of abelian groups, but $\mathbb{I}_2$ is not flat.

We are going to use a very general fact in the middle of the next proof. If N is a submodule of a module M and there is a commutative diagram

$$
\begin{array}{ccc}
M & \xrightarrow{\ f\ } & X \\
{\scriptstyle g}\downarrow & \nearrow_{h} & \\
M/N, & &
\end{array}
$$

then $(\ker f)/N = \ker h$ [since $hg(m) = f(m)$ for all $m \in M$, we have $m \in \ker f$ if and only if $g(m) \in \ker h$].

Theorem 7.5.   *The functors $\operatorname{Tor}_n^R(A, \square)$ and $\operatorname{Tor}_n^R(\square, B)$ can be computed using flat resolutions of either variable; more precisely, for all flat resolutions $\mathbf{F}$ and $\mathbf{G}$ of A and B, respectively, and for all $n \geq 0$,*

$$H_n(\mathbf{F}_A \otimes_R B) \cong \operatorname{Tor}_n^R(A, B) \cong H_n(A \otimes_R \mathbf{G}_B).$$

Remark. There is a much simpler proof of this using spectral sequences; see Corollary 10.23. ◄

Proof. It suffices to prove that $H_n(\mathbf{F}_A \otimes_R B) \cong \operatorname{Tor}_n^R(A, B)$; the other isomorphism follows by replacing R by R^{op}; that is, by Theorem 7.1.

The proof is by dimension shifting; that is, by a slow starting induction on $n \geq 0$. If

$$\to F_2 \xrightarrow{d_2} \to F_1 \xrightarrow{d_1} F_0 \to A \to 0$$

is a flat resolution, then $F_1 \otimes_R B \xrightarrow{d_1 \otimes 1} F_0 \otimes_R B \to A \otimes_R B \to 0$ is exact, and so

$$H_0(\mathbf{F}_A \otimes_R B) = \operatorname{coker}(d_1 \otimes 1) \cong A \otimes_R B \cong \operatorname{Tor}_0^R(A, B).$$

For $n = 1$, there is a commutative diagram

$$
\begin{array}{ccccc}
F_2 & \xrightarrow{d_2} & F_1 & \xrightarrow{\quad d_1 \quad} & F_0 \\
& & \searrow{\scriptstyle d_1'} & \nearrow{\scriptstyle i} & \\
& & & Y, &
\end{array}
$$

where $Y = \ker d_1$, $i: Y \to F_0$ is the inclusion, and d_1' and d_1 differ only in their target. Applying $\square \otimes_R B$ gives a commutative diagram

$$
\begin{array}{ccc}
F_1 \otimes_R B & \xrightarrow{d_1 \otimes 1} & F_0 \otimes_R B \\
\searrow{\scriptstyle d_1' \otimes 1} & & \nearrow{\scriptstyle i \otimes 1} \\
& Y. &
\end{array}
$$

Now

$$\operatorname{im}(d_1 \otimes 1) = \operatorname{im}(i \otimes 1), \tag{1}$$

because right exactness of $\square \otimes_R B$ gives $d_1' \otimes 1$ surjective. Next, consider

$$
\begin{array}{ccccc}
F_2 \otimes_R B & \xrightarrow{d_2 \otimes 1} & F_1 \otimes_R B & \xrightarrow{d_1 \otimes 1} & F_0 \otimes_R B \\
& & \downarrow{\scriptstyle \alpha} & {\scriptstyle \delta} \dashrightarrow \nearrow & \\
& & F_1 \otimes_R B / \operatorname{im}(d_2 \otimes 1) & & \\
& & \downarrow{\scriptstyle \beta} & \searrow{\scriptstyle \gamma} & \\
& & F_1 \otimes_R B / \ker(d_1 \otimes 1), & &
\end{array}
$$

where α is the natural map. Since $\operatorname{im}(d_2 \otimes 1) \subseteq \ker(d_1 \otimes 1)$, the enlargement of coset map β is surjective, while $\gamma: F_1 \otimes_R B / \ker(d_1 \otimes 1)$ is the injection of the First Isomorphism Theorem [whose image is $\mathrm{im}(d_1 \otimes 1)$]. If we define $\delta = \gamma\beta$, the general fact mentioned before the theorem says that

$$\ker\delta = (\ker(d_1 \otimes 1)/\mathrm{im}(d_2 \otimes 1) = H_1(\mathbf{F}_A \otimes_R B).$$

On the other hand, $\mathrm{im}\,\delta = \mathrm{im}\,\gamma$ (because $\delta = \gamma\beta$ and β is surjective). But $d_1 \otimes 1 = \gamma\beta\alpha$, so that $\mathrm{im}(d_1 \otimes 1) = \mathrm{im}\,\gamma$ (both β and α are surjective). Therefore, $H_1(\mathbf{F}_A \otimes_R B) \cong \mathrm{im}\,\gamma = \mathrm{im}(d_1 \otimes 1)$; but $\mathrm{im}(d_1 \otimes 1) = \mathrm{im}(i \otimes 1)$, by Eq. (1), so that $H_1(\mathbf{F}_A \otimes_R B) \cong \mathrm{im}(i \otimes 1)$.

Consider the fragment of the long exact sequence for Tor:

$$\mathrm{Tor}_1^R(F_0, B) \to \mathrm{Tor}_1^R(A, B) \to Y \otimes_R B \xrightarrow{i \otimes 1} F_0 \otimes_R B.$$

Now $\mathrm{Tor}_1^R(F_0, B) = \{0\}$, because F_0 is flat (Theorem 7.2), so that

$$\mathrm{Tor}_1^R(A, B) \cong \ker(i \otimes 1).$$

Hence, $\mathrm{Tor}_1(A, B) \cong \ker(i \otimes 1) \cong H_1(\mathbf{F}_A \otimes B)$.

For the inductive step $n \geq 1$, there is an exact sequence

$$\mathrm{Tor}_{n+1}(F_0, B) \to \mathrm{Tor}_{n+1}(A, B) \to \mathrm{Tor}_n(Y, B) \to \mathrm{Tor}_n(F_0, B).$$

Since F_0 is flat, the two ends are $\{0\}$, and $\mathrm{Tor}_{n+1}(A, B) \cong \mathrm{Tor}_n(Y, B)$. Now $\mathbf{F}' = \to F_2 \to F_1 \to Y \to 0$ is a flat resolution of Y, and so $H_n(\mathbf{F}'_Y \otimes B) \cong \mathrm{Tor}_n(Y, B)$, by the inductive hypothesis. But $H_n(\mathbf{F}'_Y \otimes B) = H_{n+1}(\mathbf{F}_A \otimes B)$: in the notation of $\mathbf{F}$, both are $(\ker d_{n+1} \otimes 1)/(\mathrm{im}\, d_{n+2} \otimes 1)$. $\quad\bullet$

Proposition 7.6. *If $(B_k)_{k \in K}$ is a family of left R-modules, then there are natural isomorphisms, for all $n \geq 0$,*

$$\mathrm{Tor}_n^R\left(A, \bigoplus_{k \in K} B_k\right) \cong \bigoplus_{k \in K} \mathrm{Tor}_n^R(A, B_k).$$

There is also an isomorphism if the direct sum is in the first variable.

Proof. The proof is by dimension shifting. The base step is Theorem 2.65, for $\mathrm{Tor}_0^R(A, \square)$ is naturally equivalent to $A \otimes \square$.

For the inductive step, choose, for each $k \in K$, a short exact sequence

$$0 \to N_k \to P_k \to B_k \to 0,$$

where P_k is projective. There is an exact sequence

$$0 \to \bigoplus_k N_k \to \bigoplus_k P_k \to \bigoplus_k B_k \to 0,$$

and $\bigoplus_k P_k$ is projective, for every direct sum of projectives is projective. There is a commutative diagram with exact rows:

$$\text{Tor}_1(A, \bigoplus_k P_k) \to \text{Tor}_1(A, \bigoplus_k B_k) \overset{\partial}{\to} A \otimes \bigoplus_k N_k \to A \otimes \bigoplus_k P_k$$

$$\bigoplus_k \text{Tor}_1(A, P_k) \to \bigoplus_k \text{Tor}_1(A, B_k) \underset{\partial'}{\to} \bigoplus_k A \otimes N_k \to \bigoplus_k A \otimes P_k,$$

where the maps in the bottom row are just the usual induced maps in each coordinate, and the maps τ and σ are the isomorphisms given by Theorem 2.65. The proof is completed by dimension shifting. •

Example 7.7.

(i) We show, for every abelian group B, that

$$\text{Tor}_1^{\mathbb{Z}}(\mathbb{I}_n, B) \cong B[n] = \{b \in B : nb = 0\}.$$

There is an exact sequence

$$0 \to \mathbb{Z} \overset{\mu_n}{\to} \mathbb{Z} \to \mathbb{I}_n \to 0,$$

where μ_n is multiplication by n. Applying $\square \otimes B$ gives exactness of

$$\text{Tor}_1(\mathbb{Z}, B) \to \text{Tor}_1(\mathbb{I}_n, B) \to \mathbb{Z} \otimes B \overset{1 \otimes \mu_n}{\to} \mathbb{Z} \otimes B.$$

Now $\text{Tor}_1(\mathbb{Z}, B) = \{0\}$, because $\mathbb{Z}$ is projective. Moreover, $1 \otimes \mu_n$ is also multiplication by n, while $\mathbb{Z} \otimes B = B$. In more detail, $\mathbb{Z} \otimes \square$ is naturally isomorphic to the identity functor on **Ab**, and so there is a commutative diagram with exact rows:

$$
\begin{array}{ccccccc}
0 & \longrightarrow & B[n] & \longrightarrow & B & \overset{\mu_n}{\longrightarrow} & B \\
 & & \downarrow & & \downarrow{\tau_B} & & \downarrow{\tau_B} \\
0 & \longrightarrow & \text{Tor}_1(\mathbb{I}_n, B) & \longrightarrow & \mathbb{Z} \otimes B & \underset{1 \otimes \mu_n}{\longrightarrow} & \mathbb{Z} \otimes B.
\end{array}
$$

By Proposition 2.71, there is an isomorphism $B[n] \cong \text{Tor}_1(\mathbb{I}_n, B)$.

(ii) We can now compute $\text{Tor}_1^{\mathbb{Z}}(A, B)$ whenever A and B are finitely generated abelian groups. By the fundamental theorem, both A and B are direct sums of cyclic groups. Since Tor commutes with direct sums, $\text{Tor}_1^{\mathbb{Z}}(A, B)$ is the direct sum of groups $\text{Tor}_1^{\mathbb{Z}}(C, D)$, where C and D are cyclic. We may assume that C and D are finite; otherwise, they are projective and $\text{Tor}_1 = \{0\}$. This calculation can be completed using part (i) and the fact that if D is a cyclic group of finite order m, then $D[n]$ is a cyclic group of order d, where $d = (m, n)$ is their gcd. ◄

Proposition 7.8. *If $\{B_i, \varphi_j^i\}$ is a direct system of left R-modules over a directed index set I, then for all right R-modules A and all $n \geq 0$, there is an isomorphism*

$$\operatorname{Tor}_n^R\big(A, \varinjlim B_i\big) \cong \varinjlim \operatorname{Tor}_n^R(A, B_i).$$

Proof. The proof is by dimension shifting. The base step is Theorem 5.27, for $\operatorname{Tor}_0(A, \square)$ is naturally isomorphic to $A \otimes \square$.

For the inductive step, choose, for each $i \in I$, a short exact sequence

$$0 \to N_i \to P_i \to B_i \to 0,$$

where P_i is projective. Since the index set is directed, Proposition 5.33 says that there is an exact sequence

$$0 \to \varinjlim N_i \to \varinjlim P_i \to \varinjlim B_i \to 0.$$

Now $\varinjlim P_i$ is flat, for every projective module is flat, and a direct limit of flat modules is flat, by Corollary 5.34. There is a commutative diagram with exact rows:

$$\operatorname{Tor}_1(A, \varinjlim P_i) \to \operatorname{Tor}_1(A, \varinjlim B_i) \overset{\partial}{\to} A \otimes \varinjlim N_i \to A \otimes \varinjlim P_i$$

$$\varinjlim \operatorname{Tor}_1(A, P_i) \to \varinjlim \operatorname{Tor}_1(A, B_i) \underset{\tilde\partial}{\to} \varinjlim A \otimes N_i \to \varinjlim A \otimes P_i,$$

where the maps in the bottom row are just the usual induced maps between direct limits, and the maps τ and σ are the isomorphisms given by Theorem 5.27. The step $n \geq 2$ is routine.   •

We can now augment Theorem 3.66.

Theorem 7.9 (Chase). *The following are equivalent for a ring R.*

(i) *Every direct product of flat right R-modules is flat.*

(ii) *For every set X, the right R-module R^X is flat.*

(iii) *Every finitely generated submodule of a free left R-module is finitely presented.*

(iv) *R is left coherent.*

Proof. The equivalence of the last three statements was proved in Theorem 3.66. It is obvious that (i) $\Rightarrow$ (ii), for R viewed as a right module over itself is flat. We complete the proof by showing that (iii) $\Rightarrow$ (i).

Let $(A_i)_{i \in I}$ be a family of flat right R-modules, and write $A = \prod_{i \in I} A_i$. Define the functor $G \colon {}_R\mathbf{Mod} \to \mathbf{Ab}$ on objects by $G(C) = \prod_i (A_i \otimes_R C)$ and on maps $f \colon C \to C'$ by $Gf \colon (a_i \otimes c) \mapsto (a_i \otimes fc)$. It is easy to see that G is an additive functor that is exact (because all the A_i are flat). Define a natural transformation $\tau \colon A \otimes_R \square \to G$, where $\tau_C \colon (\prod_i A_i) \otimes_R C \to \prod_i (A_i \otimes_R C)$ is given by $(a_i) \otimes c \mapsto (a_i \otimes c)$.

Let C be a finitely generated left R-module, and let $0 \to K \to F \to C \to 0$ be an exact sequence, where F is a finitely generated free module. There is a commutative diagram with exact rows:

$$
\begin{array}{ccccccc}
A \otimes_R K & \longrightarrow & A \otimes_R F & \longrightarrow & A \otimes_R C & \longrightarrow & 0. \\
{\scriptstyle \tau_K}\downarrow & & \downarrow{\scriptstyle \tau_F} & & \downarrow{\scriptstyle \tau_C} & & \downarrow \\
0 \longrightarrow & GK & \longrightarrow & GF & \longrightarrow & GC & \longrightarrow & 0.
\end{array}
$$

Since $GR \cong C$ and $F \cong R^n$, the additivity of G shows that τ_F is an isomorphism. Hence, if C is finitely presented; that is, if K is finitely generated, then τ_K is also a surjection. A diagram chase shows that τ_C is a surjection, and so the Five Lemma shows that τ_C is an isomorphism. Now K is a finitely generated submodule of the free module F, so that K is finitely presented, by hypothesis; therefore, τ_K is an isomorphism. It follows that the arrow $A \otimes_R K \to A \otimes_R F$ in the top row is an injection. But exactness of $0 = \mathrm{Tor}_1(A, F) \to \mathrm{Tor}_1(A, C) \to A \otimes_R K \to A \otimes_R F$ allows us to conclude that $\mathrm{Tor}_1(A, C) = \{0\}$ whenever C is finitely presented.

Consider now any finitely generated left R-module C, and let $0 \to K \to F \to C \to 0$ be an exact sequence with F finitely generated free. The family $(K_i)_{i \in I}$ of all the finitely generated submodules of K forms a direct system with $\varinjlim K_i = K$, by Example 5.32(iii), and Exercise 5.21 says that $C \cong \varinjlim (F/K_i)$. Now F/K_i is finitely presented for all i, so that $\mathrm{Tor}_1(A, F/K_i) = \{0\}$ for all i. Proposition 7.8 gives $\mathrm{Tor}_1(A, C) \cong \mathrm{Tor}_1(A, \varinjlim(F/K_i)) \cong \varinjlim \mathrm{Tor}_1(A, F/K_i) = \{0\}$. Thus, $\mathrm{Tor}_1(A, C) = \{0\}$ for every finitely generated left R-module C. Therefore, $\mathrm{Tor}_1(A, B) = \{0\}$ for every left R-module B, by Proposition 3.48, and so Theorem 7.2 says that A is flat. $\quad \bullet$

Corollary 7.10. *If R is left noetherian, then every direct product of flat right R-modules is flat.*

Proof. Every left noetherian ring is left coherent. $\quad \bullet$

There are other constructions of Tor. For example, $\mathrm{Tor}_1^{\mathbb{Z}}(A, B)$ can be defined by generators and relations. Consider all triples (a, n, b), where $a \in A$, $b \in B$, $na = 0$, and $nb = 0$; then $\mathrm{Tor}_1^{\mathbb{Z}}(A, B)$ is generated by all such triples subject to the relations (whenever both sides are defined)

$$(a + a', n, b) = (a, n, b) + (a', n, b),$$
$$(a, n, b + b') = (a, n, b) + (a, n, b'),$$
$$(ma, n, b) = (a, mn, b) = (a, m, nb).$$

For a proof of this result, and its generalization to $\operatorname{Tor}_n^R(A, B)$ for arbitrary rings R, see Mac Lane, *Homology*, pp. 150–159 and Mac Lane, "Slide and torsion products for modules," *Rendiconti del Sem. Mat.* 15 (1955), 281–309.

7.1.1 Domains

We are now going to assume that R is a domain, so that the notion of torsion submodule is defined.

Notation.   Denote $\operatorname{Frac}(R)$ by Q and Q/R by K.

Lemma 7.11.   *Let R be a domain.*

(i) *If A is a torsion R-module, then $\operatorname{Tor}_1^R(K, A) \cong A$.*

(ii) *For every R-module A, we have $\operatorname{Tor}_n^R(K, A) = \{0\}$ for all $n \geq 2$.*

(iii) *If A is a torsion-free R-module, then $\operatorname{Tor}_1^R(K, A) = \{0\}$.*

Proof.

(i) Exactness of $0 \to R \to Q \to K \to 0$ gives exactness of

$$\operatorname{Tor}_1(Q, A) \to \operatorname{Tor}_1(K, A) \to R \otimes A \to Q \otimes A.$$

Now Q is flat, by Corollary 5.35, and so $\operatorname{Tor}_1(Q, A) = \{0\}$, by Theorem 7.2. The last term $Q \otimes A = \{0\}$ because Q is divisible and A is torsion (Proposition 2.73), and so the middle map $\operatorname{Tor}_1(K, A) \to R \otimes A$ is an isomorphism.

(ii) The sequence $\operatorname{Tor}_n(Q, A) \to \operatorname{Tor}_n(K, A) \to \operatorname{Tor}_{n-1}(R, A)$ is exact. Since $n \geq 2$, we have $n - 1 \geq 1$, and so both the first and third Tors are $\{0\}$, because Q and R are flat. Thus, exactness gives $\operatorname{Tor}_n(K, A) = \{0\}$.

(iii) By Lemma 4.33(ii), there is an exact sequence $0 \to A \to V \to T \to 0$, where V is a vector space over Q and T is torsion. Since every vector space has a basis, V is a direct sum of copies of Q. Corollary 5.35 says that Q is flat, and Lemma 3.46 says that a direct sum of flat modules is flat. We conclude that V is flat. Exactness of $0 \to A \to V \to V/A \to 0$ gives exactness of $\operatorname{Tor}_2(K, V/A) \to \operatorname{Tor}_1(K, A) \to \operatorname{Tor}_1(K, V)$. Now $\operatorname{Tor}_2(K, V/A) = \{0\}$, by (ii), and $\operatorname{Tor}_1(K, V) = \{0\}$, for V is flat, and so $\operatorname{Tor}_1(K, A) = \{0\}$.   •

The next result shows why Tor is so called.

Theorem 7.12. $\mathrm{Tor}_1^R(K, A) \cong tA$ *for all R-modules A. In fact, the functor* $\mathrm{Tor}_1^R(K, \square)$ *is naturally isomorphic to the torsion functor.*

Proof. Consider the exact sequence

$$\mathrm{Tor}_2(K, A/tA) \to \mathrm{Tor}_1(K, tA) \xrightarrow{\iota_{A*}} \mathrm{Tor}_1(K, A) \to \mathrm{Tor}_1(K, A/tA),$$

where ι_{A*} is the map induced by the inclusion $i_A : tA \to A$. The first term is $\{0\}$, by Lemma 4.33(ii), and the last term is $\{0\}$, by Lemma 7.11(iii). Therefore, the map $\iota_{A*} : \mathrm{Tor}_1(K, tA) \to \mathrm{Tor}_1(K, A)$ is an isomorphism.

Let $f : A \to B$ and let $f' : tA \to tB$ be its restriction. The following diagram commutes, because $\mathrm{Tor}_1(K, \square)$ is a functor, which says that the isomorphisms ι_{A*} constitute a natural transformation.

$$\begin{array}{ccc}
\mathrm{Tor}_1(K, tA) & \xrightarrow{\iota_{A*}} & \mathrm{Tor}_1(K, A) \\
f'_* \downarrow & & \downarrow f_* \\
\mathrm{Tor}_1(K, tB) & \xrightarrow[\iota_{B*}]{} & \mathrm{Tor}_1(K, B)
\end{array}$$

Proof of naturality is left to the reader. •

Corollary 7.13. *Let R be a domain.*

(i) *For every R-module A, there is an exact sequence*

$$0 \to tA \to A \to Q \otimes_R A \to K \otimes_R A \to 0.$$

(ii) *An R-module A is torsion if and only if* $Q \otimes_R A = \{0\}$.

Proof.

(i) In the exact sequence

$$\mathrm{Tor}_1(Q, A) \to \mathrm{Tor}_1(K, A) \to R \otimes A \to Q \otimes A \to K \otimes A \to 0,$$

Q flat gives $\mathrm{Tor}_1(Q, A) = \{0\}$, Lemma 4.33(ii) gives $\mathrm{Tor}_1(K, A) \cong tA$, and $R \otimes_R A \cong A$.

(ii) Necessity is Exercise 3.33 on page 152; sufficiency follows at once from part (i). •

Another reason for the name Tor is that Tor_n^R is a torsion R-module for all $n \geq 1$.

Lemma 7.14. *If R is a domain and B is a torsion R-module, then $\operatorname{Tor}_n^R(A, B)$ is torsion for all A and for all $n \geq 0$.*

Proof. The proof is by dimension shifting. If $n = 0$, then each generator $a \otimes b$ is torsion, and so $\operatorname{Tor}_0^R(A, B)$ is torsion.

If $n = 1$, there is an exact sequence $0 \to N \to P \to A \to 0$ with P projective, and this gives exactness of

$$0 = \operatorname{Tor}_1^R(P, B) \to \operatorname{Tor}_1^R(A, B) \to N \otimes_R B.$$

Since $N \otimes_R B$ is torsion, so is its submodule $\operatorname{Tor}_1^R(A, B)$.

For the inductive step, look further out in the long exact sequence. There is exactness

$$0 = \operatorname{Tor}_{n+1}^R(P, B) \to \operatorname{Tor}_{n+1}^R(A, B) \to \operatorname{Tor}_n^R(N, B) \to \operatorname{Tor}_n^R(P, B) = 0.$$

But $\operatorname{Tor}_n^R(N, B)$ is torsion, by induction, and so $\operatorname{Tor}_{n+1}^R(A, B) \cong \operatorname{Tor}_n^R(N, B)$ is torsion. •

Theorem 7.15. *If R is a domain, then $\operatorname{Tor}_n^R(A, B)$ is a torsion module for all A, B and all $n \geq 1$.*

Proof. Let $n = 1$, and consider the special case when B is torsion-free. By Lemma 4.33(i), there is an exact sequence

$$0 \to B \to V \to T \to 0,$$

where V is a vector space over Q and $T = V/B$ is torsion. This gives an exact sequence

$$\operatorname{Tor}_2^R(A, T) \to \operatorname{Tor}_1^R(A, B) \to \operatorname{Tor}_1^R(A, V).$$

Now $\operatorname{Tor}_2^R(A, T)$ is torsion, by Lemma 7.14, while $\operatorname{Tor}_1^R(A, V) = \{0\}$, because V is flat (V is a direct sum of copies of Q). Thus, $\operatorname{Tor}_1^R(A, B)$ is a quotient of a torsion module, and hence it is torsion.

Now let B be arbitrary. Exactness of $0 \to tB \to B \to B/tB \to 0$ gives exactness of

$$\operatorname{Tor}_1^R(A, tB) \to \operatorname{Tor}_1^R(A, B) \to \operatorname{Tor}_1^R(A, B/tB).$$

The flanking terms are torsion, for tB is torsion and B/tB is torsion-free. Thus, $\operatorname{Tor}_1^R(A, B)$ is torsion, being an extension of one torsion module by another. The proof is completed by dimension shifting. •

7.1.2 Localization

We are now going to see that Tor gets along well with localization.

Proposition 7.16. *Let R and A be rings, and let $T: {}_R\text{Mod} \to {}_A\text{Mod}$ be an exact additive functor. Then T commutes with homology; that is, for every complex $(\mathbf{C}, d) \in {}_R\text{Comp}$ and for every $n \in \mathbb{Z}$, there is an isomorphism*

$$H_n(T\mathbf{C}, T d) \cong T H_n(\mathbf{C}, d).$$

Proof. Consider the commutative diagram with exact bottom row,

$$
\begin{array}{ccccc}
C_{n+1} & \xrightarrow{\ d_{n+1}\ } & C_n & \xrightarrow{\ d_n\ } & C_{n-1} \\
\scriptstyle d'_{n+1}\downarrow & & \uparrow\scriptstyle k & & \\
0 \longrightarrow \operatorname{im} d_{n+1} & \xrightarrow[\ j\ ]{} & \ker d_n & \longrightarrow & H_n(\mathbf{C}) \longrightarrow 0,
\end{array}
$$

where j, and k are inclusions and d'_{n+1} is just d_{n+1} with its target changed from C_n to $\operatorname{im} d_{n+1}$. Applying the exact functor T gives the commutative diagram with exact bottom row

$$
\begin{array}{ccccc}
TC_{n+1} & \xrightarrow{\ T d_{n+1}\ } & TC_n & \xrightarrow{\ T d_n\ } & TC_{n-1} \\
\scriptstyle T d'_{n+1}\downarrow & & \uparrow\scriptstyle Tk & & \\
0 \longrightarrow T(\operatorname{im} d_{n+1}) & \xrightarrow[\ Tj\ ]{} & T(\ker d_n) & \longrightarrow & T H_n(\mathbf{C}) \longrightarrow 0.
\end{array}
$$

On the other hand, because T is exact, we have $T(\operatorname{im} d_{n+1}) = \operatorname{im} T(d_{n+1})$ and $T(\ker d_n) = \ker(T d_n)$, so that the bottom row is

$$0 \to \operatorname{im}(T d_{n+1}) \to \ker(T d_n) \to T H_n(\mathbf{C}) \to 0.$$

By definition, $\ker(T d_n)/\operatorname{im}(T d_{n+1}) = H_n(T\mathbf{C})$, and a diagram chase, Proposition 2.70, gives $H_n(T\mathbf{C}) \cong T H_n(\mathbf{C})$. •

Localization commutes with Tor, essentially because $S^{-1}R$ is a flat R-module.

Proposition 7.17. *If S is a multiplicative subset of a commutative ring R, then for all $n \geq 0$ and all R-modules A and B, there are isomorphisms, natural in A and B,*

$$S^{-1} \operatorname{Tor}_n^R(A, B) \cong \operatorname{Tor}_n^{S^{-1}R}(S^{-1}A, S^{-1}B).$$

Proof. First consider the case $n = 0$. For a fixed R-module A, there is a natural isomorphism

$$\tau_{A,B}: S^{-1}(A \otimes_R B) \to S^{-1}A \otimes_{S^{-1}R} S^{-1}B,$$

for either is a universal solution U of the universal mapping problem

$$S^{-1}A \times S^{-1}B \longrightarrow U$$
$$f \searrow \quad \nearrow \tilde{f}$$
$$M,$$

where M is an $S^{-1}R$-module, f is $S^{-1}R$-bilinear, and $\tilde{f}$ is an $S^{-1}R$-map.

If $\mathbf{P}_B$ is a deleted projective resolution of B, then $S^{-1}(\mathbf{P}_B)$ is a deleted projective resolution of $S^{-1}B$, for localization is an exact functor that preserves projectives. Naturality of the isomorphisms $\tau_{A,B}$ gives an isomorphism of complexes

$$S^{-1}(A \otimes_R \mathbf{P}_B) \cong S^{-1}A \otimes_{S^{-1}R} S^{-1}(\mathbf{P}_B),$$

so that their homology groups are isomorphic. Since localization is an exact additive functor, Proposition 7.16 applies: for all $n \geq 0$,

$$H_n(S^{-1}(A \otimes_R \mathbf{P}_B)) \cong S^{-1}H_n(A \otimes_R \mathbf{P}_B) \cong S^{-1}\mathrm{Tor}_n^R(A, B).$$

On the other hand, since $S^{-1}(\mathbf{P}_B)$ is a deleted projective resolution of $S^{-1}B$, the definition of Tor gives

$$H_n(S^{-1}A \otimes_{S^{-1}R} S^{-1}(\mathbf{P}_B)) \cong \mathrm{Tor}_n^{S^{-1}R}(S^{-1}A, S^{-1}B).$$

Proof of naturality is left for the reader.   •

Corollary 7.18. *Let A be an R-module over a commutative ring R. If $A_{\mathfrak{m}}$ is a flat $R_{\mathfrak{m}}$-module for every maximal ideal $\mathfrak{m}$, then A is a flat R-module.*

Proof. Since $A_{\mathfrak{m}}$ is flat, Proposition 7.2 gives $\mathrm{Tor}_n^{R_{\mathfrak{m}}}(A_{\mathfrak{m}}, B_{\mathfrak{m}}) = \{0\}$ for all $n \geq 1$, for every R-module B, and for every maximal ideal $\mathfrak{m}$. But Proposition 7.17 gives $\mathrm{Tor}_n^R(A, B)_{\mathfrak{m}} = \{0\}$ for all maximal ideals $\mathfrak{m}$ and all $n \geq 1$. Finally, Proposition 4.90 shows that $\mathrm{Tor}_n^R(A, B) = \{0\}$ for all $n \geq 1$. Since this is true for all R-modules B, we have A flat.   •

Lemma 7.19. *If R is a left noetherian ring and A is a finitely generated left R-module, then there is a projective resolution $\mathbf{P}$ of A in which each P_n is finitely generated.*

Proof. Since A is finitely generated, there exist a finitely generated free left R-module P_0 and a surjective R-map $\varepsilon : P_0 \to A$. Since R is left noetherian, $\ker \varepsilon$ is finitely generated, and so there exist a finitely generated free left R-module P_1 and a surjective R-map $d_1 : P_1 \to \ker \varepsilon$. Define $D_1 : P_1 \to P_0$ as the composite id_1, where $i : \ker \varepsilon \to P_0$ is the inclusion; there is an exact sequence

$$0 \to \ker D_1 \to P_1 \xrightarrow{D_1} P_0 \xrightarrow{\varepsilon} A \to 0.$$

This construction can be iterated, for $\ker D_1$ is finitely generated, and the proof is completed by induction. (We remark that we have, in fact, constructed a *free* resolution of A, each of whose terms is finitely generated.) •

Theorem 7.20. *If R is a commutative noetherian ring, and if A and B are finitely generated R-modules, then $\operatorname{Tor}_n^R(A, B)$ is a finitely generated R-module for all $n \geq 0$.*

Remark. There is an analogous result for Ext (see Theorem 7.36). ◄

Proof. Note that Tor is an R-module because R is commutative. We prove that Tor_n is finitely generated by induction on $n \geq 0$. The base step holds, for $A \otimes_R B$ is finitely generated, by Exercise 3.13 on page 115(i). If $n \geq 0$, choose a projective resolution $\cdots \to P_1 \xrightarrow{d_1} P_0 \to A \to 0$ as in Lemma 7.19. Since $P_n \otimes_R B$ is finitely generated, so are $\ker(d_n \otimes 1_B)$ (by Proposition 3.18) and its quotient $\operatorname{Tor}_n^R(A, B)$. •

Exercises

*7.1 If R is right hereditary, prove that $\operatorname{Tor}_j^R(A, B) = \{0\}$ for all $j \geq 2$ and for all right R-modules A and B.
  Hint. Every submodule of a projective module is projective.

7.2 If $0 \to A \to B \to C \to 0$ is an exact sequence of right R-modules with both A and C flat, prove that B is flat.

*7.3 If F is flat and $\pi\colon P \to F$ is a surjection with P flat, prove that $\ker \pi$ is flat.

7.4 If A, B are finite abelian groups, prove that $\operatorname{Tor}_1^{\mathbb{Z}}(A, B) \cong A \otimes_{\mathbb{Z}} B$.

7.5 Let R be a domain with $\operatorname{Frac}(R) = Q$ and $K = Q/R$. Prove that the right derived functors of t (the torsion submodule functor) are

$$R^0 t = t, \quad R^1 t = K \otimes_R \square, \quad R^n t = 0 \quad \text{for all } n \geq 2.$$

7.6 Let k be a field, let $R = k[x, y]$, and let I be the ideal (x, y).

    (i) Prove that $x \otimes y - y \otimes x \in I \otimes_R I$ is nonzero.
        Hint. Consider $(I/I^2) \otimes (I/I^2)$.

    (ii) Prove that $x(x \otimes y - y \otimes x) = 0$, and conclude that $I \otimes_R I$ is not torsion-free.

7.7 Prove that the functor $T = \operatorname{Tor}_1^{\mathbb{Z}}(G, \square)$ is left exact for every abelian group G, and compute its right derived functors $L_n T$.

7.2   Ext

We now examine Ext more closely. As we said in the last chapter, all proper-
ties of $\operatorname{Ext}_R^n(A, \square)$ and $\operatorname{Ext}_R^n(\square, B)$ must follow from the axioms characterizing them, Theorems 6.48 and 6.64. In particular, their construction as derived
functors need not be used. As for Tor, it is possible that a proof of some
property of Ext using derived functors may be simpler than a proof from the
axioms. For example, it was very easy to prove Proposition 6.38: $\operatorname{Ext}_R^n(A, \square)$
preserves multiplications, but it is not obvious how to give a new proof of this
fact from the axioms.

We begin by showing that Ext behaves like Hom with respect to direct
sums and direct products.

Proposition 7.21.   *If* $(A_k)_{k \in K}$ *is a family of modules, then there are natural
isomorphisms, for all* $n \geq 0$,

$$\operatorname{Ext}_R^n\left(\bigoplus_{k \in K} A_k, B\right) \cong \prod_{k \in K} \operatorname{Ext}_R^n(A_k, B).$$

Proof.   The proof is by dimension shifting. The step $n = 0$ is Theorem 2.31,
because $\operatorname{Ext}_R^0(\square, B)$ is naturally isomorphic to $\operatorname{Hom}_R(\square, B)$.

For the step $n = 1$, choose, for each $k \in K$, a short exact sequence

$$0 \to L_k \to P_k \to A_k \to 0,$$

where P_k is projective. There is an exact sequence

$$0 \to \bigoplus_k L_k \to \bigoplus_k P_k \to \bigoplus_k A_k \to 0,$$

and $\bigoplus_k P_k$ is projective, for every direct sum of projectives is projective.
There is a commutative diagram with exact rows:

$$\operatorname{Hom}(\bigoplus P_k, B) \twoheadrightarrow \operatorname{Hom}(\bigoplus L_k, B) \xrightarrow{\partial} \operatorname{Ext}^1(\bigoplus A_k, B) \to \operatorname{Ext}^1(\bigoplus P_k, B)$$

$$\left\downarrow{\scriptstyle \tau} \qquad\qquad\qquad \left\downarrow{\scriptstyle \sigma} \qquad\qquad\qquad \downarrow \qquad\qquad\qquad \right. \right.$$

$$\prod \operatorname{Hom}(P_k, B) \twoheadrightarrow \prod \operatorname{Hom}(L_k, B) \xrightarrow[d]{} \prod \operatorname{Ext}^1(A_k, B) \twoheadrightarrow \prod \operatorname{Ext}^1(P_k, B),$$

where the maps in the bottom row are just the usual induced maps in each
coordinate, and the maps τ and σ are the natural isomorphisms given by The-
orem 2.31. Now $\operatorname{Ext}^1(\bigoplus P_k, B) = \{0\} = \prod \operatorname{Ext}^1(P_k, B)$, because $\bigoplus P_k$
and each P_k are projective; thus, the maps ∂ and d are surjective. This is
precisely the diagram in Proposition 2.70, and so there exists an isomorphism
$\operatorname{Ext}^1(\bigoplus A_k, B) \to \prod \operatorname{Ext}^1(A_k, B)$ making the augmented diagram commute.

We now prove the inductive step for $n \geq 1$. Look further out in the long exact sequence. There is a commutative diagram

$$\text{Ext}^n\left(\bigoplus P_k, B\right) \to \text{Ext}^n\left(\bigoplus L_k, B\right) \xrightarrow{\partial} \text{Ext}^{n+1}\left(\bigoplus A_k, B\right) \to \text{Ext}^{n+1}\left(\bigoplus P_k, B\right)$$

$$\prod \text{Ext}^n(P_k, B) \to \prod \text{Ext}^n(L_k, B) \xrightarrow{d} \prod \text{Ext}^{n+1}(A_k, B) \to \prod \text{Ext}^{n+1}(P_k, B),$$

where $\sigma \colon \text{Ext}^n(\bigoplus L_k, B) \to \prod \text{Ext}^n(L_k, B)$ is an isomorphism that exists by the inductive hypothesis. Since $n \geq 1$, all four Exts whose first variable is projective are $\{0\}$; it follows from exactness of the rows that both ∂ and d are isomorphisms. Finally, the composite $d\sigma\partial^{-1} \colon \text{Ext}^{n+1}(\bigoplus A_k, B) \to \prod \text{Ext}^{n+1}(A_k, B)$ is an isomorphism, as desired. $\bullet$

There is a dual result in the second variable.

Proposition 7.22. *If $(B_k)_{k \in K}$ is a family of modules, then there are natural isomorphisms, for all $n \geq 0$,*

$$\text{Ext}_R^n\left(A, \prod_{k \in K} B_k\right) \cong \prod_{k \in K} \text{Ext}_R^n(A, B_k).$$

Proof. The proof is by dimension shifting. The step $n = 0$ is Theorem 2.30, for $\text{Ext}^0(A, \square)$ is naturally isomorphic to the covariant functor $\text{Hom}(A, \square)$.

For the step $n = 1$, choose, for each $k \in K$, a short exact sequence

$$0 \to B_k \to E_k \to N_k \to 0,$$

where E_k is injective. There is an exact sequence $0 \to \prod_k B_k \to \prod_k E_k \to \prod_k N_k \to 0$, and $\prod_k E_k$ is injective, for every product of injectives is injective, by Proposition 3.28. The proof finishes as that of Proposition 7.21.
$\bullet$

It follows that Ext^n commutes with finite direct sums in either variable; of course, this also follows from Ext^n being an additive functor in either variable.

Remark. These last two proofs cannot be generalized by replacing direct sums by direct limits or direct products by inverse limits; the reason is that direct limits of projectives need not be projective and inverse limits of injectives need not be injective. $\blacktriangleleft$

Example 7.23.

(i) We show, for every abelian group B, that if $m \geq 2$, then

$$\text{Ext}^1_{\mathbb{Z}}(\mathbb{I}_m, B) \cong B/mB.$$

There is an exact sequence

$$0 \to \mathbb{Z} \xrightarrow{\mu_m} \mathbb{Z} \to \mathbb{I}_m \to 0,$$

where μ_m is multiplication by m. Applying $\text{Hom}(\square, B)$ gives exactness of $\text{Hom}(\mathbb{Z}, B) \xrightarrow{\mu_m^*} \text{Hom}(\mathbb{Z}, B) \to \text{Ext}^1(\mathbb{I}_m, B) \to \text{Ext}^1(\mathbb{Z}, B)$. Now $\text{Ext}^1(\mathbb{Z}, B) = \{0\}$ because $\mathbb{Z}$ is projective. Moreover, μ_m^* is also multiplication by m, while $\text{Hom}(\mathbb{Z}, B) \cong B$. More precisely, $\text{Hom}(\mathbb{Z}, \square)$ is naturally equivalent to the identity functor on **Ab**, and so there is a commutative diagram with exact rows:

$$
\begin{array}{ccccccc}
B & \xrightarrow{\mu_m} & B & \longrightarrow & B/mB & \longrightarrow & 0 \\
\tau_B \downarrow & & \tau_B \downarrow & & \downarrow & & \\
\text{Hom}(\mathbb{Z}, B) & \xrightarrow[\mu_m^*]{} & \text{Hom}(\mathbb{Z}, B) & \longrightarrow & \text{Ext}^1(\mathbb{I}_m, B) & \longrightarrow & 0.
\end{array}
$$

By Proposition 2.70, there is an isomorphism $B/mB \cong \text{Ext}^1(\mathbb{I}_m, B)$.

(ii) We can now compute $\text{Ext}^1_{\mathbb{Z}}(A, B)$ whenever A and B are finitely generated abelian groups. By the Fundamental Theorem of Finitely Generated Abelian Groups, both A and B are direct sums of cyclic groups. Since $\text{Ext}^1_{\mathbb{Z}}$ commutes with finite direct sums, $\text{Ext}^1_{\mathbb{Z}}(A, B)$ is the direct sum of groups of the form $\text{Ext}^1_{\mathbb{Z}}(C, D)$, where C and D are cyclic. We may assume that C is finite; otherwise, $C \cong \mathbb{Z}$, and $\text{Ext}^1(C, D) = \{0\}$. This calculation can be completed using part (i) and Exercise 2.29 on page 94: if D is a cyclic group of finite order m, then D/nD is a cyclic group of order $d = (m, n)$.   ◄

Definition.   Given R-modules C and A, an *extension* of A by C is a short exact sequence

$$0 \to A \xrightarrow{i} B \xrightarrow{p} C \to 0.$$

An extension is *split* if there exists an R-map $s: C \to B$ with $ps = 1_C$.

Of course, if $0 \to A \to B \to C \to 0$ is a split extension, then $B \cong A \oplus C$. The converse is false; there are nonsplit extensions with $B \cong A \oplus C$.

Proposition 7.24. If $\text{Ext}_R^1(C, A) = \{0\}$, then every extension of A by C splits.

Proof. Apply $\text{Hom}(C, \square)$ to $0 \to A \xrightarrow{i} B \xrightarrow{p} C$ to obtain an exact sequence $\text{Hom}(C, B) \xrightarrow{p_*} \text{Hom}(C, C) \xrightarrow{\partial} \text{Ext}^1(C, A)$. By hypothesis, $\text{Ext}^1(C, A) = \{0\}$, so that p_* is surjective. Hence, there exists $s \in \text{Hom}(C, B)$ with $1_C = p_*(s)$; that is, $1_C = ps$, and this says that the extension splits. $\quad \bullet$

We will soon prove the converse of Proposition 7.24.

Corollary 7.25.

(i) *A left R-module P is projective if and only if $\text{Ext}_R^1(P, B) = \{0\}$ for every R-module B.*

(ii) *A left R-module E is injective if and only if $\text{Ext}_R^1(A, E) = \{0\}$ for every left R-module A.*

Proof.

(i) If P is projective, then $\text{Ext}_R^1(P, B) = \{0\}$ for all B, by Corollary 6.58. Conversely, if $\text{Ext}_R^1(P, B) = \{0\}$ for all B, then every exact sequence $0 \to B \to X \to P \to 0$ splits, by Proposition 7.24, and so P is projective, by Proposition 3.3.

(ii) Similar to the proof of (i), but using Proposition 3.40. $\quad \bullet$

The next definition arises from Schreier's solution to the extension problem in Group Theory (see Proposition 9.12).

Definition. Given modules C and A, two extensions $\xi : 0 \to A \to B \to C \to 0$ and $\xi' : 0 \to A \to B' \to C \to 0$ of A by C are *equivalent* if there exists a map $\varphi : B \to B'$ making the following diagram commute:

$$
\begin{array}{ccccccccc}
\xi = 0 & \longrightarrow & A & \xrightarrow{i} & B & \xrightarrow{p} & C & \longrightarrow & 0 \\
& & \downarrow{1_A} & & \downarrow{\varphi} & & \downarrow{1_C} & & \\
\xi' = 0 & \longrightarrow & A & \xrightarrow{i'} & B' & \xrightarrow{p'} & C & \longrightarrow & 0.
\end{array}
$$

We denote the equivalence class of an extension ξ by $[\xi]$, and we define

$$e(C, A) = \big\{ [\xi] : \xi \text{ is an extension of } A \text{ by } C \big\}.$$

Example 7.26.   If two extensions are equivalent, then the Five Lemma shows that the map φ must be an isomorphism; it follows that equivalence is, indeed, an equivalence relation (for we can now prove symmetry). However, the converse is false: there are inequivalent extensions having isomorphic middle terms. For example, let p be an odd prime, and consider the diagram with exact rows

$$
\begin{array}{ccccccccc}
0 & \longrightarrow & K & \xrightarrow{\ i\ } & G & \xrightarrow{\ \pi\ } & Q & \longrightarrow & 0 \\
& & \downarrow{\scriptstyle 1_K} & & \downarrow & & \downarrow{\scriptstyle 1_Q} & & \\
0 & \longrightarrow & K & \xrightarrow[\ i'\ ]{} & G & \xrightarrow[\ \pi'\ ]{} & Q & \longrightarrow & 0.
\end{array}
$$

In the top row, $K = \langle a \rangle$, a cyclic group of order p, $G = \langle g \rangle$, a cyclic group of order p^2, $i \colon K \to G$ is defined by $i(a) = pg$, $Q = G/\operatorname{im} i$, and π is the natural map. In the bottom row, define $i'(a) = 2pg$; note that i' is an injection because p is odd. Suppose there is a map $\varphi \colon G \to G$ making the diagram commute. Commutativity of the first square gives $\varphi(pa) = 2pa$, and this forces $\varphi(g) = 2g$, by Exercise 7.8 on page 435. But commutativity of the second square now gives $g + \operatorname{im} i = 2g + \operatorname{im} i$, which says that $g \in \operatorname{im} i$, a contradiction. Therefore, the two extensions are not equivalent.   ◄

We are going to show that there is a bijection $\psi \colon e(C, A) \to \operatorname{Ext}^1(C, A)$. Given an extension $\xi \colon 0 \to A \to B \to C \to 0$ and the chosen projective resolution $\mathbf{P}$ of C (in the definition of Ext^n), form the diagram

$$
\begin{array}{ccccccccc}
\longrightarrow & P_2 & \xrightarrow{\ d_2\ } & P_1 & \xrightarrow{\ d_1\ } & P_0 & \longrightarrow & C & \longrightarrow & 0 \\
& \downarrow & & \downarrow{\scriptstyle \alpha_1} & & \downarrow{\scriptstyle \alpha_0} & & \downarrow{\scriptstyle 1_C} & & \\
\longrightarrow & 0 & \longrightarrow & A & \longrightarrow & B & \longrightarrow & C & \longrightarrow & 0.
\end{array}
$$

By Theorem 6.16, the Comparison Theorem, there exist dashed arrows which comprise a chain map $(\alpha_n) \colon \mathbf{P}_C \to \xi$ over 1_C. In particular, the first component $\alpha_1 \colon P_1 \to A$ satisfies $\alpha_1 d_2 = 0$; thus, $d_2^*(\alpha_1) = 0$, $\alpha_1 \in \ker d_2^*$, and α_1 is a cocycle. Define

$$
\psi \colon e(C, A) \to \operatorname{Ext}_R^1(C, A) \quad \text{by} \quad [\xi] \mapsto \operatorname{cls}(\alpha_1)
$$

[recall that $\operatorname{Ext}_R^1(C, A) = (\ker d_2^*)/(\operatorname{im} d_1^*)$ and $\operatorname{cls}(\alpha_1) = \alpha_1 + \operatorname{im} d_1^*$]. To see that ψ is well-defined, we show that ψ does not depend on the chain map $(\alpha_n) \colon \mathbf{P}_C \to \xi$ nor on the choice of extension in the equivalence class $[\xi]$.

$$
\begin{array}{ccccc}
P_2 & \xrightarrow{\ d_2\ } & P_1 & \xrightarrow{\ d_1\ } & P_0 \\
& {\scriptstyle s_1}\swarrow \ {\scriptstyle \alpha_1}\downarrow\downarrow{\scriptstyle \alpha_1'} & & {\scriptstyle s_0}\swarrow & \\
0 & \longrightarrow & A & \longrightarrow & B
\end{array}
$$

The Comparison Theorem says that if $(\alpha'_n)\colon \mathbf{P}_C \to \xi$ is another chain map over 1_C, then (α_n) and (α'_n) are homotopic: there are maps s_0 and s_1 with $\alpha'_1 - \alpha_1 = 0 \cdot s_1 + s_0 d_1 = s_0 d_1$, so that $\alpha'_1 - \alpha_1 \in \operatorname{im} d_1^*$ and $\operatorname{cls}(\alpha_1) = \operatorname{cls}(\alpha'_1)$. To see that equivalent extensions ξ and ξ' determine the same element of Ext^1, consider the diagram

$$\begin{array}{ccccccccc}
\longrightarrow & P_2 & \xrightarrow{d_2} & P_1 & \xrightarrow{d_1} & P_0 & \longrightarrow & C & \longrightarrow 0 \\
& \downarrow & & \downarrow{\alpha_1} & & \downarrow{\alpha_0} & & \downarrow{1_C} & \\
\xi = \longrightarrow & 0 & \longrightarrow & A & \longrightarrow & B & \longrightarrow & C & \longrightarrow 0 \\
& \downarrow & & \downarrow{1_A} & & \downarrow{\beta} & & \downarrow{1_C} & \\
\xi' \longrightarrow & 0 & \longrightarrow & A & \longrightarrow & B' & \longrightarrow & C & \longrightarrow 0.
\end{array}$$

Regarding the equivalence from row 2 to row 3 as a chain map over 1_C, we see that the bottom row ξ' gives the same cocycle α_1 as does ξ.

Lemma 7.27. *The function $\psi\colon e(C,A) \to \operatorname{Ext}^1_R(C,A)$, given by $[\xi] \mapsto \operatorname{cls}(\alpha_1)$, is well-defined, and if ξ is a split extension, then $\psi\colon [\xi] \mapsto 0$.*

Proof. We have just proved that ψ is is a well-defined function, for it is independent of the choices. If ξ is a split extension, there is a map $j\colon C \to B$ with $pj = 1_C$, and

$$\begin{array}{ccccccccc}
\longrightarrow & P_2 & \xrightarrow{d_2} & P_1 & \xrightarrow{d_1} & P_0 & \xrightarrow{\varepsilon} & C & \longrightarrow 0 \\
& \downarrow & & \downarrow{0} & & \downarrow{j\varepsilon} & & \downarrow{1_C} & \\
\longrightarrow & 0 & \longrightarrow & A & \longrightarrow & B & \xrightarrow{p} & C & \longrightarrow 0
\end{array}$$

is a commutative diagram with $\alpha_1 = 0$. $\bullet$

We will prove that ψ is a bijection by constructing its inverse; to each cocycle $\alpha\colon P_1 \to A$, we must find an extension of A by C. Let us begin by analyzing the diagram defining the map α_1.

Lemma 7.28. *Let $\Xi = 0 \to X_1 \xrightarrow{j} X_0 \xrightarrow{\varepsilon} C \to 0$ be an extension of a module X_1 by a module C. Given a map $h\colon X_1 \to A$, consider the diagram*

$$\begin{array}{ccccccccc}
\Xi = 0 & \longrightarrow & X_1 & \xrightarrow{j} & X_0 & \xrightarrow{\varepsilon} & C & \longrightarrow 0 \\
& & \downarrow{h} & & & & \downarrow{1_C} & \\
& & A & & & & C.
\end{array}$$

(i) *There exists a commutative diagram with exact rows that completes the given diagram:*

$$
\begin{array}{ccccccccc}
0 & \longrightarrow & X_1 & \xrightarrow{\;j\;} & X_0 & \xrightarrow{\;\varepsilon\;} & C & \longrightarrow & 0 \\
 & & {\scriptstyle h}\downarrow & & \downarrow{\scriptstyle \beta} & & \downarrow{\scriptstyle 1_C} & & \\
0 & \longrightarrow & A & \xrightarrow[\;i\;]{} & B & \xrightarrow[\;\eta\;]{} & C & \longrightarrow & 0.
\end{array}
$$

(ii) *Any two bottom rows of completed diagrams are equivalent extensions.*

Proof.

(i) Define B to be the pushout of j and h. As in Lemma 3.41, let $S \subseteq A \oplus X_0$ by $S = \{(hx_1, -jx_1) : x_1 \in X_1\}$ and define $B = (A \oplus X_0)/S$. If we define $i\colon A \to B$ by $a \mapsto (a, 0) + S$ and $\beta\colon X_0 \to B$ by $x_0 \mapsto (0, x_0) + S$, then i is an injection and the first square commutes. It is easy to check that $\eta\colon B \to C$, given by $(a, x_0) + S \mapsto \varepsilon x_0$ is well-defined, the second square commutes, and the bottom row is exact.

(ii) Let

$$
\begin{array}{ccccccccc}
0 & \longrightarrow & X_1 & \xrightarrow{\;j\;} & X_0 & \xrightarrow{\;\varepsilon\;} & C & \longrightarrow & 0 \\
 & & {\scriptstyle h}\downarrow & & \downarrow{\scriptstyle \beta'} & & \downarrow{\scriptstyle 1_C} & & \\
0 & \longrightarrow & A & \xrightarrow[\;i'\;]{} & B' & \xrightarrow[\;\eta'\;]{} & C & \longrightarrow & 0
\end{array}
$$

be a second completion of the diagram. We must define $\theta\colon B \to B'$ making the following diagram commute.

$$
\begin{array}{ccccccccc}
0 & \longrightarrow & A & \xrightarrow{\;i\;} & B & \xrightarrow{\;\eta\;} & C & \longrightarrow & 0 \\
 & & {\scriptstyle 1_A}\downarrow & & \downarrow{\scriptstyle \theta} & & \downarrow{\scriptstyle 1_C} & & \\
0 & \longrightarrow & A & \xrightarrow[\;i'\;]{} & B' & \xrightarrow[\;\eta'\;]{} & C & \longrightarrow & 0
\end{array}
$$

Now θ, given by $(a, x_0) + S \mapsto i'a + \beta'x_0$, is well-defined, and it makes the diagram commute; that is, the extensions are equivalent. •

Notation.   Denote the extension of A by C just constructed by

$$
h\Xi.
$$

Here is the dual result.

Lemma 7.29.   *Let $\Xi' = 0 \to A \to Y_0 \to Y_1 \to 0$ be an extension of A by Y_1, where A and Y_0 are modules. Given a map $k: C \to Y_1$, consider the diagram*

$$
\begin{array}{ccc}
A & & C \\
{\scriptstyle 1_A}\downarrow & & \downarrow{\scriptstyle k} \\
\end{array}
$$

$$\Xi' = 0 \longrightarrow A \longrightarrow Y_0 \xrightarrow{p} Y_1 \longrightarrow 0.$$

(i) *There exists a commutative diagram with exact rows that completes the given diagram:*

$$\Xi'k = 0 \longrightarrow A \longrightarrow B \longrightarrow C \longrightarrow 0$$
$$ \quad {\scriptstyle 1_A}\downarrow \qquad \downarrow \qquad \downarrow{\scriptstyle k}$$
$$\Xi' = 0 \longrightarrow A \longrightarrow Y_0 \xrightarrow{p} Y_1 \longrightarrow 0.$$

(ii) *Any two top rows of completed diagrams are equivalent extensions.*

Proof.   Dual to that of Lemma 7.28; in particular, construct the top row using the pullback of k and p.   •

Notation.   Denote the extension of A by C just constructed by

$$\Xi'k.$$

Theorem 7.30.   *The function $\psi: e(C, A) \to \mathrm{Ext}^1(C, A)$ is a bijection.*

Proof.   We construct the inverse $\theta: \mathrm{Ext}^1(C, A) \to e(C, A)$ of ψ. Choose a projective resolution of C,

$$\to P_2 \xrightarrow{d_2} P_1 \xrightarrow{d_1} P_0 \to C \to 0,$$

and choose a 1-cocycle $\alpha_1: P_1 \to A$. Since α_1 is a cocycle, we have $0 = d_2^*(\alpha_1) = \alpha_1 d_2$; thus, α_1 induces a homomorphism $\overline{\alpha}_1: P_1/\mathrm{im}\, d_2 \to A$ [if $x_1 \in P_1$, then $\overline{\alpha}_1: x_1 + \mathrm{im}\, d_2 \mapsto \alpha_1(x_1)$]. Let Ξ denote the extension

$$\Xi = 0 \to P_1/\mathrm{im}\, d_2 \to P_0 \to C \to 0.$$

As in the lemma, there is a commutative diagram with exact rows:

$$0 \longrightarrow P_1/\mathrm{im}\, d_2 \longrightarrow P_0 \longrightarrow C \longrightarrow 0$$
$$ \quad {\scriptstyle \overline{\alpha}_1}\downarrow \qquad \downarrow{\scriptstyle \beta} \qquad \downarrow{\scriptstyle 1_C}$$
$$0 \longrightarrow A \xrightarrow{i} B \longrightarrow C \longrightarrow 0.$$

Define $\theta: \mathrm{Ext}^1(C, A) \to e(C, A)$ using the construction in Lemma 7.28(ii):

$$\theta: \mathrm{cls}(\alpha_1) \mapsto [\overline{\alpha}_1 \Xi].$$

Let us see that θ is independent of the choice of cocycle α_1. If α_1' is another representative of the coset $\alpha_1 + \operatorname{im} d_1^*$, then there is a map $s \colon P_0 \to A$ with $\alpha_1' = \alpha_1 + sd_1$. But it is easy to see that the following diagram commutes:

$$
\begin{array}{ccccccccc}
P_2 & \xrightarrow{d_2} & P_1 & \xrightarrow{d_1} & P_0 & \longrightarrow & C & \longrightarrow & 0 \\
{\scriptstyle 0}\downarrow & & {\scriptstyle \alpha_1+sd_1}\downarrow & & \downarrow{\scriptstyle \beta+is} & & \downarrow{\scriptstyle 1_C} & & \\
0 & \longrightarrow & A & \xrightarrow{i} & B & \longrightarrow & C & \longrightarrow & 0.
\end{array}
$$

As the bottom row has not changed, $[\overline{\alpha}_1 \, \Xi] = [\overline{\alpha}_1' \, \Xi]$.

It remains to show that the composites $\psi\theta$ and $\theta\psi$ are identities. If $\operatorname{cls}(\alpha_1) \in \operatorname{Ext}^1(C, A)$, then $\theta(\operatorname{cls}(\alpha_1))$ is the bottom row of the diagram

$$
\begin{array}{ccccccccc}
P_2 & \longrightarrow & P_1 & \longrightarrow & P_0 & \longrightarrow & C & \longrightarrow & 0 \\
{\scriptstyle 0}\downarrow & & {\scriptstyle \alpha_1}\downarrow & & \downarrow{\scriptstyle \alpha_0} & & \downarrow{\scriptstyle 1_C} & & \\
0 & \longrightarrow & A & \longrightarrow & B & \longrightarrow & C & \longrightarrow & 0,
\end{array}
$$

and $\psi\theta(\operatorname{cls}(\alpha_1)) = \operatorname{cls}(\alpha_1)$ because α_1 is the first component of a chain map $\mathbf{P}_C \to \theta(\operatorname{cls}(\alpha_1))$ over 1_C; therefore, $\psi\theta$ is the identity.

For the other composite, start with an extension ξ, and then imbed it as the bottom row of a diagram, using Lemma 7.28(i).

$$
\begin{array}{ccccccccc}
0 & \longrightarrow & P/\operatorname{im} d_2 & \xrightarrow{\overline{d}_1} & P_0 & \longrightarrow & C & \longrightarrow & 0 \\
 & & {\scriptstyle \overline{\alpha}_1}\downarrow & & \downarrow & & \downarrow{\scriptstyle 1_C} & & \\
0 & \longrightarrow & A & \longrightarrow & B & \longrightarrow & C & \longrightarrow & 0
\end{array}
$$

Both ξ and $\overline{\alpha}_1 \, \Xi$ are bottom rows of such a diagram, and so Lemma 7.28(ii) shows that $[\xi] = [\alpha_1' \, \Xi]$. Hence, $\theta\psi$ is the identity, and ψ is a bijection. •

We can now prove the converse of Proposition 7.24.

Theorem 7.31.   *Every extension* $0 \to A \xrightarrow{i} B \xrightarrow{p} C \to 0$ *splits if and only if* $\operatorname{Ext}_R^1(C, A) = \{0\}$.

Proof.   Sufficiency is Proposition 7.24. For the converse, if every extension is split, then $|e(C, A)|$ is the number of equivalence classes of split extensions. But all split sequences are equivalent, by Exercise 7.9 on page 435, so that $|e(C, A)| = 1$. Therefore, $|\operatorname{Ext}_R^1(C, A)| = 1$, by Theorem 7.30, and $\operatorname{Ext}_R^1(C, A) = \{0\}$. •

Whenever meeting a homology group, we must ask what it means for it to be zero, for its elements can then be construed as being obstructions. Thus, nonzero elements of $\operatorname{Ext}_R^1(C, A)$ describe nonsplit extensions (indeed, this result is why Ext is so called).

Example 7.32. If p is a prime, then $\text{Ext}^1_{\mathbb{Z}}(\mathbb{I}_p, \mathbb{I}_p) \cong \mathbb{I}_p$, as we saw in Example 7.23(i), so that $|\text{Ext}^1_{\mathbb{Z}}(\mathbb{I}_p, \mathbb{I}_p)| = p$. It follows from Theorem 7.30 that there are p equivalence classes of extensions $0 \to \mathbb{I}_p \to B \to \mathbb{I}_p \to 0$. However, if $|B| = p^2$, then elementary Group Theory says that there are, to isomorphism, only two choices for B, namely, $B \cong \mathbb{I}_{p^2}$ or $B \cong \mathbb{I}_p \oplus \mathbb{I}_p$ (two inequivalent extensions with middle group $\mathbb{I}_{p^2}$ are displayed in Example 7.26). Therefore, if $\text{mid}(A, C)$ is the number of middle terms B in extensions of the form $0 \to A \to B \to C \to 0$, then $\text{mid}(A, C) \leq |\text{Ext}^1_{\mathbb{Z}}(C, A)|$; moreover, this inequality can be strict. ◄

Proposition 7.33.

(i) *If F is a torsion-free abelian group and T is a group of **bounded order** (i.e., $nT = \{0\}$ for some positive integer n), then $\text{Ext}^1(F, T) = \{0\}$.*

(ii) *If the torsion subgroup tG of an abelian group G is of bounded order, then tG is a direct summand of G.*

Proof.

(i) Since F is torsion-free, it is a flat $\mathbb{Z}$-module, by Theorem 4.35, so that exactness of $0 \to \mathbb{Z} \to \mathbb{Q}$ gives exactness of $0 \to \mathbb{Z} \otimes F \to \mathbb{Q} \otimes F$. Thus, $F \cong \mathbb{Z} \otimes F$ can be imbedded in a vector space V over $\mathbb{Q}$, namely, $V = \mathbb{Q} \otimes F$. Applying the contravariant functor $\text{Hom}(\square, T)$ to $0 \to F \to V \to V/F \to 0$ gives an exact sequence

$$\text{Ext}^1(V, T) \to \text{Ext}^1(F, T) \to \text{Ext}^2(V/F, T).$$

The last term is $\{0\}$, by Exercise 6.18 on page 376. Also, $\text{Ext}^1(V, T)$ is a vector space over $\mathbb{Q}$, by Proposition 6.19, so that $\text{Ext}^1(F, T)$ is divisible. Now multiplication $\mu_n \colon \mathbb{Q} \to \mathbb{Q}$ is an isomorphism, and so the induced map $\mu_n^* \colon \text{Ext}^1(F, T) \to \text{Ext}^1(F, T)$ is an isomorphism. On the other hand, multiplication $\mu'_n \colon T \to T$ is the zero map (since $nT = \{0\}$), and so the induced map $\mu'_n{}_* \colon \text{Ext}^1(F, T) \to \text{Ext}^1(F, T)$ is the zero map. But both induced maps are multiplication by n, so that $\text{Ext}^1(F, T) = \{0\}$, as desired. (We have solved Exercise 6.20 on page 377.)

(ii) To prove that the extension $0 \to tG \to G \to G/tG \to 0$ splits, it suffices to prove that $\text{Ext}^1(G/tG, tG) = \{0\}$. Since G/tG is torsion-free, this follows from part (i) and Corollary 7.31. •

Remark. Let T be a torsion abelian group. We say that T has Property S if, whenever T is the torsion subgroup of a group G, then it is a direct summand of G [in homological language, a torsion group T has Property S if and only if $\text{Ext}^1_{\mathbb{Z}}(C, T) = \{0\}$ for every torsion-free abelian group C]. We have just proved that a group T of bounded order has Property S, and it follows easily that a torsion group $B \oplus D$ has Property S if B has bounded order and D is divisible. The converse is true, and it follows from a theorem of Kulikov: a pure subgroup of bounded order is a direct summand (see Fuchs, *Infinite Abelian Groups* I, p. 118). ◄

Proposition 7.34.   *There exists an abelian group G whose torsion subgroup is not a direct summand of G; in fact, we may choose $tG = \bigoplus_p \mathbb{I}_p$, where the direct sum is over all primes p.*

Proof.   By Corollary 7.31, it suffices to prove that $\text{Ext}^1_{\mathbb{Z}}(\mathbb{Q}, \bigoplus_p \mathbb{I}_p) \neq 0$, where p ranges over all primes. In Exercise 3.31 on page 151, it is shown that $D = (\prod_p \mathbb{I}_p)/(\bigoplus_p \mathbb{I}_p)$ is a torsion-free divisible group; that is, D is a vector space over $\mathbb{Q}$. Exactness of $0 \to \bigoplus_p \mathbb{I}_p \to \prod_p \mathbb{I}_p \to D \to 0$ gives exactness of

$$\text{Hom}\Big(\mathbb{Q}, \prod_p \mathbb{I}_p\Big) \to \text{Hom}(\mathbb{Q}, D) \xrightarrow{\partial} \text{Ext}^1\Big(\mathbb{Q}, \bigoplus_p \mathbb{I}_p\Big) \to \text{Ext}^1\Big(\mathbb{Q}, \prod_p \mathbb{I}_p\Big).$$

Now $\text{Hom}(\mathbb{Q}, \prod_p \mathbb{I}_p) \cong \prod \text{Hom}(\mathbb{Q}, \mathbb{I}_p) = \{0\}$, by Theorem 2.31, while Propositions 7.22 and 7.33(i) give $\text{Ext}^1(\mathbb{Q}, \prod_p \mathbb{I}_p) \cong \prod_p \text{Ext}^1(\mathbb{Q}, \mathbb{I}_p) = \{0\}$. Hence, ∂ is an isomorphism. But $\text{Hom}(\mathbb{Q}, D) \neq \{0\}$, because D is a nonzero vector space over $\mathbb{Q}$. Therefore, $\text{Ext}^1(\mathbb{Q}, \bigoplus_p \mathbb{I}_p) \neq \{0\}$.   •

It should come as no surprise that Ext^1 does not preserve infinite direct sums. We just saw that $\text{Ext}^1(\mathbb{Q}, \bigoplus_p \mathbb{I}_p) \neq \{0\}$, while $\bigoplus_p \text{Ext}^1(\mathbb{Q}, \mathbb{I}_p) = \{0\}$. Therefore, $\text{Ext}^1(\mathbb{Q}, \oplus_p \mathbb{I}_p) \not\cong \bigoplus_p \text{Ext}^1(\mathbb{Q}, \mathbb{I}_p)$.

7.2.1 Baer Sum

If X is a set and $\psi : X \to G$ is a bijection to a group G, then there is a unique group structure on X making it a group and ψ an isomorphism [if $x, x' \in X$, then $x = \psi^{-1}(g)$ and $x' = \psi^{-1}(g')$; define $xx' = \psi^{-1}(gg')$]. In particular, Theorem 7.30 implies that there is a group structure on $e(C, A)$. It was R. Baer who made this explicit. If $f : C \to A$ and $g : C' \to A'$, define

$$f \oplus g : C \oplus C' \to A \oplus A' \quad \text{by} \quad f \oplus g : (c, c') \mapsto (fc, gc').$$

Define the **diagonal map** $\Delta_C : C \to C \oplus C$ by $\Delta_C : c \mapsto (c, c)$, and define the **codiagonal map** $\nabla_A : A \oplus A \to A$ by $\nabla_A : (a_1, a_2) \mapsto a_1 + a_2$. Let us show that if $f, g : C \to A$ are homomorphisms, then $\nabla_A(f \oplus g)\Delta_C = f + g$. If $c \in C$, then

$$\nabla_A(f \oplus g)\Delta_C : c \mapsto (c, c) \mapsto (fc, gc) \mapsto fc + gc = (f + g)c.$$

Thus, if $f, g: C \to A$, then the formula $f + g = \nabla_A(f \oplus g)\Delta_C$ describes addition in $\mathrm{Hom}(C, A)$. Now Ext^1 generalizes Hom [for $\mathrm{Hom} = \mathrm{Ext}^0$], and Baer mimicked this definition to define addition in $e(C, A)$.

Definition.   The *direct sum* of extensions

$$\xi = 0 \to A \xrightarrow{i} B \xrightarrow{p} C \to 0 \quad \text{and} \quad \xi' = 0 \to A' \xrightarrow{i'} B' \xrightarrow{p'} C' \to 0$$

is the extension

$$\xi \oplus \xi' = 0 \to A \oplus A' \xrightarrow{i \oplus i'} B \oplus B' \xrightarrow{p \oplus p'} C \oplus C' \to 0.$$

By analogy with the sum of two homomorphisms, define $[\xi] + [\xi']$ in $e(C, A)$ to be the function $e(C, A) \times e(C, A) \to e(C, A)$ given by the composite

$$([\xi], [\xi']) \mapsto [\xi \oplus \xi'] \mapsto [\nabla(\xi \oplus \xi')] \mapsto [(\nabla(\xi \oplus \xi'))\Delta].$$

Pay attention to the parentheses, for we have not proved that this operation is associative (even though it is). The reader can check that this addition is independent of the choice of representative; that is, if $[\xi] = [\xi_1]$ and $[\xi'] = [\xi_1']$, then $[\xi] + [\xi'] = [\xi_1] + [\xi_1']$.

We will need three bookkeeping formulas. Recall that if ξ is an extension of A by C, then $\psi[\xi] = \mathrm{cls}(\alpha_1)$, where $\mathbf{P}$ is a projective resolution of C and α_1 is the first component of a chain map $\mathbf{P} \to \xi$ over 1_C.

Formula I. $\psi[\xi \oplus \xi'] = \psi[\xi] \oplus \psi[\xi']$, where the right side is obtained from the direct sum $\xi \oplus \xi'$ pictured below.

$$
\begin{array}{ccccccccc}
0 & \longrightarrow & P_1 \oplus P_1' & \longrightarrow & P_0 \oplus P_0' & \longrightarrow & C \oplus C' & \longrightarrow & 0 \\
& & {\scriptstyle \alpha_1 \oplus \alpha_1'}\downarrow & & \downarrow{\scriptstyle \alpha_0 \oplus \alpha_0'} & & \downarrow{\scriptstyle 1_{C \oplus C'}} & & \\
0 & \longrightarrow & A \oplus A' & \longrightarrow & B \oplus B' & \longrightarrow & C \oplus C' & \longrightarrow & 0
\end{array}
$$

We know that $\mathrm{Ext}^1(C \oplus C', A \oplus A')$ is independent of the choice of projective resolution of the first variable; hence, if $\mathbf{P}$ and $\mathbf{P}'$ are chosen resolutions of C and C', respectively, then we may assume that $\mathbf{P} \oplus \mathbf{P}'$ is the chosen resolution of $C \oplus C'$.

Formula II.   If $[\xi] \in e(C, A)$ and $h: A \to A'$, then

$$\psi[h\xi] = h\psi[\xi] \quad \text{in} \quad e(C, A').$$

This formula follows from the diagram

$$
\begin{array}{ccccccccc}
P_2 & \longrightarrow & P_1 & \longrightarrow & P_0 & \longrightarrow & C & \longrightarrow & 0 \\
& & {\scriptstyle \alpha_1}\downarrow & & \downarrow & & \downarrow{\scriptstyle 1_C} & & \\
\xi = & & 0 \longrightarrow A & \longrightarrow & B & \longrightarrow & C & \longrightarrow & 0 \\
& & {\scriptstyle h}\downarrow & & \downarrow & & \downarrow{\scriptstyle 1_C} & & \\
h\xi = & & 0 \longrightarrow A' & \longrightarrow & E & \longrightarrow & C & \longrightarrow & 0.
\end{array}
$$

Formula III. If $[\xi] \in e(C, A)$ and $k: C' \to C$, then $[\xi k] \in e(C', A)$; let us denote the function $e(C', A) \to \operatorname{Ext}^1(C', A)$ by ψ'; that is, if $\mathbf{P}'$ is the chosen projective resolution of C', then $\psi'([\xi k]) = \operatorname{cls}(k_1')$, where k_1' is the first component of a chain map $\mathbf{P}'_{C'} \to \xi k$ over $1_{C'}$. We can now state the formula:

$$\psi'[\xi k] = \psi[\xi] k_1' \quad \text{in} \quad e(C', A).$$

Consider the diagram with middle row ξk and bottom row ξ:

$$
\begin{array}{ccccccccc}
P_2' & \xrightarrow{d_2'} & P_1' & \xrightarrow{d_1'} & P_0' & \longrightarrow & C' & \longrightarrow & 0 \\
\downarrow & & \gamma_1 \downarrow & & \downarrow \gamma_0 & & \downarrow 1_{C'} & & \\
0 & \longrightarrow & A & \longrightarrow & B' & \longrightarrow & C' & \longrightarrow & 0 \\
\downarrow & & 1_A \downarrow & & \downarrow & & \downarrow k & & \\
0 & \longrightarrow & A & \longrightarrow & E & \longrightarrow & C & \longrightarrow & 0.
\end{array}
$$

Thus, $\psi'[\xi k] = \gamma_1$. Having pictured $\psi'([\xi k])$, let us now picture $\psi[\xi] k_1'$.

$$
\begin{array}{ccccccccc}
P_2' & \longrightarrow & P_1' & \longrightarrow & P_0' & \longrightarrow & C' & \longrightarrow & 0 \\
\downarrow & & k_1' \downarrow & & \downarrow k_0' & & \downarrow k & & \\
P_2 & \longrightarrow & P_1 & \longrightarrow & P_0 & \longrightarrow & C & \longrightarrow & 0 \\
\downarrow & & \alpha_1 \downarrow & & \downarrow & & \downarrow 1_C & & \\
0 & \longrightarrow & A & \longrightarrow & B & \longrightarrow & C & \longrightarrow & 0
\end{array}
$$

Both γ_1 and $\alpha_1 k_1'$ are first components of chain maps $\mathbf{P}'_{C'} \to \xi$ over k. The Comparison Theorem says such chain maps are unique to homotopy, and so $\operatorname{cls}(\gamma_1) = \operatorname{cls}(\alpha_1 k_1')$.

This formula will be used for $[\xi \oplus \xi'] \in \operatorname{Ext}^1(C \oplus C, A \oplus A)$, where $[\xi], [\xi'] \in e(C, A)$, and the diagonal map $\Delta: C \to C \oplus C$. The appropriate diagram is

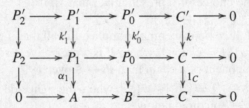

$$
\begin{array}{ccccccccc}
P_2 & \longrightarrow & P_1 & \longrightarrow & P_0 & \longrightarrow & C & \longrightarrow & 0 \\
\downarrow & & \Delta_1' \downarrow & & \downarrow \Delta_0' & & \downarrow \Delta & & \\
P_2 \oplus P_2 & \longrightarrow & P_1 \oplus P_1 & \longrightarrow & P_0 \oplus P_0 & \longrightarrow & C \oplus C & \longrightarrow & 0 \\
\downarrow & & \alpha_1 \oplus \alpha_1' \downarrow & & \downarrow & & \downarrow 1_{C \oplus C} & & \\
0 & \longrightarrow & A \oplus A & \longrightarrow & B \oplus B' & \longrightarrow & C \oplus C & \longrightarrow & 0.
\end{array}
$$

As in the proof of Formula I, we may assume that $\mathbf{P} \oplus \mathbf{P}$ is the chosen projective resolution of $C \oplus C$. Furthermore, we also know that ψ is independent of the chain map $\mathbf{P}_C \to \mathbf{P}_C \oplus \mathbf{P}_C$ over Δ. But the sequence of diagonal maps $\Delta_n: P_n \to P_n \oplus P_n$ is a chain map over Δ, and so we may take Δ_1' to be the diagonal map.

Theorem 7.35 (Baer). $e(C, A)$ *is an abelian group under* **Baer sum**, *defined by*

$$[\xi] + [\xi'] = [(\nabla(\xi \oplus \xi'))\Delta],$$

and $\psi : e(C, A) \to \text{Ext}^1_R(C, A)$ *is an isomorphism.*

Remark.   The associative law $[(h\xi)k] = [h(\xi k)]$ does hold, but we will not need it in this proof. ◄

Proof.   The formula for Baer sum defines a relation

$$\rho : e(C, A) \times e(C, A) \to e(C, A),$$

which we do not yet know is a function. After verifying Formula III, we saw that we may assume here that Δ'_1 is a diagonal map.

$$
\begin{aligned}
\psi[(\nabla(\xi \oplus \xi'))\Delta] &= \psi[\nabla(\xi \oplus \xi')]\Delta & \text{(Formula } \mathbf{III}) \\
&= (\nabla\psi[\xi \oplus \xi'])\Delta & \text{(Formula } \mathbf{II}) \\
&= \nabla(\psi[\xi] \oplus \psi[\xi'])\Delta & \text{(Formula } \mathbf{I}) \\
&= \psi[\xi] + \psi[\xi'].
\end{aligned}
$$

There are two conclusions from this computation. First, $\psi \circ \rho$ is a function,[1] so that, since ψ is a bijection, $\rho = \psi^{-1}(\psi\rho)$ is also a function. Second, $\psi([\xi] + [\xi']) = \psi[\xi] + \psi[\xi']$, so that ρ is the good addition on $e(C, A)$ making it a group and ψ an isomorphism. •

One can prove directly [without using $\text{Ext}^1(C, A)$] that $e(C, A)$ is an abelian group under Baer sum and that $e(C, A)$ repairs the loss of exactness after applying Hom to a short exact sequence (see Exercises 7.23 through 7.26 on pages 436–437). This approach has the advantage that it avoids choosing resolutions of either variable, and no projectives or injectives are required! This illustrates Mac Lane's viewpoint that the Ext functors should be defined by the axioms in Theorems 6.48 and 6.64, so that resolutions may be relegated to their proper place as aids to computation. Baer's description of Ext^1 as $e(C, A)$ has been generalized by N. Yoneda to a description of Ext^n for all $n \geq 1$. Elements of Yoneda's $\text{Ext}^n(C, A)$ are certain equivalence classes of exact sequences

$$0 \to A \to B_1 \to \cdots \to B_n \to C \to 0,$$

---

[1]Let X, Y, Z be sets, and let $\rho \subseteq X \times Y$ and $\psi \subseteq Y \times Z$ be relations. Recall that the composite $\psi \circ \rho$ is the relation

$$\psi \circ \rho = \{(x, z) : \text{there is } y \in Y \text{ with } (x, y) \in \rho \text{ and } (y, z) \in \psi\}.$$

In particular, it makes sense to consider the composite of a relation and a function.

and we add them by a generalized Baer sum (see Mac Lane, *Homology*, pp. 82–87). Thus, there is a construction of Ext^n for all $n \geq 1$ that does not use derived functors, projectives, or injectives.

Another construction of $\mathrm{Ext}_R^1(C, A)$ is given in the remark on page 512, which mimics the homological classification of group extensions in terms of *factor sets*.

Here are some localization results for Ext.

Theorem 7.36.   *If R is a commutative noetherian ring, and if A and B are finitely generated R-modules, then $\mathrm{Ext}_R^n(A, B)$ is a finitely generated R-module for all $n \geq 0$.*

Proof.   Since R is commutative, Ext is an R-module. The proof, an induction on $n \geq 0$ showing that Ext^n is finitely generated, is essentially that of Theorem 7.20 with $\square \otimes_R B$ replaced by $\mathrm{Hom}_R(\square, B)$. If $n = 0$, then $\mathrm{Hom}_R(A, B)$ is finitely generated, by Exercise 3.13 on page 115, for R is noetherian.[2] If $n \geq 1$, choose a projective resolution $\cdots \to P_1 \xrightarrow{d_1} P_0 \to A \to 0$, as in Lemma 7.19. Since $\mathrm{Hom}_R(P_n, B)$ is finitely generated, so are $\ker d_n^*$ (by Proposition 3.18) and its quotient $\mathrm{Ext}_R^n(A, B)$.   •

Lemma 7.37.   *Let S be a multiplicative subset of a commutative ring R, and let M and A be R-modules with A finitely presented. Then there is a natural $S^{-1}R$-isomorphism*

$$\tau_{A,B}: S^{-1}\mathrm{Hom}_R(A, M) \to \mathrm{Hom}_{S^{-1}R}(S^{-1}A, S^{-1}M).$$

Proof.   It suffices to construct natural isomorphisms

$$\theta_A: \mathrm{Hom}_R(A, S^{-1}M) \to \mathrm{Hom}_{S^{-1}R}(S^{-1}A, S^{-1}M)$$

and

$$\varphi_A: S^{-1}\mathrm{Hom}_R(A, M) \to \mathrm{Hom}_R(A, S^{-1}M),$$

for then we can define $\tau_A = \theta_A \varphi_A$.

Assume first that $A = R^n$ is a finitely generated free R-module. If $a_1, \ldots, a_n$ is a basis of A, then $S^{-1}A = S^{-1}R \otimes_R R^n$ is a free $S^{-1}R$-module with basis $a_1/1, \ldots, a_n/1$. The map

$$\theta_{R^n}: \mathrm{Hom}_R(A, S^{-1}M) \to \mathrm{Hom}_{S^{-1}R}(S^{-1}A, S^{-1}M),$$

given by $f \mapsto \widetilde{f}$, where $\widetilde{f}(a_i/\sigma) = f(a_i)/\sigma$, is easily seen to be a well-defined R-isomorphism.

---

[2]Exercise 3.13(iii) gives an example of a commutative ring R and finitely generated R-modules A and B for which $\mathrm{Hom}_R(A, B)$ is not a finitely generated R-module.

If, now, A is a finitely presented R-module, then there is an exact sequence

$$R^t \to R^n \to A \to 0. \tag{1}$$

Apply the contravariant functors $\operatorname{Hom}_R(\Box, M')$ and $\operatorname{Hom}_{S^{-1}R}(\Box, M')$, where $M' = S^{-1}M$ is first viewed as an R-module; we obtain a commutative diagram with exact rows

$$
\begin{array}{ccccc}
0 \longrightarrow \operatorname{Hom}_R(A, M') & \longrightarrow & \operatorname{Hom}_R(R^n, M') & \longrightarrow & \operatorname{Hom}_R(R^t, M') \\
\quad \downarrow \theta_A & & \downarrow \theta_{R^n} & & \downarrow \theta_{R^t} \\
0 \to \operatorname{Hom}_L(S^{-1}A, M') & \to & \operatorname{Hom}_L((S^{-1}R)^n, M') & \to & \operatorname{Hom}_L((S^{-1}R)^t, M'),
\end{array}
$$

where $L = S^{-1}R$. Since the vertical maps θ_{R^n} and θ_{R^t} are isomorphisms, there is a dashed arrow θ_A that must be an isomorphism, by Proposition 2.71. If $\beta \in \operatorname{Hom}_R(A, M)$, then the reader may check that

$$\theta_A(\beta) = \widetilde{\beta} \colon a/\sigma \mapsto \beta(a)/\sigma,$$

from which it follows that the isomorphisms θ_A are natural.

Construct $\varphi_A \colon S^{-1}\operatorname{Hom}_R(A, M) \to \operatorname{Hom}_R(A, S^{-1}M)$ by defining $\varphi_A \colon g/\sigma \mapsto g_\sigma$, where $g_\sigma(a) = g(a)/\sigma$. Now φ_A is well-defined, for it arises from the R-bilinear function $S^{-1}R \times \operatorname{Hom}_R(A, M) \to \operatorname{Hom}_R(A, S^{-1}M)$ given by $(r/\sigma, g) \mapsto rg_\sigma$ [for $S^{-1}\operatorname{Hom}_R(A, M) = S^{-1}R \otimes_R \operatorname{Hom}_R(A, M)$]. Observe that φ_A is an isomorphism when A is finitely generated free, and consider the commutative diagram

$$
\begin{array}{ccccc}
0 \to S^{-1}\operatorname{Hom}_R(A, M) & \to & S^{-1}\operatorname{Hom}_R(R^n, M) & \to & S^{-1}\operatorname{Hom}_R(R^t, M) \\
\quad \downarrow \varphi_A & & \downarrow \varphi_{R^n} & & \downarrow \varphi_{R^t} \\
0 \to \operatorname{Hom}_R(A, S^{-1}M) & \to & \operatorname{Hom}_R(R^n, S^{-1}M) & \to & \operatorname{Hom}_R(R^t, S^{-1}M).
\end{array}
$$

The top row is exact, for it arises from (1) by first applying the left exact contravariant functor $\operatorname{Hom}_R(\Box, M)$ and then applying the exact localization functor. The bottom row is exact, for it arises from (1) by applying the left exact contravariant functor $\operatorname{Hom}_R(\Box, S^{-1}M)$. The Five Lemma shows that φ_A is an isomorphism. $\bullet$

Example 7.38. Lemma 7.37 can be false if A is not finitely presented. For example, let $R = \mathbb{Z}$ and $S^{-1}R = \mathbb{Q}$. We claim that

$$\mathbb{Q} \otimes_\mathbb{Z} \operatorname{Hom}_\mathbb{Z}(\mathbb{Q}, \mathbb{Z}) \ncong \operatorname{Hom}_\mathbb{Q}(\mathbb{Q} \otimes_\mathbb{Z} \mathbb{Q}, \mathbb{Q} \otimes_\mathbb{Z} \mathbb{Z}).$$

The left-hand side is $\{0\}$ because $\operatorname{Hom}_\mathbb{Z}(\mathbb{Q}, \mathbb{Z}) = \{0\}$. On the other hand, the right-hand side is $\operatorname{Hom}_\mathbb{Z}(\mathbb{Q}, \mathbb{Q}) \cong \mathbb{Q}$. ◄

Proposition 7.39.   *Let R be a commutative noetherian ring, and let S be a multiplicative subset. If A is a finitely generated R-module, then there are isomorphisms, natural in A, B,*

$$S^{-1} \operatorname{Ext}_R^n(A, B) \cong \operatorname{Ext}_{S^{-1}R}^n(S^{-1}A, S^{-1}B)$$

for all $n \geq 0$ and all R-modules B.

Proof.   Since R is noetherian and A is finitely generated, Lemma 7.19 says there is a projective resolution **P** of A each of whose terms is finitely generated. By Lemma 7.37, there is a natural isomorphism

$$\tau_{A,B} \colon S^{-1} \operatorname{Hom}_R(A, B) \to \operatorname{Hom}_{S^{-1}R}(S^{-1}A, S^{-1}B)$$

for every R-module B (a finitely generated module over a noetherian ring must be finitely presented). Now $\tau_{A,B}$ gives an isomorphism of complexes

$$S^{-1}(\operatorname{Hom}_R(\mathbf{P}_A, B)) \cong \operatorname{Hom}_{S^{-1}R}(S^{-1}(\mathbf{P}_A), S^{-1}B).$$

Taking homology of the left-hand side gives

$$H_n(S^{-1}(\operatorname{Hom}_R(\mathbf{P}_A, B))) \cong S^{-1} H_n(\operatorname{Hom}_R(\mathbf{P}_A, B)) \cong S^{-1} \operatorname{Ext}_R^n(A, B),$$

because localization is an exact functor. On the other hand, homology of the right-hand side is

$$H_n(\operatorname{Hom}_{S^{-1}R}(S^{-1}(\mathbf{P}_A), S^{-1}B)) = \operatorname{Ext}_{S^{-1}R}^n(S^{-1}A, S^{-1}B),$$

because $S^{-1}(\mathbf{P}_A)$ is an $S^{-1}R$-projective resolution of $S^{-1}A$.   •

Remark.   An alternative proof of Proposition 7.39 can be given using a deleted injective resolution $\mathbf{E}^B$ in the second variable. We must still assume that A is finitely generated, in order to use Lemma 7.37, but now we use the fact, when R is noetherian, that localization preserves injectives.   ◄

Corollary 7.40.   *Let A be a finitely generated R-module over a commutative noetherian ring R. Then $A_{\mathfrak{m}}$ is a projective $R_{\mathfrak{m}}$-module for every maximal ideal $\mathfrak{m}$ if and only if A is a projective R-module.*

Proof.   Sufficiency is easy: if A is a free (or projective) R-module, then $S^{-1}A$ is a free (or projective) $S^{-1}R$-module for any multiplicative subset $S \subseteq R$, for tensor product commutes with direct sums. In particular, if A is projective, then $A_{\mathfrak{m}}$ is projective. Necessity follows from Proposition 7.39: for every R-module B and maximal ideal $\mathfrak{m}$, we have

$$\operatorname{Ext}_R^1(A, B)_{\mathfrak{m}} \cong \operatorname{Ext}_{R_{\mathfrak{m}}}^1(A_{\mathfrak{m}}, B_{\mathfrak{m}}) = \{0\},$$

because $A_{\mathfrak{m}}$ is projective. By Proposition 4.90, $\operatorname{Ext}_R^1(A, B) = \{0\}$ for all B, which says that A is projective.   •

Exercises

*7.8   **(i)**   Let G be a p-primary abelian group, where p is prime. If $(m, p) = 1$, prove that $x \mapsto mx$ is an automorphism of G.

   (ii)   If p is an odd prime and $G = \langle g \rangle$ is a cyclic group of order p^2, prove that $\varphi : x \mapsto 2x$ is the unique automorphism with $\varphi(pg) = 2pg$.

*7.9   Prove that any two split extensions of modules A by C are equivalent.

7.10   Prove that if A is an abelian group with $nA = A$ for some positive integer n, then every extension $0 \to A \to E \to \mathbb{I}_n \to 0$ splits.

*7.11   **(i)**   Find an abelian group B for which $\text{Ext}^1_{\mathbb{Z}}(\mathbb{Q}, B) \neq \{0\}$.

   (ii)   Prove that $\mathbb{Q} \otimes_{\mathbb{Z}} \text{Ext}^1_{\mathbb{Z}}(\mathbb{Q}, B) \neq \{0\}$ for the group B in (i).

   (iii)   Prove that Proposition 7.39 may be false when A is not finitely generated, even when $R = \mathbb{Z}$.

*7.12   Let E be a left R-module. Prove that E is injective if and only if $\text{Ext}^1_R(A, E) = \{0\}$ for every left R-module A.

*7.13   **(i)**   Prove that the covariant functor $E = \text{Ext}^1_{\mathbb{Z}}(G, \square)$ is right exact for every abelian group G, and compute its left derived functors $L_n E$.

   (ii)   Prove that the contravariant functor $F = \text{Ext}^1_{\mathbb{Z}}(\square, G)$ is right exact for every abelian group G, and compute its left derived functors $L_n F$. (See the footnote on page 370.)

7.14   **(i)**   If A is an abelian group with $mA = A$ for some nonzero $m \in \mathbb{Z}$, prove that every exact sequence $0 \to A \to G \to \mathbb{I}_m \to 0$ splits. Conclude that $m \, \text{Ext}^1_{\mathbb{Z}}(A, B) = \{0\} = m \, \text{Ext}^1_{\mathbb{Z}}(B, A)$.

   (ii)   If A and C are abelian groups with $mA = \{0\} = nC$, where $(m, n) = 1$, prove that every extension of A by C splits.

7.15   **(i)**   For any ring R, prove that a left R-module B is injective if and only if $\text{Ext}^1_R(R/I, B) = \{0\}$ for every left ideal I.

   Hint. Use the Baer criterion.

   (ii)   If D is an abelian group and $\text{Ext}^1_{\mathbb{Z}}(\mathbb{Q}/\mathbb{Z}, D) = \{0\}$, prove that D is divisible. The converse is true because divisible abelian groups are injective. Does this hold if we replace $\mathbb{Z}$ by a domain R and $\mathbb{Q}/\mathbb{Z}$ by $\text{Frac}(R)/R$?

7.16   Let G be an abelian group G. Prove that G is free abelian if and only if $\text{Ext}^1_{\mathbb{Z}}(G, F) = \{0\}$ for every free abelian group F.

*7.17   Let A be a torsion abelian group and let S^1 be the circle group. Prove that $\text{Ext}^1_{\mathbb{Z}}(A, \mathbb{Z}) \cong \text{Hom}_{\mathbb{Z}}(A, S^1)$.

*7.18 An abelian group W is a **Whitehead group** if $\text{Ext}^1_{\mathbb{Z}}(W, \mathbb{Z}) = \{0\}$.[3]

    (i) Prove that every subgroup of a Whitehead group is a Whitehead group.

    (ii) Prove that $\text{Ext}^1_{\mathbb{Z}}(A, \mathbb{Z}) \cong \text{Hom}_{\mathbb{Z}}(A, S^1)$ if A is a torsion group and S^1 is the circle group. Prove that if $A \neq \{0\}$ is torsion, then A is not a Whitehead group; conclude further that every Whitehead group is torsion-free.

    Hint. Use Exercise 7.17.

    (iii) Let A be a torsion-free abelian group of rank 1; i.e., A is a subgroup of $\mathbb{Q}$. Prove that $A \cong \mathbb{Z}$ if and only if $\text{Hom}_{\mathbb{Z}}(A, \mathbb{Z}) \neq \{0\}$.

    (iv) Let A be a torsion-free abelian group of rank 1. Prove that if A is a Whitehead group, then $A \cong \mathbb{Z}$.

    Hint. Use an exact sequence $0 \to \mathbb{Z} \to A \to T \to 0$, where T is a torsion group whose p-primary component is either cyclic or isomorphic to Prüfer's group of type p^∞.

    (v) (**K. Stein**). Prove that every countable[4] Whitehead group is free abelian.

    Hint. Use Exercise 3.4 on page 114, *Pontrjagin's Lemma*: if A is a countable torsion-free group and every subgroup of A having finite rank is free abelian, then A is free abelian.

7.19 We have constructed the bijection $\psi \colon e(C, A) \to \text{Ext}^1(C, A)$ using a projective resolution of C. Define a function $\psi' \colon e(C, A) \to \text{Ext}^1(C, A)$ using an injective resolution of A, and prove that ψ' is a bijection.

7.20 Consider the diagram

$$
\begin{array}{ccccccccc}
\xi_1 = & 0 & \longrightarrow & A_1 & \longrightarrow & B_1 & \longrightarrow & C_1 & \longrightarrow 0 \\
& & & \downarrow{\scriptstyle h} & & & & \downarrow{\scriptstyle k} & \\
\xi_2 = & 0 & \longrightarrow & A_2 & \longrightarrow & B_2 & \longrightarrow & C_2 & \longrightarrow 0.
\end{array}
$$

Prove that there is a map $\beta \colon B_1 \to B_2$ making the diagram commute if and only if $[h\xi_1] = [\xi_2 k]$.

7.21   **(i)** Prove, in $e(C, A)$, that $-[\xi] = [(-1_A)\xi] = [\xi(-1_C)]$.

    (ii) Generalize (i) by replacing (-1_A) and (-1_C) by μ_r for any r in the center of R.

---

[3] Dixmier proved that a locally compact abelian group A is path connected if and only if $A \cong \mathbb{R}^n \oplus \widehat{D}$, where D is a (discrete) Whitehead group and $\widehat{D}$ is its Pontrjagin dual.

[4] The question whether $\text{Ext}^1_{\mathbb{Z}}(G, \mathbb{Z}) = \{0\}$ implies G is free abelian is known as **Whitehead's problem**. S. Shelah proved that it is undecidable whether uncountable Whitehead groups must be free abelian (see Eklof, "Whitehead's problem is undecidable," *Amer. Math. Monthly* 83 (1976), 775–788).

7.22 Prove that $[\xi] = [0 \to A \xrightarrow{i} B \to C \to 0] \in e(C, A)$ has finite order if and only if there are a nonzero $m \in \mathbb{Z}$ and a map $s: B \to A$ with $si = m \cdot 1_A$.

*7.23   **(i)**   Prove that $e(C, \square): {}_R\mathbf{Mod} \to \mathbf{Ab}$ is a covariant functor if, for $h: A \to A'$, we define $h_*: e(C, A) \to e(C, A')$ by $[\xi] \mapsto [h\xi]$.

       (ii)   Prove that $e(C, \square)$ is naturally isomorphic to $\mathrm{Ext}^1_R(C, \square)$.

7.24 Consider the extension $\chi = 0 \to A' \xrightarrow{i} A \xrightarrow{p} A'' \to 0$.

       (i)   Define $D: \mathrm{Hom}_R(C, A'') \to e(C, A')$ by $k \mapsto [\chi k]$, and prove exactness of

$$\mathrm{Hom}(C, A) \xrightarrow{p_*} \mathrm{Hom}(C, A'') \xrightarrow{D} e(C, A')$$
$$\xrightarrow{i_*} e(C, A) \xrightarrow{p_*} e(C, A'').$$

       (ii)   Prove commutativity of

$$
\begin{array}{ccc}
\mathrm{Hom}(C, A'') & \xrightarrow{D} & e(C, A') \\
& \searrow{\scriptstyle \partial} & \downarrow{\scriptstyle \psi} \\
& & \mathrm{Ext}^1(C, A'),
\end{array}
$$

where ∂ is the connecting homomorphism.

7.25   **(i)**   Prove that $e(\square, A): {}_R\mathbf{Mod} \to \mathbf{Ab}$ is a contravariant functor if, for $k: C' \to C$, we define $k^*: e(C, A) \to e(C', A)$ by $[\xi] \mapsto [\xi k]$.

       (ii)   Prove that $e(\square, A)$ is naturally isomorphic to $\mathrm{Ext}^1_R(\square, A)$.

*7.26 Consider the extension $X = 0 \to C' \xrightarrow{i} C \xrightarrow{p} C'' \to 0$.

       (i)   Define $D': \mathrm{Hom}_R(C', A) \to e(C'', A)$ by $h \mapsto [hX]$, and prove exactness of

$$\mathrm{Hom}(C, A) \xrightarrow{i^*} \mathrm{Hom}(C', A) \xrightarrow{D'} e(C'', A)$$
$$\xrightarrow{p^*} e(C, A) \xrightarrow{i^*} e(C', A).$$

       (ii)   Prove commutativity of

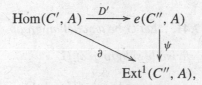

$$
\begin{array}{ccc}
\mathrm{Hom}(C', A) & \xrightarrow{D'} & e(C'', A) \\
& \searrow{\scriptstyle \partial} & \downarrow{\scriptstyle \psi} \\
& & \mathrm{Ext}^1(C'', A),
\end{array}
$$

where ∂' is the connecting homomorphism.

7.3   Cotorsion Groups

Here is a circle of ideas involving two related questions. Which abelian groups can be equipped with a topology that makes them compact topological groups? Which abelian groups A are *realizable* in the sense that $A \cong \text{Ext}^1_{\mathbb{Z}}(X, Y)$ for some abelian groups X and Y? The first question was answered by Kaplansky; the second was answered, independently, by Harrison, Nunke, and Fuchs.

We will use two results, consequences of the Künneth Formula (which we will prove later). All results in the section may be generalized to modules over Dedekind rings.

Proposition 10.86. *Given a commutative hereditary ring R and R-modules A, B, and C, there is an isomorphism*

$$\text{Ext}^1_R(\text{Tor}^R_1(A, B), C) \cong \text{Ext}^1_R(A, \text{Ext}^1_R(B, C)).$$

Proposition 10.87. *If R is a commutative hereditary ring and A, B, C are R-modules, then*

$$\text{Ext}^1_R(A \otimes_R B, C) \oplus \text{Hom}_R(\text{Tor}^R_1(A, B), C))$$
$$\cong \text{Ext}^1_R(A, \text{Hom}_R(B, C)) \oplus \text{Hom}_R(A, \text{Ext}^1_R(B, C)).$$

Notation. In this section, we abbreviate $\text{Hom}_{\mathbb{Z}}(A, B)$, $\text{Ext}^1_{\mathbb{Z}}(A, B)$, and $A \otimes_{\mathbb{Z}} B$ to, respectively, $\text{Hom}(A, B)$, $\text{Ext}^1(A, B)$, and $A \otimes B$.

We begin with a fundamental notion.

Definition. If G is an abelian group, then its *maximal divisible subgroup* is

$$dG = \langle S \subseteq G : S \text{ is divisible} \rangle.$$

We say that G is *reduced* if $dG = \{0\}$.

It is easy to see that G is reduced if and only if $\text{Hom}(\mathbb{Q}, G) = \{0\}$. It follows that if $0 \to A \to B \to C \to 0$ is exact and both A and C are reduced, then B is reduced.

Proposition 7.41.   *Let G be an abelian group.*

(i) *dG is a divisible subgroup of G.*

(ii) *The exact sequence $0 \to dG \to G \xrightarrow{\pi} G/dG \to 0$ is split, and G/dG is reduced.*

(iii) $d: G \mapsto dG$ *defines a left exact additive functor* **Ab** $\to$ **Ab**, *and* $G \mapsto$ *G/dG defines a right exact additive functor* **Ab** $\to$ **Ab**.

Proof.

(i) This result is a special case of Exercise 5.23 on page 255, but, nevertheless, we give a proof of it. We claim that if $S_1, \ldots, S_m$ are divisible subgroups, then $S_1 + \cdots + S_m$ is a divisible subgroup. If $x \in S_1 + \cdots + S_m$, then $x = s_1 + \cdots + s_m$, where $s_i \in S_i$. If $n > 0$, then there are $s_i' \in S_i$ with $s_i = n s_i'$; hence, $x = n(s_1' + \cdots + s_m')$ and $S_1 + \cdots + S_m$ is divisible. It follows that dG is divisible, for if $x \in dG$ and $n > 0$, then $x \in S_1 + \cdots + S_m$ for some divisible subgroups S_i, and so $x = nx'$ for $x' \in S_1 + \cdots + S_m \subseteq dG$.

(ii) The exact sequence splits, for divisible abelian groups are injective (Corollary 3.35). If $D \subseteq G/dG$ is a nonzero divisible subgroup, then $\pi^{-1}(D)$ is a divisible subgroup of G properly containing dG, a contradiction.

(iii) If $f: G \to H$ is a homomorphism and $S \subseteq G$ is divisible, then $f(S)$ is divisible, and so $f(S) \subseteq dH$. Hence, $f(dG) \subseteq dH$. The reader may show that if we define $df = f|dG$, then d is, indeed, a functor as stated. The proof that $G \mapsto G/dG$ gives a right exact functor is also routine. •

Recall Corollary 3.72: a subgroup $S \subseteq G$ is *pure* if $S \cap nG = nS$ for all $n > 0$; that is, if $s \in S$ and $s = ng$ for some $g \in G$ and $n > 0$, then there is $s' \in S$ with $s = ns'$. We say that a sequence $0 \to A \xrightarrow{i} B \to C \to 0$ is *pure exact* if it is exact and $i(A)$ is a pure subgroup of B. We note that if C is torsion-free, then every exact sequence $0 \to A \to B \to C \to 0$ is pure exact.

Definition.   An abelian group A is *algebraically compact* if every pure exact sequence $0 \to A \to B \to C \to 0$ splits.

The motivation for this definition comes from *Pontrjagin duality*. The category **LCA** having as objects all locally compact abelian topological groups and as morphisms $\mathrm{Hom}_c(G, H)$, all continuous homomorphisms $G \to H$, admits a duality $G \mapsto \widehat{G} = \mathrm{Hom}_c(G, S^1)$, where $S^1 = \mathbb{R}/\mathbb{Z}$ is the circle group; that is, $\widehat{\widehat{G}} = G$. Now discrete abelian groups are locally compact, and $\widehat{G}$ is compact if and only if G is discrete (see Hewitt–Ross, *Abstract Harmonic Analysis*).

An abelian group C is *cocyclic* if there is a prime p such that C is isomorphic to a subgroup of the Prüfer group $\mathbb{Z}(p^\infty)$; that is, either C is finite cyclic of order p^m for some m or $C \cong \mathbb{Z}(p^\infty)$.

Theorem 7.42.   *The following statements are equivalent for a (discrete) abelian group G.*

(i)   *G is algebraically compact.*

(ii)   *There is a compact topological group C and G is isomorphic to an algebraic direct summand of C (that is, if one forgets the topology on C, then $C \cong G \oplus B$ for some not necessarily closed subgroup B).*

(iii)   *G is a direct summand of a product of cocyclic groups.*

(iv)   *$G \cong D \oplus E$, where D is divisible and E is a direct summand of a product of finite cyclic groups.*

Proof.   See Fuchs, *Infinite Abelian Groups* I, Chapter VII. The key idea is a theorem of Łos (Fuchs, *Infinite Abelian Groups* I, p. 127) that every abelian group can be imbedded as a pure subgroup of a product of cocyclic groups   •

There is another class of abelian groups, *complete groups*, that is closely related to the class of algebraically compact groups.

Definition.   The ***n-adic topology*** on an abelian group A is the family of all cosets of subgroups $n!A$ for all $n > 0$. If $A_\omega = \bigcap_{n>0} n!A$, then A is ***metric*** if $A_\omega = \{0\}$; in this case, A is a metric space in the n-adic topology.[5] We say that A is ***complete*** if A is metric and complete as a metric space: every Cauchy sequence in A converges to a limit in A (see Example 5.19, which discusses the p-adic topology and shows that completeness corresponds to being a certain inverse limit).

There are abelian groups G that are not metric. For example, since dG is divisible, $dG = n!(dG)$ for all $n > 0$, and so $dG \subseteq G_\omega \subseteq G$. It is easy to see, for every abelian group G, that G/G_ω is metric; in particular, G/G_ω is always reduced.

Theorem 7.43 (Kaplansky).   *If A is a reduced group A, then A is complete if and only if it is algebraically compact. Hence, every reduced algebraically compact group is metric.*

Proof.   Fuchs, *Infinite Abelian Groups* I, p. 163.   •

There are reduced groups that are not metric.

---

[5]If $a, b \in A$, define $\|a - b\| = e^n$ if $a - b \in n!A$, but $a - b \notin (n+1)!A$.

Proposition 7.44.   *If T is the group with presentation*

$$T = (a, b_n : n \geq 1 \,|\, pa, a - p^n b_n : n \geq 1),$$

where p is prime, then T is a reduced torsion group with $T_\omega = \langle a + R \rangle \cong \mathbb{I}_p$.

Sketch of proof. We have $T = F/R$, where F is the free abelian group with basis $\{a, b_n : n \geq 1\}$ and $R = \langle pa, a - p^n b_n : n \geq 1 \rangle$. First, $a + R \neq 0$, lest it lead to an impossible equation in the free abelian group F. It is obvious that $a + R \in T_\omega$, and it is not difficult to show that it generates T_ω. To see that T is reduced, apply $\mathrm{Hom}(\mathbb{Q}, \square)$ to $0 \to T_\omega \to T \to T/T_\omega \to 0$; the flanking terms are $\{0\}$, hence the middle Hom is $\{0\}$, and so T is reduced.   •

Definition.   An abelian group G is ***cotorsion*** if $\mathrm{Ext}^1(\mathbb{Q}, G) = \{0\}$.

Thus, G is cotorsion if every exact sequence $0 \to G \to B \to \mathbb{Q} \to 0$ splits.

The premier example of a cotorsion group is $\mathrm{Ext}^1(X, Y)$ viewed as an abelian group. Before seeing this, let us first note some elementary properties of cotorsion groups.

Proposition 7.45.

   (i) *Every algebraically compact group is cotorsion.*

  (ii) *G is cotorsion if and only if every exact sequence* $0 \to G \to B \to X \to 0$ *with X torsion-free splits.*

 (iii) *A quotient of a cotorsion group is cotorsion.*

 (iv) *A direct summand of a cotorsion group is cotorsion.*

  (v) *A direct product of cotorsion groups is cotorsion.*

 (vi) *A torsion cotorsion group* $G = B \oplus D$, *where D is divisible and B has bounded order; that is,* $nB = \{0\}$ *for some* $n > 0$.

Proof.

   (i) If G is algebraically compact, then every pure exact sequence $0 \to G \to B \to C \to 0$ splits. Since any short exact sequence with $C = \mathbb{Q}$ is pure exact, $0 \to G \to B \to \mathbb{Q} \to 0$ always splits; that is, $\mathrm{Ext}^1(\mathbb{Q}, G) = \{0\}$.

  (ii) If X is torsion-free, there is an exact sequence $0 \to X \to \mathbb{Q} \otimes X \to C \to 0$, which gives exactness of

$$\mathrm{Ext}^1(\mathbb{Q} \otimes X, G) \to \mathrm{Ext}^1(X, G) \to \mathrm{Ext}^2(C, G).$$

Now $\text{Ext}^2_{\mathbb{Z}}(C, G) = \{0\}$, because $D(\mathbb{Z}) = 1$. Also, $\mathbb{Q} \otimes X \cong \bigoplus \mathbb{Q}$, because $\mathbb{Q} \otimes X$ is a vector space over $\mathbb{Q}$, so that

$$\text{Ext}^1(\mathbb{Q} \otimes X, G) \cong \text{Ext}^1\left(\bigoplus \mathbb{Q}, G\right) \cong \prod \text{Ext}^1(\mathbb{Q}, G) = \{0\}.$$

Therefore, $\text{Ext}^1(X, G) = \{0\}$.

(iii) If $C \to C' \to 0$ is exact, then $\text{Ext}^1(\mathbb{Q}, C) \to \text{Ext}^1(\mathbb{Q}, C') \to 0$ is exact, and $\text{Ext}^1(\mathbb{Q}, C') = \{0\}$. Thus, C' is cotorsion.

(iv) If C is cotorsion and $C = A \oplus B$, then

$$\{0\} = \text{Ext}^1(\mathbb{Q}, C) \cong \text{Ext}^1(\mathbb{Q}, A) \oplus \text{Ext}^1(\mathbb{Q}, B),$$

and so both A and B are cotorsion.

(v) This follows from $\text{Ext}^1(\mathbb{Q}, \prod_i C_i) \cong \prod_i \text{Ext}^1(\mathbb{Q}, C_i)$.

(vi) This statement follows from the result quoted in the remark on page 427: a pure subgroup of bounded order is a direct summand. •

Here is the important result.

Theorem 7.46. *An abelian group G is cotorsion if and only if*

$$G \cong D \oplus \text{Ext}^1(X, Y),$$

where X, Y, D are abelian groups and D is divisible.

Proof. If $G \cong D \oplus \text{Ext}^1(X, Y)$. then

$$\text{Ext}^1(\mathbb{Q}, G) \cong \text{Ext}^1(\mathbb{Q}, D) \oplus \text{Ext}^1(\mathbb{Q}, \text{Ext}^1(X, Y)).$$

Since D is divisible, it is injective and $\text{Ext}^1(\mathbb{Q}, D) = \{0\}$. Proposition 10.86 gives

$$\text{Ext}^1(\mathbb{Q}, \text{Ext}^1(X, Y)) \cong \text{Ext}^1(\text{Tor}_1(\mathbb{Q}, X), Y).$$

Now $\text{Tor}_1(\mathbb{Q}, X) = \{0\}$, because $\mathbb{Q}$ is flat, and so $\text{Ext}^1(\mathbb{Q}, \text{Ext}^1(X, Y)) = \{0\}$. Hence, G is cotorsion.

Conversely, let G be cotorsion. As any abelian group, $G = dG \oplus A$, where A is reduced. Since every direct summand of a cotorsion group is cotorsion, it suffices to prove that $A \cong \text{Ext}^1(X, Y)$ for some X, Y. Apply $\text{Hom}(\square, A)$ to the exact sequence

$$0 \to \mathbb{Z} \to \mathbb{Q} \to K \to 0,$$

where $K = \mathbb{Q}/\mathbb{Z}$, to obtain exactness of

$$\text{Hom}(\mathbb{Q}, A) \to \text{Hom}(\mathbb{Z}, A) \to \text{Ext}^1(K, A) \to \text{Ext}^1(\mathbb{Q}, A). \tag{1}$$

The first term vanishes because A is reduced; the last term vanishes because A is cotorsion. Therefore, $A \cong \text{Hom}(\mathbb{Z}, A) \cong \text{Ext}^1(K, A)$. •

Corollary 7.47.

(i) *For every abelian group G, $\mathrm{Ext}^1(K, G)$ and $\mathrm{Hom}(K, G)$ are reduced and cotorsion.*

(ii) *If a group A is reduced and cotorsion, then $A \cong \mathrm{Ext}^1(K, A)$.*

Proof.

(i) The proof of Theorem 7.46 shows that $\mathrm{Ext}^1(K, G)$ is cotorsion. By Proposition 10.87,

$$\mathrm{Ext}^1(\mathbb{Q} \otimes K, G) \oplus \mathrm{Hom}(\mathrm{Tor}_1(\mathbb{Q}, K), G))$$
$$\cong \mathrm{Ext}^1(\mathbb{Q}, \mathrm{Hom}(K, G)) \oplus \mathrm{Hom}(\mathbb{Q}, \mathrm{Ext}^1(K, G)).$$

The left side vanishes because $\mathbb{Q} \otimes K = \{0\}$, by Proposition 2.73, and $\mathrm{Tor}_1(\mathbb{Q}, K) = \{0\}$, because $\mathbb{Q}$ is flat, and so the right side vanishes as well. Therefore, $\mathrm{Hom}(\mathbb{Q}, \mathrm{Ext}^1(K, G)) = \{0\}$; that is, $\mathrm{Ext}^1(K, G)$ is reduced; also, $\mathrm{Ext}^1(\mathbb{Q}, \mathrm{Hom}(K, G)) = \{0\}$; that is, $\mathrm{Hom}(K, G)$ is cotorsion. Finally, we prove that $\mathrm{Hom}(K, T)$ is reduced. The Adjoint Isomorphism says

$$\mathrm{Hom}(\mathbb{Q}, \mathrm{Hom}(K, T)) \cong \mathrm{Hom}(\mathbb{Q} \otimes K, T),$$

and $\mathrm{Hom}(\mathbb{Q} \otimes K, T) = \{0\}$ because $\mathbb{Q} \otimes K = \{0\}$.

(ii) If A is reduced and cotorsion, then it was shown in Eq. (1) that $A \cong \mathrm{Ext}^1(K, A)$. •

Proposition 7.48.

(i) *If T is a reduced torsion group, then there is an exact sequence*

$$0 \to T \to \mathrm{Ext}^1(K, T) \to V \to 0,$$

where V is torsion-free divisible.

(ii) *There exist cotorsion groups that are not algebraically compact.*

Proof.

(i) Apply $\mathrm{Hom}(\square, T)$ to $0 \to \mathbb{Z} \to \mathbb{Q} \to K \to 0$ to obtain exactness of

$$\mathrm{Hom}(\mathbb{Q}, T) \succ \mathrm{Hom}(\mathbb{Z}, T) \succ \mathrm{Ext}^1(K, T) \succ \mathrm{Ext}^1(\mathbb{Q}, T) \succ \mathrm{Ext}^1(\mathbb{Z}, T).$$

The outside terms vanish because T is reduced and $\mathbb{Z}$ is projective. Now $\mathrm{Hom}(\mathbb{Z}, T) \cong T$, and $\mathrm{Ext}^1(\mathbb{Q}, T)$ is a vector space over $\mathbb{Q}$, hence is torsion-free.

(ii) Let T be the reduced torsion group in Proposition 7.44. The group $G = \text{Ext}^1(K, T)$ is reduced and cotorsion, by Corollary 7.47, and T is its torsion subgroup, by (i). Now

$$T_\omega = \bigcap_{n \geq 1} nT \subseteq \bigcap_{n \geq 1} nG = G_\omega,$$

so that $T_\omega \neq \{0\}$ implies $G_\omega \neq \{0\}$. But if A is a reduced algebraically compact group, then $A_\omega = \{0\}$, by Theorem 7.43, and so G is not algebraically compact.   •

We are going to see the etymology of cotorsion.

Proposition 7.49.   *The assignment $\eta \colon T \mapsto \text{Hom}(K, T)$ is a bijection from all isomorphism classes of divisible torsion groups to all isomorphism classes of torsion-free reduced cotorsion groups.*

Proof.   Let T be a torsion abelian group. Now $\text{Hom}(K, T)$ is reduced and cotorsion, by Corollary 7.47. Exactness of $\mathbb{Q} \to K \to 0$ gives exactness of $0 \to \text{Hom}(K, T) \to \text{Hom}(\mathbb{Q}, T)$. But $\text{Hom}(\mathbb{Q}, T)$ is torsion-free, since it is a vector space over $\mathbb{Q}$, and so its subgroup $\text{Hom}(K, T)$ is also torsion-free.

(i) η is an injection. The exact sequence

$$0 \to \text{Hom}(K, T) \to \text{Hom}(\mathbb{Q}, T) \to \text{Hom}(\mathbb{Z}, T) \to \text{Ext}^1(K, T)$$

simplifies to $0 \to \text{Hom}(K, T) \to \text{Hom}(\mathbb{Q}, T) \to T \to 0$ [note that $\text{Ext}^1(K, T) = \{0\}$ because T is divisible, hence injective]. Tensoring by K gives exactness of

$$\text{Tor}_1(K, \text{Hom}(\mathbb{Q}, T)) \to \text{Tor}_1(K, T)$$
$$\to K \otimes \text{Hom}(K, T) \to K \otimes \text{Hom}(\mathbb{Q}, T).$$

Now $\text{Hom}(\mathbb{Q}, T)$ is torsion-free divisible, for it is a vector space over $\mathbb{Q}$. Hence, the first term vanishes because $\text{Hom}(\mathbb{Q}, T)$ is flat, while the last term vanishes because K is torsion and $\text{Hom}(\mathbb{Q}, T)$ is divisible. But $\text{Tor}_1(K, T) \cong T$, by Lemma 7.11. Therefore, $T \cong K \otimes \text{Hom}(K, T)$.

Suppose that T and T' are divisible torsion. If $\eta(T) = \eta(T')$, then $\text{Hom}(K, T) \cong \text{Hom}(K, T')$, and

$$T \cong K \otimes \text{Hom}(K, T) \cong K \otimes \text{Hom}(K, T') \cong T'.$$

(ii) η is a surjection. For any group G, the group $K \otimes G$ is divisible torsion: it is torsion because K is torsion, and it is divisible because exactness of $\mathbb{Q} \to K \to 0$ gives exactness of $\mathbb{Q} \otimes G \to K \otimes G \to 0$. There is an exact sequence

$$\text{Tor}_1(K, G) \to \mathbb{Z} \otimes G \to \mathbb{Q} \otimes G \to K \otimes G \to 0.$$

If G is torsion-free, then the Tor term vanishes, and this simplifies to

$$0 \to G \to \mathbb{Q} \otimes G \to K \otimes G \to 0.$$

Applying $\text{Hom}(K, \square)$ gives exactness of

$$\text{Hom}(K, \mathbb{Q} \otimes G) \to \text{Hom}(K, K \otimes G) \to \text{Ext}^1(K, G) \to \text{Ext}^1(\mathbb{Q}, G).$$

Assume further that G is reduced cotorsion. The first term vanishes because K is torsion and $\mathbb{Q} \otimes G$ is torsion-free, and the last term vanishes because G is cotorsion. Therefore, $\text{Hom}(K, K \otimes G) \cong \text{Ext}^1(K, G)$. Since G is reduced cotorsion, $G \cong \text{Ext}^1(K, G)$, by Corollary 7.47.   •

Corollary 7.50.   *A group G is torsion-free reduced cotorsion if and only if it is a direct summand of a product of copies of p-adic integers $\mathbb{Z}_p$ for various primes p. Hence, G is algebraically compact.*

Proof.   If G is torsion-free reduced cotorsion, then there exists a torsion divisible T with $G \cong \text{Hom}(K, T)$. Now $T \cong \bigoplus_{i \in I} D_i$, where each D_i is a Prüfer group $\mathbb{Z}(p^\infty)$ for some prime p (Kaplansky, *Infinite Abelian Groups*, p. 10). The exact sequence

$$0 \to \bigoplus D_i \to \prod D_i \to X \to 0$$

splits, because $T \cong \bigoplus D_i$ is injective, and so $G \cong \text{Hom}(K, T)$ is a direct summand of $\text{Hom}(K, \prod_i D_i) \cong \prod_i \text{Hom}(K, D_i)$. But $K \cong \bigoplus_p \mathbb{Z}(p^\infty)$, and $\text{Hom}(\mathbb{Z}(p^\infty), \mathbb{Z}(p^\infty)) \cong \mathbb{Z}_p$, the p-adic integers (Fuchs, *Infinite Abelian Groups* I, p. 181).

Now $\mathbb{Z}_p \cong \text{Hom}(K, \mathbb{Z}(p^\infty))$, where $\mathbb{Z}_p$ is the group of p-adic integers, and so $\mathbb{Z}_p$ is torsion-free reduced cotorsion, by Proposition 7.49. Hence, Proposition 7.45 says that a direct product of copies of $\mathbb{Z}_p$ (for various primes) is torsion-free reduced cotorsion, as is any direct summand.

The last statement follows from Theorem 7.42.   •

Definition.   A cotorsion group G is *adjusted* if it is reduced and has no torsion-free direct summands.

We shall see that adjusted groups arise from reduced torsion groups.

Proposition 7.51.   *If G is a reduced cotorsion group, then there exists a unique adjusted subgroup A such that $G = A \oplus B$ and B is torsion-free.*

Proof.   Let $tG \subseteq H \subseteq G$ be such that $H/tG = d(G/tG)$; thus, $G/tG \cong (H/tG) \oplus (G/tG)/(H/tG)$ with $G/H \cong (G/tG)/(H/tG)$ reduced. Now G/tG is a torsion-free group, and H/tG is a pure subgroup [for $H/tG = d(G/tG)$ is even a direct summand]; therefore, $G/H \cong (G/tG)/(H/tG)$ is torsion-free. We claim that H is a direct summand of G. Consider the exact sequence $\operatorname{Hom}(\mathbb{Q}, G/H) \to \operatorname{Ext}^1(\mathbb{Q}, H) \to \operatorname{Ext}^1(\mathbb{Q}, G)$: the first term vanishes because G/H is reduced, and the last term vanishes because G is cotorsion. Therefore, $\operatorname{Ext}^1(\mathbb{Q}, H) = \{0\}$, and H is cotorsion. By Proposition 7.45(ii), $\operatorname{Ext}^1(X, H) = \{0\}$ for every torsion-free group X; in particular, $\operatorname{Ext}^1(G/H, H) = \{0\}$; that is, $G \cong H \oplus (G/H)$.

We claim that H is adjusted. If $H = S \oplus S'$ and S is torsion-free, then $S \cap tG = \{0\}$, and so $H/tG = S \oplus (S'/tG)$.[6] Since H/tG is divisible, S is divisible. We have shown that every torsion-free direct summand of H is divisible. But H is reduced, for it is a subgroup of the reduced group G, and so it has no torsion-free direct summands; that is, H is adjusted.

Finally, we prove uniqueness of H. Suppose that $H' \subseteq G$ is an adjusted direct summand with G/H' torsion-free. We claim that $H \subseteq H'$. Otherwise,

$$\frac{G}{H'} \supseteq \frac{H + H'}{H'} \cong \frac{H}{H \cap H'} \cong \frac{H/tG}{(H \cap H')/tG}.$$

But G/H' is reduced (being isomorphic to a summand of G) and the last group is divisible, being a quotient of $H/tG = d(G/tG)$. Hence, $H \subseteq H'$. Now $G = H \oplus B$ with B torsion-free, so that $H' = H \oplus (H' \cap B)$. But $H' \cap B$ is torsion-free, contradicting H' having no torsion-free summands (because it is adjusted). Therefore, $H = H'$.   •

Corollary 7.52.   *A reduced cotorsion group G is adjusted if and only if $G \cong \operatorname{Ext}^1(K, tG)$.*

Proof.   Since G is reduced cotorsion, it has a unique adjusted direct summand H with G/H divisible; the proof of Proposition 7.51 identifies H as the subgroup with $tG \subseteq H$ and $d(G/tG) = H/tG$. If G is adjusted, then $H = G$; that is, $G/tG = d(G/tG)$ is divisible.

Consider the exact sequence

$$\operatorname{Hom}(K, G/tG) \to \operatorname{Ext}^1(K, tG) \to \operatorname{Ext}^1(K, G) \to \operatorname{Ext}^1(K, G/tG).$$

---

[6]If $h \in tG \subseteq H$, then $nh = 0$ for some $n > 0$. Now $h = s + s'$, where $s \in S$ and $s' \in S'$, and $0 = nh = ns + ns'$. Hence, $ns \in S \cap S' = \{0\}$, so that $ns = 0$. But S is torsion-free, and so $s = 0$ and $h = s' \in S'$; that is, $tG \subseteq S'$.

The first term vanishes because K is torsion and G/tG is torsion-free, and the last term vanishes because G/tG is divisible (hence, injective). Therefore, $\operatorname{Ext}^1(K, tG) \cong \operatorname{Ext}^1(K, G)$. But, as any reduced cotorsion group, $G \cong \operatorname{Ext}^1(K, G)$, by Corollary 7.47(ii).

Conversely, if $G \cong \operatorname{Ext}^1(K, tG)$, then G is reduced and cotorsion, by Corollary 7.47. If $G = X \oplus Y$, where Y is torsion-free, then $tG \cap Y = \{0\}$. Hence, $G/tG \cong X/tG \oplus Y$; that is, Y is a direct summand of G/tG. Since G/tG is divisible, Y is divisble. But Y is reduced, being a subgroup of the reduced group G, and so $Y = \{0\}$. Therefore, G is adjusted. $\bullet$

Proposition 7.53. *The assignment* $\zeta : A \mapsto tA$ *is a bijection from all isomorphism classes of adjusted cotorsion groups to all reduced torsion groups.*

Proof. Let A and A' be adjusted. If $\zeta(A) = \zeta(A')$, then $tA \cong tA'$ and $\operatorname{Ext}^1(K, tA) \cong \operatorname{Ext}^1(K, tA')$. Hence, $A \cong \operatorname{Ext}^1(K, tA) \cong \operatorname{Ext}^1(K, tA') \cong A'$, by Corollary 7.52; that is, ζ is an injection.

Let T be a reduced torsion group. We claim that $A = \operatorname{Ext}^1(K, T)$ is an adjusted cotorsion group with $tA = T$. Of course, A is cotorsion. Consider the exact sequence

$$\operatorname{Hom}(\mathbb{Q}, T) \to \operatorname{Hom}(\mathbb{Z}, T) \to \operatorname{Ext}^1(K, T) \to \operatorname{Ext}^1(\mathbb{Q}, T) \to \operatorname{Ext}^1(\mathbb{Z}, T).$$

The first term vanishes because T is reduced, and the last term vanishes because $\mathbb{Z}$ is projective. The sequence simplifies to

$$0 \to T \to \operatorname{Ext}^1(K, T) \to \operatorname{Ext}^1(\mathbb{Q}, T) \to 0.$$

Since $\operatorname{Ext}^1(\mathbb{Q}, T)$ is a vector space over $\mathbb{Q}$, it is torsion-free (divisible), and so $T \cong t \operatorname{Ext}^1(K, T) = tA$. Finally, A is adjusted, by Corollary 7.52. $\bullet$

Theorem 7.54. *There is a bijection from all isomorphism classes of torsion abelian groups to all isomorphism classes of reduced cotorsion groups; it is given by*

$$T \mapsto \operatorname{Hom}(K, T) \oplus \operatorname{Ext}^1(K, T).$$

Proof. If T is torsion, then $T = dT \oplus T'$, where T' is reduced torsion, and

$$\operatorname{Hom}(K, T) \oplus \operatorname{Ext}^1(K, T) = \operatorname{Hom}(K, dT) \oplus \operatorname{Ext}^1(K, T').$$

Now $\operatorname{Hom}(K, dT)$ is torsion-free reduced cotorsion, by Proposition 7.49, while $T' \mapsto \operatorname{Ext}^1(K, T')$, being the inverse of the bijection ζ of Proposition 7.53, is an adjusted cotorsion group. $\bullet$

It follows from Theorem 7.54 that any classification of torsion groups gives a classification of reduced cotorsion groups. For example, all *countable* reduced torsion groups T are classified by Ulm's Theorem (Kaplansky, *Infinite Abelian Groups*, p. 27), and so all reduced cotorsion groups $\operatorname{Ext}^1(K, T)$ are classified for such T.

7.4 Universal Coefficients

In Chapter 6, we defined the homology groups $H_n(X)$ of a topological space X with coefficients in an abelian group A as

$$H_n(X, A) = H_n(\mathbf{S}_\bullet(X) \otimes_{\mathbb{Z}} A).$$

Of course, $H_n(X, \mathbb{Z}) = H_n(X)$, the homology group defined in Chapter 1. Similarly, we defined cohomology groups with coefficients as

$$H^n(X, A) = H_{-n}(\mathrm{Hom}_{\mathbb{Z}}(\mathbf{S}_\bullet(X), A)).$$

Your first guess is $H_n(X, A) \cong H_n(X) \otimes_{\mathbb{Z}} A$, but this is usually not the case. The next theorem allows us to compute $H_n(X, A)$ from $H_n(X)$; the corresponding result for cohomology will be given afterwards.

We will use Exercise 2.17 on page 67 in the next proof: if

$$A \xrightarrow{f} B \xrightarrow{g} C \xrightarrow{h} D \xrightarrow{k} E$$

is exact, then there is a short exact sequence

$$0 \to \mathrm{coker}\, f \xrightarrow{\alpha} C \xrightarrow{\beta} \ker k \to 0,$$

where $\alpha \colon b + \mathrm{im}\, f \mapsto gb$ and $\beta \colon c \mapsto hc$.

Theorem 7.55 (Universal Coefficient Theorem for Homology, I). *Let R be a ring, let A be a left R-module, and let $(\mathbf{K}, d)$ be a complex of flat right R-modules whose subcomplex $\mathbf{B}$ of boundaries also has all terms flat.*

For all $n \geq 0$, there is an exact sequence

$$0 \to H_n(\mathbf{K}) \otimes_R A \xrightarrow{\lambda_n} H_n(\mathbf{K} \otimes_R A) \xrightarrow{\mu_n} \mathrm{Tor}_1^R(H_{n-1}(\mathbf{K}), A) \to 0,$$

where $\lambda_n \colon \mathrm{cls}(z) \otimes a \mapsto \mathrm{cls}(z \otimes a)$, and both λ_n and μ_n are natural.

Proof. By Corollary 7.4, each term in the fundamental exact sequence

$$0 \to Z_n \xrightarrow{i_n} K_n \xrightarrow{d_n'} B_{n-1} \to 0 \tag{1}$$

is flat, where i_n is the inclusion and d_n' is obtained from the differential d_n by changing its target from K_{n-1} to $\mathrm{im}\, d_n = B_{n-1}$. Since B_{n-1} is flat, $\mathrm{Tor}_1^R(B_{n-1}, A) = \{0\}$, and we have exactness of

$$0 \to Z_n \otimes_R A \xrightarrow{i_n \otimes 1} K_n \otimes_R A \xrightarrow{d_n' \otimes 1} B_{n-1} \otimes_R A \to 0.$$

The maps $i_n \otimes 1, d_n' \otimes 1$ can be assembled to give an exact sequence of complexes

$$0 \to \mathbf{Z} \otimes_R A \xrightarrow{i \otimes 1} \mathbf{K} \otimes_R A \xrightarrow{d' \otimes 1} \mathbf{B}[-1] \otimes_R A \to 0$$

(recall that $\mathbf{B}[-1]$ is the complex obtained from $\mathbf{B}$ by reindexing: $\mathbf{B}[-1]_n = B_{n-1}$ for all n). The corresponding long exact sequence is

$$H_{n+1}(\mathbf{B}[-1] \otimes_R A) \xrightarrow{\partial_{n+1}} H_n(\mathbf{Z} \otimes_R A) \xrightarrow{(i_n \otimes 1)_*} H_n(\mathbf{K} \otimes_R A)$$

$$\to H_n(\mathbf{B}[-1] \otimes_R A) \xrightarrow{\partial_n} H_{n-1}(\mathbf{Z} \otimes_R A).$$

Since $\mathbf{Z}$ and $\mathbf{B}[-1]$ have zero differentials, we have $H_{n+1}(\mathbf{B}[-1] \otimes_R A) = B_n \otimes_R A$ and $H_n(\mathbf{Z} \otimes_R A) = Z_n \otimes_R A$, by Example 6.1(iii). Thus, we may rewrite the long exact sequence as

$$B_n \otimes_R A \xrightarrow{\partial_{n+1}} Z_n \otimes_R A \to H_n(\mathbf{K} \otimes_R A)$$

$$\to B_{n-1} \otimes_R A \xrightarrow{\partial_n} Z_{n-1} \otimes_R A.$$

By Exercise 2.17, there are short exact sequences

$$\operatorname{coker} \partial_{n+1} \overset{\alpha_n}{\rightarrowtail} H_n(\mathbf{K} \otimes_R A) \overset{\beta_n}{\twoheadrightarrow} \ker \partial_n, \tag{2}$$

where $\alpha_n : z \otimes a + \operatorname{im} \partial_{n+1} \mapsto \operatorname{cls}(i_n z \otimes a)$.

We compute the connecting homomorphism ∂_{n+1} using its definition.

$$
\begin{array}{ccc}
 & & K_{n+1} \otimes_R A \xrightarrow{d'_{n+1} \otimes 1} B_n \otimes_R A \\
 & & \quad\quad\quad\quad \downarrow{\scriptstyle d_{n+1} \otimes 1} \\
Z_n \otimes_R A & \xrightarrow{i_n \otimes 1} & K_n \otimes_R A.
\end{array}
$$

If $b \in B_n$, then $b = d_{n+1}k$ for some $k \in K_{n+1}$, and so

$$\partial_{n+1} : b \otimes a \mapsto k \otimes a \mapsto b \otimes a \to (i_n \otimes 1)^{-1}(b \otimes a).$$

Now $(i_n \otimes 1)^{-1}(b \otimes a) = b \otimes a$, where b is regarded as an element of Z_n; thus, if $j_n : B_n \to Z_n$ is the inclusion, then $\partial_{n+1} = j_n \otimes 1$ and $\partial_n = j_{n-1} \otimes 1$. Hence, exact sequence (2) is

$$\operatorname{coker}(j_n \otimes 1) \overset{\alpha_n}{\rightarrowtail} H_n(\mathbf{K} \otimes_R A) \overset{\beta_n}{\twoheadrightarrow} \ker(j_{n-1} \otimes 1), \tag{3}$$

where $\alpha_n : \operatorname{cls}(z \otimes a) + \operatorname{im}(j_n \otimes 1) \mapsto \operatorname{cls}(i_n z \otimes a)$. The reader may prove that both α_n and β_n are natural.

Consider the flanking terms in (3). Since B_n and Z_n are flat, the exact sequence $0 \to B_n \xrightarrow{j_n} Z_n \to H_n(\mathbf{K}) \to 0$ is a flat resolution of $H_n(\mathbf{K})$. Thus, $0 \to B_n \xrightarrow{j_n} Z_n \to 0$ is a deleted flat resolution of $H_n(\mathbf{K})$; after tensoring by A, its homology is given by

$$H_1 = \ker(j_{n-1} \otimes 1) = \operatorname{Tor}_1^R(H_{n-1}(\mathbf{K}), A)$$

and

$$H_0 = \operatorname{coker}(j_n \otimes 1) = \operatorname{Tor}_0^R(H_n(\mathbf{K}), A).$$

Recall Theorem 7.5: Tor can be computed using flat resolutions instead of projective resolutions. Thus, exact sequence (3) is

$$0 \to \operatorname{Tor}_0^R(H_n(\mathbf{K}), A) \to H_n(\mathbf{K} \otimes_R A) \to \operatorname{Tor}_1^R(H_{n-1}(\mathbf{K}), A) \to 0.$$

But $\operatorname{Tor}_0^R(H_n(\mathbf{K}), A) \cong H_n(\mathbf{K}) \otimes_R A$; making this isomorphism explicit, we see that the imbedding $\lambda_n : H_n(\mathbf{K}) \otimes_R A \to H_n(\mathbf{K} \otimes_R A)$ is given by $\operatorname{cls}(z) \otimes a \mapsto \operatorname{cls}(z \otimes a)$. •

Corollary 7.56 (Universal Coefficient Theorem for Homology, II). *Let R be a right hereditary ring, let A be a left R-module, and let $(\mathbf{K}, d)$ be a complex of projective right R-modules.*

(i) *For all $n \geq 0$, there is an exact sequence*

$$0 \to H_n(\mathbf{K}) \otimes_R A \xrightarrow{\lambda_n} H_n(\mathbf{K} \otimes_R A) \xrightarrow{\mu_n} \operatorname{Tor}_1^R(H_{n-1}(\mathbf{K}), A) \to 0,$$

where $\lambda_n : \operatorname{cls}(z) \otimes a \mapsto \operatorname{cls}(z \otimes a)$, and both λ_n and μ_n are natural.

(ii) *For all $n \geq 0$, the exact sequence splits[7]:*

$$H_n(\mathbf{K} \otimes_R A) \cong H_n(\mathbf{K}) \otimes_R A \oplus \operatorname{Tor}_1^R(H_{n-1}(\mathbf{K}), A).$$

Proof.

(i) Since R is right hereditary, every submodule of a projective right R-module is also projective, by Corollary 4.14. Therefore, $B_n \subseteq K_n$ is projective, hence flat, and so the hypothesis of Theorem 7.55 is satisfied.

(ii) The sequence $0 \to Z_n \otimes A \xrightarrow{i_n \otimes 1} K_n \otimes A \to B_{n-1} \otimes A \to 0$ is split exact, because exact sequence (1) splits. More precisely, $\operatorname{im}(i_n \otimes 1)$ is a direct summand of $K_n \otimes A$. There are inclusions

$$\operatorname{im}(d_{n+1} \otimes 1) \subseteq \operatorname{im}(i_n \otimes 1) \subseteq \ker(d_n \otimes 1) \subseteq K_n \otimes A.$$

By Corollary 2.24(i), $\operatorname{im}(i_n \otimes 1)$ is a direct summand of $\ker(d_n \otimes 1)$ and, by Corollary 2.24(ii), $\operatorname{im}(i_n \otimes 1) / \operatorname{im}(d_{n+1} \otimes 1)$ is a direct summand of $\ker(d_n \otimes 1) / \operatorname{im}(d_{n+1} \otimes 1) = H_n(\mathbf{K} \otimes_R A)$. Now $d_{n+1} = i_n j_n d'_{n+1}$. Using the general fact that $\operatorname{im} fg = f(\operatorname{im} g)$, we see that

$$\operatorname{im}(d_{n+1} \otimes 1) = (i_n \otimes 1) \operatorname{im}(j_n d'_{n+1} \otimes 1)$$

---

[7] The splitting need not be natural.

and
$$\operatorname{im}(j_n d'_{n+1} \otimes 1) = (j_n \otimes 1)\operatorname{im}(d'_{n+1} \otimes 1).$$

But $\operatorname{im}(d'_{n+1} \otimes 1) = B_n \otimes_R A$, because d'_{n+1} is surjective and $\square \otimes_R A$ is right exact; therefore,

$$(j_n \otimes 1)\operatorname{im}(d'_{n+1} \otimes 1) = (j_n \otimes 1)(B_n \otimes_R A) = \operatorname{im}(j_n \otimes 1).$$

Therefore,

$$\operatorname{im}(i_n \otimes 1)/\operatorname{im}(d_{n+1} \otimes 1) = \operatorname{im}(i_n \otimes 1)/(i_n \otimes 1)\operatorname{im}(j_n \otimes 1).$$

But $\operatorname{im}(i_n \otimes 1) = Z_n \otimes_R A$, so that $\operatorname{im}(i_n \otimes 1)/(i_n \otimes 1)\operatorname{im}(j_n \otimes 1) = Z_n \otimes_R A/\operatorname{im}(j_n \otimes 1) = \operatorname{coker}(j_n \otimes 1) = H_n(\mathbf{K}) \otimes_R A$. •

Corollary 7.57. *If X is a topological space and A is an abelian group, then, for all $n \geq 0$,*

$$H_n(X, A) \cong H_n(X) \otimes_{\mathbb{Z}} A \ \oplus \ \operatorname{Tor}_1^{\mathbb{Z}}(H_{n-1}(X), A).$$

Proof. Now $H_n(X) = H_n(\mathbf{S}_\bullet(X))$ and $H_n(X, A) = H_n(\mathbf{S}_\bullet(X) \otimes_{\mathbb{Z}} A)$. The Universal Coefficient Theorem applies at once, for $\mathbb{Z}$ is hereditary and every term of $\mathbf{S}_\bullet(X)$ is free abelian. •

Corollary 7.58. *If either $H_{n-1}(X)$ or A is a torsion-free abelian group, then*

$$H_n(X, A) \cong H_n(X) \otimes_{\mathbb{Z}} A.$$

Proof. Either hypothesis forces $\operatorname{Tor}_1^{\mathbb{Z}}(H_{n-1}(X), A) = \{0\}$. •

Here is the dual result.

Theorem 7.59 (Universal Coefficient Theorem for Cohomology). *Let R be a ring, let A be a left R-module, and let $(\mathbf{K}, d)$ be a complex of projective left R-modules whose subcomplex $\mathbf{B}$ of boundaries has all terms projective.*

(i) *Then, for all $n \geq 0$, there is an exact sequence*

$$0 \to \operatorname{Ext}_R^1(H_{n-1}, A) \xrightarrow{\lambda_n} H^n(\operatorname{Hom}_R(\mathbf{K}, A)) \xrightarrow{\mu_n} \operatorname{Hom}_R(H_n, A)) \to 0$$

[where H_n abbreviates $H_n(\mathbf{K})$] with both λ_n and μ_n natural.

(ii) *If R is left hereditary, then, for all $n \geq 0$, the exact sequence splits[8]:*

$$H^n(\operatorname{Hom}_R(\mathbf{K}, A)) \cong \operatorname{Hom}_R(H_n(\mathbf{K}), A) \ \oplus \ \operatorname{Ext}_R^1(H_{n-1}(\mathbf{K}), A).$$

---

[8]The splitting need not be natural.

Proof.   Adapt the proof of Theorem 7.55, using the (contravariant) functor $\operatorname{Hom}_R(\square, A)$ instead of $\square \otimes_R A$. The stronger hypothesis that boundaries be projective (instead of flat) is needed because Ext requires projective resolutions.   ●

The next result shows that the homology groups of a space determine its cohomology groups.

Corollary 7.60.   *If X is a topological space and A is an abelian group, then for all $n \geq 0$,*

$$H^n(X, A) \cong \operatorname{Hom}_{\mathbb{Z}}(H_n(X), A) \oplus \operatorname{Ext}_{\mathbb{Z}}^1(H_{n-1}(X), A).$$

Proof.   By definition,

$$H^n(X) = H_{-n}(\operatorname{Hom}_{\mathbb{Z}}(\mathbf{S}_\bullet(X), \mathbb{Z})),$$

while

$$H^n(X, A) = H_{-n}(\operatorname{Hom}_{\mathbb{Z}}(\mathbf{S}_\bullet(X), A)).$$

The Universal Coefficient Theorem for Cohomology applies at once, for $\mathbb{Z}$ is hereditary and every term of $\mathbf{S}_\bullet(X)$ is free abelian.   ●

Corollary 7.61.   *Let $\mathbf{K}$ be a complex of free abelian groups. If $H_{n-1}(\mathbf{K})$ is free or A is divisible (for example, if A is the additive group of a field of characteristic 0), then*

$$H^n(\operatorname{Hom}_{\mathbb{Z}}(\mathbf{K}, A)) \cong \operatorname{Hom}_{\mathbb{Z}}(H_n(\mathbf{K}), A).$$

Proof.   Either hypothesis forces $\operatorname{Ext}_{\mathbb{Z}}^1(H_{n-1}, A) = \{0\}$.   ●

Of course, variations on this theme are played by other hypotheses guaranteeing the vanishing of Ext^1.

Corollary 7.62.   *If $\mathbf{K}$ is a complex of vector spaces over a field k, and if V is a vector space over k, then for all $n \geq 0$,*

$$H^n(\operatorname{Hom}_k(\mathbf{K}, V)) \cong \operatorname{Hom}_k(H_n(\mathbf{K}), V).$$

In particular, $H^n(\operatorname{Hom}_k(\mathbf{K}, k)) \cong H_n(\mathbf{K})^*$, where $*$ denotes the dual space.

Proof.   As every k-module, V is injective, and so $\operatorname{Ext}_k^1(H_{n-1}, V) = \{0\}$.   ●

It is known, for any sequence of abelian groups $C_0, C_1, C_2, \ldots$, that there exists a topological space X with homology groups $H_n(X) \cong C_n$ for all n. In contrast, if the cohomology group $H^n(X, \mathbb{Z})$ is countable, then it is a direct sum of a finite group and a free abelian group [Nunke–Rotman, "Singular cohomology groups," *J. London Math Soc*, 37 (1962), 301–306].

8

Homology and Rings

We are now going to show that homology is a valuable tool in studying rings.

8.1 Dimensions of Rings

We can use Ext and Tor to define various dimensions of a ring, essentially measuring how far it is from being semisimple. We shall see, for example, that semisimple rings have dimension 0, while hereditary rings have dimension 1. The basic idea has already arisen, in the proof of Proposition 7.33: if the torsion subgroup tG of an abelian group G is of bounded order, then tG is a direct summand of G. The proof used Exercise 6.18, which says that $\text{Ext}^2_{\mathbb{Z}}(A, B) = \{0\}$ for all abelian groups A and B. Here is that proof generalized from abelian groups to modules over Dedekind rings.

Recall that if $\to P_n \xrightarrow{d_n} P_{n-1} \to \cdots \to P_1 \xrightarrow{d_1} P_0 \to 0$ is a deleted projective resolution of a left R-module A, then

$$\text{Ext}^n_R(A, B) = \frac{\ker d_n^*}{\operatorname{im} d_{n-1}^*},$$

where d_n^* is the induced map $\text{Hom}_R(P_{n-1}, B) \to \text{Hom}_R(P_n, B)$.

Proposition 8.1. *If R is left hereditary, then $\text{Ext}^n_R(A, B) = \{0\}$ for all $n \geq 2$ and for all left R-modules A, B.*

Proof. There is an exact sequence $0 \to P_1 \to P_0 \to A \to 0$ with P_0 projective. Since R is left hereditary, Corollary 4.14 says that P_1 is projective,

J.J. Rotman, *An Introduction to Homological Algebra*, Universitext, DOI 10.1007/978-0-387-68324-9_8, © Springer Science+Business Media LLC 2009

and so this short exact sequence can be viewed as a projective resolution of A in which $P_n = \{0\}$ for all $n \geq 2$. Hence, the differentials $d_n^* = 0$ for all $n \geq 2$, and $\text{Ext}_R^n(A, B) = \{0\}$ for all $n \geq 2$.   •

Proposition 8.2.   *If R is a Dedekind ring and A is a torsion-free R-module, then $\text{Ext}_R^1(A, B)$ is a divisible R-module for every R-module B.*

Proof.   By Lemma 4.33(ii) (which holds for every domain R), there is an exact sequence $0 \to A \xrightarrow{i} V \to X \to 0$, where V is a vector space over $Q = \text{Frac}(R)$. This gives rise to the exact sequence

$$\text{Ext}_R^1(V, B) \xrightarrow{i^*} \text{Ext}_R^1(A, B) \to \text{Ext}_R^2(X, B).$$

The last term is $\{0\}$, by Proposition 8.1, so that i^* is surjective. But $\text{Ext}_R^1(V, B)$ is also a vector space over Q, by Exercise 6.19 on page 376, and so it is a divisible R-module. Therefore, its image, $\text{Ext}_R^1(A, B)$, is divisible.   •

We have seen, in Proposition 7.34, that the torsion subgroup of an abelian group need not be a direct summand.

Theorem 8.3.   *Let R be a Dedekind ring, and let B be an R-module with torsion submodule $T = tB$. If there is a nonzero $r \in R$ with $rT = \{0\}$, then T is a direct summand of B.*

Proof.   We must show that $0 \to T \to B \to B/T \to 0$ splits. Since B/T is torsion-free, it suffices to prove that $\text{Ext}_R^1(A, T) = \{0\}$ whenever A is torsion-free. Now $\text{Ext}_R^1(A, T)$ is divisible, by Proposition 8.2; on the other hand, $rT = \{0\}$ implies $r\,\text{Ext}_R^1(A, T) = \{0\}$, by Proposition 6.38. It follows that $\text{Ext}_R^1(A, T) = \{0\}$, for if E is a divisible module, then $rE = E$ for all $r \neq 0$. Therefore, the short exact sequence splits.   •

We can measure how far away a module is from being projective.

Definition.   If A is a left R-module (for some ring R), then $\text{pd}_R(A) \leq n$ (pd abbreviates *projective dimension*) if there is a finite projective resolution

$$0 \to P_n \to \cdots \to P_1 \to P_0 \to A \to 0.$$

We will usually omit the subscript R. If no such finite resolution exists, then $\text{pd}(A) = \infty$; otherwise, $\text{pd}(A) = n$ if n is the length of a shortest projective resolution of A.

Example 8.4.

(i) $\mathrm{pd}(A) = 0$ if and only if A is projective.

(ii) If R is semisimple, then Proposition 4.5 gives $\mathrm{pd}(A) = 0$ for every left R-module A.

(iii) If R is left hereditary, then $\mathrm{pd}(A) \leq 1$ for every left R-module A, for Theorem 4.19 says that every submodule of a projective R-module is projective. ◀

Recall that the *nth syzygy* K_n of a module R is defined by $K_0 = \ker \varepsilon$ and $K_n = \ker d_n$ for $n \geq 1$, where $\mathbf{P} = \; \to P_n \xrightarrow{d_n} P_{n-1} \to \cdots \to P_1 \xrightarrow{d_1} P_0 \xrightarrow{\varepsilon} A \to 0$ is a projective resolution of A. Obviously, the syzygies of A depend on the choice of projective resolution of A.

Definition. Two modules A and B are ***projectively equivalent*** if there exist projective modules P and P' with $A \oplus P \cong B \oplus P'$.

It is clear that this is an equivalence relation.

Proposition 8.5. *Let $(K_n)_{n \geq 0}$ and $(K'_n)_{n \geq 0}$ be syzygies of a left R-module A defined by two projective resolutions of A.*

(i) *For each $n \geq 0$, K_n and K'_n are projectively equivalent.*

(ii) *For every left R-module B, we have $\mathrm{Ext}^1_R(K_n, B) \cong \mathrm{Ext}^1_R(K'_n, B)$.*

(iii) *For every left R-module B and every $n \geq 1$, we have $\mathrm{Ext}^{n+1}_R(A, B) \cong \mathrm{Ext}^1_R(K_{n-1}, B)$.*

Proof.

(i) This follows at once from Exercise 3.15 on page 128, the generalized Schanuel Lemma.

(ii) If P is projective, $\mathrm{Ext}^1_R(K_n \oplus P, B) \cong \mathrm{Ext}^1_R(K_n, B) \oplus \mathrm{Ext}^1_R(P, B) \cong \mathrm{Ext}^1_R(K_n, B)$. Thus, if $K_n \oplus P \cong K'_n \oplus P'$, then $\mathrm{Ext}^1_R(K_n, B) \cong \mathrm{Ext}^1_R(K'_n, B)$.

(iii) Corollary 6.55. •

As a result of Proposition 8.5, one often abuses language and speaks of *the* nth syzygy of a module, even though such a module is only defined to projective equivalence.

Proposition 8.6.   *The following are equivalent for a left R-module A.*

(i) $\text{pd}(A) \leq n$.

(ii) $\text{Ext}_R^k(A, B) = \{0\}$ *for all left R-modules B and all* $k \geq n + 1$.

(iii) $\text{Ext}_R^{n+1}(A, B) = \{0\}$ *for all left R-modules B.*

(iv) *There exists a projective resolution of A whose* $(n-1)$*st syzygy is projective.*

(v) *Every projective resolution of A has its* $(n-1)$*st syzygy projective.*

Proof.

(i) $\Rightarrow$ (ii) There is a projective resolution $0 \to P_n \to \cdots \to P_0 \to A \to 0$ with $P_k = \{0\}$ for all $k \geq n + 1$. Therefore, the induced maps $d_k^*\colon \text{Hom}_R(P_{k-1}, B) \to \text{Hom}_R(P_k, B)$ are 0 for all $k \geq n + 1$, and so $\text{Ext}_R^k(A, B) = \{0\}$ for all $k \geq n + 1$.

(ii) $\Rightarrow$ (iii) Trivial.

(iii) $\Rightarrow$ (iv) Let K_{n-1} be the $(n-1)$st syzygy of a projective resolution of A. By hypothesis, $\text{Ext}_R^{n+1}(A, B) = \{0\}$ for all B. Now $\text{Ext}_R^{n+1}(A, B) \cong \text{Ext}_R^1(K_{n-1}, B)$, by Proposition 8.5(iii), and so $\text{Ext}_R^1(K_{n-1}, B) = \{0\}$ for all B. Hence, K_{n-1} is projective, by Corollary 7.25.

(iv) $\Rightarrow$ (v) Assume that K_{n-1} and K'_{n-1} are $(n-1)$st syzygies arising from two projective resolutions of A. By Proposition 8.5(i), there are projective modules P and P' with $K_{n-1} \oplus P \cong K'_{n-1} \oplus P'$. But if K_{n-1} is projective, then K'_{n-1} is a direct summand of the projective module $K_{n-1} \oplus P$ and, hence, is projective.

(v) $\Rightarrow$ (i) If $\cdots \to P_1 \to P_0 \to A \to 0$ is a projective resolution of A, then

$$0 \to K_{n-1} \to P_{n-1} \to \cdots \to P_1 \to P_0 \to A \to 0$$

is an exact sequence, where K_{n-1} is the $(n-1)$st syzygy. But if K_{n-1} is projective, then this sequence is a projective resolution of A, and this says that $\text{pd}(A) \leq n$.   •

We now introduce a notation that will soon be simplified.

Definition.   If R is a ring, then its *left projective global dimension* is

$$\ell\text{pD}(R) = \sup\{\text{pd}(A) : A \in \text{obj}(_R\textbf{Mod})\}.$$

Corollary 8.7. $\ell\mathrm{pD}(R) \leq n$ *if and only if* $\mathrm{Ext}_R^{n+1}(A, B) = \{0\}$ *for all left R-modules A and B.*

Proof.   Immediate from Proposition 8.6(iii).   •

Example 8.8.

(i) If R is semisimple, then $\ell\mathrm{pD}(R) = 0$.

(ii) If R is left hereditary, then $\ell\mathrm{pD}(R) \leq 1$.   ◄

All of this discussion can be repeated using injective modules instead of projectives; we merely state the definitions and results. The fundamental reason this can be done is Theorem 6.67: Ext does not depend on the variable being resolved.

Definition.   If B is a left R-module (for some ring R), then $\mathrm{id}_R(B) \leq n$ (id abbreviates *injective dimension*) if there is a finite injective resolution

$$0 \to B \to E^0 \to E^1 \to \cdots \to E^n \to 0.$$

We will usually omit the subscript R. If no such finite resolution exists, then $\mathrm{id}(B) = \infty$; otherwise, $\mathrm{id}(B) = n$ if n is the length of a shortest injective resolution of B.

Example 8.9.

(i) $\mathrm{id}(B) = 0$ if and only if B is injective.

(ii) If R is semisimple, then Proposition 4.5 gives $\mathrm{id}(B) = 0$ for every left R-module B.

(iii) If R is left hereditary, then $\mathrm{id}(B) \leq 1$ for every left R-module B, for Theorem 4.19 says that every submodule of an injective R-module is injective.   ◄

Recall that the *nth cosyzygy* V^n of a module B is defined by $V^0 = \mathrm{coker}\,\eta$ and $V^n = \mathrm{coker}\,d^{n-1}$ for $n \geq 1$, where $\mathbf{E} = 0 \to B \xrightarrow{\eta} E^0 \xrightarrow{d^0} E^1 \xrightarrow{d^1} E^2 \to \cdots$ is an injective resolution of B. Obviously, the cosyzygies of B depend on the choice of injective resolution of B.

Definition.   Two modules V and W are *injectively equivalent* if there exist injective modules E and E' with $V \oplus E \cong W \oplus E'$.

It is obvious that this is an equivalence relation.

Proposition 8.10. *Let* $(V^n)_{n \geq 0}$ *and* $(V'^n)_{n \geq 0}$ *be cosyzygies of a left R-module B defined by two injective resolutions of B.*

(i) *For each* $n \geq 0$, V^n *and* V'^n *are injectively equivalent.*

(ii) *For every left R-module A, we have* $\mathrm{Ext}^1_R(A, V^n) \cong \mathrm{Ext}^1_R(A, V'^n)$.

(iii) *For every left R-module A and every* $n \geq 1$, *we have* $\mathrm{Ext}^{n+1}_R(A, B) \cong \mathrm{Ext}^1_R(A, V^{n-1})$.

As a result of Proposition 8.10, one often abuses language and speaks of *the* nth cosyzygy of a module, even though such a module is only defined to injective equivalence.

Proposition 8.11. *The following are equivalent for a left R-module B.*

(i) $\mathrm{id}(B) \leq n$.

(ii) $\mathrm{Ext}^k_R(A, B) = \{0\}$ *for all left R-modules A and all* $k \geq n + 1$.

(iii) $\mathrm{Ext}^{n+1}_R(A, B) = \{0\}$ *for all left R-modules A.*

(iv) *There exists an injective resolution of B whose* $(n-1)$st *cosyzygy is injective.*

(v) *Every injective resolution of B has its* $(n-1)$st *cosyzygy injective.*

Definition. If R is a ring, then its *left injective global dimension* is

$$\ell\mathrm{iD}(R) = \sup\{\mathrm{id}(B) : B \in \mathrm{obj}(_R\mathbf{Mod})\}.$$

Corollary 8.12. $\ell\mathrm{iD}(R) \leq n$ *if and only if* $\mathrm{Ext}^{n+1}_R(A, B) = \{0\}$ *for all left R-modules A and B.*

Example 8.13.

(i) If R is semisimple, then $\ell\mathrm{iD}(R) = 0$.

(ii) If R is left hereditary and B is a left R-module, then Example 8.9(iii) shows that $\ell\mathrm{iD}(R) \leq 1$. ◄

We now combine Corollaries 8.7 and 8.12.

Theorem 8.14.   *For any ring R, we have $\ell\mathrm{pD}(R) = \ell\mathrm{iD}(R)$.*

Proof.   Both dimensions are characterized by $\mathrm{Ext}_R^{n+1}(A, B) = \{0\}$ for all left R-modules A and B.   •

In light of Theorem 8.14, we now simplify earlier notation.

Definition.   If R is a ring, then its *left global dimension* $\ell\mathrm{D}(R)$ is the common value of $\ell\mathrm{pD}(R)$ and $\ell\mathrm{iD}(R)$.

We now see why, in Chapter 4, we first discussed semisimple rings and then hereditary rings; semisimple rings have global dimension 0 while hereditary rings, having global dimension 1, are only one step removed from semisimple rings. Hilbert's Syzygy Theorem, which we will soon prove, states that if k is a field, then $k[x_1, \ldots, x_n]$ has global dimension n.

All of the results just proved for left modules hold, mutatis mutandis, for right modules. The *right global dimension* $r\mathrm{D}(R)$ is the common value of $r\mathrm{pD}(R)$ and $r\mathrm{iD}(R)$. Of course, if R is a commutative ring, then we speak of the *global dimension* $\mathrm{D}(R)$, dropping ℓ and r.

Recall that a ring R is left semisimple if it is a direct sum of minimal left ideals. The Wedderburn–Artin Theorem says that a ring R is left semisimple if and only if it is isomorphic to a direct product of matrix algebras over division rings; it follows that these rings are also *right* semisimple. Therefore, $\ell\mathrm{D}(R) = 0 = r\mathrm{D}(R)$.

The first example of a ring for which the left and right global dimensions differ was given by Kaplansky ["On the dimension of rings and modules X," *Nagoya Math. J.* 13 (1958), 85–88] who exhibited a ring R with $\ell\mathrm{D}(R) = 1$ and $r\mathrm{D}(R) = 2$. Jategaonkar ["A counterexample in ring theory and homological algebra," *J. Algebra* 12 (1966), 97–105] proved that if $1 \leq m \leq n \leq \infty$, then there exists a ring R with $\ell\mathrm{D}(R) = m$ and $r\mathrm{D}(R) = n$. The same phenomenon, but with n finite, can be found in Fossum–Griffith–Reiten, *Trivial Extensions of Abelian Categories*, pp. 74–75.

On the positive side, we shall prove that $\ell\mathrm{D}(R) = r\mathrm{D}(R)$ when R is left and right noetherian. Jensen ["Homological dimensions of rings with countably generated ideals," *Math Scand.* 18 (1966), 97–105] proved that if all one-sided ideals (i.e., all left ideals and all right ideals) of a ring R are countably generated, then $|\ell\mathrm{D}(R) - r\mathrm{D}(R)| \leq 1$. This, in turn, was generalized by Osofsky ["Upper bounds of homological dimension," *Nagoya Math. J.* 32 (1968), 315–322] who showed that if every one-sided ideal of R can be generated by at most $\aleph_n$ elements, then $|\ell\mathrm{D}(R) - r\mathrm{D}(R)| \leq n + 1$; thus, Set Theory makes its presence known in these results. Indeed, if R is the direct product of countably many fields, then Osofsky (*Homological Dimension of Modules*, p. 60) proved that $\mathrm{D}(R) \leq 2$, and that $\mathrm{D}(R) = 2$ if and only if the Continuum Hypothesis holds. Another example involving Set

Theory is the ring E of all entire functions (a function $f: \mathbb{C} \to \mathbb{C}$ is ***entire*** if it is holomorphic at every $z \in \mathbb{C}$). Now E is a Bézout ring, and every Bézout ring R of cardinality $\aleph_m$ has global dimension $D(R) \le m + 2$ (Osborne, *Basic Homological Algebra*, p. 357). In particular, assuming the Continuum Hypothesis, $D(E) \le 3$.

Lemma 8.15. *A left R-module is injective if and only if* $\operatorname{Ext}_R^1(R/I, B) = \{0\}$ *for all left ideals I.*

Proof. If B is injective, then $\operatorname{Ext}_R^1(A, B) = \{0\}$ for every left R-module A. Conversely, apply $\operatorname{Hom}_R(\square, B)$ to $0 \to I \to R \to R/I \to 0$ to obtain exactness of $\operatorname{Hom}_R(R, B) \to \operatorname{Hom}_R(I, B) \to \operatorname{Ext}_R^1(R/I, B) = 0$. The result now follows from Theorem 3.30, Baer's Criterion, for every map $I \to B$ can be extended to R.   •

It is natural to ask whether there is an analog of Lemma 8.15 to test for projectivity. The obvious candidates do not work (but see Exercise 8.11 on page 467). If we assume that $\operatorname{Ext}_R^1(A, R/I) = \{0\}$ for all I, then Exercise 8.1 on page 466 shows, when R is Dedekind, that we may conclude only that A is torsion-free. If we assume that A satisfies $\operatorname{Ext}_R^1(A, I) = \{0\}$ for all ideals I, then this, too, is not enough to force A to be projective. Indeed, when $R = \mathbb{Z}$ and $I \ne \{0\}$, then we are assuming $\operatorname{Ext}_{\mathbb{Z}}^1(A, \mathbb{Z}) = \{0\}$ (for nonzero ideals here are principal and, hence, are isomorphic to $\mathbb{Z}$), and we are posing ***Whitehead's problem***: if A is an abelian group with $\operatorname{Ext}_{\mathbb{Z}}^1(A, \mathbb{Z}) = \{0\}$, is A free abelian? (See Exercise 7.18 on page 436.) It is not difficult to prove that every countable subgroup of A is free. However, Shelah ["Infinite abelian groups, Whitehead problem, and some constructions," *Israel Math. J.* 18 (1974), 243–256] proved that Whitehead's problem is undecidable: the statement "$\operatorname{Ext}_{\mathbb{Z}}^1(A, \mathbb{Z}) = \{0\}$ and $|A| = \aleph_1$ implies A is free" and its negation are each consistent with the ZFC axioms of Set Theory.

We now develop some ways to compute global dimension. The next result shows that left global dimension is determined by finitely generated modules; indeed, it is determined by cyclic modules.

Theorem 8.16 (Auslander). *For any ring R,*

$$\ell D(R) = \sup\{\operatorname{pd}(R/I) : I \text{ is a left ideal}\}.$$

Proof. (**Matlis**) If $\sup_I\{\operatorname{pd}(R/I)\} = \infty$, we are done, and so we may assume that $\operatorname{pd}(R/I) \le n$ for all left ideals I; thus, $\operatorname{Ext}_R^{n+1}(R/I, B) = \{0\}$ for every left R-module B. Now Theorem 8.14 says $\ell D(R) = \ell iD(R)$, so that it suffices to prove $\operatorname{id}(B) \le n$ for every B. Take an injective resolution of B with $(n-1)$st cosyzygy V^{n-1}. By Proposition 8.10(iii), $\{0\} = \operatorname{Ext}_R^{n+1}(R/I, B) \cong \operatorname{Ext}_R^1(R/I, V^{n-1})$ for all left ideals I. But Lemma 8.15 gives V^{n-1} injective and, therefore, $\operatorname{id}(B) \le n$.   •

We now introduce another notion of dimension arising from flat resolutions and Tor, for von Neumann regular rings will then appear naturally.

Definition. If A is a right R-module, then $\text{fd}_R(A) \leq n$ (fd abbreviates *flat dimension*) if there is a finite flat resolution

$$0 \to F_n \to \cdots \to F_1 \to F_0 \to A \to 0.$$

We will usually omit the subscript R. If no such finite resolution exists, then $\text{fd}(A) = \infty$; otherwise, define $\text{fd}(A) = n$ if n is the length of a shortest flat resolution of A.

A right R-module A is flat if and only if $\text{fd}(A) = 0$.

Definition. If $\to F_n \xrightarrow{d_n} F_{n-1} \to \cdots \to F_1 \xrightarrow{d_1} F_0 \xrightarrow{\varepsilon} A \to 0$ is a flat resolution of A, define its *nth syzygy*[1] by $Y_0 = \ker \varepsilon$ and $Y_n = \ker d_n$ for $n \geq 1$.

Of course, the nth syzygy of A depends on the flat resolution. Unfortunately, Schanuel's Lemma does not hold for flat resolutions (for the Comparison Theorem may not apply). For example, two flat resolutions of the abelian group $\mathbb{Q}/\mathbb{Z}$ are

$$0 \to \mathbb{Z} \to \mathbb{Q} \to \mathbb{Q}/\mathbb{Z} \to 0 \quad \text{and} \quad 0 \to S \to F \to \mathbb{Q}/\mathbb{Z} \to 0,$$

where F is a free abelian group (of infinite rank) mapping onto $\mathbb{Q}/\mathbb{Z}$ (recall Corollary 3.51: an abelian group is flat if and only if it is torsion-free). Now $\mathbb{Z} \oplus F$ is free abelian, but it is not isomorphic to $\mathbb{Q} \oplus S$, for $\mathbb{Q}$ is not projective. Still, we should be able to link flat dimension to Tor, for Theorem 7.5 says that Tor can be computed using flat resolutions.

Proposition 8.17. *The following four statements are equivalent for a right R-module A.*

(i) $\text{fd}(A) \leq n$.

(ii) $\text{Tor}_k^R(A, B) = \{0\}$ *for all left R-modules B and all $k \geq n + 1$.*

(iii) $\text{Tor}_{n+1}^R(A, B) = \{0\}$ *for all left R-modules B.*

(iv) *Every flat resolution of A has its $(n-1)$st syzygy flat.*

---

[1] We denote syzygies by Y_n because the Greek word *syzygy* means *yoke*.

Proof.

(i) $\Rightarrow$ (ii) By Theorem 7.5, we may compute $\mathrm{Tor}_k^R(A, B)$ using any flat resolution of A; in particular, we may use the given flat resolution whose kth term $F_k = \{0\}$ for all $k \geq n + 1$. Therefore, $\mathrm{Tor}_k^R(A, B) = \{0\}$ for all $k \geq n + 1$, because all the differentials $d_k \otimes_R 1_B$ are 0.

(ii) $\Rightarrow$ (iii) Trivial.

(iii) $\Rightarrow$ (iv) Assume that $\mathrm{Tor}_{n+1}^R(A, B) = \{0\}$ for all left R-modules B. If Y_{n-1} is the $(n-1)$st syzygy of A, then $\mathrm{Tor}_{n+1}^R(A, B) \cong \mathrm{Tor}_1^R(Y_{n-1}, B)$, by Corollary 6.23. Hence, $\mathrm{Tor}_1^R(Y_{n-1}, B) = \{0\}$ for all B, so that Y_{n-1} flat, by Theorem 7.2.

(iv) $\Rightarrow$ (i) Analogous to the proof of Proposition 8.6.   •

Remark. The proof of (iii) $\Rightarrow$ (iv) just given could have been used in the proof of Proposition 8.6, thus avoiding projective equivalence there.   ◀

Definition. The *right weak dimension* of a ring R is defined by

$$\mathrm{rwD}(R) = \sup\{\mathrm{fd}(A) : A \in \mathrm{obj}(\mathbf{Mod}_R)\}.$$

Proposition 8.18. $\mathrm{rwD}(R) \leq n$ *if and only if* $\mathrm{Tor}_{n+1}^R(A, B) = \{0\}$ *for all right R-modules A and all left R-modules B.*

Proof. Immediate from Proposition 8.17.   •

Definition. The *left weak dimension* of a ring R is defined by

$$\ell\mathrm{wD}(R) = \sup\{\mathrm{fd}(B) : B \in \mathrm{obj}(_R\mathbf{Mod})\}.$$

Theorem 8.19. *For any ring R, we have $\ell\mathrm{wD}(R) = r\mathrm{wD}(R)$.*

Proof. We can prove the left versions of Propositions 8.17 and 8.18, obtaining the same formula for $\ell\mathrm{wD}(R)$ and $r\mathrm{wD}(R)$.   •

Definition. The *weak dimension* $\mathrm{wD}(R)$ of a ring R is the common value of $\ell\mathrm{wD}(R)$ and $r\mathrm{wD}(R)$.

In the beginning, it is a nuisance that $A \otimes_R B$ and $\mathrm{Tor}_n^R(A, B)$ are hybrids in that each requires A to be a right R-module and B to be a left R-module. However, we now see that weak dimension requires no left/right distinction as does global dimension; that is, $\mathrm{wD}(R) = \mathrm{wD}(R^{\mathrm{op}})$. This fact will soon be exploited.

Example 8.20.

(i) If R is a ring, then $\mathrm{wD}(R) = 0$ if and only if every R-module is flat. In light of Theorem 4.9, we have $\mathrm{wD}(R) = 0$ if and only if R is von Neumann regular.

(ii) If R is a ring, then $\mathrm{wD}(R) \leq 1$ if and only if every submodule of a flat R-module is flat. In Corollary 4.36, we saw that a domain R has $\mathrm{wD}(R) \leq 1$ if and only if R is a Prüfer ring. No "intrinsic" description of all rings of weak dimension 1 is known.

(iii) Let R be a ring with $\mathrm{wD}(R) \leq 1$. Theorem 4.32 shows that R is left semihereditary if and only if it is left coherent. Similarly, R is right semihereditary if and only if it is right coherent. ◄

The next result explains why wD is called "weak dimension."

Proposition 8.21. *Let R be a ring.*

(i) *For every right R-module A, we have $\mathrm{fd}(A) \leq \mathrm{pd}(A)$.*

(ii) $\mathrm{wD}(R) \leq \min\{\ell\,\mathrm{D}(R), r\,\mathrm{D}(R)\}.$

Proof.

(i) The inequality obviously holds if $\mathrm{pd}(A) = \infty$, and so we may assume that $\mathrm{pd}(A) \leq n$; that is, there is a projective resolution

$$0 \to P_n \to \cdots \to P_0 \to A \to 0.$$

Since projective modules are flat, this is a flat resolution of A showing that $\mathrm{fd}(A) \leq n$. A similar argument works for left R-modules.

(ii) This follows at once from part (i). •

The inequality in Proposition 8.21 can be strict, for there are von Neumann regular rings that are not semisimple. In Corollary 8.28, we shall see that the inequality in Proposition 8.21 is an equality when R is both left and right noetherian.

Corollary 8.22. *If $\mathrm{Ext}_R^{n+1}(A, B) = \{0\}$ for all left R-modules A and B (or for all right R-modules A and B), then $\mathrm{Tor}_{n+1}^R(C, D) = \{0\}$ for all right R-modules C and left R-modules D.*

Proof. The Ext condition says that $\ell\,\mathrm{D}(R) \leq n$, while the Tor condition says that $\mathrm{wD}(R) \leq n$. Thus, this is just a restatement of Proposition 8.21. •

We now understand why $\mathrm{Tor}_2^R(A, B) = \{0\}$ for hereditary rings R.

Proposition 8.23.   *If S is a multiplicative subset of a commutative ring R, then*
$$\mathrm{wD}(S^{-1}R) \leq \mathrm{wD}(R).$$

Proof.   We may assume that $\mathrm{wD}(R) = n < \infty$. By Proposition 8.18, it suffices to show that $\mathrm{Tor}_{n+1}^{S^{-1}R}(A, B) = \{0\}$ for all $S^{-1}R$-modules A and B. Now there are R-modules M, N with $A \cong S^{-1}M$ and $B \cong S^{-1}N$ as $S^{-1}R$-modules, by Corollary 4.79(ii), and so Proposition 7.17 gives

$$\begin{aligned}
\mathrm{Tor}_{n+1}^{S^{-1}R}(A, B) &= \mathrm{Tor}_{n+1}^{S^{-1}R}(S^{-1}M, S^{-1}N) \\
&\cong S^{-1}\,\mathrm{Tor}_{n+1}^{R}(M, N) \\
&= \{0\}. \quad \bullet
\end{aligned}$$

Lemma 8.24.   *A left R-module B is flat if and only if $\mathrm{Tor}_1^R(R/I, B) = \{0\}$ for every right ideal I.*

Proof.   Proposition 3.58 says that a left R-module B is flat if and only if the sequence $0 \to I \otimes_R B \xrightarrow{i \otimes 1} R \otimes_R B$ is exact for every right ideal I, where $i : I \to R$ is the inclusion. If B is flat, $1 \otimes i$ is an injection, and exactness of

$$0 = \mathrm{Tor}_1^R(R, B) \to \mathrm{Tor}_1^R(R/I, B) \to I \otimes_R B \xrightarrow{i \otimes 1} R \otimes_R B$$

shows that $\mathrm{Tor}_1^R(R/I, B) = \{0\}$. The converse is obvious. $\bullet$

Weak dimension, as global dimension, is determined by finitely generated modules; indeed, it is even determined by cyclic modules.

Theorem 8.25.   *For any ring R, we have*
$$\begin{aligned}
\mathrm{wD}(R) &= \sup\{\mathrm{fd}(R/I) : I \text{ is a right ideal}\} \\
&= \sup\{\mathrm{fd}(R/J) : J \text{ is a left ideal}\}.
\end{aligned}$$

Proof.   The proof is the same as that of Theorem 8.16, using Lemma 8.24 instead of Lemma 8.15. $\bullet$

We can now complete the proof of Theorem 4.32.

Corollary 8.26.   *Every left ideal of a ring R is flat if and only if every submodule of a flat left R-module is flat.*

Proof.   Sufficiency is clear, for every left ideal is a submodule of R.

Conversely, if every left ideal I is flat, then $0 \to I \to R \to R/I \to 0$ is a flat resolution of R/I, which shows that $\mathrm{fd}(R/I) \leq 1$. By Theorem 8.25, we have $\mathrm{wD}(R) \leq 1$; hence, if S is a submodule of a flat module B, then $\mathrm{fd}(B/S) \leq 1$. There is a flat resolution $\to B \xrightarrow{\nu} B/S \to 0$, where ν is the natural map. By Proposition 8.17, the 0th syzygy Y_0 is flat. But $Y_0 = \ker \nu = S$, and so S is flat. $\bullet$

Let us return to dimensions. For noetherian rings, weak dimension is nothing new.

Theorem 8.27. *Let R be a right noetherian ring.*

(i) *Let A be a finitely generated [2] right R-module, then $\mathrm{fd}(A) = \mathrm{pd}(A)$.*

(ii) $\mathrm{wD}(R) = r\,\mathrm{D}(R)$.

(iii) *If R is left noetherian, then $\mathrm{wD}(R) = \ell\,\mathrm{D}(R)$.*

Proof.

(i) As we saw in the proof of Proposition 8.21, $\mathrm{fd}(A) \leq \mathrm{pd}(A)$. For the reverse inequality, we may assume that $\mathrm{fd}(R) = n < \infty$, and we must show that $\mathrm{pd}(A) \leq n$. A projective resolution $\to P_k \to P_{k-1} \to \cdots \to P_0 \to A \to 0$ in which each P_k is finitely generated, as in Lemma 7.19, is also a flat resolution, so that Proposition 8.17(iv) says that the $(n-1)$st syzygy Y_{n-1} is flat; that is,

$$0 \to Y_{n-1} \to P_{n-1} \to \cdots \to P_0 \to A \to 0 \qquad (1)$$

is a flat resolution. Since R is right noetherian, Corollary 3.57 says that Y_{n-1} is projective. But all the P_i are projective; hence, sequence (1) is a projective resolution of A, and so $\mathrm{pd}(A) \leq n$.

(ii) Statements (ii) and (iii) follows from part (i), for both weak and global dimension are determined by dimensions of finitely generated modules.   •

Osofsky (*Homological Dimension of Modules*, p. 57) generalized Theorem 8.27 as follows. If every right ideal of a ring R can be generated by at most $\aleph_n$ elements (where we agree that $\aleph_{-1}$ means *finite*), then $r\,\mathrm{D}(R) \leq \mathrm{wD}(R) + n + 1$.

Corollary 8.28 (Auslander). *If R is both left and right noetherian, then $\ell\,\mathrm{D}(R) = r\,\mathrm{D}(R)$.*

Proof. In this case, $\ell\,\mathrm{D}(R) = \mathrm{wD}(R) = r\,\mathrm{D}(R)$.   •

---

[2] We must assume that A is finitely generated; for example, $\mathbb{Q}$ is a flat $\mathbb{Z}$-module, so that $\mathrm{fd}(\mathbb{Q}) = 0$, but $\mathrm{pd}(\mathbb{Q}) = 1$.

Exercises

*8.1 Let R be a Dedekind ring. Prove that an R-module A is torsion-free if and only if $\text{Ext}^1_R(A, R/I) = \{0\}$ for every nonzero ideal I.
Hint. Use Theorem 8.3.

8.2   **(i)**   If R is quasi-Frobenius, prove that $\ell D(R) = 0$ or ∞.

   (ii)   Prove that $D(\mathbb{I}_n) = \infty$ if n is not squarefree.

   (iii)   Let G be a finite group and let k be a field whose characteristic divides $|G|$; prove that $\ell D(kG) = \infty$.

*8.3 If M is a left R-module with $\text{pd}(M) = n < \infty$, prove that there is a free R-module F with $\text{Ext}^n_R(M, F) \neq \{0\}$.

*8.4   **(i)**   If I is a left ideal in a ring R, prove that either R/I is projective or $\text{pd}(R/I) = 1 + \text{pd}(I)$ (we agree that $1 + \infty = \infty$).

   (ii)   If $0 \to M' \to M \to M'' \to 0$ is exact and if two of the modules have finite projective (or injective) dimension, prove that the third module has finite projective (or injective) dimension as well.

*8.5 Let $0 \to M' \to M \to M'' \to 0$ be an exact sequence of left R-modules for some ring R. Prove each of the following using the long exact Ext sequence.

   (i)   If $\text{pd}(M') < \text{pd}(M)$, prove that $\text{pd}(M'') = \text{pd}(M)$.

   (ii)   If $\text{pd}(M') > \text{pd}(M)$, prove that $\text{pd}(M'') = \text{pd}(M') + 1$.

   (iii)   If $\text{pd}(M') = \text{pd}(M)$, prove that $\text{pd}(M'') \leq \text{pd}(M') + 1$.

*8.6 Let $0 \to M' \to M \to M'' \to 0$ be an exact sequence of left R-modules for some ring R.

   (i)   Prove that if $\text{pd}(M') = n < \infty$ and $\text{pd}(M'') \leq n$, then $\text{pd}(M) = n$.

   (ii)   Prove that $\text{pd}(M) \leq \max\{\text{pd}(M'), \text{pd}(M'')\}$. Moreover, if $\text{pd}(M') = 1 + \text{pd}(M'')$ and the short exact sequence is not split, then the inequality is an equality.

*8.7 Let $0 \to M' \to M \to M' \to 0$ be an exact sequence of left R-modules for some ring R.

   (i)   Prove that $\text{pd}(M'') \leq 1 + \max\{\text{pd}(M), \text{pd}(M')\}$.
   Hint. Use the long exact Ext sequence.

   (ii)   If $0 \to M' \to M \to M'' \to 0$ is exact and M projective, prove that either all three modules are projective or that $\text{pd}(M'') = 1 + \text{pd}(M')$.
   Hint. Use the long exact Ext sequence.

*8.8 Let $0 \to M' \to M \to M'' \to 0$ be an exact sequence of right R-modules for some ring R. Prove that if $\text{fd}(M') = n < \infty$ and $\text{fd}(M'') \leq n$, then $\text{fd}(M) = n$.

*8.9 Given a family of left R-modules $(A_k)_{k \in K}$, prove that

$$\mathrm{pd}\Big(\bigoplus_{k \in K} A_k\Big) = \sup_{k \in K}\{\mathrm{pd}(A_k)\}.$$

Conclude that if $\ell \mathrm{D}(R) = \infty$, then there exists a module A with $\mathrm{pd}(A) = \infty$. [A priori, there might exist a ring R with $\mathrm{pd}(A) < \infty$ for every left R-module A, yet $\ell \mathrm{D}(R) = \infty$ because there is a sequence of left R-modules A_n with $\sup_n\{\mathrm{pd}(A_n)\} = \infty$.]

8.10 Let R be a Dedekind ring with fraction field Q. Prove that an R-module B is injective if and only if $\mathrm{Ext}_R^1(Q/R, B) = \{0\}$.

*8.11 If R is a Dedekind ring, prove that an R-module P is projective if and only if $\mathrm{Ext}_R^1(P, F) = \{0\}$ for every free R-module F.

8.12 If R is a left coherent ring, prove that $\mathrm{wD}(R) \leq 1$ if and only if R is left hereditary. Conclude that if R is left and right coherent, then $\mathrm{wD}(R) \leq 1$ if and only if R is left and right hereditary.

8.13 Prove that every left ideal of a ring R is flat if and only if every right ideal is flat.

8.2 Hilbert's Syzygy Theorem

If R is a (not necessarily commutative) ring, let $R[x]$ denote the polynomial ring in which the indeterminate x commutes with all the coefficients in R. Since $R[x_1, \ldots, x_{n+1}] = \big(R[x_1, \ldots, x_n]\big)[x_{n+1}]$, the polynomial ring $R[x_1, \ldots, x_n]$ consists of polynomials in which the indeterminates commute with each other as well as with the coefficients in R. Hilbert's Syzygy Theorem states that $\ell \mathrm{D}(R[x]) = 1 + \ell \mathrm{D}(R)$ (actually, Hilbert proved only the special case when $R = \mathbb{C}[x_1, \ldots, x_n]$).

If M is a left R-module, write

$$M[x] = R[x] \otimes_R M.$$

Since $R[x]$ is the free R-module with basis $\{1, x, x^2, \ldots\}$, and since tensor product commutes with direct sums, we may view the underlying R-module of $M[x]$ as a direct sum: $M[x] = \bigoplus_{i \in \mathbb{N}} M_i$, where $M_i \cong M$. Thus, elements of $M[x]$ are "vectors" $(x^i \otimes m_i)$, where $i \geq 0$, $m_i \in M$, and almost all $m_i = 0$. One may also regard these elements as "polynomials" with coefficients in M.

If an abelian group M is an R-module, we may use the subscript R, writing $\mathrm{pd}_R(M)$ to denote its projective dimension. Now $M[x]$ is an R-module as well as an $R[x]$-module, and so both $\mathrm{pd}_R(M[x])$ and $\mathrm{pd}_{R[x]}(M[x])$ are defined.

Lemma 8.29. *For every left R-module M, we have*

$$\mathrm{pd}_R(M) = \mathrm{pd}_{R[x]}(M[x]).$$

Proof. It suffices to prove, for any $n \geq 0$, that if one dimension is finite and $\leq n$, then the other is also finite and $\leq n$.

If $\mathrm{pd}_R(M) \leq n$, then there is a projective resolution

$$0 \to P_n \to \cdots \to P_0 \to M \to 0.$$

Since $R[x]$ is a flat right R-module ($R[x]$ is a free R-module), there is an exact sequence of left $R[x]$-modules

$$0 \to R[x] \otimes_R P_n \to \cdots \to R[x] \otimes_R P_0 \to R[x] \otimes_R M \to 0.$$

But $R[x] \otimes_R P_i$ is $R[x]$-projective for all i, and so $\mathrm{pd}_{R[x]}(M[x]) \leq n$.

If $\mathrm{pd}_{R[x]}(M[x]) \leq n$, then there is an $R[x]$-projective resolution

$$0 \to Q_n \to \cdots \to Q_0 \to M[x] \to 0. \qquad (1)$$

View this sequence as an R-exact sequence. Now each Q_i is a direct summand of a free $R[x]$-module F_i; that is, $F_i = Q_i \oplus S_i$ for some $R[x]$-module S_i. But F_i is also a free R-module, and so Q_i is projective as an R-module. Thus, the exact sequence (1) is also an R-projective resolution of $M[x] \cong \bigoplus_{i \in \mathbb{N}} M_i$, a direct sum of countably many copies of M. But $\mathrm{pd}_R(M[x]) = \mathrm{pd}_R(M)$, by Exercise 8.9 on the previous page.   •

Corollary 8.30. *If $\ell \mathrm{D}(R) = \infty$, then $\ell \mathrm{D}(R[x]) = \infty$.*

Proof. If $\ell \mathrm{D}(R) = \infty$, there exists an R-module M with $\mathrm{pd}(M) = \infty$, by Exercise 8.9. But $\mathrm{pd}_{R[x]}(M[x]) = \infty$, by Lemma 8.29, and so $\ell \mathrm{D}(R[x]) = \infty$.   •

Lemma 8.31. *If M is a left $R[x]$-module, then there is an $R[x]$-exact sequence*

$$0 \to M[x] \to M[x] \xrightarrow{e} M \to 0,$$

where $e \colon M[x] \to M$ by $x^i \otimes m_i \mapsto x^i m_i$.

Proof. Clearly, e is a surjective $R[x]$-map, and there is an $R[x]$-exact sequence $0 \to \ker e \to M[x] \to M \to 0$. It suffices to prove that $M[x] \cong \ker e$ as left $R[x]$-modules. Define $f \colon M[x] \to \ker e$ by

$$f \colon \sum x^i \otimes m_i \mapsto \sum x^i (1 \otimes x - x \otimes 1) m_i.$$

It is routine to see that f is an $R[x]$-map with im $f \subseteq \ker e$. To see that f is an isomorphism, we write its formula in more detail:

$$\sum_{i=0}^{k} x^i \otimes m_i \mapsto 1 \otimes xm_0 + \sum_{i=1}^{k} x^i \otimes (xm_i - m_{i-1}) - x^{k+1} \otimes m_k.$$

If $\sum x^i \otimes m_i \in \ker f$, then

$$0 = -m_k = xm_k - m_{k-1} = \cdots = xm_1 - m_0,$$

so that each $m_i = 0$; hence, f is injective. If $\sum_{i=0}^{k} x^i \otimes v_i \in \ker e$, then $\sum_{i=0}^{k} x^i v_i = 0$ in M, and we can solve the equations

$$-v_0 = xm_0, \ v_1 = xm_1 - m_0, \ \ldots, \ v_k = -m_{k-1}$$

recursively to show that f is surjective. •

Corollary 8.32. *For every ring R, we have*

$$\ell D(R[x]) \leq 1 + \ell D(R).$$

Proof. Let us agree that $\infty = \infty + 1$, so that Corollary 8.30 lets us assume that $\ell D(R) = n < \infty$. Take a left $R[x]$-module M, and view it as a left R-module. By Lemma 8.29, we have $\mathrm{pd}_{R[x]}(M[x]) = \mathrm{pd}_R(M) \leq n$. Apply Exercise 8.7 on page 466 to the $R[x]$-exact sequence $0 \to M[x] \to M[x] \to M \to 0$ in Lemma 8.31; we obtain $\mathrm{pd}_{R[x]}(M) \leq 1 + \mathrm{pd}_{R[x]}(M[x]) \leq 1 + n$.
•

We are now going to prove the reverse inequality: if $\ell D(R) = n$, then $\ell D(R[x]) \geq 1 + \ell D(R)$. Note that we have already proven this in the infinite-dimensional case in Corollary 8.30.

Given a left R-module M, we had to create a left $R[x]$-module $M[x]$. The reverse direction is easier, for a left $R[x]$-module N can be viewed as a left R-module by forgetting the action of x. The proper context in which to view the upcoming discussion is that of ***change of rings*** (we will elaborate on this circle of ideas when discussing spectral sequences). If $\varphi \colon R \to R^*$ is a ring homomorphism, then every left R^*-module M^* acquires a left R-module structure via the formula

$$rm^* = \varphi(r)m^*, \qquad r \in R \text{ and } m^* \in M^*.$$

For example, the inclusion $\varphi \colon k \to k[x]$ that takes $a \in k$ to the constant polynomial a views each left $k[x]$-module as a left k-module by forgetting the action of x. Every R^*-map $f^* \colon M^* \to N^*$ can also be viewed as an R-map:

$$f^*(rm^*) = f^*(\varphi(r)m^*) = \varphi(r)f^*(m^*) = rf^*(m^*).$$

Proposition 8.33. *Every ring homomorphism $\varphi\colon R \to R^*$ defines an exact additive functor $U\colon {}_{R^*}\mathbf{Mod} \to {}_{R}\mathbf{Mod}$. Moreover, if φ is surjective, then $U_*\colon \mathrm{Hom}_{R^*}(M^*, N^*) \to \mathrm{Hom}_R(UM^*, UN^*)$, given by $f^* \mapsto Uf^*$, is an isomorphism for all left R^*-modules M^* and N^*.*

Proof. It is easy to see that U is an additive functor; U is exact because, for any R^*-map f^*, both f^* and Uf^* are equal as additive functions and, hence, they have the same kernel and the same image. Equality of the underlying functions of f^* and Uf^* shows that U_* is an injection, for every R^*-map is an R-map. If φ is surjective, then U_* is a surjection, for every R-map $g\colon M^* \to N^*$ is an R^*-map: if $r \in R$ and $r^* = \varphi(r)$, then

$$g(r^*m^*) = g(\varphi(r)m^*) = g(rm^*) \quad [\text{for } \varphi(r)m^* = rm^*]$$
$$= rg(m^*) = \varphi(r)g(m^*) = r^*g(m^*).$$

Therefore, $g \in \mathrm{Hom}_{R^*}(M^*, N^*)$ and $g = Ug$. $\quad\bullet$

Since the underlying additive functions of an R^*-map f^* and its corresponding R-map Uf^* are the same, one usually identifies them and writes $\mathrm{Hom}_{R^*}(M^*, N^*) = \mathrm{Hom}_R(M^*, N^*)$.

Let $\ell\,\mathrm{D}(R) = n$. We will prove that $\ell\,\mathrm{D}(R[x]) \geq 1 + \ell\,\mathrm{D}(R) = 1 + n$ by exhibiting a pair of left $R[x]$-modules V and W with $\mathrm{Ext}^{n+1}_{R[x]}(V, W) \neq \{0\}$. The next result will help us do this.

Definition. Let R be a ring. An element $x \in Z(R)$ is **regular** on a left R-module M if the multiplication $M \to M$, given by $m \mapsto xm$, is an injection; that is, if $m \in M$ and $xm = 0$ imply $m = 0$.

If $M = R$, then x is regular on R if x is not a zero-divisor.

Theorem 8.34 (Rees). *Let R be a ring, let $x \in Z(R)$ be a nonzero element that is neither a unit nor a zero-divisor, and let $R^* = R/xR$. If M is a left R-module and x is regular on M, then there is an isomorphism*

$$\mathrm{Ext}^n_{R^*}(L^*, M/xM) \cong \mathrm{Ext}^{n+1}_R(L^*, M)$$

for every R^*-module L^* and every $n \geq 0$.

Proof. Since a left R^*-module is merely a left R-module annihilated by x, the quotient M/xM is also a left R^*-module; thus, $\mathrm{Ext}^n_{R^*}(L^*, M/xM)$ makes sense for every left R^*-module L^*. The natural map $R \to R^*$ gives a change of rings functor $U\colon {}_{R^*}\mathbf{Mod} \to {}_{R}\mathbf{Mod}$, so that UL^* is a left R-module and $\mathrm{Ext}^{n+1}_R(UL^*, M)$ also makes sense. However, we shall write $\mathrm{Ext}^{n+1}_R(L^*, M)$ as in the statement instead of the more accurate $\mathrm{Ext}^{n+1}_R(UL^*, M)$.

Recall Theorem 6.64, the axioms characterizing the contravariant Ext functors. If $(G^n\colon {}_{R^*}\mathbf{Mod} \to \mathbf{Ab})_{n\geq 0}$ is a sequence of additive contravariant functors satisfying the following three axioms:

(i) every short exact sequence gives a long exact sequence having natural connecting homomorphisms;

(ii) there exists a left R^*-module M^* with $G^0 \cong \operatorname{Hom}_{R^*}(\square, M^*)$;

(iii) $G^n(P^*) = \{0\}$ for all free R^*-modules P^* and all $n \geq 1$ (we have used the remark after Theorem 6.64 to write *free* instead of *projective*),

then

$$G^n \cong \operatorname{Ext}^n_{R^*}(\square, M^*) \text{ for all } n \geq 0.$$

Set $M^* = M/xM$, and define $G^n \colon {}_{R^*}\mathbf{Mod} \to \mathbf{Ab}$, for all $n \geq 0$, by

$$G^n = \operatorname{Ext}^{n+1}_R(\square, M).$$

It is clear that this sequence of functors satisfies axiom (i). Let us verify axiom (ii). Exactness of the sequence of R-modules $0 \to M \xrightarrow{\mu} M \to M/xM \to 0$, where $\mu \colon m \mapsto xm$, gives exactness of

$$\operatorname{Hom}_R(L^*, M) \to \operatorname{Hom}_R(L^*, M/xM) \xrightarrow{\partial} \operatorname{Ext}^1_R(L^*, M) \xrightarrow{\mu_*} \operatorname{Ext}^1_R(L^*, M).$$

We claim that $\operatorname{Hom}_R(L^*, M) = \{0\}$. If $u \in L^*$, then $xu = 0$, because L^* is a left R^*-module. Hence, if $f \colon L^* \to M$ is an R-map, then $xf(u) = f(xu) = f(0) = 0$. But $f(u) \in M$, so that $\mu \colon M \to M$ being an injection gives $f(u) = 0$; hence, $f = 0$. Thus, ∂ is injective.

We claim that $\mu_* = 0$. On the one hand, μ_* is multiplication by x. On the other hand, if $\nu \colon L^* \to L^*$ is multiplication by x, then $\nu = 0$ (because L^* is a left R^*-module), and so the induced map $\nu^* = 0$. But $\mu_* = \nu^*$, for both are multiplication by x; hence, $\mu_* = 0$. Thus, ∂ is surjective.

We conclude that the connecting homomorphism ∂ is a natural isomorphism $\operatorname{Hom}_R(L^*, M/xM) \to \operatorname{Ext}^1_R(L^*, M)$. But the change of rings map $R \to R^*$ is surjective, and so $\operatorname{Hom}_R(L^*, M/xM) = \operatorname{Hom}_{R^*}(L^*, M/xM)$, by Proposition 8.33. Thus, $G^0 = \operatorname{Ext}^1_R(\square, M)$ and $\operatorname{Hom}_{R^*}(\square, M/xM)$ are naturally isomorphic.

We now prove (the modified) axiom (iii): $G^n(F^*) = \operatorname{Ext}^{n+1}_R(F^*, M) = \{0\}$ for all free left R^*-modules F^* and all $n \geq 1$. Choose a basis of F^*, and let Q be the free left R-module with the same basis. In more detail, $F^* = \bigoplus_{k \in K} R^* u_k = \bigoplus_{k \in K} (R/xR) u_k$, and $Q = \bigoplus_{k \in K} R u_k$. If $\lambda \colon Q \to Q$ is multiplication by x, then the hypothesis that x is a central nonzero divisor in R shows that λ is an injective R-map; moreover, $\operatorname{coker} \lambda = Q/xQ = \left(\bigoplus_{k \in K} R u_k \right) / x \left(\bigoplus_{k \in K} R u_k \right) = \bigoplus_{k \in K} (R/xR) u_k = F^*$. Thus, there is an R-exact sequence $0 \to Q \xrightarrow{\lambda} Q \to F^* \to 0$. In the exact sequence

$$\operatorname{Ext}^n_R(Q, M) \to \operatorname{Ext}^{n+1}_R(F^*, M) \to \operatorname{Ext}^{n+1}_R(Q, M),$$

the flanking Exts are $\{0\}$ because Q is a free R-module. Hence, $G^n(F^*) = \text{Ext}_R^{n+1}(F^*, M/xM) = \{0\}$ for all $n \geq 1$ Therefore, $(G^n)_{n \geq 0}$ satisfies the axioms, and $G^n \cong \text{Ext}_{R*}^n(\Box, M/xM)$. But $G^n = \text{Ext}_R^{n+1}(\Box, M)$, and this gives the desired isomorphism. •

Corollary 8.35. *Let R be a ring, and let $x \in Z(R)$ be a nonzero element that is neither a unit nor a zero-divisor. If $R^* = R/xR$, and $\ell D(R^*) = n < \infty$, then*

$$\ell D(R) \geq \ell D(R^*) + 1.$$

Proof. Assume that L^* is a left R^*-module with $\text{pd}_{R*}(L^*) = n$. By Exercise 8.3 on page 466, there is a free left R^*-module F^* with $\text{Ext}_{R*}^n(L^*, F^*) \neq \{0\}$. Define Q as the free left R-module having the same basis as F^* (as in the proof of Theorem 8.34), so that $Q/xQ \cong F^*$. By Theorem 8.34,

$$\text{Ext}_{R*}^n(L^*, F^*) \cong \text{Ext}_R^{n+1}(L^*, Q).$$

This says that $\text{pd}_R(L^*) \geq n + 1$, and so $\ell D(R) \geq n + 1 = 1 + \ell D(R^*)$. •

Theorem 8.36. *For any ring R,*

$$\ell D(R[x]) = 1 + \ell D(R).$$

Proof. Corollary 8.32 gives $\ell D(R[x]) \leq 1 + \ell D(R)$, while Corollary 8.30 gives the reverse inequality if $\ell D(R) = \infty$. Assume that $\ell D(R) = n < \infty$. If $R = R[x]$, then $R^* = R/xR = R[x]/xR[x] = R$; hence, Corollary 8.35 applies to give $\ell D(R[x]) \geq 1 + \ell D(R)$. •

Theorem 8.37 (Hilbert Theorem on Syzygies). *If k is a (not necessarily commutative) ring and $k[x_1, \ldots, x_n]$ is the polynomial ring in which the indeterminates commute with each other and with the coefficients in k, then*

$$\ell D(k[x_1, \ldots, x_n]) = \ell D(k) + n.$$

In particular, if k is a field, then $D(k[x_1, \ldots, x_n]) = n$.

Proof. Induction on $n \geq 1$ using Theorem 8.36. •

If $\varphi \colon R \to R^*$ is a ring map, then change of rings says that every left R^*-module M can be viewed as a left R-module; we now compare its projective dimensions over the two rings.

Proposition 8.38. *Let $R^* = R/(x)$, let $x \in Z(R)$ not be a zero-divisor, and let M be a left R-module with x regular on M.*

(i) *If $\mathrm{pd}_R(M) = n < \infty$, then $\mathrm{pd}_{R*}(M/xM) \leq n - 1$.*

(ii) *If $\mathrm{pd}_{R*}(M/xM) = n < \infty$, then $\mathrm{pd}_R(M) \geq n + 1$.*

Proof.

(i) Since $\mathrm{pd}(M) = n < \infty$, we have $\mathrm{Ext}_R^{n+1}(L, M) = \{0\}$ for all left R-modules L; in particular, $\mathrm{Ext}_R^{n+1}(L^*, M) = \{0\}$ for all left R^*-modules L^* (by change of rings, L^* can be viewed as a left R-module). By the Rees Theorem, $\mathrm{Ext}_{R*}^n(L^*, M/xM) \cong \mathrm{Ext}_R^{n+1}(L^*, M)$. Therefore, $\mathrm{Ext}_{R*}^n(L^*, M/xM) = \{0\}$ for all L^*, and $\mathrm{pd}_{R*}(M/xM) \leq n - 1$.

(ii) If $\mathrm{pd}_{R*}(M/xM) = n < \infty$, then there is a left R^*-module L^* with $\mathrm{Ext}_{R*}^n(L^*, M/xM) \neq \{0\}$. We have $\mathrm{Ext}_R^{n+1}(L^*, M) \neq \{0\}$, by the Rees Theorem, and so $\mathrm{pd}_R(M) \geq n + 1$. •

Here is another change of rings theorem, simpler than the theorem of Rees, which also yields Corollary 8.35. Note that the hypothesis does not involve regularity.

Proposition 8.39 (Kaplansky).   *Let R be a ring, let $x \in Z(R)$ not be a unit or a zero-divisor, let $R^* = R/(x)$, and let M^* be a left R^*-module. If $\mathrm{pd}_{R*}(M^*) = n < \infty$, then $\mathrm{pd}_R(M^*) = n + 1$.*

Proof. We note that a left R^*-module is merely a left R-module M with $xM = \{0\}$.

The proof is by induction on $n \geq 0$. If $n = 0$, then M^* is a projective left R^*-module. Since x is not a zero-divisor, there is an exact sequence of left R-modules

$$0 \to R \xrightarrow{x} R \to R^* \to 0,$$

so that $\mathrm{pd}_R(R^*) \leq 1$. Now M^*, being a projective left R^*-module, is a direct summand of a free left R^*-module F^*. But $\mathrm{pd}_R(F^*) \leq 1$, for it is a direct sum of copies of R^*, and so $\mathrm{pd}_R(M^*) \leq 1$. Finally, if $\mathrm{pd}_R(M^*) = 0$, then M^* would be a projective left R-module; but this contradicts Exercise 3.2 on page 114, for $xM = \{0\}$. Therefore, $\mathrm{pd}_R(M^*) = 1$.

If $n \geq 1$, then there is an exact sequence of left R^*-modules

$$0 \to K^* \to F^* \to M^* \to 0 \tag{2}$$

with F^* free. Now $\mathrm{pd}_{R*}(K^*) = n - 1$, so induction gives $\mathrm{pd}_R(K^*) = n$.

If $n = 1$, then $\mathrm{pd}_R(K^*) = 1$ and $\mathrm{pd}_R(K^*) \leq 2$, by Exercise 8.5 on page 466. There is an exact sequence of left R-modules

$$0 \to L \to F \to M^* \to 0 \qquad (3)$$

with F free. Since $xM^* = \{0\}$, we have $xF \subseteq \ker(F \to M^*) = L$, and this gives an exact sequence of R^*-modules (each term is annihilated by x)

$$0 \to L/xF \to F/xF \to M^* \to 0.$$

Thus, $\mathrm{pd}_{R*}(L/xF) = \mathrm{pd}_{R*}(M^*) - 1 = 0$, because F/xF is a free R^*-module, so the exact sequence of R^*-modules

$$0 \to xF/xL \to L/xL \to L/xF \to 0$$

splits. Since $M^* \cong F/L \cong xF/xL$, we see that M^* is a direct summand of L/xL. Were L a projective left R-module, then L/xL and, hence, M^* would be projective left R^*-modules, contradicting $\mathrm{pd}_{R*}(M^*) = 1$. Exact sequence (3) shows that $\mathrm{pd}_R(M^*) = 1 + \mathrm{pd}_R(L) \geq 2$, and so $\mathrm{pd}_R(M^*) = 2$.

Finally, assume that $n \geq 2$. Exact sequence (2) gives $\mathrm{pd}_{R*}(K^*) = n - 1 > 1 \geq \mathrm{pd}_R(F^*)$; hence, Exercise 8.5 on page 466 gives

$$\mathrm{pd}_R(M^*) = \mathrm{pd}_R(K^*) + 1 = n + 1. \quad \bullet$$

The following theorem compares global dimensions (in case $A = R[x]$, it only gives the inequality $\ell\,\mathrm{D}(R) \leq \ell\,\mathrm{D}(R[x])$). Note that the hypothesis of the theorem makes sense, for if a ring R is a subring of a ring A, then A is an (R, R)-bimodule.

Theorem 8.40 (McConnell–Roos).   *Let A be a ring that is a faithfully flat[3] right R-module, where R is a subring of A with $\ell\,\mathrm{D}(R) = n < \infty$. If either*

(i) *A is a projective left R-module or*

(ii) *A is a flat left R-module and R is left noetherian,*

then $\ell\,\mathrm{D}(R) \leq \ell\,\mathrm{D}(A)$.

Remark.   K. R. Goodearl, ["Global dimension of differential operator rings," *Proc. AMS* 45 (1974), 315–322] gives an example showing that one must assume that $\ell\,\mathrm{D}(R)$ is finite. He displays a commutative ring R with $\mathrm{D}(R) = \infty$ and a differential ring $R[\theta]$ with $r\,\mathrm{D}(R[\theta]) = 1$.   ◄

---

[3]Recall that a right R-module A is *faithfully flat* if it is flat and $A \otimes_R X = \{0\}$ implies $X = \{0\}$.

Proof. We claim, for every left R-module M, that $\varphi : M \to A \otimes_R M$, defined by $m \mapsto 1 \otimes m$, is an injection. Exactness of $0 \to \ker \varphi \to M \xrightarrow{\varphi} A \otimes_R M$ gives exactness of $0 \to A \otimes_R \ker \varphi \to A \otimes_R M \xrightarrow{1 \otimes \varphi} A \otimes_R (A \otimes_R M)$, because A is flat. The reader may show that multiplication in the ring A gives an R-map $\mu : A \otimes_R (A \otimes_R M) \to A \otimes_R M$ with $a \otimes a' \otimes \mapsto aa' \otimes m$. Now the composite $\mu(1 \otimes \varphi) = 1_{A \otimes M}$, and so $1 \otimes \varphi$ is an injection. Therefore, $\{0\} = \ker(1 \otimes \varphi) = A \otimes \ker \varphi$; since A is faithfully flat, $\ker \varphi = \{0\}$, and so φ is an injection.

(i) Choose a left R-module M with $\mathrm{pd}_R(M) = n$. There is an exact sequence

$$0 \to M \xrightarrow{\varphi} A \otimes_R M \to C \to 0,$$

where $C = \mathrm{coker}\, \varphi$. Now $\mathrm{pd}_R(C) \leq n$ (for $\ell \mathrm{D}(R) = n$), so that Exercise 8.6 on page 466 gives $\mathrm{pd}_R(A \otimes_R M) = n$. We claim that $\mathrm{pd}_A(A \otimes_R M) \geq n$ [of course, $A \otimes_R M$ is a left A-module]; this will suffice to prove $\ell \mathrm{D}(A) \geq n = \ell \mathrm{D}(R)$.

Assume that A is R-projective. Every free left A-module F is a direct sum of copies of A; as each A is R-projective, we see that F, too, is R-projective. Now any A-projective Q is an A-direct summand of a free A-module F and, hence, is also an R-direct summand of F; thus, Q is R-projective. If $\mathrm{pd}_A(A \otimes_R M) = d < n$, then there is an A-projective resolution

$$0 \to Q_d \to \cdots \to Q_0 \to A \otimes_R M \to 0.$$

But we have just seen that A-projectives are R-projective, so that this is also an R-projective resolution of $A \otimes_R M$. Thus, $\mathrm{pd}_R(A \otimes_R M) \leq d < n$, a contradiction.

(ii) Since R is left noetherian, Theorem 8.27 says that $\ell \mathrm{D}(R) = \mathrm{wD}(R)$. Choose a left R-module X with $\mathrm{fd}_R(X) = n$, and let $\varphi' : X \to A \otimes_R X$ send $x \mapsto 1 \otimes x$. There is an exact sequence

$$0 \to X \xrightarrow{\varphi'} A \otimes_R X \to C' \to 0,$$

where $C = \mathrm{coker}\, \varphi'$. Now $\mathrm{fd}_R(C') \leq n$ (for $\mathrm{wD}(R) = n$), so that Exercise 8.8 on page 466 gives $\mathrm{fd}_R(A \otimes_R X) = n$. Now $A \otimes_R X$ is a left A-module, and we claim that $\mathrm{pd}_A(A \otimes_R X) \geq n$; this will suffice to prove $\ell \mathrm{D}(A) \geq n = \ell \mathrm{D}(R)$.

Every left A-module B is also a left R-module; we claim that if B is a flat left A-module, then it is also a flat left R-module. Both $\square \otimes_R B$ and $\square \otimes_R (A \otimes_A B)$ are functors $_R\mathbf{Mod} \to \mathbf{Ab}$, and they are naturally isomorphic. Since A is a flat left R-module, however, the latter functor is exact, being the composite of exact functors. Hence, $\square \otimes_R B$ is exact and $_R B$ is R-flat.

If $\mathrm{pd}_A(A \otimes_R X) = d < n$, then there is an A-projective resolution

$$0 \to B_d \to \cdots \to B_0 \to A \otimes_R X \to 0,$$

which is also an A-flat resolution. Since A-flats are R-flat, this is also an R-flat resolution of $A \otimes_R X$. Thus, $\mathrm{fd}_R(A \otimes_R X) \le d < n$, a contradiction.   •

8.3   Stably Free Modules

If k is a field, then Theorem 4.100, the Quillen–Suslin Theorem, shows that finitely generated projective $k[x_1, \ldots, x_n]$-modules are free. However, the proof uses a theorem of Serre saying that projective $k[x_1, \ldots, x_n]$-modules are stably free (a module P is *stably free* if there exist finitely generated free modules F and F' with $F' \cong P \oplus F$). Of course, finitely generated free modules are stably free. We prove Serre's theorem in this section. The basic idea of this proof is due to Borel, Serre, and Swan; our exposition provides details of the sketch given by Kaplansky in *Commutative Rings*, pp. 134–135.

Definition.   A module M *has* **FFR** (a *finite free resolution*) of *length* $\le n$ if M is finitely generated and there is an exact sequence

$$0 \to F_n \to F_{n-1} \to \cdots \to F_0 \to M \to 0$$

in which each F_i is a finitely generated free module.

It is redundant to say that a module M having FFR is finitely generated (it is even finitely presented, by Corollary 3.13); we have made this explicit in the definition only for emphasis.

The next proposition shows why FFR is relevant here.

Proposition 8.41.   *A finitely generated projective left R-module P has* FFR *if and only if P is stably free.*

Proof.   If P is stably free, then P is finitely generated and there is a finitely generated free module F with $P \oplus F$ free. Hence, P has FFR of length ≤ 1, for $0 \to F \to P \oplus F \to P \to 0$ is exact. Conversely, assume that P has FFR: there is a free resolution $0 \to F_n \to F_{n-1} \to \cdots \to F_0 \to P \to 0$ with each F_i finitely generated. We prove that P is stably free by induction on the length $n \ge 0$. If $n = 0$, then there is an exact sequence $0 \to F_0 \to P \to 0$

with F_0 finitely generated free. Exactness gives $F_0 \cong P$, so that P is free and, hence, stably free. For the inductive step, assume that there is a free resolution $0 \to F_{n+1} \to F_n \to \cdots F_1 \to F_0 \to P \to 0$ with each F_i finitely generated. If K is the 0th syzygy, we may factor this resolution into two exact sequences:

$$0 \to F_{n+1} \to \cdots F_1 \to K \to 0 \quad \text{and} \quad 0 \to K \to F_0 \to P \to 0.$$

The first exact sequence shows that K has FFR of length $\leq n$. Since P is projective, the short exact sequence splits, $F_0 \cong P \oplus K$, and so K is finitely generated projective. By induction, K is stably free; that is, there is a finitely generated free module Q with $K \oplus Q$ finitely generated free. Thus, P is stably free, for $P \oplus (K \oplus Q) \cong (P \oplus K) \oplus Q \cong F_0 \oplus Q$.   •

Lemma 8.42.   *If a module M has a projective resolution*

$$0 \to P_n \to \cdots \to P_2 \xrightarrow{d_2} P_1 \xrightarrow{d_1} P_0 \xrightarrow{\varepsilon} M \to 0$$

in which each P_i is stably free, then M has FFR of length $\leq n + 1$.

Proof.   We do an induction on $n \geq 0$. If $n = 0$, then $\varepsilon \colon P_0 \to M$ is an isomorphism. Since $M \cong P_0$ is stably free, there are finitely generated free modules F_0 and F_1 with $F_0 \cong M \oplus F_1$, and so there is an exact sequence $0 \to F_1 \to F_0 \to M \to 0$, as desired. Let $n > 0$. Since P_0 is stably free, there is a finitely generated free module F with $P_0 \oplus F$ finitely generated free. There is an exact sequence

$$0 \to P_n \to \cdots \to P_2 \xrightarrow{d_2'} P_1 \oplus F \xrightarrow{d_1 \oplus 1_F} P_0 \oplus F \xrightarrow{\varepsilon'} M \to 0,$$

where $d_2' \colon p_2 \mapsto (d_2, p_2, 0)$ and $\varepsilon' \colon (p_0, f) \mapsto \epsilon(p_0)$. Now $\ker \varepsilon'$ has a stably free resolution with $n - 1$ terms, and so it has FFR of length n, by induction. Splicing this FFR for $\ker \varepsilon'$ with the short exact sequence $0 \to \ker \varepsilon' \to P_0 \oplus F \to M \to 0$ (see Exercise 2.6 on page 65) shows that M has FFR of length $\leq n + 1$.   •

Proposition 8.43.   *Let $0 \to M' \to M \to M'' \to 0$ be an exact sequence of left R-modules, where R is left noetherian. If two of the modules have FFR, then so does the third.*

Proof.   Since modules having FFR are finitely generated, two of the modules in the short exact sequence are finitely generated; as R is left noetherian, the third module is finitely generated as well. The noetherian hypothesis also allows us to use Lemma 7.19: there are free resolutions of M' and M'' each of whose terms is finitely generated.

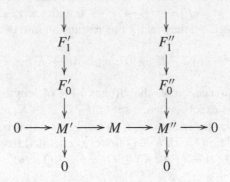

By the Horseshoe Lemma (Proposition 6.24), we may insert a free resolution $\to F_1 \to F_0 \to M \to 0$ between them; note that all $F_i \cong F_i' \oplus F_i''$ are finitely generated free. For each $n \geq 0$, there is an exact sequence of syzygies: $0 \to K_n' \to K_n \to K_n'' \to 0$. If any of these syzygies, say, K_n, is stably free, then the truncated resolution $0 \to K_n \to F_n \to \cdots \to F_0 \to M \to 0$ is a finite resolution each of whose terms is stably free (finitely generated free modules F_i are stably free!). Lemma 8.42 applies to show that M has FFR.

By hypothesis, two of $\{M', M, M''\}$ have FFR of length $\leq n$. Assume first that one of these two is M''. By Exercise 8.16 on page 484, K_n'' is stably free, as is one of the other syzygies. The exact sequence of syzygies now splits (for stably free modules are projective). In this case, the third syzygy is also stably free, by Example 4.92(iv), for its complement is stably free. As above, M' and M have FFR, by Lemma 8.42.

The remaining case assumes that M' and M have FFR of length $\leq n$, so that K_n' and K_n are stably free, by Exercise 8.16. Now splice the short exact sequence $0 \to K_n' \to K_n \to K_n'' \to 0$ and the truncated resolution $0 \to K_n'' \to F_n'' \to \cdots \to F_0'' \to M'' \to 0$ to obtain a resolution of M'' by stably free modules:

$$0 \to K_n' \to K_n \to F_n'' \to \cdots \to F_0'' \to M'' \to 0.$$

Lemma 8.42 applies to show that M'' has FFR.   •

Definition.   A *family* $\mathfrak{F}$ is a subclass of $\mathrm{obj}(_R\mathbf{Mod})$ such that whenever an exact sequence $0 \to M' \to M \to M'' \to 0$ has two terms in $\mathfrak{F}$, the third term also lies in $\mathfrak{F}$.

Proposition 8.43 says that if R is left noetherian, then the class of all FFR left R-modules is a family.

Lemma 8.44.   *Every intersection of families of left R-modules is a family.*

Proof. Let $\mathfrak{F}^* = \bigcap_\alpha \mathfrak{F}_\alpha$, where each $\mathfrak{F}_\alpha$ is a family. If $0 \to M' \to M \to M'' \to 0$ is an exact sequence having two terms in $\mathfrak{F}^*$, then these two terms lie in every $\mathfrak{F}_\alpha$. Since each $\mathfrak{F}_\alpha$ is a family, the third term must lie in every $\mathfrak{F}_\alpha$, and so the third term lies in $\mathfrak{F}^*$.   •

Definition. In light of Lemma 8.44, we may define $\mathfrak{F}(\mathcal{X})$, the *family generated by* a subclass $\mathcal{X} \subseteq \mathrm{obj}(_R\mathbf{Mod})$, as the intersection of all the families containing $\mathcal{X}$.

If $\mathcal{X} \subseteq \mathrm{obj}(_R\mathbf{Mod})$, define an $\mathcal{X}$-*child* to be a module occurring in a short exact sequence whose other two terms lie in $\mathcal{X}$, and define

$$\mathfrak{C}(\mathcal{X}) \text{ to be the class of all } \mathcal{X}\text{-children.}$$

We claim that $\mathfrak{C}(\mathcal{X})$ contains $\mathcal{X}$. If $\mathcal{X} = \varnothing$, there is nothing to prove. If $M \in \mathcal{X}$, then exactness of $0 \to M \xrightarrow{1_M} M \to 0 \to 0$ shows that $\{0\} \in \mathfrak{C}(\mathcal{X})$, while M and $\{0\}$ being terms in this same short exact sequence shows that $M \in \mathfrak{C}(\mathcal{X})$; that is, $\mathcal{X} \subseteq \mathfrak{C}(\mathcal{X})$. Define an ascending chain of subclasses:

$$\mathfrak{C}^0(\mathcal{X}) = \mathcal{X}; \quad \mathfrak{C}^{n+1}(\mathcal{X}) = \mathfrak{C}(\mathfrak{C}^n(\mathcal{X})).$$

The union $\bigcup_{n=0}^\infty \mathfrak{C}^n(\mathcal{X})$ consists of all the descendants of $\mathcal{X}$.

Lemma 8.45. *If $\mathcal{X}$ is a subclass of $\mathrm{obj}(_R\mathbf{Mod})$, then $\bigcup_{n=0}^\infty \mathfrak{C}^n(\mathcal{X}) = \mathfrak{F}(\mathcal{X})$, the family generated by $\mathcal{X}$.*

Proof. It is clear that every family $\mathfrak{F}$ containing $\mathcal{X}$ must contain $\mathfrak{C}(\mathcal{X})$ and $\mathfrak{C}^n(\mathcal{X})$ for all n; that is, $\bigcup_{n=0}^\infty \mathfrak{C}^n(\mathcal{X}) \subseteq \mathfrak{F}$ for all $\mathfrak{F}$. Hence, $\bigcup_{n=0}^\infty \mathfrak{C}^n(\mathcal{X}) \subseteq \bigcap_\mathfrak{F} \mathfrak{F} = \mathfrak{F}(\mathcal{X})$.

For the reverse inclusion, it suffices to prove that $\bigcup_{n=0}^\infty \mathfrak{C}^n(\mathcal{X})$ is a family containing $\mathcal{X}$. Let $0 \to M' \to M \to M'' \to 0$ be an exact sequence having two terms in the union. There exists $n \geq 0$ with $\mathfrak{C}^n(\mathcal{X})$ containing these two terms, and so the third term lies in $\mathfrak{C}^{n+1}(\mathcal{X})$. Thus, the union is a family.   •

Corollary 8.46. *If R is left noetherian and $\mathcal{X}$ is a class of left R-modules each of whose members has FFR, then every member of the family $\mathfrak{F}(\mathcal{X})$ generated by $\mathcal{X}$ has FFR.*

Proof. If $M \in \mathfrak{F}(\mathcal{X})$, there is a smallest number $n \geq 0$ with $M \in \mathfrak{C}^n(\mathcal{X})$; we prove that M has FFR by induction on n. If $n = 0$, then $M \in \mathcal{X}$, and so M has FFR, by hypothesis. If $n > 0$, then there is a short exact sequence whose other two terms lie in $\mathfrak{C}^{n-1}(\mathcal{X})$. By induction, these two terms have FFR, and so M has FFR, by Proposition 8.43.   •

The next theorem is the main result about families. We will state it now so that the reader may see that Serre's Theorem follows quickly from it.

Theorem 8.47. *Let R be a commutative noetherian ring. If every finitely generated R-module has FFR, then every finitely generated $R[x]$-module has FFR.*

Theorem 8.48 (Serre). *If k is a field, then every finitely generated projective $k[x_1, \ldots, x_n]$-module is stably free.*

Proof. We begin by proving, by induction on $n \geq 1$, that every finitely generated $k[x_1, \ldots, x_n]$-module has FFR. If $n = 1$, then $k[x]$ is a PID, and every $k[x]$-module has FFR of length ≤ 1. If $n > 1$, then $k[x_1, \ldots, x_n]$ is noetherian, by the Hilbert Basis Theorem. By induction, every finitely generated R-module has FFR, where $R = k[x_1, \ldots, x_n]$, and so Theorem 8.47 says that every finitely generated $R[x_{n+1}]$-module has FFR (of course, $R[x_{n+1}] = k[x_1, \ldots, x_{n+1}]$). In particular, finitely generated projective $k[x_1, \ldots, x_{n+1}]$-modules have FFR, and so they are stably free, by Proposition 8.41. •

We prepare a lemma for the proof of Theorem 8.47.

Definition. If M is an R-module, where R is commutative, then a subset $X \subseteq M$ is *scalar closed* if $x \in X$ implies that $rx \in X$ for all $r \in R$.

Every submodule of a module M is scalar closed;

$$\text{Zer}(R) = \{r \in R : r = 0 \text{ or } r \text{ is a zero-divisor}\}$$

is an example of a scalar closed subset (of R) that is not a submodule.

Definition. Let $X \subseteq M$ be a scalar closed subset. The *annihilator* of $x \in X$ is

$$\text{ann}(x) = \{r \in R : rx = 0\},$$

the *annihilator* of X is

$$\text{ann}(X) = \{r \in R : rx = 0 \text{ for all } x \in X\},$$

and

$$\mathcal{A}(X) = \{\text{ann}(x) : x \in X \text{ and } x \neq 0\}.$$

Note that $\text{ann}(x)$ and $\text{ann}(X)$ are ideals.

Lemma 8.49. *Let R be a commutative noetherian ring, let M be a nonzero finitely generated R-module, and let $X \subseteq M$ be a nonempty scalar closed subset.*

  (i) *An ideal I maximal among $\mathcal{A}(X)$ is a prime ideal.*

(ii) *There is a descending chain*

$$M = M_0 \supseteq M_1 \supseteq M_2 \supseteq \cdots \supseteq M_n = \{0\}$$

whose factor modules $M_i/M_{i+1} \cong R/\mathfrak{p}_i$ *for prime ideals* $\mathfrak{p}_i$.

Proof.

(i) Since R is noetherian, the nonempty set $\mathcal{A}(X)$ contains a maximal element, by Proposition 3.16; call it $I = \text{ann}(x)$. Suppose that a, b are elements in R with $ab \in I$ and $b \notin I$; that is, $abx = 0$ but $bx \neq 0$. Thus, $\text{ann}(bx) \supseteq I + Ra \supseteq I$. If $a \notin I$, then $\text{ann}(bx) \supseteq I + Ra \supsetneq I$. But $bx \in X$ because X is scalar closed, so that $\text{ann}(bx) \in \mathcal{A}(X)$, which contradicts the maximality of $I = \text{ann}(x)$. Therefore, $a \in I$ and I is a prime ideal.

(ii) Since R is left noetherian, it has the maximum condition on left ideals. Thus, the nonempty set $\mathcal{A}(M)$ has a maximal element, say, $\mathfrak{p}_1 = \text{ann}(x_1)$, which is prime, by part (i). Define $M_1 = \langle x_1 \rangle$, and note that $M_1 \cong R/\text{ann}(x_1) = R/\mathfrak{p}_1$. Now repeat this procedure. Let $\mathfrak{p}_2 = \text{ann}(x_2 + M_1)$ be a maximal element of $\mathcal{A}(M/M_1)$, so that $\mathfrak{p}_2$ is prime, and define $M_2 = \langle x_2, x_1 \rangle$. Note that $\{0\} \subsetneq M_1 \subsetneq M_2$ and that $M_2/M_1 \cong R/\text{ann}(x_2 + M_1) = R/\mathfrak{p}_2$. By Proposition 3.18, the module M has ACC, and so this process terminates, say, with $M^* \subseteq M$. We must have $M^* = M$, however, lest the process continue for another step. Now reindex the subscripts to get the desired statement. •

Here is the proof of Theorem 8.47.

Proof. Let $\mathcal{X}$ be the class of all finitely generated *extended* $R[x]$-modules M; that is, $M \cong R[x] \otimes_R B$ for some finitely generated R-module B. By hypothesis, B has FFR: there is an R-exact sequence

$$0 \to F_m \to \cdots \to F_1 \to F_0 \to B \to 0$$

in which all F_i are finitely generated free R-modules. Since $R[x]$ is a flat R-module, tensoring this sequence by $R[x]$ yields an $R[x]$-exact sequence. But each $R[x] \otimes_R F_i$ is a free $R[x]$-module, and so M has FFR. By Corollary 8.46, every module in $\mathcal{F}(\mathcal{X})$ has FFR. Thus, our task is to prove that every finitely generated $R[x]$-module M lies in $\mathcal{F} = \mathcal{F}(\mathcal{X})$.

We begin by normalizing M. Suppose that $\text{ann}(M) \cap R \neq \{0\}$. Let $m \in M$ be nonzero, and let $\text{ann}(m)$ be its annihilator; note that $\text{ann}(m) \cap R \supseteq \text{ann}(M) \cap R \neq \{0\}$. If we write $I = \text{ann}(m) \cap R$, then $R/I \cong \langle m \rangle_R$, the R-submodule of M generated by m. Since $R[x]$ is a flat R-module, there is an exact sequence

$$0 \to R[x] \otimes_R I \to R[x] \to R[x] \otimes_R \langle m \rangle_R \to 0. \tag{1}$$

Corollary 3.59 says that $R[x] \otimes_R I \cong R[x]I$, so that $R[x]I \neq \{0\}$. Now $R[x]/R[x]I \cong R[x] \otimes_R \langle m \rangle_R$ is a cyclic submodule of M, say, $\langle m_1 \rangle$. Thus, $\langle m_1 \rangle$ is extended, for $\langle m_1 \rangle \cong R[x] \otimes_R \langle m \rangle_R$, and so $\langle m_1 \rangle \in \mathcal{X} \subseteq \mathcal{F}$. Exactness of (1) implies $\operatorname{ann}(m_1) \cong R[x] \otimes_R I \cong R[x]I$, so that $\operatorname{ann}(m_1) \cap R \neq \{0\}$. This argument can be applied to $M/\langle m_1 \rangle$: there is $m_2 + \langle m_1 \rangle \in M/\langle m_1 \rangle$ with $\operatorname{ann}(m_2 + \langle m_1 \rangle) \cap R \neq \{0\}$ and with $\langle m_1, m_2 \rangle/\langle m_1 \rangle \in \mathcal{X}$. It follows that $\langle m_1, m_2 \rangle \in \mathcal{F}$ and $\operatorname{ann}(\langle m_1, m_2 \rangle \cap R \neq \{0\}$. This process must stop, by Proposition 3.15, for M has ACC. We conclude that if $\operatorname{ann}(M) \cap R \neq \{0\}$, then $M \in \mathcal{F}$.

By Lemma 8.49(ii), there is a descending chain

$$M = M_0 \supseteq M_1 \supseteq M_2 \supseteq \cdots \supseteq M_n = \{0\}$$

whose factor modules $M_i/M_{i+1} \cong R[x]/\mathfrak{p}_i$ for prime ideals $\mathfrak{p}_i$. It thus suffices, by induction on n, to show that $M = R[x]/\mathfrak{p} \in \mathcal{F}$. Our normalization allows us to assume that $\operatorname{ann}(R[x]/\mathfrak{p}) \cap R = \mathfrak{p} \cap R = \{0\}$. But $\mathfrak{p} \cap R$ is a prime ideal in R, so that R and, hence, $R[x]$ are domains. Choose a nonzero $f(x) \in \mathfrak{p} \subseteq R[x]$, and consider the exact sequence $0 \to (f) \to \mathfrak{p} \to \mathfrak{p}/(f)$. Now $(f) \cong R[x]$, since $R[x]$ is a domain, while $\operatorname{ann}(\mathfrak{p}/(f)) \neq \{0\}$ [for it contains $f(x)$]. Thus, both (f), $\mathfrak{p}/(f) \in \mathcal{F}$, so that $\mathfrak{p} \in \mathcal{F}$. Finally, $R[x]/\mathfrak{p} \in \mathcal{F}$, for both $R[x]$, $\mathfrak{p} \in \mathcal{F}$.   •

We will need the following proposition in the next section.

Lemma 8.50 (Prime Avoidance).   *Let* $\mathfrak{p}_1, \ldots, \mathfrak{p}_n$ *be prime ideals in a commutative ring* R. *If* J *is an ideal with* $J \subseteq \mathfrak{p}_1 \cup \cdots \cup \mathfrak{p}_n$, *then* J *is contained in some* $\mathfrak{p}_i$.

Proof.   The proof is by induction on $n \geq 1$, and the base step is trivially true. For the inductive step, let $J \subseteq \mathfrak{p}_1 \cup \cdots \cup \mathfrak{p}_{n+1}$, and define

$$D_i = \mathfrak{p}_1 \cup \cdots \cup \widehat{\mathfrak{p}_i} \cup \cdots \cup \mathfrak{p}_{n+1}.$$

We may assume that $J \subsetneq D_i$ for all i, for otherwise the inductive hypothesis can be invoked to complete the proof. Hence, for each i, there exists $a_i \in J$ with $a_i \notin D_i$; since $J \subseteq D_i \cup \mathfrak{p}_i$, we must have $a_i \in \mathfrak{p}_i$. Consider the element

$$b = a_1 + a_2 \cdots a_{n+1}.$$

Now $b \in J$ because all the a_i are. We claim that $b \notin \mathfrak{p}_1$. Otherwise, $a_2 \cdots a_{n+1} = b - a_1 \in \mathfrak{p}_1$; but $\mathfrak{p}_1$ is a prime ideal, and so $a_i \in \mathfrak{p}_1$ for some $i \geq 2$. This is a contradiction, for $a_i \in \mathfrak{p}_1 \subseteq D_i$ and $a_i \notin D_i$. Therefore, $b \notin \mathfrak{p}_i$ for any i, contradicting $J \subseteq \mathfrak{p}_1 \cup \cdots \cup \mathfrak{p}_n$.   •

Proposition 8.51.   *If R is a commutative noetherian ring, then there are finitely many prime ideals $\mathfrak{p}_1, \ldots, \mathfrak{p}_n$ with*

$$\mathrm{Zer}(R) = \{r \in R : r = 0 \text{ or } r \text{ is a zero-divisor}\} \subseteq \mathfrak{p}_1 \cup \cdots \cup \mathfrak{p}_n.$$

Proof.   In Lemma 8.49(i), which applies because $\mathrm{Zer}(R)$ is scalar closed, we proved that any ideal I that is a maximal member of the family $\mathcal{A}(X) = (\mathrm{ann}(x))_{x \in \mathrm{Zer}(R)}$ is prime (maximal members exist because R is noetherian). Let $(\mathfrak{p}_\alpha)_{\alpha \in A}$ be the family of all such maximal members. If x is a zero-divisor, then there is a nonzero $r \in R$ with $rx = 0$; that is, $x \in \mathrm{ann}(r)$ [of course, $r \in \mathrm{Zer}(R)$]. It follows that every zero-divisor x lies in some $\mathfrak{p}_\alpha$, and so $\mathrm{Zer}(R) \subseteq \bigcup_{\alpha \in A} \mathfrak{p}_\alpha$. It remains to prove that we may choose the index set A to be finite.

Each $\mathfrak{p}_\alpha = \mathrm{ann}(x_\alpha)$ for some $x_\alpha \in X$; let S be the submodule of M generated by all the x_α. Since R is noetherian and M is finitely generated, the submodule S is generated by finitely many of the x_α; say, $S = \langle x_1, \ldots, x_n \rangle$. We claim that $\mathrm{ann}(X) \subseteq \mathfrak{p}_1 \cup \ldots \cup \mathfrak{p}_n$, and it suffices to prove that $\mathfrak{p}_\alpha \subseteq \mathfrak{p}_1 \cup \ldots \cup \mathfrak{p}_n$ for all α. Now $\mathfrak{p}_\alpha = \mathrm{ann}(x_\alpha)$, and $x_\alpha \in S$; hence,

$$x_\alpha = r_1 x_1 + \cdots + r_n x_n$$

for $r_i \in R$. If $a \in \mathfrak{p}_1 \cap \cdots \cap \mathfrak{p}_n = \mathrm{ann}(x_1) \cap \cdots \cap \mathrm{ann}(x_n)$, then $ax_i = 0$ for all i and so $ax_\alpha = 0$. Therefore,

$$\mathfrak{p}_1 \cap \cdots \cap \mathfrak{p}_n \subseteq \mathrm{ann}(x_\alpha) = \mathfrak{p}_\alpha.$$

But $\mathfrak{p}_\alpha$ is prime, so that $\mathfrak{p}_i \subseteq \mathfrak{p}_\alpha$ for some i,[4] and this contradicts the maximality of $\mathfrak{p}_i$.   •

Exercises

*8.14 Consider the change of rings functors $U: {}_R\mathbf{Mod} \to {}_{R'}\mathbf{Mod}$ and $V: \mathbf{Mod}_R \to \mathbf{Mod}_{R'}$ arising from a ring map $\varphi: R' \to R$. Prove that the function $B \otimes_R A \to VB \otimes_{R'} UA$, given by

$$\sum_i b_i \otimes a_i \text{ (in } B \otimes_R R) \mapsto \sum_i b_i \otimes a_i \text{ (in } VB \otimes_{R'} UA),$$

is a well-defined injective $\mathbb{Z}$-map.

Hint. The relations defining $VB \otimes_{R'} UA$ include all ordered pairs $(b\varphi(r), a) - (b, \varphi(r)b)$, which are special cases of the relations $(br, a) - (b, rb)$ in $B \otimes_R A$.

---

[4]Assume that $I_1 \cap \cdots \cap I_n \subseteq \mathfrak{p}$, where $\mathfrak{p}$ is prime. If $I_i \not\subseteq \mathfrak{p}$ for all i, then there are $u_i \in I_i$ with $u_i \notin \mathfrak{p}$. But $u_1 \cdots u_n \in I_1 \cap \cdots \cap I_n \subseteq \mathfrak{p}$; since $\mathfrak{p}$ is prime, some $u_i \in \mathfrak{p}$, a contradiction.

8.15 Let $\mathfrak{F}$ be a family, and let $M \in \mathfrak{F}$. If $M' \cong M$, prove that $M' \in \mathfrak{F}$.

*8.16 Let R be left noetherian, and let M be a left R-module having FFR of length $\leq n$. Prove that the nth syzygy of every free resolution of M, each of whose terms is finitely generated, is stably free.

Hint. Use Exercise 3.15 on page 128, the generalized Schanuel Lemma.

8.17 Let S be a multiplicative set in a commutative ring R. If an R-module M has FFR, prove that the $S^{-1}R$-module $S^{-1}M$ has FFR and that $\chi(S^{-1}M) = \chi(M)$. [The *Euler characteristic* $\chi(M)$ is defined in Exercise 3.16 on page 129.]

8.4   Commutative Noetherian Local Rings

This section discusses the theorems of Auslander, Buchsbaum, and Serre about regular local rings. We are now going to focus on commutative noetherian local rings, the main results being that such rings have finite global dimension if and only if they are *regular local* rings (regular local rings arise quite naturally in Algebraic Geometry in describing nonsingular points on varieties), and that they are unique factorization domains. Let us begin with a localization result.

Proposition 8.52.   *Let R be a commutative noetherian[5] ring.*

(i) *If A is a finitely generated R-module, then*

$$\mathrm{pd}(A) = \sup_{\mathfrak{m}}\{\mathrm{pd}(A_{\mathfrak{m}})\},$$

where $\mathfrak{m}$ ranges over all the maximal ideals of R.

(ii)

$$\mathrm{D}(R) = \sup_{\mathfrak{m}}\{\mathrm{D}(R_{\mathfrak{m}})\},$$

where $\mathfrak{m}$ ranges over all the maximal ideals of R.

Proof.

(i) We first prove that $\mathrm{pd}(A) \geq \mathrm{pd}(A_{\mathfrak{m}})$ for every maximal ideal $\mathfrak{m}$. If $\mathrm{pd}(A) = \infty$, there is nothing to prove, and so we may assume that $\mathrm{pd}(A) = n < \infty$. Thus, there is an R-projective resolution

$$0 \to P_n \to P_{n-1} \to \cdots \to P_0 \to A \to 0.$$

---

[5]Part (ii) of this proposition may be false if R is not noetherian. For example, an infinite Boolean ring R is not semisimple, and so $\mathrm{D}(R) > 0$. On the other hand, $R_{\mathfrak{m}} \cong \mathbb{F}_2$ is a field for every maximal ideal $\mathfrak{m}$, and so $\sup_{\mathfrak{m}}\{\mathrm{D}(R_{\mathfrak{m}})\} = 0$.

Since $R_\mathfrak{m}$ is a flat R-module, by Theorem 4.80,

$$0 \to R_\mathfrak{m} \otimes_R P_n \to R_\mathfrak{m} \otimes_R P_{n-1} \to \cdots \to R_\mathfrak{m} \otimes_R P_0 \to A_\mathfrak{m} \to 0$$

is an $R_\mathfrak{m}$-projective resolution of $A_\mathfrak{m}$, and so $\mathrm{pd}(A_\mathfrak{m}) \leq n$. (This implication does not need the hypothesis that R is noetherian nor that A is finitely generated.)

For the reverse inequality, it suffices to assume that $\sup_\mathfrak{m}\{\mathrm{pd}(A_\mathfrak{m})\} = n < \infty$. Since R is noetherian, Theorem 8.27(i) says that $\mathrm{pd}(A) = \mathrm{fd}(A)$. Now $\mathrm{pd}(A_\mathfrak{m}) \leq n$ if and only if $\mathrm{Tor}_{n+1}^{R_\mathfrak{m}}(A_\mathfrak{m}, B_\mathfrak{m}) = \{0\}$ for all $R_\mathfrak{m}$-modules $B_\mathfrak{m}$, by Proposition 8.17. However, Proposition 7.17 gives an isomorphism $\mathrm{Tor}_{n+1}^{R_\mathfrak{m}}(A_\mathfrak{m}, B_\mathfrak{m}) \cong \left(\mathrm{Tor}_{n+1}^R(A, B)\right)_\mathfrak{m}$. Therefore, Proposition 4.90(i) gives $\mathrm{Tor}_{n+1}^R(A, B) = \{0\}$. We conclude that $n \geq \mathrm{pd}(A)$.

(ii) This follows at once from part (i), for $\mathrm{D}(R) = \sup_A\{\mathrm{pd}(A)\}$, where A ranges over all finitely generated (even cyclic) R-modules, by Theorem 8.16.   •

We now set up notation that will be used in the rest of this section.

Notation.   We denote a commutative noetherian local ring by R, by $(R, \mathfrak{m})$, or by $(R, \mathfrak{m}, k)$, where $\mathfrak{m}$ is its unique maximal ideal and k is its *residue field* $k = R/\mathfrak{m}$.

Theorem 8.16 allows us to compute the global dimension of a ring R as the supremum of the projective dimensions of its cyclic modules. When R is a local ring, there is a dramatic improvement; global dimension is determined by the projective dimension of one cyclic module: the residue field k, as we shall see in Theorem 8.55.

Lemma 8.53.   *Let $(R, \mathfrak{m}, k)$ be a local ring. If M is a finitely generated R-module, then $\mathrm{pd}(M) \leq n$ if and only if $\mathrm{Tor}_{n+1}^R(M, k) = \{0\}$.*

Proof.   Assume that $\mathrm{pd}(M) \leq n$. By Proposition 8.21(i), we have $\mathrm{fd}(M) \leq \mathrm{pd}(M)$, so that $\mathrm{Tor}_{n+1}^R(M, B) = \{0\}$ for every R-module B. In particular, $\mathrm{Tor}_{n+1}^R(M, k) = \{0\}$.

We prove the converse by induction on $n \geq 0$. For the base step $n = 0$, we must prove that $\mathrm{Tor}_1^R(M, k) = \{0\}$ implies $\mathrm{pd}(M) = 0$; that is, M is projective. By Theorem 4.62, there is a projective cover: an exact sequence

$$0 \to N \xrightarrow{\ i\ } F \xrightarrow{\ \varphi\ } M \to 0$$

with F a finitely generated free R-module and $N \subseteq \mathfrak{m}F$. Since $\text{Tor}_1^R(M, k) = \{0\}$, the sequence

$$0 \to N \otimes_R k \xrightarrow{i \otimes 1} F \otimes_R k \xrightarrow{\varphi \otimes 1} M \otimes_R k \to 0$$

is exact. Tensor $0 \to \mathfrak{m} \to R \to k \to 0$ by N; right exactness gives a natural isomorphism

$$\tau_N : N \otimes_R k \to N/\mathfrak{m}N;$$

if $n \in N$ and $b \in k$, then $\tau_N : n \otimes b \mapsto n + \mathfrak{m}N$. There is a commutative diagram

$$
\begin{array}{ccc}
0 \longrightarrow N \otimes_R k & \xrightarrow{i \otimes 1} & F \otimes_R k \\
\tau_N \downarrow & & \downarrow \tau_F \\
N/\mathfrak{m}N & \xrightarrow{\bar{i}} & F/\mathfrak{m}F,
\end{array}
$$

where $\bar{i} : n + \mathfrak{m}N \mapsto n + \mathfrak{m}F$. Since $i \otimes 1$ is an injection, so is $\bar{i}$. But $N \subseteq \mathfrak{m}F$ says that the map $\bar{i}$ is the zero map. Thus, $N/\mathfrak{m}N = \{0\}$, so that $N = \mathfrak{m}N$. Hence, $N = \{0\}$, by Nakayama's lemma (Corollary 4.51) (which applies because finitely generated modules over a noetherian ring are finitely presented). Therefore, $\varphi : F \to M$ is an isomorphism, and so M is free.

For the inductive step, we must prove that if $\text{Tor}_{n+2}^R(M, k) = \{0\}$, then $\text{pd}(M) \le n + 1$. Take a projective resolution $\mathbf{P}$ of M, and let Ω_n be its nth syzygy. Since $\mathbf{P}$ must also be a flat resolution of M, we have $Y_n = \Omega_n$ (where Y_n denotes the nth syzygy of $\mathbf{P}$ viewed as a flat resolution of M). By Corollary 6.23, $\text{Tor}_{n+2}^R(M, k) \cong \text{Tor}_1^R(Y_n, k)$. The base step shows that $Y_n = \Omega_n$ is free, and this gives $\text{pd}(M) \le n + 1$, by Lemma 8.6.   •

Corollary 8.54.   *Let $(R, \mathfrak{m}, k)$ be a local ring. If M is a finitely generated R-module, then*

$$\text{pd}(M) = \sup\{i : \text{Tor}_i^R(M, k) \ne \{0\}\}.$$

Proof.   Let $n = \sup\{i : \text{Tor}_i^R(M, k) \ne \{0\}\}$. Then $\text{pd}(M) \le n - 1$; since $\text{pd}(M) \not< n$, we have $\text{pd}(M) = n$.   •

Theorem 8.55.   *Let $(R, \mathfrak{m}, k)$ be a local ring.*

(i) $\text{D}(R) \le n$ *if and only if* $\text{Tor}_{n+1}^R(k, k) = \{0\}$.

(ii) $\text{D}(R) = \text{pd}(k)$.

Proof.

(i) If $D(R) \leq n$, then Lemma 8.53 applies at once to give $\mathrm{Tor}^R_{n+1}(k, k) = \{0\}$. Conversely, if $\mathrm{Tor}^R_{n+1}(k, k) = \{0\}$, Lemma 8.53 gives $\mathrm{pd}(k) \leq n$. Now $\mathrm{Tor}^R_{n+1}(M, k) = \{0\}$ for every R-module M, by Proposition 8.17. In particular, if M is finitely generated, then Lemma 8.53 gives $\mathrm{pd}(M) \leq n$. Finally, $D(R) = \sup_M \{\mathrm{pd}(M)\}$, where M ranges over all finitely generated (even cyclic) R-modules, by Proposition 8.52. Hence, $D(R) \leq n$.

(ii) Immediate from part (i) and Corollary 8.54. $\bullet$

Definition. A *prime chain* of *length* n in a commutative ring R is a strictly decreasing chain of prime ideals

$$\mathfrak{p}_0 \supsetneq \mathfrak{p}_1 \supsetneq \cdots \supsetneq \mathfrak{p}_n.$$

If R is a commutative ring, then its *Krull dimension*, $\dim(R)$, is the length n of a longest prime chain in R.

Let k be a field and let $R = k[x_1, \ldots, x_n]$ be the polynomial ring. If $\mathfrak{p}_{i-1} = (x_1, \ldots, x_i)$, then $\mathfrak{p}_i$ is a prime ideal, for $R/\mathfrak{p}_{i-1} \cong k[x_{i+1}, \ldots, x_n]$ is a domain, and

$$\mathfrak{p}_0 \supsetneq \mathfrak{p}_1 \supsetneq \cdots \supsetneq \mathfrak{p}_{n-1} \supsetneq (0)$$

is a prime chain of length n. It turns out that this prime chain has maximal length, so that $\dim(k[x_1, \ldots, x_n]) = n$.

We cite some results of Commutative Algebra that do not use homology. If $(R, \mathfrak{m}, k)$ is a local ring, then elements $x_1, \ldots, x_d$ in $\mathfrak{m}$ comprise a *minimal set of generators* (no proper subset of them generates $\mathfrak{m}$) if their cosets mod $\mathfrak{m}^2$ form a basis of the k-vector space $\mathfrak{m}/\mathfrak{m}^2$ (Rotman, *Advanced Modern Algebra*, Proposition 11.165). It follows that any two minimal sets of generators of $\mathfrak{m}$ have the same number of elements, namely,

$$V(R) = \dim_k(\mathfrak{m}/\mathfrak{m}^2).$$

There is always an inequality $\dim(R) \leq V(R)$ (*Advanced Modern Algebra*, Corollary 11.166).

Proposition 8.56. *Let $(R, \mathfrak{m}, k)$ be a noetherian local ring. If $x \in \mathfrak{m} - \mathfrak{m}^2$, then $(R^*, \mathfrak{m}^*, k)$ is a local ring, where $R^* = R/(x)$ and $\mathfrak{m}^* = \mathfrak{m}/(x)$, and*

$$V(R) = V(R^*) + 1.$$

Proof. Let $\{y_1^*, \ldots, y_t^*\}$ be a minimal generating set of $\mathfrak{m}^*$, and let $y_i^* = y_i + \mathfrak{m}$. It is clear that $\{x, y_1, \ldots, y_t\}$ generates $\mathfrak{m}$, and we now show that it is a minimal generating set; that is, their cosets mod $\mathfrak{m}$ form a basis of $\mathfrak{m}/\mathfrak{m}^2$.

If $rx + \sum_i r_i y_i \in \mathfrak{m}^2$, where $r_i, r \in R$, then we must show that each term lies in $\mathfrak{m}^2$; that is, all $r_i, r \in \mathfrak{m}$. Passing to R^*, we have $\sum_i r_i^* y_i^* \in (\mathfrak{m}^*)^2$ [where $*$ denotes coset mod (x)] for $r^* x^* = 0$. But $\{y_1^*, \ldots, y_t^*\}$ is a basis of $\mathfrak{m}^*/(\mathfrak{m}^*)^2$, so that $r_i^* \in \mathfrak{m}^*$ and $r_i \in \mathfrak{m}$ for all i. Therefore, $rx \in \mathfrak{m}^2$. But $x \notin \mathfrak{m}^2$, and so $r \in \mathfrak{m}$, as desired.   •

Definition.   A local ring $(R, \mathfrak{m}, k)$ is *regular* of *dimension n* if it is noetherian and
$$n = \dim(R) = V(R).$$

It is clear that every field is a regular local ring of dimension 0, and it is easy to see that every local PID is a regular local ring of dimension 1. Regular local rings must be domains (Rotman, *Advanced Modern Algebra*, Proposition 11.172); it follows that if p is a prime, $R = \mathbb{I}_{p^2}$ is a noetherian local ring that is not regular.

Example 8.57.   Regular local rings arise in connection with nonsingular points on varieties. In more detail, let k be an algebraically closed field, and let $I \subseteq k[X]$ be a set of polynomials, where $k[X]$ abbreviates $k[x_1, \ldots, x_m]$. We regard each $f(X) \in k[X]$ as a k-valued function, and we define the *variety* of I to be
$$\mathrm{Var}(I) = \{a \in k^m : f(a) = 0 \text{ for all } f \in I\}.$$
Given a subset $A \subseteq k^m$, define
$$\mathrm{Id}(A) = \{f(X) \in k[X] : f(a) = 0 \text{ for all } a \in A\}.$$
The *coordinate ring* of A is
$$k[A] = \{f|A : f(X) \in k[X]\}.$$
Now $\mathrm{Id}(A)$ is always an ideal in $k[X]$, and $k[A] \cong k[X]/\mathrm{Id}(A)$. A variety V is *irreducible* if its coordinate ring $k[V]$ is a domain; that is, if $\mathrm{Id}(V)$ is a prime ideal (it is common usage to assume, as part of the definition, that varieties are irreducible). Let $k(V) = \mathrm{Frac}(k[V])$.

If V is a variety and $a \in V$, then the *local ring of V at a* is the localization
$$\mathcal{O}_{a,V} = k[V]_{\mathrm{Id}(a)},$$
where $\mathrm{Id}(a) = \{f/g \in k[V] : f(a) = 0 \text{ and } g(a) \neq 0\}$. Now $\mathcal{O}_{a,V}$ is a local ring with maximal ideal $\mathfrak{m}_{a,V}$ the localization of $\mathrm{Id}(a)$, and $\mathcal{O}_{a,V}/\mathfrak{m}_{a,V} \cong k$. We can define formal partial derivatives $\partial f/\partial x_i$ for every $f \in k[V]$, which allows us to define the *tangent space* $T_{a,V}$ of V at a. There is an isomorphism $T_{a,V}^* \cong \mathfrak{m}_{a,V}/\mathfrak{m}_{a,V}^2$ of vector spaces over k, where $T_{a,V}^*$ is the dual space of $T_{a,V}$, and so $\dim_k(\mathfrak{m}_{a,V}/\mathfrak{m}_{a,V}^2) = \dim(T_{a,V})$. We say a is a *nonsingular*

point if the tangent space at a has the expected number of linearly independent tangents. There are several ways to express this algebraically, but all of them say that $a \in V$ is nonsingular if and only if $(\mathcal{O}_{a,V}, \mathfrak{m}_{a,V}, k)$ is a regular local ring. ◄

Serre and Auslander–Buchsbaum proved, independently, that R is regular if and only if $D(R)$ is finite, in which case $D(R) = \dim(R) = V(R)$. The proof that if $(R, \mathfrak{m}, k)$ is regular, then $D(R) = \dim(R) = V(R)$ is not too difficult.

Definition. Let R be a commutative ring and let M be an R-module. A sequence $x_1, \ldots, x_n$ in R is an **M-regular sequence** if x_1 is regular on M (i.e., the multiplication map $M \to M$, given by $m \mapsto xm$, is an injection), x_2 is regular on $M/(x_1)M$, x_3 is regular on $M/(x_1, x_2)M$, $\cdots$, x_n is regular on $M/(x_1, \ldots, x_{n-1})M$. If $M = R$, then $x_1, \ldots, x_n$ is also called an R-*sequence*.

For example, if $R = k[x_1, \ldots, x_n]$ is a polynomial ring over a field k, then it is easy to see that $x_1, \ldots, x_n$ is an R-sequence.

Proposition 8.58. *A noetherian local ring $(R, \mathfrak{m}, k)$ is regular if and only if $\mathfrak{m}$ is generated by an R-sequence $x_1, \ldots, x_d$. Moreover, in this case,*

$$d = V(R).$$

Proof. Rotman, *Advanced Modern Algebra*, Proposition 11.173. ●

Lemma 8.59. *Let $(R, \mathfrak{m}, k)$ be a local ring, let M be a finitely generated R-module, and let $x \in \mathfrak{m}$ be regular on M. If $\mathrm{pd}(M) = n < \infty$, then $\mathrm{pd}(M/xM) = n + 1$.*

Proof. Since x is regular on M, there is an exact sequence

$$0 \to M \xrightarrow{x} M \to M/xM \to 0,$$

where the first map is multiplication by x. There is a long exact sequence arising from applying $\square \otimes_R k$; consider the fragment for $i > n + 1$:

$$0 = \mathrm{Tor}_i^R(M, k) \to \mathrm{Tor}_i^R(M/xM, k) \to \mathrm{Tor}_{i-1}^R(M, k) = 0$$

[since $i - 1 > n = \mathrm{pd}(M)$, the outside terms vanish, by Lemma 8.53]. Thus, $\mathrm{Tor}_i^R(M/xM, k) = \{0\}$ for all $i > n + 1$, and so $\mathrm{pd}(M/xM) \le n + 1$ [we are using Theorem 8.27(i): since R is noetherian and M is finitely generated, $\mathrm{fd}(M) = \mathrm{pd}(M)$].

Now consider the fragment of the long exact sequence for $i = n + 1$:

$$0 = \mathrm{Tor}^R_{n+1}(M, k) \to \mathrm{Tor}^R_{n+1}(M/xM, k) \to \mathrm{Tor}^R_n(M, k) \xrightarrow{x} \mathrm{Tor}^R_n(M, k).$$

Since $x \in \mathfrak{m}$, multiplication by x annihilates $k = R/\mathfrak{m}$, and hence multiplication by x is the zero map on $\mathrm{Tor}^R_n(M, k)$. Exactness shows that the map $\mathrm{Tor}^R_{n+1}(M/xM, k) \to \mathrm{Tor}^R_n(M, k)$ is an isomorphism. But $\mathrm{pd}(M) = n$ gives $\mathrm{Tor}^R_n(M, k) \neq \{0\}$, by Corollary 8.54. Hence, $\mathrm{Tor}^R_{n+1}(M/xM, k) \neq \{0\}$, and $\mathrm{pd}(M/xM) = n + 1$.   •

Proposition 8.60.   *If $(R, \mathfrak{m}, k)$ is a regular local ring, then $\mathrm{D}(R)$ is finite; in fact,*

$$D(R) = V(R) = \dim(R).$$

Proof.   Since R is regular, $\mathfrak{m}$ can be generated by an R-sequence $x_1, \ldots, x_d$, by Proposition 8.58. But Lemma 8.59 applied to the modules R, $R/(x_1)$, $R/(x_1, x_2), \cdots, R/(x_1, \ldots, x_d) = R/\mathfrak{m} = k$ shows that $\mathrm{pd}(k) = d$. Hence, $d = V(R) = \dim(R)$, by Proposition 8.58. On the other hand, Theorem 8.55 gives $d = \mathrm{pd}(k) = D(R)$.   •

The converse of Proposition 8.60: a noetherian local ring of finite global dimension is regular, is more difficult to prove. The following proof is essentially that in Lam, *Lectures on Modules and Rings*, Chapter 2, § 5F.

In proving that $\mathrm{D}(R)$ finite implies R regular, we cannot assume that R is a domain (though this will turn out to be true); hence, we must deal with zero-divisors.

Proposition 8.61.   *Let $(R, \mathfrak{m}, k)$ be a local ring.*

(i) *If $\mathfrak{m} - \mathfrak{m}^2$ consists of zero-divisors, then there is a nonzero $a \in R$ with $a\mathfrak{m} = \{0\}$.*

(ii) *If $0 < D(R) = n < \infty$, then there exists a nonzero-divisor $x \in \mathfrak{m} - \mathfrak{m}^2$.*

Proof.

(i) By Proposition 8.51, there are prime ideals $\mathfrak{p}_1, \ldots, \mathfrak{p}_n$ with

$$\mathfrak{m} - \mathfrak{m}^2 \subseteq \mathrm{Zer}(R) \subseteq \mathfrak{p}_1 \cup \cdots \cup \mathfrak{p}_n.$$

If we can show that

$$\mathfrak{m} \subseteq \mathfrak{p}_1 \cup \cdots \cup \mathfrak{p}_n, \tag{1}$$

then Prime Avoidance, Lemma 8.50, gives $\mathfrak{m} \subseteq \mathfrak{p}_i$ for some i. But $\mathfrak{p}_i = \mathrm{ann}(a)$ for some $a \in \mathfrak{m}$, so that $a\mathfrak{m} = \{0\}$, as desired.

To verify Eq. (1), it suffices to prove $\mathfrak{m}^2 \subseteq \mathfrak{p}_1 \cup \cdots \cup \mathfrak{p}_n$. Now $\mathfrak{m} \neq \mathfrak{m}^2$, by Nakayama's Lemma (we may assume that $\mathfrak{m} \neq \{0\}$, for the result is trivially true otherwise), and so there exists $x \in \mathfrak{m} - \mathfrak{m}^2 \subseteq \mathfrak{p}_1 \cup \cdots \cup \mathfrak{p}_n$. Let $y \in \mathfrak{m}^2$. For every integer $s \geq 1$, we have $x + y^s \in \mathfrak{m} - \mathfrak{m}^2 \subseteq \mathfrak{p}_1 \cup \cdots \cup \mathfrak{p}_n$; that is, $x + y^s \in \mathfrak{p}_j$ for some $j = j(s)$. By the pigeonhole principle, there are an integer j and integers $s < t$ with $x + y^s, x + y^t \in \mathfrak{p}_j$. Subtracting, $y^s(1 - y^{t-s}) \in \mathfrak{p}_j$. But $1 - y^{t-s}$ is a unit [if $u \in \mathfrak{m}$, then $1 - u$ is a unit; otherwise, $(1 - u)$ is a proper ideal, $(1 - u) \subseteq \mathfrak{m}$, $1 - u \in \mathfrak{m}$, and $1 \in \mathfrak{m}$]. Since $\mathfrak{p}_j$ is a prime ideal, $y \in \mathfrak{p}_j$.

(ii) **(Griffith)** In light of (i), it suffices to show there is no nonzero $a \in \mathfrak{m}$ with $a\mathfrak{m} = \{0\}$. If, on the contrary, such an a exists and if $\mu \colon R \to R$ is given by $\mu \colon r \mapsto ar$, then $\mathfrak{m} \subseteq \ker \mu$. This inclusion cannot be strict; if $b \in \ker \mu$, then $ab = 0$, but if $b \notin \mathfrak{m}$, then b is a unit (for Rb is not contained in the maximal ideal), and so $ab \neq 0$. Hence, $\mathfrak{m} = \ker \mu$. Thus, $k = R/\mathfrak{m} = R/\ker \mu \cong \operatorname{im} \mu = Ra$; that is, $k \cong Ra$. Consider the exact sequence $0 \to Ra \to R \to R/Ra \to 0$; by Exercise 8.4 on page 466, either $\operatorname{pd}(R/Ra) = \operatorname{pd}(Ra) + 1 = \operatorname{pd}(k) + 1$ or $\operatorname{pd}(R/Ra) = 0$. In the first case, $\operatorname{pd}(R/Ra) = \operatorname{pd}(k) + 1 > \operatorname{pd}(k)$, contradicting Theorem 8.55(ii) [which says that $\operatorname{pd}(k) = D(R)$]. In the second case, $0 = \operatorname{pd}(Ra) = \operatorname{pd}(k)$, contradicting $\operatorname{pd}(k) = D(R) > 0$.   •

Theorem 8.62 (Serre–Auslander–Buchsbaum).   *A noetherian local ring* $(R, \mathfrak{m}, k)$ *is regular if and only if* $D(R)$ *is finite; in fact,*

$$D(R) = V(R) = \dim(R).$$

Proof.   Necessity is Proposition 8.60. We prove the converse by induction on $D(R) = n \geq 0$. If $n = 0$, then R is semisimple. Since R is commutative, it is the direct product of finitely many fields; since R is local, it is a field, and hence it is regular.

If $n \geq 1$, then Proposition 8.61(ii) says that $\mathfrak{m} - \mathfrak{m}^2$ contains a nonzero-divisor x. Now $(R^*, \mathfrak{m}^*, k)$ is a local ring, where $R^* = R/(x)$ and $\mathfrak{m}^* = \mathfrak{m}/(x)$. Since x is not a zero-divisor, it is regular on $\mathfrak{m}$; since $\operatorname{pd}_R(\mathfrak{m}) < \infty$, Proposition 8.38(i) gives $\operatorname{pd}_{R^*}(\mathfrak{m}/x\mathfrak{m}) < \infty$.

Consider a short exact sequence of R-modules

$$0 \to k \xrightarrow{\alpha} B \to C \to 0 \tag{2}$$

in which $\alpha(1) = e$, where $e \in B - \mathfrak{m}B$. Now the coset $e + \mathfrak{m}B$ is part of a basis of the k-vector space $B/\mathfrak{m}B$, and so there is a k-map $\beta \colon B/\mathfrak{m}B \to k$ with $\beta(e + \mathfrak{m}B) = 1$. The composite $\pi \colon B \xrightarrow{\text{nat}} B/\mathfrak{m}B \xrightarrow{\beta} k$ shows that the exact sequence (2) splits, for $\pi\alpha = 1_k$. In particular, this applies when $B = \mathfrak{m}/x\mathfrak{m}$ and k is the cyclic submodule generated by $e + x\mathfrak{m}$. Thus, k is a direct summand of $\mathfrak{m}/x\mathfrak{m}$. It follows that $\mathrm{pd}_{R*}(k) < \infty$, so that Proposition 8.39 gives $\mathrm{pd}_{R*}(k) = \mathrm{pd}_R(k) - 1$. Theorem 8.55 now gives

$$D(R^*) = \mathrm{pd}_{R*}(k) = n - 1.$$

By induction, R^* is a regular local ring and $\dim(R^*) = n - 1$. Hence, there is a prime chain of length $n - 1$ in R^*

$$\mathfrak{p}_0^* \supsetneq \mathfrak{p}_1^* \supsetneq \cdots \supsetneq \mathfrak{p}_{n-1}^* = (0)$$

[we have $\mathfrak{p}_{n-1}^* = (0)$ because R^* is a domain (being regular) and (0) is a prime ideal]. Taking inverse images gives a prime chain in R:

$$\mathfrak{p}_0 \supsetneq \mathfrak{p}_1 \supsetneq \cdots \supsetneq \mathfrak{p}_{n-1} = (x).$$

Were (x) a minimal prime ideal, then every element in it would be nilpotent, by Proposition 4.76. Since x is not a zero-divisor, it is not nilpotent, and so there is a prime ideal $\mathfrak{q} \subsetneq (x)$. Hence, $\dim(R) \geq n$.

Since $x \in \mathfrak{m} - \mathfrak{m}^2$, Proposition 8.56 says that $V(R) = 1 + V(R^*) = 1 + \dim(R^*) = n$. Therefore,

$$n = V(R) \geq \dim(R) \geq n$$

[the inequality $V(R) \geq \dim(R)$ always being true], and $\dim(R) = V(R)$; that is, R is a regular local ring of dimension n.   •

Corollary 8.63. *If S is a multiplicative subset of a regular local ring, then $S^{-1}R$ is also a regular local ring. In particular, if $\mathfrak{p}$ is a prime ideal in R, then $R_{\mathfrak{p}}$ is regular.*

Proof. Theorem 8.27(ii) says that $D(R) = \mathrm{wD}(R)$ in this case. But Proposition 8.23 says that $\mathrm{wD}(S^{-1}R) \leq \mathrm{wD}(R)$. Therefore,

$$D(S^{-1}R) = \mathrm{wD}(S^{-1}R) \leq \mathrm{wD}(R) = D(R) < \infty.$$

It follows from Corollary 4.74 that $S^{-1}R$ is a local ring; therefore, $S^{-1}R$ is regular, by Theorem 8.62. The second statement follows: if $S = R - \mathfrak{p}$, then $R_{\mathfrak{p}}$ is a local ring, and so the Serre–Auslander–Buchsbaum Theorem says that $R_{\mathfrak{p}}$ is regular.   •

There are several proofs that regular local rings are unique factorization domains; most use the notion of *depth* that was used in the original proof of Auslander and Buchsbaum. If M is a finitely generated R-module, then its **depth** is defined by

$$\mathrm{depth}(M) = \text{length of a maximal regular } M\text{-sequence}.$$

The depth of M was originally called its *codimension* because of the following result of Auslander and Buchsbaum.

Theorem (Codimension Theorem).   *Let* $(R, \mathfrak{m})$ *be a noetherian local ring, and let* M *be a finitely generated* R-*module with* $\mathrm{pd}(M) < \infty$. *Then*

$$\mathrm{pd}(M) + \mathrm{depth}(M) = \mathrm{depth}(R).$$

In particular, if R *is regular, then* $\mathrm{depth}(R) = \mathrm{D}(R)$ *and*

$$\mathrm{pd}(M) + \mathrm{depth}(M) = \mathrm{D}(R).$$

Proof.   Rotman, *Advanced Modern Algebra*, Proposition 11.181.   •

M. Nagata proved, using the result of Serre–Auslander–Buchsbaum, that if one knew that every regular local ring R with $\mathrm{D}(R) = 3$ is a unique factorization domain, then this is so for every regular local ring [see "A general theory of algebraic geometry over Dedekind rings II," *Amer. J. Math.* 80 (1958), 382–420].

Corollary 8.64.   *If* $(R, \mathfrak{m})$ *is a noetherian local ring with* $\mathrm{D}(R) = 3$ *and* $\mathfrak{p} \neq \mathfrak{m}$ *is a prime ideal in* R, *then*

$$\mathrm{pd}(\mathfrak{p}) \leq 1.$$

Proof.   By hypothesis, there exists $x \in \mathfrak{m} - \mathfrak{p}$. Now x is regular on $R/\mathfrak{p}$: if $x(r + \mathfrak{p}) = \mathfrak{p}$, then $xr \in \mathfrak{p}$ and $r \in \mathfrak{p}$, for $x \notin \mathfrak{p}$ and $\mathfrak{p}$ is prime. Therefore, $\mathrm{depth}(R/\mathfrak{p}) \geq 1$ and so $\mathrm{pd}(R/\mathfrak{p}) \leq 2$, by the Codimension Theorem. But there is an exact sequence $0 \to \mathfrak{p} \to R \to R/\mathfrak{p} \to 0$, which shows that $\mathrm{pd}(\mathfrak{p}) \leq 1$.   •

Theorem 8.65 (Auslander–Buchsbaum).   *Every regular local ring* R *is a unique factorization domain.*

Proof.   A standard result of Commutative Algebra is that a domain R is a unique factorization domain if every *minimal* nonzero prime ideal $\mathfrak{p}$ (there is no nonzero prime ideal $\mathfrak{q}$ with $\mathfrak{q} \subsetneq \mathfrak{p}$) is principal. If $\mathrm{D}(R) = 3$, then Corollary 8.64 gives $\mathrm{pd}(\mathfrak{p}) \leq 1$. Auslander and Buchsbaum showed that $\mathrm{pd}(\mathfrak{p}) = 1$ gives a contradiction, so that $\mathrm{pd}(\mathfrak{p}) = 0$; that is, $\mathfrak{p}$ is a projective, hence free, R-module (Corollary 4.16). But any ideal in a domain R that is free as an R-module must be principal.   •

Another proof, not using Nagata's difficult proof, is based on the following criterion.

Proposition 8.66.   *If* R *is a noetherian domain for which every finitely generated* R-*module has* FFR, *then* R *is a unique factorization domain.*

Proof.   This is Theorem 184 in Kaplansky, *Commutative Rings*. The proof uses a criterion for a domain to be a unique factorization domain (his Theorem 179), which involves showing that if a commutative ring R has the property that every finitely generated R-module has FFR, then so does $R[x]$. But this is just our Theorem 8.47. •

Unique factorization for regular local rings follows easily from this last proposition. If $D(R) = n$, then every finitely generated R-module M has a projective resolution $0 \to P_n \to \cdots \to P_0 \to M \to 0$ in which each P_i is finitely generated (Lemma 7.19). But (finitely generated) projective R-modules are free [Theorem 4.57], and so every finitely generated R-module has FFR.

9

Homology and Groups

Applications of homology to Group Theory are usually called *Cohomology of Groups*. The history of this subject is quite interesting. Its algebraic origins can be found in the early 1900s, with Schur's work on *projective representations* [homomorphisms of groups to PGL(n, k)] in the first decade and in Schreier's work on extensions of groups in 1926. Its topological origins lie in the discovery, by Hurewicz in the 1930s, that if X is a connected *aspherical space* (the higher homotopy groups of X are all trivial), then all the homology and cohomology groups of X are determined by the fundamental group $\pi = \pi_1(X)$. But it was a theorem of Hopf in 1944, about actions of fundamental groups, that led Eilenberg and Mac Lane to define and develop the basic ideas of Cohomology of Groups. This mixed parentage (which explains why groups in the early papers are always denoted by π instead of by G) is a reflection of a deep relationship between Group Theory and Algebraic Topology.

9.1 Group Extensions

Exactness of a sequence of nonabelian groups,

$$\to G_{n+1} \xrightarrow{d_{n+1}} G_n \xrightarrow{d_n} G_{n-1} \to,$$

is defined just as it is for abelian groups: $\operatorname{im} d_{n+1} = \ker d_n$ for all n. Of course, each $\ker d_n$ is a normal subgroup of G_n.

J.J. Rotman, *An Introduction to Homological Algebra*, Universitext,
DOI 10.1007/978-0-387-68324-9_9, © Springer Science+Business Media LLC 2009

Definition.   If K and Q are groups, then an **extension** of K by Q is a short exact sequence $1 \to K \xrightarrow{i} E \xrightarrow{p} Q \to 1$.

Unless we say otherwise, we will assume that the map $i: K \to E$ is the inclusion. The notation K and Q reminds us of kernel and quotient. Henceforth, we shall denote the elements of K by $a, b, c, \dots$ and the elements of Q by $x, y, z, \dots$.

A group E having a normal subgroup K can be "factored" into K and E/K. The **extension problem** is the inverse question: find all possible extensions of a given ordered pair of groups (K, Q). In other words, to what extent can E be recovered from a normal subgroup K and the quotient $Q = E/K$? For example, we know that $|E| = |K||Q|$ if E is finite. O. Schreier ["Über die Erweiterung von Gruppen, I," *Monatsh. Math. Phys.* 34 (1926), 165–180; "Über die Erweiterung von Gruppen, II," *Abh. Math. Sem. Hamburg* 4 (1926), 321–346] solved the extension problem for groups by constructing all possible multiplication tables for E. Even though the proof of Schreier's Theorem consists of manipulating and organizing long series of elementary calculations, his results are still strong enough to yield a proof of the Schur–Zassenhaus Lemma. Our discussion in this section displays the origins of several definitions of Homological Algebra, but, more importantly, it gives interpretations of low-dimensional cohomology groups.

Remark.   We must point out that Schreier's solution of the extension problem does not allow us, given K and Q, to determine the number of nonisomorphic middle groups E of extensions $1 \to K \to E \to Q \to 1$. It is not easy to recognize whether two multiplication tables of a group of order n arise from a given group E; after all, there are $n!$ different lists of the elements of E, each of which gives a multiplication table. If E' is another group of order n, the problem of determining whether or not E and E' are isomorphic is essentially the problem of comparing two families of multiplication tables, one for E and one for E', to see if there is a pair of tables that coincide. ◀

The significance of the extension problem arises from the Jordan–Hölder Theorem. Assume that a group E has a composition series; say,

$$E = K_0 \ge K_1 \ge K_2 \ge \cdots \ge K_{n-1} \ge K_n = \{1\},$$

with simple factor groups $Q_1, \dots, Q_n$, where $Q_i = K_{i-1}/K_i$ for all $i \ge 1$. Since $K_n = \{1\}$, we have $Q_n = K_{n-1}$. If we could solve the extension problem, then K_{n-2} could be found from the extension $1 \to K_{n-1} \to K_{n-2} \to Q_{n-1} \to 1$; that is, from Q_n, Q_{n-1}. Iterating, E could be recaptured from $Q_n, Q_{n-1}, \dots, Q_1$. Since all finite simple groups are classified (the proof having been completed in 2005), all finite groups could be surveyed if we could solve the extension problem.[1]

The definition of extension makes sense for any, possibly nonabelian, group K, but, to keep hypotheses uniform, we assume in our discussion that K is abelian, even when this assumption is not needed. As usual, we write abelian groups K additively. The group E containing K is allowed to be nonabelian, and it, too, is written additively (it would be confusing to do otherwise). The group Q will always be written multiplicatively. Proposition 9.1(iii) gives a reason for using this mixture of additive and multiplicative notation.

Definition. If $0 \to K \to E \xrightarrow{p} Q \to 1$ is an extension, then a *lifting* is a function $\ell \colon Q \to E$, not necessarily a homomorphism, with $p\ell = 1_Q$. We assume further that $\ell(1) = 0$.

If K is a subgroup of a group E, then a right *transversal* of K (or a *complete system of coset representatives* of K) is a subset $T \subseteq E$ consisting of exactly one element from each right coset $K + t$ of K. We normalize T by assuming that $t = 0$ is chosen for the coset K.

Given a right transversal, we can construct a lifting. For each $x \in Q$, surjectivity of p provides $\ell(x) \in E$ with $p\ell(x) = x$; thus, the function $x \mapsto \ell(x)$ is a lifting if we choose $\ell(1) = 0$. Conversely, given a lifting, we claim that $\ell(Q)$ is a right transversal of K. If $K + e$ is a coset, then $p(e) \in Q$; say, $p(e) = x$; hence, $p(e - \ell(x)) = 1$, so that $e - \ell(x) \in K$ and $K + e = K + \ell(x)$. Thus, every coset has a representative in $\ell(Q)$. Finally, we must show that $\ell(Q)$ does not contain two elements in the same coset. If $K + \ell(x) = K + \ell(y)$, then there is $a \in K$ with $a + \ell(x) = \ell(y)$. Apply p to this equation; since $p(a) = 1$, we have $x = y$ and so $\ell(x) = \ell(y)$.

The *automorphism group* $\mathrm{Aut}(E)$ of a group E is the group whose elements are all the isomorphisms of E with itself and whose operation is composition. An automorphism φ is *inner* if it is a conjugation; that is, there is $c \in E$ with $\varphi(e) = c + e - c$ for all $e \in E$ (in additive notation). An automorphism of E is *outer* if it is not inner. The subset $\mathrm{Inn}(E) \subseteq \mathrm{Aut}(E)$ consisting of all the inner automorphisms of E is a normal subgroup of $\mathrm{Aut}(E)$; the quotient $\mathrm{Aut}(E)/\mathrm{Inn}(E)$ is denoted by $\mathrm{Out}(E)$ and is called the *outer automorphism group*.

---

[1] Alas, these remarks are not practical. Besche–Eick–O'Brien ["The groups of order at most 2000," *Electron. Res. Announc. Amer. Math. Soc.* 7 (2001), 1–4] have shown that there are 56,092 nonisomorphic groups of order 2^8, and 10,494,213 groups of order 2^9. Besche–Eick–O'Brien ["A millenium project: constructing small groups," *Internat. J. Algebra Comput.*, 12 (2002), 623–644] show that there are 49,487,365,422 groups of order 2^{10}.

Proposition 9.1. *Let* $0 \to K \to E \overset{p}{\to} Q \to 1$ *be an extension of an abelian group* K *by a group* Q, *and let* $\ell: Q \to E$ *be a lifting.*

(i) *For every* $x \in Q$, *conjugation* $\theta_x: K \to K$, *defined by*

$$\theta_x: a \mapsto \ell(x) + a - \ell(x),$$

is independent of the choice of lifting $\ell(x)$ *of* x.

(ii) *The function* $\theta: Q \to \text{Aut}(K)$, *defined by* $x \mapsto \theta_x$, *is a homomorphism.*

(iii) K *is a left* $\mathbb{Z}Q$-*module with scalar multiplication given by*

$$xa = \theta_x(a) = \ell(x) + a - \ell(x).$$

Proof.

(i) Suppose that ℓ' is another lifting, so that $p\ell'(x) = x$ for all $x \in Q$. There is $b \in K$ with $\ell'(x) = \ell(x) + b$ [for $-\ell(x) + \ell'(x) \in \ker p = K$]. Therefore,

$$\ell'(x) + a - \ell'(x) = \ell(x) + b + a - b - \ell(x)$$
$$= \ell(x) + a - \ell(x),$$

because a and b commute in the abelian group K.

(ii) Now $\theta_x(a) \in K$, for $K \lhd E$, so that $\theta_x: K \to K$; also, $\theta_x \in \text{Aut}(K)$, because conjugations are automorphisms. Let us see that $\theta: Q \to \text{Aut}(K)$ is a homomorphism. If $x, y \in Q$ and $a \in K$, then

$$\theta_x(\theta_y(a)) = \theta_x(\ell(y) + a - \ell(y)) = \ell(x) + \ell(y) + a - \ell(y) - \ell(x),$$

while

$$\theta_{xy}(a) = \ell(xy) + a - \ell(xy).$$

But $\ell(x) + \ell(y)$ and $\ell(xy)$ are both liftings of xy, and so $\theta_x \theta_y = \theta_{xy}$ follows from part (i).

(iii) Parts (i) and (ii). ●

The homomorphism $\theta: Q \to \text{Aut}(K)$ in Proposition 9.1(ii) tells "how" K is normal in E. For example, let K be a cyclic group of order 3 and $Q = \langle x \rangle$ be cyclic of order 2. If $E = K \times Q$, then E is abelian and K lies in the center $Z(E)$. In this case, $\ell(x) + a - \ell(x) = a$ for all $a \in K$, and $\theta_x = 1_K$. On the other hand, if $E = S_3$ and $K = A_3$ (which does not lie in the center of S_3), then conjugating $(1\ 2\ 3)$ by $\ell(x) = (1\ 2)$ gives $(1\ 3\ 2)$; thus, $\theta_x \neq 1_K$.

In Proposition 2.1, we saw that if R is a ring, then an abelian group K is a left R-module if and only if there is a ring homomorphism $\varphi \colon R \to \text{End}(K)$ with $\varphi(r) \colon a \mapsto ra$ for all $a \in K$.[2] In the special case $R = \mathbb{Z}Q$, we note that $\varphi \colon \mathbb{Z}Q \to \text{End}(K)$ is completely determined by its restriction $\theta \colon Q \to \text{Aut}(K)$ [because $\text{Aut}(K)$ is the group of (two-sided) units in $\text{End}(K)$]. In down-to-earth language, if we know xa for all $x \in Q$ and $a \in K$, then $(\sum m_x x)a = \sum_x m_x(xa)$. Consequently, we use the following abbreviations.

Definition. If Q is a group and K is an abelian group, then a *Q-module* is a left $\mathbb{Z}Q$-module. We will abbreviate $\text{Hom}_{\mathbb{Z}Q}(A, B)$ to $\text{Hom}_Q(A, B)$ and $M \otimes_{\mathbb{Z}Q} N$ to $M \otimes_Q N$ when A, B, N are left $\mathbb{Z}Q$-modules and M is a right $\mathbb{Z}Q$-module.

Exercise 9.3 on page 503 shows that every left $\mathbb{Z}Q$-module K is also a right $\mathbb{Z}Q$-module if one defines ax to be $x^{-1}a$, where $x \in Q$ and $a \in K$. Thus, if K and L are Q-modules, we can always adjust them so that $K \otimes_Q L$ is defined.

An abelian group K can be a Q-module in many ways. In particular, if K happens to be a Q-module, then the $\mathbb{Z}Q$-action arising from conjugation, as in Proposition 9.1, may not be the same as the given Q action. We give a name to those extensions for which these two $\mathbb{Z}Q$ actions coincide.

Definition. Let K be a Q-module. An extension $0 \to K \to E \to Q \to 1$ *realizes the operators* if, for all $x \in Q$ and $a \in K$, we have

$$xa = \ell(x) + a - \ell(x).$$

Definition. A Q-module K is *trivial* if $xa = a$ for all $x \in Q$ and all $a \in K$.

Proposition 9.2. *Let K be a Q-module, and let $0 \to K \to E \to Q \to 1$ be an extension that realizes the operators. Then K is a trivial Q-module if and only if $K \subseteq Z(E)$.*

Proof. Since the extension realizes the operators, $xa = \ell(x) + a - \ell(x)$ for all $x \in Q$ and $a \in K$. If K is a trivial Q-module, then $xa = a$, so that a commutes with $\ell(x)$ for all x. A general element of E has the form $b + \ell(y)$, where $b \in K$ and $y \in Q$; hence, $a + b + \ell(y) = b + a + \ell(y) = b + \ell(y) + a$ (for K is abelian), and so $a \in Z(E)$.

Conversely, if $a \in Z(E)$, then it commutes, in particular, with every $\ell(x)$, and so $xa = \ell(x) + a - \ell(x) = a$. $\bullet$

---

[2] Exercise 9.12 on page 513 puts this into the context of adjoint functors.

9.1.1 Semidirect Products

The simplest extension of K by Q is a *semidirect product*.

Definition.   An extension $0 \to K \to E \overset{p}{\to} Q \to 1$ is *split* if there is a homomorphism $j \colon Q \to E$ with $pj = 1_Q$. The middle group E in a split extension is called a *semidirect product* of K by Q and is denoted by $K \rtimes Q$.

Thus, an extension is split if and only if there is a lifting, namely, j, that is also a homomorphism.

Proposition 9.3.   *The following conditions are equivalent for an additive group E having a normal abelian subgroup K with $E/K \cong Q$.*

   (i) *E is a semidirect product of K by Q.*

  (ii) *There is a subgroup $C \subseteq E$ (called a **complement** of K) with $C \cong Q$, $K \cap C = \{0\}$, and $K + C = E$.*

 (iii) *Each $e \in E$ has a unique expression $e = a + x$, where $a \in K$ and $x \in C$.*

Proof.   A routine adaptation of Proposition 2.20.   •

Example 9.4.

   (i) An abelian group E is a semidirect product if and only if it is a direct product (usually called a direct sum), for every subgroup of an abelian group is normal. Thus, cyclic groups of prime power order are *not* semidirect products, for they cannot be a direct sum of two proper subgroups.

  (ii) A direct product $K \times Q$ is a semidirect product of K by Q (and also of Q by K). A semidirect product is so called because a direct product E of K and C requires, in addition to $K + C = E$ and $K \cap C = \{0\}$, that both subgroups K and Q be normal. For example, $E = S_3$ is a semidirect product of $K = \langle \sigma \rangle$ by $Q = \langle \tau \rangle$, where $\sigma = (1\ 2\ 3)$ and $\tau = (1\ 2)$. Note that K is a normal subgroup of order 3, but that Q, a subgroup of order 2, is not normal. It follows that the nonabelian group S_3 is not the direct product $K \times Q$, for this last group is abelian.

This example also shows that complements need not be unique. For example, S_3 is the semidirect product of K by $\langle \tau' \rangle$, where τ' is any transposition in S_3. However, any two complements in a semidirect product E of K by Q are isomorphic, for every complement Q of K is isomorphic to E/K.

(iii) The **dihedral group** D_8 of order 8 is a semidirect product in two ways: $D_8 \cong \mathbb{I}_4 \rtimes \mathbb{I}_2$ and $D_8 \cong \mathbf{V} \rtimes \mathbb{I}_2$, where $\mathbf{V}$ is the four-group. The other nonabelian group of order 8, the quaternion group $\mathbf{Q}$, is not a semidirect product (see Exercise 9.7 on page 503). ◄

We now construct semidirect products.

Theorem 9.5. *Given a group Q and a Q-module K, there exists a split extension $0 \to K \to K \rtimes Q \to Q \to 1$ that realizes the operators. The elements of $K \rtimes Q$ are all ordered pairs $(a, x) \in K \times Q$, and its operation is*

$$(a, x) + (b, y) = (a + xb, xy).$$

Remark. The operation looks more natural in multiplicative notation;

$$(ax)(by) = a(xbx^{-1})xy. \quad ◄$$

Proof. We begin by proving that $K \rtimes Q$ is a group. For associativity,

$$[(a, x) + (b, y)] + (c, z) = (a + xb, xy) + (c, z)$$
$$= (a + xb + (xy)c, (xy)z).$$

On the other hand,

$$(a, x) + [(b, y) + (c, z)] = (a, x) + (b + yc, yz)$$
$$= (a + x(b + yc), x(yz)).$$

Of course, $(xy)z = x(yz)$, because of associativity in Q. The first coordinates are also equal: since K is a Q-module,

$$x(b + yc) = xb + x(yc) = xb + (xy)c,$$

and so the operation is associative. The identity element is $(0, 1)$, for

$$(0, 1) + (a, x) = (0 + 1a, 1x) = (a, x),$$

and the inverse of (a, x) is $(-x^{-1}a, x^{-1})$, for

$$(-x^{-1}a, x^{-1}) + (a, x) = (-x^{-1}a + x^{-1}a, x^{-1}x) = (0, 1).$$

Therefore, $K \rtimes Q$ is a group.

Define a function $p : K \rtimes Q \to Q$ by $p : (a, x) \mapsto x$. Since the only "twist" occurs in the first coordinate, p is a surjective homomorphism with $\ker p = \{(a, 1) : a \in K\}$. If we define $i : K \to K \rtimes Q$ by $i : a \mapsto (a, 1)$, then

$$0 \to K \xrightarrow{i} K \rtimes Q \xrightarrow{p} Q \to 1$$

is an extension. The function $j: Q \to K \rtimes Q$, defined by $j: x \mapsto (0, x)$, is a homomorphism, for $(0, x) + (0, y) = (0, xy)$. Now $pjx = p(0, x) = x$, so that $pj = 1_Q$, and the extension splits. Finally, $K \rtimes Q$ realizes the operators: if $x \in Q$, then every lifting of x has the form $\ell(x) = (b, x)$ for some $b \in K$, and

$$
\begin{aligned}
(b, x) + (a, 1) - (b, x) &= (b + xa, x) + (-x^{-1}b, x^{-1}) \\
&= (b + xa + x(-x^{-1}b), xx^{-1}) \\
&= (b + xa - b, 1) \\
&= (xa, 1). \quad \bullet
\end{aligned}
$$

Theorem 9.6. *Let K be an abelian group. If a group E is a semidirect product of K by a group Q, then there is a Q-module structure on K so that $E \cong K \rtimes Q$.*

Proof. Regard E as a group having subgroups K and Q [so we may write x instead of $\ell(x)$] with $K \lhd E$ and Q a complement of K. If $a \in K$ and $x \in Q$, define

$$
xa = x + a - x.
$$

By Proposition 9.3(iii), each $e \in E$ has a unique expression as $e = a + x$, where $a \in K$ and $x \in Q$. It follows that $\varphi: E \to K \rtimes Q$, defined by $\varphi: a + x \mapsto (a, x)$, is a bijection. We now show that φ is an isomorphism.

$$
\begin{aligned}
\varphi((a + x) + (b + y)) &= \varphi(a + x + b + (-x + x) + y) \\
&= \varphi(a + (x + b - x) + x + y) \\
&= (a + xb, x + y).
\end{aligned}
$$

The definition of addition in $K \rtimes Q$ now gives

$$
\begin{aligned}
(a + xb, x + y) &= (a, x) + (b, y) \\
&= \varphi(a + x) + \varphi(b + y). \quad \bullet
\end{aligned}
$$

Exercises

In the first two exercises, the group K need not be abelian; in all other exercises, it is assumed to be abelian.

   *9.1 Let E be a group of order mn, where $(m, n) = 1$. Prove that a normal subgroup K of order m has a complement in E if and only if there exists a subgroup $C \subseteq E$ of order n. (Kernels in this exercise may not be abelian groups.)

*9.2 (*Baer*). Call a group E *injective*[3] in **Groups** if it solves the obvious universal mapping problem: for every group G and every (not necessarily abelian) subgroup $S \subseteq G$, every homomorphism $f : S \to E$ can be extended to G:

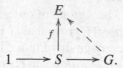

Prove that E is injective if and only if $E = \{1\}$.

Hint. Let A be free with basis $\{x, y\}$, and let B be the semidirect product $B = A \rtimes \langle z \rangle$, where z is an element of order 2 that acts on A by $zxz = y$ and $zyz = x$.

*9.3   (i)   Let K be a Q-module, where Q is a group. Prove that K is also a right $\mathbb{Z}Q$-module if one defines ax to be $x^{-1}a$, where $x \in Q$ and $a \in K$.

      (ii) If a Q-module K is made into a right $\mathbb{Z}Q$-module, as in part (i), give an example showing that K is not a $(\mathbb{Z}Q, \mathbb{Z}Q)$-bimodule.

9.4 Give an example of a split extension $0 \to K \to G \xrightarrow{p} Q \to 1$ in **Groups** for which there does not exist a homomorphism $q : G \to K$ with $qi = 1_K$. Compare with Exercise 2.8.

*9.5 Let $0 \to B \to A \to \mathbb{I}_p \to 0$ be an exact sequence of finite abelian p-groups, where p is prime. If B is cyclic, prove that either A is cyclic or the sequence splits.

*9.6 If $G = K \rtimes Q$ and $Q \subseteq N \subseteq G$, prove that $N = (N \cap K) \rtimes Q$.
Hint. Adapt the proof of Corollary 2.24.

*9.7 Prove that **Q**, the group of quaternions, is not a semidirect product.
Hint. The quaternion group **Q** is the subgroup of order 8,

$$\mathbf{Q} = \{I, A, A^2, A^3, B, BA, BA^2, BA^3\}$$
$$= \langle A \rangle \cup B\langle A \rangle \subseteq \mathrm{GL}(2, \mathbb{C}),$$

where $I = \begin{bmatrix} 1 & 0 \\ 0 & 1 \end{bmatrix}$, $A = \begin{bmatrix} 0 & 1 \\ -1 & 0 \end{bmatrix}$, and $B = \begin{bmatrix} 0 & i \\ i & 0 \end{bmatrix}$. Note that $A^2 = -I$ is the unique element of order 2 and that $Z(\mathbf{Q}) = \langle -I \rangle$.

9.8 If K and Q are solvable groups, prove that a semidirect product of K by Q is also solvable.

---

[3]The term *injective* had not yet been coined when R. Baer, who introduced the notion of injective module, proved this result. After recognizing that injective groups are duals of free groups, he jokingly called such groups *fascist groups*, and he was delighted to have proved that they are trivial.

9.9 Let K be an abelian group, let Q be a group, and let $\theta : Q \rightarrow$ Aut(K) be a homomorphism. Prove that $K \rtimes Q \cong K \times Q$ if and only if θ is the trivial map; that is, $\theta_x = 1_K$ for all $x \in Q$.

*9.10   **(i)** If K is cyclic of prime order p, prove that Aut(K) is cyclic of order $p - 1$.

     (ii) Let G be a group of order pq, where $p > q$ are primes. If $q \nmid (p - 1)$, prove that G is cyclic. Conclude, for example, that every group of order 15 is cyclic.

*9.11   **(i)** Prove that Aut$(S_3) \cong$ GL$(2, 2) \cong S_3$.

     (ii) Prove that if G is a group, then Aut$(G) = \{1\}$ if and only if $|G| \leq 2$. Conclude that every abelian group of order > 2 has an outer automorphism.

     (iii) Prove that D_8 has an outer automorphism.
     Hint. $D_8 = \langle a, b \rangle$, where $a^4 = 1 = b^2$ and $bab = a^{-1}$. Define $\varphi : D_8 \rightarrow D_8$ by $\varphi(a) = a^3$ and $\varphi(b) = b$.

     (iv) Prove that $\mathbf{Q}$ has an outer automorphism.
     Hint. Show that Aut$(\mathbf{Q}) \cong S_4$ and Inn$(\mathbf{Q}) \cong \mathbf{V}$.

9.1.2 General Extensions and Cohomology

We now solve the extension problem. In light of our discussion of semidirect products, it is reasonable to refine the problem by assuming that K is a Q-module and then to seek all those extensions E realizing the operators. One way to describe a group E is to give an addition table for it; that is, to list all its elements $a_1, a_2, \ldots$ and all sums $a_i + a_j$. Indeed, this is how we constructed semidirect products: elements are the ordered pairs (a, x) with $a \in K$ and $x \in Q$, and addition is $(a, x) + (b, y) = (a + xb, xy)$.

Suppose an extension $0 \rightarrow K \rightarrow E \rightarrow Q \rightarrow 1$ is given. If $\ell : Q \rightarrow E$ is a lifting (ℓ need not be a homomorphism), then im ℓ is a transversal[4] of K in E. The group E is the disjoint union of the cosets of K, so that every element can be expressed uniquely as $a + \ell x$. If $x, y \in Q$, then $\ell(xy)$ and $\ell x + \ell y$ represent the same coset of K, and so

$$\ell x + \ell y = f(x, y) + \ell(xy) \quad \text{for some } f(x, y) \in K. \tag{1}$$

Definition. Given an extension $0 \rightarrow K \rightarrow E \rightarrow Q \rightarrow 1$ and a lifting $\ell : Q \rightarrow E$, a *factor set*[5] (or *cocycle*) is a function $f : Q \times Q \rightarrow K$ such that, for all $x, y \in Q$,
$$\ell(x) + \ell(y) = f(x, y) + \ell(xy).$$

---

[4] Since K is a normal subgroup, each right coset Kx is equal to the left coset xK, and so it makes no difference whether one chooses a right transversal or a left one.

[5] If we switch to multiplicative notation, we see that a factor set occurs in the factorization $\ell(x)\ell(y) = f(x, y)\ell(xy)$.

Of course, a factor set depends on the choice of lifting ℓ. When E is a split extension, there exists a lifting that is a homomorphism, and the corresponding factor set is identically 0. Therefore, we can regard a factor set as the obstruction to a lifting ℓ being a homomorphism; that is, a factor set describes how an extension differs from being a split extension.

Proposition 9.7. *Let Q be a group, K a Q-module, and $0 \to K \to E \to Q \to 1$ an extension realizing the operators. If $\ell: Q \to E$ is a lifting and $f: Q \times Q \to K$ is its corresponding factor set, then*

   (i) *for all $x, y \in Q$,*
$$f(1, y) = 0 = f(x, 1);$$

   (ii) *the **cocycle identity**[6] holds: for all $x, y, z \in Q$, we have*
$$f(x, y) + f(xy, z) = xf(y, z) + f(x, yz).$$

Proof. Set $x = 1$ in the equation that defines $f(x, y)$,
$$\ell(x) + \ell(y) = f(x, y) + \ell(xy),$$

to see that $\ell(y) = f(1, y) + \ell(y)$ [since $\ell(1) = 0$ is part of the definition of lifting], and hence $f(1, y) = 0$. Setting $y = 1$ gives the other equation in (i).

The cocycle identity follows from associativity in E. For all $x, y, z \in Q$, we have
$$[\ell(x) + \ell(y)] + \ell(z) = f(x, y) + \ell(xy) + \ell(z)$$
$$= f(x, y) + f(xy, z) + \ell(xyz).$$

On the other hand,
$$\ell(x) + [\ell(y) + \ell(z)] = \ell(x) + f(y, z) + \ell(yz)$$
$$= xf(y, z) + \ell(x) + \ell(yz)$$
$$= xf(y, z) + f(x, yz) + \ell(xyz). \quad \bullet$$

It is more interesting that the converse is true. The next result generalizes the construction of $K \rtimes Q$ in Proposition 9.5.

---

[6]Written as an alternating sum, $f(x, y) = xf(y, z) - f(xy, z) + f(x, yz)$, this identity is reminiscent of the formulas describing geometric cycles as described in Section 1.1.

Theorem 9.8.   *Given a group Q and a Q-module K, then a function $f: Q \times Q \to K$ is a factor set if and only if it satisfies the cocycle identity: for all $x, y, z \in Q$,*

$$xf(y, z) - f(xy, z) + f(x, yz) - f(x, y) = 0,$$

and, for all $x, y \in Q$,

$$f(1, y) = 0 = f(x, 1).$$

More precisely, if f satisfies these two identities, then there is an extension $0 \to K \to E \to Q \to 1$ realizing the operators, and there is a lifting $\ell: Q \to E$ whose corresponding factor set is f.

Proof.   Necessity is Proposition 9.7. For the converse, define E to be the set of all ordered pairs (a, x) in $K \times Q$ equipped with the operation

$$(a, x) + (b, y) = (a + xb + f(x, y), xy)$$

(if f is identically 0, then $E = K \rtimes Q$). The proof that E is a group is similar to the proof of Proposition 9.5. The cocycle identity is used to prove associativity, the identity is $(0, 1)$, and the inverse of (a, x) is

$$-(a, x) = (-x^{-1}a - x^{-1}f(x, x^{-1}), x^{-1}).$$

Define $p: E \to Q$ by $p: (a, x) \mapsto x$. Because the only "twist" occurs in the first coordinate, it is easy to see that p is a surjective homomorphism with $\ker p = \{(a, 1) : a \in K\}$. If we define $i: K \to E$ by $i: a \mapsto (a, 1)$, then we have an extension $0 \to K \xrightarrow{i} E \xrightarrow{p} Q \to 1$.

To see that this extension realizes the operators, we must show, for every lifting ℓ, that $xa = \ell(x) + a - \ell(x)$ for all $a \in K$ and $x \in Q$. Now $\ell(x) = (b, x)$ for some $b \in K$ and

$$
\begin{aligned}
\ell(x) + (a, 1) - \ell(x) &= (b, x) + (a, 1) - (b, x) \\
&= (b + xa, x) + (-x^{-1}b - x^{-1}f(x, x^{-1}), x^{-1}) \\
&= (b + xa + x[-x^{-1}b - x^{-1}f(x, x^{-1})] + f(x, x^{-1}), 1) \\
&= (xa, 1).
\end{aligned}
$$

Finally, we show that f is the factor set determined by some lifting ℓ. Define $\ell(x) = (0, x)$ for all $x \in Q$. The factor set F determined by ℓ is defined by

$$
\begin{aligned}
F(x, y) &= \ell(x) + \ell(y) - \ell(xy) \\
&= (0, x) + (0, y) - (0, xy) \\
&= (f(x, y), xy) + (-(xy)^{-1}f(xy, (xy)^{-1}), (xy)^{-1}) \\
&= (f(x, y) + xy[-(xy)^{-1}f(xy, (xy)^{-1})] \\
&\qquad + f(xy, (xy)^{-1}), xy(xy)^{-1}) \\
&= (f(x, y), 1). \quad \bullet
\end{aligned}
$$

Definition. Given a group Q, a Q-module K, and a factor set f, denote the group constructed in Theorem 9.8 by $\mathbf{Gr}(K; Q, f)$, and denote the extension of K by Q constructed there by

$$\mathbf{XGr}(K, Q, f) = 0 \to K \to \mathbf{Gr}(K, Q, f) \to Q \to 1.$$

The next result shows that we have found all the extensions of a Q-module K by a group Q.

Theorem 9.9. *Let Q be a group, let K be a Q-module, and let $0 \to K \to E \to Q \to 1$ be an extension realizing the operators. Then there exists a factor set $f: Q \times Q \to K$ with*

$$E \cong \mathbf{Gr}(K, Q, f).$$

Proof. Let $\ell: Q \to E$ be a lifting, and let $f: Q \times Q \to K$ be the corresponding factor set: that is, for all $x, y \in Q$, we have

$$\ell(x) + \ell(y) = f(x, y) + \ell(xy).$$

Since E is the disjoint union of the cosets, $E = \bigcup_{x \in Q} K + \ell(x)$, each $e \in E$ has a unique expression $e = a + \ell(x)$ for $a \in K$ and $x \in Q$. Uniqueness implies that the function $\varphi: E \to \mathbf{Gr}(K, Q, f)$, given by

$$\varphi: e = a + \ell(x) \mapsto (a, x),$$

is a well-defined bijection. We now show that φ is an isomorphism.

$$
\begin{aligned}
\varphi(a + \ell(x) + b + \ell(y)) &= \varphi(a + \ell(x) + b - \ell(x) + \ell(x) + \ell(y)) \\
&= \varphi(a + xb + \ell(x) + \ell(y)) \\
&= \varphi(a + xb + f(x, y) + \ell(xy)) \\
&= (a + xb + f(x, y), xy) \\
&= (a, x) + (b, y) \\
&= \varphi(a + \ell(x)) + \varphi(b + \ell(y)). \quad \bullet
\end{aligned}
$$

Remark. Note that if $a \in K$, then $\varphi(a) = \varphi(a + \ell(1)) = (a, 1)$, and, if $x \in Q$, then $\varphi(\ell(x)) = (0, x)$. This would not be so had we chosen a lifting ℓ with $\ell(1) \neq 0$. ◄

We have described all extensions in terms of factor sets, but a factor set depends on a choice of lifting.

Lemma 9.10.   *Given a group Q, a Q-module K, an extension $0 \to K \to E \to Q \to 1$ realizing the operators, and liftings ℓ and ℓ' giving factor sets f and f', respectively, there exists a function $h\colon Q \to K$ with $h(1) = 0$ and, for all $x, y \in Q$,*

$$f'(x, y) - f(x, y) = xh(y) - h(xy) + h(x).$$

Proof.   For each $x \in Q$, both $\ell(x)$ and $\ell'(x)$ lie in the same coset of K in E, and so there exists an element $h(x) \in K$ with

$$\ell'(x) = h(x) + \ell(x).$$

Since $\ell(1) = 0 = \ell'(1)$, we have $h(1) = 0$. The main formula is derived as follows:

$$\begin{aligned} \ell'(x) + \ell'(y) &= [h(x) + \ell(x)] + [h(y) + \ell(y)] \\ &= h(x) + xh(y) + \ell(x) + \ell(y), \end{aligned}$$

because E realizes the operators. The equations continue,

$$\begin{aligned} \ell'(x) + \ell'(y) &= h(x) + xh(y) + f(x, y) + \ell(xy) \\ &= h(x) + xh(y) + f(x, y) - h(xy) + \ell'(xy). \end{aligned}$$

By definition, f' satisfies $\ell'(x) + \ell'(y) = f'(x, y) + \ell'(xy)$. Therefore,

$$f'(x, y) = h(x) + xh(y) + f(x, y) - h(xy),$$

and so

$$f'(x, y) - f(x, y) = xh(y) - h(xy) + h(x). \quad \bullet$$

Definition.   Given a group Q and a Q-module K, a function $g\colon Q \times Q \to K$ is called a ***coboundary*** if there exists a function $h\colon Q \to K$ with $h(1) = 0$ such that, for all $x, y \in Q$,

$$g(x, y) = xh(y) - h(xy) + h(x).$$

The term *coboundary* arises because its formula is an alternating sum analogous to the formula for geometric boundaries in Chapter 1.

We have just shown that if f and f' are factor sets of an extension G that arise from different liftings, then $f' - f$ is a coboundary.

Definition.   Given a group Q and a Q-module K, define

$$Z^2(Q, K) = \{\text{all factor sets } f\colon Q \times Q \to K\}$$

and

$$B^2(Q, K) = \{\text{all coboundaries } g\colon Q \times Q \to K\}.$$

Proposition 9.11. *Given a group Q and a Q-module K, then $Z^2(Q, K)$ is an abelian group with operation pointwise addition,*

$$f + f': (x, y) \mapsto f(x, y) + f'(x, y),$$

and $B^2(Q, K)$ is a subgroup of $Z^2(Q, K)$.

Proof. Pointwise addition is an (associative) operation on Z^2, for Theorem 9.8 implies that the sum of two factor sets is a factor set. Now the zero function, $f(x, y) = 0$ for all $x, y \in Q$, is a factor set (of the semidirect product), and $-f$ set is a factor set if f is (using Theorem 9.8 again), so that Z^2 is a group.

Note that $B^2 \subseteq Z^2$, for if $g(x, y) = xh(y) - h(xy) + h(x)$, then $g(1, y) = 0 = g(x, 1)$ [this uses the hypothesis that $h(1) = 0$] and g satisfies the cocycle identity (a routine calculation); moreover, B^2 is nonempty, for the zero function is a coboundary. To see that B^2 is a subgroup of Z^2, it now suffices to prove it is closed under subtraction. But if g, g' are coboundaries, then they are factor sets (because $B^2 \subseteq Z^2$), and so $g - g' \in B^2$, by Lemma 9.10. •

The following quotient group suggests itself.

Definition. The *second cohomology group* is defined by

$$H^2(Q, K) = Z^2(Q, K)/B^2(Q, K).$$

Definition. Given a group Q and a Q-module K, two extensions of K by Q realizing the operators are called **equivalent** if there are factor sets f and f' of each so that $f' - f$ is a coboundary.

The notion of equivalence of extensions of modules that arose in Chapter 6 first arose in the context of group extensions.

Proposition 9.12. *Given a group Q and a Q-module K, two extensions of K by Q realizing the operators are equivalent if and only if there exists an isomorphism $\gamma: E \to E'$ making the following diagram commute:*

$$
\begin{array}{ccccccccc}
0 & \longrightarrow & K & \overset{i}{\longrightarrow} & E & \overset{p}{\longrightarrow} & Q & \longrightarrow & 1 \\
& & \downarrow{1_K} & & \downarrow{\gamma} & & \downarrow{1_Q} & & \\
0 & \longrightarrow & K & \underset{i'}{\longrightarrow} & E' & \underset{p'}{\longrightarrow} & Q & \longrightarrow & 1.
\end{array}
$$

Remark. A diagram chase shows that any homomorphism γ making the diagram commute is necessarily an isomorphism. ◄

Proof. Assume that the two extensions are equivalent. We begin by setting up notation. Let $\ell\colon Q \to E$ and $\ell'\colon Q \to E'$ be liftings, and let f, f' be the corresponding factor sets; that is, for all x, $y \in Q$, we have

$$\ell(x) + \ell(y) = f(x, y) + \ell(xy),$$

with a similar equation for f' and ℓ'. Equivalence means that there is a function $h\colon Q \to K$ with $h(1) = 0$ and

$$f(x, y) - f'(x, y) = xh(y) - h(xy) + h(x)$$

for all x, $y \in Q$. Since $E = \bigcup_{x \in Q} K + \ell(x)$ is a disjoint union, each $e \in E$ has a unique expression $e = a + \ell(x)$ for $a \in K$ and $x \in Q$; similarly, each $e' \in E'$ has a unique expression $e' = a + \ell'(x)$.

Define $\gamma\colon E \to E'$ by

$$\gamma(a + \ell(x)) = a + h(x) + \ell'(x).$$

This function makes the diagram commute. If $a \in K$, then

$$\gamma(a) = \gamma(a + \ell(1)) = a + h(1) + \ell'(1) = a;$$

furthermore,

$$p'\gamma(a + \ell(x)) = p'(a + h(x) + \ell'(x)) = x = p(a + \ell(x)).$$

Finally, γ is a homomorphism:

$$\gamma\big([a + \ell(x)] + [b + \ell(y)]\big) = \gamma(a + xb + f(x, y) + \ell(xy))$$
$$= a + xb + f(x, y) + h(xy) + \ell'(xy),$$

while

$$\gamma(a + \ell(x)) + \gamma(b + \ell(y)) = \big(a + h(x) + \ell'(x)\big) + \big(b + h(y) + \ell'(y)\big)$$
$$= a + h(x) + xb + xh(y) + f'(x, y) + \ell'(xy)$$
$$= a + xb + \big(h(x) + xh(y) + f'(x, y)\big) + \ell'(xy)$$
$$= a + xb + f(x, y) + h(xy) + \ell'(xy).$$

We have used the given equation for $f' - f$ [remember that the terms other than $\ell'(xy)$ all lie in the abelian group K, and so they may be rearranged]. Therefore, the diagram commutes.

Conversely, assume that there exists a homomorphism γ making the diagram commute; thus, $\gamma(a) = a$ for all $a \in K$ and

$$x = p(\ell(x)) = p'\gamma(\ell(x))$$

for all $x \in Q$. It follows that $\gamma\ell\colon Q \to E'$ is a lifting. Applying γ to the equation $\ell(x) + \ell(y) = f(x, y) + \ell(xy)$, which defines the factor set f, we see that γf is the factor set determined by the lifting $\gamma \ell$. But $\gamma f(x, y) = f(x, y)$ for all $x, y \in Q$, because $f(x, y) \in K$. Therefore, f is also a factor set of the second extension. On the other hand, if f' is any other factor set for the second extension, then Lemma 9.10 shows that $f' - f \in B^2$; that is, the extensions are equivalent. •

Remark. We have seen, in Example 7.26, that there can be two inequivalent extensions of K by Q with isomorphic middle groups. Since both extensions in that example have abelian middle groups, each K is a trivial Q-module, and so both extensions realize the operators. ◄

The next theorem summarizes the calculations in this section.

Theorem 9.13 (Schreier). *Let Q be a group, let K be a Q-module, and let $e(Q, K)$ denote the family of all the equivalence classes of extensions of K by Q realizing the operators. There is a bijection*

$$\varphi: H^2(Q, K) \to e(Q, K)$$

that takes 0 to the class of the split extension.

Proof. Denote the equivalence class of an extension

$$0 \to K \to E \to Q \to 1$$

by $[E]$. Define $\varphi: H^2(Q, K) \to e(Q, K)$ by

$$\varphi: f + B^2 \mapsto [\mathbf{XGr}(K, Q, f)],$$

where f is a factor set of the extension and the target extension is that constructed in Theorem 9.8.

First, φ is a well-defined injection: f and g are factor sets with $f + B^2 = g + B^2$ if and only if $[\mathbf{XGr}(K, Q, f)] = [\mathbf{XGr}(K, Q, g)]$, by Proposition 9.12. To see that φ is a surjection, let $[E] \in e(Q, K)$. By Theorem 9.9 and the remark following it, $[E] = [\mathbf{XGr}(K, Q, f)]$ for some factor set f, and so $[E] = \varphi(f + B^2)$. Finally, the zero factor set corresponds to the semidirect product. •

Corollary 9.14. *If Q is a group, K is a Q-module, and $H^2(Q, K) = \{0\}$, then every extension of K by Q realizing the operators splits. Thus, if $0 \to K \to E \to Q \to 1$ is an extension realizing the operators, then $E \cong K \rtimes Q$.*

Proof. By the theorem, $e(Q, K) = \operatorname{im} \varphi = \{[\varphi(0)]\}$; that is, every extension of K by Q realizing the operators is equivalent to the split extension. In this case, the middle group E of an extension is a semidirect product $K \rtimes Q$. •

Remark. Schreier's approach to extensions can be modified to give another construction of $\mathrm{Ext}^1_R(C, A)$ for any ring R. Given an extension $0 \to A \to B \to C \to 0$ of left R-modules, choose a lifting $\ell \colon C \to B$ with $\ell 0 = 0$. Each element of B has a unique expression of the form $a + \ell c$, and, when we try to add two elements, a factor set emerges:

$$(a + \ell c) + (a' + \ell c') = a + a' + f(c, c') + \ell(c + c'). \tag{2}$$

Here, $f \colon C \times C \to A$ satisfies the identities

(i) $f(c, 0) = 0 = f(0, c')$,

(ii) $f(c', c'') - f(c + c', c'') + f(c, c' + c'') - f(c, c') = 0$,

(iii) $f(c, c') = f(c', c)$.

The second and third identities arise, respectively, from associativity and commutativity of addition. If we define addition on $\widetilde{B} = A \times C$ by Eq. (1), then $\widetilde{B}$ is an abelian group because f satisfies (i), (ii), (iii). To ensure that $\widetilde{B}$ is a left R-module, we return to the left R-module B. Define a function $g \colon R \times C \to A$ by $g(r, c) = r\ell c - \ell(rc)$. Additional identities arise from the module axioms:

(i) $g(1, c) = 0 = g(r, 0)$;

(ii) $rg(s, c) = g(rs, c) - g(r, sc)$;

(iii) $g(r + s, c) + f(rc, sc) = g(r, c) + g(s, c)$;

(iv) $g(r, c + c') + f(rc, rc') = g(r, c) + g(r, c')$.

The ordered pair (f, g) conveys all the necessary data to make $\widetilde{B}$ into a left R-module if one defines scalar multiplication by

$$r(a + \ell c) = ra + \ell(rc)$$

(remember that A and C are left R-modules). The set of all such (f, g) forms an abelian group $Z(C, A)$, where each coordinate acts via pointwise addition, and choosing a second lifting $\ell' \colon C \to A$ determines a subgroup $B(C, A) \subseteq Z(C, A)$. Obviously, the resolution and the boundary formula are more complicated than those for groups, but they are simple for $\mathbb{Z}$-modules when g can be forgotten. ◀

Exercises

*9.12 Let U: **Rings** $\to$ **Groups** be the functor assigning to each ring R its group of (two-sided) units $U(R)$; let F: **Groups** $\to$ **Rings** be the functor assigning to each group G its integral group ring $\mathbb{Z}G$ and to each group homomorphism $\varphi\colon G \to H$ the ring homomorphism $F(\varphi)\colon \mathbb{Z}G \to \mathbb{Z}H$, defined by $\sum_{x\in G} m_x x \mapsto \sum_{x\in G} m_x \varphi(x)$. Prove that (F, U) is an adjoint pair of functors.

*9.13 Let Q be a group and let K be a Q-module. Prove that any two split extensions of K by Q realizing the operators are equivalent.

9.14 Let Q be abelian, let K be a Q-module, and let $A(Q, K)$ be the subset of $H^2(Q, K)$ consisting of all $[0 \to K \to E \to Q \to 1]$ with E abelian.

   (i) Prove that $A(Q, K)$ is a subgroup of $H^2(Q, K)$.

   (ii) Prove that $A(Q, K) \cong \operatorname{Ext}^1_{\mathbb{Z}}(Q, K)$.

9.15 The ***generalized quaternion group*** $\mathbf{Q}_n$, for $n \geq 3$, is the subgroup of $\mathrm{GL}(2, \mathbb{C})$ generated by $A = \left[\begin{smallmatrix} 0 & \omega \\ \omega & 0 \end{smallmatrix}\right]$ and $B = \left[\begin{smallmatrix} 0 & 1 \\ -1 & 0 \end{smallmatrix}\right]$, where ω is a primitive 2^{n-1}th root of unity. Note that $|\mathbf{Q}_n| = 2^n$ and that

$$A^{2^{n-1}} = 1, \quad BAB^{-1} = a^{-1}, \quad \text{and} \quad B^2 = A^{2^{n-2}}.$$

   (i) Prove that B is the unique element of order 2, $\dot{Z}(\mathbf{Q}_n) = \langle B \rangle$, and that $\mathbf{Q}_n$ is not a semidirect product.

   (ii) Prove that $\mathbf{Q}_n$ is a ***central extension*** (i.e., θ is trivial) of $\mathbb{I}_2$ by $D_{2^{n-1}}$.

   (iii) Using factor sets, give a proof of the existence of $\mathbf{Q}_n$.

9.16 If p is an odd prime, prove that every group G of order $2p$ is a semidirect product of $\mathbb{I}_p$ by $\mathbb{I}_2$, and conclude that either G is cyclic or $G \cong D_{2p}$.

9.17   **(i)** Let T be the subgroup of $\mathrm{GL}(2, \mathbb{C})$ generated by $\left[\begin{smallmatrix} \omega & 0 \\ 0 & \omega^2 \end{smallmatrix}\right]$ and $\left[\begin{smallmatrix} 0 & i \\ i & 0 \end{smallmatrix}\right]$, where $\omega = e^{2\pi i/3}$ is a primitive cube root of unity. Prove that $|T| = 12$.

   (ii) Prove that T has a presentation

$$(a, b \mid a^6 = 1, b^2 = a^3 = (ab)^2).$$

   (iii) Prove that $T \cong \mathbb{I}_3 \rtimes \mathbb{I}_4$.
   Hint. Let $K = \langle u \rangle \cong \mathbb{I}_3$, let $Q = \langle x \rangle \cong \mathbb{I}_4$, and make K into a Q-module by $xu = 2u$, $x(2u) = u$, and $x^2 u = u$. In $K \rtimes Q$, define $a = (2u, x^2)$ and $b = (0, x)$.

   (iv) Prove that every group G of order 12 is isomorphic to exactly one of the following five groups:

$$\mathbb{I}_{12}, \quad \mathbf{V} \times \mathbb{I}_3, \quad A_4, \quad S_3 \times \mathbb{I}_2, \quad T.$$

9.1.3 Stabilizing Automorphisms

The Schur–Zassenhaus Lemma, Theorem 9.43, gives a condition guaranteeing that $H^2(Q, K) = \{0\}$: if Q and K are finite groups whose orders are relatively prime. In this case, the middle group E of an extension is a semidirect product. If C, C' are complements of K in E, then $C \cong C'$ (for both are isomorphic to $E/K \cong Q$). The Schur–Zassenhaus Lemma goes on to say that C and C' are conjugate subgroups. Let us examine conjugacy.

We begin with a computational lemma. Let Q be a group, let K be a Q-module, and let $0 \to K \to E \to Q \to 1$ be a split extension. Choose a lifting $\ell \colon Q \to E$, so that every element $e \in E$ has a unique expression of the form

$$e = a + \ell x,$$

where $a \in K$ and $x \in Q$.

Definition.   An automorphism φ of a group E *stabilizes* an extension $0 \to K \xrightarrow{i} E \xrightarrow{p} Q \to 1$ if the following diagram commutes:

$$
\begin{array}{ccccccccc}
0 & \longrightarrow & K & \xrightarrow{i} & E & \xrightarrow{p} & Q & \longrightarrow & 1 \\
 & & \downarrow{\scriptstyle 1_K} & & \downarrow{\scriptstyle \varphi} & & \downarrow{\scriptstyle 1_Q} & & \\
0 & \longrightarrow & K & \xrightarrow{i} & E & \xrightarrow{p} & Q & \longrightarrow & 1.
\end{array}
$$

The set of all stabilizing automorphisms of an extension of K by Q, where K is a Q-module, is a group under composition; it is denoted by

$$\mathrm{Stab}(Q, K).$$

We shall see, in Corollary 9.17, that $\mathrm{Stab}(Q, K)$ does not depend on the extension.

Proposition 9.15.   *Let Q be a group, let K be a Q-module, and let*

$$0 \to K \to E \xrightarrow{p} Q \to 1$$

be an extension. If $\ell \colon Q \to E$ is a lifting, then every stabilizing automorphism $\varphi \colon E \to E$ has the form

$$\varphi(a + \ell x) = a + d(x) + \ell x,$$

where $d(x) \in K$ is independent of the choice of lifting ℓ. Moreover, this formula defines a stabilizing automorphism if and only if, for all $x, y \in Q$, the function $d \colon Q \to K$ satisfies

$$d(xy) = d(x) + x d(y).$$

Proof. If φ is stabilizing, then $\varphi(a) = a$, for all $a \in K$, and $p\varphi = p$. To use the second constraint on φ, suppose that $\varphi(\ell x) = d(x) + \ell y$ for some $d(x) \in K$ and $y \in Q$. Then

$$x = p(\ell x) = p\varphi(\ell x) = p(d(x) + \ell y) = y;$$

that is, $x = y$. Therefore,

$$\varphi(a + \ell x) = \varphi(a) + \varphi(\ell x) = a + d(x) + \ell x.$$

To see that the formula for d holds, we first show that d is independent of the choice of lifting. Suppose that $\ell' : Q \to G$ is another lifting, so that $\varphi(\ell' x) = d'(x) + \ell' x$ for some $d'(x) \in K$. Now there is $k(x) \in K$ with $\ell' x = k(x) + \ell x$, for $p\ell' x = x = p\ell x$. Therefore,

$$
\begin{aligned}
d'(x) &= \varphi(\ell' x) - \ell' x \\
&= \varphi(k(x) + \ell x) - \ell' x \\
&= k(x) + d(x) + \ell x - \ell' x \\
&= d(x),
\end{aligned}
$$

because $k(x) + \ell x - \ell' x = 0$.

There is a factor set $f : Q \to K$ with $\ell x + \ell y = f(x, y) + \ell(xy)$ for each $x, y \in Q$. We compute $\varphi(\ell x + \ell y)$ in two ways.

On the one hand,

$$
\begin{aligned}
\varphi(\ell x + \ell y) &= \varphi(f(x, y) + \ell(xy)) \\
&= \varphi(f(x, y)) + \varphi\ell(xy) \\
&= f(x, y) + \varphi\ell(xy) \qquad \text{for } f(x, y) \in K \\
&= f(x, y) + d(xy) + \ell(xy).
\end{aligned}
$$

On the other hand,

$$
\begin{aligned}
\varphi(\ell x + \ell y) &= \varphi(\ell x) + \varphi(\ell y) \\
&= d(x) + \ell x + d(y) + \ell y \\
&= d(x) + xd(y) + f(x, y) + \ell(xy).
\end{aligned}
$$

After canceling $\ell(xy)$ from the right, all terms lie in the abelian group K; now cancel $f(x, y)$ from both sides to obtain

$$d(xy) = d(x) + xd(y).$$

The proof of the converse: if $\varphi(a + \ell x) = a + d(x) + \ell x$ (where d satisfies the given identity), then φ is a stabilizing isomorphism, is left to the reader. •

We give a name to functions like d.

Definition.   Let Q be a group and let K be a Q-module. A *derivation* (or *crossed homomorphism*) is a function $d: Q \to K$ such that

$$d(xy) = xd(y) + d(x).$$

The set of all derivations, $\text{Der}(Q, K)$, is an abelian group under pointwise addition. If K is a trivial Q-module, then $\text{Der}(Q, K) = \text{Hom}(Q, K)$.

If d is a derivation, then $d(1) = 0$, for $d(1 \cdot 1) = 1d(1) + d(1)$.

Proposition 9.15 can be restated. If $0 \to K \to E \to Q \to 1$ is an extension with lifting $\ell: G \to E$, then $\ell(x) = (d(x), x)$, where $d: Q \to K$, and ℓ is a homomorphism if and only if d is a derivation.

Recall that $\text{Stab}(Q, K)$ denotes the group of all the stabilizing automorphisms of an extension of K by Q.

Corollary 9.16.   *Let Q be a group, K a Q-module, and $0 \to K \to E \to Q \to 1$ an extension. The function $\sigma: \varphi \mapsto d$, where $\varphi(\ell x) = d(x) + \ell x$, is an isomorphism*

$$\sigma: \text{Stab}(Q, K) \to \text{Der}(Q, K).$$

Proof.   If φ is a stabilizing automorphism and $\ell: Q \to E$ is a lifting, then Proposition 9.15 says that $\varphi(a + \ell x) = a + d(x) + \ell x$, where d is a derivation. This proposition further states that d is independent of the choice of lifting; that is, σ is a well-defined function $\text{Stab}(Q, K) \to \text{Der}(Q, K)$. The reader can easily check that σ is a homomorphism.

We now show that σ is an isomorphism. If $d \in \text{Der}(Q, K)$, define $\varphi: E \to E$ by $\varphi(a + \ell x) = a + d(x) + \ell x$. Now φ is stabilizing, by Proposition 9.15, and $d \mapsto \varphi$ is inverse to σ.   •

It is not obvious from its definition that $\text{Stab}(Q, K)$ is abelian, for its operation is composition.

Corollary 9.17.   *If Q is a group and K is a Q-module, then $\text{Stab}(Q, K)$ is an abelian group that does not depend on the extension of K by Q used to define it.*

Proof.   By Corollary 9.16, $\text{Stab}(Q, K) \cong \text{Der}(Q, K)$; hence, $\text{Stab}(Q, K)$ is abelian because $\text{Der}(Q, K)$ is. Moreover, $\text{Der}(Q, K)$ is defined without referring to any extension of K by Q.   •

Example 9.18. If Q is a group and K is a Q-module, then a function $d_0: Q \to K$ of the form $d_0(x) = xa_0 - a_0$, where $a_0 \in K$, is a derivation:

$$
\begin{aligned}
d_0(x) + xd_0(y) &= xa_0 - a_0 + x(ya_0 - a_0) \\
&= xa_0 - a_0 + xya_0 - xa_0 \\
&= xya_0 - a_0 \\
&= d_0(xy).
\end{aligned}
$$

If the action of Q on K is conjugation, say, $xa = \ell x + a - \ell x$, then

$$
xa_0 - a_0 = \ell x + a_0 - \ell x - a_0;
$$

that is, $xa_0 - a_0$ is the commutator of x and a_0 (in multiplicative notation, $xa_0 - a_0$ becomes $xa_0 x^{-1} a_0^{-1}$). ◄

Definition. A derivation $d_0: Q \to K$ of the form $d_0(x) = xa_0 - a_0$, where $a_0 \in K$, is called a ***principal derivation***. The set of all principal derivations is denoted by

$$
\mathrm{PDer}(Q, K).
$$

$\mathrm{PDer}(Q, K)$ is a subgroup of $\mathrm{Der}(Q, K)$, because $(xa - a) - (xb - b) = x(a - b) - (a - b)$.

Lemma 9.19. *Let* $0 \to K \to E \to Q \to 1$ *be an extension, and let* $\ell: Q \to E$ *be a lifting.*

(i) *A function* $\varphi: E \to E$ *is an inner stabilizing automorphism by some* $a_0 \in K$ *if and only if*

$$
\varphi(a + \ell x) = a + xa_0 - a_0 + \ell x.
$$

(ii) $\qquad \mathrm{Stab}(Q, K)/\mathrm{Inn}(Q, K) \cong \mathrm{Der}(Q, K)/\mathrm{PDer}(Q, K),$

where $\mathrm{Inn}(Q, K) = \mathrm{Inn}(E) \cap \mathrm{Stab}(Q, K)$.

Proof.

(i) If we write $d(x) = xa_0 - a_0$, then $\varphi(a + \ell x) = a + d(x) + \ell x$. But d is a (principal) derivation, and so φ is a stabilizing automorphism, by Proposition 9.15. Finally, φ is conjugation by $-a_0$, for

$$
-a_0 + (a + \ell x) + a_0 = -a_0 + a + xa_0 + \ell x = \varphi(a + \ell x).
$$

Conversely, assume that φ is a stabilizing conjugation. That φ is stabilizing says that $\varphi(a + \ell x) = a + d(x) + \ell x$; that φ is conjugation by $a_0 \in K$ says that $\varphi(a + \ell x) = a_0 + a + \ell x - a_0$. But $a_0 + a + \ell x - a_0 = a_0 + a - xa_0 + \ell x$, so that $d(x) = a_0 - xa_0$.

(ii) We prove that $\sigma(\text{Inn}(Q, K)) = \text{PDer}(Q, K)$, where $\sigma: \text{Stab}(Q, K) \to \text{Der}(Q, K)$ is the isomorphism of Corollary 9.16: if $\varphi \in \text{Stab}(Q, K)$, then $\varphi(a + \ell x) = a + d(x) + \ell x$ and $\sigma: \varphi \mapsto d$. If $\varphi \in \text{Inn}(Q, K)$, then $\varphi(a + \ell x) = a + xa_0 - a_0 + \ell x$, by part (i), so that $\sigma(\varphi) = d$ with $d(x) = xa_0 - a_0$; hence, $\sigma(\varphi) \in \text{PDer}(Q, K)$ and $\sigma(\text{Inn}(Q, K)) \subseteq \text{PDer}(Q, K)$. For the reverse inclusion, if $d_0 \in \text{PDer}(Q, K)$, define $\varphi_0: E \to E$ by $\varphi_0(a + \ell x) = a + d_0(x) + \ell x$. Now $\varphi_0 \in \text{Inn}(Q, K)$, by part (i), and so $d_0 = \sigma(\varphi_0)$. •

Definition.   If Q is a group and K is a Q-module, define

$$H^1(Q, K) = \text{Der}(Q, K)/\text{PDer}(Q, K).$$

Corollary 9.20.   *For every group Q and Q-module K,*

$$H^1(Q, K) \cong \text{Stab}(Q, K)/\text{Inn}(Q, K).$$

Proof.   Immediate from the definition of $H^1(Q, K)$ and Lemma 9.19(iii). •

Proposition 9.21.   *Let $0 \to K \to E \to Q \to 1$ be a split extension, and let C and C' be complements of K in E. If $H^1(Q, K) = \{0\}$, then C and C' are conjugate.*

Proof.   Since E is a semidirect product, there are liftings $\ell: Q \to E$, with image C, and $\ell': Q \to E$, with image C', that are homomorphisms. Thus, the factor sets f and f' determined by each of these liftings are identically zero, and so $f' - f = 0$. But Lemma 9.10 says that there exists $h: Q \to K$, namely, $h(x) = \ell'x - \ell x$, with

$$0 = f'(x, y) - f(x, y) = xh(y) - h(xy) + h(x);$$

thus, h is a derivation. Since $H^1(Q, K) = \{0\}$, h is a principal derivation: there is $a_0 \in K$ with

$$\ell'x - \ell x = h(x) = xa_0 - a_0$$

for all $x \in Q$. Since addition in E satisfies $\ell'x - a_0 = -xa_0 + \ell'x$, we have

$$\ell x = a_0 - xa_0 + \ell'x = a_0 + \ell'x - a_0.$$

But $\text{im } \ell = C$ and $\text{im } \ell' = C'$, and so C and C' are conjugate via a_0. •

9.2   Group Cohomology

If A and B are left R-modules, for some ring R, then the abelian group $\text{Hom}_R(A, B)$ is usually not an R-module unless A or B is a bimodule (see Proposition 2.54). Similarly, if A is a right R-module and B is a left R-module, then $A \otimes_R B$ is only an abelian group (see Proposition 2.51). However, module structures are available when $R = \mathbb{Z}G$.

Definition. Let G be a group, and let A and B be left $\mathbb{Z}G$-modules. The *diagonal action* on $\text{Hom}_G(A, B)$ is given by

$$(g\varphi)(a) = g\varphi(g^{-1}a),$$

where $g \in G$, $\varphi \colon A \to B$, and $a \in A$.

If M is a right $\mathbb{Z}G$-module and B is a left $\mathbb{Z}G$-module, the *diagonal action* on $M \otimes_G B$ is given by

$$g(m \otimes b) = gm \otimes gb,$$

where $g \in G$, $m \in M$, and $b \in B$. Note that if M is G-trivial, then diagonal action is $g(m \otimes b) = m \otimes gb$.

Exercise 9.18 on page 557 asks you to prove that diagonal action makes $\text{Hom}_G(A, B)$ and $M \otimes_G B$ into G-modules.

Consider the formulas that have arisen in the Section 9.1.

$$
\begin{aligned}
\text{factor set:} \quad & 0 = xf(y, z) - f(xy, z) + f(x, yz) - f(x, y) \\
\text{coboundary:} \quad & f(x, y) = xh(y) - h(xy) + h(x) \\
\text{derivation:} \quad & 0 = xd(y) - d(xy) + d(x) \\
\text{principal derivation:} \quad & d(x) = xa_0 - a_0
\end{aligned}
$$

A pattern suggests that the next equation is

$$0 = xa_0 - a_0.$$

Definition. If G is a group (we now denote groups by G instead of by Q) and K is a G-module, then the submodule of *fixed points* is defined by

$$K^G = \{a \in K : xa = a \text{ for all } x \in G\}.$$

It is easy to see that K^G is a G-trivial submodule; indeed, it is the unique maximal G-trivial submodule of K. If $\varphi \colon K \to L$ is a G-map and $a \in K^G$, then $xa = a$ for all $x \in G$, and so $\varphi(xa) = \varphi(a)$. Since φ is a G-map, $\varphi(xa) = x\varphi(a)$, and so $\varphi(a) \in L^G$. Define $\varphi^G = \varphi|K^G$.

Definition.   The *fixed-point functor* $\text{Fix}^G \colon {}_{\mathbb{Z}G}\mathbf{Mod} \to {}_{\mathbb{Z}G}\mathbf{Mod}$ is defined by $\text{Fix}^G(K) = K^G$ and $\text{Fix}^G(\varphi) = \varphi^G = \varphi|K^G$.

It is easy to see that Fix^G is an additive functor.

Proposition 9.22.   *If $\mathbb{Z}$ is viewed as a G-trivial module, then*

$$\text{Fix}^G \cong \text{Hom}_G(\mathbb{Z}, \square),$$

and, hence, Fix^G is left exact.

Proof.   Define $\tau_K \colon \text{Hom}_G(\mathbb{Z}, K) \to K^G$ by $f \mapsto f(1)$. Now $f(1) \in L^G$: if $x \in G$, then $x f(1) = f(x \cdot 1) = f(1)$, because $\mathbb{Z}$ is G-trivial. To see that τ_K is an isomorphism, we display its inverse. If $a \in K^G$, there is a $\mathbb{Z}$-map $f_a \colon \mathbb{Z} \to K$ with $f_a(1) = a$. Since $xa = a$ for all $x \in G$, it follows that f_a is a G-map, and $a \mapsto f_a$ is the inverse of τ_K. The reader may check naturality: the following diagram commutes.

$$
\begin{array}{ccc}
\text{Hom}_G(\mathbb{Z}, K) & \xrightarrow{\ \tau_K\ } & K^G \\
{\scriptstyle \varphi_*}\big\downarrow & & \big\downarrow{\scriptstyle \varphi^G} \\
\text{Hom}_G(\mathbb{Z}, L) & \xrightarrow[\ \tau_L\ ]{} & L^G
\end{array}
\qquad \bullet
$$

Definition.   If G is a group and K is a G-module, then the *cohomology groups* of G with coefficients in K are

$$H^n(G, K) = \text{Ext}^n_{\mathbb{Z}G}(\mathbb{Z}, K),$$

where $\mathbb{Z}$ is viewed as a trivial G-module.

Having defined group cohomology $H^n(G, K)$ as the right derived functors of $\text{Fix}^G \cong \text{Hom}_{\mathbb{Z}G}(\mathbb{Z}, \square)$, we are now obliged to show that these groups coincide with Schreier's groups when $n = 1$ and $n = 2$.

As $\text{Ext}(\mathbb{Z}, \square)$ is computed with a G-projective resolution of $\mathbb{Z}$, let us begin by mapping $\mathbb{Z}G$ onto $\mathbb{Z}$.

Proposition 9.23.   *There is a G-exact sequence*

$$0 \to \mathcal{G} \to \mathbb{Z}G \xrightarrow{\ \epsilon\ } \mathbb{Z} \to 0,$$

where $\epsilon \colon \mathbb{Z}G \to \mathbb{Z}$ is defined by $\sum_{x \in G} m_x x \mapsto \sum_{x \in G} m_x$. The function ϵ is a ring map as well as a G-map, and $\ker \epsilon = \mathcal{G}$ is a two-sided ideal in $\mathbb{Z}G$.

Proof.   We can calculate directly that ϵ is a G-map, but let us be fancy and use the functor $F \colon \mathbf{Groups} \to \mathbf{Rings}$ (in Exercise 9.12 on page 513) assigning to each group G its integral group ring $\mathbb{Z}G$. The trivial group homomorphism $\varphi \colon G \to \{1\}$ induces a ring map $F\varphi \colon \mathbb{Z}G \to \mathbb{Z}\{1\} = \mathbb{Z}$, namely, $F\varphi = \epsilon \colon \sum m_x x \mapsto \sum m_x$. Since ϵ is a ring homomorphism, $\mathcal{G} = \ker \epsilon$ is a two-sided ideal in $\mathbb{Z}G$. $\bullet$

The map $\epsilon \colon \mathbb{Z}G \to \mathbb{Z}$ is important because of the special role played by the G-trivial module $\mathbb{Z}$.

Definition.   The map $\epsilon \colon \mathbb{Z}G \to \mathbb{Z}$, given by $\sum m_x x \mapsto \sum m_x$, is called the *augmentation*, and $\mathcal{G} = \ker \epsilon$ is called the *augmentation ideal*.

Lemma 9.24.   *The additive group of the augmentation ideal $\mathcal{G}$ is the free abelian group with basis $\{x - 1 : x \in G^\times\}$, where $G^\times = \{x \in G : x \neq 1\}$.*

Proof.   If $u = \sum_{x \in G} m_x x \in \mathbb{Z}G$, then $u \in \ker \epsilon$ if and only if $\sum_{x \in G} m_x = 0$. Therefore, $u = u - \left(\sum_{x \in G} m_x\right)1 = \sum_{x \in G^\times} m_x(x - 1)$. Thus, $\mathcal{G}$ is additively generated by all $x - 1$. Suppose that $\sum_{x \in G^\times} m_x(x - 1) = 0$. Then we have $\sum_{x \in G^\times} m_x x - \left(\sum_{x \in G^\times} m_x\right)1 = 0$. But, as an abelian group, $\mathbb{Z}G$ is free with basis $\{x \in G\}$. Hence, $m_x = 0$ for all $x \in G^\times$.   •

The next result shows that the hybrid $\mathrm{Der}(G, A)$ (G is a group and A is a G-module) may be viewed as an ordinary Hom between G-modules.

Proposition 9.25.   *There is a natural isomorphism*

$$\tau \colon \mathrm{Hom}_G(\mathcal{G}, \square) \to \mathrm{Der}(G, \square);$$

the maps $\tau_A \colon \mathrm{Hom}_G(\mathcal{G}, A) \to \mathrm{Der}(G, A)$ are given by $\tau_A(f) = f'$, where $f' \colon G \to A$ is given by $x \mapsto f(x - 1)$ for all $x \in G$.

Proof.   It is routine to check that if $f \colon \mathcal{G} \to A$ is a G-map, then f' is a derivation and that τ_A is a homomorphism.

We construct the inverse of τ_A. If $d \in \mathrm{Der}(G, A)$, define $\tilde{d} \colon \mathcal{G} \to A$, where $\tilde{d}(x - 1) = d(x)$ (Lemma 9.24 shows that $\tilde{d}$ is a well-defined $\mathbb{Z}$-map, for $\mathcal{G}$ is the free abelian group with basis $\{x - 1 : x \in G^\times\}$). Since d is a derivation, it is easy to see that $\tilde{d}$ is a G-map. Define $\sigma_A \colon \mathrm{Der}(G, A) \to \mathrm{Hom}_G(\mathcal{G}, A)$ by $\sigma_A \colon d \mapsto \tilde{d}$. The reader may prove that both composites $\sigma_A \tau_A$ and $\tau_A \sigma_A$ are identities, and that τ is natural.   •

We now compute the cohomology groups of a finite cyclic group.

Lemma 9.26.   *Let $G = \langle x \rangle$ be a finite cyclic group of order k, and define elements D and N of $\mathbb{Z}G$ by $D = x - 1$ and $N = 1 + x + x^2 + \cdots + x^{k-1}$. Then*

$$\to \mathbb{Z}G \xrightarrow{D} \mathbb{Z}G \xrightarrow{N} \mathbb{Z}G \xrightarrow{D} \mathbb{Z}G \xrightarrow{\epsilon} \mathbb{Z} \to 0$$

is a G-free resolution of $\mathbb{Z}$, where the maps alternate being multiplication by D and multiplication by N.

Proof. Since $\mathbb{Z}G$ is commutative here, the maps D and N are G-maps (we abuse notation by denoting the maps multiplication by D and N by D and N, respectively). Since $ND = DN = x^k - 1 = 0$, the composites $ND = 0 = DN$, while if $u \in \mathbb{Z}G$, then

$$\epsilon(Du) = \epsilon\big((x-1)u\big) = \epsilon(x-1)\epsilon(u) = 0,$$

because ϵ is a ring map. Thus, we have a complex, and it only remains to prove exactness.

We have already noted that ϵ is surjective. Now $\ker \epsilon = \mathcal{G} = \operatorname{im} D$, by Lemma 9.24, and so we have exactness at the zeroth step.

We show that $\ker D \subseteq \operatorname{im} N$. If $u = \sum_{i=0}^{k-1} m_i x^i$, then

$$(x-1)u = (m_{k-1} - m_0) + (m_0 - m_1)x + \cdots + (m_{k-2} - m_{k-1})x^{k-1}.$$

Hence, if $u \in \ker D$, then $Du = (x-1)u = 0$, and $m_{k-1} = m_0 = m_1 = \cdots = m_{k-2}$; thus, $u = m_0 N \in \operatorname{im} N$.

We show that $\ker N \subseteq \operatorname{im} D$. If $u = \sum_{i=0}^{k-1} m_i x^i \in \ker N$, then $0 = \epsilon(Nu) = \epsilon(N)\epsilon(u) = k\epsilon(u)$, so that $\epsilon(u) = \sum_{i=0}^{k-1} m_i = 0$; this is used in showing

$$u = -D\big(m_0 + (m_0 + m_1)x + \cdots + (m_0 + \cdots + m_{k-1})x^{k-1}\big) \in \operatorname{im} D. \quad \bullet$$

Theorem 9.27. *Let G be a finite cyclic group. If A is a G-module, define $_NA = \{a \in A : Na = 0\}$. Then, for all $n \geq 1$,*

$$H^0(G, A) = A^G,$$
$$H^{2n-1}(G, A) = {}_NA/DA,$$
$$H^{2n}(G, A) = A^G/NA.$$

Proof. Apply $\operatorname{Hom}_G(\mathbb{Z}, \square)$ to the resolution in Lemma 9.26, and take homology. In more detail, if $d_{2n+1} = D$ and $d_{2n} = N$ for $n \geq 0$, then

$$\ker N^* = {}_NA, \quad \operatorname{im} N^* = NA, \quad \ker D^* = A^G, \quad \operatorname{im} D^* = DA,$$

where N^* and D^* are the induced maps. The formulas follow from the definition: $H^m(G, A) = \ker d_{m+1}^*/\operatorname{im} d_m^*$. $\quad \bullet$

Corollary 9.28. *If $G = \{1\}$, then $H^n(G, A) = \{0\}$ for all $n > 0$ and all G-modules A.*

Proof. This follows at once from the theorem. $\quad \bullet$

Corollary 9.29. *Let G be a finite cyclic group of order k. If A is a G-trivial module, then, for all $n \geq 1$,*

$$H^0(G, A) = A,$$
$$H^{2n-1}(G, A) = {}_k A = \{a \in A : ka = 0\},$$
$$H^{2n}(G, A) = A/kA.$$

In particular, if A is the G-trivial module $\mathbb{Z}$, then

$$H^0(G, \mathbb{Z}) = \mathbb{Z}, \quad H^{2n-1}(G, \mathbb{Z}) = \{0\}, \quad H^{2n}(G, \mathbb{Z}) = \mathbb{I}_k.$$

Corollary 9.30. *If G is a finite cyclic group, then the global dimension $\mathrm{D}(\mathbb{Z}G) = \infty$.*

Proof. We have $\mathrm{pd}(\mathbb{Z}) = \infty$, for $H^{2n}(G, \mathbb{Z}) = \mathrm{Ext}^{2n}_{\mathbb{Z}G}(\mathbb{Z}, \mathbb{Z}) \neq \{0\}$ for all n, by Corollary 9.29. •

Remark. A group G for which there exists a positive integer d such that $H^n(G, A) \cong H^{n+d}(G, A)$ for all $n \geq 1$ and all G-modules A is said to have *periodic cohomology*. We have just seen that finite cyclic groups have periodic cohomology. Other examples arise topologically as groups acting freely as orientation-preserving homeomorphisms of spheres (see Adem–Milgram, *Cohomology of Finite Groups*, p. 143). It can be proved that a finite group G has periodic cohomology if and only if its Sylow p-subgroups are cyclic, for all odd primes p, while its Sylow 2-subgroups are either cyclic or generalized quaternion (see Brown, *Cohomology of Groups*, VI, §9, or Adem–Milgram, IV, §6). For example, $G = \mathrm{SL}(2, 5)$, the group of all unimodular 2×2 matrices over $\mathbb{F}_5$, has periodic cohomology: it is a group of order $120 = 8 \cdot 3 \cdot 5$; its Sylow 3-subgroups and Sylow 5-subgroups are cyclic (for they have prime order), and its Sylow 2-subgroups are isomorphic to the quaternions. ◄

The next corollary is a key lemma in Class Field Theory.

Definition. If G is a finite cyclic group and A is a finite G-module, then the *Herbrand quotient* is

$$h(A) = |H^2(G, A)|/|H^1(G, A)|.$$

[Note that $h(A)$ is defined, for A finite implies finiteness of both $H^2(G, A)$ and $H^1(G, A)$, by Theorem 9.27.]

Corollary 9.31 (Herbrand). *If $G = \langle x \rangle$ is a finite cyclic group of order k, and if A is a finite G-module, then $h(A) = 1$.*

Proof. (**Hoechsmann**) There are exact sequences

$$0 \to {}_N A \to A \xrightarrow{N} NA \to 0 \quad \text{and} \quad 0 \to \ker D \to A \xrightarrow{D} DA \to 0,$$

where $D = x - 1$ and $N = 1 + x + x^2 + \cdots + x^{k-1}$. We claim that $\ker D = A^G$. If $a \in \ker D$, then $(x - 1)a = 0$, then $xa = a$ and, by induction, $x^m a = a$ for all $m \geq 1$. Since $G = \langle x \rangle$, we have $a \in A^G$, and so $\ker D \subseteq A^G$. The reverse inclusion is clear, for if $a \in A^G$, then $xa = a$. Thus, $|{}_N A||NA| = |A| = |A^G||DA|$, and so Theorem 9.27 gives

$$|H^1(G, A)| = |{}_N A/DA| = |A^G/NA| = |H^2(G, A)|. \quad \bullet$$

The next two results are used in Number Theory.

Theorem 9.32.   *Let E/k be a Galois extension with Galois group $G = \text{Gal}(E/k)$. The multiplicative group $E^\times$ is a kG-module, and*

$$H^1(G, E^\times) = \{0\}.$$

Remark.   This theorem is one of the first results in what is called *Galois Cohomology*. Another early result is that $H^n(G, E) = \{0\}$ for all $n \geq 1$, where E (in contrast to $E^\times$) is the additive group of the Galois extension (this result follows easily from the Normal Basis Theorem; see Jacobson, *Basic Algebra* I, p. 283). ◄

Proof. If $c: G \to E^\times$ is a 1-cocycle, denote $c(\sigma)$ by c_σ. In multiplicative notation, the cocycle condition is the identity $\sigma(c_\tau)c_{\sigma\tau}^{-1}c_\sigma = 1$ for all $\sigma, \tau \in G$; that is,

$$\sigma(c_\tau) = c_{\sigma\tau}c_\sigma^{-1}. \tag{1}$$

For $e \in E^\times$, define

$$b = \sum_{\tau \in G} c_\tau \tau(e).$$

By Dedekind's lemma on the independence of characters (Rotman, *Advanced Modern Algebra*, p. 220), there is some $e \in E^\times$ with $b \neq 0$. For such an element e, we have, using Eq. (1),

$$\sigma(b) = \sum_{\tau \in G} \sigma(c_\tau)\sigma\tau(e)$$

$$= \sum_{\tau \in G} c_{\sigma\tau}c_\sigma^{-1}\sigma\tau(e)$$

$$= c_\sigma^{-1} \sum_{\tau \in G} c_{\sigma\tau}\sigma\tau(e)$$

$$= c_\sigma^{-1} \sum_{\omega \in G} c_\omega \omega(e)$$

$$= c_\sigma^{-1} b.$$

Hence, $c_\sigma = b\sigma(b)^{-1}$, c is a coboundary, and $H^1(G, E^\times) = \{0\}$. ●

The next corollary describes the elements of norm 1 in a cyclic extension; it is so called because it was Theorem 90 in an 1897 treatise of Hilbert on Number Theory. The result itself is due to Kummer.

Corollary 9.33 (Hilbert's Theorem 90). *Let E/k be a Galois extension whose Galois group $G = \mathrm{Gal}(E/k)$ is cyclic, say, with generator σ. If $u \in E^\times$, then $Nu = 1$ if and only if there is $v \in E^\times$ with*

$$u = \sigma(v)v^{-1}.$$

Proof. By Theorem 9.27, we have $H^1(G, E^\times) = \ker N / \mathrm{im}\, D$, where N is the norm (remember that $E^\times$ is a multiplicative group) and $De = \sigma(e)e^{-1}$. Theorem 9.32 gives $H^1(G, E^\times) = \{0\}$, so that $\ker N = \mathrm{im}\, D$. Hence, if $u \in E^\times$, then $Nu = 1$ if and only if there is $v \in E^\times$ with $u = \sigma(v)v^{-1}$. ●

9.3 Bar Resolutions

Do the low-dimensional cohomology groups agree with Schreier's groups?

Notation. Let B_0 be the free G-module with basis the symbol [ ] (so that $B_0 \cong \mathbb{Z}G$). If $n \geq 1$, define B_n to be the free G-module with basis G^n, the cartesian product of n copies of G. We shall denote the elements of G^n by $[x_1 \mid \ldots \mid x_n]$ instead of by $(x_1, \ldots, x_n)$.

Lemma 9.34. *The sequence of G-modules*

$$B_3 \xrightarrow{d_3} B_2 \xrightarrow{d_2} B_1 \xrightarrow{d_1} B_0 \xrightarrow{\epsilon} \mathbb{Z} \to 0$$

is exact; it is the beginning of a G-free resolution of the G-trivial module $\mathbb{Z}$, where

$$d_3[x \mid y \mid z] = x[y \mid z] - [xy \mid z] + [x \mid yz] - [x \mid y],$$
$$d_2[x \mid y] = x[y] - [xy] + [x],$$
$$d_1[x] = x[\,] - [\,],$$
$$\epsilon[\,] = 1.$$

Proof. We have defined each of d_3, d_2, and d_1 on bases of free modules, and so each extends to a G-map. If the sequence is exact, then it can be completed to a G-free resolution of $\mathbb{Z}$ by splicing it to a G-free resolution of $\ker d_3$. We prove that $\operatorname{im} d_2 \subseteq \ker d_1$ and $\operatorname{im} d_3 \subseteq \ker d_2$. One could prove the reverse inclusions now, but it is not routine (the definition of the resolution will be completed, and its exactness will be proved, in Proposition 9.37).

$$
\begin{aligned}
d_1 d_2 [x \mid y] &= d_1(x[y] - [xy] + [x]) \\
&= x d_1[y] - d_1[xy] + d_1[x]) \\
&= x(y - 1) - (xy - 1) + (x - 1) = 0
\end{aligned}
$$

(the equation $d_1 x[y] = x d_1[y]$ holds because d_1 is a G-map).

$$
\begin{aligned}
d_2 d_3 [x \mid y \mid z] &= d_2(x[y \mid z] - [xy \mid z] + [x \mid yz] - [x \mid y]) \\
&= x d_2[y \mid z] - d_2[xy \mid z] + d_2[x \mid yz] - d_2[x \mid y] \\
&= x(y[z] - [yz] + [y])\ ^{\cdot}- (xy[z] - [xyz] + [xy]) \\
&\quad + (x[yz] - [xyz] + [x]) - (x[y] - [xy] + [x]) = 0. \quad \bullet
\end{aligned}
$$

Corollary 9.35. $\operatorname{Ext}^1_{\mathbb{Z}G}(\mathbb{Z}, K) \cong \operatorname{Der}(G, K)/\operatorname{PDer}(G, K)$; *that is, the* Ext *version of* $H^1(G, K)$ *coincides with Schreier's cohomology group* $H^1(G, K)$.

Proof. We compute $\operatorname{Ext}^1_{\mathbb{Z}G}(\mathbb{Z}, K)$ by applying the contravariant functor $\operatorname{Hom}_G(\square, K)$ to the exact sequence in Lemma 9.34:

$$
\operatorname{Hom}_G(B_2, K) \xleftarrow{d_2^*} \operatorname{Hom}_G(B_1, K) \xleftarrow{d_1^*} \operatorname{Hom}_G(B_0, K);
$$

by definition, $\operatorname{Ext}^1(\mathbb{Z}, K) = \ker d_2^* / \operatorname{im} d_1^*$.

There is no loss in generality in identifying a G-map $g \in \operatorname{Hom}_G(B_1, K)$ with its restriction to the basis G^2; hence, $g(x[y]) = xg[y]$ for all $x, y \in G$. Moreover, we may extend any function $\delta \colon G \to K$ to a G-map $B_1 \to K$ by $[y] \mapsto \delta(y)$; in particular, we may regard $\operatorname{Der}(G, K) \subseteq \operatorname{Hom}_G(B_1, K)$.

If $g \colon G \to K$ lies in $\ker d_2^*$, then

$$
\begin{aligned}
0 &= (d_2^* g)[x \mid y] \\
&= g d_2[x \mid y] \\
&= g(x[y] - [xy] + [x]) \\
&= xg[y] - g[xy] + g[x].
\end{aligned}
$$

Thus, $g[xy] = xg[y] + g[x]$, g is a derivation, and $\ker d_2^* \subseteq \operatorname{Der}(G, K)$. For the reverse inclusion, take $\delta \in \operatorname{Der}(G, K)$. Then

$$
\begin{aligned}
(d_2^* \delta)[x \mid y] &= \delta d_2[x \mid y] \\
&= \delta(x[y] - [xy] + [x]) \\
&= x\delta[y] - \delta[xy] + \delta[x] = 0,
\end{aligned}
$$

because δ is (the restriction of) a G-map. Hence, $\ker d_2^* = \mathrm{Der}(G, K)$.

Let us compute $\mathrm{im}\, d_1^*$. If $t \in \mathrm{Hom}_G(B_0, K)$, then $t[\ ] = a_0 \in K$. Now

$$(d_1^* t)[x] = t d_1[x] = t(x[\ ] - [\ ]) = xt[\ ] - t[\ ] = xa_0 - a_0,$$

because t is (the restriction of) a G-map. Thus, $d_1^* t$ is a principal derivation; that is, $\mathrm{im}\, d_1^* \subseteq \mathrm{PDer}(G, K)$. For the reverse inclusion, let $p: G \to K$ be a principal derivation: $p(x) = xb_0 - b_0$ for some $b_0 \in K$. Now define $u: B_0 \to K$ by $u[\ ] = b_0$, so that $p = d_1^* u$ and $\mathrm{PDer}(G, K) \subseteq \mathrm{im}\, d_1^*$.

Therefore, $\mathrm{Ext}^1(\mathbb{Z}, K) = \mathrm{Der}(G, K)/\mathrm{PDer}(G, K)$, which is the definition of Schreier's $H^1(G, K)$. $\quad\bullet$

Let us now consider $\mathrm{Ext}^2(G, K) = \ker d_3^* / \mathrm{im}\, d_2^*$. If $f: G^2 \to K$ lies in $\ker d_3^*$, then $0 = d_3^* f = f d_3$. Hence, for all $x, y, z \in G$, we have

$$\begin{aligned}
0 &= f d_3[x \mid y \mid z] \\
&= f(x[y \mid z] - [xy \mid z] + [x \mid yz] - [x \mid y]) \\
&= xf[y \mid z] - f[xy \mid z] + f[x \mid yz] - f[x \mid y],
\end{aligned}$$

the equation $f(x[y \mid z]) = xf[y \mid z]$ holding because f is (the restriction of) a G-map. Thus, f satisfies the cocycle identity; it would be a factor set if $f[1 \mid y] = 0 = f[x \mid 1]$ for all $x, y \in G$. Alas, we do not know this. If f lies in $\mathrm{im}\, d_2^*$, then there is some $h: G \to K$ with $f = d_2^* h = h d_2$. Thus,

$$\begin{aligned}
f[x \mid y] &= h d_2[x \mid y] \\
&= h(x[y] - [xy] + [x]) \\
&= xh[y] - h[xy] + h[x];
\end{aligned}$$

the equation $h(x[y]) = xh[y]$ holding because h is (the restriction of) a G-map. Now f is almost a coboundary, but it may not be one because we cannot guarantee $h(1) = 0$. Thus, $\mathrm{Ext}^2(\mathbb{Z}, K)$ does not quite coincide with Schreier's cohomology group $H^2(G, K)$. That these two groups are isomorphic is shown in Theorem 9.39. The key idea is that we can choose any projective resolution of $\mathbb{Z}$ to compute the groups $\mathrm{Ext}^n_{\mathbb{Z}G}(\mathbb{Z}, K)$.

Before we investigate other resolutions of $\mathbb{Z}$, we must say that there is a group-theoretic interpretation of $H^3(Q, K)$. Given an extension with a non-abelian kernel N, say, $1 \to N \to G \to Q \to 1$, the normality of N need not give a homomorphism $Q \to \mathrm{Aut}(N)$; instead, there is a homomorphism $\theta: Q \to \mathrm{Out}(N)$, called a *coupling*, where $\mathrm{Out}(N) = \mathrm{Aut}(N)/\mathrm{Inn}(N)$. Not every homomorphism $\theta: Q \to \mathrm{Out}(N)$ actually corresponds to a coupling arising from some extension, and the group $H^3(Q, Z(N)) = \{0\}$ if and only if every such θ can be realized by some extension (see Robinson, *A Course in the Theory of Groups*, Section 11.4). For $n \geq 4$, Robinson says, "While group-theoretic interpretations of the cohomology groups in dimensions greater than 3 are known, no really convincing applications to group theory have been made." On the other hand, higher cohomology groups will be used to define the *cohomological dimension* cd(G) of a group G [cd$(G) \leq n$ if $H^q(G, A) = \{0\}$ for all G-modules A and all $q > n$].

There are two constant themes in Homological Algebra: *low-dimensional* homology groups should have interesting interpretations (we do not expect an interpretation of H^{1409}); homology groups should be amenable to computation. We have just interpreted $H^n(G, A)$ for $n \leq 3$; we are now going to construct explicit G-free resolutions of the G-trivial module $\mathbb{Z}$. After all, since homology groups are independent of projective resolutions, choosing an explicit resolution may help in computing them. One consequence of this technical interlude will be a topological interpretation of cohomology groups, but let us begin by completing the definition of the resolution in Lemma 9.34.

Definition. Let G be a group. The *bar resolution* of $\mathbb{Z}$ is the sequence

$$\mathbf{B}(G) = \to B_2 \xrightarrow{d_2} B_1 \xrightarrow{d_1} B_0 \xrightarrow{\epsilon} \mathbb{Z} \to 0,$$

where B_0 is the free G-module on the single generator $[\ ]$, $\epsilon: B_0 \to \mathbb{Z}$ is the augmentation, B_n is the free G-module, for $n \geq 1$, with basis all symbols $[x_1 \mid x_2 \mid \cdots \mid x_n]$, where $x_i \in G$, and $d_n: B_n \to B_{n-1}$ is given by

$$d_n: [x_1 \mid \cdots \mid x_n] \mapsto x_1[x_2 \mid \cdots \mid x_n]$$
$$+ \sum_{i=1}^{n-1} (-1)^i [x_1 \mid \cdots \mid x_i x_{i+1} \mid \cdots \mid x_n]$$
$$+ (-1)^n [x_1 \mid \cdots \mid x_{n-1}].$$

Remark. Just as the G-module $B_0(G) \cong \mathbb{Z}G$ can be viewed as the free $\mathbb{Z}$-module with basis $\{x[\ ]: x \in G\}$, so, too, can $B_n(G)$ be viewed as the free $\mathbb{Z}$-module with basis $\{x[x_1 \mid x_2 \mid \cdots \mid x_n]: x, x_1, \ldots, x_n \in G\}$. ◄

The low-dimensional part of the bar resolution does agree with the sequence in Lemma 9.34.

$$d_1: [x] \mapsto x[\ ] - [\ ];$$
$$d_2: [x \mid y] \mapsto x[y] - [xy] + [x];$$
$$d_3: [x \mid y \mid z] \mapsto x[y \mid z] - [xy \mid z] + [x \mid yz] - [x \mid y].$$

There are actually two bar resolutions: the bar resolution just defined, which we have already found lacking; the *normalized bar resolution*, which will easily show that $H^2(G, K)$ coincides with Schreier's second cohomology group.

It is not obvious that the bar resolution is a complex, let alone that it is an exact sequence; we prove this by comparing $\mathbf{B}(G)$ to a resolution familiar to algebraic topologists.

Definition.   Let G be a group. The *homogeneous resolution* $\mathbf{P}(G)$ of $\mathbb{Z}$ (or *standard resolution*) is the sequence

$$\mathbf{P}(G) = \to P_2 \xrightarrow{\partial_2} P_1 \xrightarrow{\partial_1} P_0 \xrightarrow{\epsilon} \mathbb{Z} \to 0$$

in which each P_n is the free abelian group with basis all $(n+1)$-tuples of elements of G made into a G-module by defining

$$x(x_0, x_1, \ldots, x_n) = (xx_0, xx_1, \ldots, xx_n).$$

In particular, P_0 is the free abelian group with basis $\{(y) : y \in G\}$, made into a G-module by $x(y) = (xy)$. The map $\epsilon \colon P_0 \to \mathbb{Z}$, given by $\epsilon \colon \sum m_y(y) \mapsto \sum m_y$, is (essentially) the augmentation.[7] Define $\partial_n \colon P_n \to P_{n-1}$, whenever $n \geq 1$, by

$$\partial_n \colon (x_0, x_1, \ldots, x_n) \mapsto \sum_{i=0}^{n} (-1)^i (x_0, \ldots, \widehat{x_i}, \ldots, x_n).$$

It is clear that $\mathbf{P}(G)$ is a complex, for its differentials are essentially those arising in Algebraic Topology (see Proposition 1.1). Finally, each P_n is a free G-module with basis $\{(1, x_1, \ldots, x_n) : x_i \in G \text{ for all } i\}$, as the reader may check.

Proposition 9.36.   *The homogeneous resolution $\mathbf{P}(G)$ is a G-free resolution of the G-trivial module $\mathbb{Z}$.*

Proof.   To prove exactness of $\mathbf{P}(G)$, it suffices, by Proposition 6.15 and Exercise 6.10 on page 339, to construct a contracting homotopy; that is, $\mathbb{Z}$-maps

$$\leftarrow P_2 \xleftarrow{s_1} P_1 \xleftarrow{s_0} P_0 \xleftarrow{s_{-1}} \mathbb{Z}$$

such that $\epsilon s_{-1} = 1_{\mathbb{Z}}$ and $\partial_{n+1}s_n + s_{n-1}\partial_n = 1_{P_n}$ for all $n \geq 0$. Define $s_{-1} \colon \mathbb{Z} \to P_0$ by $m \mapsto m(1)$, where the 1 in the parentheses is the identity element of the group G; for $n \geq 0$, define $s_n \colon P_n \to P_{n+1}$ by

$$s_n \colon (x_0, x_1, \ldots, x_n) \mapsto (1, x_0, x_1, \ldots, x_n).$$

Here are the computations. First, $\epsilon s_{-1}(m) = \epsilon(m(1)) = m$. If $n \geq 0$, then

$$\partial_{n+1} s_n(x_0, \ldots, x_n) = \partial_{n+1}(1, x_0, \ldots, x_n)$$

$$= (x_0, \ldots, x_n) + \sum_{i=0}^{n} (-1)^{i+1}(1, x_0, \ldots, \widehat{x_i}, \ldots, x_n)$$

---

[7] Actually, the map ϵ is the composite of the isomorphism $P_0 \to \mathbb{Z}G$, given by $\sum m_y(y) \mapsto \sum m_y y$, with the augmentation $\mathbb{Z}G \to \mathbb{Z}$.

[the range of summation has been rewritten because x_i sits in the $(i + 1)$st position in $(1, x_0, \ldots, x_n)$]. On the other hand,

$$s_{n-1}\partial_n(x_0, \ldots, x_n) = s_{n-1} \sum_{j=0}^{n} (-1)^j (x_0, \ldots, \widehat{x}_j, \ldots, x_n)$$

$$= \sum_{j=0}^{n} (-1)^j (1, x_0, \ldots, \widehat{x}_j, \ldots, x_n).$$

It follows that $(\partial_{n+1}s_n + s_{n-1}\partial_n)(x_0, \ldots, x_n) = (x_0, \ldots, x_n)$.   •

Proposition 9.37.   *The bar resolution $\mathbf{B}(G)$ is a G-free resolution of $\mathbb{Z}$.*

Proof.   For each $n \geq 0$, define $\tau_n \colon P_n \to B_n$ by

$$\tau_n \colon (x_0, \ldots, x_n) \mapsto x_0[x_0^{-1}x_1 \mid x_1^{-1}x_2 \mid \cdots \mid x_{n-1}^{-1}x_n],$$

and define $\sigma_n \colon B_n \to P_n$ by

$$\sigma_n \colon [x_1 \mid \cdots \mid x_n] \mapsto (1, x_1, x_1x_2, x_1x_2x_3, \ldots, x_1x_2 \cdots x_n).$$

It is routine to check that τ_n and σ_n are inverse, and so each τ_n is an isomorphism. The reader can also check that $\tau = (\tau_n \colon \mathbf{P}(G) \to \mathbf{B}(G))_{n \geq 0}$ is a chain map; that is, the following diagram commutes:

$$
\begin{array}{ccc}
P_n & \xrightarrow{\ \tau_n\ } & B_n \\
{\scriptstyle\partial_n}\downarrow & & \downarrow{\scriptstyle d_n} \\
P_{n-1} & \xrightarrow[\ \tau_{n-1}\ ]{} & B_{n-1}.
\end{array}
$$

We now can see that $\mathbf{B}(G)$ is a complex. Since $d_n = \tau_{n-1}\partial_n\tau_n^{-1}$, we have

$$d_n d_{n+1} = (\tau_{n-1}\partial_n\tau_n^{-1})(\tau_n\partial_{n+1}\tau_{n+1}^{-1}) = \tau_{n-1}\partial_n\partial_{n+1}\tau_{n+1}^{-1} = 0.$$

Finally, Exercise 6.2 on page 338 shows that both complexes have the same homology groups. By Proposition 9.36, the complex $\mathbf{P}(G)$ is an exact sequence, so that all of its homology groups are $\{0\}$. It follows that all of the homology groups of $\mathbf{B}(G)$ are $\{0\}$ and, hence, $\mathbf{B}(G)$ is an exact sequence.   •

Although we now know that the bar resolution is, in fact, a resolution, we have seen that it is inadequate to prove that Schreier's second cohomology group is $\mathrm{Ext}^2(\mathbb{Z}, \square)$. We now introduce another resolution by modifying $\mathbf{B}(G)$. Define $U_n \subseteq B_n$ to be the submodule generated by all $[x_1 \mid \cdots \mid x_n]$ having at least one $x_i = 1$. It is easy to check that $d_n(U_n) \subseteq U_{n-1}$, for if $x_i = 1$, then every term in the expression for $d_n[x_1 \mid \cdots \mid x_n]$ involves x_i except two (those involving $x_{i-1}x_i$ and x_ix_{i+1}) that cancel. Hence, $\mathbf{U}(G)$ is a subcomplex of $\mathbf{B}(G)$.

Definition. The *normalized bar resolution* $\mathbf{B}^\star(G)$ is the quotient complex

$$\mathbf{B}^\star(G) = \mathbf{B}(G)/\mathbf{U}(G) = \; \to B_2^\star \xrightarrow{d_2^\star} B_1^\star \xrightarrow{d_1^\star} B_0^\star \xrightarrow{\epsilon} \mathbb{Z} \to 0.$$

Note that $B_0^\star = B_0$ (for $U_0 = \{0\}$) and, when $n \geq 1$, $B_n^\star = B_n/U_n$ is the free G-module with basis all cosets $[x_1 \mid \cdots \mid x_n]^\star = [x_1 \mid \cdots \mid x_n] + U_n$ in which all $x_i \neq 1$. Moreover, each differential $d_n^\star$ has the same formula as the map d_n in the bar resolution except that all symbols $[x_1 \mid \cdots \mid x_n]$ now occur with stars; in particular, $[x_1 \mid \cdots \mid x_n]^\star = 0$ if some $x_i = 1$.

Theorem 9.38. *The normalized bar resolution $\mathbf{B}^\star(G)$ is a G-free resolution of $\mathbb{Z}$.*

Proof. To prove exactness of $\mathbf{B}^\star(G)$, it suffices, by Proposition 6.15 and Exercise 6.10 on page 339, to construct a contracting homotopy

$$\leftarrow B_2^\star \xleftarrow{t_1} B_1^\star \xleftarrow{t_0} B_0^\star \xleftarrow{t_{-1}} \mathbb{Z},$$

where each t_n is a $\mathbb{Z}$-map. Define $t_{-1} \colon \mathbb{Z} \to B_0^\star$ by $t_{-1} \colon m \mapsto m[\,]$. To define t_n for $n \geq 0$, we take advantage of the fact that t_n need only be a $\mathbb{Z}$-map. Since $\mathbb{Z}G$ is a free abelian group, $B_n^\star$ is also a free abelian group, with basis $\{x[x_1 \mid \cdots \mid x_n]^\star : x, x_i \in G$ and $x_i \neq 1\}$ ($x = 1$ is allowed), and it suffices to define t_n on this basis; moreover, freeness allows us to choose the values without restriction. Thus, for $n \geq 0$, we define $t_n \colon B_n^\star \to B_{n+1}^\star$ by

$$t_n \colon x[x_1 \mid \cdots \mid x_n]^\star \mapsto [x \mid x_1 \mid \cdots \mid x_n]^\star.$$

That we have constructed a contracting homotopy is routine; the reader may check that $\epsilon t_{-1} = 1_\mathbb{Z}$ and, for $n \geq 0$, that $d_{n+1}^\star t_n + t_{n-1} d_n^\star = 1_{B_n^\star}$. •

Theorem 9.39. $\mathrm{Ext}^2_{\mathbb{Z}G}(\mathbb{Z}, K)$ *coincides with Schreier's second cohomology group* $H^2(G, K)$.

Proof. When we use the normalized bar resolution $\mathbf{B}^\star(G)$, the identities $f(1, y) = 0 = f(x, 1)$ and $h(1) = 0$ that were lacking in the calculation on page 527 are now present. •

Proposition 9.40. *If G is a finite group of order m, then $mH^n(G, K) = \{0\}$ for all $n \geq 1$ and all G-modules K.*

Proof. Use the bar resolution $\mathbf{B}(G)$. If $f \colon B_n \to K$, define $g \colon B_{n-1} \to K$ by

$$g[x_1 \mid \cdots \mid x_{n-1}] = \sum_{x \in G} f[x_1 \mid \ldots \mid x_{n-1} \mid x];$$

g is well-defined because G is finite and K is abelian. Now sum the cocycle identity over all $x = x_{n+1} \in G$:

$$(d_n f)[x_1 \mid \cdots \mid x_n] = x_1 f[x_2 \mid \cdots \mid x_n]$$
$$+ \sum_{i=1}^{n-2} (-1)^i f[x_1 \mid \cdots \mid x_i x_{i+1} \mid \cdots \mid x_n]$$
$$+ (-1)^{n-1} f[x_1 \mid \cdots \mid x_n x]$$
$$+ (-1)^n f[x_1 \mid \cdots \mid x_{n-1}].$$

In the next-to-last term, as x varies over G, so does $x_n x$. Therefore, if $d_n f = 0$, we have

$$0 = x_1 g[x_2 \mid \cdots \mid x_n] + \sum_{i=1}^{n-2} (-1)^i g[x_1 \mid \cdots \mid x_i x_{i+1} \mid \cdots \mid x]$$
$$+ (-1)^{n-1} g[x_1 \mid \cdots \mid x_{n-1}]$$
$$+ m(-1)^n f[x_1 \mid \cdots \mid x_n]$$

(the last term is independent of x). Hence,

$$0 = g d_{n-1} + m(-1)^n f,$$

and so $mf = \pm g d_{n-1} = d_{n-1}^* g$ is a coboundary.   •

We will give a second proof of this in Corollary 9.89.

Corollary 9.41. *If G is a finite group and A is a finitely generated G-module, then $H^n(G, A)$ is finite for all $n \geq 0$.*

Proof. Since G is finite, the group ring $\mathbb{Z}G$ is a finitely generated abelian group; it follows that finitely generated G-modules, for example, A and the terms B_n in the bar resolution, are also finitely generated abelian groups. Now $\mathrm{Hom}_G(B_n, A) \subseteq \mathrm{Hom}_{\mathbb{Z}}(B_n, A)$. Since the latter is finitely generated as an abelian group, so are $\mathrm{Hom}_G(B_n, A)$, $\ker d_n^*$, and $H^n(G, A)$. If $|G| = m$, then $m H^n(G, A) = \{0\}$, by Proposition 9.40. Therefore, $H^n(G, A)$ is finite, for it is a finitely generated abelian group of finite exponent.   •

We now apply Schreier's theorem.

Theorem 9.42 (Zassenhaus). *Let G be a finite group of order mn, where $(m, n) = 1$. If K is an abelian normal subgroup of order n, then G is a semidirect product of K by G/K, and any two complements of K are conjugate.*

Proof. Define $Q = G/K$. Note that $|Q| = |G|/|K| = mn/n = m$. By Corollary 9.14 and Proposition 9.21, it suffices to prove that $H^2(Q, K) = \{0\} = H^1(Q, K)$. For every $q \geq 0$, we know that $mH^q(Q, K) = \{0\}$, by Proposition 9.40. Since $(m, n) = 1$ and K is a finite abelian group of order n, the multiplication map $\mu_m \colon K \to K$, given by $a \mapsto ma$, is an automorphism. Hence, the induced map $(\mu_m)_*$, which is also multiplication by m, is an automorphism of $H^q(Q, K)$. Therefore, $H^q(Q, K) = \{0\}$. •

The hypothesis that K be abelian can be removed.

Theorem 9.43 (Schur–Zassenhaus). *Let G be a finite group of order mn, where $(m, n) = 1$. If K is a normal subgroup of order n, then G is a semidirect product of K by G/K, and any two complements of K are conjugate.*

Sketch of Proof. The proof that G is a semidirect product is a series of normalizations, eventually reaching the case K abelian (see Rotman, *An Introduction to the Theory of Groups*, p. 190).

Proving conjugacy of complements is much more difficult. We first prove that complements are conjugate if either K or Q is a solvable group (see Robinson, *A Course in the Theory of Groups*, p. 255). Since $|Q|$ and $|K|$ are relatively prime, at least one of K or Q has odd order. The deep Feit-Thompson Theorem, which says that every group of odd order is solvable, now completes the proof. •

The homogeneous resolution $\mathbf{P}(G)$ suggests a connection between group cohomology and cohomology of topological spaces, which we now sketch.

Definition. Let X be a topological space, and let $\mathrm{Aut}(X)$ be the group of all homeomorphisms of X with itself. We say that a group G *operates on* X if there is a homomorphism $G \to \mathrm{Aut}(X)$.

When a group G operates on a space X, we may regard each $g \in G$ as a homeomorphism of X; moreover, there are identities

$$g_1(g_2 x) = (g_1 g_2)x \qquad \text{and} \qquad 1x = x \tag{1}$$

for all $g_1, g_2 \in G$ and $x \in X$.

Definition. A group G *operates without fixed points* on a space X if $gx = x$ for some $x \in X$ implies $g = 1$.

Proposition 9.44. *If a group G operates on a space X, then the singular complex*

$$\mathbf{S}_\bullet(X) = \to S_n(X) \xrightarrow{\partial_n} S_{n-1}(X) \xrightarrow{\partial_{n-1}} S_{n-2}(X) \to$$

is a complex of G-modules. Moreover, if G operates without fixed points, then each $S_n(X)$ is a free G-module.

Proof.   Recall from Chapter 1 that $S_n(X)$ is the free abelian group with basis all singular n-simplexes; that is, all continuous $T : \Delta_n \to X$, where $\Delta_n = [v_0, v_1, \ldots, v_n]$ is the standard n-simplex. Since G operates on X, we may regard any $g \in G$ as a homeomorphism of X, so that gT is also an n-simplex. The identities in Eq. (1) show that $S_n(X)$ is a G-module. Also, in the notation of Section 1.3, if $T^i = T\epsilon_i^n$ is the ith face of T, then $(gT)^i = gT\epsilon_i^n = g(T^i)$, and this shows that ∂ is a G-map.

If $x \in X$, then its orbit is the equivalence class $\{gx : g \in G\}$; the orbit space X/G is the space of all orbits. Choose a transversal $X_0 \subseteq X$; that is, a subset consisting of exactly one element from each orbit. Call an n-simplex T *basic* if $Tv_0 \in X_0$. We claim that $S_n(X)$ is a free G-module with basis all basic n-simplexes. These simplexes do generate $S_n(X)$, for if $\sigma : \Delta_n \to X$, then $\sigma v_0 = gx_0$ for some $x_0 \in X_0$ and $g \in G$ (because X is the union of all the orbits); hence, $g^{-1}\sigma$ is basic, and $\sigma = g(g^{-1}\sigma)$.

To see that the family of all basic n-simplexes is a G-basis of $S_n(X)$, we first show that if $gT_1 = hT_2$, where T_1, T_2 are basic, then $g = h$ and $T_1 = T_2$. For suppose that $T_1 v_0 \neq T_2 v_0$; then, as these elements of X lie in distinct orbits, $gT_1 v_0 \neq hT_2 v_0$, a contradiction. Therefore, $T_1 v_0 = T_2 v_0 = x$, say. But $gx = hx$ implies that $g^{-1}h$ fixes x; since G operates without fixed points, $g = h$. Finally, since g is a homeomorphism, $T_1 = T_2$. Finally, suppose that $\sum_j \alpha_j T_j = 0$, where the T_j are distinct, and $\alpha_j = \sum_k m_{jk} g_k \in \mathbb{Z}G$, where the g_k are distinct. Then $\sum_{j,k} m_{jk}(g_k T_j) = 0$. Hence, all the simplexes $g_k T_j$ are distinct, so that each $m_{jk} = 0$ and, therefore, each $\alpha_j = 0$.   •

Definition.   A topological space X is *acyclic* if $H_0(X) \cong \mathbb{Z}$ and $H_n(X) = \{0\}$ for all $n \geq 1$.

Theorem 9.45.   *If a group G operates on an acyclic space X without fixed points, then the singular complex $\mathbf{S}_\bullet(X)$ is a deleted G-free resolution of the G-trivial module $\mathbb{Z}$.*

Proof.   We know that $\mathbf{S}_\bullet(X)$ is a complex of G-free modules. That X is acyclic says that $\to S_2(X) \xrightarrow{d_2} S_1(X) \xrightarrow{d_1} S_0(X) \to 0$ is exact at each $n \geq 1$ and, hence, $\to S_2(X) \xrightarrow{d_2} S_1(X) \to \operatorname{coker} d_1 \to 0$ is exact. But, since X is acyclic, $\operatorname{coker} d_1 = H_0(X) \cong \mathbb{Z}$.   •

Let A be a G-module. Now $H^\bullet(X; A)$ is defined as the homology of the complex $\operatorname{Hom}_G(\mathbf{S}_\bullet(X), A)$. If X is acyclic and G operates on X without fixed points, then $H^\bullet(G, A)$ is the homology of $\operatorname{Hom}_G(\mathbf{S}_\bullet(X), A)$, by Theorem 9.45. Suppose now that G also acts *properly* on X; that is, each $x \in X$

lies in some open set U with $gU \cap U = \varnothing$ for all $g \in G^{\times}$ (this hypothesis implies that G operates without fixed points). Then there is an isomorphism of complexes (see Mac Lane, *Homology*, pp. 135–136)

$$\text{Hom}_{\mathbb{Z}}(\mathbf{S}_{\bullet}(X/G), A) \cong \text{Hom}_G(\mathbf{S}_{\bullet}(X), A)$$

whenever A is G-trivial and X/G is the orbit space of X. But isomorphic complexes have isomorphic homology: for all $n \geq 0$,

$$H^n(X/G, A) \cong H^n(G, A). \tag{2}$$

The next step is to exhibit a space X satisfying all the conditions above. Given a group G, there exists an ***Eilenberg–Mac Lane space*** $K(G, 1)$ that is path-connected, that is *aspherical* (its nth homotopy groups vanish for all $n > 1$), and whose fundamental group $\pi_1(K(G, 1)) \cong G$ (see Adem–Milgram, *Cohomology of Finite Groups*, Chapter II, Brown, *Cohomology of Groups*, §1.4, or Spanier, *Algebraic Topology*, §8.1). The desired space BG, called the ***classifying space*** of G, is the universal covering space of $K(G, 1)$; the space BG is acyclic, G acts properly on BG, and $BG/G \approx K(G, 1)$. It follows from Eq. (2) that if A is a G-trivial module, then

$$H^n(K(G, 1), A) \cong H^n(G, A);$$

the cohomology of an abstract group G coincides with the cohomology of a certain topological space.

9.4 Group Homology

The groups $H^n(G, A)$ are obtained by applying the functor $\text{Hom}(\square, A)$ to the bar resolution $\mathbf{B}(G)$, obtaining $H^n(G, A) = \text{Ext}^n_{\mathbb{Z}G}(\mathbb{Z}, A)$. If we were topologists, we would also study *homology groups* by applying $\square \otimes_G A$ to $\mathbf{B}(G)$ [tensoring here is well-defined, by Exercise 9.3 on page 503, because left $\mathbb{Z}G$-modules [e.g., $B_n(G)$] can also be viewed as right $\mathbb{Z}G$-modules].

Definition. Let G be a group, let A be a G-module, and let $\mathbb{Z}$ be the integers viewed as a trivial G-module. The ***homology groups*** of G are

$$H_n(G, A) = \text{Tor}_n^{\mathbb{Z}G}(\mathbb{Z}, A).$$

Proposition 9.46. *If A is a G-module, then there is a natural isomorphism*

$$\eta_A \colon H_0(G, A) = \mathbb{Z} \otimes_G A \to A/\mathcal{G}A$$

given by

$$m \otimes a \mapsto ma + \mathcal{G}A.$$

Proof. By definition, $H_0(G, A) = \text{Tor}_0^{\mathbb{Z}G}(\mathbb{Z}, A) = \mathbb{Z} \otimes_G A$. Exactness of $0 \to \mathcal{G} \to \mathbb{Z}G \to \mathbb{Z} \to 0$ gives exactness of

$$\mathcal{G} \otimes_G A \to \mathbb{Z}G \otimes_G A \to \mathbb{Z} \otimes_G A \to 0.$$

Now $\mathbb{Z}G \otimes_G A \to A$, given by $\sum_x m_x x \otimes a \mapsto \sum_x m_x a$, is a surjection, and, under this map, $\text{im}\,\mathbb{Z} \otimes_G A$ goes into $\mathcal{G}A$. Hence, the map $\eta_A : m \otimes a \mapsto ma + \mathcal{G}A$ is an isomorphism. •

It is easy to see that $A/\mathcal{G}A$ is G-trivial; indeed, it is the largest G-trivial quotient of A. We often denote $A/\mathcal{G}A$ by A_G, in analogy with A^G, the largest G-trivial submodule of A.

Example 9.47. Suppose that E is a semidirect product of an abelian group A by a group G. Recall that $[G, A]$ is the subgroup generated by all commutators of the form $[x, a] = xax^{-1}a^{-1}$, where $x \in G$ and $a \in A$. If we write commutators additively, then

$$[x, a] = (x + a - x) - a = xa - a = (x - 1)a.$$

Hence, $\mathcal{G}A = [G, A]$ and $A/\mathcal{G}A = A/[G, A]$ here. ◄

We compute the homology groups of a finite cyclic group $G = \langle x \rangle$ of order k (the reader should compare this with Theorem 9.27, the cohomology groups of G). Define elements D and N of $\mathbb{Z}G$ by $D = x - 1$ and $N = 1 + x + x^2 + \cdots + x^{k-1}$. Lemma 9.26 gives a free G-resolution of $\mathbb{Z}$:

$$\to \mathbb{Z}G \xrightarrow{D} \mathbb{Z}G \xrightarrow{N} \mathbb{Z}G \xrightarrow{D} \mathbb{Z}G \xrightarrow{\epsilon} \mathbb{Z} \to 0, \tag{1}$$

where the maps alternate between being multiplication by D and multiplication by N.

Theorem 9.48. *If G is a cyclic group of finite order and A is a G-module, then, for $n \geq 1$,*

$$H_0(G, A) = A/\mathcal{G}A,$$
$$H_{2n-1}(G, A) = A^G/NA,$$
$$H_{2n}(G, A) = {}_N A/\mathcal{G}A,$$

where ${}_N A = \{a \in A : Na = 0\}$.

Proof. Apply $\square \otimes_G A$ to the resolution of $\mathbb{Z}$ in Eq. (1). After identifying $\mathbb{Z}G \otimes_G A$ with A via $\gamma \otimes a \mapsto \gamma a$, we have $\ker(D \otimes 1) = \mathcal{G}A$, $\text{im}(D \otimes 1) = \mathcal{G}A$, $\ker(N \otimes 1) = {}_N A$, and $\text{im}(N \otimes 1) = NA$. •

Corollary 9.49. *If G is a finite cyclic group of order k and A is a trivial G-module, then, for all $n \geq 1$,*

$$H_0(G, A) = A,$$
$$H_{2n-1}(G, A) = A/kA,$$
$$H_{2n}(G, A) = {}_kA.$$

In particular,

$$H_0(G, \mathbb{Z}) = \mathbb{Z}, \quad H_{2n-1}(G, \mathbb{Z}) = \mathbb{I}_k, \quad H_{2n}(G, \mathbb{Z}) = \{0\}.$$

Proof. Since A is G-trivial, we have $A^G = A$ and $\mathcal{G}A = \{0\}$ [for $Da = (x-1)a = 0$ because $xa = a$]. •

Corollary 9.50. *If G is a finite cyclic group and A is a G-module, then, for all $n \geq 1$,*

$$H_{2n-1}(G, A) = H^{2n}(G, A),$$
$$H_{2n}(G, A) = H^{2n-1}(G, A).$$

Proof. Theorems 9.48 and 9.27. •

We will understand Corollary 9.50 better once we introduce the Tate groups (see Proposition 9.105).

Let us now compute low-dimensional homology groups of not necessarily cyclic groups.

Lemma 9.51.

(i) *The connecting homomorphism $\partial \colon H_1(G, \mathbb{Z}) \to H_0(G, \mathcal{G})$ is an isomorphism.*

(ii) *For any group G, there is an isomorphism $H_1(G, \mathbb{Z}) \cong \mathcal{G}/\mathcal{G}^2$.*

(iii) *An explicit formula for an isomorphism is $\mathrm{cls}(z) \mapsto -z + \mathcal{G}^2$.*

Proof.

(i) The long exact sequence arising from $0 \to \mathcal{G} \to \mathbb{Z}G \xrightarrow{\varepsilon} \mathbb{Z} \to 0$ ends with

$$H_1(G, \mathbb{Z}G) \rightarrowtail H_1(G, \mathbb{Z}) \xrightarrow{\partial} H_0(G, \mathcal{G}) \rightarrowtail H_0(G, \mathbb{Z}G) \xrightarrow{\epsilon_*} H_0(G, \mathbb{Z}) \rightarrowtail 0.$$

Now $H_1(G, \mathbb{Z}G) = \{0\}$ because $\mathbb{Z}G$ is projective, so that ∂ is an injection. Proposition 9.46 with $A = \mathcal{G}$ gives $H_0(G, \mathcal{G}) \cong \mathcal{G}/\mathcal{G}^2$. Now $H_0(G, \mathbb{Z}) \cong \mathbb{Z}$, because $\mathbb{Z}$ is G-trivial, so that ϵ_* is essentially a map

$\mathbb{Z} \to \mathbb{Z}$. But every nonzero map $\mathbb{Z} \to \mathbb{Z}$ is an injection; as ϵ_* is a surjection, it is nonzero, and so ϵ_* is injective. Exactness of the homology sequence says that ϵ_* is injective if and only if ∂ is surjective. Hence, ∂ is an isomorphism.

(ii) Now $H_0(G, \mathcal{G}) = \mathbb{Z} \otimes_G \mathcal{G}$. We conclude from Proposition 9.46 that if $m \in \mathbb{Z}$ and $\gamma \in \mathcal{G}$, then the composite $\varphi = \eta_\mathcal{G} \partial \colon H_1(G, \mathbb{Z}) \to H_0(G, \mathcal{G}) \to \mathcal{G}/\mathcal{G}^2$ is an isomorphism, where $\eta_\mathcal{G} \colon m \otimes \gamma \mapsto m\gamma + \mathcal{G}^2$.

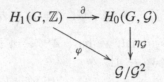

(iii) The isomorphism $\varphi \colon H_1(G, \mathbb{Z}) \to \mathcal{G}/\mathcal{G}^2$ is equal to $\eta_\mathcal{G} \partial$, so that we can give an explicit formula once we have a formula for the connecting homomorphism ∂. We give a formula for $\partial = j^{-1} D_1 r^{-1}$ using the bar resolution and the Horseshoe Lemma, Proposition 6.24.

$$
\begin{array}{ccccccccc}
0 & \longrightarrow & B_2 & \longrightarrow & B_2 \oplus B_1 & \overset{r}{\longrightarrow} & B_1 & \longrightarrow & 0 \\
& & \downarrow{d_2} & & \downarrow{D_1} & & \downarrow{d_1} & & \\
0 & \longrightarrow & B_1 & \overset{j}{\longrightarrow} & B_1 \oplus B_0 & \overset{q}{\longrightarrow} & B_0 & \longrightarrow & 0 \\
& & \downarrow{d_1'} & & \downarrow{D_0} & \overset{\sigma}{\nwarrow} & \downarrow{\varepsilon} & & \\
0 & \longrightarrow & \mathcal{G} & \overset{i}{\longrightarrow} & \mathbb{Z}G & \overset{\varepsilon}{\longrightarrow} & \mathbb{Z} & \longrightarrow & 0
\end{array}
$$

The third column is the bar resolution **B**; the first column is the resolution of $\mathcal{G}$ obtained by truncating **B**; thus, $d_1' \colon B_1 \to \mathcal{G}$ is the map $[x] \mapsto x - 1$. Recall that the horizontal maps into and out of the direct sums are just injections and projections. As in the proof of the Horseshoe Lemma, the boundary homomorphism D_0 is given by $(b_1, b_0) \mapsto id_1' b_1 + \sigma b_0$, where $\epsilon\sigma = \epsilon$; obviously, we may take $\sigma = 1_{\mathbb{Z}G}$ (for $B_0 = \mathbb{Z}G$). Thus, $D_0 \colon (b_1, b_0) \mapsto id_1' b_1 + b_0$, and so $\ker D_0 = \{(b_1, b_0) : id_1' b_1 = -b_0\}$.

Using the proof of the Horseshoe Lemma again, the construction of D_1 involves a map $\tau \colon B_1 \to \ker D_0$ with $q'\tau = d_1'$, where $q' = q | \ker D_0$.

$$
\begin{array}{ccccccccc}
0 & \longrightarrow & B_2 & \longrightarrow & B_2 \oplus B_1 & \overset{r}{\longrightarrow} & B_1 & \longrightarrow & 0 \\
& & \downarrow{d_2'} & & \downarrow{D_1'} & \overset{\tau}{\nwarrow} & \downarrow{d_1'} & & \\
0 & \longrightarrow & \ker d_1 & \overset{j'}{\longrightarrow} & \ker D_0 & \overset{q'}{\longrightarrow} & \mathcal{G} & \longrightarrow & 0
\end{array}
$$

Here, d_2' and d_1' differ from d_2 and d_1, respectively, only in their targets; in particular, $d_2' \colon [x \mid y] \mapsto x[y] - [xy] + [x] \in \ker d_1$. For a basis element $[x]$ of B_1 (where $x \in G$), the map τ must satisfy $\tau[x] = (\beta_1, \beta_0) \in \ker D_0$, so that $id_1\beta_1 = -\beta_0$. Also, $q'\tau[x] = (\beta_1, \beta_0) = \beta_0$, so that $\beta_0 = -(x - 1)$. We can construct such a map τ. Since B_1 is the free module with basis all $[x]$, there exists a map τ with $\tau[x] = (-[x], x-1)$ for $x \in G$; thus, $\tau b_1 = (-b_1, d_1b_1)$ and $q'\tau = d_1'$. Define $D_1 \colon B_2 \oplus B_1 \to B_1 \oplus B_0$ by $(b_2, b_1) \mapsto jd_2b_2 + \tau b_1$; that is,

$$D_1 \colon (b_2, b_1) \mapsto (jd_2b_2 - b_1, d_1b_1)$$

(the map $D_1' \colon B_2 \oplus B_1 \to \ker D_0$ differs from D_1 only in its target). We can now give a formula for the boundary homomorphism ∂. If z is a 1-cycle, then

$$\begin{aligned}
\partial(\mathrm{cls}\, z) &= \mathrm{cls}(j^{-1}D_1r^{-1}z) \\
&= \mathrm{cls}(j^{-1}D_1(0, z)) \\
&= \mathrm{cls}(j^{-1}(-z, 0)) \\
&= -\mathrm{cls}\, z \in H_0(G, \mathcal{G}).
\end{aligned}$$

The isomorphism $\eta_G \colon H_0(G, \mathcal{G}) = \mathbb{Z} \otimes_G \mathcal{G} \to \mathcal{G}/\mathcal{G}^2$ is $1 \otimes z \mapsto z + \mathcal{G}^2$, by Proposition 9.46. Therefore, if z is a 1-cycle, then

$$\varphi\, \mathrm{cls}(z) = \eta_G \partial\, \mathrm{cls}(z) = \eta_G(-\mathrm{cls}(z)) = -z + \mathcal{G}^2. \quad \bullet$$

Theorem 9.52. *Let G' denote the commutator subgroup of a group G. There is an isomorphism*

$$\theta_G \colon H_1(G, \mathbb{Z}) \to G/G'.$$

If $H_1(G, \mathbb{Z})$ is computed with the bar resolution, then a formula for θ_G is

$$\theta_G \colon \mathrm{cls}\left(\sum_x m_x[x]\right) \mapsto \prod_x x^{-m_x} G',$$

where $\sum_x m_x = 0$.

Proof. Since $H_1(G, \mathbb{Z}) \cong \mathcal{G}/\mathcal{G}^2$, by Lemma 9.51, it is enough to prove $\mathcal{G}/\mathcal{G}^2 \cong G/G'$. Define $\lambda \colon G \to \mathcal{G}/\mathcal{G}^2$ by

$$\lambda \colon x \mapsto (x - 1) + \mathcal{G}^2.$$

To see that λ is a homomorphism, note that

$$xy - 1 - (x - 1) - (y - 1) = (x - 1)(y - 1) \in \mathcal{G}^2,$$

so that

$$\begin{aligned}
\lambda(xy) &= xy - 1 + \mathcal{G}^2 \\
&= (x - 1) + (y - 1) + \mathcal{G}^2 \\
&= x - 1 + \mathcal{G}^2 + y - 1 + \mathcal{G}^2 \\
&= \lambda(x) + \lambda(y).
\end{aligned}$$

Since $\mathcal{G}/\mathcal{G}^2$ is abelian, $G' \subseteq \ker \lambda$; therefore, λ induces a homomorphism $\lambda' : G/G' \to \mathcal{G}/\mathcal{G}^2$, namely, $xG' \mapsto x - 1 + \mathcal{G}^2$.

We show that λ' is an isomorphism by constructing its inverse. Recall Lemma 9.24: $\mathcal{G}$ is a free abelian group with basis all $x - 1$, where $x \in G^\times$. It follows that there is a (well-defined) $\mathbb{Z}$-homomorphism $\mu : \mathcal{G} \to G/G'$ with

$$\mu : x - 1 \mapsto xG'.$$

If $\mathcal{G}^2 \subseteq \ker \mu$, then μ induces a $\mathbb{Z}$-homomorphism $\mu' : \mathcal{G}/\mathcal{G}^2 \to G/G'$, which, obviously, is the inverse of λ', and this will complete the proof.

If $u \in \mathcal{G}^2$, then

$$u = \left(\sum_{x \neq 1} m_x(x - 1)\right)\left(\sum_{y \neq 1} n_y(y - 1)\right)$$

$$= \sum_{x,y} m_x n_y (x - 1)(y - 1)$$

$$= \sum_{x,y} m_x n_y \big((xy - 1) - (x - 1) - (y - 1)\big).$$

Therefore, $\mu(u) = \prod_{x,y} (xyx^{-1}y^{-1})^{m_x n_y} G' = G'$, and so $u \in \ker \mu$, as desired.

By Lemma 9.51(iii), the isomorphism $H_1(G, \mathbb{Z}) \to \mathcal{G}/\mathcal{G}^2$ is just cls $z \mapsto -z + \mathcal{G}^2$; in particular, $\mathrm{cls}([x] - [1]) \mapsto -x + 1 + \mathcal{G}^2$. Therefore, if $z = \sum_x m_x[x]$ is a 1-cycle, that is, $\sum_x m_x = 0$, then the composite $H_1(G, \mathbb{Z}) \to \mathcal{G}/\mathcal{G}^2 \to G/G'$ sends $\mathrm{cls}(\sum_x m_x[x]) \mapsto \prod_x x^{-m_x} G'$. •

The Universal Coefficient Theorem gives a nice description of $H_1(G, A)$ (see Robinson, *A Course in the Theory of Groups*, p. 342, for a proof not using Universal Coefficients).

Proposition 9.53.   *If G is a group and A is a trivial [8] G-module, then there is a natural isomorphism*

$$H_1(G, A) \cong H_1(G, \mathbb{Z}) \otimes_{\mathbb{Z}} A \cong (G/G') \otimes_{\mathbb{Z}} A.$$

Proof.   Let **B** be the bar resolution of the trivial G-module $\mathbb{Z}$, and let $\mathbf{K} = \mathbf{B} \otimes_G \mathbb{Z}$. By Exercise 9.20 on page 558, we may view **K** as a complex of free $\mathbb{Z}$-modules. By the Universal Coefficient Theorem for homology (Theorem 7.55), there is an exact sequence

$$0 \to H_1(\mathbf{K}) \otimes_{\mathbb{Z}} A \xrightarrow{\lambda} H_1(\mathbf{K} \otimes_{\mathbb{Z}} A) \xrightarrow{\mu} \mathrm{Tor}_1^{\mathbb{Z}}(H_0(\mathbf{K}), A) \to 0$$

---

[8]This result may be false if A is not G-trivial.

with both λ and μ natural. By Proposition 9.46, we have

$$H_0(\mathbf{K}) = H_0(G, \mathbb{Z}) = \mathbb{Z}_G = \mathbb{Z},$$

because $\mathbb{Z}$ is G-trivial; thus, $\text{Tor}_1^{\mathbb{Z}}(H_0(\mathbf{K}), A) = \text{Tor}_1^{\mathbb{Z}}(\mathbb{Z}, A) = \{0\}$. Therefore, $\lambda \colon H_1(\mathbf{K}) \otimes_{\mathbb{Z}} A \to H_1(\mathbf{K} \otimes_{\mathbb{Z}} A)$ is a natural isomorphism. Finally, the associative law for tensor product, Proposition 2.57, gives

$$\mathbf{K} \otimes_{\mathbb{Z}} A = (\mathbf{B} \otimes_G \mathbb{Z}) \otimes_{\mathbb{Z}} A \cong \mathbf{B} \otimes_G (\mathbb{Z}) \otimes_{\mathbb{Z}} A,$$

where $(\mathbb{Z}) \otimes_{\mathbb{Z}} A$ is a G-module via $x(n \otimes a) = xn \otimes a = n \otimes a$ (because $\mathbb{Z}$ is G-trivial). Thus, $(\mathbb{Z}) \otimes_{\mathbb{Z}} A$ is G-trivial. Now $\mathbb{Z} \otimes_{\mathbb{Z}} A \cong A$ as abelian groups; but this is a G-isomorphism because A is G-trivial. Hence, $\mathbf{K} \otimes_{\mathbb{Z}} A \cong \mathbf{B} \otimes_G A$ and $H_1(\mathbf{K} \otimes_G A) = H_1(\mathbf{B} \otimes_G A) = H_1(G, A)$.

The last isomorphism follows from Theorem 9.52.   •

9.4.1 Schur Multiplier

The *Schur multiplier*[9] is a subtle invariant of a group that turns out to be very useful.

Definition.   The *Schur multiplier* of a group G is $H_2(G, \mathbb{Z})$.

We are going to prove *Hopf's formula*: if a group G has a presentation $G = F/R$, where F is a free group and R is the normal subgroup of relations, then $H_2(G, \mathbb{Z}) \cong (R \cap F)/[F, R]$.

Proposition 9.54.   *If G is a free group with basis X, then its augmentation ideal $\mathcal{G}$ is a free G-module with basis $X - 1 = \{x - 1 : x \in X\}$.*

Remark.   Recall Lemma 9.24: the augmentation ideal $\mathcal{G}$ is a free abelian group with basis $X - 1$.   ◄

Proof.   The formulas

$$xy - 1 = (x - 1) + x(y - 1) \quad \text{and} \quad x^{-1} - 1 = -x^{-1}(x - 1)$$

show that if $g = x_1^{e_1} \cdots x_n^{e_n}$, then $g - 1$ is a G-linear combination of $X - 1$. Therefore, $X - 1$ generates $\mathcal{G}$ as a G-module.

To see that $\mathcal{G}$ is freely generated by $X - 1$, we complete the diagram

---

[9]The multiplier is often called the *multiplicator*, which is a transliteration from German.

where i is the inclusion, A is a G-module, φ is a function, and $\widetilde{\varphi}$ is a G-map (uniqueness of $\widetilde{\varphi}$ follows from $\mathcal{G}$ being generated by $X - 1$). Since $\operatorname{Hom}_G(\mathcal{G}, A) \cong \operatorname{Der}(G, A)$, by Proposition 9.25, we seek a derivation. Consider the extension $0 \to A \to E \xrightarrow{\pi} G \to 1$; since G is free, this extension splits, so we may assume that E consists of all ordered pairs $(a, g) \in A \times G$ and $\pi(a, g) = g$. Define a function $\ell \colon X \to E$ by $\ell x = (\varphi(x - 1), x)$. As G is free on X, the function ℓ extends to a homomorphism $L \colon G \to E$, say, $L(g) = (dg, g)$. By Proposition 9.15, $d \colon G \to A$ is a derivation. Now the isomorphism of Proposition 9.25 yields a G-map $\widetilde{\varphi} \colon \mathcal{G} \to A$, namely, $\widetilde{\varphi}(g - 1) = dg$. Since $\ell x = Lx = (dx, x) = (\varphi(x - 1), x)$, we have $\widetilde{\varphi}(x - 1) = \varphi(x - 1)$, which shows that $\widetilde{\varphi}$ extends φ.   •

Corollary 9.55.   *If G is a free group, then*
$$H_n(G, A) = \{0\} = H^n(G, A)$$
for all $n \geq 2$ and all G-modules A.

Proof.   The sequence $0 \to \mathcal{G} \to \mathbb{Z}G \to \mathbb{Z} \to 0$ is a G-free resolution of $\mathbb{Z}$.

•

In Corollary 9.28, we saw that all the higher cohomology groups of the trivial group $G = \{1\}$ vanish; the next corollary shows that the homology groups vanish as well.

Corollary 9.56.   *If $G = \{1\}$, then*
$$H_n(G, A) = \{0\} = H^n(G, A)$$
for all $n \geq 1$ and all G-modules A.

Proof.   If $G = \{1\}$, then $\mathbb{Z}G \cong \mathbb{Z}$ and $\mathcal{G} = \{0\}$. Thus, $0 \to \mathbb{Z}G \to \mathbb{Z} \to 0$ is a G-free resolution of $\mathbb{Z}$.   •

Let G be a group, let F be a free group (with basis X), and let $\pi \colon F \to G$ be a surjection with kernel R. As every subgroup of a free group is free (Nielsen–Schreier Theorem), R is free, say, with basis Y. Recall (see Exercise 9.12 on page 513): the homomorphism π induces a surjective ring map $\mathbb{Z}F \to \mathbb{Z}G$; namely, $\pi_* \colon \sum m_f f \mapsto \sum m_f \pi(f)$. Hence, $\ker \pi_*$ is a two-sided ideal in $\mathbb{Z}F$, and $\mathbb{Z}F / \ker \pi_* \cong \mathbb{Z}G$.

Definition.   If G is a group, F is a free group (with basis X), and $\pi \colon F \to G$ is a surjection with kernel R, then the ***relation ideal***, denoted by $\mathfrak{R}$, is $\ker \pi_*$.

Beware! We denote the relation ideal by $\mathfrak{R}$, and not by $\mathcal{R}$, for $\mathfrak{R}$ is a two-sided ideal in $\mathbb{Z}F$; in contrast, the augmentation ideal $\mathcal{R}$ of R is merely a two-sided ideal in $\mathbb{Z}R \subseteq \mathbb{Z}F$. However, the next lemma shows that the relation ideal $\mathfrak{R}$ is the two-sided ideal in $\mathbb{Z}F$ generated by the augmentation ideal $\mathcal{R}$.

Lemma 9.57. *Let F be a free group with basis X, and let R be a normal subgroup of F with basis Y. Then the relation ideal $\mathfrak{R}$ is the free (left or right) F-module with basis $Y - 1 = \{y - 1 : y \in Y\}$. Moreover, $\mathfrak{R}$ is the two-sided ideal in $\mathbb{Z}F$ generated by the augmentation ideal $\mathcal{R}$ of R.*

Proof. Clearly, $Y - 1 \subseteq \ker \pi_*$. Choose a left transversal T of R in F:

$$F = \bigcup_{t \in T} tR.$$

If $\alpha \in \mathbb{Z}F$, then $\alpha = \sum_{i,j} m_{ij} t_i r_j$, where $t_i r_j \in t_i R$ and $m_{ij} \in \mathbb{Z}$. If $\alpha \in \mathfrak{R}$, then

$$0 = \pi_* \alpha = \sum_{i,j} m_{ij} \pi(t_i),$$

where the $\pi(t_i)$ are distinct elements of G. Therefore,

$$\alpha = \alpha - 0 = \sum_i \left(\sum_j m_{ij} t_i (r_j - 1) \right)$$

is an F-linear combination of elements of the form $r - 1$ with $r \in R$. However, the proof of Proposition 9.54 shows that each such $r - 1$ is an R-linear combination, a fortiori, an F-linear combination, of $Y - 1$. Thus, $\mathfrak{R}$ is generated as an F-module by $Y - 1$. The same argument, using $F = \bigcup_{t \in T} Rt$, shows that $\mathfrak{R}$ is also generated by $Y - 1$ as a right $\mathbb{Z}F$-module.

To see that $Y - 1$ freely generates $\mathfrak{R}$, assume that $\sum \alpha_k (y_k - 1) = 0$, where $\alpha_k \in \mathbb{Z}F$. It is easy to see that $\alpha_k = \sum_p t_p \beta_{pk}$, where $\beta_{pk} \in \mathbb{Z}R$. Now the coset representatives $\{t_p\}$ are independent over $\mathbb{Z}R$ ($0 = \sum_{p,q} t_p m_{pq} r_q$ implies that each $m_{pq} = 0$, since all $t_p r_q$ are distinct), from which it follows that $\sum_k \beta_{pk}(y_k - 1) = 0$ for each p. The problem has been reduced to Proposition 9.54, for R is free with basis Y. Again, using the transversal as right coset representatives proves that $\mathfrak{R}$ is the free right $\mathbb{Z}F$-module with basis $Y - 1$.

The second statement follows from Lemma 9.24: the augmentation ideal of R is the free abelian group with basis $\{y - 1 : x \in Y\}$. •

Lemma 9.58. *Let G be a group, let R be a normal subgroup of a free group F with $F/R \cong G$, and let M be a free left F-module with basis W.*

(i) *$M/\mathfrak{R}M$ is a free left G-module with basis $\{w + \mathfrak{R}M : w \in W\}$. A similar statement holds for $M'/M'\mathfrak{R}$ when M' is a free right F-module.*

(ii) *$M/\mathcal{F}M$ is a free abelian group with basis $\{w + \mathcal{F}M : w \in W\}$. A similar statement holds for $M'/M'\mathcal{F}$ when M' is a free right F-module.*

Proof.

(i) First of all, we let G act on $M/\mathfrak{R}M$ by

$$g(m + \mathfrak{R}M) = fm + \mathfrak{R}M, \quad \text{where } \pi(f) = g.$$

This action is well-defined, for if $\pi(f_1) = \pi(F)$, then $f_1 - f \in$ $\ker \pi_* = \mathfrak{R}$ and $(f_1 - f)m \in \mathfrak{R}M$. Since $M = \bigoplus_{w \in W}(\mathbb{Z}F)w$ and $\mathfrak{R} = \bigoplus_{w \in W}\mathfrak{R}w$, it follows that $M/\mathfrak{R}M \cong \bigoplus_{w \in W}(\mathbb{Z}F)w/\mathfrak{R}w \cong$ $\bigoplus(\mathbb{Z}F/\mathfrak{R})w$. The last module is G-free, for $\mathbb{Z}F/\mathfrak{R} \cong \mathbb{Z}G$.

(ii) Specialize the argument in (i) to the case $G = \{1\}$. The (trivial) map $F \to G$ induces the ring map $\mathbb{Z}F \to \mathbb{Z}G = \mathbb{Z}$ whose kernel is $\mathcal{F}$, the augmentation ideal of F, and $\mathcal{F}$ now plays the role of $\mathfrak{R}$ in part (i).   •

Recall that if $A \subseteq B \subseteq M$ are submodules, then there is a surjection $M/A \to M/B$, defined by $m + A \mapsto m + B$. If also $M \subseteq M'$, then $M/B \subseteq M'/B$, and the composite $M/A \to M/B \to M'/B$ is called *enlargement of coset*.

Theorem 9.59 (Gruenberg). *Let $1 \to R \to F \to G \to 1$ be an exact sequence of groups, where F is free with basis X and R is free with basis Y.*

(i) *For all $n \geq 1$, $P_{2n} = \mathfrak{R}^n/\mathfrak{R}^{n+1}$ is the G-free module with basis*

$$\{(y_1 - 1) \cdots (y_n - 1) + \mathfrak{R}^{n+1} : y_i \in Y\},$$

and $P_{2n-1} = \mathfrak{R}^{n-1}\mathcal{F}/\mathfrak{R}^n\mathcal{F}$ is the G-free module with basis

$$\{(y_1 - 1) \cdots (y_{n-1} - 1)(x - 1) + \mathfrak{R}^n\mathcal{F} : y_i \in Y \text{ and } x \in X\}.$$

(ii) *There is a G-free resolution of $\mathbb{Z}$,*

$$\to P_2 \xrightarrow{d_2} P_1 \xrightarrow{d_1} \mathbb{Z}G \xrightarrow{\epsilon} \mathbb{Z} \to 0,$$

where the maps $d_k \colon P_k \to P_{k-1}$ are enlargements of coset.

Proof.

(i) Since $\mathcal{F}$ is F-free on $X - 1$ and $\mathfrak{R}$ is F-free on $Y - 1$, iterated use of Exercise 9.25 on page 559 shows that the F-modules $\mathfrak{R}^{n-1}\mathcal{F}$ and $\mathfrak{R}^n$ are free with bases $\{(y_1 - 1) \cdots (y_{n-1} - 1)(x - 1) : y_i \in Y, x \in X\}$ and $\{(y_1 - 1) \cdots (y_n - 1) : y_j \in Y\}$, respectively. Lemma 9.58 now applies, showing that P_{2n-1} and P_{2n} are free G-modules with the stated bases.

(ii) Let us describe the maps in more detail. Since $\mathfrak{R}^{n+1} \subseteq \mathfrak{R}^n \mathcal{F}$, the enlargement of coset map $d_{2n} \colon P_{2n} \to P_{2n-1}$ is

$$d_{2n} \colon \mathfrak{R}^n / \mathfrak{R}^{n+1} \to \mathfrak{R}^n / \mathfrak{R}^n \mathcal{F} \to \mathfrak{R}^{n-1} \mathcal{F} / \mathfrak{R}^n \mathcal{F}.$$

The Third Isomorphism Theorem gives

$$\text{im } d_{2n} \colon \mathfrak{R}^n / \mathfrak{R}^n \mathcal{F} \quad \text{and} \quad \ker d_{2n} = \mathfrak{R}^n \mathcal{F} / \mathfrak{R}^{n+1}.$$

Since $\mathfrak{R}^{n+1} \mathcal{F} \subseteq \mathfrak{R}^{n+1}$, the map $d_{2n+1} \colon P_{2n+1} \to P_{2n}$ is

$$d_{2n+1} = \mathfrak{R}^n \mathcal{F} / \mathfrak{R}^{n+1} \to \mathfrak{R}^n \mathcal{F} / \mathfrak{R}^{n+1} \to \mathfrak{R}^n / \mathfrak{R}n + 1.$$

Our calculation in part (i) now gives exactness at all terms in the sequence with the possible exception of $\mathcal{F} / \mathfrak{R} \mathcal{F} = P_1 \xrightarrow{d_1} \mathbb{Z}G \xrightarrow{\epsilon} \mathbb{Z}$. Let us interpret $\mathfrak{R}^0 = \mathbb{Z}F$, so that $P_0 = \mathfrak{R}^0 / \mathfrak{R}^1 = \mathbb{Z}F / \mathfrak{R} \cong \mathbb{Z}G$; hence, $d_1 \colon x - 1 + \mathfrak{R}\mathcal{F} \mapsto x - 1 + \mathfrak{R} \mapsto \pi x - 1$. Thus, im d_1 is the augmentation ideal $\mathcal{G} = \ker \epsilon$, and the proof is complete. ●

Definition. Given a presentation of a group G as F/R, where F is free, the G-free resolution of $\mathbb{Z}$ in Theorem 9.59 is called the ***Gruenberg resolution***.

Example 9.60. If $G = \langle g \rangle$ is a cyclic group of order k, then one presentation of G has $F = \langle x \rangle$, $R = \langle x^k \rangle$, and $\pi(x) = g$. Let us check that the Gruenberg resolution of $\mathbb{Z}$ in this case is the resolution of Lemma 9.26. The module P_{2n} is free on one generator $x_{2n} = (x^k - 1)^n + \mathfrak{R}^{n+1}$, and P_{2n-1} is free on one generator $x_{2n-1} = (x^k - 1)^{n+1}(x-1) + \mathfrak{R}^n \mathcal{F}$. Under the enlargement of coset map $P_{2n+1} \to P_{2n}$, we have $x_{2n+1} = (x^k - 1)^n (x-1) + \mathfrak{R}^n \mathcal{F} \mapsto (x^k - 1)^n (x - 1) + \mathfrak{R}^{n+1}$. But $(x^k - 1)^n (x - 1) = (x^k - 1)^{n-1}(x^k - 1)(x - 1) = (x - 1)x_{2n}$; that is, if $D = x - 1$, then the map $P_{2n+1} \to P_{2n}$ is just μ_D, multiplication by D. The map $P_{2n} \to P_{2n-1}$ takes $x_{2n} = (x^k-1)^n + \mathfrak{R}^{n+1} \mapsto (x^k-1)^n + \mathfrak{R}^n \mathcal{F}$. But $(x^k - 1)^n = (x - 1)(x^k - 1)^{n-1}(1 + x + \cdots + x^{k-1})$. Hence, if $N = 1 + x + \cdots + x^{k-1}$, then $x_{2n} \mapsto Nx_{2n-1}$ in P_{2n-1}, and the map $P_{2n} \to P_{2n-1}$ is multiplication by N. ◄

In Lemma 9.51 and Theorem 9.52, we saw that $\mathcal{G} / \mathcal{G}^2 \cong G/G'$. Thus, there is a relationship between group ring constructions and group constructions. Here is another such.

Lemma 9.61. *In the notation of the Gruenberg resolution, there are isomorphisms of abelian groups,*

$$\mathfrak{R} / \mathfrak{R} \mathcal{F} \cong R/R' \quad \text{and} \quad (\mathfrak{R}\mathcal{F} + \mathcal{F}\mathfrak{R}) / \mathfrak{R}\mathcal{F} \cong [F, R]/R'.$$

Proof. Now $\mathfrak{R}/\mathfrak{R}\mathcal{F}$ is free abelian with basis $\{y - 1 + \mathfrak{R}\mathcal{F} : y \in Y\}$, by Lemma 9.58, while R/R' is the free (multiplicative) abelian group with basis $\{yR' : y \in Y\}$. Thus, there is an isomorphism of abelian groups, $\theta : \mathfrak{R}/\mathfrak{R}\mathcal{F} \to R/R'$, given by $\theta : y - 1 + \mathfrak{R}\mathcal{F} \mapsto yR'$. If $r = y_1^{e_1} \cdots y_n^{e_n}$, what is $\theta(r - 1 + \mathfrak{R}\mathcal{F})$? The identities

$$(uv - 1) - (u - 1) - (v - 1) = (u - 1)(v - 1)$$

and

$$-(u^{-1} - 1) - (u - 1) = (u - 1)(u^{-1} - 1),$$

together with $\mathfrak{R}^2 \subseteq \mathfrak{R}\mathcal{F}$, show that $\theta(r - 1 + \mathfrak{R}\mathcal{F}) = rR'$.

Now restrict the isomorphism θ to $(\mathfrak{R}\mathcal{F} + \mathcal{F}\mathfrak{R})/\mathfrak{R}\mathcal{F}$, the subgroup generated by all $(f - 1)(r - 1) + \mathfrak{R}\mathcal{F}$, where $r \in R$. The identity

$$(f - 1)(r - 1) = ([f, r] - 1) + ([f, r] - 1)(rf - 1) + (r - 1)(f - 1)$$

shows that $(f - 1)(r - 1) + \mathfrak{R}\mathcal{F} = [f, r] - 1 + \mathfrak{R}\mathcal{F}$. Therefore,

$$\theta\big((f - 1)(r - 1) + \mathfrak{R}\big[\begin{smallmatrix} 0 & 1 \\ -1 & 0 \end{smallmatrix}\big]\big) = [f, r]R'. \quad \bullet$$

Remark. Actually, more is true. The multiplicative abelian group R/R' is a G-module if one defines $g(rR') = frf^{-1}R'$, where $\pi(f) = g$, and the isomorphism in the lemma is now a G-isomorphism, for

$$\theta(g(r - 1) + \mathfrak{R}\mathcal{F}) = \theta(f(r - 1) + \mathfrak{R}\mathcal{F}) \quad \text{(see Lemma 9.58)}$$
$$= \theta(f(r - 1) + f(r - 1)(f^{-1} - 1) + \mathfrak{R}\mathcal{F})$$
$$= \theta(frf^{-1} - 1 + \mathfrak{R}\mathcal{F})$$
$$= frf^{-1}R' = g(rR') = g\theta(r - 1 + \mathfrak{R}\mathcal{F}). \quad \blacktriangleleft$$

We need one more elementary lemma before we prove Hopf's formula.

Lemma 9.62. *Let G be a group, and let R be a normal subgroup of a free group F with $F/R \cong G$. If M is an F-module, then*

$$\mathbb{Z} \otimes_G (M/\mathfrak{R}M) \cong M/\mathcal{F}M.$$

Proof. A G-module A may be regarded as an F-module annihilated by $\mathfrak{R}$; moreover, if $\pi f = g$ (where $\pi : F \to G$), then $ga = fa$ for all $a \in A$. Therefore, $(g - 1)a = (f - 1)a$, which implies that $\mathcal{G}A = \mathcal{F}A$. In particular,

$$\mathcal{G}(M/\mathfrak{R}M) = \mathcal{F}(M/\mathfrak{R}M) = \mathcal{F}M/\mathfrak{R}M.$$

By Proposition 9.46,

$$\mathbb{Z} \otimes_G (M/\mathfrak{R}M) \cong (M/\mathfrak{R}M)/\mathcal{G}(M/\mathfrak{R}M)$$
$$= (M/\mathfrak{R}M)/(\mathcal{F}M/\mathfrak{R}M) \cong M/\mathcal{F}M. \quad \bullet$$

Theorem 9.63 (Hopf's Formula). *Let G be a group, and let R be a normal subgroup of a free group F with $F/R \cong G$. Then*

$$H_2(G, \mathbb{Z}) \cong (R \cap F')/[F, R],$$

and so $(R \cap F')/[F, R]$ depends only on G and not on the choice of F and R.

Proof. Apply $\mathbb{Z} \otimes_G \square$ to the Gruenberg resolution

$$\to \mathfrak{R}F/\mathfrak{R}^2 F \xrightarrow{d_3} \mathfrak{R}/\mathfrak{R}^2 \xrightarrow{d_2} F/\mathfrak{R}F \to;$$

using Lemma 9.62, obtain the complex

$$\to \mathfrak{R}F/F\mathfrak{R}F \xrightarrow{\Delta_3} \mathfrak{R}/F\mathfrak{R} \xrightarrow{\Delta_3} F/F^2 \to$$

(the maps Δ_n are still enlargements of coset). Now

$$H_2(G, \mathbb{Z}) = \ker \Delta_2 / \operatorname{im} \Delta_3$$
$$= (\mathfrak{R}F^2)/(F\mathfrak{R} + \mathfrak{R}F)$$
$$= \ker(\mathfrak{R}/(F\mathfrak{R} + \mathfrak{R}F) \to F/F^2),$$

the last arrow being enlargement of coset. But $F/F^2 \cong F/F'$ and, by Lemma 9.61,

$$\frac{\mathfrak{R}}{F\mathfrak{R} + \mathfrak{R}F} \cong \frac{\mathfrak{R}/\mathfrak{R}F}{(F\mathfrak{R} + \mathfrak{R}F)/\mathfrak{R}F}$$
$$\cong \frac{R/R'}{[F, R]/R'} \cong R/[F, R].$$

We conclude that

$$H_2(G, \mathbb{Z}) \cong \ker(R/[F, R] \to F/F') = (R \cap F')/[F, R],$$

for the reader may check commutativity of the diagram

$$
\begin{array}{ccc}
\mathfrak{R}/(F\mathfrak{R} + \mathfrak{R}F) & \longrightarrow & F/F^2 \\
\downarrow & & \downarrow \\
R/[F, R] & \longrightarrow & F/F'.
\end{array} \quad \bullet
$$

Example 9.64. Let us compute $H_2(\mathbf{V}, \mathbb{Z})$, where $\mathbf{V} = \mathbb{I}_2 \oplus \mathbb{I}_2$ is the four-group, using the presentation

$$\mathbf{V} = (x_1, x_2 \mid x_1^2, x_2^2, [x_1, x_2]).$$

Here, F is free with basis $\{x_1, x_2\}$; the normal subgroup R generated by the relations turns out to have five generators, three of which are the displayed relations. [In contrast to free abelian groups, a basis of a subgroup can have larger cardinal than a basis of the big group. In fact, the commutator subgroup of a free group of rank 2 is a free group of infinite rank (Rotman, *An Introduction to the Theory of Groups*, Theorem 11.48).]

We are going to use Hopf's formula to show that $H_2(\mathbf{V}, \mathbb{Z}) \cong \mathbb{I}_2$. Set $K = F/[F, R]$. Since F/R is abelian, $F' \subseteq R$ and $R \cap F' = F'$. Therefore,

$$K' = F'/[F, R] = (R \cap F')/[F, R] \cong H_2(G, \mathbb{Z}).$$

Define $a = x[F, R]$ and $b = y[F, R]$, so that $K = \langle a, b \rangle$. If $L = \langle [a, b] \rangle$, then clearly $L \subseteq K'$; on the other hand, $L \lhd K$ and K/L is abelian, being generated by two commuting elements, so that $L \supseteq K'$. Therefore, $K' = L$ is cyclic with generator $[a, b]$.

We claim that $[a, b]^2 = 1$. Observe first that $F' \subseteq R$ implies that $[F', F] \subseteq [F, R]$, from which it follows that $K' \subseteq Z(K)$, the center of K; furthermore, $b^2 \in Z(K)$ (for $y^2 \in R$ and $[y^2, f] \in [R.F]$ for all $f \in F$). Hence,

$$b^2 = ab^2a^{-1} = (aba^{-1})^2 = ([a, b]b)^2$$
$$= [a, b]^2 b^2, \quad \text{since } [a, b] \in Z(K).$$

Canceling b^2 gives $[a, b]^2 = 1$, and so $|K'| \le 2$.

It remains to show that $K' \neq \{1\}$; that is, $K = F/[F, R]$ is not abelian. Consider the group $\mathbf{Q}$ of quaternions of order 8, with presentation

$$\mathbf{Q} = (x, y \mid x^2 = y^2, xyx = y).$$

Recall that $\mathbf{Q}$ has the following properties:

(i) if S is the normal subgroup of F generated by $x^2 y^{-2}$ and $xyxy^{-1}$, then $\mathbf{Q}$ is generated by $c = xS$ and $d = yS$;

(ii) $\langle c^2 \rangle$ has order 2;

(iii) $Z(\mathbf{Q}) = \langle c^2 \rangle = \mathbf{Q}'$.

It suffices to show that if $[F, R] \subseteq S$, then $F/[F, R]$ maps onto the nonabelian group $F/S = \mathbf{Q}$, for then $F/[F, R] = K$ is not abelian and $K' \neq \{1\}$. To see that $[F, R] \subseteq S$, it is enough to prove $[\bar{r}, \bar{f}] = 1$ in $\mathbf{Q}$ for every $f \in F$ and $r = x^2, y^2,$ or $[x, y]$, where bar means coset mod S. But this is true, for $\bar{x}^2 = c^2 = \bar{y}^2$ lie in $Z(\mathbf{Q})$, and $[\bar{x}, \bar{y}] \in \mathbf{Q}' = Z(\mathbf{Q})$.

Recall that if p is a prime, then an ***elementary abelian group*** of order p^n is the direct product of n copies of $\mathbb{I}_p$. Using spectral sequences, Theorem 10.55 generalizes this example by showing that if G is an elementary abelian p-group of order p^n, then $H_2(G, \mathbb{Z})$ is elementary abelian of order $p^{n(n-1)/2}$. We sketch another proof of this in the next example. ◀

Example 9.65. The multiplier of any abelian group G can be computed with *exterior algebra*.

Definition. If R is a commutative ring and G is an R-module, then its ***exterior square*** is

$$G \wedge G = (G \otimes_R G)/W,$$

where W is the submodule generated by all $\{g \otimes g : g \in G\}$.

Theorem. *If G is an abelian group, then $H_2(G, \mathbb{Z}) \cong G \wedge G$.*

Sketch of proof. Given a presentation $1 \to R \to F \to G \to 1$ of G, we have $F' \subseteq R$ because G is abelian. Now $H_2(G, \mathbb{Z}) \cong (R \cap F')/[F, R]) = F'/[F, R]$, by Hopf's Formula. But $\theta : (F/R) \times (F/R) \to F'/[F, R]$, given by $(f_1R, f_2R) \mapsto [f_1, f_2][F, R]$, is a bilinear function with $\theta(fR, fR) = 0$ for all $f \in F$. Thus, θ induces a homomorphism $\Theta : G \wedge G \to H_2(G, \mathbb{Z})$, and one then proves that Θ is an isomorphism by constructing its inverse (see Robinson, *A Course in the Theory of Groups*, p. 348). •

If P is a free R-module of rank n, then $P \wedge P$ is a free R-module of rank $\binom{n}{2}$ (see Rotman, *Advanced Modern Algebra*, p. 749). An elementary abelian p-group E of order p^n can be viewed as an n-dimensional vector space over $\mathbb{F}_p$. It follows that $H_2(E, \mathbb{Z})$ is an elementary abelian p-group of order $p^{n(n-1)/2}$. ◄

Definition. A group G is ***finitely presented*** if it has a finite presentation

$$(x_1, \ldots, x_n \mid y_1, \ldots, y_r).$$

There exist finitely generated groups that are not finitely presented. Obviously, there are only countably many finitely presented groups, but B. H. Neumann proved that there are uncountably many finitely generated groups (Rotman, *An Introduction to the Theory of Groups*, Theorem 11.73).

Notation. If A is a finitely generated abelian group, let

$$\rho(A) = \text{rank}(A) = \text{rank}(A/tA),$$

and let

$$d(A) = \text{the smallest cardinal of a generating set of } A.$$

Here are some facts about finitely generated abelian groups (which are easily verified using the Fundamental Theorem; in fact, (iii) and (iv) are true for arbitrary abelian groups).

(i) If A is a finitely generated abelian group, then $\rho(A) \leq d(A)$. Moreover, $\rho(A) = d(A)$ if and only if A is free abelian, while $\rho(A) = 0$ if and only if A is finite.

(ii) If A and A' are finitely generated abelian groups with A' free abelian, then $d(A \oplus A') = d(A) + d(A')$. This equality is not generally true: if $A = \mathbb{I}_2$ and $B = \mathbb{I}_3$, then $A \oplus B \cong \mathbb{I}_6$; thus, $d(A \oplus B) = 1$ and $d(A) + d(B) = 2$.

(iii) $\rho(A) = \dim_{\mathbb{Q}}(\mathbb{Q} \otimes A)$.

(iv) If $0 \to A' \to A \to A'' \to 0$ is exact, then

$$\rho(A) = \rho(A') + \rho(A'').$$

Lemma 9.66. *Let a group G have a presentation $(x_1, \ldots, x_n \mid y_1, \ldots, y_r)$, so that $G \cong F/R$, where F is free on $\{x_1, \ldots, x_n\}$ and R is the normal subgroup of F generated by $y_1, \ldots, y_r$. Then $R/[F, R]$ is a finitely generated abelian group, and*
$$d\big(R/[F, R]\big) \leq r.$$

Proof. Since $R \lhd F$, we see that $[F, R]$ is a normal subgroup of R containing $[R, R] = r'$. Therefore, $R/[F, R]$ is an abelian group. It suffices to show that $R/[F, R]$ is generated by the cosets of the ys. Every element of R is a product of elements of the form fsf^{-1}, where s lies in the subgroup generated by the ys. But $fsf^{-1}s^{-1} \in [F, R]$, so that $fsf^{-1} \equiv s \mod [F, R]$, and the result follows. ●

Theorem 9.67. *If a group G has a presentation $(x_1, \ldots, x_n \mid y_1, \ldots, y_r)$, then $H_2(G, \mathbb{Z})$ is finitely generated with $d(H_2(G, \mathbb{Z})) \leq r$ and*

$$n - r \leq \rho(G/G') - d(H_2(G, \mathbb{Z})).$$

Proof. Let F be free on $\{x_1, \ldots, x_n\}$, and let R be the normal subgroup of F generated by $\{y_1, \ldots, y_r\}$. We will use the descending chain of normal subgroups of F:

$$F \supseteq F'R \supseteq R \supseteq R \cap F' \supseteq [F, R].$$

By Lemma 9.66, $R/[F, R]$ is a finitely generated abelian group with at most r generators. There is an exact sequence

$$0 \to (R \cap F')/[F, R] \to R/[F, R] \to R/(R \cap F') \to 0. \qquad (2)$$

By Hopf's formula,

$$d(H_2(G, \mathbb{Z})) = d((R \cap F')/[F, R]) \leq r$$

(recall Corollary 4.15: if A is an abelian group with r generators, then every subgroup of A can be generated by r or fewer elements). Since F/F' is free abelian and $R/(R \cap F') \cong F'R/F' \subseteq F/F'$, the group $R/(R \cap F')$ is free abelian; hence, the exact sequence (2) splits. Using fact (ii), the additivity of d when one summand is free,

$$d(R/[F, R]) = d(R/(R \cap F')) + d((R \cap F')/[F, R])$$
$$= d(F'R/F') + d(H_2(G, \mathbb{Z})).$$

There is another exact sequence

$$0 \to F'R/F' \to F/F' \to F/F'R \to 0,$$

and fact (iv) gives $\rho(F/F') = \rho(F'R/F') + \rho(F/F'R)$. Since F/F' and its subgroup $F'R/F'$ are free abelian, however, we have

$$d(F'R/F') = d(F/F') - \rho(F'/F'R) = n - \rho(F/F'R).$$

We conclude that

$$r \geq d(R/[F, R]) = n - \rho(F/F'R) + d(H_2(G, \mathbb{Z})).$$

This completes the proof, because $F/F'R \cong G/G'$ (for $F'R$ is the subgroup of F mapping onto G' under the map $F \to G$). •

Definition.   A finite presentation $(x_1, \ldots, x_n \mid y_1, \ldots, y_r)$ is **balanced** if $n = r$. A group is **balanced** if it has a balanced presentation.

Corollary 9.68.   *If a group G is balanced, then*

$$d(H_2(G, \mathbb{Z})) \leq \rho(G/G').$$

In particular, if G is a finite balanced group, then $H_2(G, \mathbb{Z}) = \{0\}$.

It follows that $H_2(G, \mathbb{Z}) = \{0\}$ for every finite cyclic group G. On the other hand, we have seen that $H_2(\mathbf{V}, \mathbb{Z}) \neq \{0\}$, and so there is no balanced presentation of the four-group.

The converse of Corollary 9.68 is false; see Swan, "Minimal resolutions for finite groups," *Topology* 4 (1965), 193–208, for a counterexample. It is not known whether every p-group with a trivial Schur multiplier has a balanced presentation.

The next result shows that a finite group usually has more relations than generators.

Corollary 9.69. *If G is a finite group with a presentation having n generators and r relations, then $d(H_2(G, \mathbb{Z})) \le r - n$. Hence, $n \le r$.*

Given a group G, regard the cyclic group $\mathbb{I}_p$ of prime order p as a trivial G-module. Since multiplication by p is the zero map on $\mathbb{I}_p$, so is the induced map on $H_n(G, \mathbb{I}_p)$ for every n. Thus, $pH_n(G, \mathbb{I}_p) = \{0\}$, and so $H_n(G, \mathbb{I}_p)$ is a vector space over $\mathbb{F}_p$; let $\delta_n(G)$ be its dimension. By Exercise 9.24 on page 558, if G is a finite p-group, then $\delta_1(G) = d(G)$, the minimal number of generators of G. In view of Theorem 9.67, it is reasonable to expect that $\delta_2(G)$ somehow involves the number of relators of G.

Theorem (Golod–Šafarevič).   *If G is a finite p-group, then*

$$\delta_2(G) > \tfrac{1}{4}\delta_1(G)^2.$$

Proof.   Golod-Šafarevič, "On the class field tower," *Izv. Akad. Nauk SSSR* 28 (1964), 261–272. See Gruenberg, *Cohomological Topics in Group Theory*, p. 104. There is a proof of this inequality without Homological Algebra in Herstein, *Noncommutative Rings*, Chapter 8.   •

Thus, a criterion exists to determine whether a p-group G is finite. Golod and Šafarevič constructed a finitely generated p-group G violating their inequality, and concluded that G is an infinite finitely generated p-group. Burnside had proved 50 years earlier that a finitely generated subgroup of $GL(n, \mathbb{C})$ all of whose elements have finite order must be finite (it follows that the group of Golod–Šafarevič has no faithful finite-dimensional complex representation). **Burnside's problem** asks whether a finitely generated group B of **exponent** e (i.e., $x^e = 1$ for all $x \in B$) must be finite (there is no uniform bound on orders of elements in the group of Golod–Šafarevič). In 1968, Adjan and Novikov proved that the answer is negative when e is odd and sufficiently large; their proof is over 300 pages long. A. Ol'shanskii found a more elegant proof ["On the Novikov-Adyan theorem," *Mat. Sb.* 118 (1982); English translation: *Math USSR-Sb* 46 (1983), 203–236]. The solution of the Burnside problem was completed by S. V. Ivanov "The free Burnside groups of sufficiently large exponents," *Internat. J. Math* 4 (1994), 1–308, who showed that there exist infinite finitely generated groups of exponent $2^k m$ for all $k \ge 48$ and all odd $m \ge 1$. It is an open question whether a finitely presented group of finite exponent must be finite.

We are now going to see that the Schur multiplier arises as a cohomology group (indeed, this was Schur's original investigation ["Über die Darstellungen der endlichen Gruppen durch gebrochene lineare Substitutionen," *J. Reine Angew. Math.* 127 (1904), 20–50, "Untersuchungen über die Darstellungen der endlichen Gruppen durch gebrochene lineare Substitutionen," *J. Reine Angew. Math.* 132 (1907), 85–137]).

Definition.   An exact sequence $0 \to A \to E \to G \to 1$ of groups is a *central extension* of a group G if $A \subseteq Z(E)$. A *universal central extension* is a central extension $0 \to M \to U \to G \to 1$ for which there exists a commutative diagram whenever the bottom row is a central extension:

$$\begin{array}{ccccccccc} 0 & \longrightarrow & M & \longrightarrow & U & \longrightarrow & G & \longrightarrow & 1 \\ & & \downarrow & & \downarrow & & \downarrow{\scriptstyle 1_G} & & \\ 0 & \longrightarrow & A & \longrightarrow & E & \longrightarrow & G & \longrightarrow & 1. \end{array}$$

If G is a finite group, then G has a universal central extension if and only if G is *perfect*; that is, $G = G'$, in which case $M \cong H_2(G, \mathbb{Z})$ (see Milnor, *Introduction to Algebraic K-Theory*, pp 43–46). For example, nonabelian simple groups are perfect. What if G is not perfect?

Definition.   A *stem extension* of a group G is a central extension

$$1 \to A \to E \to G \to 1$$

in which $A \subseteq E'$ (Schur called such groups E "representation groups").
  A *stem cover* is a stem extension of the form

$$0 \to H_2(G, \mathbb{Z}) \to S \to G \to 1.$$

We have seen that $H_2(\mathbf{V}, \mathbb{Z}) \cong \mathbb{I}_2$, and the two nonabelian groups of order 8, namely, D_8 and $\mathbf{Q}$, give stem covers of $\mathbf{V}$. We will need two facts.

  (i) Each stem extension is a homomorphic image (a *homomorphism* here is an ordered triple of homomorphisms making the diagram commute) of of some stem cover (this is the analog of the diagrammatic property of the universal central extension when G is not perfect). (See Gruenberg, *Cohomological Topics in Group Theory*, p. 213.)

  (ii) If $1 \to A \to E \to G \to 1$ is a central extension in which A is divisible, then there exists a homomorphism of every stem cover of G into it. (See Gruenberg, *Cohomological Topics in Group Theory*, p. 216.)

  Representation Theory deals with homomorphisms $G \to \mathrm{GL}(n, k)$ where k is a field.

Definition.   The *projective general linear group* is

$$\mathrm{PGL}(n, k) = \mathrm{GL}(n, k)/Z(\mathrm{GL}(n, k)).$$

Note that $Z(\mathrm{GL}(n, k))$, the center of the matrix group $\mathrm{GL}(n, k)$, consists of all the nonzero scalar matrices, and so it is isomorphic to $k^{\times}$, the multiplicative group of k of nonzero elements of k.
  A *projective representation* of G is a homomorphism $G \to \mathrm{PGL}(n, k)$.

Now we prefer a representation of G, that is, a homomorphism $G \to$ $GL(n, k)$, but we may only have a projective representation. The next theorem says that a projective representation of a group G can be replaced by an "honest" representation for a price: G must be replaced by a stem cover of G.

Theorem 9.70 (Schur). *Let $\tau: G \to PGL(n, k)$ be a projective representation, where k is algebraically closed. If $0 \to M \to S \xrightarrow{\pi} G \to 1$ is a stem cover, then τ arises from a representation T of S; that is, there exist a homomorphism T and a commutative diagram*

$$\begin{array}{ccc} S & \xrightarrow{\pi} & G \\ {\scriptstyle T}\downarrow & & \downarrow{\scriptstyle \tau} \\ GL(n, k) & \longrightarrow & PGL(n, k). \end{array}$$

Sketch of Proof. By Lemma 7.29, there is a commutative diagram

$$\begin{array}{ccccccccc} 1 & \longrightarrow & k^\times & \longrightarrow & E & \longrightarrow & G & \longrightarrow & 1 \\ & & {\scriptstyle 1_{k^\times}}\downarrow & & {\scriptstyle \sigma}\downarrow & & \downarrow{\scriptstyle \tau} & & \\ 1 & \longrightarrow & k^\times & \longrightarrow & GL(n, k) & \longrightarrow & PGL(n, k) & \longrightarrow & 1. \end{array}$$

The top row is a central extension. Since k is algebraically closed, the group $k^\times$ is divisible (every element has an nth root for every $n > 0$), and so fact (ii) above gives a commutative diagram

$$\begin{array}{ccccccccc} 1 & \longrightarrow & H_2(G, \mathbb{Z}) & \longrightarrow & S & \longrightarrow & G & \longrightarrow & 1 \\ & & \downarrow & & {\scriptstyle \rho}\downarrow & & \downarrow{\scriptstyle 1_G} & & \\ 1 & \longrightarrow & k^\times & \longrightarrow & E & \longrightarrow & G & \longrightarrow & 1. \end{array}$$

The desired representation is $T = \sigma\rho: S \to GL(n, k)$. ●

Consider the diagram

$$\begin{array}{ccc} S & \xrightarrow{\pi} & G \\ {\scriptstyle T}\downarrow & {\scriptstyle L}\swarrow & \downarrow{\scriptstyle \tau} \\ GL(n, k) & \longrightarrow & PGL(n, k), \end{array}$$

where T is a representation of G arising from a projective representation τ. For each $x \in G$, choose a lifting $L(x) \in GL(n, k)$ of $\tau(x)$. A straightforward computation gives a function $f: G \times G \to k^\times$ with

$$L(xy) = f(x, y)L(x)L(y)$$

(we see now why such a *function* f is called a multiplier), and, using the fact that $k^\times$ is a trivial G-module, we can see that f is a factor set (this is why factor sets are so called). Moreover, if another set of liftings $L'(x)$ of $\tau(x)$ is chosen, then there is a factor set f' satisfying $L'(xy) = f'(x, y)L'(x)L'(y)$, and $f^{-1}f'$ is a coboundary (written multiplicatively). In other words, when k is algebraically closed, each projective representation τ gives an element of the *cohomology group* $H^2(G, k^{\times})$. Thus, it appears that there may be a connection between homology and cohomology.

We begin with a mixed identity, a variant of Lemma 3.55.

Lemma 9.71. *Let R and S be rings, and let $_RA$, $_RB_S$, C_S be modules, where A is finitely presented and C is injective. Then there is a natural isomorphism*

$$\mathrm{Hom}_S(B, C) \otimes_R A \cong \mathrm{Hom}_S(\mathrm{Hom}_R(A, B), C).$$

Proof. Note that the hypothesis makes the Homs into modules, so that the terms in the statement make sense. For $f \otimes a \in \mathrm{Hom}_S(B, C) \otimes_R A$, define $\sigma_A(f \otimes a) \in \mathrm{Hom}_S(\mathrm{Hom}_R(A, B), C)$ by $\sigma_A(f \otimes a) \colon g \mapsto f(g(a))$. It is straightforward to check, for any left R-module A, that $\sigma_A \colon \mathrm{Hom}_S(B, C) \otimes_R A \to \mathrm{Hom}_S(\mathrm{Hom}_R(A, B), C)$ is a homomorphism, natural in A. Moreover, if A is finitely generated free, then σ_A is an isomorphism.

As A is finitely presented, there is an exact sequence $F_1 \to F_0 \to A \to 0$ with F_1, F_0 finitely generated free. Consider the diagram in which we denote Hom_S and Hom_R by h_S and h_R, respectively.

$$
\begin{array}{ccccccc}
h_S(B, C) \otimes_R F_1 & \longrightarrow & h_S(B, C) \otimes_R F_0 & \longrightarrow & h_S(B, C) \otimes_R A & \to & 0 \\
\downarrow{\scriptstyle\sigma_{F_1}} & & \downarrow{\scriptstyle\sigma_{F_0}} & & \downarrow{\scriptstyle\sigma_A} & & \\
h_S(h_R(F_1, B), C) & \to & h_S(h_R(F_0, B), C) & \to & h_S(h_R(A, B), C) & \to & 0
\end{array}
$$

The top row is exact because tensor product is right exact; the bottom row is exact because $\mathrm{Hom}_R(\square, C)$ is (contravariant) exact (for C is injective) and $\mathrm{Hom}_R(\square, B)$ is (contravariant) left exact. The diagram commutes because of the naturality of σ. Finally, σ_{F_1} and σ_{F_0} are isomorphisms, because F_1, F_0 are finitely generated free, and so the Five Lemma says that σ_A is an isomorphism. •

Lemma 9.72. *Let R and S be rings with R left noetherian, and let $_RA$, $_RB_S$, C_S be modules, where A is finitely generated and C is injective. Then, for all $n \geq 0$, there are isomorphisms*

$$\mathrm{Tor}_n^R(\mathrm{Hom}_S(B, C), A) \cong \mathrm{Hom}_S(\mathrm{Ext}_R^n(A, B), C).$$

Proof. Since R is left noetherian and A is finitely generated, A is finitely presented. Thus, Lemma 9.71 applies to give a natural isomorphism

$$\mathrm{Hom}_S(B, C) \otimes_R A \cong \mathrm{Hom}_S(\mathrm{Hom}_R(A, B), C).$$

Let $\mathbf{P}_A = \rightarrow P_i \rightarrow P_{i-1} \rightarrow \cdots \rightarrow P_0 \rightarrow 0$ be a deleted projective resolution of A with each P_i finitely generated. Naturality of the isomorphism $\operatorname{Hom}_S(B, C) \otimes_R P_i \cong \operatorname{Hom}_S(\operatorname{Hom}_R(P_i, B), C)$ gives isomorphic complexes. Take homology; there are isomorphisms for all $n \geq 0$,

$$H_n(\operatorname{Hom}_S(B, C) \otimes_R \mathbf{P}_A) \cong \operatorname{Hom}_S(H_n(\operatorname{Hom}_R(\mathbf{P}_A, B)), C),$$

for H_n commutes with the contravariant exact (because C is injective) functor $\operatorname{Hom}_S(\square, C)$. The left-hand side is $\operatorname{Tor}_n^R(\operatorname{Hom}_S(B, C), A)$, and the right-hand side is $\operatorname{Hom}_S(\operatorname{Ext}_R^n(A, B), C)$. •

Theorem 9.73 (Duality Theorem). *For every finite group G, for every G-module B, and for all $n \geq 0$, there are isomorphisms*

$$H^n(G, B)^* \cong H_n(G, B^*),$$

where $B^* = \operatorname{Hom}_{\mathbb{Z}}(B, \mathbb{Q}/\mathbb{Z})$.

Proof. By Lemma 9.72, given rings R and S, there are isomorphisms for all $n \geq 0$,

$$\operatorname{Tor}_n^R(\operatorname{Hom}_S(B, C), A) \cong \operatorname{Hom}_S(\operatorname{Ext}_R^n(A, B), C).$$

Set $S = \mathbb{Z}$ and $R = \mathbb{Z}G$; note that $\mathbb{Z}G$ is left noetherian because G is finite. Setting $A = \mathbb{Z}$ and $C = \mathbb{Q}/\mathbb{Z}$ (which is $\mathbb{Z}$-injective), the isomorphism becomes

$$\operatorname{Tor}_n^{\mathbb{Z}G}(B^*, \mathbb{Z}) \cong \operatorname{Ext}_{\mathbb{Z}G}^n(\mathbb{Z}, B)^*.$$

Now $\operatorname{Tor}_n^{\mathbb{Z}G}(B^*, \mathbb{Z}) \cong \operatorname{Tor}_n^{\mathbb{Z}G}(\mathbb{Z}, B^*) = H_n(G, B^*)$, by Exercise 9.23 on page 558, while $\operatorname{Ext}_{\mathbb{Z}G}^n(\mathbb{Z}, B)^* = H^n(G, B)^*$. •

The next result explains why $H_2(G, \mathbb{Z})$ is called the *multiplier* (it is isomorphic to H^2 whose elements are essentially factor sets).

Theorem 9.74. *If G is a finite group, then*

$$H_2(G, \mathbb{Z}) \cong H^2(G, \mathbb{Q}/\mathbb{Z}),$$

where the abelian group $\mathbb{Q}/\mathbb{Z}$ is viewed as a trivial G-module.

Proof. Exactness of $0 \rightarrow \mathbb{Z} \rightarrow \mathbb{Q} \rightarrow \mathbb{Q}/\mathbb{Z} \rightarrow 0$ gives exactness of

$$H^2(G, \mathbb{Q}) \rightarrow H^2(G, \mathbb{Q}/\mathbb{Z}) \rightarrow H^3(G, \mathbb{Z}) \rightarrow H^3(G, \mathbb{Q}).$$

We claim that $H^2(G, \mathbb{Q}) = \{0\} = H^3(G, \mathbb{Q})$. Multiplication by $|G|$ is an isomorphism of $\mathbb{Q}$, and so its induced map on $H^n(G, \mathbb{Q})$, which is multiplication by $|G|$, is also an isomorphism. But $m H^n(G, \mathbb{Q}) = \{0\}$, by Proposition 9.40, so that $H^n(G, \mathbb{Q}) = \{0\}$. We conclude that

$$H^2(G, \mathbb{Q}/\mathbb{Z}) \cong H^3(G, \mathbb{Z}).$$

A similar argument gives an isomorphism in homology:

$$H_3(G, \mathbb{Q}/\mathbb{Z}) \cong H_2 G, \mathbb{Z}).$$

Applying the Duality Theorem gives

$$H^3(G, \mathbb{Z})^* \cong H_3(G, \mathbb{Z}^*).$$

But $H^3(G, \mathbb{Z})$ is finite, so that $H^3(G, \mathbb{Z})^* \cong H^3(G, \mathbb{Z})$. Also, $\mathbb{Z}^* \cong \mathbb{Q}/\mathbb{Z}$, so that $H^3(G, \mathbb{Z}) \cong H_3(G, \mathbb{Q}/\mathbb{Z})$. Assembling all the isomorphisms gives $H^2(G, \mathbb{Q}/\mathbb{Z}) \cong H_2(G, \mathbb{Z})$.   •

Exercises

*9.18   **(i)**   Let A and B be G-modules. Prove that $\mathrm{Hom}_{\mathbb{Z}}(A, B)$ is a G-module under diagonal action:

$$(g\varphi)(a) = g\varphi(g^{-1}a)$$

for all $\varphi \colon A \to B$, $g \in G$, and $a \in A$. Moreover, prove that $\mathrm{Hom}_G(A, B) = \mathrm{Hom}_{\mathbb{Z}}(A, B)^G$.

(ii)   Let A be a right $\mathbb{Z}G$-module, and let B be a (left) G-module. Prove that $A \otimes \mathbb{Z}B$ is a G-module under diagonal action:

$$g(a \otimes b) = ga \otimes ga$$

for all $g \in G$, $a \in A$, and $b \in B$.

9.19 Let G and Q be groups, and let $\varphi \colon \mathbb{Z}G \to \mathbb{Z}Q$ be a ring homomorphism.

(i)   Prove that if K is a Q-module, then φ equips K with the structure of a G-module (which we denote by $^{\varphi}K$).

Hint. See Proposition 2.1: if $\sigma \colon \mathbb{Z}Q \to \mathrm{End}(K)$, then $\varphi\sigma \colon \mathbb{Z}G \to \mathrm{End}(K)$.

(ii)   If G and Q are groups with isomorphic group rings, $\mathbb{Z}G \cong \mathbb{Z}Q$, prove that G and Q have the same homology and the same cohomology: for every G-module K, $H_n(G, {}^{\varphi}K) \cong H_n(Q, K)$ and $H^n(G, {}^{\varphi}K) \cong H^n(Q, K)$.

(iii)   If G and Q are abelian groups with isomorphic group rings, prove that $G \cong Q$.

Hint. $H_1(G, \mathbb{Z}) \cong H_1(Q, \mathbb{Z})$.

There exist nonisomorphic finite groups whose integral group rings are isomorphic [see M. Hertweck, "A counterexample to the isomorphism problem for integral group rings," *Annals Math* 154 (2001), 115–138].

*9.20 Let G be a group and, for $n \geq 1$, let B_n be the free G-module with basis $X = \{[x_0|\cdots|x_n] : x_i \in G\}$ (B_n is the nth term of the bar resolution). Prove that $(B_n)_G = B_n/\mathcal{G}B_n$ is the free $\mathbb{Z}$-module with basis X.

9.21 Give an example of a group G and a G-module A for which Proposition 9.53 is false; that is, $H_1(G, A) \not\cong H_1(G, \mathbb{Z}) \otimes_\mathbb{Z} A$.

9.22 Let G be a group, let k be a commutative ring, and let kG be the group algebra.

   (i) If A is a left kG-module, prove, for all $n \geq 0$, that

$$\mathrm{Ext}^n_{kG}(k, A) \cong \mathrm{Ext}^n_{\mathbb{Z}G}(\mathbb{Z}, A).$$

   Conclude that the cohomology groups $H^n(G, A)$ do not depend on the coefficient ring k.

   (ii) If A is a left kG-module, prove, for all $n \geq 0$, that

$$\mathrm{Tor}_n^{kG}(k, A) \cong \mathrm{Tor}_n^{\mathbb{Z}G}(\mathbb{Z}, A).$$

   Conclude that the homology groups $H_n(G, A)$ do not depend on the coefficient ring k.

*9.23   **(i)** Prove that $\varphi : \mathbb{Z}G \to (\mathbb{Z}G)^{\mathrm{op}}$, defined by $\varphi : \sum_x m_x x \mapsto \sum_x m_x x^{-1}$, is a ring isomorphism.

   (ii) Let A be a left $\mathbb{Z}G$-module and B be a right $\mathbb{Z}G$-module. Prove that $\mathrm{Tor}_n^{\mathbb{Z}G}(B, A) \cong \mathrm{Tor}_n^{\mathbb{Z}G}(A, B)$, where A, B are viewed [as in part (i)] as right and left $\mathbb{Z}G$-modules, respectively, in the second Tor. Compare with Theorem 7.1.

*9.24 For a group G and integer $m > 0$, view $\mathbb{I}_m$ as a trivial G-module. Prove that $H_1(G, \mathbb{I}_m) \cong G/G'G^m$, where G' is the commutator subgroup and G^m is the subgroup generated by all mth powers.
Hint. Consider the exact sequence of G-trivial modules

$$0 \to \mathbb{Z} \xrightarrow{m} \mathbb{Z} \to \mathbb{I}_m \to 0.$$

Remark. The *Frattini subgroup* $\Phi(G)$ of a group G is the intersection of its maximal subgroups. If p is a prime and G is a finite p-group, then $\Phi(G) = G'G^p$. The *Burnside Basis Theorem* says that $G/\Phi(G)$ is a vector space over $\mathbb{F}_p$ whose dimension is the cardinal of a minimal set of generators of G (see Rotman, *An Introduction to the Theory of Groups*, pp. 123–124). Compare this with the following. Recall that the *Jacobson radical* $J(R)$ of a ring R is the intersection of its maximal left ideals. If $(R, \mathfrak{m}, k)$ is a noetherian local ring, then $J(R) = \mathfrak{m}$, and $\mathfrak{m}/\mathfrak{m}^2$ is a vector space over k whose dimension is the cardinal of a minimal set of generators of $\mathfrak{m}$.   ◄

*9.25 Let F be a free group. If I and J are two-sided ideals in $\mathbb{Z}F$, which are free F-modules on U and V, respectively, prove that IJ is a free F-module with basis $UV = \{uv : u \in U, v \in V\}$.

9.26 If G is a group, prove that $P_n \cong \bigotimes^{n+1} \mathbb{Z}G$, where P_n is the nth term in the homogeneous resolution $\mathbf{P}(G)$ and

$$\overset{n}{\bigotimes} \mathbb{Z}G = \mathbb{Z}G \otimes_{\mathbb{Z}} \mathbb{Z}G \otimes_{\mathbb{Z}} \cdots \otimes_{\mathbb{Z}} \mathbb{Z}G,$$

the tensor product of $\mathbb{Z}G$ with itself n times.

9.27 If G is a finite cyclic group, prove, for all G-modules A and for all $n \geq 1$, that $H^n(G, A) \cong H_{n+1}(G, A)$.

9.28 Let G be a finite cyclic group, and let $0 \to A \to B \to C \to 0$ be an exact sequence of G-modules.

    (i) Prove that there is an *exact hexagon*:

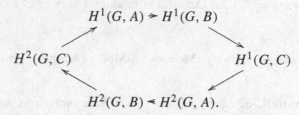

    (ii) Prove that if the Herbrand quotient is defined for two of the modules A, B, C [that is, both $H^1(G, M)$ and $H^2(G, M)$ are finite, where $M = A, B,$ or C], then it is defined for the third one, and

$$h(B) = h(A)h(C).$$

9.29 If $R = \mathbb{Z}[x]/(x^k - 1)$, prove that $\mathrm{D}(R) = \infty$ (where D is global dimension).

9.30 **(Barr–Rinehart)** For a group G, define $\widetilde{H}^n(G, A) = \mathrm{Ext}^n_{\mathbb{Z}G}(\mathcal{G}, A)$, where A is a G-module and $\mathcal{G}$ is the augmentation ideal. Prove that $\widetilde{H}^0(G, A) \cong \mathrm{Der}(G, A)$ and $\widetilde{H}^n(G, A) \cong H^{n+1}(G, A)$ for $n \geq 1$.

9.5 Change of Groups

Both the homology functors $H_q(G, \square) = \mathrm{Tor}^{\mathbb{Z}G}_q(\mathbb{Z}, \square)$ and the cohomology functors $H^q(G, \square) = \mathrm{Ext}^q_{\mathbb{Z}G}(\mathbb{Z}, \square)$ are covariant functors $_{\mathbb{Z}G}\mathbf{Mod} \to \mathbf{Ab}$; in this section, we shall see that they are also functors of the first variable. Interesting maps arise from group homomorphisms. We shall see that an exact sequence $1 \to S \to G \to G/S \to 1$ of groups gives an exact cohomology sequence $0 \to H^1(G/S, K^S) \to H^1(G, K) \to H^1(S, K)$, and this sequence will be used in §9.8 to prove a theorem of Gaschütz that finite nonabelian p-groups have outer automorphisms.

Definition. A G-module A is *G-acyclic* if $H^n(G, A) = \{0\}$ for all $n \geq 1$.

Of course, injective G-modules A are G-acyclic, but, since $H^n(G, A) = \text{Ext}_G^n(\mathbb{Z}, A)$, it is very likely that there are others [for injectivity is equivalent to $\text{Ext}_G^n(C, A) = \{0\}$ for every G-module C, not just for $C = \mathbb{Z}$].

We discussed the notion of *change of rings* in Chapter 8: if $\varphi: R' \to R$ is a ring homomorphism, then every left R-module A acquires a left R'-module structure by $r'a = \varphi(r')a$, and every R-map is an R'-map. Consider the following special case of the change of rings functor $U = U_\varphi: {}_R\textbf{Mod} \to {}_{R'}\textbf{Mod}$.

Definition. A group homomorphism $f: G' \to G$ induces a ring homomorphism $\mathbb{Z}G' \to \mathbb{Z}G$, also denoted by f, namely, $f: \sum n_{x'} x' \mapsto \sum n_{x'} f(x')$. By change of rings, every G-module A is a G'-module: if $x' \in G'$ and $a \in A$, then $x'a = f(x')a$. Denote a G-module A viewed as a G'-module by

$$UA = {}_fA,$$

and call $U = U_f: {}_{\mathbb{Z}G}\textbf{Mod} \to {}_{\mathbb{Z}G'}\textbf{Mod}$ a *change of groups functor*.

Lemma 9.75. *Let $f: G' \to G$ be a homomorphism, and let $U: {}_{\mathbb{Z}G}\textbf{Mod} \to {}_{\mathbb{Z}G'}\textbf{Mod}$ be the corresponding change of groups functor.*

(i) *If $\mathbf{P}$ is a G-acyclic complex, then $U\mathbf{P}$ is a G'-acyclic complex.*

(ii) *Let $S \subseteq G$, and let $f: S \to G$ be the inclusion. If P is a projective G-module, then UP is a projective S-module. Moreover, if $\mathbf{P}$ is a G-projective resolution of a G-module A, then $U\mathbf{P}$ is an S-projective resolution of UA.*

Proof.

(i) Proposition 8.33 says that $U: {}_{\mathbb{Z}G}\textbf{Mod} \to {}_{\mathbb{Z}G'}\textbf{Mod}$ is an exact additive functor, and so U preserves exact sequences.

(ii) If T is a right transversal of S in G, then G is the disjoint union $\bigcup_{t \in T} St$; thus, every $x \in G$ has a unique expression of the form $x = st$, where $s \in S$ and $t \in T$. As an S-module, $\mathbb{Z}G$ is free, for $\mathbb{Z}G = \bigoplus_{t \in T} (\mathbb{Z}S)t$ is a direct sum of S-modules. It follows that if P is a projective S-module, then $UP = {}_fP$ is a projective S-module. The second statement now follows from part (i). •

Recall that if S is a subgroup of a group G and A is an S-module, then $\text{Hom}_S(\mathbb{Z}G, A)$ is a (left) G-module by *diagonal action*: if $y \in G$ and $g: \mathbb{Z}G \to A$, define

$$yg: x \mapsto yg(y^{-1}x).$$

In particular, if $S = \{1\}$, then $\mathbb{Z}S = \mathbb{Z}$ and $\mathrm{Hom}_{\mathbb{Z}}(\mathbb{Z}G, A)$ is a G-module.

If S is a subgroup of a group G, we regard $\mathbb{Z}G$ as a $(\mathbb{Z}G, \mathbb{Z}S)$-bimodule, as usual, by letting S act as right multiplication. If A is an S-module, then $\mathbb{Z}G \otimes_S A$ is a G-module with *diagonal action*:

$$y(x \otimes a) = (yx) \otimes ya.$$

In particular, if $S = \{1\}$, then $\mathbb{Z}G \otimes_{\mathbb{Z}} A$ is a G-module.

Proposition 9.76 (Eckmann–Shapiro Lemma). *Let G be a group, let $S \subseteq G$ be a subgroup, and let A be an S-module.*

(i) $H^n(S, A) \cong H^n(G, \mathrm{Hom}_S(\mathbb{Z}G, A))$ *for all $n \geq 0$.*

(ii) $H_n(S, A) \cong H_n(G, \mathbb{Z}G \otimes_S A)$ *for all $n \geq 0$.*

Proof.

(i) If $\to P_1 \to P_0 \to \mathbb{Z} \to 0$ is a G-free resolution, then

$$H^n(G, \mathrm{Hom}_S(\mathbb{Z}G, A)) = H^n(\mathrm{Hom}_G(\mathbf{P}, \mathrm{Hom}_S(\mathbb{Z}G, A))).$$

By the adjoint isomorphism, we have, for all i,

$$\mathrm{Hom}_G(P_i, \mathrm{Hom}_S(\mathbb{Z}G, A)) \cong \mathrm{Hom}_S(P_i \otimes_G \mathbb{Z}G, A) \cong \mathrm{Hom}_S(U P_i, A),$$

where $U : {}_{\mathbb{Z}G}\mathbf{Mod} \to {}_{\mathbb{Z}S}\mathbf{Mod}$ is the change of groups functor. But $U\mathbf{P}$ is an S-projective resolution of $\mathbb{Z}$, by Lemma 9.75(ii). Thus, there is an isomorphism of complexes

$$\mathrm{Hom}_S(\mathbf{P}, A) \cong \mathrm{Hom}_G(U\mathbf{P}, \mathrm{Hom}_S(\mathbb{Z}G, A)),$$

and isomorphisms in homology: $H^n(S, A) \cong H^n(G, \mathrm{Hom}_S(\mathbb{Z}G, A))$.

(ii) By definition, $H_n(G, \mathbb{Z}G \otimes_S A) = H_n(\mathbf{P} \otimes_G (\mathbb{Z}G \otimes_S A))$. Since $\mathbb{Z}G$ is a $(\mathbb{Z}G, \mathbb{Z}S)$-bimodule, the associativity of tensor product (Proposition 2.57) gives $P_i \otimes_G (\mathbb{Z}G \otimes_S A) \cong P_i \otimes_S A$ for all i. But $U\mathbf{P}$ is an S-projective resolution of $\mathbb{Z}$, by Lemma 9.75(ii), and $H_n(U\mathbf{P} \otimes_S A) = H_n(S, A)$. •

Definition. If G is a group, then a G-module of the form $\mathrm{Hom}_{\mathbb{Z}}(\mathbb{Z}G, A)$, where A is an abelian group, is called a ***coinduced module***. A G-module of the form $\mathbb{Z}G \otimes_{\mathbb{Z}} A$ is called an ***induced module***.

Note that every free G-module F is induced, for if X is a basis of F, then $F = \mathbb{Z}G \otimes_{\mathbb{Z}} A$, where A is the free abelian group with basis X.

Proposition 9.77.

(i) *Every coinduced G-module A is G-acyclic:* $H^n(G, A) = \{0\}$ *for all $n \geq 1$; moreover, every G-module can be imbedded in a coinduced module.*

(ii) *Every induced module B satisfies $H_n(G, B) = \{0\}$ for all $n \geq 1$; moreover, every G-module is a quotient of an induced module.*

Proof.

(i) If $S = \{1\}$, the Eckmann–Shapiro Lemma gives

$$H^n(G, \operatorname{Hom}_{\mathbb{Z}}(\mathbb{Z}G, A)) \cong H^n(\{1\}, A).$$

But $H^n(\{1\}, A) = \{0\}$ for all $n \geq 1$, by Corollary 9.28. The proof of Theorem 3.38 gives an imbedding $\varphi \colon A \to \operatorname{Hom}_{\mathbb{Z}}(\mathbb{Z}G, A)$ [if $a \in A$, define $\varphi(a) \colon x \mapsto xa$].

(ii) The vanishing of $H_n(G, B)$ when B is induced is the dual of that just given for cohomology. Moreover, the map $\mathbb{Z}G \otimes_{\mathbb{Z}} A \to A$, given by $x \otimes a \mapsto xa$, is a surjection.   •

We can now give another, equivalent, description of group cohomology and group homology.

Theorem 9.78.

(i) *Given a group G, the sequence $(H^n(G, \square))_{n \geq 0}$ is the cohomological extension of Fix^G for which $H^n(G, A) = \{0\}$ for all coinduced G-modules A and all $n \geq 1$.*

(ii) *Given a group G, the sequence $(H_n(G, \square))_{n \geq 0}$ is the homological extension of $\mathbb{Z} \otimes_G \square$ for which $H_n(G, B) = \{0\}$ for all induced G-modules B and all $n \geq 1$.*

Proof.

(i) By Proposition 9.22, we have $\operatorname{Fix}^G \cong \operatorname{Hom}_G(\mathbb{Z}, \square)$, while Proposition 9.77 says that $_G\mathbf{Mod}$ has enough co-$\mathcal{Y}$-objects, where $\mathcal{Y}$ is the class of all coinduced G-modules. Hence, $(R^n\operatorname{Fix}^G)_{n \geq 0}$ is a cohomological extension of Fix^G. On the other hand, $(\operatorname{Ext}^n(G, \square))_{n \geq 0}$ is also a cohomological extension of Fix^G. Therefore, the uniqueness assertion of Corollary 6.50 gives the result.

(ii) Similar to part (i); Corollary 6.35 applies, where $\mathcal{X}$ is the class of all induced G-modules.   •

When G is finite, coinduced modules and induced modules coincide.

Proposition 9.79. *If G is finite and A is an abelian group, then there is a natural G-isomorphism*

$$\theta : \operatorname{Hom}_{\mathbb{Z}}(\mathbb{Z}G, A) \to \mathbb{Z}G \otimes_{\mathbb{Z}} A$$

given by $\theta : f \mapsto \sum_{x \in G} x \otimes f(x)$.

Remark. Of course, finiteness of G is needed for θ to be defined.

We can adapt the proof to show that if $S \subseteq G$ is a subgroup of finite index and A is an S-module, then there is a G-isomorphism $\operatorname{Hom}_S(\mathbb{Z}G, A) \to \mathbb{Z}G \otimes_S A$. ◄

Proof. It is obvious that $\theta(f + g) = \theta(f) + \theta(g)$. To see that θ is a G-map, we use the diagonal actions. Let $y \in G$ and $f : \mathbb{Z}G \to A$; then

$$\theta(yf) = \sum_{x \in G} x \otimes (yf)(x) = \sum_{x \in G} x \otimes yf(y^{-1}x).$$

Change variables: $z = y^{-1}x$, and note that as x varies over G, so does z. Hence,

$$\sum_{x \in G} x \otimes yf(y^{-1}x) = \sum_{z \in G} yz \otimes yf(z) = y \sum_{z \in G} z \otimes f(z) = y\theta(f).$$

We show that θ is injective. Now $\mathbb{Z}G \otimes_{\mathbb{Z}} A = \bigoplus_{x \in G} A_x$, where A_x is the subgroup $\{x \otimes a : a \in A\}$. If $\theta(f) = \sum_{x \in G} x \otimes f(x) = 0$, then $f(x) = 0$ for all $x \in G$, and so $f = 0$. Finally, we show that θ is surjective. Let $u = \sum_{x \in G} x \otimes a_x \in \mathbb{Z}G \otimes_{\mathbb{Z}} A$. Since $\mathbb{Z}G$ is the free abelian group with basis G, there is a $\mathbb{Z}$-map $f : \mathbb{Z}G \to A$ with $f(x) = a_x$ for all $x \in G$, and $\theta(f) = u$.

To prove naturality, we must show that the following diagram commutes for every map $\varphi : A \to B$ of abelian groups.

$$
\begin{array}{ccc}
\operatorname{Hom}_{\mathbb{Z}}(\mathbb{Z}G, A) & \xrightarrow{\varphi_*} & \operatorname{Hom}_{\mathbb{Z}}(\mathbb{Z}G, B) \\
\Big\downarrow{\scriptstyle \theta} & & \Big\downarrow{\scriptstyle \theta} \\
\mathbb{Z}G \otimes_{\mathbb{Z}} A & \xrightarrow{1 \otimes \varphi} & \mathbb{Z}G \otimes_{\mathbb{Z}} A
\end{array}
$$

If $f : \mathbb{Z}G \to A$, then going clockwise, $f \mapsto \varphi f \mapsto \sum x \otimes (\varphi)f(x)$; going counterclockwise, $f \mapsto \sum x \otimes f(x) \mapsto \sum x \otimes \varphi(f(x))$. •

Corollary 9.80. *Let S be a subgroup of a finite group G, and let A be a G-module.*

(i) $\operatorname{Hom}_{\mathbb{Z}}(\mathbb{Z}G, A)$ *is S-coinduced.*

(ii) *If S is normal, then* $\mathrm{Hom}_{\mathbb{Z}}(\mathbb{Z}G, Z)^S \cong \mathrm{Hom}_{\mathbb{Z}}(\mathbb{Z}(G/S), A)$*, and so* $\mathrm{Hom}_{\mathbb{Z}}(\mathbb{Z}G, Z)^S$ *is G/S-coinduced.*

Proof.

(i) The following isomorphisms hold because Hom commutes with finite direct sums in either variable.

$$\mathrm{Hom}_{\mathbb{Z}}(\mathbb{Z}G, A) \cong \mathrm{Hom}_{\mathbb{Z}}(\oplus \mathbb{Z}S, A)$$
$$\cong \oplus \mathrm{Hom}_{\mathbb{Z}}(\mathbb{Z}S, A)$$
$$\cong \mathrm{Hom}_{\mathbb{Z}}(\mathbb{Z}S, \oplus A).$$

(ii) Since G is finite, it suffices to show that $(\mathbb{Z}G \otimes_{\mathbb{Z}} A)^S$ is G/S-induced. As each $u \in (\mathbb{Z}G \otimes_{\mathbb{Z}} A)^S$ has a unique expression of the form $u = \sum_{x \in G} n_x x$, the result will follow if $n_{sx} = n_x$ for all $x \in G$ and $s \in S$ (that is, if the coefficients are constant on cosets of S). Since $s^{-1}u = u$, we have

$$u = \sum n_x x = \sum n_x s^{-1} x = \sum n_{sx} x.$$

Uniqueness of expression gives $n_x = n_{sx}$, as desired.   •

9.5.1 Restriction and Inflation

We now show that group cohomology and group homology are functorial in the first variable G.

Definition.   Let G', G be groups and let $\alpha: G' \to G$ be a homomorphism. If A' is a G'-module and $f: A \to A'$ is a $\mathbb{Z}$-map, then we call (α, f) a *cocompatible pair* if $f: {}_\alpha A \to A'$ is a G'-map, where ${}_\alpha A$ denotes A made into a G'-module as on page 560; that is,

$$f((\alpha\, x')a) = x' f(a).$$

Define a category **Pairs*** with objects all ordered pairs (G, A) (where G is a group and A is a G-module), with morphisms $(G, A) \to (G', A')$ being cocompatible pairs (α, f), and with composition $(\beta, g)(\alpha, f) = (\alpha\beta, gf)$ [of course, $1_{(G,A)} = (1_G, 1_A)$]. Note that composition in **Pairs*** makes sense: if $\beta: G'' \to G'$ and $g: {}_\beta A' \to A''$, then $gf: A \to A''$ is a $\mathbb{Z}$-map and $gf(\alpha\beta(x'')a) = g((\beta x'')(fa)) = x'' gf(a)$.

Given a cocompatible pair $(\alpha, f): (G, A) \to (G', A')$ [so $\alpha: G' \to G$ and $f: {}_\alpha A \to A'$], let $U: {}_{\mathbb{Z}G}\mathbf{Mod} \to {}_{\mathbb{Z}G'}\mathbf{Mod}$ be the change of groups functor, let $\mathbf{P}'$ be a G'-projective resolution of $\mathbb{Z}$, and let $\mathbf{P}$ be a G-projective resolution of $\mathbb{Z}$. Since U is a functor, there is a map

$$\mathrm{Hom}_G(\mathbf{P}, A) \to \mathrm{Hom}_{G'}(U\mathbf{P}, UA) = \mathrm{Hom}_{G'}(U\mathbf{P}, {}_\alpha A).$$

The G'-complex $U\mathbf{P}$ is acyclic, by Lemma 9.75(i), and so the Comparison Theorem (Theorem 6.16) gives a chain map $\tau(\alpha)\colon \mathbf{P}' \to U\mathbf{P}$ over $1_\mathbb{Z}$ that is unique to homotopy. Apply the (contravariant) functor $\operatorname{Hom}_{G'}(\square, {}_\alpha A)$ to obtain a chain map

$$\tau^{\#}(\alpha)\colon \operatorname{Hom}_{G'}(U\mathbf{P}, {}_\alpha A) \to \operatorname{Hom}_{G'}(\mathbf{P}', {}_\alpha A).$$

Next, $f\colon {}_\alpha A \to A'$ induces a chain map

$$f_{\#}\colon \operatorname{Hom}_{G'}(\mathbf{P}', {}_\alpha A) \to \operatorname{Hom}_{G'}(\mathbf{P}', A').$$

Thus, we have

$$\operatorname{Hom}_G(\mathbf{P}, A) \xrightarrow{U} \operatorname{Hom}_{G'}(U\mathbf{P}, {}_\alpha A) \xrightarrow{f_{\#}} \operatorname{Hom}_{G'}(\mathbf{P}', {}_\alpha A) \xrightarrow{\tau^{\#}(\alpha)} \operatorname{Hom}_{G'}(\mathbf{P}', A'),$$

and we define

$$(\alpha, f)^*\colon H^n(G, A) \to H^n(G', A')$$

to be the map in homology induced by this composite. If $\zeta \in \operatorname{Hom}_G(P_n, A)$ is a cocycle, where $\mathbf{P} = \to P_n \to P_{n-1} \to \cdots \to P_0 \to \mathbb{Z} \to 0$, then

$$(\alpha, f)^*\colon \operatorname{cls}(\zeta) \mapsto \operatorname{cls}(f\zeta\tau(\alpha)).$$

There is a "ladder" theorem for these induced maps.

Proposition 9.81. *Let $\alpha\colon G' \to G$ be a group homomorphism, and consider a commutative diagram*

$$
\begin{array}{ccccccccc}
0 & \longrightarrow & A & \longrightarrow & B & \longrightarrow & C & \longrightarrow & 0 \\
& & \downarrow{\scriptstyle(\alpha, f)} & & \downarrow{\scriptstyle(\alpha, g)} & & \downarrow{\scriptstyle(\alpha, h)} & & \\
0 & \longrightarrow & A' & \longrightarrow & B' & \longrightarrow & C' & \longrightarrow & 0,
\end{array}
$$

where the top row is an exact sequence of G-modules, the bottom row is an exact sequence of G'-modules, and the vertical arrows are cocompatible. Then there is a commutative diagram with exact rows

$$
\begin{array}{ccccccc}
\longrightarrow H^n(G, C) & \xrightarrow{\delta} & H^{n+1}(G, A) & \longrightarrow & H^{n+1}(G, B) & \longrightarrow & H^{n+1}(G, C) \\
\downarrow{\scriptstyle(\alpha, h)^*} & & \downarrow{\scriptstyle(\alpha, f)^*} & & \downarrow{\scriptstyle(\alpha, g)^*} & & \downarrow{\scriptstyle(\alpha, h)^*} \\
\longrightarrow H^n(G', C') & \xrightarrow{\delta'} & H^{n+1}(G', A') & \longrightarrow & H^{n+1}(G', B') & \longrightarrow & H^{n+1}(G', C').
\end{array}
$$

Proof. The proof is left to the reader. $\bullet$

Here are three important special cases of induced maps $(\alpha, f)^*$.

Definition. Let S be a subgroup of a group G, let A be a G-module, and let α be the inclusion $i: S \to G$. Then $(i, 1_A)$ is a cocompatible pair: if $x' \in S$, then $i(\alpha(x')a) = x'a = x'i(a)$. The induced homomorphism $(i, 1_A)^*$ is called **restriction** and is denoted by

$$\mathrm{Res}^n: H^n(G, A) \to H^n(S, A)$$

(it is customary to write A instead of $_iA$).

If ζ is an n-cocycle, then $\mathrm{Res}^n: \mathrm{cls}(z) \mapsto \mathrm{cls}(\zeta\tau(\alpha))$. The homogeneous resolution $\mathbf{P}(S) = (P_n(S))$ of $\mathbb{Z}$ is a subcomplex of the homogeneous resolution $\mathbf{P}(G)$ of $\mathbb{Z}$, and we may choose $\tau(\alpha)$ to be the inclusion $\mathbf{P}(S) \to \mathbf{P}(G)$. Thus,

$$\mathrm{Res}^n: \mathrm{cls}(\zeta) \mapsto \mathrm{cls}(\zeta i), \tag{1}$$

where $\zeta: P_n(G) \to A$ and ζi is the restriction $\zeta | P_n(S)$.

Definition. Given a normal subgroup S of a group K', let $\pi: K' \to K'/S$ be the natural map. If B' is a K'-module, view $(B')^S$ as a (K'/S)-module by $(k'S)b = k'b$ for all $k' \in K'$ and $b \in (B')^S$; this action is well-defined, for if $k'S = k''S$, then $k'' = k's$ for some $s \in S$, and so $k''b = k'sb = k'b$. If $f: A^S \to A$ is the inclusion, then (π, f) is a cocompatible pair: if $b \in (B')^S$, then $f(\pi(k')b) = (k'S)b = k'b = k'f(b)$.

The induced homomorphism $(\pi, f)^*: H^n(K'/S, (B')^S) \to H^n(K', B')$ is called **inflation**. In more customary notation, with $K' = G$ and $B' = A$, we have

$$\mathrm{Inf}^n: H^n(G/S, A^S) \to H^n(G, A).$$

The homogeneous resolution $\mathbf{P}(G/S)$ of $\mathbb{Z}$ is a quotient complex of the homogeneous resolution $\mathbf{P}(G)$ of $\mathbb{Z}$, and we may choose $\tau(\alpha)$ to be the natural map. If ζ is an n-cocycle, then $\mathrm{Inf}^n: H^n(G/S, A^S) \to H^n(G, A)$ is given by

$$\mathrm{Inf}^n: \mathrm{cls}(\zeta) \mapsto \mathrm{cls}(\zeta\pi), \tag{2}$$

where $\zeta\pi(x_0, \dots, x_n) = \zeta(x_0S, \dots, x_nS)$. In words, evaluate $\zeta\pi(x_0, \dots, x_n)$ by letting its value depend, for all coordinates j, only on the coset x_jS; thus, ζ has been "inflated" from a function $(G/S) \times \cdots \times (G/S) \to A$ to a function $G \times \cdots \times G \to A$ by letting it be constant on cosets of S.

Conjugation gives another important example.

Definition. Let $G' = G$ and let $\alpha: G \to G$ be γ_y, conjugation by y, where $y \in G$; that is, $\alpha = \gamma_y: x \mapsto yxy^{-1}$. If A is a G-module, define $f =$

$f_y \colon A \to A$ to be the $\mathbb{Z}$-map $f_y(a) = y^{-1}a$. Now (γ_y, f_y) is a cocompatible pair: if $x \in G$, then

$$f_y((\gamma_y x)a) = f_y(yxy^{-1}a) = y^{-1}yxy^{-1}a = xy^{-1}a = xf_y(a).$$

If $\mathbf{P}(G)$ is the homogeneous resolution of $\mathbb{Z}$, then the terms in $U\mathbf{P}$ are $_\alpha P_n(G)$, where $\alpha(x)u = yxy^{-1}u$ for all $u \in P_n(G)$. Thus, if $v = (x_0, \dots, x_n)$ is a basis element of $P_n(G)$, then $xv = (xx_0, \dots, xx_n)$; however, if we view $(x_0, \dots, x_n)$ as an element of $_\alpha P_n(G)$, then

$$x \cdot (x_0, \dots, x_n) = yxy^{-1}(x_0, \dots, x_n) = (yxy^{-1}x_0, \dots, yxy^{-1}x_n).$$

Define $\tau \colon \mathbf{P}(G) \to U\mathbf{P}(G)$ by $\tau(x_0, \dots, x_n) = (yx_0, \dots, yx_n)$. Note that $\tau \colon P_n(G) \to {}_\alpha P_n(G)$ is a G-map:

$$\tau(x(x_0, \dots, x_n)) = \tau(xx_0, \dots, xx_n) = (yxx_0, \dots, yxx_n),$$

and

$$
\begin{aligned}
x \cdot \tau(x_0, \dots, x_n) &= yxy^{-1}\tau_n(x_0, \dots, x_n) \\
&= yxy^{-1}(yx_0, \dots, yx_n) \\
&= (yxy^{-1}yx_0, \dots, yxy^{-1}yx_n) \\
&= (yxx_0, \dots, yxx_n).
\end{aligned}
$$

It is easy to see that τ is a chain map over $1_{\mathbb{Z}}$. Thus, the map $(\gamma_y, f_y)^*$ induced by conjugation γ_y and $f_y \colon a \mapsto y^{-1}a$ is

$$\mathrm{cls}(\zeta) \mapsto \mathrm{cls}(f_y \zeta \tau) = \mathrm{cls}(y^{-1}\zeta y). \tag{3}$$

If $S \lhd G$, then we can use conjugation to make $H^n(S, A)$ into a (G/S)-module.

Lemma 9.82. *Let G be a group and let A be a G-module.*

(i) *If $y \in G$ and $f \colon a \mapsto y^{-1}a$, then $(\gamma_y, f_y)^* \colon \mathrm{cls}(\zeta) \mapsto \mathrm{cls}(y^{-1}\zeta y)$ is the identity map on $H^n(G, A)$.*

(ii) *If S is a normal subgroup of G, then $H^n(S, A)$ is a G/S-module. Moreover, if A is a trivial G-module, then $H^n(S, A)$ is a trivial G/S-module.*

Proof.

(i) We have just seen that $\gamma_y(\mathrm{cls}(\zeta)) = \mathrm{cls}(y^{-1}\zeta y)$. But

$$
\begin{aligned}
y^{-1}\zeta y(x_0, \dots, x_n) &= (y^{-1}\zeta)(yx_0, \dots, yx_n) \\
&= \zeta(y^{-1}yx_0, \dots, y^{-1}yx_n) \\
&= \zeta(x_0, \dots, x_n).
\end{aligned}
$$

Thus, γ_y fixes every $\mathrm{cls}(\zeta)$, and so γ_y is the identity on $H^n(G, A)$.

(ii) If S is any (not necessarily normal) subgroup of G, and if $\rho_y = \gamma_y | S$, then $\rho_y \colon S \to ySy^{-1}$. If A is a (ySy^{-1})-module, then $f_y \colon A \to A$ is given by $f_y(a) = y^{-1}a$, and (ρ_y, f_y) is a cocompatible pair, because $\rho_y = \alpha | S$. Of course, $(\rho_y, f_y)^* \colon H^n(ySy^{-1}, A) \to H^n(S, A)$. If now $S \lhd G$, then $ySy^{-1} = S$ and $(\rho_y, f_y)^* \colon H^n(S, A) \to H^n(S, A)$. Define a (G/S)-action on $H^n(S, A)$ by

$$(yS)\,\mathrm{cls}(\zeta) = (\rho_y, f_y)^*\,\mathrm{cls}(\zeta)$$

for all $y \in G$ and $\mathrm{cls}(\zeta) \in H^n$. If $yS = y'S$, then $y' = ys$ for some $s \in S$, and $(\rho_{y'}, f_{y'})^* = (\rho_{ys}, f_{ys})^* = (\rho_y, f_y)^*(\rho_s, f_s)^*$. But $(\rho_s, f_s)^*$ is the identity, by (i), and so $(\rho_{y'}, f_{y'})^* = (\rho_y, f_y)^*$. Hence, this action is well-defined, and $H^n(S, A)$ is a (G/S)-module.

If A is a trivial G-module, then

$$(yS)\,\mathrm{cls}(\zeta) = (\rho_y, f_y)^*\,\mathrm{cls}(\zeta) = \mathrm{cls}(y^{-1}\zeta y) = \mathrm{cls}(\zeta),$$

and so yS acts trivially. $\bullet$

Corollary 9.83.   *If S is a normal subgroup of a group G and if A is a G-module, then, for all $n \geq 0$,*

$$\mathrm{im}\,\mathrm{Res}^n \subseteq H^n(S, A)^{G/S}.$$

Proof.   If $\zeta \in \mathrm{Der}(G, A)$, then $\mathrm{Res}^n(\mathrm{cls}(\zeta)) = \mathrm{cls}(\zeta | S)$. Hence, for $y \in G$, we have $(\rho_y, f_y)^*\,\mathrm{cls}(\zeta | S) = \mathrm{cls}(y^{-1}(\zeta | S)y)$. But

$$
\begin{aligned}
y^{-1}(\zeta | S)y(s_0, \ldots, s_n) &= y^{-1}\zeta(ys_0, \ldots, ys_n) \\
&= \zeta(y^{-1}ys_0, \ldots, y^{-1}ys_n) \\
&= \zeta(s_0, \ldots, s_n).   \quad \bullet
\end{aligned}
$$

Theorem 9.84.   *Let S be a normal subgroup of a group G, and let A be a G-module.*

(i) *There is an exact sequence*

$$0 \to H^1(G/S, A^S) \xrightarrow{\mathrm{Inf}^1} H^1(G, A) \xrightarrow{\mathrm{Res}^1} H^1(S, A)^{G/S} \xrightarrow{d} H^2(G/S, A^S).$$

(ii) *If $q \geq 1$ and $H^i(S, A) = \{0\}$ for all $1 \leq i \leq q - 1$, then there is an exact sequence*

$$0 \to H^q(G/S, A^S) \xrightarrow{\mathrm{Inf}^q} H^q(G, A) \xrightarrow{\mathrm{Res}^q} H^q(S, A)^{G/S}.$$

(iii) *If $q \geq 1$ and $H^i(S, A) = \{0\}$ for all $1 \leq i \leq q - 1$, then*

$$\mathrm{Inf}^i \colon H^i(G/S, A^S) \to H^i(G, A)$$

is an isomorphism for all $i \leq q - 1$.

Remark. In Theorem 10.53, which uses spectral sequences, we shall see that the exact sequence can be lengthened to a five-term sequence that ends with

$$\rightarrow H^1(S, A)^{G/S} \xrightarrow{d} H^2(G/S, A^S) \xrightarrow{\text{Inf}^2} H^2(G, A).$$

We do not prove exactness involving the map $d: H^1(S, A)^G \rightarrow H^2(G/S, A^S)$ (indeed, we do not even define d here), but there is a proof in Suzuki, *Group Theory* I, pp. 214–216. We do prove exactness of the rest of the sequence, however, using a proof of Serre, *Local Fields*, pp. 125–126. ◄

Proof.

(i) Corollary 9.83 says that $\text{im Res}^1 \subseteq H^1(S, A)^{G/S}$, and so it suffices to prove exactness when the third term is $H^1(S, A)$. We use the interpretation of H^1 as Der/PDer (formally, we are using the normalized bar resolution).

Inf^1 is an injection. If $\zeta: G/S \rightarrow A^S$ is a derivation and $\dot{\zeta}: G \rightarrow A$ is the function obtained from ζ by making it constant on cosets of S, then $\text{Inf}^1: \text{cls}(\zeta) \mapsto \text{cls}(\dot{\zeta})$. If $\text{cls}(\zeta) \in \ker \text{Inf}^1$, then $\dot{\zeta}$ is a principal derivation; that is, there is $a \in A$ with $\dot{\zeta}(x) = xa - a$ for all $x \in G$. Since $\dot{\zeta}$ is constant on cosets of S, we have $xsa - a = xa - a$ for all $s \in S$ and $x \in G$; hence, $xsa = xa$. If $x = 1$, then $sa = a$; that is, $a \in A^S$. It follows that ζ is a principal derivation, and $\text{cls}(\zeta) = 0$.

$\text{im Inf}^1 \subseteq \ker \text{Res}^1$. If $Z: G \rightarrow A$ is a derivation, then $\text{Res}^1 \text{cls}(Z) = \text{cls}(Z|S)$. In particular, if $\text{cls}(Z) = \text{cls}(\dot{\zeta}) \in \text{im Inf}^1$, then Z is constant on the coset S. But $Z(1) = 0$ (because every derivation sends 1 to 0), and so $Z = 0$.

$\ker \text{Res}^1 \subseteq \text{im Inf}^1$. If $\text{cls}(Z) \in \ker \text{Res}^1$, then $\text{cls}(Z|S) = 0$; that is, $Z|S$ is a principal derivation. Thus, there is $a \in A$ with $Z(s) = sa - a$ for all $s \in S$. Replacing Z by $Z - \delta$, where δ is the principal derivation $g \mapsto ga - a$, we may assume that $Z(s) = 0$ for all $s \in S$. But Z is constant on cosets of S, for the definition of derivation gives $Z(gs) = gZ(s) + Z(g) = Z(g)$. Now $\zeta: G/S \rightarrow A$, defined by $\zeta(gS) = Z(g)$, is a well-defined derivation, and $Z = \dot{\zeta} \in \text{im Inf}^1$.

(ii) We prove the result by induction on $q \geq 1$. The base step $q = 1$ is just part (i). Now the proof of Theorem 3.38 gives an imbedding $\varphi: A \rightarrow A^* = \text{Hom}_{\mathbb{Z}}(\mathbb{Z}G, A)$ [if $a \in A$, define $\varphi(a): x \mapsto xa$]; thus, there is an exact sequence of G-modules

$$0 \rightarrow A \rightarrow A^* \rightarrow C \rightarrow 0 \qquad (4)$$

with A^* G-coinduced. Note that A^* is S-induced, by Corollary 9.80(i), and that $(A^*)^S$ is G/S-coinduced, by Corollary 9.80(ii).

By hypothesis, $H^1(S, A) = \{0\}$ (for $q \geq 2$ now), and so there is exactness of

$$0 \to A^S \to (A^*)^S \to C^S \to 0. \tag{5}$$

Consider the commutative diagram

$$
\begin{array}{ccccccc}
0 & \longrightarrow & H^{q-1}(G/S, C^S) & \xrightarrow{\text{Inf}} & H^{q-1}(G, C) & \xrightarrow{\text{Res}} & H^{q-1}(S, C) \\
& & \downarrow{\scriptstyle \delta'} & & \downarrow{\scriptstyle \delta} & & \downarrow{\scriptstyle \delta''} \\
0 & \longrightarrow & H^q(G/S, A^S) & \xrightarrow{\text{Inf}} & H^q(G, A) & \xrightarrow{\text{Res}} & H^q(S, A).
\end{array}
$$

The first column occurs in the long exact sequence arising from applying $H^\bullet(G, \square)$ to exact sequence (5), while the other two columns arise by applying $H^\bullet(G, \square)$ and $H^\bullet(S, \square)$, respectively, to exact sequence (4). Now δ' is an isomorphism, for $(A^*)^S$ is G/S-coinduced; δ is an isomorphism because A^* is G-coinduced; δ'' is an isomorphism because A^* is S-coinduced. Since C (instead of A) also satisfies the inductive hypothesis (with $i \leq q - 2$), the top row of the diagram is exact. Finally, commutativity gives the bottom row exact as well.

(iii) This follows from the exactness of the top row of the diagram in part (ii) and $H^{q-1}(S, C) = \{0\}$.   •

There is a similar discussion for group homology.

Definition. Let G', G be groups and let $\alpha : G \to G'$ be a homomorphism. If A' is a G'-module and $f : A \to A'$ is a $\mathbb{Z}$-map, then we call (α, f) a *compatible pair* if $f : A \to {}_\alpha A'$ is a G-map, where ${}_\alpha A'$ denotes A' made into a G-module as on page 560; that is,

$$f(xa) = (\alpha \, x) f(a).$$

Define the category **Pairs**$_*$ with objects all pairs (G, A) (with G a group and A a G-module), with morphisms $(G, A) \to (G', A')$ being compatible pairs (α, f), and with composition $(\beta, g)(\alpha, f) = (\beta\alpha, gf)$ [of course, $1_{(G,A)} = (1_G, 1_A)$]. Composition in **Pairs**$_*$ makes sense: if $\beta : G' \to G''$ and $g : A' \to {}_\beta A''$, then $gf : A \to A''$ is a $\mathbb{Z}$-map and $gf(xa) = g((\alpha x)(fa)) = \beta\alpha(x)gf(a)$.

Given a compatible pair $(\alpha, f) : (G, A) \to (G', A')$ (so that $\alpha : G \to G'$ and $f : A \to {}_\alpha A'$), let $U : {}_{\mathbb{Z}G'}\mathbf{Mod} \to {}_{\mathbb{Z}G}\mathbf{Mod}$ be the change of groups functor, let $\mathbf{P}$ be a G-projective resolution of $\mathbb{Z}$, and let $\mathbf{P}'$ be a G'-projective resolution of $\mathbb{Z}$. The G-complex $U\mathbf{P}'$ is acyclic, by Lemma 9.75(i), and so the Comparison Theorem gives a chain map $\tau_\#(\alpha) : U\mathbf{P}' \to \mathbf{P}$ over $1_{\mathbb{Z}}$ that is unique to homotopy. Apply the functor $\square \otimes_G {}_\alpha A$ to obtain a chain map

$$\tau_\#(\alpha) : U\mathbf{P}' \otimes_G {}_\alpha A \to \mathbf{P} \otimes_G {}_\alpha A.$$

Next, $f: A \to {}_\alpha A'$ induces a chain map

$$f_{\#}: \mathbf{P} \otimes_G A \to \mathbf{P} \otimes_G {}_\alpha A'.$$

Thus, $f_{\#}\tau_{\#}(\alpha): U\mathbf{P}' \otimes_G A \to \mathbf{P} \otimes_G {}_\alpha A'$, and we define

$$(\alpha, f)_*: H_n(G, A) \to H_n(G', A')$$

to be the map in homology induced by this composite. If $\zeta \in P_n \otimes_G A$ is a cocycle, where $\mathbf{P} = \to P_n \to P_{n-1} \to \cdots \to P_0 \to \mathbb{Z} \to 0$, then

$$(\alpha, f)_*: \ \mathrm{cls}(\zeta) \mapsto \mathrm{cls}(f\zeta\tau_n(\alpha)).$$

There is a "ladder" theorem for these induced maps.

Proposition 9.85. *Let $\alpha: G' \to G$ be a group homomorphism, and consider a commutative diagram*

$$
\begin{array}{ccccccccc}
0 & \longrightarrow & A & \longrightarrow & B & \longrightarrow & C & \longrightarrow & 0 \\
& & \downarrow{\scriptstyle(\alpha,f)} & & \downarrow{\scriptstyle(\alpha,g)} & & \downarrow{\scriptstyle(\alpha,h)} & & \\
0 & \longrightarrow & A' & \longrightarrow & B' & \longrightarrow & C' & \longrightarrow & 0,
\end{array}
$$

where the top row is an exact sequence of G-modules, the bottom row is an exact sequence of G'-modules, and the vertical arrows are compatible. Then there is a commutative diagram with exact rows

$$
\begin{array}{ccccccc}
\to H_n(G', C') & \overset{\delta}{\to} & H_{n-1}(G', A') & \to & H_{n-1}(G', B') & \to & H_{n-1}(G', C') \\
\downarrow{\scriptstyle(\alpha,h)_*} & & \downarrow{\scriptstyle(\alpha,f)_*} & & \downarrow{\scriptstyle(\alpha,g)_*} & & \downarrow{\scriptstyle(\alpha,h)_*} \\
\to H_n(G, C) & \overset{\delta'}{\to} & H_{n-1}(G, A) & \to & H_{n-1}(G, B) & \to & H_{n-1}(G, C).
\end{array}
$$

Proof.   The computations are left to the reader.   ●

If $\alpha: S \to G$ is the inclusion of a subgroup and $f: A \to A$ is the identity 1_A, then $(\alpha, f)_*$ is called **restriction**

$$\mathrm{Res}_n: H_n(S, A) \to H_n(G, A).$$

If $S \triangleleft G$ is a normal subgroup, $\alpha: G \to G/S$ is the natural map, and $f: A \to A_S = A/\mathcal{S}A$ is the natural map, then $(\alpha, f)_*$ is called **coinflation**:

$$\mathrm{Coinf}_n: H_n(G, A) \to H_n(G/S, A_S).$$

Conjugation gives maps that show that if $S \triangleleft G$, then $H_*(S, A)$ is a (G/S)-module; moreover, there is an exact sequence in homology analogous to the sequence in Theorem 9.84.

9.6   Transfer

If G is a group and S is a subgroup S, then we have constructed restriction maps $\text{Res}^n\colon H^n(G, A) \to H^n(S, A)$; moreover, if S is a normal subgroup, then there are inflation maps $\text{Inf}^n\colon H^n(G/S, A^S) \to H^n(G, A)$. Similarly, there are maps in homology: corestriction $\text{Cor}_n\colon H_n(G, A) \to H_n(S, A)$ and coinflation $\text{Coinf}_n\colon H_n(G, A) \to H_n(G/S, A_S)$. When G is a group and $S \subseteq G$ is a subgroup of finite index, Eckmann constructed *transfer* maps in the reverse direction: $\text{Tr}^n\colon H^n(G, A) \to H^n(S, A)$ and $\text{Tr}_n\colon H_n(G, A) \to H_n(G, A)$. Transfer does *not* arise via change of groups, for it is not a functor defined on **Pairs*** or **Pairs***.

Recall that if G is a finite group and A is a G-module, then the *norm* $N\colon A \to A$ is defined by $a \mapsto \sum_{x \in G} xa$. We now generalize this.

Definition.   If G is a group, $S \subseteq G$ is a subgroup of finite index n, and A is a G-module, then the **norm**[10] $N_{G/S}\colon A \to A$ is defined by

$$N_{G/S}(a) = \sum_{i=1}^{n} t_i a,$$

where $\{t_1, \ldots, t_n\}$ is a left transversal of S in G; that is, $G = \bigcup_{i=1}^{n} t_i S$. If $S = \{1\}$, we continue to use the simpler notation

$$N_{G/\{1\}} = N.$$

Lemma 9.86.   *Let S be a subgroup of finite index in a group G, and let A be a G-module.*

(i) *The norm $N_{G/S}$ is independent of the choice of left transversal of S in G.*

(ii) $N_{G/S}\colon A^S \to A^G$; *that is,* $\text{im}\, N_{G/S} \subseteq A^G$.

(iii) $N_{G/S}\colon \text{Fix}^S \to \text{Fix}^G$ *is a natural transformation making the following diagram commute for all G-maps $f\colon A \to B$.*

$$
\begin{array}{ccc}
A^S & \xrightarrow{\ f_* \ } & B^S \\
\Big\downarrow{\scriptstyle N} & & \Big\downarrow{\scriptstyle N} \\
A^G & \xrightarrow[\ f_* \ ]{} & B^G
\end{array}
$$

---

[10]Some authors call $N_{G/S}$ the **trace**.

Proof.

(i) Let $T = \{t_1, \ldots, t_m\}$ and $T' = \{t'_1, \ldots, t'_m\}$ be left transversals of S in G. Now $t'_i = t_i s_i$ for each i, where $s_i \in S$. Hence, if $a \in A^S$,

$$\sum_i t'_i a = \sum_i t_i s_i a = \sum_i t_i a.$$

Thus, $N_{G/S}(a)$ does not depend on the choice of transversal.

(ii) If T is a left transversal of S in G and $x \in G$, then $xT = \{xt : t \in T\}$ is also a left transversal: since multiplication by x is a bijection, $G = x(\bigcup_{t \in T} tS) = \bigcup_{t \in T} xtS$. Now

$$x N_{G/S}(a) = x \sum_{t \in T} ta = \sum_{t \in T} (xt)a = N_{G/S}(a),$$

the last equality holding because $N_{G/S}(a)$ does not depend on the transversal. Therefore, $N_{G/S}(a) \in A^G$.

(iii) A routine computation. •

Proposition 9.87. *Let S be a subgroup of finite index in a group G, and let A be a G-module. There exist unique homomorphisms*

$$\mathrm{Tr}^n \colon H^n(S, A) \to H^n(G, A),$$

called **transfer** *(or* **corestriction**), *such that*

(i) $\mathrm{Tr}^0 = N_{G/S} \colon A^S \to A^G$,

(ii) *for every exact sequence $0 \to A \xrightarrow{i} B \xrightarrow{p} C \to 0$ of G-modules, there is a commutative diagram*

$$
\begin{array}{ccccccc}
H^n(S, B) & \xrightarrow{p_*} & H^n(S, C) & \xrightarrow{\partial_S} & H^{n+1}(S, A) & \xrightarrow{i_*} & H^{n+1}(S, B) \\
\downarrow{\scriptstyle \mathrm{Tr}_n} & & \downarrow{\scriptstyle \mathrm{Tr}_n} & & \downarrow{\scriptstyle \mathrm{Tr}_{n+1}} & & \downarrow{\scriptstyle \mathrm{Tr}_{n+1}} \\
H^n(G, B) & \xrightarrow[p_*]{} & H^n(G, C) & \xrightarrow[\partial_G]{} & H^{n+1}(G, A) & \xrightarrow[i_*]{} & H^{n+1}(G, B).
\end{array}
$$

Proof. Both $(H^n(S, \square))$ and $(H^n(G, \square))$ are cohomological ∂-functors that vanish on injectives (recall that every injective G-module is an injective S-module), and $N_{G/S} \colon H^0(S, \square) \to H^0(G, \square)$ is a natural transformation. Thus, Theorem 6.51 applies at once. •

Theorem 9.88. *If S is a subgroup of G of finite index m and A is a G-module, then $\operatorname{Tr}^n \operatorname{Res}^n \colon H^n(G, A) \to H^n(G, A)$ is multiplication by m.*

Proof. Let T be a left transversal of S in G. If $n = 0$, then $\operatorname{Tr}^0 \colon A^S \to A^G$ sends $a \mapsto \sum_{t \in T} ta$, while $\operatorname{Res}^0 \colon A^G \to A^S$ is the inclusion. Thus, the composite $\operatorname{Tr}_0 \operatorname{Res}_0$ is multiplication by m, for A^G is G-trivial. The result now follows from Theorem 6.36, for if $\tau_n \colon H_n(G, A) \to H_n(G, A)$ is multiplication by m, then both (τ_n) and $(\operatorname{Res}_n \operatorname{Tr}_n)$ are morphisms of cohomological ∂-functors agreeing in degree 0.   •

Corollary 9.89 (= Proposition 9.40). *If G is a finite group of order m, then $mH^n(G, A) = \{0\}$ for all $n \geq 0$ and all finitely generated G-modules A.*

Proof. By Theorem 9.88, if S is a subgroup of G of finite index, then the composite
$$H^n(G, A) \xrightarrow{\operatorname{Res}^n} H^n(S, A) \xrightarrow{\operatorname{Tr}^n} H^n(G, A)$$
is multiplication by $[G : S]$. This is true, in particular, when G is finite, $S = \{1\}$, and $[G : S] = |G|$. But $H^n(\{1\}, A) = \{0\}$, and so multiplication by $|G|$ is 0.   •

If p is a prime, then a ***Sylow p-subgroup*** of a finite group G, denoted by G_p, is a maximal p-subgroup of G. It is known that such subgroups exist for every prime divisor of $|G|$ and that $|G_p|$ is the largest power of p dividing $|G|$; that is, $([G : G_p], p) = 1$. In general, Sylow p-subgroups are not unique (but any two such are conjugate); in the special case when G is abelian, however, G_p is unique and is called the ***p-primary component***.

Recall Corollary 9.41: if G is a finite group and A is a finitely generated G-module, then $H^n(G, A)$ is finite for all $n \geq 0$; hence, there is a primary decomposition $H^n(G, A) = \bigoplus_p H^n(G, A)_p$.

Corollary 9.90. *Let G be a finite group, and let A be a G-module.*

(i) $\operatorname{Res}^n \colon H^n(G, A) \to H^n(G_p, A)$ *is injective on the p-primary component $H^n(G, A)_p$ for every $n \geq 0$ and for every prime $p \mid |G|$.*

(ii) *There is an injection $\theta \colon H^n(G, A) \to \bigoplus_p H^n(G_p, A)$.*

(iii) *If $H^n(G_p, A) = \{0\}$ for all Sylow p-subgroups, then $H^n(G, A) = \{0\}$.*

Proof.

(i) If $u \in H^n(G, A)_p$, then $p^e u = 0$ for some $e \geq 0$. Let $[G : G_p] = q$. Now $\operatorname{Tr}^n \operatorname{Res}^n(u) = qu$, by Theorem 9.88. If $\operatorname{Res}^n(u) = 0$, then $qu = \operatorname{Tr}^n(\operatorname{Res}^n u) = 0$. Since $(q, p^e) = 1$ (because G_p is a Sylow p-subgroup), there are integers s and t with $1 = sq + tp^e$, and so $u = squ + tp^e u = 0$.

(ii) Since $|G|H^n(G, A) = \{0\}$, the order of any element in $H^n(G, A)$ is a divisor of $|G|$. It follows that the primary decomposition of $H^n(G, A)$ involves only the prime divisors of $|G|$:

$$H^n(G, A) = \bigoplus_{p\|G|} H^n(G, A)_p.$$

Thus, if $u \in H^n(G, A)$, then $u = (u_p)$ for $u_p \in H^n(G, A)_p$. Denote $\mathrm{Res}^n\colon H^n(G, A) \to H^n(G_p, A)$ by f_p, and consider the map $\theta\colon H^n(G, A) \to \bigoplus_p H^n(G_p, A)$ given by $\theta\colon u \mapsto (f_p u)$. If $u \in \ker\theta$, then $f_p u = 0$ for all p. But $f_p(u_p) = 0$ implies $u_p = 0$, by (i), and so θ is an injection.

(iii) Immediate from (ii). ●

Let us now consider homology.

Definition. If G is a group, $S \subseteq G$ is a subgroup of finite index n, and A is a G-module, then the **conorm** $v_{G/S}\colon A \to A$ is defined by

$$v_{G/S}(a) = \sum_i t_i^{-1}a,$$

where $\{t_1, \ldots, t_n\}$ is a left transversal of S in G; that is, $G = \bigcup_{i=1}^n t_i S$.

If $x \in G$, then each xt_i, as any element of G, lies in a unique left coset, say, $xt_i \in t_{j(i)}S$; that is,

$$xt_i = t_{j(i)}s_{i,j(i)} \tag{1}$$

for some $s_{i,j(i)} \in S$ and $j(i) \in \{1, \ldots, n\}$. It is easy to see that j is a permutation of $\{1, \ldots, n\}$; it follows that

$$v_{G/S}(a) = \sum_i t_{j(i)}^{-1}a. \tag{2}$$

Lemma 9.91. *Let S be a subgroup of finite index in a group G, and let A be a G-module.*

(i) *The conorm $v_{G/S}$ is independent of the choice of right transversal of S in G.*

(ii) $v_{G/S}\colon A_G \to A_S$.

(iii) $v_{G/S}: \square_G \to \square_S$ *is a natural transformation making the following diagram commute for all G-maps* $f: A \to B$.

$$
\begin{array}{ccc}
A_G & \xrightarrow{f_*} & B_G \\
{\scriptstyle v}\downarrow & & \downarrow{\scriptstyle v} \\
A_S & \xrightarrow[f_*]{} & B_S
\end{array}
$$

Proof. We prove (ii), leaving the other parts to the reader. Since $A_G = A/\mathcal{G}A$, it suffices to prove that $v_{G/S}(\mathcal{G}A) \subseteq \mathcal{S}A$, where $\mathcal{S}$ is the augmentation ideal of S. A typical generator of $\mathcal{G}$ is $x - 1$, for $x \in G$ and

$$
\begin{aligned}
v_{G/S}(x-1)a &= \sum_i t_{j(i)}^{-1}(x-1)a && \text{[Eq. (2)]} \\
&= \Big(\sum_i t_{j(i)}^{-1}x\Big)a - Na && \Big(\text{where } N = \sum_i t_{j(i)}^{-1}\Big) \\
&= \Big(\sum_i s_{i,j(i)} t_i^{-1}\Big)a - Na && \text{[Eq. (1)]} \\
&= \sum_i (s_{i,j(i)} - 1)(t_i^{-1}a) \in \mathcal{S}A. \quad \bullet
\end{aligned}
$$

We record a consequence of this computation.

Corollary 9.92. *Let S be a subgroup of finite index in a group G, and let A be a G-module. If $a \in A$, then* [*using the notation in* Eq. (1)]

$$
\sum_i t_{j(i)}^{-1}(x-1)a + \mathcal{S}A = \sum_i (s_{i,j(i)} - 1)(t_i^{-1}a) + \mathcal{S}A \quad \text{in } A_S.
$$

Proposition 9.93. *Let S be a subgroup of finite index in a group G, and let A be a G-module. There exist unique homomorphisms*

$$
\mathrm{Tr}_n: H_n(G, A) \to H_n(S, A),
$$

called **transfer**, *such that*

(i) $\mathrm{Tr}_0: H_0(G, A) \to H_0(S, A)$ *is given by* $a + \mathcal{G}A \mapsto v_{G/S}(a) + \mathcal{S}A$, *where $\mathcal{S}$ is the augmentation ideal of S,*

(ii) *for every exact sequence* $0 \to A \xrightarrow{i} B \xrightarrow{p} C \to 0$ *of G-modules, there is a commutative diagram*

$$
\begin{array}{ccccccc}
H_n(G, B) & \xrightarrow{p_*} & H_n(G, C) & \xrightarrow{\partial_G} & H_{n-1}(G, A) & \xrightarrow{i_*} & H_{n-1}(G, B) \\
{\scriptstyle \mathrm{Tr}_n}\downarrow & & \downarrow{\scriptstyle \mathrm{Tr}_n} & & \downarrow{\scriptstyle \mathrm{Tr}_{n-1}} & & \downarrow{\scriptstyle \mathrm{Tr}_{n-1}} \\
H_n(S, B) & \xrightarrow[p_*]{} & H_n(S, C) & \xrightarrow[\partial_S]{} & H_{n-1}(S, A) & \xrightarrow[i_*]{} & H_{n-1}(S, B).
\end{array}
$$

Proof.   By Theorem 6.36, there exists a unique morphism between homological ∂-functors that annihilates projectives.   •

Theorem 9.94.   *If S is a subgroup of G of finite index m and A is a G-module, then $\mathrm{Res}_n \mathrm{Tr}_n \colon H_n(G, A) \to H_n(G, A)$ is multiplication by m.*

Proof.   As in the proof of Theorem 9.88.   •

Corollary 9.95.   *If G is a finite group of order m, then $mH_n(G, A) = \{0\}$ for all $n \geq 0$ and all finitely generated G-modules A.*

Proof.   As in the proof of Corollary 9.89.   •

Corollary 9.96.   *Let G be a finite group, and let A be a G-module. If $H_n(G_p, A) = \{0\}$ for all Sylow p-subgroups, then $H_n(G, A) = \{0\}$.*

Proof.   As in the proof of Corollary 9.90.   •

Remark.   Propositions 9.87 and 9.93 prove that the transfer maps Tr^n and Tr_n exist, but it is not so obvious how to use them for computation.

Given a ring map $\varphi \colon R \to S$, change of rings discusses how to view S-modules as R-modules; that is, it gives a functor $_S\mathbf{Mod} \to {_R}\mathbf{Mod}$. We now construct a functor $_R\mathbf{Mod} \to {_S}\mathbf{Mod}$ in the reverse direction; that is, we show how to use φ to view R-modules as S-modules. Regard S as a right R-module by defining

$$s.r = s\varphi(r).$$

It is easy to see that S is an (S, R)-bimodule, so that if A is a left R-module, then $S \otimes_R A$ is a left S-module. Indeed, the reader may show that $A \mapsto S \otimes_R A$ is functorial. We refer the reader to Brown, *Cohomology of Groups* III, §9, to see how this general situation applies to transfer.   ◄

Transfer maps are so called because they generalize the *transfer* in Group Theory. Let $S \subseteq G$ be a subgroup of finite index; say, $[G : S] = n < \infty$. An $n \times n$ **monomial matrix over** S is an $n \times n$ permutation matrix in which all entries equal to 1 are replaced by elements of S. It is easy to check that the usual matrix product of monomial matrices is defined (nonzero entries never need to be added), it is a monomial matrix, and the set $\mathrm{Mon}_n(S)$ of all $n \times n$ monomial matrices over S is a group. Monomial matrices arise in the following context. If $T = \{t_1, \ldots, t_n\}$ is a left transversal of S in G, then Eq. (1) on page 575 shows, for any $x \in G$, that there is a permutation j of $\{1, \ldots, n\}$ with $xt_i = t_{j(i)}s_{i,j(i)}$, and so $[s_{i,j(i)}] \in \mathrm{Mon}_n(S)$.

Definition.   Let S be a subgroup of finite index n in a group G. If $\{t_1, \ldots, t_n\}$ is a left transversal of S in G, then the **transfer** $V_{G \to S} \colon G \to S/S'$ is defined by

$$x \mapsto \det[s_{i,j(i)}]S' = \prod_{i=1}^{n} s_{i,j(i)} S' = \prod_{i=1}^{n} t_{j(i)}^{-1} x t_i S',$$

where S' is the commutator subgroup of S.

Transfer $V_{G \to S} \colon G \to S/S'$ is a homomorphism whose definition is independent of the choice of left transversal of S in G (Rotman, *An Introduction to the Theory of Groups*, Theorem 7.45). Since S/S' is abelian, $G' \subseteq \ker V_{G \to S}$, and one usually views the transfer as a map $G/G' \to S/S'$. Since $G/G' \cong H_1(G, \mathbb{Z})$, by Theorem 9.52, the transfer can be viewed as a map $H_1(G, \mathbb{Z}) \to H_1(S, \mathbb{Z})$, and it is natural to wonder whether it is related to Tr_1. The next theorem is due to B. Eckmann, "Cohomology of groups and transfer," *Annals Math.* 58 (1953), 481–493.

Theorem 9.97 (Eckmann).   *If G is a group and S is a subgroup of finite index m, then* $\mathrm{Tr}_1 \colon H_1(G, \mathbb{Z}) \to H_1(S, \mathbb{Z})$ *is the transfer* $V_{G \to S}$.

Proof.   Consider the commutative diagram of S-modules with vertical maps inclusions:

$$
\begin{array}{ccccccccc}
0 & \longrightarrow & \mathcal{G} & \longrightarrow & \mathbb{Z}G & \longrightarrow & \mathbb{Z} & \longrightarrow & 0 \\
 & & \lambda\uparrow & & \uparrow & & \| & & \\
0 & \longrightarrow & S & \longrightarrow & \mathbb{Z}S & \longrightarrow & \mathbb{Z} & \longrightarrow & 0.
\end{array}
$$

Taking homology gives the bottom two rows of the following diagram.

$$
\begin{array}{ccccc}
0 \longrightarrow & G/G' = H_1(G, \mathbb{Z}) & \xrightarrow{\partial_G} & H_0(G, \mathcal{G}) = \mathcal{G}/\mathcal{G}^2 & \\
 & \mathrm{Tr}_1\big\downarrow & & \big\downarrow \mathrm{Tr}_0 & \\
0 \longrightarrow & S/S' = H_1(S, \mathbb{Z}) & \xrightarrow{\Delta} & H_0(S, \mathcal{G}) = \mathcal{G}/\mathcal{G}S & \\
 & =\big\uparrow & & \big\uparrow \lambda_* & \\
0 \longrightarrow & H_1(S, \mathbb{Z}) & \xrightarrow[\partial_S]{} & H_0(S, S) = S/S^2 &
\end{array}
$$

The top row arises from the G-exact sequence $0 \to \mathcal{G} \to \mathbb{Z}G \to \mathbb{Z} \to 0$. Recall Proposition 9.46: we may identify $H_0(G, \mathcal{G})$ with $\mathcal{G}/\mathcal{G}^2$; recall Theorem 9.52: the connecting homomorphism $\partial_G \colon H_1(G, \mathbb{Z}) \to H_0(G, \mathcal{G}) = \mathcal{G}/\mathcal{G}^2$, given by $\mathrm{cls}([x] - 1) \mapsto x - 1 + \mathcal{G}^2$, is an isomorphism. All the rows are exact, for the preceding terms $H_1(G, \mathbb{Z}G)$ and $H_1(S, \mathbb{Z}G)$ are $\{0\}$, because $\mathbb{Z}G$ is projective, even free, as a G-module and as an S-module. Finally, the diagram commutes; in particular, the top square commutes, by Proposition 9.93:

$$\mathrm{Tr}_0 \, \partial_G = \Delta \, \mathrm{Tr}_1 . \tag{3}$$

We compute. Let $G = \bigcup_i t_i S$. Using the notation in Eqs. (1) and (2),

$$
\begin{aligned}
\mathrm{Tr}_0 \, \partial_G (xG') &= \nu_{G/S}(x - 1 + \mathcal{G}^2) \\
&= \sum_i t_{j(i)}^{-1} (x - 1) + \mathcal{SG} \\
&= \sum_i (s_{i,j(i)} - 1) t_i^{-1} + \mathcal{SG} \qquad \text{by Corollary 9.92} \\
&= \sum_i (s_{i,j(i)} - 1) + \mathcal{SG},
\end{aligned}
$$

the last equation holding because $(s-1)t_i^{-1} - (s-1) = (s-1)(t_i^{-1} - 1) \in \mathcal{SG}$ for all $s \in S$.

By definition, $V_{G \to S}(xG') = \prod_i s_{i,j(i)} S'$. Hence,

$$
\begin{aligned}
\Delta V_{G \to S}(xG') &= \Delta \Big(\prod_i s_{i,j(i)} S' \Big) \\
&= \lambda_* \partial_S \Big(\prod_i s_{i,j(i)} S' \Big) \\
&= \lambda_* \Big(\sum_i s_{i,j(i)} - 1 \Big) + \mathcal{S}^2 \\
&= \sum_i (s_{i,j(i)} - 1) + \mathcal{SG}.
\end{aligned}
$$

Therefore, $\mathrm{Tr}_0 \, \partial_G(xG') = \Delta V_{G \to S}(xG')$. But $\mathrm{Tr}_0 \, \partial_G = \Delta \, \mathrm{Tr}_1$, by Eq. (3), so that $\Delta \, \mathrm{Tr}_1 = \Delta V_{G \to S}$. Since Δ is an injection, we have $\mathrm{Tr}_1 = V_{G \to S}$. •

Exercises

*9.31 Consider the commutative diagram of modules

$$
\begin{array}{ccccccc}
A & \xrightarrow{\Delta} & B & \xrightarrow{d} & C & \longrightarrow & D \\
& & \downarrow{f} & & \uparrow{g} & & \\
& & B' & \xrightarrow{\alpha} & C' & &
\end{array}
$$

in which $d\Delta = 0$, f is surjective, and g is injective.

(i) Prove that $\bar{d} \colon B/\operatorname{im} \Delta \to C$, given by $b + \operatorname{im} \Delta \mapsto db$, is a well-defined map with $\ker \bar{d} = \ker d / \operatorname{im} \Delta$.

(ii)   Prove that $\varphi: B/\operatorname{im}\Delta \to C'$, given by $b + \operatorname{im}\Delta \mapsto \alpha f b$, is well-defined.

(iii)   Using surjectivity of f, prove that $\ker \bar{d} \cong \ker \alpha$.

(iv)   As in the first three parts, prove that $\operatorname{coker}\bar{d} \cong \operatorname{coker}\alpha$.

9.32 If $T \subseteq S \subseteq G$ are subgroups of finite index, use Theorem 9.97 to prove

$$V_{G \to T} = V_{S \to T} V_{G \to S}.$$

9.7   Tate Groups

If G is a finite group and $0 \to A \to B \to C \to 0$ is an exact sequence of G-modules, then the long exact homology and cohomology sequences can be spliced to form a doubly infinite long exact sequence. This will allow us to do dimension shifting in both directions, enabling us to define change of groups maps in homology as well as in cohomology.

Recall that

$$H_0(G, A) = A_G = A/\mathcal{G}A \quad \text{and} \quad H^0(G, A) = A^G,$$

where $\mathcal{G}$ is the augmentation ideal of G.

Lemma 9.98.   *Let G be a finite group, and let A be a G-module.*

(i)   *If $N: A \to A$ is the norm map, then $\mathcal{G}A \subseteq \ker N$, $\operatorname{im} N \subseteq A^G$, and $\alpha_A: H_0(G, A) \to H^0(G, A)$, given by*

$$\alpha_A: a + \mathcal{G}A \mapsto Na,$$

is a well-defined natural map.

(ii)   *If $_N A = \ker N = \{a \in A : Na = 0\}$, then*

$$\ker \alpha_A = {}_N A/\mathcal{G}A \quad \text{and} \quad \operatorname{coker}\alpha_A = A^G/NA;$$

that is, there is an exact sequence

$$0 \to {}_N A/\mathcal{G}A \to H_0(G, A) \xrightarrow{\alpha_A} H^0(G, A) \to A^G/NA \to 0.$$

Proof.

(i)   By Lemma 9.24, $\mathcal{G}$ is generated by all $y-1$ for $y \in G$. If $(y-1)a \in \mathcal{G}A$, then $N(y-1)a = 0$, for $N(y-1) = \sum_x xy - \sum_x x = 0$. That $Na \in \operatorname{im} N$ is Lemma 9.86 [a simpler proof here is $yNa = (\sum_x yx)a = Na$]. It follows that α_A is a well-defined map whose values lie in A^G.

To prove naturality, we must show that the following diagram commutes for $f: A \to B$.

$$
\begin{array}{ccc}
H_0(G, A) & \xrightarrow{\ f_* \ } & H_0(G, B) \\
\alpha_A \downarrow & & \downarrow \alpha_B \\
H^0(G, A) & \xrightarrow[\ f_* \ ]{} & H^0(G, B).
\end{array}
$$

Going clockwise, $a + \mathcal{G}A \mapsto fa + \mathcal{G}B \mapsto Nfa$; going counterclockwise, $a + \mathcal{G}A \mapsto Na \mapsto f(Na)$.

(ii) Both equalities are easily verified. •

Proposition 9.99. *If G is a finite group and A is an induced G-module, then*

$$
{}_N A / \mathcal{G}A = \{0\} = A^G / NA
$$

and $\alpha_A : H_0(G, A) \to H^0(G, A)$ is a natural isomorphism.

Proof. We know that $\mathcal{G}A \subseteq {}_N A$ is always true. We prove the reverse inclusion when A is induced; that is, $A = \mathbb{Z}G \otimes_{\mathbb{Z}} B$ for some abelian group B. If $a \in {}_N A$, then a has a unique expression $a = \sum_{x \in G} x \otimes b_x$, where $b_x \in B$, and $Na = 0$. Since $Nx = N$ for every $x \in G$, we have

$$
0 = N\left(\sum_x x \otimes b_x \right) = \sum_x Nx \otimes b_x = \sum_x N \otimes b_x
$$
$$
= N\left(\sum_x 1 \otimes b_x \right) = N\left(1 \otimes \sum_x b_x \right).
$$

If $b \in B$ and $N(1 \otimes b) = 0$, then $b = 0$, for the x-component of $N(1 \otimes b)$ in $\mathbb{Z}G \otimes B$ is $0 = x \otimes b$. It follows that $\sum_x b_x = 0$ and $\sum_x 1 \otimes b_x = 1 \otimes \sum_x b_x = 0$. Therefore,

$$
a = a - \sum_x 1 \otimes b_x = \sum_x (x - 1) \otimes b_x \in \mathcal{G}A,
$$

for $\mathcal{G}$ is generated by all $x - 1$.

We know that $NA \subseteq A^G$ is always true. We prove the reverse inclusion when $A = \mathbb{Z}G \otimes_{\mathbb{Z}} B$. If $a \in A^G$, then a has a unique expression $a = \sum_{x \in G} x \otimes b_x$ and $ya = a$ for all $y \in G$. Thus, $\sum yx \otimes b_x = \sum x \otimes b_x$. It follows that $b_{yx} = b_x$ for all y, x. Therefore, all b_x are equal, say, $b_x = b$ for all x, and so $a = \sum x \otimes b_x = \sum x \otimes b = N(1 \otimes b) \in NA$.

Finally, α_A is natural, by Lemma 9.98, and it is an isomorphism because $\ker \alpha_A = \{0\} = \operatorname{coker} \alpha$. •

Definition. If G is a finite group and A is a G-module, then

$$\widehat{H}^q(G, A) = \begin{cases} H^q(G, A) & \text{if } q \geq 1, \\ A^G/NA & \text{if } q = 0, \\ {}_NA/\mathcal{G}A & \text{if } q = -1, \\ H_{q-1}(G, A) & \text{if } q \leq -2. \end{cases}$$

We will soon see that the Tate groups arise naturally as the cohomology groups of a doubly infinite complex.

Proposition 9.100. *If G is a finite group and A is an induced G-module, then*

$$\widehat{H}^q(G, A) = \{0\}$$

for all $q \in \mathbb{Z}$.

Proof. Since G is finite, $\widehat{H}^q(G, A) = \{0\}$ for all $q \geq 1$ and all $q \leq -2$, by Proposition 9.77. If $q = -1, 0$, then $\widehat{H}^q(G, A) = \{0\}$, by Proposition 9.99. •

Theorem 9.101. *If G is a finite group and $0 \to A \to B \to C \to 0$ is an exact sequence of G-modules, then there is a long exact sequence*

$$\to \widehat{H}^q(G, A) \to \widehat{H}^q(G, B) \to \widehat{H}^q(G, C) \to \widehat{H}^{q-1}(G, A) \to .$$

Proof. Since $\widehat{H}^q(G, A) = H^q(G, A)$ for all $q \geq 1$, we have exactness of the sequence for all such q; similarly, since $\widehat{H}^q(G, A) = H_{q-1}(G, A)$ for all $q \leq -2$, we have exactness of the sequence for all such q.

Let us prove exactness for $q = -1$ and $q = 0$. By naturality of the maps α_A, there is a commutative diagram

$$
\begin{array}{ccccccccc}
H_1(G, C) & \xrightarrow{\partial} & H_0(G, A) & \to & H_0(G, B) & \to & H_0(G, C) & \longrightarrow & 0 \\
& & \downarrow{\alpha_A} & & \downarrow{\alpha_B} & & \downarrow{\alpha_C} & & \\
0 & \longrightarrow & H^0(G, A) & \to & H^0(G, B) & \to & H^0(G, C) & \xrightarrow{\delta} & H^1(G, A).
\end{array}
$$

The version of the Snake Lemma in Exercise 6.5 on page 338 gives an exact sequence

$$\ker \alpha_A \to \ker \alpha_B \to \ker \alpha_C \to \operatorname{coker} \alpha_A \to \operatorname{coker} \alpha_B \to \operatorname{coker} \alpha_C;$$

that is, there is an exact sequence

$$\widehat{H}^{-1}(G, A) \to \widehat{H}^{-1}(G, B) \to \widehat{H}^{-1}(G, C)$$
$$\to \widehat{H}^0(G, A) \to \widehat{H}^0(G, B) \to \widehat{H}^0(G, C).$$

The reader is asked to prove exactness at $\widehat{H}^{-1}(G, A)$ and $\widehat{H}^0(G, C)$ in Exercise 9.36 on page 595. •

Definition.   Let G be a finite group and let $\mathbb{Z}$ be viewed (as usual) as a trivial G-module. A ***complete resolution*** is an exact sequence $\mathbf{X}$

where each X_q is finitely generated G-free, ϵ is surjective, η is injective, and $d = \eta\epsilon$.

We will use the next lemma to prove the existence of complete resolutions.

Lemma 9.102.   *If*

$$\to F_{n+1} \to F_n \to F_{n-1} \to$$

is an exact sequence of finitely generated free abelian groups, then

$$\to \operatorname{Hom}(F_{n-1}, \mathbb{Z}) \to \operatorname{Hom}(F_n, \mathbb{Z}) \to \operatorname{Hom}(F_{n+1}, \mathbb{Z}) \to$$

is also an exact sequence of finitely generated free abelian groups.

Proof.   Factor the original exact sequence into short exact sequences:

$$\longrightarrow F_{n+1} \longrightarrow F_n \longrightarrow F_{n-1} \longrightarrow$$
$$\searrow \quad \nearrow \qquad \searrow \quad \nearrow$$
$$K_n \qquad\qquad K_{n-1}.$$

Since every subgroup of a free abelian group is free abelian, hence projective, each exact sequence $0 \to K_n \to F_n \to K_{n-1} \to 0$ splits. The result follows, for $\operatorname{Hom}_{\mathbb{Z}}(\square, \mathbb{Z})$ preserves split exact sequences.   •

Proposition 9.103.   *Every finite group G has a complete resolution* $\mathbf{X}$.

Proof.   Since G is finite, there is a G-free resolution of $\mathbb{Z}$

$$\to P_1 \to P_0 \to \mathbb{Z} \to 0$$

in which each P_i is finitely generated. Define the ***dual M^**** of a G-module M by
$$M^* = \operatorname{Hom}_{\mathbb{Z}}(M, \mathbb{Z})$$

[as usual, M^* is a G-module with diagonal action: $xf : m \mapsto xf(x^{-1}m)$]. Since $\mathbb{Z}^* = \mathbb{Z}$, Lemma 9.102 gives an exact sequence

$$0 \to \mathbb{Z} \to P_0^* \to P_1^* \to$$

in which each P_i^* is a finitely generated G-free module. Splicing these two sequences together gives a doubly infinite exact sequence

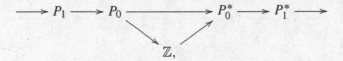

which is a complete resolution once we relabel: for all $q \geq 0$, set

$$X_q = P_q \quad \text{and} \quad X_{-q} = P_{q+1}^*. \quad \bullet$$

Remark.   If G is finite, then every complete resolution $\mathbf{X}$ is of the form just constructed (see Brown, *Cohomology of Groups*, p. 133). ◄

Proposition 9.104.   *If G is a finite group and $\mathbf{X}$ is the complete resolution constructed in* Proposition 9.103, *then for all $q \in \mathbb{Z}$,*

$$\widehat{H}^q(G, A) \cong H^q(\text{Hom}_G(\mathbf{X}, A)).$$

Proof.   The left half of $\mathbf{X}$ gives a deleted G-free resolution of $\mathbb{Z}$

$$\mathbf{L} = \ \to P_1 \to P_0 \to 0,$$

and so $H^q(\text{Hom}_G(\mathbf{L}, A)) = \text{Ext}_{\mathbb{Z}G}^q(\mathbb{Z}, A) = H^q(G, A)$ for all $q \geq 0$. Therefore, $H^q(\text{Hom}_G(\mathbf{X}, A)) = H^q(G, A)$ for all $q \geq 1$ (this is not true for $q = 0$ because $X_{-1} \neq \{0\} = L_{-1}$).

We now treat the right half of $\mathbf{X}$. The reader may check that if M is any finitely generated free G-module, then there is a natural G-isomorphism $\sigma\colon M \otimes_G A \to \text{Hom}_{\mathbb{Z}}(M^*, A)$, where M^* is the dual of M, defined as follows: if $f \in M^* = \text{Hom}_{\mathbb{Z}}(M, \mathbb{Z})$, then

$$\sigma(m \otimes a)\colon f \mapsto f(m)a.$$

By Proposition 9.99, the composite $\tau\colon M \otimes_G A \to \text{Hom}_G(M^*, A)$,

$$(M \otimes_{\mathbb{Z}} A)_G \xrightarrow{\ \alpha_A\ } (M \otimes_{\mathbb{Z}} A)^G \xrightarrow{\ \sigma\ } \text{Hom}_{\mathbb{Z}}(M^*, A)^G,$$
$$\underbrace{\phantom{(M \otimes_{\mathbb{Z}} A)_G \xrightarrow{\ \alpha_A\ } (M \otimes_{\mathbb{Z}} A)^G \xrightarrow{\ \sigma\ } \text{Hom}}}_{\tau}$$

is a natural isomorphism [recall that $M \otimes_G A = (M \otimes_{\mathbb{Z}} A)_G$ and $\text{Hom}_{\mathbb{Z}}(M^*, A)^G = \text{Hom}_G(M^*, A)$]. Naturality of τ implies that τ gives an isomorphism of complexes $\mathbf{L} \otimes_G A \cong \text{Hom}_G(\mathbf{L}^*, A)$. Since $X_{-q} = P_{q+1}^*$, it follows, for all $q \geq 1$, that

$$H^{-q-1}(\text{Hom}_G(\mathbf{X}, A)) = H^{-q-1}(\text{Hom}_G(\mathbf{L}^*, A))$$
$$\cong H_q(\mathbf{L} \otimes_G A) = \text{Tor}_q^{\mathbb{Z}G}(\mathbb{Z}, A) = H_q(G, A).$$

Therefore, $H^q(\mathrm{Hom}_G(\mathbf{X}, A)) \cong H_q(G, A)$ for all $q \leq -2$.

It remains to prove $H^q(\mathrm{Hom}_G(\mathbf{X}, A)) \cong \widehat{H}^q(G, A)$ for $q = -1, 0$; that is, $H^{-1}(\mathrm{Hom}_G(\mathbf{X}, A)) = {}_N A/\mathcal{G}A$ and $H^0(\mathrm{Hom}_G(\mathbf{X}, A)) = A^G/NA$. The reader may show that the following diagram commutes:

$$
\begin{array}{ccc}
\mathrm{Hom}_G(P_{-1}, A) & \xrightarrow{\;d^*\;} & \mathrm{Hom}_G(P_0, A) \\
{\scriptstyle=}\big\downarrow & & \big\uparrow{\scriptstyle\varepsilon^*} \\
\mathrm{Hom}_G(P_0^*, A) & & \\
{\scriptstyle\tau^{-1}}\big\downarrow & & \\
P_0 \otimes_G A & & \\
{\scriptstyle\varepsilon\otimes 1}\big\downarrow & & \\
A_G & \xrightarrow{\;\alpha_A\;} & A^G
\end{array}
$$

[if $u \otimes a \in P_0 \otimes_G A$ and $v \in P_0$, then both clockwise and counterclockwise composites send $u \otimes a$ to the map $f \in \mathrm{Hom}_G(P_0, A)$ with $f \colon v \mapsto \varepsilon(Nv)\varepsilon(u)$]. The result now follows from Exercise 9.31 on page 579, for $\widehat{H}^{-1}(G, A) = \ker \alpha_A$ and $\widehat{H}^0(G, A) = \mathrm{coker}\,\alpha_A$. $\quad\bullet$

We now give another proof of Corollary 9.50 that explains why indices are off by 1.

Proposition 9.105 (= Corollary 9.50). *If G is a finite cyclic group and A is a G-module, then, for all $n \geq 1$,*

$$H_{2n-1}(G, A) = H^{2n}(G, A),$$
$$H_{2n}(G, A) = H^{2n-1}(G, A).$$

Proof. There is a complete resolution

$$
\xrightarrow{\;N\;} \mathbb{Z}G \xrightarrow{\;D\;} \mathbb{Z}G \xrightarrow{\qquad N \qquad} \mathbb{Z}G \xrightarrow{\;D\;} \mathbb{Z}G \xrightarrow{\;N\;}
$$
$$
\searrow{\scriptstyle\varepsilon} \qquad \nearrow{\scriptstyle\eta}
$$
$$
\mathbb{Z},
$$

where $D \colon x \mapsto x - 1$, ε is the augmentation, and $\eta \colon 1 \mapsto N$. The result now follows by recalling that $H_q(G, A) = \widehat{H}^{-q-1}(G, A)$ for all $q \geq 1$. $\quad\bullet$

There are products defined in Tate cohomology.

Theorem. *Let G be a finite group, and let A, B be G-modules. There exists a unique family of homomorphisms*

$$\widehat{H}^p(G, A) \otimes_{\mathbb{Z}} \widehat{H}^q(G, B) \to \widehat{H}^{p+q}(G, A \otimes B)$$

for every ordered pair $(p, q) \in \mathbb{Z} \times \mathbb{Z}$ (denoted by $a \otimes b \mapsto a.b$) that satisfies the following conditions:

(i) *they are functorial in A and in B;*

(ii) *when $p = 0 = q$, they are induced by the obvious product $A^G \otimes B^G \to (A \otimes B)^G$;*

(iii) *if $0 \to A' \to A \to A'' \to 0$ is an exact sequence of G-modules, and if*

$$0 \to A' \otimes B \to A \otimes B \to A'' \otimes B \to 0$$

is an exact sequence, then

$$(\delta a'').b = \delta(a''.b) \in \widehat{H}^{p+q+1}(G, A' \otimes B),$$

where $a'' \in \widehat{H}^p(G, A'')$ and $b \in \widehat{H}^q(G, B)$;

(iv) *if $0 \to B' \to B \to B'' \to 0$ is an exact sequence of G-modules and if*

$$0 \to A \otimes B' \to A \otimes B \to A \otimes B'' \to 0$$

is exact, then

$$a.(\delta b'') = (-1)^p \delta(a.b'') \in \widehat{H}^{p+q+1}(G, A \otimes B)$$

for all $a \in \widehat{H}^p(G, A)$ and $b'' \in \widehat{H}^q(G, B'')$.

Proof. See any of the following: Brown, *Cohomology of Groups*, Chapter V; Cassels–Fröhlich, *Algebraic Number Theory*, p. 105; Evens, *The Cohomology of Groups*, Chapter 3; Serre, *Local Fields*, p. 139; or Weiss, *Cohomology of Groups*, Chapter 4. •

The special case $A = k = B$, where k is a commutative ring viewed as a trivial G-module, is most interesting. Now $\widehat{H}^*(G, k) = \bigoplus_{q \geq 0} \widehat{H}^q(G, k)$, called the **cohomology ring of** G **over** k, is a graded ring with multiplication **cup product**, where if $u \in \widehat{H}^q(G, k)$ and $v \in \widehat{H}^q(G, k)$, then

$$u \cup v = \mu^*(u.v),$$

where $\mu: k \otimes k \to k$ is the multiplication in the ring k, $u.v \in \widehat{H}^{p+q}(G, k \otimes k)$, and $\mu^*: \widehat{H}^{p+q}(G, k \otimes k) \to \widehat{H}^{p+q}(G, k)$. This added structure has important applications. For example, cup product is used to prove the **Integral Duality Theorem**: $H_{p-1}(G, \mathbb{Z}) \cong \widehat{H}^p(G, \mathbb{Z})$ for all $p \in \mathbb{Z}$. It is a theorem of Evens and Venkov, independently, that if G is a finite group and k is a noetherian commutative ring on which G acts trivially, then the cohomology ring $\widehat{H}^*(G, k)$ is a noetherian k-algebra (see Evens, *The Cohomology of Groups*, p. 92).

9.8   Outer Automorphisms of p-Groups

If G is a group of order p, where p is prime, then $\mathrm{Aut}(G) \cong \mathbb{I}_{p-1}$, by Exercise 9.10 on page 504 and, of course, every automorphism except 1_G is outer (because G is abelian). We are now going to prove a theorem of Gaschütz, "Nichtabelsche p-Gruppen besitzen äussere p-Automorphismen," *J. Algebra* 4 (1966), 1–2: if G is a finite p-group with $|G| > p$, then G has an outer automorphism of order p (it is not obvious that G has any outer automorphisms!).

Lemma 9.106 (Gaschütz).   *Let p be a prime, let G be a finite p-group, and let A be a G-module that is also a finite p-group. If $H^1(G, A) = \{0\}$, then $H^1(S, A) = \{0\} = H^2(S, A)$ for all subgroups $S \subseteq G$.*

Remark.   Gaschütz proved, given the hypotheses, that $H^q(S, A) = \{0\}$ for all $q \geq 1$ and all subgroups $S \subseteq G$.   ◄

Proof.   We prove the lemma by induction on $|G| \geq 1$. The base step is true, by Corollary 9.28. Assume now that $|G| > 1$, and choose a maximal subgroup $M \subseteq G$. Note that $M \lhd G$ and $|G/M| = p$.[11] By Theorem 9.84(i), there is an exact sequence

$$0 \to H^1(G/M, A^M) \to H^1(G, A) \to H^1(M, A)^{G/M} \to H^2(G/M, A^M).$$

Now $H^1(G, A) = \{0\}$, by hypothesis, so that $H^1(G/M, A^M) = \{0\}$, by exactness. Recall that if B is a finite J-module, where J is a finite cyclic group, then its *Herbrand quotient* is $h(B) = |H^2(J, B)|/|H^1(J, B)|$; Corollary 9.31 says that $h(B) = 1$. As A^M is a finite G/M-module, we have $h(A^M) = 1$, and so

$$H^2(G/M, A^M) = \{0\}.$$

The exact sequence now gives

$$H^1(M, A)^{G/M} = \{0\}.$$

Since $H^1(M, A)$ is finite (Corollary 9.41) of p-power order (Proposition 9.40) acted on by the p-group G/M, there must be a fixed point if $H^1(M, A) \neq \{0\}$ (Suzuki, *Group Theory* I, p. 87); we conclude that $H^1(M, A) = \{0\}$.

Now let S be any proper subgroup of G. There is a maximal subgroup M containing S. By induction, $H^1(S, A) = \{0\} = H^2(S, A)$, and it remains to show that $H^2(G, A) = \{0\}$. By Theorem 9.84(ii), there is an exact sequence

$$0 \to H^2(G/M, A^M) \to H^2(G, A) \to H^2(M, A)^{G/M}.$$

We have already seen that $H^2(G/M, A^M) = \{0\}$, while $H^2(M, A)^{G/M} = \{0\}$ because $H^2(M, A) = \{0\}$. Exactness now gives $H^2(G, A) = \{0\}$.   •

---

[11]Rotman, *An Introduction to the Theory of Groups*, Theorem 5.40.

As is common in the Theory of Groups, many proofs proceed by a series of reductions. We follow the proof in Gruenberg, *Cohomological Topics in Group Theory*, pp. 110–115. In contrast to earlier notation in this chapter, all groups and subgroups will now be written multiplicatively.

Lemma 9.107. *Let p be a prime, let G be a finite p-group with $|G| > p$, and let A be a normal abelian subgroup. If $H^1(G/A, A) \neq \{0\}$, then there exists an outer automorphism of G of order p.*

Proof. Let $\varphi \colon G \to G$ be an automorphism that stabilizes the extension $1 \to A \to G \to G/A \to 1$. Now $H^1(G/A, A) \cong \mathrm{Stab}(G, A)/\mathrm{Inn}(G, A)$, by Corollary 9.20, so that the hypothesis gives such a φ that is not an inner automorphism. Moreover, $|G/A|H^1(G/A, A) = \{0\}$, by Proposition 9.40, and so every φ has order some power of p.   •

Lemma 9.108. *Let p be a prime, and let G be a finite p-group with $|G| > p$. If there is a maximal subgroup M with $Z(M) \subseteq Z(G)$, then there exists an outer automorphism of G of order p.*

Proof. We have noted earlier that $M \lhd G$ and $|G/M| = p$. Now $M \neq \{1\}$ because $|G| > p$, and so $Z(M) \neq \{1\}$.[12] In particular, $Z(G) \neq \{1\}$. Now there exists a homomorphism $f \colon G \to Z(M)$ with $M = \ker f$; for example, let f be the composite of the natural map $G \to G/M$ with a map taking a generator of G/M to an element of order p in $Z(M)$. Define $\varphi \colon G \to G$ by

$$\varphi(x) = xf(x).$$

Note that φ fixes M pointwise, for if $x \in M$, then $x \in \ker f$, and so $\varphi(x) = xf(x) = x$. Conversely, if $\varphi(x) = x$, then $xf(x) = x$, $f(x) = 1$, and $x \in \ker f = M$. It follows that $\varphi \neq 1_G$. Now φ is a homomorphism:

$$\varphi(xy) = xyf(xy) = xyf(x)f(y) = xf(x)yf(y) = \varphi(x)\varphi(y),$$

for $f(x) \in Z(M) \subseteq Z(G)$. The map φ is injective: if $1 = \varphi(x) = xf(x)$, then $x = f(x)^{-1} \in Z(M) \subseteq M$; hence, $\varphi(x) = x$ and $x = 1$. Since G is finite, $\varphi \in \mathrm{Aut}(G)$. Now $\varphi^p(x) = xf(x)^p = x$, for $\mathrm{im}\, f$ is cyclic of order p, and so $\varphi^p = 1_G$. Therefore, φ is an automorphism of order p.

If φ is inner, there is $g \in G$ with $\varphi(x) = gxg^{-1}$ for all $x \in G$. But φ fixes M pointwise, so that if $x \in M$, then $x = \varphi(x) = gxg^{-1}$; that is, $g \in C_G(M)$, the *centralizer of M in G*. If $g \in M$, then $g \in C_G(M) \cap M = Z(M) \subseteq Z(G)$, which says that conjugation by g is 1_G, contradicting $\varphi \neq 1_G$. Therefore, $g \notin M$. By maximality, $G = \langle g \rangle M$. But this also implies $g \in Z(G)$: every

---

[12]Ibid., Theorem 5.41(i): if H is a nontrivial normal subgroup of a finite p-group G, then $H \cap Z(G) \neq \{1\}$.

$z \in G$ has the form $z = g^i m$ for some $m \in M$ and $i \geq 0$, and so g commutes with z. Again, $g \in Z(G)$, so that conjugation by g is 1_G in this case as well, a contradiction. Therefore, φ is outer. $\bullet$

There are finite groups that have no outer automorphisms. For example, if $n \neq 2, 6$, then every automorphism of the symmetric group S_n is inner.[13]

Theorem 9.109 (Gaschütz). *If p is a prime and G is a finite p-group with $|G| > p$, then there exists an outer automorphism of G of order p.*[14]

Proof. Let A be a maximal abelian normal subgroup of G. We may assume that $H^1(G/A, A) = \{0\}$, by Lemma 9.107, and so Lemma 9.106 gives $H^2(G/A, A) = \{0\}$; by the Schur-Zassenhaus Theorem, $G = A \rtimes Q$, where $Q \cong G/A$.

Choose a maximal subgroup M containing Q; by Lemma 9.108, we may assume that $Z(M) \not\subseteq Z(G)$. Write

$$B = A \cap M \quad \text{and} \quad C = A \cap Z(M).$$

We are going to prove that B is cyclic. Now $M \triangleleft G$ and $|G/M| = p$.[15] Since $Z(M) \not\subseteq Z(G)$, there is $y \in Z(M)$ with $y \notin Z(G)$. Hence, $G = AM$, because $G = A \rtimes Q$ and $M \supseteq Q$. If $y \in A$, then y commutes with all elements in $Q \subseteq M$ and in A, giving $y \in Z(G)$, a contradiction. Thus, $y \notin A$ and, hence, $y \notin C = A \cap Z(M)$. Replacing y by a suitable pth power if necessary, we may assume that $y^p \in C$. Hence, $|\langle y, C \rangle / C| = p$, and so $S = \langle y, C \rangle A/A$ is cyclic with generator $s = yA$ of order p. The normal subgroup A is an S-module, where s acts as conjugation by y. We claim that $A^S \neq A$; otherwise, $y \in C_G(A) = A$, the *centralizer of A in G*;[16] that is, if $gag^{-1} = a$ for all $a \in A$, then $g \in A$, contradicting $y \notin A$. Now $B = A \cap M \subseteq A^S$, for $y \in Z(M)$ fixes M pointwise. For the reverse inclusion, it is obvious that $A^S \subseteq A$. If $A^S \not\subseteq M$, then maximality of M gives $G = A^S M$, so that $y \in Z(G)$, a contradiction. Therefore, $A^S = B$. Since $S = A\langle y, C \rangle / A \subseteq G/A$, Lemma 9.106 gives $H^2(S, A) = \{0\}$. But $H^2(S, A) = A^S/NA$, by Theorem 9.27, where $N = 1 + s + \cdots + s^{p-1}$, so that $A^S = NA$. Thus, $A/B = A/(A \cap M) \cong AM/M = G/M$, which is

---

[13] Ibid., p. 158.

[14] If one merely wants the existence of outer automorphisms (not necessarily of order p), then the hypothesis can be weakened to "$|G| > 2$." If p is a prime and $|G| = p$, then G is cyclic, $|\operatorname{Aut}(G)| = p - 1$, and there are no automorphisms of order p; if $|G| = 2$, then $\operatorname{Aut}(G) = \{1\}$ and there are no outer automorphisms at all.

[15] Ibid., Theorem 5.40: if M is a maximal subgroup of a finite p-group G, then $M \triangleleft G$ and $|G/M| = p$.

[16] Ibid., Theorem 5.41(ii): if A is a maximal abelian normal subgroup of a finite p-group G, then $C_G(A) = A$

cyclic of order p. Hence, $A = \langle a, B \rangle$, where $a \in A$, $a^p \in B$, and

$$B = A^S = NA = \langle Na, NB \rangle = \langle Na, B^p \rangle = \langle Na \rangle,$$

because $B^p \subseteq \Phi(B)$.[17] We have shown that B is cyclic.

Suppose that every abelian normal subgroup of G is cyclic. Since $G = A \rtimes Q$, we have $M = (A \cap M) \rtimes Q = B \rtimes Q$, by Exercise 9.6 on page 503. Note that $C = A \cap Z(M) \lhd G$, for it is the intersection of two normal subgroups. We claim that $B \neq C$. Now

$$\frac{B}{C} = \frac{A \cap M}{A \cap Z(M)} = \frac{A \cap M}{(A \cap M) \cap Z(M)} \cong \frac{(A \cap M)Z(M)}{Z(M)} = \frac{BZ(M)}{Z(M)}.$$

If $B = C$, then $BZ(M) = Z(M)$, so that $B \subseteq Z(M)$ and Exercise 9.6 on page 503 gives $Z(M) = B \rtimes (Q \cap Z(M)) = B \times (Q \cap Z(M))$. But $Z(M)$, B, and $Q \cap Z(M)$ are nontrivial groups, the latter being nontrivial because $Q \subseteq M$. This cannot be, for cyclic p-groups are indecomposable. Thus, $B/C \neq \{1\}$, so that $Z(G/C) \cap B/C \neq \{1\}$, by [If H is a nontrivial normal subgroup of a finite p-group G, then $H \cap Z(G) \neq \{1\}$. In particular, $Z(G) \neq \{1\}$,[18], and so there is a nontrivial $xC \in B/C$ commuting with every $gC \in G/C$; that is, $x \in B$, $x \notin C$, and $xgx^{-1}g^{-1}C = C$ for all $g \in G$. Thus, $\langle x, C \rangle \lhd G$; since $\langle x, C \rangle$ is abelian, it is cyclic, by hypothesis. Now

$$\frac{\langle x, C \rangle Z(M)}{C} \cong \frac{\langle x, C \rangle}{C} \times \frac{Z(M)}{C},$$

for $\langle x, C \rangle \subseteq B$ implies $\langle x, C \rangle \cap Z(M) \subseteq B \cap Z(M) = C$. But both factors on the right-hand side are nontrivial: $\langle x, C \rangle \neq C$, because $x \notin C$; $Z(M) \neq C$, because $y \in Z(M)$, but $y \notin C$. The indecomposability of cyclic p-groups has been contradicted again, and we conclude that not all abelian normal subgroups of G can be cyclic.

We may now assume that G contains a noncyclic abelian normal subgroup, say, D. There is a maximal abelian normal subgroup A containing D, and it is not cyclic (for every subgroup of a cyclic group is cyclic). Knowing that A is not cyclic, Exercise 9.5 on page 503 says that the exact sequence $0 \to B \to A \to \mathbb{I}_p \to 0$ splits: $A = B \times \langle a \rangle$, where $a^p = 1$ (for $A/B \cong G/M$ is cyclic of order p). We have seen that $B = \langle Na \rangle$ if $A = \langle a, B \rangle$ and $a^p \in B$ (recall that $N = 1 + s + \cdots + s^{p-1}$, where $S = \langle s \rangle$ and $s^p = 1$). Now $sa = ba^i$ for some $b \in B$ and $i \geq 1$, for $sa \in A$; since $a^p = 1$, we must have $b^p = 1$. Iterating, $s^j a = b^{1+i+\cdots+j} a^{i^j}$. In particular,

---

[17] Ibid., Theorem 5.48: if G is a finite p-group, then $\Phi(G) = G^p G'$, where $G^p = \langle g^p : g \in G \rangle$ and G' is the commutator subgroup, and Theorem 5.47: the Frattini subgroup $\Phi(G)$ consists of all the *nongenerators* of G; that is, if $G = \langle X, \Phi(G) \rangle$, then $G = \langle X \rangle$.

[18] Ibid., Theorem 5.41(i).

if $j = p$, then $s^p = 1$ gives $a = s^p a = b^{1+i+\cdots+i^p} a^{i^p}$. By Fermat's Theorem, $i^p \equiv i \bmod p$, so that $a^{1-i} \in B \cap \langle a \rangle = \{1\}$. Since $a^p = 1$, we have $i \equiv 1 \bmod p$, and so $sa = ba$ and $s^j a = b^j a$. Hence,

$$Na = a(ba)(b^2 a) \cdots (b^{p-1} a) = b^{p(p-1)/2} a^p = b^{p(p-1)/2}.$$

If p is odd, then $Na = b^{p(p-1)/2} = 1$ (because $b^p = 1$), so that $B = \langle Na \rangle = \{1\}$ and $A = \langle a \rangle$ has order p, contradicting our hypothesis that A is not cyclic. Therefore, $p = 2$, $|B| = 2$, and $A \cong \mathbf{V}$. Now $N_G(A) = G$, the *normalizer of A in G*, because $A \lhd G$, and $A = C_G(A)$;[19] that is, if $gag^{-1} = a$ for all $a \in A$, then $g \in A$. Therefore, [If $A \subseteq G$, then $C_G(A) \lhd N_G(A)$, and there is an imbedding $N_G(A)/C_G(A) \to \operatorname{Aut}(A)$[20] gives G/A imbedded in $\operatorname{Aut}(A)$. But $A \cong \mathbf{V}$ and $\operatorname{Aut}(\mathbf{V}) \cong S_3$. Since G/A is a 2-group and $|S_3| = 6$, we have $|G/A| = 2$ and $|G| = 8$.

There are only five groups of order 8: three abelian ones, $\mathbf{Q}$, and D_8, and Exercise 9.11 on page 504 shows that each of these has an outer automorphism of order 2. •

We mention another nice result.

Theorem. *If G is a finite group, then G is nilpotent if and only if whenever A is a finite G-module for which $\widehat{H}^n(G, A) = \{0\}$ for some $n \in \mathbb{Z}$, then $\widehat{H}^q(G, A) = \{0\}$ for all $q \in \mathbb{Z}$.*

Proof. K. Hoechsmann, P. Roquette, and H. A. Zassenhaus, "A cohomological characterization of finite nilpotent groups," *Arch. Mat.* 19 (1968), 225–244. •

9.9   Cohomological Dimension

We have interpreted $H^n(G, A)$ for $n = 0, 1, 2, 3$, but are higher-dimensional groups of any value? The following definition is reminiscent of global dimension of rings.

Definition. A group G has *cohomological dimension* $\leq n$, denoted by $\operatorname{cd}(G) \leq n$, if

$$H^q(G, A) = \{0\}$$

for all $q > n$ and all G-modules A. Define $\operatorname{cd}(G) = \infty$ if no such integer n exists.

We say that $\operatorname{cd}(G) = n$ if $\operatorname{cd}(G) \leq n$ and it is not true that $\operatorname{cd}(G) \leq n - 1$; that is, $\operatorname{cd}(G) \leq n$ and $H^n(G, A) \neq \{0\}$ for some G-module A.

---

[19]Ibid., Theorem .41(ii): if A is a maximal abelian normal subgroup of a finite p-group G, then $C_G(A) = A$.

[20]Ibid., Theorem 7.1(i).

Example 9.110.

(i) If $G = \{1\}$, then $\mathrm{cd}(G) = 0$, which is merely a restatement of Corollary 9.28. We will prove the converse in Corollary 9.113.

(ii) If G is a finite cyclic group of order $k > 1$, then $\mathrm{cd}(G) = \infty$, as we see from Corollary 9.30 with $A = \mathbb{Z}$.

(iii) If $G \cong \mathbb{Z}$ is an infinite cyclic group, then $\mathrm{cd}(G) \leq 1$, by Theorem 9.55. If $\mathrm{cd}(G) = 0$, then $H^1(G, A) = \{0\}$ for all modules A. In particular, $H^1(G, \mathbb{Z}) = \mathrm{Der}(G, \mathbb{Z})/\mathrm{PDer}(G, \mathbb{Z}) \cong \mathrm{Hom}(\mathbb{Z}, \mathbb{Z}) \neq \{0\}$. Hence, $\mathrm{cd}(G) = 1$.

(iv) If $G \neq \{1\}$ is free, then the argument just given in (iii) for $\mathbb{Z}$ applies here, so that $\mathrm{cd}(G) = 1$.

(v) If G is a free abelian group of finite rank n, then $\mathrm{cd}(G) = n$ (see Theorem 10.57). ◄

Proposition 9.111. *If G is a group and $S \subseteq G$ is a subgroup, then*

$$\mathrm{cd}(S) \leq \mathrm{cd}(G).$$

Proof. We may assume that $\mathrm{cd}(G) = n < \infty$. If $m > n$, there is a $\mathbb{Z}S$-module A with $H^m(S, A) \neq \{0\}$, and the Eckmann–Shapiro Lemma, Proposition 9.76, applies to give $H^m(G, \mathrm{Hom}_{\mathbb{Z}S}(\mathbb{Z}G, A)) \cong H^m(S, A) \neq \{0\}$, contradicting $\mathrm{cd}(G) = n$. •

Corollary 9.112. *A group G of finite cohomological dimension is torsion-free*; *that is, G has no elements of finite order (other than 1).*

Proof. The statement follows from Proposition 9.111, for if S is a finite cyclic subgroup with $|S| > 1$, then $\mathrm{cd}(S) = \infty$. •

Corollary 9.113. *A group $G = \{1\}$ if and only if $\mathrm{cd}(G) = 0$.*

Proof. If $G = \{1\}$, then $\mathrm{cd}(G) = 0$, by Example 9.110(i). Conversely, if $\mathrm{cd}(G) = 0$, then Proposition 9.111 gives $\mathrm{cd}(S) = 0$ for every cyclic subgroup $S \subseteq G$. By Example 9.110(ii), all $S = \{1\}$, and so $G = \{1\}$. •

Are there groups G with $\mathrm{cd}(G) = 1$ that are not free?

Theorem 9.114 (Stallings). *Let G be a finitely presented group, and let $\mathbb{F}_2 G$ be the group algebra of G over $\mathbb{F}_2$. If $H^1(G, \mathbb{F}_2 G)$ has more than two elements, then G is a free product, $G = H * K$, where $H \neq \{1\}$ and $K \neq \{1\}$ (free product is the coproduct in* **Groups**).

Proof. J. Stallings, "On torsion-free groups with infinitely many ends," *Annals Math.* 88 (1968), 312–334 •

As a consequence, he proves the following corollaries.

Corollary 9.115. *If G is a finitely generated group with $\mathrm{cd}(G) = 1$, then G is free.*

The next corollary answers a question of Serre, "Sur la dimension cohomologique des groupes profinis," *Topology* 3 (1965), 413–420.

Corollary 9.116. *If G is a torsion-free finitely generated group having a free subgroup S of finite index, then G is free.*

Proof. Serre ["Cohomologie des groupes discrets," *Annals Math. Studies* 70 (1971), 77-169] proved that if H is a subgroup of finite index in a torsion-free group G, then $\mathrm{cd}(G) = \mathrm{cd}(H)$. Now S is finitely generated[21] (a subgroup of a finitely generated group need not be finitely generated).[22] Hence, $\mathrm{cd}(S) \leq 1$, so that S is free, by Corollary 9.115. By Serre's Theorem, $\mathrm{cd}(G) = 1$; hence G is free, by Corollary 9.115. •

Swan showed that both corollaries remain true if we remove the hypothesis that G be finitely generated.

Theorem 9.117 (Stallings–Swan). *A torsion-free group having a free subgroup of finite index must be free.*

Proof. R. G. Swan, "Groups of cohomological dimension 1," *J. Algebra* 12 (1969), 585–610. •

There are interesting groups of cohomological dimension 2; we merely mention some results whose proofs can be found in Gruenberg, *Cohomological Topics in Group Theory*, Chapter 8.

---

[21]Rotman, *Advanced Modern Algebra*, Corollary 5.91: if G is a finitely generated group, then every subgroup of finite index is also finitely generated

[22]Ibid., Corollary 5.90: if G is a free group of rank 2, then its commutator subgroup G' is free of infinite rank.

Theorem (Berstein). *If $G = \varinjlim_{i \in I} G_i$, where I is a countable directed set, then*

$$\mathrm{cd}(G) \leq 1 + \sup_{i \in I}\{\mathrm{cd}(G_i)\}.$$

Proof.   I. Berstein, "On the dimension of modules and algebras IX; direct limits," *Nagoya Math. J.* 13 (1958), 83–84.   •

Corollary. $\mathrm{cd}(\mathbb{Q}) = 2$, *where $\mathbb{Q}$ is the group of rationals.*

Proof.   Since $\mathbb{Q} = \varinjlim\langle 1/n \rangle$, we have $\mathrm{cd}(\mathbb{Q}) \leq 2$, by Berstein's Theorem. However, $\mathrm{cd}(\mathbb{Q}) \neq 1$ because $\mathbb{Q}$ is not free.   •

Theorem.   *If $*_{i \in I} G_i$ denotes the free product of groups G_i, then*

$$\mathrm{cd}(*_{i \in I} G_i) = \sup_{i \in I}\{\mathrm{cd}(G_i)\}.$$

Proof.   Gruenberg, *Cohomological Topics in Group Theory*, pp. 138–140.   •

Theorem (Lyndon).   *If G is a group having only one defining relation r, that is not a proper power (that is, $r \neq w^h$ for $h > 1$), then $\mathrm{cd}(G) \leq 2$.*

Proof.   Gruenberg, *Cohomological Topics in Group Theory*, pp. 129–130.   •

Gruenberg also mentions a theorem of Papakyriakopoloulos, "On Dehn's lemma and the asphericity of knots," *Annals Math.* 66 (1957), 1–26: if K is a tame knot, then $\mathrm{cd}(\pi_1(\mathbb{R}^3 - K)) \leq 2$, and the knot is trivial if and only if the fundamental group is infinite cyclic.

Exercises

9.33   **(i)**   Give an example of a induced G-module that is not injective.

(ii)   Give an example of an induced G-module that is not projective.

*9.34   The ring $\mathcal{L}_{\mathbb{Z}}[x]$ of all *Laurent polynomials* over $\mathbb{Z}$ in one indeterminate consists of all formal sums $\sum_{i=k}^{n} m_i x^i$, where $m_i \in \mathbb{Z}$ and $k \leq n$ are (possibly negative) integers.

(i)   Prove that $\mathcal{L}_{\mathbb{Z}}[x] \cong \mathbb{Z}G$, where $G \cong \mathbb{Z}$.

(ii)   If $S = \{x^k : k \geq 0\}$, prove that $\mathcal{L}_{\mathbb{Z}}[x] \cong S^{-1}\mathbb{Z}[x]$.

(iii)   If G is the free abelian group with basis $\{x_1, \ldots, x_n\}$, define the ring $\mathcal{L}_{\mathbb{Z}}[x_1, \ldots, x_n]$ of *Laurent polynomials* over $\mathbb{Z}$ in n indeterminates to be $S^{-1}\mathbb{Z}[x_1, \ldots, x_n]$. Prove that $\mathcal{L}_{\mathbb{Z}}[x_1, \ldots, x_n] \cong \mathbb{Z}G$.

(iv) Prove that $\mathrm{cd}(\mathbb{Z}G) \leq n+1$ when G is free abelian of rank n.
Hint. Use Hilbert's Syzygy Theorem.

9.35 Let G be a group. If B and A are G-modules, make $A \otimes_{\mathbb{Z}} B$ into a G-module with diagonal action:

$$g(b \otimes a) = (gb) \otimes (ga).$$

If A is a G-module, let A_0 be its underlying abelian group. Prove that $\mathbb{Z}G \otimes_{\mathbb{Z}} A_0 \cong \mathbb{Z}G \otimes_{\mathbb{Z}} A$ as G-modules.
Hint. Define $f : \mathbb{Z}G \otimes_{\mathbb{Z}} A_0 \to \mathbb{Z}G \otimes_{\mathbb{Z}} A$ by $g \otimes a \mapsto g \otimes ga$.

*9.36 Let G be a finite group and let $0 \to A \to B \to C \to 0$ be an exact sequence of G-modules. Find a formula for the boundary maps, and prove exactness of

$$\widetilde{H}^{-2}(G, C) \to \widetilde{H}^{-1}(G, A) \to \widetilde{H}^{-1}(G, B)$$

and

$$\widetilde{H}^{0}(G, B) \to \widetilde{H}^{0}(G, C) \to \widetilde{H}^{1}(G, A).$$

9.10 Division Rings and Brauer Groups

Brauer groups, which are useful in studying division rings, turn out to be cohomology groups. We begin by discussing some standard definitions and examples, and we state some important theorems without proof (the reader is referred to my book *Advanced Modern Algebra*). Our ultimate goal in this section is illustrate again the value of Homological Algebra, this time by constructing a (noncommutative) division ring of characteristic $p > 0$.

Definition. A *division algebra* over a field k is a division ring regarded as an algebra over its center k.

Example 9.118. The most familiar example of a noncommutative division ring is the *quaternions*, the four-dimensional $\mathbb{R}$-algebra $\mathbb{H}$ with basis $1, i, j, k$, such that

$$i^2 = j^2 = k^2 = -1,$$

$$ij = k = -ji, \quad jk = i = -kj, \quad ki = j = -ik.$$

It is routine to check that $\mathbb{H}$ is an $\mathbb{R}$-algebra with center $\mathbb{R} \cdot 1$. To see that it is a division ring, define the *conjugate* of $h = a + bi + cj + dk \in \mathbb{H}$ to be $\overline{h} = a - bi - cj - dk$. Then $h\overline{h} = a^2 + b^2 + c^2 + d^2$, so that if $h \neq 0$, then $h\overline{h} \neq 0$. If $h \neq 0$, define

$$h^{-1} = \overline{h}/h\overline{h} = \overline{h}/(a^2 + b^2 + c^2 + d^2).$$

If k is a field such that $a^2 + b^2 + c^2 + d^2 = 0$ if and only if $a = b = c = d = 0$, then these formulas show that the four-dimensional k-algebra with basis $1, i, j, k$ is also a division ring (of course, this construction does not apply when $k = \mathbb{C}$). ◀

W. R. Hamilton discovered the quaternions in 1843, and F. G. Frobenius, in 1880, proved that the only division algebras over $\mathbb{R}$ are $\mathbb{R}$, $\mathbb{C}$, and $\mathbb{H}$ (see Theorem 9.128). No other examples of noncommutative division rings were known until *cyclic algebras* were found in the early 1900s, by J. M. Wedderburn and by L. E. Dickson. In 1932, A. A. Albert found an example of a *crossed product algebra* that is not a cyclic algebra, and S. A. Amitsur ("On central division algebras," *Israel J. Math.* 12 (1972), 408–420) found an example of a noncommutative division ring that is not a crossed product algebra.

Definition.   A k-algebra A over a field k is **central simple** if it is finite-dimensional, simple (no two-sided ideals other than A and $\{0\}$), and its center $Z(A) = k$.

Example 9.119.

(i) Every division algebra Δ that is finite-dimensional over its center k is a central simple k-algebra. The quaternions $\mathbb{H}$ is a central simple $\mathbb{R}$-algebra, and every field is a central simple algebra over itself.

(ii) If k is a field, then $\mathrm{Mat}_n(k)$ is a central simple k-algebra, where $\mathrm{Mat}_n(k)$ is the k-algebra of all $n \times n$ matrices with entries in k.

(iii) If A is a central simple k-algebra, then its opposite algebra A^{op} is also a central simple k-algebra. ◀

Theorem 9.120.   *Let A be a central simple k-algebra. If B is a simple k-algebra, then $A \otimes_k B$ is a central simple $Z(B)$-algebra. In particular, if B is a central simple k-algebra, then $A \otimes_k B$ is a central simple k-algebra.*

Proof.   Advanced Modern Algebra, Theorem 9.112. •

It is not generally true that the tensor product of simple k-algebras is again simple; we must pay attention to the centers. In fact, a tensor product of division algebras need not be a division algebra, as we see in the next example.

Example 9.121.   The eight-dimensional $\mathbb{R}$-algebra $\mathbb{C} \otimes_{\mathbb{R}} \mathbb{H}$ is also a four-dimensional $\mathbb{C}$-algebra: a basis is

$$1 = 1 \otimes 1, \quad 1 \otimes i, \quad 1 \otimes j, \quad 1 \otimes k.$$

The reader can prove that the vector space isomorphism $\mathbb{C} \otimes_{\mathbb{R}} \mathbb{H} \to \mathrm{Mat}_2(\mathbb{C})$ with $1 \otimes 1 \mapsto \left[\begin{smallmatrix} 1 & 0 \\ 0 & 1 \end{smallmatrix}\right] = I$, $1 \otimes i \mapsto \left[\begin{smallmatrix} i & 0 \\ 0 & -i \end{smallmatrix}\right]$, $1 \otimes j \mapsto \left[\begin{smallmatrix} 0 & 1 \\ -1 & 0 \end{smallmatrix}\right]$, $1 \otimes k \mapsto \left[\begin{smallmatrix} 0 & i \\ i & 0 \end{smallmatrix}\right]$ is an isomomorphism of $\mathbb{C}$-algebras. ◄

The next theorem puts the existence of the isomorphism in Example 9.121 into the context of central simple algebras.

Notation. If A is an algebra over a field k, then we write $[A : k] = \dim_k(A)$.

Theorem 9.122. *Let k be a field and let A be a central simple k-algebra.*

(i) *If $\overline{k}$ is the algebraic closure of k, then there is an integer n with*

$$\overline{k} \otimes_k A \cong \mathrm{Mat}_n(\overline{k}).$$

(ii) *If A is a central simple k-algebra, then there is an integer n with*

$$[A : k] = n^2.$$

Proof. *Advanced Modern Algebra*, Theorem 9.114. ●

The division ring $\mathbb{H}$ of quaternions is a central simple $\mathbb{R}$-algebra, and so its dimension $[\mathbb{H} : \mathbb{R}]$ must be a square (it is 4). Moreover, since $\mathbb{C}$ is algebraically closed, Theorem 9.122 gives $\mathbb{C} \otimes_{\mathbb{R}} \mathbb{H} \cong \mathrm{Mat}_2(\mathbb{C})$ (Example 9.121 displays an explicit isomorphism).

Definition. A *splitting field* for a central simple k-algebra A is a field extension E/k for which there exists an integer n such that $E \otimes_k A \cong \mathrm{Mat}_n(E)$.

Theorem 9.122 says that the algebraic closure $\overline{k}$ of a field k is a splitting field for every central simple k-algebra A. There always exists a splitting field that is a finite extension of k.

Theorem 9.123. *If D is a division algebra over a field k and E is a maximal subfield of D, then E is a splitting field for D; that is, $E \otimes_k D \cong \mathrm{Mat}_s(E)$, where $s = [D : E] = [E : k]$.*

Proof. *Advanced Modern Algebra*, Theorem 9.118. ●

Corollary 9.124. *If D is a division algebra over a field k, then all maximal subfields have the same degree over k.*

This corollary can be illustrated by Example 9.121. The division algebra $\mathbb{H}$ of quaternions is a four-dimensional $\mathbb{R}$-algebra, and a maximal subfield must have degree 2 over $\mathbb{R}$. And so it is, for $\mathbb{C}$ is a maximal subfield.

Recall that a *unit* in a noncommutative ring A is an element having a two-sided inverse in A.

Theorem 9.125 (Skolem–Noether).   *Let A be a central simple k-algebra over a field k, and let B and B' be isomorphic simple k-subalgebras of A. If $\psi : B \to B'$ is an isomorphism, then there exists a unit $u \in A$ with $\psi(b) = ubu^{-1}$ for all $b \in B$.*

Proof.   *Advanced Modern Algebra*, Corollary 9.121.   •

Theorem 9.126 (Wedderburn).   *Every finite division ring D is a field.*

Proof. (**van der Waerden**) Let $Z = Z(D)$, and let E be a maximal subfield of D. If $d \in D$, then $Z(d)$ is a subfield of D and, hence, there is a maximal subfield E_d containing $Z(d)$. By Corollary 9.124, all maximal subfields have the same degree, hence have the same order, and hence are isomorphic [two finite fields are isomorphic if and only if they have the same order (it is not generally true that maximal subfields of a division algebra are isomorphic)]. For every $d \in D$, the Skolem–Noether theorem says there is $x_d \in D$ with $E_d = x_d E x_d^{-1}$. Therefore, $D = \bigcup_x x E x^{-1}$, and so

$$D^{\times} = \bigcup_x x E^{\times} x^{-1}.$$

If E is a proper subfield of D, then $E^{\times}$ is a proper subgroup of $D^{\times}$. But this equation contradicts a standard exercise of group theory: if H is a proper subgroup of a finite group G, then $G \neq \bigcup_{x \in G} x H x^{-1}$. Therefore, $D^{\times} = E^{\times}$, $D = E$, and D is commutative.   •

Lemma 9.127.   *If Δ is a division algebra over a field k, then a subfield $E \subseteq \Delta$ is a maximal subfield if and only if every $a \in \Delta$ commuting with each $e \in E$ must lie in E.*

Proof.   *Advanced Modern Algebra*, Lemma 9.117.   •

Theorem 9.128 (Frobenius).   *If D is a noncommutative finite-dimensional real division algebra, then $D \cong \mathbb{H}$.*

Proof. If E is a maximal subfield of D, then $[D : E] = [E : \mathbb{R}] \leq 2$. If $[E : \mathbb{R}] = 1$, then $[D : \mathbb{R}] = 1^2 = 1$ and $D = \mathbb{R}$. Hence, $[E : \mathbb{R}] = 2$ and $[D : \mathbb{R}] = 4$. If $[E : \mathbb{R}] = 2$, then $E \cong \mathbb{C}$; let us identify E with $\mathbb{C}$. Now complex conjugation is an automorphism of E, so that the Skolem–Noether theorem gives $x \in D$ with $\bar{z} = xzx^{-1}$ for all $z \in E$. In particular, $-i = xix^{-1}$. Hence,

$$x^2 i x^{-2} = x(-i)x^{-1} = -xix^{-1} = i,$$

and so x^2 commutes with i. Therefore, $x^2 \in C_D(E) = E$, by Lemma 9.127, and so $x^2 = a + bi$ for $a, b \in \mathbb{R}$. But

$$a + bi = x^2 = xx^2x^{-1} = x(a + bi)x^{-1} = a - bi,$$

so that $b = 0$ and $x^2 \in \mathbb{R}$. If $x^2 > 0$, then there is $t \in \mathbb{R}$ with $x^2 = t^2$. Now $(x + t)(x - t) = 0$ gives $x = \pm t \in \mathbb{R}$, contradicting $-i = xix^{-1}$. Therefore, $x^2 = -r^2$ for some real r. The element j, defined by $j = x/r$, satisfies $j^2 = -1$ and $ji = -ij$. The list $1, i, j, ij$ is linearly independent over $\mathbb{R}$: if $a + bi + cj + dij = 0$, then $(-di - c)j = a + ib \in \mathbb{C}$. Since $j \notin \mathbb{C}$ (lest $x \in \mathbb{C}$), we must have $-di - c = 0 = a + bi$. Hence, $a = b = 0 = c = d$. Since $[D : \mathbb{R}] = 4$, the list $1, i, j, ij$ is a basis of D. It is now routine to see that if we define $k = ij$, then $ki = j = -ik$, $jk = i = -kj$, and $k^2 = -1$, and so $D \cong \mathbb{H}$. $\bullet$

Brauer introduced the Brauer group to study division rings. Since construction of division rings was notoriously difficult, he considered the wider class of central simple algebras. Theorem 9.131 shows the success of this approach.

Definition. Two central simple k-algebras A and B are *similar*, denoted by $A \sim B$, if there are integers n and m with

$$A \otimes_k \mathrm{Mat}_n(k) \cong B \otimes_k \mathrm{Mat}_m(k).$$

By the Wedderburn theorem, $A \cong \mathrm{Mat}_n(\Delta)$ for a unique division algebra Δ over k, and we shall see that $A \sim B$ if and only if they determine the same division algebra.

Theorem 9.123 can be extended from division algebras to central simple algebras.

Theorem 9.129. *Let A be a central simple k-algebra over a field k, so that A is isomorphic to a ring of matrices over Δ, a division algebra over k. If E is a maximal subfield of Δ, then E splits A; that is, there are an integer n and an isomorphism*

$$E \otimes_k A \cong \mathrm{Mat}_n(E).$$

Proof. *Advanced Modern Algebra*, Theorem 9.127. $\bullet$

Definition. If $[A]$ denotes the equivalence class of a central simple k-algebra A under similarity, define the *Brauer group* $\mathrm{Br}(k)$ to be the set

$$\mathrm{Br}(k) = \big\{ [A] : A \text{ is a central simple } k\text{-algebra} \big\}$$

with binary operation

$$[A][B] = [A \otimes_k B].$$

Theorem 9.130. $\mathrm{Br}(k)$ *is an abelian group for every field k. Moreover, if* $A \cong \mathrm{Mat}_n(\Delta)$ *for a division algebra* Δ, *then* Δ *is central simple and* $[A] = [\Delta]$ *in* $\mathrm{Br}(k)$.

Proof.   *Advanced Modern Algebra*, Theorem 9.128. The operation is well-defined, the identity is $[k]$, and $[A]^{-1} = [A^{\mathrm{op}}]$.   •

The next theorem shows the significance of the Brauer group.

Theorem 9.131.   *If k is a field, then there is a bijection from* $\mathrm{Br}(k)$ *to the family* $\mathcal{D}$ *of all isomorphism classes of finite-dimensional division algebras over k, and so*

$$|\mathrm{Br}(k)| = |\mathcal{D}|.$$

Therefore, there exists a noncommutative division ring, finite-dimensional over its center k, if and only if $\mathrm{Br}(k) \neq \{0\}$.

Proof.   Define a function $\varphi \colon \mathrm{Br}(k) \to \mathcal{D}$ by setting $\varphi([A])$ to be the isomorphism class of Δ, where $A \cong \mathrm{Mat}_n(\Delta)$. Note that Theorem 9.130 shows that $[A] = [\Delta]$ in $\mathrm{Br}(k)$. Let us see that φ is well-defined. If $[\Delta] = [\Delta']$, then $\Delta \sim \Delta'$, so there are integers n and m with $\Delta \otimes_k \mathrm{Mat}_n(k) \cong \Delta' \otimes_k \mathrm{Mat}_m(k)$. Hence, $\mathrm{Mat}_n(\Delta) \cong \mathrm{Mat}_m(\Delta')$. By the uniqueness in the Wedderburn–Artin theorems, $\Delta \cong \Delta'$ (and $n = m$). Therefore, $\varphi([\Delta]) = \varphi([\Delta'])$.

Clearly, φ is surjective, for if Δ is a finite-dimensional division algebra over k, then the isomorphism class of Δ is equal to $\varphi([\Delta])$. To see that φ is injective, suppose that $\varphi([\Delta]) = \varphi([\Delta'])$. Then, $\Delta \cong \Delta'$, which implies $\Delta \sim \Delta'$.   •

Example 9.132.

(i) If k is an algebraically closed field, then $\mathrm{Br}(k) = \{0\}$ (Theorem 9.122).

(ii) If k is a finite field, then $\mathrm{Br}(k) = \{0\}$ (Wedderburn's Theorem 9.126).

(iii) If $k = \mathbb{R}$, then Frobenius's Theorem 9.128 shows that $\mathrm{Br}(\mathbb{R}) \cong \mathbb{I}_2$, for its only nonzero element is $[\mathbb{H}]$.   ◀

Proposition 9.133.   *If E/k is a field extension, then there is a homomorphism*

$$f_{E/k} \colon \mathrm{Br}(k) \to \mathrm{Br}(E)$$

given by $[A] \mapsto [E \otimes_k A]$.

Proof. If A and B are central simple k-algebras, then $E \otimes_k A$ and $E \otimes_k B$ are central simple E-algebras, by Theorem 9.120. If $A \sim B$, then $E \otimes_k A \sim E \otimes_k B$ as E-algebras, by Exercise 9.42 on page 606. It follows that the function $f_{E/k}$ is well-defined. Finally, $f_{E/k}$ is a homomorphism, because

$$(E \otimes_k A) \otimes_E (E \otimes_k B) \cong (E \otimes_E E) \otimes_k (A \otimes_k B) \cong E \otimes_k (A \otimes_k B),$$

by associativity of tensor product.   •

Definition.   If E/k is a field extension, then the ***relative Brauer group***, $\mathrm{Br}(E/k)$, is the kernel of homomorphism $f_{E/k} \colon \mathrm{Br}(k) \to \mathrm{Br}(E)$:

$$\mathrm{Br}(E/k) = \ker f_{E/k} = \big\{ [A] \in \mathrm{Br}(k) : A \text{ is split by } E \big\}.$$

Corollary 9.134.   *For every field k, we have*

$$\mathrm{Br}(k) = \bigcup_{E/k \text{ finite}} \mathrm{Br}(E/k).$$

Proof.   This follows at once from Theorem 9.129.   •

We now show that the Brauer group is related to cohomology. Suppose that V is a vector space over a field E having basis $\{u_\sigma : \sigma \in G\}$ for some set G, so that each $v \in V$ has a unique expression as an E-linear combination $v = \sum_\sigma a_\sigma u_\sigma$ for $a_\sigma \in E$. For a function $\mu \colon V \times V \to V$, with $\mu(u_\sigma, u_\tau)$ denoted by $u_\sigma u_\tau$, define ***structure constants*** $g_\alpha^{\sigma,\tau} \in E$ by

$$u_\sigma u_\tau = \sum_{\alpha \in G} g_\alpha^{\sigma,\tau} u_\alpha.$$

To have the associative law, we must have $u_\sigma(u_\tau u_\omega) = (u_\sigma u_\tau)u_\omega$; expanding this equation, the coefficient of each u_β is

$$\sum_\alpha g_\alpha^{\sigma,\tau} g_\beta^{\alpha,\omega} = \sum_\gamma g_\gamma^{\tau,\omega} g_\beta^{\sigma,\gamma}.$$

Let us simplify these equations. Let G be a group and suppose that $g_\alpha^{\sigma,\tau} = 0$ unless $\alpha = \sigma\tau$; that is, $u_\sigma u_\tau = f(\sigma, \tau)u_{\sigma\tau}$, where $f(\sigma, \tau) = g_{\sigma\tau}^{\sigma,\tau}$. The function $f \colon G \times G \to E^\times$, given by $f(\sigma, \tau) = g_{\sigma\tau}^{\sigma,\tau}$, satisfies the following equation for all $\sigma, \tau, \omega \in G$:

$$f(\sigma, \tau)f(\sigma\tau, \omega) = f(\tau, \omega)f(\sigma, \tau\omega),$$

an equation reminiscent of the cocycle identity written in multiplicative notation. This is why factor sets enter into the next definition.

Let E/k be a Galois extension with $\mathrm{Gal}(E/k) = G$, and let $f: G \times G \to E^{\times}$ be a factor set: in multiplicative notation,

$$f(\sigma, 1) = 1 = f(1, \tau) \quad \text{for all } \sigma, \tau \in G,$$

and, if we denote the action of $\sigma \in G$ on $a \in E^{\times}$ by a^{σ}, then

$$f(\sigma, \tau) f(\sigma\tau, \omega) = f(\tau, \omega)^{\sigma} f(\sigma, \tau\omega).$$

Definition. Given a Galois extension E/k with Galois group $G = \mathrm{Gal}(E/k)$ and a factor set $f: G \times G \to E^{\times}$, define the **crossed product algebra** (E, G, f) to be the vector space over E having as a basis the set of all symbols $\{u_{\sigma} : \sigma \in G\}$ and multiplication

$$(a u_{\sigma})(b u_{\tau}) = a b^{\sigma} f(\sigma, \tau) u_{\sigma\tau}$$

for all $a, b \in E$. If G is a cyclic group, then the crossed product algebra (E, G, f) is called a **cyclic algebra**.

Since every element in (E, G, f) has a unique expression of the form $\sum a_{\sigma} u_{\sigma}$, the definition of multiplication extends by linearity to (E, G, f). We note two special cases:

$$u_{\sigma} b = b^{\sigma} u_{\sigma};$$
$$u_{\sigma} u_{\tau} = f(\sigma, \tau) u_{\sigma\tau}.$$

Proposition 9.135. *If E/k is a Galois extension with Galois group $G = \mathrm{Gal}(E/k)$, and if $f: G \times G \to E^{\times}$ is a factor set, then (E, G, f) is a central simple k-algebra that is split by E.*

Proof. Denote (E, G, f) by A. First, we show that A is a k-algebra. To prove that A is associative, it suffices to prove that

$$a u_{\sigma} (b u_{\tau} c u_{\omega}) = (a u_{\sigma} b u_{\tau}) c u_{\omega},$$

where $a, b, c \in E$. Using the definition of multiplication,

$$\begin{aligned}
a u_{\sigma} (b u_{\tau} c u_{\omega}) &= a u_{\sigma} (b c^{\tau} f(\tau, \omega) u_{\tau\omega}) \\
&= a \big(b c^{\tau} f(\tau, \omega)\big)^{\sigma} f(\sigma, \tau\omega) u_{\sigma\tau\omega} \\
&= a b^{\sigma} c^{\sigma\tau} f(\tau, \omega)^{\sigma} f(\sigma, \tau\omega) u_{\sigma\tau\omega}.
\end{aligned}$$

We also have

$$\begin{aligned}
(a u_{\sigma} b u_{\tau}) c u_{\omega} &= a b^{\sigma} f(\sigma, \tau) u_{\sigma\tau} c u_{\omega} \\
&= a b^{\sigma} f(\sigma, \tau) c^{\sigma\tau} f(\sigma\tau, \omega) u_{\sigma\tau\omega} \\
&= a b^{\sigma} c^{\sigma\tau} f(\sigma, \tau) f(\sigma\tau, \omega) u_{\sigma\tau\omega}.
\end{aligned}$$

The cocycle identity shows that multiplication in A is associative.

That u_1 is the unit in A follows from our assuming that factor sets are normalized:

$$u_1 u_\tau = f(1, \tau) u_{1\tau} = u_\tau \quad \text{and} \quad u_\sigma u_1 = f(\sigma, 1) u_{\sigma 1} = u_\sigma.$$

We have shown that A is a ring. We claim that $ku_1 = \{au_1 : a \in k\}$ is the center $Z(A)$. If $a \in E$, then $u_\sigma a u_1 = a^\sigma u_\sigma$. If $a \in k = E^G$, then $a^\sigma = a$ for all $\sigma \in G$, and so $k \subseteq Z(A)$. For the reverse inclusion, suppose that $z = \sum_\sigma a_\sigma u_\sigma \in Z(A)$. For any $b \in E$, we have $zbu_1 = bu_1 z$. But

$$zbu_1 = \sum a_\sigma u_\sigma b u_1 = \sum a_\sigma b^\sigma u_\sigma.$$

On the other hand,

$$bu_1 z = \sum b a_\sigma u_\sigma.$$

For every $\sigma \in G$, we have $a_\sigma b^\sigma = b a_\sigma$, so that if $a_\sigma \neq 0$, then $b^\sigma = b$. If $\sigma \neq 1$ and $H = \langle \sigma \rangle$, then $E^H \neq E^{\{1\}} = E$ (see *Advanced Modern Algebra*, Theorem 4.33), and so there exists $b \in E$ with $b^\sigma \neq b$. We conclude that $z = a_1 u_1$. For every $\sigma \in G$, the equation $(a_1 u_1) u_\sigma = u_\sigma (a_1 u_1)$ gives $a_1^\sigma = a_1$, and so $a_1 \in E^G = k$. Therefore, $Z(A) = ku_1$.

We now show that A is simple. Observe first that each u_σ is invertible, for its inverse is $f(\sigma^{-1}, \sigma)^{-1} u_{\sigma^{-1}}$ (remember that $\text{im } f \subseteq E^\times$, so that its values are nonzero). Let I be a nonzero two-sided ideal in A, and choose a nonzero $y = \sum_\sigma c_\sigma u_\sigma \in I$ of shortest length; that is, y has the smallest number of nonzero coefficients. Multiplying by $(c_\sigma u_\sigma)^{-1}$ if necessary, we may assume that $y = u_1 + c_\tau u_\tau + \cdots$. Suppose that $c_\tau \neq 0$. Since $\tau \neq 1_E$, there is $a \in E$ with $a^\tau \neq a$. Now I contains $ay - ya = b_\tau u_\tau + \cdots$, where $b_\tau = c_\tau (a - a^\tau) \neq 0$. Hence, I contains $y - c_\tau b_\tau^{-1}(ay - ya)$, which is shorter than y (it involves u_1 but not u_τ). We conclude that y must have length 1; that is, $y = c_\sigma u_\sigma$. But y is invertible, and so $I = A$ and A is simple.

Finally, Theorem 9.129 says that A is split by K, where K is any maximal subfield of A. The reader may show, using Lemma 9.127, that $Eu_1 \cong E$ is a maximal subfield. •

In light of Proposition 9.135, it is natural to expect a connection between relative Brauer groups and cohomology.

Theorem. *Let E/k be a Galois extension with $G = \text{Gal}(E/k)$. There is an isomorphism $H^2(G, E^\times) \to \text{Br}(E/k)$ with $\text{cls} f \mapsto [(G, E, f)]$.*

Remark. The usual proofs of this theorem are rather long. Each of the items: the isomorphism is a well-defined function; it is a homomorphism; it is injective; it is surjective, must be checked, and the proofs are computational.

For example, see the proof in Herstein, *Noncommutative Rings*, pp. 110–116; there is a less computational proof in Serre, *Local Fields*, pp. 164–167, using the method of *descent*. ◄

What is the advantage of this isomorphism? In Corollary 9.134, we saw that $\mathrm{Br}(k) = \bigcup_{E/k \text{ finite}} \mathrm{Br}(E/k)$.

Corollary 9.136.   *Let k be a field.*

(i) *The Brauer group $\mathrm{Br}(k)$ is a torsion group.*

(ii) *If A is a central simple k-algebra, then there is an integer n so that the tensor product of A with itself r times, where r is the order of $[A]$ in $\mathrm{Br}(k)$, is a matrix algebra: $A \otimes_k A \otimes_k \cdots \otimes_k A \cong \mathrm{Mat}_n(k)$.*

Sketch of proof.

(i) $\mathrm{Br}(k)$ is the union of the relative Brauer groups $\mathrm{Br}(E/k)$, where E/k is finite. It can be shown that $\mathrm{Br}(k)$ is the union of those $\mathrm{Br}(E/k)$ for which E/k is a Galois extension. We may now invoke Proposition 9.40, which says that $|G| H^2(G, E^\times) = \{0\}$.

(ii) Tensor product is the binary operation in the Brauer group.   •

Remark.   It is proved, using Class Field Theory, that $\mathrm{Br}(\mathbb{Q}_p) \cong \mathbb{Q}/\mathbb{Z}$, where $\mathbb{Q}_p$ is the field of p-adic numbers. Moreover, there is an exact sequence

$$0 \to \mathrm{Br}(\mathbb{Q}) \to \mathrm{Br}(\mathbb{R}) \oplus \sum_p \mathrm{Br}(\mathbb{Q}_p) \to \mathbb{Q}/\mathbb{Z} \to 0.$$

In a series of deep papers in the 1930s, $\mathrm{Br}(k)$ was computed for the most interesting fields k of algebraic number theory, by Albert, Brauer, Hasse, and Noether. ◄

Recall Theorem 9.131: there exists a noncommutative division k-algebra over a field k if and only if $\mathrm{Br}(k) \neq \{0\}$.

The following notion of *norm* arises in Algebraic Number Theory.

Definition.   If E/k is a (finite) Galois extension with Galois group $G = \mathrm{Gal}(E/k)$, then the **norm** $N \colon E^\times \to k^\times$ is given by

$$N(u) = \prod_{\sigma \in G} \sigma(u).$$

Corollary 9.137. *Let k be a field. If there is a cyclic Galois extension E/k such that the norm $N: E^\times \to k^\times$ is not surjective, then there exists a non-commutative k-division algebra.*

Sketch of proof. If G is a finite cyclic group, then Theorem 9.27 gives

$$H^2(G, E^\times) = (E^\times)^G / \operatorname{im} N = k^\times / \operatorname{im} N.$$

Therefore, $\operatorname{Br}(E/k) \neq \{0\}$ if N is not surjective, and so $\operatorname{Br}(k) \neq \{0\}$. $\quad\bullet$

If k is a finite field and E/k is a finite extension, then it follows from Wedderburn's theorem on finite division rings that the norm $N: E^\times \to k^\times$ is surjective.

Theorem 9.138. *If p is a prime, then there exists a noncommutative division algebra of characteristic $p > 0$.*

Proof. If k is a field of characteristic p, it suffices to find a cyclic extension E/k for which the norm $N: E^\times \to k^\times$ is not surjective; that is, we must find some $z \in k^\times$ that is not a norm.

If p is an odd prime, let $k = \mathbb{F}_p(x)$. Since p is odd, $t^2 - x$ is a separable irreducible polynomial, and so $E = k(\sqrt{x})$ is a Galois extension of degree 2. If $u \in E$, then there are polynomials $a, b, c \in \mathbb{F}_p[x]$ with $u = (a + b\sqrt{x})/c$. Moreover,

$$N(u) = (a^2 - b^2 x)/c^2.$$

We claim that $x^2 + x$ is not a norm. Otherwise,

$$a^2 - b^2 x = c^2(x^2 + x).$$

Since $c \neq 0$, the polynomial $c^2(x^2 + x) \neq 0$, and it has even degree. On the other hand, if $b \neq 0$, then $a^2 - b^2 x$ has odd degree, and this is a contradiction. If $b = 0$, then $u = a/c$; since $a^2 = c^2(x^2 + x)$, we have $c^2 \mid a^2$, hence $c \mid a$, and so $u \in \mathbb{F}_p[x]$ is a polynomial. But it is easy to see that $x^2 + x$ is not the square of a polynomial. We conclude that $N: E^\times \to k^\times$ is not surjective.

Here is an example in characteristic 2. Let $k = \mathbb{F}_2(x)$, and let $E = k(\alpha)$, where α is a root of $f(t) = t^2 + t + x + 1$ [$f(t)$ is irreducible and separable; its other root is $\alpha + 1$]. As before, each $u \in E$ can be written in the form $u = (a + b\alpha)/c$, where $a, b, c \in \mathbb{F}_2[x]$. Of course, we may assume that x is not a divisor of all three polynomials a, b, and c. Moreover,

$$N(u) = \big((a + b\alpha)(a + b\alpha + b)\big)/c^2 = \big(a^2 + ab + b^2(x + 1)\big)/c^2.$$

We claim that x is not a norm. Otherwise,

$$a^2 + ab + b^2(x + 1) = c^2 x. \tag{1}$$

Now $a(0)$, the constant term of a, is either 0 or 1. Consider the four cases arising from the constant terms of a and b; that is, evaluate Eq. (1) at $x = 0$. We see that $a(0) = 0 = b(0)$; that is $x \mid a$ and $x \mid b$. Hence, $x^2 \mid a^2$ and $x^2 \mid b^2$, so that Eq. (1) has the form $x^2 d = c^2 x$, where $d \in \mathbb{F}_2[x]$. Dividing by x gives $xd = c^2$, which forces $c(0) = 0$; that is, $x \mid c$, and this is a contradiction.   •

For further discussion of the Brauer group, see Gille and Szamuely, *Central Simple Algebras and Galois Cohomology*, Janusz, *Algebraic Number Theory*, and Reiner, *Maximal Orders*.

Exercises

9.37 Prove that $\mathbb{H} \otimes_\mathbb{R} \mathbb{H} \cong \mathrm{Mat}_4(\mathbb{R})$ as $\mathbb{R}$-algebras.

Hint. Every simple left artinian ring, e.g., $\mathbb{H} \otimes_\mathbb{R} \mathbb{H}$, is isomorphic to $\mathrm{Mat}_n(\Delta)$ for some $n \geq 1$ and some division ring Δ.

9.38 We have given one isomorphism $\mathbb{C} \otimes_\mathbb{R} \mathbb{H} \cong \mathrm{Mat}_2(\mathbb{C})$ in Example 9.121. Describe all possible isomorphisms between these two algebras.

Hint. Use the Skolem–Noether Theorem.

9.39 Prove that $\mathbb{C} \otimes_\mathbb{R} \mathbb{C} \cong \mathbb{C} \times \mathbb{C}$ as $\mathbb{R}$-algebras.

9.40 (i) Let $\mathbb{C}(x)$ and $\mathbb{C}(y)$ be function fields. Prove that $R = \mathbb{C}(x) \otimes_\mathbb{C} \mathbb{C}(y)$ is isomorphic to a subring of $\mathbb{C}(x, y)$. Conclude that R has no zero-divisors.

    (ii) Prove that $\mathbb{C}(x) \otimes_\mathbb{C} \mathbb{C}(y)$ is not a field.

    Hint. Show that R is isomorphic to the subring of $\mathbb{C}(x, y)$ consisting of polynomials of the form $f(x, y)/g(x)h(y)$.

9.41 Let A be a central simple k-algebra. If A is split by a field E, prove that A is split by any field extension E' of E.

*9.42 Let E/k be a field extension. If A and B are central simple k-algebras with $A \sim B$, prove that $E \otimes_k A \sim E \otimes_k B$ as central simple E-algebras.

9.43 Prove that $\mathrm{Mat}_2(\mathbb{H}) \cong \mathbb{H} \otimes_\mathbb{R} \mathrm{Mat}_2(\mathbb{R})$ as $\mathbb{R}$-algebras.

9.44 (i) Let A be a four-dimensional vector space over $\mathbb{Q}$, and let $1, i, j, k$ be a basis. Prove that A is a division algebra over $\mathbb{Q}$ if we define 1 to be the identity and

$$i^2 = -1, \qquad j^2 = -2, \qquad k^2 = -2,$$
$$ij = k, \qquad jk = 2i, \qquad ki = j,$$
$$ji = -k, \qquad kj = -2i, \qquad ik = -j.$$

   (ii)   Prove that $\mathbb{Q}(i)$ and $\mathbb{Q}(j)$ are nonisomorphic maximal sub-
          fields of A.

9.45 Let D be the $\mathbb{Q}$-subalgebra of $\mathbb{H}$ having basis $1, i, j, k$.

   (i)   Prove that D is a division algebra over $\mathbb{Q}$.
          Hint. Compute the center $Z(D)$.

   (ii)   For any pair of nonzero rationals p and q, prove that D has
          a maximal subfield isomorphic to $\mathbb{Q}(\sqrt{-p^2 - q^2})$.
          Hint. Compute $(pi + qj)^2$.

9.46 **(Dickson)** If D is a division algebra over a field k, then each $d \in D$
   is algebraic over k. Prove that $d, d' \in D$ are conjugate in D if and
   only if $\mathrm{irr}(d, k) = \mathrm{irr}(d', k)$ [$\mathrm{irr}(d, k)$ is the polynomial in $k[x]$ of
   least degree having d as a root].
   Hint. Use the Skolem–Noether theorem.

9.47 Prove that if A is a central simple k-algebra with $A \sim \mathrm{Mat}_n(k)$, then
   $A \cong \mathrm{Mat}_m(k)$ for some integer m.

9.48 Show that the structure constants in the crossed product (E, G, f)
   are

$$g_\alpha^{\sigma, \tau} = \begin{cases} f(\sigma, \tau) & \text{if } \alpha = \sigma\tau, \\ 0 & \text{otherwise.} \end{cases}$$

10

Spectral Sequences

Given a map $f\colon A \to M$, where A is a submodule of a module B, when can f be extended to a map $B \to M$? Applying $\mathrm{Hom}(\square, M)$ to the exact sequence $0 \to A \xrightarrow{i} B \to B/A \to 0$, where i is the inclusion, gives exactness of

$$\mathrm{Hom}(B, M) \xrightarrow{i^*} \mathrm{Hom}(A, M) \xrightarrow{\partial} \mathrm{Ext}^1(B/A, M).$$

Now there exists a map $g\colon B \to M$ extending f if and only if $f = gi = i^*(g) \in \mathrm{im}\, i^* = \ker \partial$. Thus, we may regard $\partial(f) \in \mathrm{Ext}^1(B/A, M)$ as an *obstruction*, for f can be extended to B if and only if $\partial(f) = 0$. For example, if $\mathrm{Ext}^1(B/A, M) = \{0\}$, then every map $f\colon A \to M$ can be extended to B. But what value would the notion of obstruction have if we could not compute Ext^1? The basic reason for the success of Homological Algebra is that homology groups can often be computed. So far, our most useful techniques have involved dimension shifting and "ladder" diagrams arising from the naturality of connecting homomorphisms. But many problems resist solution by routine application of the axioms for homology. For example, we had to be clever to prove that Tor and Ext are independent of the variable being resolved.

The most important functors in Homological Algebra are Hom, tensor, and their derived functors Ext and Tor, each of which involves two variables. More precisely, Ext and Tor involve resolutions of each variable. Considering two complexes simultaneously leads to *bicomplexes*, and a bicomplex M yields a complex—its *total complex,* $\mathrm{Tot}(M)$. Computing the homology of $\mathrm{Tot}(M)$ involves several steps: there are two *filtrations*, each of which determines a *spectral sequence*, and both spectral sequences *converge* to $H_n(\mathrm{Tot}(M))$.

Of course, the reader must digest these new ideas in order to apply them, but it is worth the effort.

J.J. Rotman, *An Introduction to Homological Algebra*, Universitext, DOI 10.1007/978-0-387-68324-9_10, © Springer Science+Business Media LLC 2009

10.1 Bicomplexes

Almost all spectral sequences arise from bicomplexes, and so we begin discussing them now.

Definition. A *graded module* is an indexed[1] family $M = (M_p)_{p \in \mathbb{Z}}$ of R-modules (for some ring R). Graded modules M are often denoted by $M_\bullet$.

Example 10.1.

(i) If $(\mathbf{C}, d) = \to C_{n+1} \xrightarrow{d_{n+1}} C_n \xrightarrow{d_n} C_{n-1} \to$ is a complex, then $\mathbf{C} = (C_n)_{n \in \mathbb{Z}}$ is a graded module.

(ii) If $(\mathbf{C}, d)$ is a complex, then the family $H_\bullet(\mathbf{C}) = (H_p(\mathbf{C}))_{p \in \mathbb{Z}}$ of its homology modules is a graded module. ◄

Definition. Let M and N be graded modules, and let $a \in \mathbb{Z}$. A *graded map of degree* a, denoted by $f \colon M \to N$, is a family of homomorphisms $f = (f_p \colon M_p \to N_{p+a})_{p \in \mathbb{Z}}$. The *degree* of f is a, and we denote it by $\deg(f) = a$.

Example 10.2.

(i) If $(\mathbf{C}, d)$ is a complex, then its differential $d \colon \mathbf{C} \to \mathbf{C}$, given by $d = (d_n \colon C_n \to C_{n-1})_{n \in \mathbb{Z}}$, is a graded map of degree -1.

(ii) If $f \colon \mathbf{C} \to \mathbf{C}'$ is a chain map, then $f = (f_n \colon C_n \to C'_n)_{n \in \mathbb{Z}}$ is a graded map of degree 0.

(iii) If $f, g \colon \mathbf{C} \to \mathbf{C}'$ are homotopic chain maps, then a homotopy $s = (s_n \colon C_n \to C'_{n+1})_{n \in \mathbb{Z}}$ is a graded map of degree $+1$. ◄

Proposition 10.3. *If $M \xrightarrow{f} N \xrightarrow{g} P$ are graded maps of degree a and b, respectively, then their composite gf is a graded map of degree $a + b$.*

Proof. Since $f_p \colon M_p \to N_{p+a}$ and $g_{p+a} \colon N_{p+a} \to P_{p+a+b}$, we have $gf \colon M_p \to P_{p+a+b}$, and so gf has degree $a + b$. •

---

[1] Other index sets do arise, but we assume that $\mathbb{Z}$ is the index set for graded modules.

Graded modules (over a given ring R) and graded maps form a category. If $f: M \to N$ is a graded map of degree a, then $f \in \prod_p \operatorname{Hom}(M_p, N_{p+a})$, and so

$$\operatorname{Hom}(M, N) = \bigcup_{a \in \mathbb{Z}} \left(\prod_{p \in \mathbb{Z}} \operatorname{Hom}(M_p, N_{p+a}) \right).$$

Let us define exactness in the category of graded modules. If $M' = (M'_p)_{p \in \mathbb{Z}}$ and $M = (M_p)_{p \in \mathbb{Z}}$ are graded modules, then M' is a **submodule** of M if $M'_p \subseteq M_p$ for all p; thus, inclusions are graded maps of degree 0. If $M' \subseteq M$, then the **quotient** is $M/M' = (M_p/M'_p)_{p \in \mathbb{Z}}$, and the natural map $M \to M/M'$ is a graded map of degree 0. If $f: M \to N$ is a graded map of degree a, then there is an inclusion $\ker f = (\ker f_p) \to (M_p)$. On the other hand, we want $\operatorname{im} f \subseteq N$, so that its pth term $(\operatorname{im} f)_p \subseteq N_p$. Thus, we must define

$$\operatorname{im} f = (\operatorname{im} f_{p-a})_{p \in \mathbb{Z}} \subseteq (N_p)_{p \in \mathbb{Z}}.$$

Of course, **exactness** of $A \xrightarrow{f} B \xrightarrow{g} C$ means that $\operatorname{im} f = \ker g$; that is, $\operatorname{im} f_{p-a} = \ker g_p$ for all $p \in \mathbb{Z}$.

Given a short exact sequence $0 \to \mathbf{C}' \xrightarrow{i} \mathbf{C} \xrightarrow{\pi} \mathbf{C}'' \to 0$ of complexes, the long exact sequence of homology modules is sometimes called an *exact triangle*:

$$
\begin{array}{ccc}
H_\bullet(\mathbf{C}') & \xrightarrow{\quad i_* \quad} & H_\bullet(\mathbf{C}) \\
 & {\scriptstyle \partial} \nwarrow \quad \swarrow {\scriptstyle \pi_*} & \\
 & H_\bullet(\mathbf{C}''). &
\end{array}
$$

If we regard each vertex as a graded module, then the arrows are, indeed, graded maps: i_* and π_* have degree 0, and the connecting homomorphism ∂ has degree -1. More generally, given an exact triangle of graded modules and graded maps,

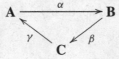

with α, β, γ having degree a, b, c, respectively, we can reconstruct the long exact sequence from which it comes. Choose some $p \in \mathbb{Z}$, and go right and left from A_p:

$$\to B_{p-b-c} \xrightarrow{\beta} C_{p-c} \xrightarrow{\gamma} A_p \xrightarrow{\alpha} B_{p+a} \xrightarrow{\beta} C_{p+a+b} \xrightarrow{\gamma} A_{p+a+b+c} \to .$$

Definition.   A *bigraded module* is a doubly indexed familty

$$M = (M_{p,q})_{(p,q) \in \mathbb{Z} \times \mathbb{Z}}$$

of R-modules. Bigraded modules M are often denoted by $M_{\bullet\bullet}$.

Just as one may picture a graded module as the x-axis with a module sitting on each integer point, so we may picture a bigraded module as the plane with a module sitting on each lattice point.

Definition. Let M and N be bigraded modules, and let $(a, b) \in \mathbb{Z} \times \mathbb{Z}$. A *bigraded map of bidegree* (a, b), denoted by $f: M \to N$, is a family of homomorphisms $f = (f_{p,q}: M_{p,q} \to N_{p+a,q+b})_{(p,q)\in\mathbb{Z}\times\mathbb{Z}}$. The *bidegree* of f is (a, b), and we denote it by $\deg(f) = (a, b)$.

As in Proposition 10.3, bidegrees add when composing bigraded maps: if $f: M \to N$ and $g: N \to P$ are bigraded maps of bidegrees (a, b) and (a', b'), respectively, then $gf: M \to P$ has bidegree $(a + a', b + b')$.

All bigraded modules (over a fixed ring R) and bigraded maps form a category. If $f: M \to N$ is a bigraded map of bidegree (a, b), then $f \in \prod_{(p,q)} \mathrm{Hom}(M_{p,q}, N_{p+a,q+b})$, and so

$$\mathrm{Hom}(M, N) = \bigcup_{(a,b)} \Big(\prod_{(p,q)} \mathrm{Hom}(M_{p,q}, N_{p+a,q+b}) \Big).$$

Let us define exactness in the category of bigraded modules. If $M' = (M'_{p,q})$ and $M = (M_{p,q})$ are bigraded modules, then M' is a *submodule* of M if $M'_{p,q} \subseteq M_{p,q}$ for all (p, q); inclusions are bigraded maps of bidegree $(0, 0)$. If $M' \subseteq M$, then the *quotient* is $M/M' = (M_{p,q}/M'_{p,q})$, and the natural map $M \to M/M'$ is a bigraded map of bidegree $(0, 0)$. If $f: M \to N$ is a bigraded map of bidegree (a, b), then $\ker f = (\ker f_{p,q}) \subseteq M$. On the other hand, $\mathrm{im}\, f \subseteq N$, so that its (p, q)th term $(\mathrm{im}\, f)_{p,q} \subseteq N_{p,q}$. Thus, we must define

$$\mathrm{im}\, f = (\mathrm{im}\, f_{p-a,q-b}) \subseteq (N_{p,q}).$$

Of course, *exactness* of $A \xrightarrow{f} B \xrightarrow{g} C$ means that $\mathrm{im}\, f = \ker g$; that is, $\mathrm{im}\, f_{p-a,q-b} = \ker g_{p,q}$ for all $(p, q) \in \mathbb{Z} \times \mathbb{Z}$.

Consider a triangle $(A, B, C, \alpha, \beta, \gamma)$ of bigraded modules and bigraded maps,

$$A \xrightarrow{\ \ \alpha\ \ } B$$
$$\gamma \searrow \quad \swarrow \beta$$
$$C,$$

which is exact at each vertex: $\ker \alpha = \mathrm{im}\, \gamma$, $\ker \beta = \mathrm{im}\, \alpha$, and $\ker \gamma = \mathrm{im}\, \beta$. This exact triangle is really a host of long exact sequences, one for each $(p, q) \in \mathbb{Z} \times \mathbb{Z}$. If α, β, γ have bidegree $(a, a'), (b, b'), (c, c')$, respectively, go left and right of $A_{p,q}$ to obtain the exact sequence

$$\to C_{p-c,q-c'} \xrightarrow{\gamma} A_{p,q} \xrightarrow{\alpha} B_{p+a,q+a'} \xrightarrow{\beta} C_{p+a+b,q+a'+b''} \to .$$

Conversely, given a family of such long exact sequences, one for each (p, q), we can define an exact triangle from which they come.

Equipping a bigraded module with differentials gives a *bicomplex*.

Definition.   A *bicomplex* (or *double complex*) is an ordered triple (M, d', d''), where $M = (M_{p,q})$ is a bigraded module, $d', d'' : M \to M$ are differentials of bidegree $(-1, 0)$ and $(0, -1)$, respectively (so that $d'd' = 0$ and $d''d'' = 0$), and

$$d'_{p,q-1}d''_{p,q} + d''_{p-1,q}d'_{p,q} = 0.$$

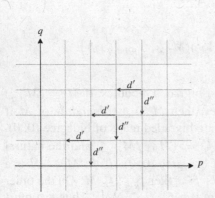

Fig. 10.1   Bicomplex.

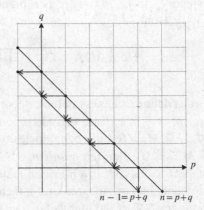

Fig. 10.2   Total complex.

As any bigraded module, a bicomplex M may be pictured as a family of modules in the pq-plane with $M_{p,q}$ sitting on the lattice point (p, q) (see Fig. 10.1). For each p, q, the differential $d'_{p,q} : M_{p,q} \to M_{p-1,q}$ points to the left, and the differential $d''_{p,q} : M_{p,q} \to M_{p,q-1}$ points down; thus, the rows $M_{*,q}$ and the columns $M_{p,*}$ are complexes. The identity $d'_{p,q-1}d''_{p,q} = -d''_{p-1,q}d'_{p,q}$ says that each square of the diagram *anticommutes*.

Example 10.4.   Let $M = (M_{p,q})$ be a bigraded module, and assume that there are bigraded maps $d' : M \to M$ of bidegree $(-1, 0)$ and $d'' : M \to M$ of bidegree $(0, -1)$ making the rows and columns of M complexes. If M is a commutative diagram, then we can make it into a bicomplex with a *sign change*.

Define $\Delta''_{p,q} = (-1)^p d''_{p,q}$. Changing sign does not affect kernels and images, and so $\Delta'' \Delta'' = 0$. Finally,

$$
\begin{aligned}
d'_{p,q-1}\Delta''_{p,q} + \Delta''_{p-1,q}d'_{p,q} &= (-1)^p d'_{p,q-1}d''_{p,q} + (-1)^{p-1}d''_{p-1,q}d'_{p,q} \\
&= (-1)^p\left(d'_{p,q-1}d''_{p,q} - d''_{p-1,q}d'_{p,q}\right) \\
&= 0.
\end{aligned}
$$

Therefore, (M, d', Δ'') is a bicomplex. ◄

Before giving some examples of bicomplexes, let us explain the significance of the anticommutativity equation $d'd'' + d''d' = 0$.

Definition. If M is a bicomplex, then its ***total complex***, denoted by $\mathrm{Tot}(M)$, is the complex with nth term

$$
\mathrm{Tot}(M)_n = \bigoplus_{p+q=n} M_{p,q}
$$

and with differentials $D_n \colon \mathrm{Tot}(M)_n \to \mathrm{Tot}(M)_{n-1}$ given by

$$
D_n = \sum_{p+q=n} (d'_{p,q} + d''_{p,q})
$$

(see Fig. 10.2).

Lemma 10.5. *If M is a bicomplex, then $(\mathrm{Tot}(M), D)$ is a complex.*

Proof. The summands of $\mathrm{Tot}(M)_n$ are the modules $M_{p,q}$ lying on the $45°$ line $p + q = n$ in the pq-plane. Note that $\mathrm{im}\, d'_{p,q} \subseteq M_{p-1,q}$ and $\mathrm{im}\, d''_{p,q} \subseteq M_{p,q-1}$; in either case, the sum of the indices is $p + q - 1 = n - 1$, and so $\mathrm{im}\, D \subseteq \mathrm{Tot}(M)_{n-1}$. We show that D is a differential.

$$
\begin{aligned}
DD &= \sum_{p,q}(d' + d'')(d' + d'') \\
&= \sum d'd' + \sum(d'd'' + d''d') + \sum d''d'' \\
&= 0,
\end{aligned}
$$

because each of the summands is 0. •

Spectral sequences arose as a method of computing the homology of the total complex $\mathrm{Tot}(M)$.

Recall that if $\mathcal{A}, \mathcal{B}, \mathcal{C}$ are categories, a function $T \colon \mathcal{A} \times \mathcal{B} \to \mathcal{C}$ is a *bifunctor* if

(i) $T(A, \square) \colon \mathcal{B} \to \mathcal{C}$ is a functor for each $A \in \mathrm{obj}(\mathcal{A})$,

(ii) $T(\square, B): \mathcal{A} \to \mathcal{C}$ is a functor for each $B \in \mathrm{obj}(\mathcal{B})$,

(iii) for each pair of morphisms $f: A' \to A$ in $\mathcal{A}$ and $g: B' \to B$ in $\mathcal{B}$, there is a commutative diagram

$$
\begin{array}{ccc}
T(A', B') & \xrightarrow{\; T(A',g) \;} & T(A', B) \\
{\scriptstyle T(f,B')} \downarrow & & \downarrow {\scriptstyle T(f,B)} \\
T(A, B') & \xrightarrow[\; T(A,g) \;]{} & T(A, B).
\end{array}
$$

We saw, in Exercise 2.35 on page 96, that $\square \otimes_R \square: \mathbf{Mod}_R \times {}_R\mathbf{Mod} \to \mathbf{Ab}$ and $\mathrm{Hom}_R(\square, \square): \mathbf{Mod}_R \times \mathbf{Mod}_R \to \mathbf{Ab}$ are bifunctors (if the definition of bifunctor is modified so that T can be covariant or contravariant in either variable). Exercise 10.2 on the facing page says that the derived functors $\mathrm{Tor}_n^R(\square, \square)$ and $\mathrm{Ext}_R^n(\square, \square)$ are also bifunctors.

Example 10.6.

(i) Let R be a ring, and let

$$
\mathbf{A} = \to A_p \xrightarrow{\; \Delta'_p \;} A_{p-1} \to \cdots \to A_0 \to 0
$$

and

$$
\mathbf{B} = \to B_q \xrightarrow{\; \Delta''_q \;} B_{q-1} \to \cdots \to B_0 \to 0
$$

be *positive complexes* of right R-modules; that is, $A_p = \{0\}$ for negative p and $B_q = \{0\}$ for negative q. Define (M, d', d'') by

$$
M_{p,q} = A_p \otimes_R B_q, \quad d'_{p,q} = \Delta'_p \otimes 1_{B_Q}, \quad \text{and} \quad d''_{p,q} = (-1)^p 1_{A_p} \otimes \Delta''_q.
$$

Since tensor is a bifunctor, the diagram consisting of the bigraded module M and the arrows $\Delta'_p \otimes 1_{B_Q}$ and $1_{A_p} \otimes \Delta''_q$ commutes. Incorporating the sign $(-1)^p$ in vertical arrows gives a bicomplex, as in Example 10.4. This bicomplex is concentrated in the first quadrant.

Definition. A *first quadrant bicomplex* is a bicomplex $(M_{p,q})$ for which $M_{p,q} = \{0\}$ whenever p or q is negative.

The total complex here is called the *tensor product of complexes* and is denoted by $\mathrm{Tot}(M) = \mathbf{A} \otimes_R \mathbf{B}$. Thus,

$$
(\mathbf{A} \otimes_R \mathbf{B})_n = \bigoplus_{p+q=n} A_p \otimes_R B_q,
$$

and $D_n \colon (\mathbf{A} \otimes_R \mathbf{B})_n \to (\mathbf{A} \otimes_R \mathbf{B})_{n-1}$ is given by

$$D_n \colon a_p \otimes b_q \mapsto \Delta' a_p \otimes b_q + (-1)^p a_p \otimes \Delta''_q b_q.$$

(ii) Let $A = A_R$ and $B = {}_R B$ be modules, and let

$$\mathbf{P}_A = \to P_p \xrightarrow{\Delta'_p} P_{p-1} \to \cdots \to P_0 \to 0$$

and

$$\mathbf{Q}_B = \to Q_q \xrightarrow{\Delta''_q} Q_{q-1} \to \cdots \to Q_0 \to 0$$

be deleted projective resolutions. A special case of the bicomplex in part (i) is $\mathbf{P}_A \otimes_R \mathbf{Q}_B$.

(iii) The ***Eilenberg-Zilber Theorem*** (see Rotman, *An Introduction to Algebraic Topology*, p. 266) states that if X and Y are topological spaces, then

$$H_n(X \times Y) \cong H_n(\mathbf{S}_\bullet(X) \otimes_{\mathbb{Z}} \mathbf{S}_\bullet(Y)),$$

where $\mathbf{S}_\bullet(X)$ is the singular complex of X.

(iv) Let $M = (M_{p,q})$ be a bigraded module, and let (M, d', d'') be a bicomplex. The ***transpose*** of M is (M^t, δ', δ''), where $M^t_{p,q} = M_{q,p}$ for all p, q, $\delta''_{p,q} = d'_{q,p}$, and $\delta'_{p,q} = d''_{q,p}$. Then (M^t, δ', δ'') is a bicomplex; that is, (M^t, d'', d') is a bicomplex. Moreover, the total complexes are identical: $\mathrm{Tot}(M^t)_n = \mathrm{Tot}(M)_n$ for all n, and $D^t_n = \sum(\delta' + \delta'') = \sum(d' + d'') = D_n$. Thus,

$$\mathrm{Tot}(M^t) = \mathrm{Tot}(M). \quad \blacktriangleleft$$

There are other interesting examples of bicomplexes. In particular, there are also *third quadrant bicomplexes*, but let us first discuss spectral sequences.

Exercises

10.1 If (M, d', d'') is a bicomplex of left R-modules, and if $F \colon {}_R\mathbf{Mod} \to {}_S\mathbf{Mod}$ is an additive functor, prove that (FM, Fd', Fd'') is a bicomplex of left S-modules.

*10.2   (i) For a ring R and a fixed $k \geq 0$, prove that $\mathrm{Tor}^R_k(\square, \square)$ is a bifunctor.

      (ii) For a ring R and a fixed $k \geq 0$, prove that $\mathrm{Ext}^k_R(\square, \square)$ is a bifunctor.

10.2   Filtrations and Exact Couples

One method of analyzing a group uses *normal series*. Recall that a sequence of subgroups (G_i) of a group G is a **normal series** if

$$G = G_0 \supseteq G_1 \supseteq G_2 \supseteq \cdots \supseteq G_n = \{1\},$$

where each $G_{i+1} \lhd G_i$. The **factor groups** of this normal series are

$$G_0/G_1, \; G_1/G_2, \; \ldots, \; G_{n-1}/G_n.$$

The factor groups of a normal series of a group G may not determine G. For example, if $\mathbf{V} = \langle a \rangle \oplus \langle b \rangle$ is the four-group, then $\mathbf{V} \supseteq \langle a \rangle \supseteq \{0\}$ is a normal series with factor groups $\mathbb{I}_2, \mathbb{I}_2$; if $\mathbb{I}_4 = \langle x \rangle$, then $\mathbb{I}_4 \supseteq \langle 2x \rangle \supseteq \{0\}$ is a normal series whose factor groups are also $\mathbb{I}_2, \mathbb{I}_2$. However, some information about a group G can be gleaned from its factor groups: for example, if G is finite, then $|G| = \prod_i |G_i/G_{i+1}|$.

Let us generalize this notion to modules.

Definition.   A *filtration* of a module M is a family $(M_p)_{p \in \mathbb{Z}}$ of submodules of M such that

$$\cdots \subseteq M_{p-1} \subseteq M_p \subseteq M_{p+1} \subseteq \cdots.$$

The **factor modules** of this filtration form the graded module $(M_p/M_{p-1})_{p \in \mathbb{Z}}$.

One can define filtrations of objects in any abelian category. In particular, a filtration of a complex $\mathbf{C}$ is a family of subcomplexes $(F^p \mathbf{C})_{p \in \mathbb{Z}}$ with

$$\cdots \subseteq F^{p-1}\mathbf{C} \subseteq F^p\mathbf{C} \subseteq F^{p+1}\mathbf{C} \subseteq \cdots.$$

Of course, the factors are $\cdots, F^p\mathbf{C}/F^{p-1}\mathbf{C}, \; F^{p+1}\mathbf{C}/F^p\mathbf{C}, \; \cdots$. In more detail, a filtration of $\mathbf{C}$ is a commutative diagram such that, for each n, the nth column is a filtration of C_n.

$$
\begin{array}{ccccccc}
\longrightarrow & C_{n+1} & \longrightarrow & C_n & \longrightarrow & C_{n-1} & \longrightarrow \\
& \uparrow & & \uparrow & & \uparrow & \\
\longrightarrow & F^p_{n+1} & \longrightarrow & F^p_n & \longrightarrow & F^p_{n-1} & \longrightarrow \\
& \uparrow & & \uparrow & & \uparrow & \\
\longrightarrow & F^{p-1}_{n+1} & \longrightarrow & F^{p-1}_n & \longrightarrow & F^{p-1}_{n-1} & \longrightarrow
\end{array}
$$

Remark. We do not insist that filtrations always be ascending, nor do we insist that they always be descending (in which case, the submodules should be re-indexed); either option is a filtration. In Group Theory, one also uses ascending series; for example, the *central series* $\{1\} \subseteq Z(G) \subseteq Z^2(G) \subseteq \cdots \subseteq G$ of a group G is ascending. ◀

Filtrations may have only finitely many terms (if $M_0 \subseteq M_1 \subseteq \cdots \subseteq M_n$, define $M_i = M_0$ for all $i \leq 0$ and $M_j = M_n$ for all $j \geq n$); moreover, the "endpoints" (if there are any) of a filtration of M need not be $\{0\}$ or M; that is, neither $\{0\}$ nor M must equal M_p for some p.

Here are two very important filtrations of the total complex.

Example 10.7. Let (M, d', d'') be a bicomplex.

Definition. The *first filtration* of $\text{Tot}(M)$ is given by

$$\left({}^I F^p \, \text{Tot}(M) \right)_n = \bigoplus_{i \leq p} M_{i,n-i}$$

$$= \cdots \oplus M_{p-2,q+2} \oplus M_{p-1,q+1} \oplus M_{p,q}.$$

Fig. 10.3 should make this clear. The term of ${}^I F^p \, \text{Tot}(M)$ of degree n is the direct sum of those $M_{i,n-i}$ lying to the left of the vertical line.

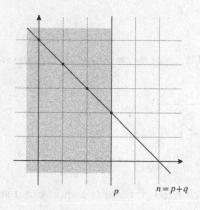

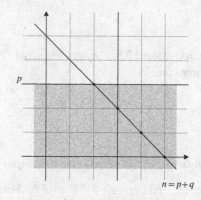

Fig. 10.3 First filtration.   **Fig. 10.4** Second filtration.

Let us check that $({}^{I}F^{p})_{n \geq 0}$ is a subcomplex of $\operatorname{Tot}(M)$.

$$D_{i,n-i}M_{i,n-i} = (d'_{i,n-i} + d''_{i,n-i})M_{i,n-i} \subseteq d'M_{i,n-i} + d''M_{i,n-i}$$

$$\subseteq M_{i-1,n-i} \oplus M_{i,n-i-1}$$

$$\subseteq ({}^{I}F^{p}\operatorname{Tot}(M))_{n-1}.$$

Definition. The *second filtration* of $\operatorname{Tot}(M)$ (see Fig. 10.4) is given by

$$\left({}^{II}F^{p}\operatorname{Tot}(M)\right)_{n} = \bigoplus_{j \leq p} M_{n-j,j}$$

$$= \cdots \oplus M_{q-1,p-2} \oplus M_{q+1,p-1} \oplus M_{q,p}.$$

The term of ${}^{II}F^{p}\operatorname{Tot}(M)$ of degree n is the direct sum of those $M_{i,n-i}$ lying below the horizontal line. One checks, as for the first filtration, that $({}^{II}F^{p})_{n \geq 0}$ is a subcomplex of $\operatorname{Tot}(M)$. We shall return to this example in §10.4. ◄

There are several ways to introduce spectral sequences. We think that using *exact couples* is the simplest way; for other discussions, see Mac Lane, *Homology*, XI.1 and XI.3, or McCleary, *User's Guide to Spectral Sequences*.

Definition. An *exact couple* is a 5-tuple $(D, E, \alpha, \beta, \gamma)$, where D and E are bigraded modules, α, β, γ are bigraded maps, and there is exactness at each vertex: $\ker \alpha = \operatorname{im} \gamma$, $\ker \beta = \operatorname{im} \alpha$, and $\ker \gamma = \operatorname{im} \beta$.

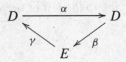

Proposition 10.8. *Every filtration* $(F^{p}\mathbf{C})_{p \in \mathbb{Z}}$ *of a complex* $\mathbf{C}$ *determines an exact couple*

$$
\begin{array}{ccc}
D & \xrightarrow{\alpha \ (1,-1)} & D \\
& \nwarrow_{\gamma \ (-1,0)} \quad \swarrow_{\beta \ (0,0)} & \\
& E &
\end{array}
$$

whose bigraded maps have the displayed bidegrees.

Proof. Abbreviate $F^{p}\mathbf{C}$ to F^{p}. For each fixed p, there is a short exact sequence of complexes,

$$0 \to F^{p-1} \xrightarrow{j^{p-1}} F^{p} \xrightarrow{\nu^{p}} F^{p}/F^{p-1} \to 0$$

(where j^{p-1} is the inclusion and ν^p is the natural map) that gives rise to the long exact sequence

$$\to H_n(F^{p-1}) \xrightarrow{\alpha} H_n(F^p) \xrightarrow{\beta} H_n(F^p/F^{p-1}) \xrightarrow{\gamma}$$

$$H_{n-1}(F^{p-1}) \xrightarrow{\alpha} H_{n-1}(F^p) \xrightarrow{\beta} H_{n-1}(F^p/F^{p-1}) \to,$$

where $\alpha = j_*^{p-1}$, $\beta = \nu_*^p$, and $\gamma = \partial$. Introduce q by writing $n = p + q$.

$$H_{p+q}(F^{p-1}) \xrightarrow{\alpha} H_{p+q}(F^p) \xrightarrow{\beta} H_{p+q}(F^p/F^{p-1}) \xrightarrow{\gamma}$$

$$H_{p+q-1}(F^{p-1}) \xrightarrow{\alpha} H_{p+q-1}(F^p) \xrightarrow{\beta} H_{p+q-1}(F^p/F^{p-1}).$$

There are two types of homology groups: homology of a subcomplex F^p or F^{p-1} and homology of a quotient complex F^p/F^{p-1}. Define

$$D = (D_{p,q}), \quad \text{where } D_{p,q} = H_{p+q}(F^p),$$
$$E = (E_{p,q}), \quad \text{where } E_{p,q} = H_{p+q}(F^p/F^{p-1}).$$

With this notation, the long exact sequence is, for fixed q,

$$\longrightarrow D_{p-1,q+1} \xrightarrow[(1,-1)]{\alpha} D_{p,q} \xrightarrow[(0,0)]{\beta} E_{p,q} \xrightarrow[(-1,0)]{\gamma} D_{p-1,q} \longrightarrow.$$

Therefore, $(D, E, \alpha, \beta, \gamma)$ is an exact couple with the displayed bidegrees.   •

Notation. By universal agreement, everyone writes $n = p + q$.

Every exact couple determines another exact couple, but let us first introduce an important notion.

Definition. A *differential bigraded module* is an ordered pair (M, d), where M is a bigraded module and $d: M \to M$ is a bigraded map with $dd = 0$.

If (M, d) is a differential bigraded module, where d has bidegree (a, b), then its *homology* $H(M, d)$ is the bigraded module whose p, q term is

$$H(M, d)_{p,q} = \frac{\ker d_{p,q}}{\operatorname{im} d_{p-a,q-b}}.$$

A bicomplex (M, d', d'') gives rise to two differential bigraded modules, namely, (M, d') and (M, d''). However, $(M, d' + d'')$ is not a differential bigraded module, for $d' + d'': M \to M$ is not a bigraded map.

A reader eager to begin applying spectral sequences can skim the routine proof of exactness in the next proposition.

Proposition 10.9. *If $(D, E, \alpha, \beta, \gamma)$ is an exact couple, then $d^1 = \beta\gamma$ is a differential $d^1\colon E \to E$, and there is an exact couple $(D^2, E^2, \alpha^2, \beta^2, \gamma^2)$, called the **derived couple**, with $E^2 = H(E, d^1)$.*

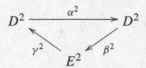

Remark. See Exercise 10.4 on page 623 for an explicit formula for d^1 when the original exact couple arises from a filtration of a complex.

Given the bidegrees of α, β, γ, there are formulas for the bidegrees of $d^1, \alpha^2, \beta^2, \gamma^2$, which we will state after the proof. ◄

Proof. To make the proof more concrete, we are going to assume that the bidegrees in the given exact couple are those that have arisen in Proposition 10.8: let α, β, γ have respective bidegrees $(1, -1)$, $(0, 0)$, $(-1, 0)$.

Define a bigraded map $d^1\colon E \to E$ by $d^1 = \beta\gamma$; the composite $\beta\gamma$ makes sense, but it is in the "wrong order." Note that $\gamma\beta = 0$, because the original couple is exact, and so d^1 is a differential: $d^1 d^1 = \beta(\gamma\beta)\gamma = 0$. Since bidegrees add, the bidegree of d^1 is $(-1, 0)$.

Define $E^2 = H(E, d^1)$. Thus, $E^2_{p,q} = \ker d^1_{p,q} / \operatorname{im} d^1_{p+1,q}$.

Define $D^2 = \operatorname{im}\alpha \subseteq D$. Thus, $D^2_{p,q} = \operatorname{im}\alpha_{p-1,q+1} \subseteq D_{p,q}$.

We now define the bigraded maps. Define $\alpha^2\colon D^2 \to D^2$ to be the restriction $\alpha|D^2$; that is, $\alpha^2 = \alpha i$, where $i\colon D^2 \to D$ is the inclusion. Since inclusions have bidegree $(0, 0)$, α^2 has bidegree $(1, -1)$, the same bidegree as that of α. If $x \in D^2_{p,q}$, then $x = \alpha u$ (for $u \in D_{p+1,q-1}$), and

$$\alpha^2_{p,q}\colon x = \alpha u \mapsto \alpha x = \alpha\alpha u.$$

Define $\beta^2\colon D^2 \to E^2$ as follows. If $y \in D^2_{p,q}$, then $y = \alpha v$ (for $v \in D_{p+1,q-1}$), and βv is a cycle [for $d^1\beta v = \beta(\gamma\beta)v = 0$]. Since $v = \alpha^{-1}y$, we set

$$\beta^2(y) = \operatorname{cls}(\beta\alpha^{-1}y).$$

To see that β^2 is well-defined (i.e., that β^2 does not depend on the choice v of the preimage $\alpha^{-1}y$), we must show that if $y = \alpha v'$, then $\operatorname{cls}(\beta v') = \operatorname{cls}(\beta v)$. Now $v' - v \in \ker\alpha = \operatorname{im}\gamma$, so that $v' - v = \gamma w$ for some $w \in E$, and hence $\beta(v' - v) = \beta\gamma w = d^1 w$ is a boundary. Note that β^2 has bidegree $(-1, 1)$.

We now define $\gamma^2\colon E^2 \to D^2$. Let $\operatorname{cls}(z) \in E^2_{p,q}$, so that $z \in E_{p,q}$ and $d^1 z = \beta\gamma z = 0$. Hence, $\gamma z \in \ker\beta = \operatorname{im}\alpha$, so that $\gamma z \in \operatorname{im}\alpha = D^2$; displaying subscripts, $\gamma_{p,q} z \in D_{p-1,q}$. Define γ^2 by

$$\gamma^2\colon \operatorname{cls}(z) \mapsto \gamma z.$$

Now γ^2 does not depend on the choice of cycle: if $w \in \operatorname{im} d^1_{p+1,q}$ is a boundary, then $w = d^1x = \beta\gamma x$, and so $\gamma w = (\gamma\beta)\gamma x = 0$. Note that γ^2 has bidegree $(-1, 0)$, the same bidegree as that of γ.

It remains to prove exactness. Since all the maps are well-defined, there is no reason to display subscripts. First of all, adjacent composites are 0.

$$\beta^2\alpha^2 : x = \alpha u \mapsto \alpha\alpha u \mapsto \operatorname{cls}(\beta\alpha^{-1}\alpha u) = \operatorname{cls}(\beta\alpha u) = 0.$$

$$\gamma^2\beta^2 : x = \alpha u \mapsto \operatorname{cls}(\beta u) \mapsto \gamma\beta u = 0.$$

$$\alpha^2\gamma^2 : \operatorname{cls}(z) \mapsto \gamma z \mapsto \alpha\gamma z = 0.$$

We have verified the inclusions of the form im $\subseteq$ ker. Here are the reverse inclusions.

ker $\alpha^2 \subseteq$ **im** γ^2. If $x \in \ker \alpha^2$, then $\alpha x \in D^2$ and $\alpha x = 0$. Hence, $x \in \ker \alpha = \operatorname{im} \gamma$, so that $x = \gamma y$ for some $y \in E$. Now $x \in \operatorname{im} \alpha = \ker \beta$, and $0 = \beta x = \beta\gamma y = d^1 y$. Thus, y is a cycle, and $x = \gamma y = \gamma^2 \operatorname{cls}(y) \in \operatorname{im} \gamma^2$.

ker $\beta^2 \subseteq$ **im** α^2. If $x \in \ker \beta^2$, then $x \in D^2 = \operatorname{im} \alpha$ and $\beta^2 x = 0$. Thus, $x = \alpha u$ and $0 = \beta^2 x = \operatorname{cls}(\beta\alpha^{-1}\alpha u) = \operatorname{cls}(\beta u)$. Hence, $\beta u \in \operatorname{im} d^1$; that is, $\beta u = d^1 w = \beta\gamma w$ for some $w \in E$. Now $u - \gamma w \in \ker \beta = \operatorname{im} \alpha = D^2$, and $\alpha^2(u - \gamma w) = \alpha u - \alpha\gamma w = \alpha u = x$. Therefore, $x \in \operatorname{im} \alpha^2$.

ker $\gamma^2 \subseteq$ **im** β^2. If $\operatorname{cls}(z) \in \ker \gamma^2$, then $\gamma^2 \operatorname{cls}(z) = \gamma z = 0$. Thus, $z \in \ker \gamma = \operatorname{im} \beta$, so that $z = \beta v$ for some $v \in D$. Hence, $\beta^2(\alpha v) = \operatorname{cls}(\beta\alpha^{-1}\alpha v) = \operatorname{cls}(\beta v) = \operatorname{cls}(z)$, and $\operatorname{cls}(z) \in \operatorname{im} \beta^2$. •

Remark. It is easy to check that if the maps α, β, γ have bidegrees (a, a'), (b, b'), and (c, c'), respectively, then the bigraded maps α^2, β^2, γ^2 have bidegrees (a, a'), $(b-a, b'-a')$, and (c, c'), respectively. Thus, only the bidegree of β changes. ◄

Definition. Define the *rth derived couple* of an exact couple $(D, E, \alpha, \beta, \gamma)$ inductively: its $(r + 1)$st derived couple $(D^{r+1}, E^{r+1}, \alpha^{r+1}, \beta^{r+1}, \gamma^{r+1})$ is the derived couple of $(D^r, E^r, \alpha^r, \beta^r, \gamma^r)$, the *rth derived couple*.

Corollary 10.10. *Let* $(D, E, \alpha, \beta, \gamma)$ *be the exact couple arising from a filtration* (F^p) *of a complex* **C**.

Then the constituents of the rth derived couple have the following properties:

(i) *the bigraded maps* $\alpha^r, \beta^r, \gamma^r$ *have bidegrees* $(1, -1)$, $(1 - r, r - 1)$, $(-1, 0)$, *respectively*;

(ii) *the diffential* d^r *has bidegree* $(-r, r-1)$, *and it is induced by* $\beta\alpha^{-r+1}\gamma$;

(iii) $E_{p,q}^{r+1} = \ker d_{p,q}^r / \operatorname{im} d_{p+r,q-r+1}^r$;

(iv) $D_{p,q}^r = \operatorname{im}(\alpha_{p-1,q+1})(\alpha_{p-2,q+2}) \cdots (\alpha_{p-r+1,q+r-1})$; *in particular, for the exact couple in* Proposition 10.8,

$$D_{p,q}^r = \operatorname{im}(j^{p-1}j^{p-2} \cdots j^{p-r+1})_* : H_n(F^{p-r+1}) \to H_n(F^p).$$

Proof.   All parts are easy inductions on $r \geq 1$, but the last statement needs more explanation. Recall that $D_{p,q}^2 = \alpha_{p-1,q+1}D_{p-1,q+1}$. Focus on the first subscript: $D_{p,q}^2 = \alpha_{p-1,\#}D_{p-1,\#}$, and so $D_{p-1,\#}^2 = \alpha_{p-2,\#}D_{p-2,\#}$. Now $D_{p,q}^3 = \alpha_{p-1,\#}^2 D_{p-1,\#}^2 = \alpha_{p-1,\#}\alpha_{p-2,\#}D_{p-2,\#}$. The result follows by induction. The final statement merely reminds us that $\alpha_{p,q} = j_*^p : H_{p+q}(F^p) \to H_{p+q}(F^{p+1})$ and that $j_*^{p-1}j_*^{p-2} \cdots j_*^{p-r+1} = (j^{p-1}j^{p-2} \cdots j^{p-r+1})_*$.   •

Definition.   A *spectral sequence* is a sequence $(E^r, d^r)_{r \geq 1}$ of differential bigraded modules such that $E^{r+1} = H(E^r, d^r)$ for all r.

If an exact couple $(D, E, \alpha, \beta, \gamma)$ is relabeled as $(D^1, E^1, \alpha^1, \beta^1, \gamma^1)$, then every exact couple yields a spectral sequence. We have proved the following result.

Theorem 10.11.   *Every filtration of a complex yields a spectral sequence as described in* Corollary 10.10.

Proof.   A filtration gives an exact couple, as in Proposition 10.8, and the E^r terms of its derived couples form a spectral sequence.   •

Spectral sequences were invented in the 1940s, independently, by J. Leray and R. C. Lyndon. Leray was a German prisoner of war from 1940 through 1945, during World War II. He was an applied mathematician, but, because he did not want the Nazis to exploit his expertise, he did abstract work in Algebraic Topology there. Lyndon invented spectral sequences, in his Ph.D. dissertation, in order to compute the cohomology groups of finite abelian groups. Exact couples were introduced by the algebraic topologist W. S. Massey, "Exact couples in algebraic topology, I, II," *Annals Math.* (2) 56 (1952), 363–396, "Exact couples in algebraic topology, III, IV, V," *Annals Math.* (2) 57 (1953), 248–286,

In practice, most spectral sequences do arise from *bounded* filtrations of complexes (but see Exercise 10.5 on the facing page). The reader meeting a more general spectral sequence is referred to another text treating spectral sequences (e.g., Mac Lane, *Homology*), but we maintain that understanding what we are doing here is sufficient preparation for a more general case.

Exercises

10.3 Let $(E^r, d^r)_{r \geq 0}$ be a spectral sequence. If there is an integer s with $E_{p,q}^s = \{0\}$ for some (p, q), prove that $E_{p,q}^r = \{0\}$ for all $r \geq s$.

*10.4 Let $(F^p C)_{p \in \mathbb{Z}}$ be a filtration of a complex $\mathbf{C}$, and let the corresponding exact couple be $(D, E, \alpha, \beta, \gamma)$ (as in Proposition 10.8). Prove that the differential $d_{p,q}^1 : E_{p,q} \to E_{p-1,q}$ is the connecting homomorphism

$$H_{p+q}(F^p/F^{p-1}) \to H_{p+q-1}(F^{p-1}/F^{p-2})$$

arising from $0 \to F^{p-1}/F^{p-2} \to F^p/F^{p-2} \to F^p/F^{p-1} \to 0$.

Hint. Recall that $d_{p,q}^1 : E_{p,q} \to E_{p-1,q}$ is the composite

$$H_{p+q}(F^p/F^{p-1}) \xrightarrow{\gamma_{p,q}} H_{p+q-1}(F^{p-1})$$
$$\xrightarrow{\beta_{p-1,q}} H_{p+q-1}(F^{p-1}/F^{p-2}),$$

where $\gamma_{p,q}$ is the connecting homomorphism and $\beta_{p-1,q}$ is the map induced by the natural map $F^{p-1} \to F^{p-1}/F^{p-2}$.

*10.5 Consider the commutative diagram in which the top row is exact,

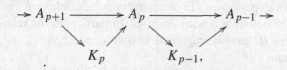

where $\operatorname{im}(A_{p+1} \to A_p) = K_p = \ker(A_p \to A_{p-1})$. For each p and module C, the exact sequence $0 \to K_p \to A_p \to K_{p-1} \to 0$ gives the long exact sequence

$$\operatorname{Tor}_q(C, K_p) \xrightarrow{\beta} \operatorname{Tor}_q(C, A_p) \xrightarrow{\gamma} \operatorname{Tor}_q(C, K_{p-1}) \xrightarrow{\alpha} \operatorname{Tor}_{q-1}(C, K_p).$$

If $D_{p,q} = \operatorname{Tor}_q(C, K_p)$ and $E_{p,q} = \operatorname{Tor}_q(C, A_p)$, prove that $\alpha, \beta,$ and γ are bigraded maps with respective bidegrees $(1, -1)$, $(0, 0)$, and $(-1, 0)$ and that $(D, E, \alpha, \beta, \gamma)$ is an exact couple. (Notice that this exact couple does not arise from a filtration.)

10.3   Convergence

We pause for an elementary interlude.

Definition.   If M is a module, then a *subquotient* of M is a module of the form M'/M'', where $M'' \subseteq M' \subseteq M$.

This definition can easily be modified to pertain to graded modules or bigraded modules.

Example 10.12.

  (i) If M has a filtration $(F^p M)$, then each factor module $F^P M / F^{p-1} M$ is a subquotient of M.

 (ii) The nth homology $H_n(\mathbf{C})$ of a complex $\mathbf{C} = (C_n)$ is a subquotient of C_n.

(iii) Each of the following properties of an abelian group E is inherited by all its subquotients: $E = \{0\}$; E is finite; E is finitely generated; E is torsion; E is p-primary; E is cyclic; $mE = \{0\}$ for some $m \in \mathbb{Z}$. ◄

Theorem 10.11 shows that a filtration of a complex $\mathbf{C}$ gives a spectral sequence. There are two obvious questions. What is the connection between the E^r terms of the spectral sequence (which are, after all, homology modules) and the homology of $H_\bullet(\mathbf{C})$? Can we use this connection to compute $H_\bullet(\mathbf{C})$?

If $\{E^r, d^r\}$ is a spectral sequence, then $E^2 = H_\bullet(E^1, d^1)$ is a subquotient of E^1: hence, $E^2 = Z^2/B^2 = $ cycles/boundaries, where

$$B^2 \subseteq Z^2 \subseteq E^1.$$

By the Correspondence Theorem, every submodule of E^2 is equal to S/B^2 for a unique submodule S with $B^2 \subseteq S$. In particular, the relative cycles Z^3 and boundaries B^3 may be regarded as quotients $B^3/B^2 \subseteq Z^3/B^2 \subseteq Z^2/B^2 = E^2$, so that

$$B^2 \subseteq B^3 \subseteq Z^3 \subseteq Z^2 \subseteq E^1.$$

More generally, for every r, there is a chain

$$B^2 \subseteq \cdots \subseteq B^r \subseteq Z^r \subseteq \cdots \subseteq Z^2 \subseteq E^1.$$

Definition.   Given a spectral sequence $\{E^r, d^r\}$, define $Z^\infty = \bigcap_r Z^r$ and $B^\infty = \bigcup_r B^r$ (an ascending union of submodules is a submodule). Then $B^\infty \subseteq Z^\infty$, and the *limit term* of the spectral sequence is the bigraded module E^∞ defined by

$$E_{p,q}^\infty = Z_{p,q}^\infty / B_{p,q}^\infty.$$

In Mac Lane's informal language, Z^r consists of the elements that "live till stage r" and B^r consists of the elements that "bound by stage r." The module Z^∞ consists of those elements that "live forever," while B^∞ consists of those elements that "eventually bound." In this sense, the E^r terms of a spectral sequence approximate the limit term.

Lemma 10.13.   *Let* $\{E^r, d^r\}$ *be a spectral sequence.*

(i) $E^{r+1} = E^r$ *if and only if* $Z^{r+1} = Z^r$ *and* $B^{r+1} = B^r$.

(ii) *If* $E^{r+1} = E^r$ *for all* $r \geq s$, *then* $E^s = E^\infty$.

Proof.

(i) If X/Y is a subquotient of Z, then $Y \subseteq X \subseteq Z$, and so $X/Y = Z$ if and only if $Y = \{0\}$ and $X = Z$. If $E^{r+1} = Z^{r+1}/B^{r+1} = E^r$, then $B^{r+1} = \{0\}$ in $E^r = Z^r/B^r$; that is, $B^{r+1} = B^r$. Hence, $E^{r+1} = Z^{r+1}/B^{r+1} = Z^{r+1}/B^r = E^r = Z^r/B^r$, so that $Z^{r+1} = Z^r$. The converse is obvious.

(ii) If $E^r = E^{r+1}$ for $r \geq s$, then $Z^s = Z^r$ for all $r \geq s$; hence, $Z^s = \bigcap_{r \geq s} Z^r = Z^\infty$. Also, $B^s = B^r$ for all $r \geq s$; hence, $B^s = \bigcup_{r \geq s} B^r = B^\infty$. Therefore,

$$E^s = Z^s/B^s = Z^\infty/B^\infty = E^\infty. \quad \bullet$$

Given a filtration (F^p) of a complex $\mathbf{C}$ with inclusions $i^p \colon F^p \to \mathbf{C}$, then $i_*^p \colon H_\bullet(F^p) \to H_\bullet(\mathbf{C})$. Since $F^p \subseteq F^{p+1}$, we have $\operatorname{im} i_*^p \subseteq \operatorname{im} i_*^{p+1}$; that is, $(\operatorname{im} i_*^p)$ is a filtration of $H_\bullet(\mathbf{C})$.

Definition.   If $(F^p\mathbf{C})$ is a filtration of a complex $\mathbf{C}$ and $i^p \colon F^p \to \mathbf{C}$ are inclusions, define

$$\Phi^p H_n(\mathbf{C}) = \operatorname{im} i_*^p.$$

We call $(\Phi^p H_n(\mathbf{C}))$ the *induced filtration* of $H_n(\mathbf{C})$.

Given a filtration of a complex $\mathbf{C}$, the induced filtration is the obvious filtration on the homology $H_\bullet(\mathbf{C})$. We are more interested in the factor modules of the induced filtration of $H_\bullet(\mathbf{C})$ than we are in the filtration. The spectral sequence (E^r) of a filtration of a complex obviously determines E^∞; thus, if

$$E_{p,q}^\infty \cong \Phi^p H_{p+q}/\Phi^{p-1} H_{p+q} \text{ for all } p, q,$$

then the spectral sequence determines the factor modules of $H_\bullet(\mathbf{C})$.

To obtain interesting results, it is not enough to have $E^\infty \cong \Phi^p/\Phi^{p-1}$; some restriction on the filtration of a complex is necessary. For example, if F^p is the zero subcomplex for all p, then $E_{p,q} = H_{p+q}(F^p/F^{p-1}) = H_{p+q}(\mathbf{0}) = \{0\}$. It follows that $E^r_{p,q} = \{0\}$ for all r, and so the subquotient $E^\infty_{p,q} = \{0\}$. The induced filtration $\Phi^p H_{p+q} = \{0\}$ for all p, and so $\Phi^p H_{p+q}/\Phi^{p-1}H_{p+q} = \{0\}$. Thus, $E^\infty_{p,q} \cong \Phi^p H_{p+q}/\Phi^{p-1}H_{p+q}$, but an isomorphism $\{0\} \cong \{0\}$ is useless.

The following condition on a filtration often holds.

Definition. A filtration $(F^p M)$ of a graded module $M = (M_n)$ is **bounded** if, for each n, there exist integers $s = s(n)$ and $t = t(n)$ such that

$$F^s M_n = \{0\} \qquad \text{and} \qquad F^t M_n = M_n.$$

If $\{F^p\}$ is a bounded filtration of a complex $\mathbf{C}$, then the induced filtration on homology is also bounded, and with the same bounds. More precisely, we know that if $i^p: F^p \to \mathbf{C}$ is the inclusion, then $\Phi^p H_n = \operatorname{im} i^p_*$, where $i^p_*: H_n(F^p) \to H_n(\mathbf{C})$. Since $F^s = 0$ and $F^t = \mathbf{C}$, we have $\Phi^s H_n = \{0\}$ and $\Phi^t H_n = H_n$. Thus, for each n, there is a finite chain (as in Group Theory),

$$\{0\} = \Phi^s H_n \subseteq \Phi^{s+1} H_n \subseteq \cdots \subseteq \Phi^t H_n = H_n.$$

Of course, $\Phi^i H_n = \{0\}$ for all $i \le s$, and $\Phi^j H_n = H_n$ for all $j \ge t$.

Definition. A spectral sequence $(E^r, d^r)_{r \ge 1}$ **converges** to a graded module H, denoted by

$$E^2_{p,q} \underset{p}{\Rightarrow} H_n,$$

if there is some *bounded* filtration $(\Phi^p H_n)$ of H with

$$E^\infty_{p,q} \cong \Phi^p H_n/\Phi^{p-1} H_n$$

for all n (we remind the reader of the notational convention $n = p + q$).

The significance of the subscript p (called the **filtration degree**) in the notation $E^2_{p,q} \underset{p}{\Rightarrow} H_n$ is that the isomorphism involves the limit term $E^\infty_{p,q}$, which may not be isomorphic to $E^\infty_{q,p}$.

Theorem 10.14. *Let* $(F^p \mathbf{C})_p$ *be a bounded filtration of a complex* $\mathbf{C}$*, and let* $(E^r, d^r)_{r \ge 1}$ *be the spectral sequence of Theorem 10.11 [so that the induced filtration* $(\Phi^p H)$ *is bounded with the same bounds* $s(n)$ *and* $t(n)$ *as* $(F^p \mathbf{C})$*]. Then*

(i) *for each p, q, we have $E^\infty_{p,q} = E^r_{p,q}$ for large r (depending on p, q),*

(ii) $E^2_{p,q} \underset{p}{\Rightarrow} H_n(\mathbf{C})$.

Proof.

(i) If p is "large"; that is, $p > t(n)$, then $F^{p-1} = F^p$, and $F^p/F^{p-1} = 0$. By definition, $E_{p,q} = H_{p+q}(F^p/F^{p-1})$, and so $E_{p,q} = \{0\}$. Since $E_{p,q}^r$ is a subquotient of $E_{p,q}$, we have $E_{p,q}^r = \{0\}$ for all r. Similarly, if p is "small"; that is, $p < s(n)$, then $F^p = 0$, and $E_{p,q}^r = \{0\}$ for all r.

How can we say anything at all about $E_{p,q}^r$? The one thing we do know is that its differential d^r has bidegree $(-r, r-1)$. Focus on first subscripts. For any fixed (p, q), $d^r(E_{p,q}^r) \subseteq E_{p-r,\#}^r$. For large r, the index $p - r$ is small, and so $E_{p-r,\#}^r = \{0\}$. Hence, $\ker d_{p,q}^r = E_{p,q}^r$. Let us compute $E_{p,q}^{r+1} = \ker d_{p,q}^r / \operatorname{im} d_{p+r,\#}^r$. Now $\operatorname{im} d_{p+r,\#}^r = \{0\}$, because the domain of $d_{p+r,\#}^r$ is $E_{p+r,\#}^r = \{0\}$ when r is large. Therefore, $E_{p,q}^{r+1} = \ker d_{p+r,\#}^r / \{0\} = E_{p,q}^r / \{0\} = E_{p,q}^r$ for large r (depending on p, q). Thus, the p, q term of $E_{p,q}^r$ is constant for large r, which says that $E_{p,q}^\infty = E_{p,q}^r$, by Lemma 10.13.

(ii) We continue focusing on the first index in the subscript by writing $\#$ for every second index. Consider the exact sequence obtained from the rth derived couple:

$$D_{p+r-2,\#}^r \xrightarrow{\ \alpha^r\ } D_{p+r-1,\#}^r \xrightarrow{\ \beta^r\ } E_{p,q}^r \xrightarrow{\ \gamma^r\ } D_{p-1,q}^r. \tag{1}$$

The indices arise from the bidegrees displayed in Corollary 10.10(i): α^r has bidegree $(1, -1)$, β^r has bidegree $(1-r, r-1)$, and γ^r has bidegree $(-1, 0)$; as in Corollary 10.10(iv), the module

$$D_{p,q}^r = \operatorname{im}(j^{p-1} j^{p-2} \cdots j^{p-r+1})_* : H_n(F^{p-r+1}) \to H_n(F^p).$$

Replacing p first by $p + r - 1$ and then by $p + r - 2$, we have

$$D_{p+r-1,\#}^r = \operatorname{im}(j^{p+r-2} \cdots j^p)_* \subseteq H_n(F^{p+r-1})$$

and

$$D_{p+r-2,\#}^r = \operatorname{im}(j^{p+r-3} \cdots j^{p-1})_* \subseteq H_n(F^{p+r-2}).$$

For large r, $F^{p+r-1} = F^t = \mathbf{C}$, and the composite $j^{p+r-2} \cdots j^p$ of inclusions is just the inclusion $i^p : F^p \to \mathbf{C}$. Therefore, $D_{p+r-1,\#}^r = \operatorname{im} i_*^p = \Phi^p H_n$. Similarly, $D_{p+r-2,\#}^r = \Phi^{p-1} H_n$ for large r. Therefore, we may rewrite exact sequence (1) as

$$\Phi^{p-1} H_n(\mathbf{C}) \to \Phi^p H_n(\mathbf{C})) \to E_{p,q}^r \to D_{p-1,q}^r,$$

where the first arrow is inclusion. If $D_{p-1,q}^r = \{0\}$ for large r, then

$$\Phi^p H_n(\mathbf{C})/\Phi^{p-1} H_n(\mathbf{C}) \cong E_{p,q}^r = E_{p,q}^\infty,$$

and we are done. But $D_{p-1,q}^r = \operatorname{im} H_n(F^{p-r}) \to H_n(F^{p-1})$, which is $\{0\}$ because $F^{p-1-r} = 0$ for large r. $\quad\bullet$

10.4   Homology of the Total Complex

Given a bicomplex (M, d', d''), recall the two filtrations of $\mathrm{Tot}(M)$ in Example 10.7. The **first filtration** of $\mathrm{Tot}(M)$ is $(^{\mathrm{I}}F^p)$ [see Fig. 10.5], where

$$^{\mathrm{I}}F^p \, \mathrm{Tot}(M)_n = \bigoplus_{i \le p} M_{i, n-i}$$

$$= \cdots \oplus M_{p-2, q+2} \oplus M_{p-1, q+1} \oplus M_{p,q},$$

is the subcomplex of $\mathrm{Tot}(M)$ whose nth term is the direct sum of those $M_{i, n-i}$ lying to the left of the vertical line.

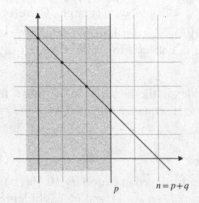

Fig. 10.5   First filtration.   **Fig. 10.6**   Second filtration.

The **second filtration** of $\mathrm{Tot}(M)$ is $(^{\mathrm{II}}F^p)$ [see Fig. 10.6], where

$$^{\mathrm{II}}F^p \, \mathrm{Tot}(M)_n = \bigoplus_{j \le p} M_{n-j, j}$$

$$= \cdots \oplus M_{q-1, p-2} \oplus M_{q+1, p-1} \oplus M_{q,p},$$

is the subcomplex of $\mathrm{Tot}(M)$ whose nth term is the direct sum of those $M_{i, n-i}$ lying below the horizontal line. **Warning**: the index p signals the first index, but we are using it to restrict the second index. This does not affect the definition of $^{\mathrm{II}}F^p$.

Lemma 10.15.   *If* $\mathrm{Tot}(M)$ *is the total complex of a bicomplex* (M, d', d''), *then the second filtration* $^{\mathrm{II}}F^p$ *of* $\mathrm{Tot}(M)$ *is equal to the first filtration of* $\mathrm{Tot}(M^t)$, *where* $\mathrm{Tot}(M^t)$ *arises from the transposed bicomplex* (M^t, d'', d').

Proof.   Recall Example 10.6(iv): the transpose M^t is defined by $M^t_{p,q} = M_{q,p}$, and so $\bigoplus_{j \le p} M_{n-j, j} = \bigoplus_{j \le p} M^t_{j, n-j}$.   •

Theorem 10.16.   *Let M be a first quadrant bicomplex, and let $^I E^r$ and $^{II} E^r$ be the spectral sequences determined by the first and second filtrations of $\text{Tot}(M)$.*

(i) *The first and second filtrations are bounded.*

(ii) *For all p, q, we have $^I E_{p,q}^\infty = {}^I E_{p,q}^r$ and $^{II} E_{p,q}^\infty = {}^{II} E_{p,q}^r$ for large r (depending on p, q).*

(iii) $^I E_{p,q}^2 \underset{p}{\Rightarrow} H_n(\text{Tot}(M))$ *and* $^{II} E_{p,q}^2 \underset{p}{\Rightarrow} H_n(\text{Tot}(M))$.

Proof.   The bounds for either filtration are $s(n) = -1$ and $t(n) = n$, and so statements (ii) and (iii) for $^I E$ follow from Theorem 10.14. Since $\text{Tot}(M^t) = \text{Tot}(M)$, where M^t is the transpose, and since the second filtration of $\text{Tot}(M)$ equals the first filtration of M^t, we have $^{II} E_{p,q}^\infty = {}^{II} E_{p,q}^r$ for large r and $^{II} E_{p,q}^2 \underset{p}{\Rightarrow} H_n(\text{Tot}(M^t)) = H_n(\text{Tot}(M))$.   ●

We have created notational monsters, $^I E_{p,q}^r$ and $^{II} E_{p,q}^r$ (they could be worse; one corner is still undecorated), but the next proposition will replace them by something less ugly.

We can compute $^I E^2$, the E^2-term of the spectral sequence arising from the first filtration $^I F^p$ of $\text{Tot}(M)$. Let us drop the prescript I in this discussion. As in the proof of Proposition 10.8, $E_{p,q} = H_n(F^p/F^{p-1})$, where

$$(F^p)_n = \cdots \oplus M_{p-2,q+2} \oplus M_{p-1,q+1} \oplus M_{p,q},$$
$$(F^{p-1})_n = \cdots \oplus M_{p-2,q+2} \oplus M_{p-1,q+1}.$$

Hence, the nth term of F^p/F^{p-1} is $M_{p,q}$. The differential $(F^p/F^{p-1})_n \to (F^p/F^{p-1})_{n-1}$ is

$$\overline{D}_n : a_n + (F^{p-1})_n \mapsto D_n a_n + (F^{p-1})_{n-1},$$

where $a_n \in (F^p)_n$; we have just seen that we may assume $a_n \in M_{p,q}$. Now $D_n a_n = (d'_{p,q} + d''_{p,q})a_n \in M_{p-1,q} \oplus M_{p,q-1}$. But $M_{p-1,q} \subseteq (F^{p-1})_n$, so that $D_n a_n \equiv d''_{p,q} a_n \bmod (F^{p-1})_{n-1}$. Thus, only d'' survives in F^p/F^{p-1}. More precisely, since $n = p + q$,

$$H_n(F^p/F^{p-1}) = \frac{\ker \overline{D}_n}{\operatorname{im} \overline{D}_{n+1}} \cong \frac{\ker d''_{p,q}}{\operatorname{im} d''_{p,q+1}} = H_q(M_{p,*}),$$

where $M_{p,*}$ is the pth column of M viewed as a complex with differential $\overline{d''}$ induced by d''. We have constructed a new bigraded module whose (p, q) term is $H_q(M_{p,*}, \overline{d''})$. Note that elements of $H_q(M_{p,*})$ have the form $\operatorname{cls} z$, where $z \in M_{p,q}$ and $d'' z = 0$.

For each fixed q, the qth row $H''(M)_{*,q}$ of $H''(M)$,

$$\dots, H_q(M_{p+1,*}), \ H_q(M_{p,*}), \ H_q(M_{p-1,*}), \dots, \tag{1}$$

can be made into a complex if we define $\overline{d'}_p \colon H_q(M_{p,*}) \to H_q(M_{p-1,*})$ by

$$\overline{d'}_p \colon \operatorname{cls}(z) \mapsto \operatorname{cls}(d'_{p,q}z),$$

where $z \in \ker d''_{p,q}$. There is a new bigraded module whose (p, q) term, denoted by $H'_p H''_q(M)$, is the pth homology of the qth row (1). Note that elements of $H'_p H''_q(M)$ have the form $\operatorname{cls}(\overline{d'}z)$, where $z \in M_{p,q}$ and $d''z = 0$.

Definition. If (M, d', d'') is a bicomplex, its *first iterated homology* is the bigraded module whose (p, q) term is $H'_p H''_q(M)$.

The first iterated homology of a bicomplex (M, d', d'') can be computed. Taking homology of pth columns gives a bigraded module whose (p, q) term is $H_q(M_{p,*})$; taking homology of the qth rows gives the first iterated homology. Before giving examples, we prove that $H'_p H''_q(M)$ is the $E^2_{p,q}$-term of the first spectral sequence.

Proposition 10.17. *If M is a first quadrant bicomplex, then*

$$^I E^1_{p,q} = H_q(M_{p,*})$$

and

$$^I E^2_{p,q} = H'_p H''_q(M) \underset{p}{\Rightarrow} H_n(\operatorname{Tot}(M)).$$

Proof. In our discussion leading up to the definition of the first iterated homology, we saw that F^p/F^{p-1} is the complex $(M_{p,*}, \overline{d''})$. Now Exercise 10.4 on page 623 identifies the differential $\overline{d''}$ with d^1, the connecting homomorphism arising from $0 \to F^{p-1}/F^{p-2} \to F^p/F^{p-2} \to F^p/F^{p-1} \to 0$.. Proposition 10.8 gives $^I E^1_{p,q} = H_{p+q}(F^p/F^{p-1})$, while our discussion above shows that $H_{p+q}(F^p/F^{p-1}) = H_q(M_{p,*})$. Therefore, $^I E^1_{p,q} = H_q(M_{p,*})$. and the elements of $^I E^1_{p,q}$ have the form $\operatorname{cls}(z)$, where $z \in M_{p,q}$ and $d''z = 0$.

We omit the prescript I for the rest of this proof. It only remains to prove that the first iterated homology is the E^2-term of the spectral sequence arising from the first filtration. We show that $d^1_{p,q} \colon E^1_{p,q} \to E^1_{p-1,q}$ takes $\operatorname{cls} z \mapsto \operatorname{cls}(\overline{d'}z) \in H'_p H''_q(M)$. As $d^1 \colon H_{p+q}(F^p/F^{p-1}) \to H_{p+q-1}(F^{p-1}/F^{p-2})$ is the connecting homomorphism, it arises from the diagram

$$
\begin{array}{ccccc}
M_{p-1,q+1} \oplus M_{p,q} & \xrightarrow{\ \pi\ } & M_{p,q} & \longrightarrow & 0 \\
& & \downarrow{\scriptstyle \overline{D}} & & \\
0 \longrightarrow M_{p-1,q} & \xrightarrow{\ i\ } & M_{p,q-1} \oplus M_{p-1,q}, & &
\end{array}
$$

where $\overline{D}: (a_{p-1,q}, a_{p,q}) \mapsto (d''a_{p-1,q} + d'a_{p,q}, d''a_{p,q})$. Let $z \in M_{p,q}$ be a cycle; that is, $d''_{p,q}z = 0$. Choose $\pi^{-1}z = (0, z)$, so that $\overline{D}(0, z) = (d'_{p,q}z, 0)$. Then,

$$d^1 \operatorname{cls}(z) = \operatorname{cls}(i^{-1}\overline{D}\pi^{-1}z) = \operatorname{cls}(\overline{d'}z) \in H'_p H''_q(M). \quad \bullet$$

We know that both the first and second spectral sequences converge to $\operatorname{Tot}(M)$; let us now compute the E^1- and E^2-terms from the second filtration.

By Lemma 10.15, the second filtration of $\operatorname{Tot}(M)$ is the first filtration of $\operatorname{Tot}(M^t)$ [M^t is the transposed bicomplex (M^t, δ', δ''), where $\delta' = d''$ and $\delta'' = d'$]. Thus, taking homology of pth columns of M^t gives a bigraded module whose (p, q) term is $H_q(M^t_{p,*})$; taking homology of the qth rows gives the second iterated homology.

Let us rewrite $H'_p H''_q(M^t)$ in terms of M. Of course, the terms $M^t{}_{p,q} = M_{q,p}$. Thus, the pth column of M^t is the pth row of M, and the (p, q) term is $H_q(M_{*,p})$, the homology of $M_{*,p}$ with respect to $\delta'' = d'$:

$$^{\mathrm{II}}E^1_{p,q} = H_q(M_{*,p}).$$

For each fixed q, the qth row of this bicomplex is the homology with respect to $\delta' = d''$ of the qth column:

$$\cdots, H_q(M_{*,p+1}), \ H_q(M_{*,p}), \ H_q(M_{*,p-1}), \cdots. \tag{2}$$

Taking homology gives a new bigraded module with (p, q) term is $H''_p H'_q(M)$, the pth homology of the qth column (2).

Definition. If (M, d', d'') is a bicomplex, its *second iterated homology* is the bigraded module whose (p, q) term is $H''_p H'_q(M)$.

We have proved the analog of Proposition 10.17 for the second filtration.

Proposition 10.18. *If M is a first quadrant bicomplex, then*

$$^{\mathrm{II}}E^1_{p,q} = H_q(M_{*,p})$$

and

$$^{\mathrm{II}}E^2_{p,q} = H''_p H'_q(M) \underset{p}{\Rightarrow} H_n(\operatorname{Tot}(M)).$$

Informally, the first iterated homology computes $^{\mathrm{I}}E^2$ by taking homology of the columns and then taking homology of the rows; the second iterated homology computes $^{\mathrm{II}}E^2$ by taking homology of the rows and then taking homology of the columns.

Even though both the first and second spectral sequences converge to $H_n(\text{Tot}(M))$, one should not expect that $^IE^\infty_{p,q} \cong \; ^{II}E^\infty_{p,q}$. The induced filtrations on H_n from $(^IF^p)$ and from $(^{II}F^p)$ need not be the same, and so the factor modules of H_n arising from them may be different.

Here is a nice way to view convergence that I learned from Dave Benson. A spectral sequence is a sequence $E^1, E^2, \ldots$ of differential bigraded modules, and so each E^r can be regarded as a *page*; thus, approximating the (p, q) term of the limit E^∞ involves focusing on the (p, q) position as we turn the pages. The E^2 page is important, for we can compute it with iterated homology and, of course, $E^\infty_{p,q}$ is a subquotient of $E^2_{p,q}$. Moreover, the differential $d^r: E^r_{p,q} \to E^r_{p-r,q+r-1}$ has bidegree $(-r, r-1)$, and it may be viewed as a family of arrows each having slope $(1-r)/r$.

We now illustrate computing $H_\bullet(\text{Tot}(M))$ using iterated homology.

Example 10.19.   A commutative diagram

$$
\begin{array}{ccc}
C & \xleftarrow{\; f \;} & D \\
{\scriptstyle j}\downarrow & & \downarrow{\scriptstyle g} \\
A & \xleftarrow{\; i \;} & B
\end{array}
$$

is a bigraded module if we define $M_{0,0} = A$, $M_{1,0} = B$, $M_{0,1} = C$, $M_{1,1} = D$, and $M_{p,q} = \{0\}$ for all other (p, q); using the sign trick, we replace g by $-g$ and obtain a first quadrant bicomplex.

We compute $H_\bullet(\text{Tot}(M))$ in the special case in which B, C are submodules of A and j, i are their inclusions. The total complex is

$$
\text{Tot}(M) : \; 0 \to D \xrightarrow{(f,-g)} C \oplus B \xrightarrow{j+i} A \to 0,
$$

where $(f, -g): d \mapsto (fd, -gd) \in C \oplus B$ and $j + i: (c, b) \mapsto jc + ib$. Now $H_2 = \ker(f, -g)$. Thus,

$$
H_2(\text{Tot}(M)) = \ker(f, -g) = \{d \in D : fd = 0 = gd\} = \ker f \cap \ker g.
$$

Also,

$$
H_0(\text{Tot}(M)) = \text{coker}(j + i) = A/\text{im}(j + i) = A/(B + C).
$$

Computing $H_1(\text{Tot}(M))$ is not so simple, and we will use spectral sequences to do it. We display the bigraded module $H_q(M_{p,*})$ as a 2×2 matrix $\begin{bmatrix} h_{10} & h_{11} \\ h_{00} & h_{01} \end{bmatrix}$ [note that $\ker(-g) = \ker g$ and $\text{coker}(-g) = \text{coker}\, g$]:

$$
[^IE^1_{p,q}] = [H_q(M_{p,*})] = \begin{bmatrix} \text{coker}\, j & \text{coker}\, g \\ \ker j & \ker g \end{bmatrix}.
$$

Let us compute the first iterated homology: f induces a map of the top row, namely, its restriction $f' : \ker g \to \ker j$, and i induces a map of the bottom row, namely, $\bar{i} : \operatorname{coker} g \to \operatorname{coker} j$, defined by $b + \operatorname{im} g \mapsto ib + \operatorname{im} j$. Thus,

$$[{}^{I}E_{p,q}^{2}] = [H_{p}' H_{q}''(M)] = \begin{bmatrix} \operatorname{coker} \bar{i} & \ker \bar{i} \\ \operatorname{coker} f' & \ker f' \end{bmatrix}.$$

Since M is a first quadrant bicomplex, the first filtration of $H_1 = H_1(\operatorname{Tot}(M))$ is $\Phi^{-1}H_1 = \{0\} \subseteq \Phi^0 H_1 \subseteq \Phi^1 H_1 = H_1$. Now ${}^{I}E_{0,1} = \ker j = \{0\}$, and so all the subquotients ${}^{I}E_{0,1}^{r} = \{0\}$; hence, ${}^{I}E_{0,1}^{\infty} = \{0\}$. Since ${}^{I}E_{p,q}^{r} \Rightarrow H_n$, we have $\{0\} = {}^{I}E_{0,1}^{\infty} = \Phi^0 H_1$. Therefore, ${}^{I}E_{1,0}^{\infty} = \Phi^1 H_1 / \Phi^0 H_1 = \Phi^1 H_1 = H_1$. If $r \geq 1$, then ${}^{I}E_{1,0}^{r+1} = \ker d_{1,0}^{r} / \operatorname{im} d_{r+1,1-r}^{r}$. But both $d_{1,0}^{r}$ and $d_{r+1,1-r}^{r}$ are identically 0 [the target of the first map and the domain of the second map are both $\{0\}$ because they lie outside the 2×2 square]. Hence, ${}^{I}E_{0,1}^{2} = {}^{I}E_{0,1}^{3} = \cdots = {}^{I}E_{0,1}^{\infty}$. Therefore,

$$H_1 = {}^{I}E_{0,1}^{2} = \ker \bar{i}.$$

Let us simplify this. By definition, $H_1 = \ker(j+i)/\ker(f, -g)$. Define $\varphi : \ker(j+i) \to \operatorname{im} j \cap \operatorname{im} i$ by $\varphi : (c, b) \mapsto jc$, where $jc + ib = 0$ (note that $jc = -ib \in \operatorname{im} j \cap \operatorname{im} i$). It is easy to see that φ is surjective and that $\operatorname{im}(f, -g) \subseteq \ker \varphi$; hence, φ induces a surjection

$$\widetilde{\varphi} : H_1 = \frac{\ker(j+i)}{\operatorname{im}(f, -g)} \to \operatorname{im} j \cap \operatorname{im} i.$$

If $v : \operatorname{im} j \cap \operatorname{im} i \to (\operatorname{im} j \cap \operatorname{im} i)/\operatorname{im} jf$ is the natural map, then $v\widetilde{\varphi}$ is a surjection $H_1 \to (\operatorname{im} j \cap \operatorname{im} i)/\operatorname{im} jf$. Finally, we claim that $v\widetilde{\varphi}$ is an isomorphism. If $(c, b) + \operatorname{im}(f, -g) \in \ker v\widetilde{\varphi}$, then $jc \in \operatorname{im} jf$; that is, $jc = jfd = igd = ib'$ for some $d \in D$ and $b' \in B$ [note that $jc = -ib$, because $(c, b) \in \ker(j+i)$, so that $ib' = jc = -ib$; since i is injective, $b' = -b$]. Since j and i are injections, $c = fd$ and $gd = b' = -b$, and so $(c, b) = (fd, -gd) \in \operatorname{im}(f, -g)$. Therefore,

$$H_1(\operatorname{Tot}(M)) = (\operatorname{im} j \cap \operatorname{im} i)/\operatorname{im} jf.$$

To compute the second iterated homology, we first transpose M:

$$M^{t} = \begin{array}{ccc} B & \overset{-g}{\longleftarrow} & D \\ {\scriptstyle i}\downarrow & & \downarrow{\scriptstyle f} \\ A & \underset{j}{\longleftarrow} & C. \end{array}$$

The second iterated homology is

$$[{}^{II}E_{p,q}^{2}] = [H_{p}'' H_{q}'(M)] = \begin{bmatrix} \operatorname{coker} \bar{j} & \ker \bar{j} \\ \operatorname{coker} g' & \ker g' \end{bmatrix},$$

which is different than the first iterated homology. The reader should check that the second iterated homology gives the same $H_1(\operatorname{Tot}(M))$. ◀

We now use Example 10.19.

Proposition 10.20.

  (i) *If R is a ring with right ideal I and left ideal J, then*

$$\operatorname{Tor}_1^R(R/I, R/J) \cong (I \cap J)/IJ.$$

  (ii) *If R is a local ring (not necessarily commutative or noetherian) with maximal ideal $\mathfrak{m}$, then*

$$\operatorname{Tor}_1^R(k, k) \cong \mathfrak{m}/\mathfrak{m}^2,$$

where $k = R/\mathfrak{m}$.

Proof.

  (i) Let $0 \to I \xrightarrow{i} R \xrightarrow{\nu} R/I \to 0$ be exact, where i is the inclusion and ν is the natural map, and let $j \colon J \to R$ be an inclusion. As in Example 10.19, we compute $H_1(\operatorname{Tot}(M))$, where $\operatorname{Tot}(M)$ is the total complex arising from the top rectangle of the diagram

$$
\begin{array}{ccc}
R \otimes_R J & \xleftarrow{\,i \otimes 1\,} & I \otimes_R J \\
{\scriptstyle 1_R \otimes j}\big\downarrow & & \big\downarrow{\scriptstyle 1_I \otimes j} \\
R \otimes_R R & \xleftarrow{\,i \otimes 1\,} & I \otimes_R R \\
{\scriptstyle 1_R \otimes \nu}\big\downarrow & & \big\downarrow{\scriptstyle 1_I \otimes \nu} \\
R \otimes_R (R/J) & \xleftarrow[\,i \otimes 1\,]{} & I \otimes (R/J).
\end{array}
$$

Both $1 \otimes i$ and $1 \otimes j$ are injections, so that $H_1(\operatorname{Tot}(M)) \cong \ker \overline{i \otimes 1}$, by Example 10.19, where $\overline{i \otimes 1} \colon \operatorname{coker} 1_I \otimes j \to \operatorname{coker} 1_R \otimes j$. But $\ker \overline{i \otimes 1} = \operatorname{Tor}_1^R(R/I, R/J)$ [because $\operatorname{Tor}_1^R(R, R/J) = \{0\}$], and so

$$H_1(\operatorname{Tot}(M)) \cong \operatorname{Tor}_1^R(R/I, R/J).$$

To complete the proof, let us simplify the diagram by applying the natural isomorphisms $A \otimes_R R \cong A$ and $R \otimes_R B \cong B$ to it, where A is a right R-module and B is a left R-module.

$$
\begin{array}{ccc}
J & \longleftarrow & IJ \\
\big\downarrow & & \big\downarrow \\
R & \longleftarrow & I
\end{array}
$$

The reason we started this proof with the more complicated version (with tensors) was to enable us to identify H_1 as Tor_1^R. It is easy to check that the natural isomorphisms give rise to a chain map

$$
\begin{array}{ccccccccc}
0 & \longrightarrow & I \otimes_R IJ & \longrightarrow & (R \otimes_R J) \oplus (I \otimes_R R) & \longrightarrow & R \otimes_R R & \longrightarrow & 0 \\
 & & \downarrow{\scriptstyle \rho} & & \downarrow{\scriptstyle \sigma} & & \downarrow{\scriptstyle \tau} & & \\
0 & \longrightarrow & IJ & \longrightarrow & J \oplus I & \longrightarrow & R & \longrightarrow & 0.
\end{array}
$$

Now σ, τ are isomorphisms, so that ρ is as well, by the Five Lemma. Therefore, these two complexes are isomorphic, hence have the same homology. In particular, $\text{Tor}_1^R(R/I, R/J) \cong H_1 \cong (I \cap J)/IJ$.

(ii) Immediate from part (i): take $I = \mathfrak{m} = J$ so that $R/I = k = R/J$. •

As a general rule, the more 0s occurring in an E_2-term, the simpler is the spectral sequence. The following situation arises often.

Definition. A spectral sequence (E^r, d^r) *collapses* on the p-axis if $E_{p,q}^2 = \{0\}$ for all $q \neq 0$; a spectral sequence (E^r, d^r) *collapses* on the q-axis if $E_{p,q}^2 = \{0\}$ for all $p \neq 0$.

Thus, a spectral sequence collapses if the only nonzero modules on its E^2 page lie on one of the axes.

Proposition 10.21. *Let (E^r, d^r) be a first quadrant spectral sequence, and let $E_{p,q}^2 \underset{p}{\Rightarrow} H_n(\text{Tot}(M))$.*

(i) *If (E^r, d^r) collapses on either axis, then $E_{p,q}^\infty \cong E_{p,q}^2$ for all p, q.*

(ii) *If (E^r, d^r) collapses on the p-axis, then $H_n(\text{Tot}(M)) \cong E_{n,0}^2$; if (E^r, d^r) collapses on the q-axis, then $H_n(\text{Tot}(M)) \cong E_{0,n}^2$.*

Proof.

(i) Assume that (E^r, d^r) collapses on the p-axis (the argument when the spectral sequence collapses on the q-axis is similar), and choose $r \geq 2$. First of all, $E_{p,q}^r = \{0\}$ for all $r \geq 2$ and $q \neq 0$, because $E_{p,q}^r$ is a subquotient of $E_{p,q}^2 = \{0\}$. Now $E_{p,0}^{r+1} = \ker d_{p,0}^r / \operatorname{im} d_{p+r,-r+1}^r$. Picturing E^r as the rth page of the spectral sequence, the differential $d^r: E^r \to E^r$ is a family of arrows each having nonzero slope $(1-r)/r$. Now $d_{p,0}^r = 0$, because its target is off the axis, hence is $\{0\}$; thus, $\ker d_{p,0}^r = E_{p,0}^r$. Also, $d_{p+r,-r+1}^r = \{0\}$, because its domain is off the axis, and so $\operatorname{im} d_{p+r,-r+1}^r = \{0\}$. Therefore, $E_{p,0}^{r+1} = E_{p,0}^r / \{0\} = E_{p,0}^r$, and Lemma 10.13 gives $E^\infty = E^2$.

(ii) The induced filtration on $H_n = H_n(\text{Tot}(M))$ is

$$\{0\} = \Phi^{-1}H_n \subseteq \Phi^0 H_n \subseteq \cdots \subseteq \Phi^{n-1}H_n \subseteq \Phi^n H_n = H_n.$$

If the spectral sequence collapses on the p-axis, then $\{0\} = E^2_{p,q}$ for all $p \leq n - 1$. But $E^\infty_{p,q} = E^2_{p,q}$, by part (i), so that $\{0\} = E^\infty_{p,q} = \Phi^p H_n / \Phi^{p-1} H_n$ for all $p \leq n - 1$. Hence, $\{0\} = \Phi^{-1}H_n = \Phi^0 H_n = \cdots = \Phi^{n-1}H_n$ and $H_n = \Phi^n H_n / \Phi^{n-1} H_n \cong E^2_{n,0}$. A similar argument can be given when the spectral sequence collapses on the q-axis.   •

Here is a proof, using spectral sequences, of Theorem 6.32: Tor is independent of the variable resolved. What is more natural than to resolve both variables simultaneously? Recall our earlier notation:

$$\text{Tor}^R_n(A, B) = H_n(\mathbf{P}_A \otimes_R B) \quad \text{and} \quad \text{tor}^R_n(A, B) = H_n(A \otimes_R \mathbf{Q}_B),$$

where $\mathbf{P}_A$ is a deleted projective resolution of A and $\mathbf{Q}_B$ is a deleted projective resolution of B.

Theorem 10.22. *Let $\mathbf{P}_A$ and $\mathbf{Q}_B$ be deleted projective resolutions of a right R-module A and a left R-module B, then $\text{Tor}^R_n(A, B) \cong \text{tor}^R_n(A, B)$; in fact, for all $n \geq 0$,*

$$H_n(\mathbf{P}_A \otimes B) \cong H_n(\mathbf{P}_A \otimes \mathbf{Q}_B) \cong H_n(A \otimes \mathbf{Q}_B).$$

Proof. Let (M, d', d'') be the bicomplex in Example 10.6(i) whose total complex is $\mathbf{P}_A \otimes \mathbf{Q}_B$. Now E^1 is the bigraded module whose (p, q) term is $H''_q(M_{p,*})$, the qth homology of the pth column

$$M_{p,*} = \to P_p \otimes Q_{q+1} \to P_p \otimes Q_q \to P_p \otimes Q_{q-1} \to .$$

Since P_p is projective, hence flat, this sequence is exact at every $q > 0$; hence, $H_q(M_{p,*}) = \{0\}$ for $q > 0$, while $H_0(M_{p,*}) = \text{coker}(P_p \otimes Q_1 \to P_p \otimes Q_{0.})$. But $Q_1 \to Q_0 \to B \to 0$ is exact, so that right exactness of tensor gives $H_0 = P_p \otimes B$. To sum up,

$$^I E^1_{p,q} = \begin{cases} \{0\} & \text{if } q > 0, \\ P_p \otimes B & \text{if } q = 0. \end{cases}$$

Therefore,

$$^I E^2_{p,q} = H'_p H''_q(M) = \begin{cases} \{0\} & \text{if } q > 0, \\ H_p(\mathbf{P}_A \otimes B) & \text{if } q = 0. \end{cases}$$

Thus, this spectral sequence collapses,[2] and Proposition 10.21 gives

$$H_n(\mathbf{P}_A \otimes \mathbf{Q}_B) = H_n(\mathrm{Tot}(M)) \cong {}^{\mathrm{I}}E_{n,0}^2 \cong H_n(\mathbf{P}_A \otimes B).$$

A similar argument applies to the second iterated homology, using the fact that each Q_q is projective and hence is flat. Hence,

$${}^{\mathrm{II}}E_{p,q}^2 = H_p'' H_q'(M) = \begin{cases} \{0\} & \text{if } q > 0, \\ H_p(A \otimes \mathbf{Q}_B) & \text{if } q = 0. \end{cases}$$

Thus, this spectral sequence also collapses, and

$$H_n(\mathbf{P}_A \otimes \mathbf{Q}_B) \cong H_n(A \otimes \mathbf{Q}_B). \quad \bullet$$

An immediate consequence is Theorem 7.5.

Corollary 10.23. *The functors* $\mathrm{Tor}_n^R(A, \square)$ *and* $\mathrm{Tor}_n^R(\square, B)$ *can be computed using flat resolutions of either variable; more precisely, for all deleted flat resolutions* $\mathbf{F}_A$ *and* $\mathbf{G}_B$ *of A and B and for all $n \geq 0$,*

$$H_n(\mathbf{F}_A \otimes_R B) \cong \mathrm{Tor}_n^R(A, B) \cong H_n(A \otimes_R \mathbf{G}_B).$$

Proof. The only property of the resolutions $\mathbf{P}_A$ and $\mathbf{Q}_B$ that was used in the proof of Theorem 10.22 is that all the terms P_p and Q_q are flat. $\bullet$

There is a similar proof showing that Ext is independent of the variable resolved, but we must first return briefly to bicomplexes to set up the appropriate notation. When it comes to visualizing bicomplexes, we must recognize that the integer points on the x-axis increase as we go to the right, in contrast to the usual notation for a complex. To conform to the geometric picture of bicomplexes, let us now write a complex $\mathbf{C}$ "backwards":

$$\leftarrow C_{-2} \xleftarrow{d_{-1}} C_{-1} \xleftarrow{d_0} C_0 \xleftarrow{d_1} C_1 \xleftarrow{d_2} C_2 \leftarrow .$$

Thus, a deleted projective resolution lives on the positive side of the x-axis, while a deleted injective resolution lives on the negative side.

Example 10.24.

(i) Given two complexes $(\mathbf{A}, \Delta')$ and $(\mathbf{B}, \Delta'')$, let us make a bicomplex whose terms are $\mathrm{Hom}(A_p, B_q)$ and whose arrows are induced maps.

---

[2] We could have concluded that the spectral sequence collapses from the calculation of $H_q''(M_{p,*})$; that is, $E_{p,q} = \{0\}$ for all $q > 0$. The extra information we now have is that $E_{p,0}^2 = \mathrm{Tor}_p(A, B)$.

Now $(\Delta'_p)^*$: $\mathrm{Hom}(A_{p-1}, B_q) \to \mathrm{Hom}(A_p, B_q)$, because of the contravariance of $\mathrm{Hom}(\square, B_q)$, so defining $M_{p,q} = \mathrm{Hom}(A_p, B_q)$ gives $(\Delta'_p)^*$ bidegree $(1, 0)$ instead of $(-1, 0)$. To conform to the definition of bicomplex, redefine the bigrading using a sign-changing trick.

$$M_{p,q} = \mathrm{Hom}(A_{-p}, B_q),$$
$$d'_{p,q} = (\Delta'_{-p+1})^*: \mathrm{Hom}(A_{-p}, B_q) \to \mathrm{Hom}(A_{-p+1}, B_q),$$
$$d''_{p,q} = (-1)^{p+q+1}(\Delta''_q)_*: \mathrm{Hom}(A_{-p}, B_q) \to \mathrm{Hom}(A_{-p}, B_{q-1}).$$

We have forced the differentials d' and d'' to have bidegrees $(-1, 0)$ and $(0, -1)$, respectively. Without the sign in d'', the diagram commutes, because Hom is a bifunctor; with the sign, (M, d', d'') is a bicomplex.

The natural candidate for the total complex is $\mathbf{Hom}(\mathbf{A}, \mathbf{B})$, whose term of degree n is

$$\mathbf{Hom}(\mathbf{A}, \mathbf{B})_n = \prod_i \mathrm{Hom}(A_i, B_{i+n}),$$

the maps of degree n as defined earlier. Set $M_{p,q} = \mathrm{Hom}(A_{-p}, B_q)$, and recall that $n = p + q$. Then we have

$$\mathbf{Hom}(\mathbf{A}, \mathbf{B})_n = \prod_p \mathrm{Hom}(A_{-p}, B_{-p+n})$$
$$= \prod_{p+q=n} \mathrm{Hom}(A_{-p}, B_q)$$
$$= \prod_{p+q=n} M_{p,q}.$$

[In contrast to $\mathrm{Tot}(M) = \bigoplus_{p+q=n} M_{p,q}$, this complex involves direct products.] Make the graded module $\mathbf{Hom}(\mathbf{A}, \mathbf{B})$ into a complex by defining its differential D_n: $\mathbf{Hom}(\mathbf{A}, \mathbf{B})_n \to \mathbf{Hom}(\mathbf{A}, \mathbf{B})_{n-1}$ as follows. Consider a map $f: \mathbf{A} \to \mathbf{B}$ of degree n as a sequence of maps $f_{-p}: A_{-p} \to B_q$, and define

$$D_n f_{-p} = (d'_{p,q} + d''_{p,q}) f_{-p}$$
$$= (\Delta'_{-p+1})^* + (-1)^{p+q+1}(\Delta''_q)_*) f_{-p}$$
$$= f_{-p}\Delta'_{-p+1} + (-1)^{p+q+1}\Delta''_q) f_{-p}.$$

A routine check shows that $DD = 0$ [the sign $(-1)^{p+q+1}$ in the definition of D is needed in this computation].

(ii) When $\mathbf{A}$ is a positive complex and $\mathbf{B}$ is a *negative complex* (that is, $B_q = \{0\}$ for positive q), the bicomplex M in part (i) has $M_{p,q} = \{0\}$ whenever p or q is positive.

Definition.   A *third quadrant bicomplex* (or *cohomology bicomplex*)
is a bicomplex $(M_{p,q})$ for which $M_{p,q} = \{0\}$ whenever p or q is positive.

For example, if $\mathbf{P}_A$ is a deleted projective resolution of a module A and
$\mathbf{E}^B$ is a deleted injective resolution of a module B, then the bicomplex
M in part (i) with $M_{p,q} = \text{Hom}(A_{-p}, E_{-q})$ is third quadrant.

For any negative integer $-n$, the line in the pq-plane with equation
$p + q = -n$ intersects the third quadrant in only finitely many lattice
points (if $n > 0$, this line does not meet the third quadrant at all). Since
there are now only finitely many nonzero $M_{p,q}$ with $p + q = -n$, the
direct product $\prod_{p+q=-n} M_{p,q}$ is a direct sum:

$$\text{Hom}(\mathbf{P}_A, \mathbf{E}^B)_{-n} = \bigoplus_{p+q=-n} \text{Hom}(P_{-p}, E_{-q}).$$

Therefore, the complex is the total complex:

$$\text{Hom}(\mathbf{P}_A, \mathbf{E}^B) = \text{Tot}(M).$$

As usual, we eliminate negative indices by raising them. For third quadrant bicomplexes, write

$$M^{p,q} = M_{-p,-q}.$$

[Once this is done, the relabeled third quadrant looks like the first quadrant. However, the differentials, which originally pointed left and down,
for they had bidegrees $(-1, 0)$ and $(0, -1)$, now point right and up.]

Consider the first filtration of $\text{Tot}(M)$ when M is a third quadrant bicomplex.

$$({}^{I}F^{-p})_{-n} = \bigoplus_{i \leq -p} M_{i,-n-i}$$
$$= M_{-n,0} \oplus \cdots \oplus M_{-p,-n+p}$$

and

$$({}^{I}F^{-p+1})_{-n} = \bigoplus_{i \leq -p-1} M_{i,-n-i}$$
$$= M_{-n,0} \oplus \cdots \oplus M_{-p,-n+p} \oplus M_{-p+1,-n+p-1}.$$

Thus,

$$\{0\} = F^{-n-1} \subseteq F^{-n} \subseteq F^{-n+1} \subseteq \cdots \subseteq F^0 = \text{Tot}(M).$$

If we lower indices and change their sign, we have

$$\{0\} = F_{n+1} \subseteq F_n \subseteq F_{n-1} \subseteq \cdots \subseteq F_0 = \mathrm{Tot}(M);$$

that is, the filtration so labeled is a *decreasing* filtration (which is acceptable). Similarly, lowered indices on the second filtration give another decreasing filtration of $\mathrm{Tot}(M)$.

For third quadrant bicomplexes, we change the signs of p, q, and n. Thus, we write $H_{-n}(\mathrm{Tot}(M)) = H^n(\mathrm{Tot}(M))$, call it the nth *cohomology module* and, for $1 \le r \le \infty$,

$$d^r_{-p,-q} = d_r^{p,q}, \quad {}^{\mathrm{I}}E^r_{-p,-q} = {}^{\mathrm{I}}E_r^{p,q}, \text{ and } \quad {}^{\mathrm{II}}E^r_{-p,-q} = {}^{\mathrm{II}}E_r^{p,q}. \quad \blacktriangleleft$$

Lemma 10.25. *If M is a third quadrant bicomplex, then the first and second filtrations are bounded.*

Proof.   The bounds are $s(n) = -n - 1$ and $t(n) = 0$.   •

Here is the third quadrant version of Proposition 10.21, so that the reader can see index raising.

Proposition 10.26. *Let (E_r, d_r) be a third quadrant spectral sequence, and let $E_2^{p,q} \underset{p}{\Rightarrow} H^n(\mathrm{Tot}(M))$.*

(i) *If (E_r, d_r) collapses on either axis, then $E_\infty^{p,q} = E_2^{p,q}$ for all p, q.*

(ii) *If (E_r, d_r) collapses on the p-axis, then $H^n(\mathrm{Tot}(M)) \cong E_2^{n,0}$;
    if (E_r, d_r) collapses on the q-axis, then $H^n(\mathrm{Tot}(M)) \cong E_2^{0,n}$.*

Proof.

(i) This proof can be copied verbatim from that of Proposition 10.21(i).

(ii) The induced filtration on $H^n = H^n(\mathrm{Tot}(M))$ is

$$\{0\} = \Psi_{n+1}H^n \subseteq \Psi_n H^n \subseteq \cdots \subseteq \Psi_{-1}H^n \subseteq \Psi_0 H^n = H^n.$$

Suppose the spectral sequence collapses on the p-axis; if $p < -n$, then $q = -n + p \ne 0$ and $E_\infty^{p,q} = E_2^{p,q} = \{0\} = \Psi_p H^n / \Psi_{p-1}H^n$. Hence, $\{0\} = \Psi_{n+1}H^n = \Psi_n H^n = \cdots = \Psi_1 H^n$, and $H^n = \Psi_0 H^n / \Psi_1 H^n \cong E_2^{n,0}$. A similar argument can be given when the spectral sequence collapses on the q-axis.   •

We can prove that Ext is independent of the variable resolved in the style of the proof of Theorem 10.22. Recall that we defined $\mathrm{Ext}^n(A, B) = H^n(\mathrm{Hom}(A, \mathbf{E}^B))$ and $\mathrm{ext}^n(A, B) = H^n(\mathrm{Hom}(\mathbf{P}_A, B))$.

Theorem 10.27.   *Let A and B be left R-modules, let $\mathbf{P}_A$ be a deleted projective resolution of A, and let $\mathbf{E}^B$ be a deleted injective resolution of B. Then $\mathrm{Ext}_R^n(A, B) \cong \mathrm{ext}_R^n(A, B)$ for all $n \geq 0$: in fact,*

$$H^n(\mathrm{Hom}(A, \mathbf{E}^B)) \cong H^n(\mathrm{Hom}_R(\mathbf{P}_A, \mathbf{E}^B)) \cong H^n(\mathrm{Hom}(\mathbf{P}_A, B)).$$

When a spectral sequence collapses, it is very easy to compute the homology of its total complex. Here is another simple example.

Proposition 10.28.   *Let M be a first quadrant bicomplex for which E^2 consists of two columns: there is $t > 0$ with $E_{p,q}^2 = \{0\}$ for $p \neq 0$ and $p \neq t$. Then*

(i) $E^2 = E^3 = \cdots = E^t$,

(ii) $E^{t+1} = E^\infty$,

(iii) *there are exact sequences*

$$0 \to E_{0,n}^{t+1} \to H_n(\mathrm{Tot}(M)) \to E_{t,n-t}^{t+1} \to 0,$$

(iv) *there is an exact sequence*

$$\to H_{n+1} \to E_{t,n+1-t}^2 \xrightarrow{d^r} E_{0,n}^2 \to H_n \to E_{t,n-t}^2 \xrightarrow{d^t} E_{0,n-1}^2 \to .$$

Proof.   The differential $d^r : E^r \to E^r$ has bidegree $(-r, r-1)$; that is, d^r goes r steps left and $r-1$ steps up. It follows that $d^r = 0$ for $r \neq t$, since it either begins or ends at $\{0\}$. Lemma 10.13 applies to prove (i) and (ii).

Consider Fig. 10.7 on the following page, the picture of the E^2-plane. Since M is first quadrant, we know that $E_{p,q}^2 \underset{p}{\Rightarrow} H_n(\mathrm{Tot}(M))$; that is,

$$\{0\} = \Phi^{-1}H_n \subseteq \Phi^0 H_n \subseteq \cdots \subseteq \Phi^n H_n = H_n$$

and

$$E_{p,q}^\infty \cong \Phi^p H_n / \Phi^{p-1} H_n \quad \text{for all } p, q.$$

It follows that $\Phi^0 H_n = E_{0,n}^\infty \subseteq H_n$. Now $\Phi^1 H_n / \Phi^0 H_n = E_{1,n-1}^\infty = \{0\}$, so that $\Phi^0 H_n = \Phi^1 H_n$. Indeed, $\Phi^0 H_n = \Phi^1 H_n = \cdots = \Phi^{t-1} H_n$. When we reach t, however, $\Phi^t H_n / \Phi^{t-1} H_n \cong E_{t,n-t}^\infty$, and then all remains constant again: $\Phi^t H_n = \Phi^{t+1} H_n = \cdots = \Phi^n H_n = H_n$. Hence, $H_n / E_{0,n}^\infty \cong E_{t,n-t}^\infty$: there is an exact sequence $0 \to E_{0,n}^\infty \to H_n \to E_{t,n-t}^\infty \to 0$. In light of (ii), this may be rewritten as

$$0 \to E_{0,n}^{t+1} \to H_n \to E_{t,n-t}^{t+1} \to 0, \tag{3}$$

which is (iii).

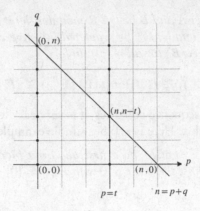

Fig. 10.7   Two columns.

There is an exact sequence obtained from $d^t = d^t_{t,n+1-t}$:

$$0 \to \ker d^t \to E^t_{t,n+1-t} \xrightarrow{d^t} E^t_{0,n} \to \operatorname{coker} d^t \to 0.$$

By definition, $E^t_{t,n+1-t} = \ker d^t_{t,n+1-t} / \operatorname{im} d^t_{2t,n+2-2t} = \ker d^t_{t,n+1-t}$, and $E^{t+1}_{0,n} = \ker d^t_{0,n} / \operatorname{im} d^t_{t,n+1-t} = E^t_{0,n} / \operatorname{im} d^t_{t,n+1-t} = \operatorname{coker} d^t_{t,n+1-t}$. The sequence may thus be rewritten as

$$0 \to E^{t+1}_{t,n+1-t} \to E^t_{t,n+1-t} \xrightarrow{d^t} E^t_{0,n} \to E^{t+1}_{0,n} \to 0. \tag{4}$$

Splicing Eqs. (3) and (4) together, remembering that $E^t = E^2$, yields the desired exact sequence

$$\to H_{n+1} \to E^2_{t,n+1-t} \xrightarrow{d^t} E^2_{0,n} \to H_n \to E^2_{t,n-t} \to . \quad \bullet$$

Note that it is possible that the terms E^r of a spectral sequence remain constant for a while before moving. Thus, it is not true in general that $E^r = E^{r+1}$ implies $E^r = E^\infty$.

The proposition simplifies when the two columns are adjacent. There is a third quadrant version of Proposition 10.28 (whose statement is left to the reader); however, we do state both versions for adjacent columns. This corollary applies, in particular, to the 2×2 rectangle in Example 10.19.

Corollary 10.29.

(i) *Let M be a first quadrant bicomplex for which E^2 consists of two adjacent columns: $E^2_{p,q} = \{0\}$ for all $p \neq 0, 1$. For each n, there is an exact sequence*

$$0 \to E^2_{0,n} \to H_n(\operatorname{Tot}(M)) \to E^2_{1,n-1} \to 0.$$

(ii) *Let M be a third quadrant bicomplex for which E_2 consists of two adjacent columns: $E_2^{p,q} = \{0\}$ for all $p \neq 0, 1$. For each n, there is an exact sequence*

$$0 \to E_2^{1,n-1} \to H^n(\mathrm{Tot}(M)) \to E_2^{0,n} \to 0.$$

We can now give a "poor man's version" of the Universal Coefficient Theorem (it mentions neither splitting nor naturality).

Corollary 10.30. *Let R be a commutative hereditary ring, and let $\mathbf{C}$ be a positive complex of free R-modules. If A is a R-module, then, for all $n \geq 0$, there are exact sequences*

$$0 \to H_n(\mathbf{C}) \otimes_R A \to H_n(\mathbf{C} \otimes_R A) \to \mathrm{Tor}_1^R(H_{n-1}(\mathbf{C}), A) \to 0$$

and

$$0 \to \mathrm{Ext}_R^1(H_{n-1}(\mathbf{C}), A) \to H^n(\mathrm{Hom}_R(\mathbf{C}, A)) \to \mathrm{Hom}_R(H_n(\mathbf{C}), A) \to 0.$$

Proof. For the first sequence, let $0 \to P_1 \to P_0 \to A \to 0$ be a projective resolution of A (which exists because R is hereditary), and apply Corollary 10.29 to the bicomplex M with $M_{p,q} = C_q \otimes_R E^p$.

For the second sequence, let $0 \to A \to E^0 \to E^1 \to 0$ be an injective resolution of A (R is hereditary), and apply Corollary 10.29 to the third quadrant bicomplex M with $M^{p,q} = \mathrm{Hom}_R(C_p, E^q)$. •

Here is a useful byproduct of convergence: the *five-term exact sequence*. We have already seen the basic idea of its proof: many differentials d^r in a spectral sequence are zero because their bidegree forces either their domain or their target to be $\{0\}$.

Theorem 10.31 (Homology Five-Term Exact Sequence). *Assume that (E^r, d^r) is a first quadrant spectral sequence, so that $E_{p,q}^2 \underset{p}{\Rightarrow} H_n(\mathrm{Tot}(M))$.*

(i) *For each n, there is a surjection $E_{0,n}^2 \to E_{0,n}^\infty$; dually, there is an injection $E_{n,0}^\infty \to E_{n,0}^2$.*

(ii) *For each n, there is an injection $E_{0,n}^\infty \to H_n(\mathrm{Tot}(M))$; dually, there is a surjection $H_n(\mathrm{Tot}(M)) \to E_{n,0}^\infty$.*

(iii) *There is an exact sequence*

$$H_2(\mathrm{Tot}(M)) \to E_{2,0}^2 \xrightarrow{d^2} E_{0,1}^2 \to H_1(\mathrm{Tot}(M)) \to E_{1,0}^2 \to 0.$$

Remark. The maps in (i) are called *edge homomorphisms*. ◄

Proof.

(i) By definition, $E_{0,n}^3 = \ker d_{0,n}^2 / \operatorname{im} d_{2,n-1}^2$. Since E^2 is first quadrant, $d_{0,n}^2 \colon E_{0,n}^2 \to E_{-2,n+1}^2 = \{0\}$ is the zero map. Therefore, $\ker d_{0,n}^2 = E_{0,n}^2$, and there is a surjection $E_{0,n}^2 \to E_{0,n}^3$. This argument can be repeated for each $d^r \colon E^r \to E^r$, using only the fact that d^r has bidegree $(-r, r-1)$. There is thus a chain of surjections $E_{0,n}^2 \to E_{0,n}^3 \to \cdots \to E_{0,n}^r$. This completes the argument, for $E_{0,n}^\infty = E_{0,n}^r$ for large r.

Dually, there is an injection $E_{n,0}^2 \to E_{n,0}^3$, for

$$E_{n,0}^3 = \ker d_{n,0}^2 / \operatorname{im} d_{n+2,-1}^2 = \ker d_{n,0}^2 \subseteq E_{n,0}^2$$

($d_{n+2,-1}^2 = 0$ because E^2 is first quadrant). Thus, there is an injection $E_{n,0}^3 \to E_{n,0}^2$. Now iterate; all $\operatorname{im} d_{n,0}^r = \{0\}$ because $\operatorname{im} d_{n,0}^r \subseteq E_{n+r,-1}^r = \{0\}$.

(ii) The definition of convergence yields

$$E_{0,n}^\infty = \Phi^0 H_n / \Phi^{-1} H_n = \Phi^0 H_n \subseteq H_n$$

and

$$E_{n,0}^\infty = \Phi^n H_n / \Phi^{n-1} H_n = H_n / \Phi^{n-1} H_n.$$

(iii) There is an exact sequence arising from d^2:

$$0 \to \ker d_{2,0}^2 \to E_{2,0}^2 \xrightarrow{d^2} E_{0,1}^2 \to \operatorname{coker} d_{2,0}^2 \to 0. \tag{5}$$

Now $\ker d_{2,0}^2 = \ker d_{2,0}^2 / \operatorname{im} d_{4,-1}^2 = E_{2,0}^3$. Iterating this argument gives $\ker d_{2,0}^2 = E_{2,0}^r$ for large r, and hence $\ker d_{2,0}^2 = E_{2,0}^\infty$. Dually,

$$\operatorname{coker} d_{2,0}^2 = E_{0,1}^2 / \operatorname{im} d_{2,0}^2 = \ker d_{0,1}^2 / \operatorname{im} d_{2,0}^2 = E_{0,1}^3.$$

Iteration gives $\operatorname{coker} d_{2,0}^2 = E_{0,1}^\infty$. Exact sequence (5) now reads

$$0 \to E_{2,0}^\infty \to E_{2,0}^2 \xrightarrow{d^2} E_{0,1}^2 \to E_{0,1}^\infty \to 0.$$

The surjection $H_2(\operatorname{Tot}(M)) \to E_{2,0}^\infty$ in (iv) gives exactness of

$$H_2(\operatorname{Tot}(M)) \to E_{2,0}^2 \xrightarrow{d^2} E_{0,1}^2 \to E_{0,1}^\infty \to 0. \tag{6}$$

Finally, the equations

$$H_1 = \Phi^1 H_1, \quad E_{1,0}^\infty = \Phi^1 H_1 / \Phi^0 H_1, \quad E_{0,1}^\infty = \Phi^0 H_1 / \Phi^{-1} H_1 = \Phi^0 H_1$$

combine to give

$$H_1/E_{0,1}^\infty \cong E_{1,0}^\infty;$$

otherwise said, there is a short exact sequence

$$0 \to E_{0,1}^\infty \to H_1(\mathrm{Tot}(M)) \to E_{1,0}^\infty \to 0.$$

Splicing this to the exact sequence (6) yields exactness of

$$H_2(\mathrm{Tot}(M)) \to E_{2,0}^2 \xrightarrow{d^2} E_{0,1}^2 \xrightarrow{\alpha} H_1(\mathrm{Tot}(M)) \to E_{1,0}^\infty \to 0,$$

where α is the composite $E_{0,1}^2 \to E_{0,1}^\infty \to H_1(\mathrm{Tot}(M))$.

If $r \geq 2$, then both $d_{1,0}^r$ and $d_{r+1,r-1}^r$ are zero maps, for $\mathrm{im}\, d_{1,0}^r \subseteq E_{1-r,r-1}^r = \{0\}$ and the domain of $d_{r+1,1-r}^r \subseteq E_{r+1,1-r}^r = \{0\}$. But $E_{1,0}^{r+1} = \ker d_{1,0}^r / \mathrm{im}\, d_{r+1,1-r}^r = E_{1,0}^r$, so that $E_{1,0}^2 = E_{1,0}^r$. Finally, $E_{1,0}^\infty = E_{1,0}^r$ for large r, so that $E_{1,0}^\infty = E_{1,0}^2$. Substituting this in the last exact sequence completes the proof.   •

Here is a variation of the five-term exact sequence whose proof is left to the reader.

Corollary 10.32. *If (E^r, d^r) is a first quadrant spectral sequence and there is $n > 0$ with $E_{p,q}^2 = \{0\}$ for all $q < n$, then there is an exact sequence*

$$H_{n+1}(\mathrm{Tot}(M)) \to E_{n+1,0}^2 \xrightarrow{d^{n+1}} E_{0,n}^2 \to H_n(\mathrm{Tot}(M)) \to E_{n,0}^2 \to 0.$$

The proofs for cohomology are dual to the proofs for homology.

Theorem 10.33 (Cohomology Five-Term Exact Sequence). *Assume that (E_r, d_r) is a third quadrant spectral sequence, so that $E_2^{p,q} \underset{p}{\Rightarrow} H^n(\mathrm{Tot}(M))$. Then there is an exact sequence*

$$0 \to E_2^{1,0} \to H^1(\mathrm{Tot}(M)) \to E_2^{0,1} \xrightarrow{d_2} E_2^{2,0} \to H^2(\mathrm{Tot}(M)).$$

Corollary 10.34. *If (E_r, d_r) is a third quadrant spectral sequence and there is $n > 0$ with $E_2^{p,q} = \{0\}$ for all $q < n$, then there is an exact sequence*

$$0 \to E_2^{n,0} \to H^n(\mathrm{Tot}(M)) \to E_2^{0,n} \xrightarrow{d_{n+1}} E_2^{n+1,0} \to H^{n+1}(\mathrm{Tot}(M)).$$

We cannot appreciate these exact sequences in this generality. After giving the Lyndon–Hochschild–Serre spectral sequence, we will see that we have lengthened the exact sequence in Theorem 9.84.

Exercises

10.6 (**Mapping Theorem**) Let (E^r, d^r) and (E''^r, d''^r) be spectral sequences. A *map of spectral sequences* is a family of bigraded maps $f = (f^r \colon E^r \to E''^r))$, each of bidegree $(0, 0)$, such that, for all r, we have $d''^r f^r = f^r d^r$ and f^{r+1} is the map induced by f^r in homology; that is, since $E^{r+1} = H(E^r, d^r)$, we have $f^{r+1} \colon \operatorname{cls}(z) \mapsto \operatorname{cls}(f^r z)$.

    (i)   Prove that if $f = (f^r \colon (E^r, d^r) \to (E''^r, d''^r))$ is a map of spectral sequences for which f^t is an isomorphism for some t, then f^r is an isomorphism for all $r \geq t$.

    (ii)   Let E^2 and E'^2 be either first or third quadrant spectral sequences. If $f = (f^r \colon (E^r, d^r) \to (E''^r, d''^r))$ is a map of spectral sequences for which f^t is an isomorphism for some t, prove that f induces an isomorphism $E^\infty \cong E'^\infty$.

10.7   (i)   If $\mathbf{P}$ is a projective resolution of a right R-module A and $\mathbf{Q}$ is a projective resolution of a left R-module B, prove that $\mathbf{P} \otimes_R \mathbf{Q}$ is a projective resolution of $A \otimes_R B$.

    (ii)   Use part (i) to redo Exercise 9.34 on page 594: if G is a free abelian group of rank r, then $\operatorname{cd}(G) \leq r$.

10.8 Prove that there is an analog of the adjoint isomorphism for complexes: if R and S are rings, $\mathbf{A}$ is a complex of right R-modules, $\mathbf{B}$ is a complex of (R, S)-bimodules, and $\mathbf{C}$ is a complex of right S-modules, then

$$\operatorname{Hom}_S(\mathbf{A} \otimes_R \mathbf{B}, \mathbf{C}) \cong \operatorname{Hom}_R(\mathbf{A}, \operatorname{Hom}_S(\mathbf{B}, \mathbf{C})).$$

*10.9   (i)   For fixed $k \geq 0$ and $(\mathbf{A}, d) \in \mathbf{Comp}(\mathbf{Mod}_R)$, $(\mathbf{C}, \delta) \in \mathbf{Comp}(_R\mathbf{Mod})$, prove that there is a bicomplex (M, d', d'') with $M_{p,q} = \operatorname{Tor}_k^R(A_p, B_q)$, $d'_{p,q} = (d_p \otimes 1_{C_q})_*$, and $d''_{p,q} = (-1)^p (1_{A_p} \otimes \delta_q)_*$. We denote $\operatorname{Tot}(M)$ by

$$\operatorname{Tor}_k^R(\mathbf{A}, \mathbf{C}).$$

[See Example 10.6(i).]

    (ii)   Prove, as in part (i), that there is a bicomplex with $(p.q)$ term $\operatorname{Ext}_R^k(A_{-p}, C_q)$; its total complex is denoted by

$$\operatorname{Ext}_R^k(\mathbf{A}, \mathbf{C}).$$

[See Example 10.24(i).]

*10.10 **(i)** If R is a ring and $0 \to \mathbf{A}' \to \mathbf{A} \to \mathbf{A}'' \to 0$ is an exact sequence in $\mathbf{Comp}(\mathbf{Mod}_R)$, prove, for any complex $\mathbf{C}$ of left R-modules, that there is an exact sequence in $\mathbf{Comp}(\mathbb{Z})$:

$$\to \mathrm{Tor}_1^R(\mathbf{A}', \mathbf{C}) \to \mathrm{Tor}_1^R(\mathbf{A}, \mathbf{C}) \to \mathrm{Tor}_1^R(\mathbf{A}'', \mathbf{C})$$
$$\to \mathbf{A}' \otimes_R \mathbf{C} \to \mathbf{A} \otimes_R \mathbf{C} \to \mathbf{A}'' \otimes_R \mathbf{C} \to 0.$$

(ii) **Definition.** A complex $\mathbf{A}$ *flat* if $0 \to \mathbf{C}' \xrightarrow{i} \mathbf{C}$ is exact in $\mathbf{Comp}(\mathbf{Mod}_R)$, then $0 \to \mathbf{A} \otimes_R \mathbf{C}' \xrightarrow{1 \otimes i} \mathbf{A} \otimes_R \mathbf{C}$ is exact in $\mathbf{Comp}(\mathbf{Ab})$.

Prove that the following statements are equivalent.

(i) $\mathbf{A}$ is a complex having all terms flat.
(ii) $\mathrm{Tor}_1^R(\mathbf{A}, \mathbf{C}) = 0$ for every $\mathbf{C}$; i.e., every term of the complex $\mathrm{Tor}_1^R(\mathbf{A}, \mathbf{C})$ is $\{0\}$.
(iii) $\mathbf{A}$ is flat.[3]

*10.11 Let $\mathbf{A}$ be a complex of right R-modules, and let $\mathbf{C}$ be a complex of left R-modules. If $\mathcal{A}$ has zero differentials and all its terms are flat, and if $H_\bullet(\mathbf{C})$ is viewed as a complex having zero differentials, prove that

$$H_n(\mathbf{A} \otimes_R \mathbf{C}) = (\mathbf{A} \otimes_R H_\bullet(\mathbf{C}))_n.$$

10.5 Cartan–Eilenberg Resolutions

We are going to see that if $\mathcal{A}$ is an abelian category with enough projectives (or injectives), then $\mathbf{Comp}(\mathcal{A})$, the category of all chain complexes over $\mathcal{A}$, also has enough projectives (or injectives). A *Cartan–Eilenberg resolution* is a special kind of projective (or injective) resolution of a complex in $\mathbf{Comp}(\mathcal{A})$, and we shall see that there are enough of them as well.

Definition. Let $\mathcal{A}$ be an abelian category, and let $(\mathbf{C}, d)$ be a complex in $\mathcal{A}$. The *fundamental exact sequences* of $\mathbf{C}$ are, for all $n \in \mathbb{Z}$:

$$0 \to Z_n \to C_n \xrightarrow{d'} B_{n-1} \to 0;$$
$$0 \to B_n \to Z_n \to H_n \to 0.$$

In $_R\mathbf{Mod}$, the morphism $d' : C_n \to B_{n-1}$ is d with its target changed from C_{n-1} to B_{n-1}; thus, $d = id'$, where $i : B \to C$ is the inclusion. In a general abelian category, there exists a factorization $d = id'$ with i monic and d' epic.

---

[3]Compare this with Theorem 10.42; characterizing flat complexes is very much simpler than characterizing projective complexes or injective complexes.

Definition. A complex $(\mathbf{C}, d)$ in an abelian category $\mathcal{A}$ is *split* if all its fundamental exact sequences are split.

Recall Example 6.1(iv): if $f \colon A \to B$ is a morphism, then

$$\Sigma^k(f)$$

is the complex with f concentrated in degrees $(k, k-1)$; that is, A is the term of degree k, B is the term of degree $k-1$, all other terms are 0, and f is the kth differential.

Lemma 10.35. *If P is a projective object in an abelian category $\mathcal{A}$ and $k \in \mathbb{Z}$, then $\Sigma^k(1_P)$ is projective in $\mathbf{Comp}(\mathcal{A})$.*

Proof. Consider the diagram in $\mathbf{Comp}(\mathcal{A})$:

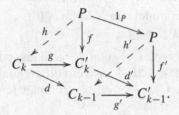

Here, g and g' are parts of an epic chain map $\mathbf{C} \to \mathbf{C}'$, so that each of them is epic in $\mathcal{A}$ (Exercise 5.66 on page 322). Since P is projective, there is $h \colon P \to C_k$ with $gh = f$. Define $h' \colon P \to C_{k-1}$ by $h' = dh$. All the faces of the prism commute, with the possible exception of the triangle on the right. In particular, $(h, h') \colon \Sigma^k(1_P) \to \mathbf{C}$ is a chain map. It remains to prove that $g'h' = f'$. But $g'h' = g'dh = d'gh = d'f = f'$. •

Proposition 10.36. *If $(\mathbf{C}, d)$ is split, then*

$$\mathbf{C} \cong \bigoplus_{k \in \mathbb{Z}} \Sigma^k(\delta_k),$$

where $\delta_k \colon B_k(\mathbf{C}) \to Z_k(\mathbf{C})$ is the inclusion and $\Sigma^k(\delta_k)$ is the complex having δ_k concentrated in degrees $(k, k-1)$.

Proof. As usual, we invoke the Metatheorem on page 316, so that we may assume that $\mathcal{A} = \mathbf{Ab}$; that is, we may assume our objects have elements. Write $B_k = B_k(\mathbf{C})$ and $Z_k = Z_k(\mathbf{C})$. Since $(\mathbf{C}, d)$ is split, there are split exact sequences $0 \to Z_k \to C_k \to B_{k-1} \to 0$, for each k, and Proposition 2.20 gives morphisms,

$$0 \to Z_k \underset{q}{\overset{i}{\rightleftarrows}} C_k \underset{j}{\overset{p}{\rightleftarrows}} B_{k-1} \to 0,$$

where p is merely d with its target changed from C_{k-1} to B_{k-1}, and

$$qi = 1_{Z_k}, \quad pj = 1_{B_{k-1}}, \quad pi = 0, \quad qj = 0, \quad \text{and} \quad iq + jp = 1_{C_k}.$$

Now $\Sigma^k(i)$ is a direct summand of $\Sigma^k(1_{C_k})$ for all k, by Exercise 10.15 on page 655.

The nth term of $\bigoplus_{k\in\mathbb{Z}} \Sigma^k(\delta_k)$ is $Z_n \oplus B_{n-1}$, and the nth differential is $\Delta_n: (z_n, b_{n-1}) \mapsto (\delta_n b_{n-1}, 0)$. Define $\theta_n: C_n \to Z_n \oplus B_{n-1}$ by $\theta: c_n \mapsto (qc_n, pc_n)$. Note that each θ is an isomorphism, for its inverse is $(z, b) \mapsto iz + jb$. Finally, $\theta = (\theta_n)$ is a chain map because the following diagram commutes, as the reader may check using the explicit formulas:

$$
\begin{array}{ccc}
C_n & \xrightarrow{\quad d \quad} & C_{n-1} \\
\theta \downarrow & & \downarrow \theta \\
Z_n \oplus B_{n-1} & \xrightarrow{\quad \Delta \quad} & Z_{n-1} \oplus B_{n-2}.
\end{array} \quad \bullet
$$

Corollary 10.37. *If $\mathcal{A}$ is an abelian category and $\mathbf{C}$ is a split complex each of whose terms is projective in $\mathcal{A}$, then $\mathbf{C}$ is projective in $\mathbf{Comp}(\mathcal{A})$.*

Proof. Proposition 10.36 gives $\mathbf{C} \cong \bigoplus_{k\in\mathbb{Z}} \Sigma^k(\delta_k)$, where $\delta_k: B_k \to Z_k$ is the inclusion. As $\mathbf{C}$ is a split complex, the exact sequence $0 \to Z_k \xrightarrow{\iota_k} C_k \to B_{k-1} \to 0$ splits; hence, $\Sigma^k(\iota_k)$ is a direct summand of $\Sigma^k(1_{C_k})$, by Exercise 10.15 on page 655. Since C_k is projective, $\Sigma^k(1_{C_k})$ is projective in $\mathbf{Comp}(\mathcal{A})$, by Lemma 10.35 and so its direct summand $\Sigma^k(\iota_k)$ is also projective in $\mathbf{Comp}(\mathcal{A})$. Similarly, splitting of $0 \to B_k \xrightarrow{\delta_k} Z_k \to H_k \to 0$ implies projectivity of $\Sigma^k(\delta_k)$, and so $\mathbf{C}$ is projective, being a direct sum of projectives. $\bullet$

We are going to prove the converse of Corollary 10.37.

Notation. If $p \in \mathbb{Z}$ and $(\mathbf{A}, d)$ is a complex, define $\mathbf{A}[p]$ to be the complex whose indices have been shifted by p: its nth term is A_{n+p} and its nth differential is d_{n+p}.

For example, if $(\mathbf{C}, d)$ is a complex with cycles $\mathbf{Z}$ and boundaries $\mathbf{B}$, then

$$0 \to \mathbf{Z} \to \mathbf{C} \xrightarrow{d'} \mathbf{B}[-1] \to 0$$

is an exact sequence of complexes, where d'_n is just d_n with its target C_{n-1} replaced by B_{n-1}.

It is obvious that

$$H_n(\mathbf{A}[p]) = H_{n+p}(\mathbf{A}).$$

Definition. If $f: (\mathbf{A}, d) \to (\mathbf{C}, \delta)$ is a chain map, then its ***mapping cone***[4] is the complex **cone**(f) whose term of degree n is **cone**$(f)_n = A_{n-1} \oplus C_n$ and whose differential $D_n: \mathbf{cone}(f)_n \to \mathbf{cone}(f)_{n-1}$ is

$$D_n : (a_{n-1}, c_n) \mapsto (-d_{n-1}a_{n-1}, \delta_n c_n - f_{n-1}a_{n-1});$$

that is, if (a_{n-1}, c_n) is viewed as a column vector,

$$D_n \begin{bmatrix} a_{n-1} \\ c_n \end{bmatrix} = \begin{bmatrix} -d_{n-1} & 0 \\ -f_{n-1} & \delta_n \end{bmatrix} \begin{bmatrix} a_{n-1} \\ c_n \end{bmatrix}.$$

It is routine to check that **cone**(f) is a complex.

Lemma 10.38. *Let $f: (\mathbf{A}, d) \to (\mathbf{C}, \delta)$ be a chain map. There is an exact sequence of complexes*

$$0 \to \mathbf{C} \xrightarrow{\ i\ } \mathbf{cone}(f) \xrightarrow{\ j\ } \mathbf{A}[-1] \to 0,$$

where $i: c \mapsto (0, c)$ and $j: (a, c) \mapsto -a$. The connecting homomorphism $H_n(\mathbf{A}[-1]) \to H_{n-1}(\mathbf{C})$ in the corresponding long exact sequence is the induced map $f_*: H_{n-1}(\mathbf{A}) \to H_{n-1}(\mathbf{C})$.

Proof. Exactness of the sequence of complexes is a routine calculation. Since $\partial = j^{-1}Di^{-1}$, we have

$$\partial: a \mapsto (-a, c) \mapsto \big(-d(-a), \delta c - f(-a)\big) \mapsto fa. \quad \bullet$$

Example 10.39. Let $f: \mathbf{A} \to \mathbf{C}$ be a chain map. Construct a two-column bicomplex $M = M_{p,q}$ with terms

$$M_{p,q} = \begin{cases} C_q & \text{if } p = 0, \\ A_q & \text{if } p = 1, \\ \{0\} & \text{if } p \geq 2, \end{cases}$$

with vertical maps the differentials in $\mathbf{A}$, $\mathbf{C}$, respectively, and with horizontal maps $-f_q: A_q \to C_q$. It is easy to see that

$$\mathrm{Tot}(M) = \mathbf{cone}(f),$$

and that the exact sequence of Lemma 10.38 is the exact sequence which occurs in Corollary 10.29. ◄

---

[4] See Gelfand–Manin, *Methods of Homological Algebra*, p. 27, for the geometric origin of this cone construction.

Lemma 10.40. *Let $\mathcal{A}$ be an abelian category, and let $(\mathbf{C}, d)$ be a complex in $\mathcal{A}$.*

(i) **cone**(1) *is an acyclic complex, where* 1 *is the identity chain map on* $\mathbf{C}$.

(ii) *If there exists a morphism* $s: \mathbf{C} \to \mathbf{C}$ *of degree* $+1$ *with* $d = dsd$ *(that is, $d_n = d_n s_{n-1} d_n$ for all n), then* $\mathbf{C}$ *is split.*

(iii) **cone**(1) *is a split complex.*

Proof.

(i) Since the connecting homomorphisms in the long exact sequence arising from $0 \to \mathbf{C} \to \mathbf{cone}(1) \to \mathbf{C}[-1] \to 0$ are identities, Exercise 2.16 on page 67 applies to show that $H_n(\mathbf{cone}(1)) = \{0\}$ for all n; that is, **cone**(1) is acyclic.

(ii) As usual, we may assume that $\mathcal{A} = \mathbf{Ab}$. If $b \in B_{n-1}$, then $b = dc$ for some $c \in C_n$. If we define $t = s|B_{n-1}: B_{n-1} \to C_n$, then the sequence $0 \to Z_n \to C_n \to B_{n-1} \to 0$ splits, for $dtb = dsb = dsdc = dc = b$; that is, $d't = 1_{B_{n-1}}$. To see that $0 \to B_n \to Z_n \xrightarrow{v} H_n \to 0$ splits, define $u: H_n \to Z_n$ by $u: \mathrm{cls}(z) \mapsto z - dsz$. We claim that u is well-defined. If $b \in B_n$, then $b = dc$ for some $c \in C_{n+1}$, and $b - dsb = dc - dsdc = dc - dc = 0$ (because $dsd = d$). Hence, if $\mathrm{cls}(z) = \mathrm{cls}(z')$, then $z' = z + b$ for some $b \in B_n$ and $z' - dsz' = z + b - ds(z + b) = z - dsz$. The map u is a splitting: the map $v: Z_n \to H_n$ is is given by $z \mapsto \mathrm{cls}(z)$, and the composite $vu: \mathrm{cls}(x) \mapsto z - dsz \mapsto \mathrm{cls}(z - dsz) = \mathrm{cls}(z)$, because dsz is a boundary.

(iii) Define $s: \mathbf{cone}(1) \to \mathbf{cone}(1)$ by $s: (c_{n-1}, c_n) \mapsto (-c_n, 0)$. A routine calculation shows that $D = DsD$, where D is the differential of **cone**(1), so that (ii) shows that **cone**(1) is split. ●

Definition. A chain map $f: (\mathbf{A}, d) \to (\mathbf{C}, \delta)$ is a *quasi-isomorphism* (or *weak equivalence*) if all of its induced maps $f_*: H_n(\mathbf{A}) \to H_n(\mathbf{C})$ are isomorphisms.

Corollary 10.41. *A chain map f is a quasi-isomorphism if and only if* **cone**(f) *is acyclic.*

Proof. Since **cone**(f) is acyclic, all of its homology groups are 0, and so all of the connecting homomorphisms are isomorphisms. ●

Theorem 10.42. *If $\mathcal{A}$ is an abelian category, then a complex $(\mathbf{C}, d)$ is projective (or injective) in $\mathbf{Comp}(\mathcal{A})$ if and only if it is split with every term C_n projective (or injective) in $\mathcal{A}$.*

Proof. Sufficiency is Corollary 10.37. Conversely, if $\mathbf{C}$ is projective, then it is easy to see that $\mathbf{C}[-1]$ is also projective; hence, the exact sequence of complexes $0 \to \mathbf{C} \to \mathbf{cone}(1) \to \mathbf{C}[-1] \to 0$ is split, where 1 is the identity chain map on $\mathbf{C}$. By Lemma 10.40, $\mathbf{cone}(1)$ is a split complex, and so its direct summand $\mathbf{C}$ is split, by Exercise 10.16 on page 655.

We now show that each term C_n is projective in $\mathcal{A}$. Given an object A', let $\varrho^n(A')$ be the complex with A' concentrated in degree n; given a morphism $f: C_n \to A'$, define a chain map $F = (F_i): \mathbf{C} \to \varrho^n(A')$, where $F_n = f$ and all other $F_i = 0$. Similarly, if $g: A \to A'$ is an epimorphism and $\varrho^n(A)$ is the complex with A concentrated in degree n, then there is a chain map $G = (G_i): \varrho^n(A) \to \varrho^n(A')$, where $G_n = g$ and all other $G_i = 0$. Since $\mathbf{C}$ is projective, there is a chain map $H: \mathbf{C} \to \varrho^n(A)$ with $GH = F$. It follows that $gh = f$, and so C_n is projective.

$$\begin{array}{ccc} & C_n & \\ h \nearrow & \downarrow f & \\ A \xrightarrow{g} A' & & \end{array} \qquad \begin{array}{ccc} & \mathbf{C} & \\ H \nearrow & \downarrow F & \\ \varrho^n(A) \xrightarrow{G} \varrho^n(A') & & \end{array}$$

This argument dualizes when $\mathbf{C}$ is injective. •

Theorem 10.43. *If an abelian category $\mathcal{A}$ has enough projectives (or enough injectives), then so does $\mathbf{Comp}(\mathcal{A})$.*

Proof. Let $\mathbf{C} = \to C_n \xrightarrow{d_n} C_{n-1} \xrightarrow{d_{n-1}} C_{n-2} \to$ be a complex in $\mathcal{A}$. For each n, there exists a projective P_n and an epic $g_n: P_n \to C_n$. Consider the following chain map $G_n: \Sigma^n(1_{P_n}) \to \mathbf{C}$:

$$\begin{array}{ccccccccc} \to & 0 & \to & P_n & \xrightarrow{1_{P_n}} & P_n & \to & 0 & \to \\ & \downarrow & & \downarrow{g_n} & & \downarrow{d_n g_n} & & \downarrow & \\ \to & C_{n+1} & \to & C_n & \xrightarrow{d_n} & C_{n-1} & \to & C_{n-2} & \to \end{array}$$

Now $\Sigma = \bigoplus_{n \in \mathbb{Z}} \Sigma^n(1_{P_n})$ is projective in $\mathbf{Comp}(\mathcal{A})$, by Lemma 10.35, and $G = \bigoplus G_n: \Sigma \to \mathbf{C}$ is an epimorphism. •

Remark. The full subcategories $\mathbf{Comp}_{\geq 0}(\mathcal{A})$ of all positive complexes and $\mathbf{Comp}^{\leq 0}(\mathcal{A})$ of all negative complexes also have enough projectives (or injectives) when $\mathcal{A}$ does. ◄

The following proposition is not needed in our discussion of projective complexes, but this is a convenient place to put it.

Proposition 10.44. *Let R be a left hereditary ring, and let $(\mathbf{A}, d)$ be a complex of left R-modules. Then there exist a complex $\mathbf{P}$ having all terms projective and a quasi-isomorphism $f : \mathbf{P} \to \mathbf{A}$.*

Proof. For fixed n, there exist a projective left R-module V_n and a surjective $\varphi_n : V_n \to Z_n$, the n-cycles in A_n. Consider the commutative diagram

$$
\begin{array}{ccc}
W_n & \xrightarrow{\ \alpha_n\ } & V_n \\
\end{array}
$$

Here, $W_n = \varphi^{-1}(B_n) \subseteq V_n$, and $\varphi'_n : W_n \to B_n$ is the restriction $\varphi_n | W_n$ (with target changed to B_n); moreover, $\alpha_n : W_n \to V_n$, $j_n : B_n \to Z_n$, and $i_n : Z_n \to A_n$ are inclusions, while $d'_{n+1} : A_{n+1} \to B_n$ differs from d_{n+1} only in its target, so that $d_{n+1} = i_n j_n d'_{n+1}$. Now W_n is projective, because every submodule of a projective is projective since R is left hereditary. As φ'_n is surjective, there exists $\varphi''_n : W_n \to A_{n+1}$ making the diagram commute.
Define

$$\mathbf{X}_n = \Sigma^{n+1}(\alpha_n) = 0 \to W_n \xrightarrow{\ \alpha_n\ } V_n \to 0,$$

the complex concentrated in degrees $n + 1$ and n. Note that

$$H_q(\mathbf{X}_n) = \begin{cases} H_n(\mathbf{A}) & \text{if } q = n, \\ \{0\} & \text{otherwise.} \end{cases}$$

Define

$$\mathbf{P} = \bigoplus_i \mathbf{X}_i = \to V_{n+1} \oplus W_n \xrightarrow{\ \delta_{n+1}\ } V_n \oplus W_{n-1} \to,$$

where $\delta_{n+1} : (v_{n+1}, w_n) \mapsto (\alpha_n w_n, 0)$, and define $f = (f_n) : \mathbf{P} \to \mathbf{A}$, where $f_n : V_n \oplus W_{n-1}$ is given by $(v_n, w_{n-1}) \mapsto i_n \varphi_n v_n$. It follows from Exercise 6.9 on page 339 that $H_n(\mathbf{P}) = H_n(\bigoplus_i \mathbf{X}_i) = H_n(\mathbf{X}_n) = H_n(\mathbf{A})$; moreover, f_* is an isomorphism. Hence, f is a quasi-isomorphism. $\bullet$

Definition. Let $\mathbf{C}$ be a complex in $\mathbf{Comp}(\mathcal{A})$, where $\mathcal{A}$ is an abelian category. A *Cartan–Eilenberg projective resolution* (or a *proper projective resolution*) of $\mathbf{C}$ is an exact sequence in $\mathbf{Comp}(\mathcal{A})$,

$$\to \mathbf{M}_{\bullet,q} \to \cdots \to \mathbf{M}_{\bullet,1} \to \mathbf{M}_{\bullet,0} \to \mathbf{C} \to 0,$$

such that the following sequences in $\mathcal{A}$ are projective resolutions for each p:

(i) $\to M_{p,1} \to M_{p,0} \to C_p \to 0;$

(ii) $\to Z_{p,1} \to Z_{p,0} \to Z_p(\mathbf{C}) \to 0;$

(iii) $\to B_{p,1} \to B_{p,0} \to B_p(\mathbf{C}) \to 0;$

(iv) $\to H_{p,1} \to H_{p,0} \to H_p(\mathbf{C}) \to 0.$

Our notation is a bit misleading; every term with a double subscript is a projective object in $\mathcal{A}$.

There is a dual notion of *Cartan–Eilenberg injective resolution*.

A Cartan–Eilenberg projective resolution can be viewed as a large commutative diagram M in $\mathcal{A}$ (which would be a bicomplex if we invoked the sign trick of Example 10.4). For each p, the pth column $M_{p,*}$ is a deleted projective resolution of C_p; for each q, the qth row $M_{*,q}$ is a complex each of whose terms is projective. Now the fundamental exact sequences of the qth row $M_{*,q}$,

$$0 \to Z_{p,q} \to M_{p,q} \to B_{p-1,q} \to 0$$
$$0 \to B_{p,q} \to Z_{p,q} \to H_{p,q} \to 0,$$

are split, because $B_{p-1,q}$ and $H_{p,q}$ are projective. Therefore, each row $M_{*,q}$ is projective in $\mathbf{Comp}(\mathcal{A})$, by Theorem 10.42, and so a Cartan–Eilenberg projective resolution of a complex $\mathbf{C}$ is, indeed, a projective resolution of $\mathbf{C}$ in the category $\mathbf{Comp}(\mathcal{A})$.

Theorem 10.45.   *If $\mathcal{A}$ is an abelian category with enough projectives (or injectives), then every complex $\mathbf{C}$ in $\mathbf{Comp}(\mathcal{A})$ has a Cartan–Eilenberg projective (or injective) resolution.*

Proof.   Let $\mathbf{C} = \to C_2 \xrightarrow{d_2} C_1 \xrightarrow{d_1} C_0 \xrightarrow{d_0} C_{-1} \to$ be a complex. For each $p \in \mathbb{Z}$, there are fundamental exact sequences

$$0 \to B_p \to Z_p \to H_p \to 0 \quad \text{and} \quad 0 \to Z_p \to C_p \to B_{p-1} \to 0.$$

Choose projective resolutions $B_{p,*}$ and $H_{p,*}$ of B_p and H_p, respectively; by the Horseshoe Lemma, Proposition 6.24, there is a projective resolution $Z_{p,*}$ of Z_p so that $0 \to B_{p,*} \to Z_{p,*} \to H_{p,*} \to 0$ is an exact sequence of complexes. Using the Horseshoe Lemma again, there is a projective resolution $M_{p,*}$ of C_p so that $0 \to Z_{p,*} \to M_{p,*} \to B_{p-1,*} \to 0$ is an exact sequence of complexes. For each p, define chain maps $d_{p,q} : M_{p,q} \to M_{p-1,q}$ as the composite

$$d_{p,q} : M_{p,q} \to B_{p-1,q} \to Z_{p-1,q} \to M_{p-1,q}. \tag{1}$$

We have a commutative two-dimensional diagram whose columns are the projective resolutions $M_{p,\bullet}$ of C_p.

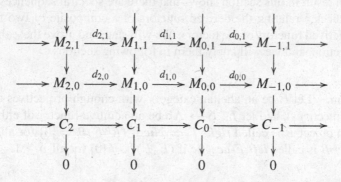

We have constructed a Cartan-Eilenberg projective resolution. ●

Exercises

10.12 Let $C = (C_{p,q})$ and $D = (D_{p,q})$ be first quadrant bicomplexes, and let $f = (f_{p,q}\colon C_{p,q} \to D_{p,q})$ be a map of bicomplexes. If, for all $p \geq 0$, we have $f_{p,*}\colon C_{p,*} \to D_{p,*}$ a quasi-isomorphism, prove that the map $\mathrm{Tot}(C) \to \mathrm{Tot}(D)$ induced by f is a quasi-isomorphism.

10.13 Let M be a first quadrant or third quadrant bicomplex all of whose rows (or all of whose columns) are exact. Prove that $\mathrm{Tot}(M)$ is acyclic.

10.14 Let $\mathbf{P} = \varrho^0(\mathbb{Z})$, the complex of abelian groups having $\mathbb{Z}$ concentrated in degree 0. Prove, without using Proposition 10.42, that $\mathbf{P}$ is not projective in **Comp(Ab)**.

*10.15 If $0 \to A' \xrightarrow{\delta} A \to A'' \to 0$ is a split exact sequence in an abelian category $\mathcal{A}$, prove that $\Sigma^k(\delta)$ is a direct summand of $\Sigma^k(1_A)$, where $\Sigma^k(\delta)$ is the complex with δ concentrated in degrees $(k, k-1)$.

*10.16 Let $\mathbf{C} = \mathbf{C}' \oplus \mathbf{C}''$ be a direct sum of complexes. If $\mathbf{C}$ is split, prove that $\mathbf{C}'$ is also split.

10.17 A complex $(\mathbf{C}, d)$ *split exact* if it is split and acyclic. Prove that if $(\mathbf{C}, d)$ is acyclic and $0 \to Z_n \to C_n \to B_{n-1} \to 0$ splits for all n, then $0 \to B_n \to Z_n \to H_n \to 0$ splits for all n; i.e., $(\mathbf{C}, d)$ is split.

10.18   (i)   Prove that every contractible complex $\mathbf{C}$ is split exact.

(ii)   Prove, using induction, that a positive or a negative complex $\mathbf{C}$ which is split exact is contractible.

10.6   Grothendieck Spectral Sequences

The main result in this section shows that there are spectral sequences, due to Grothendieck,[5] relating the derived functors of a composite of two functors and the derived functors of the factors. We will state and prove these theorems in this section, and we will apply them in following sections.

Definition.   Let $\mathcal{B}$ be an abelian category with enough projectives (or with enough injectives), and let $F \colon \mathcal{B} \to \mathbf{Ab}$ be an additive functor of either variance. An object B is called *right F-acyclic* if $(R^p F)B = \{0\}$ for all $p \geq 1$. An object B is called *left F-acyclic* if $(L_p F)B = \{0\}$ for all $p \geq 1$.

Example 10.46.

(i) If $F = \mathrm{Hom}_R(A, \square)$, then every injective R-module E is right F-acyclic because $\mathrm{Ext}_R^p(A, E) = \{0\}$ for all $p \geq 1$.

(ii) If $F = A \otimes_R \square$, then every projective R-module P is left F-acyclic because $\mathrm{Tor}_p^R(A, P) = \{0\}$ for all $p \geq 1$. In fact, every flat R-module B is left F-acyclic.

(iii) If R is a domain with $Q = \mathrm{Frac}(R)$, and if $F = Q \otimes_R \square$, then every R-module B is left F-acyclic, for $\mathrm{Tor}_p^R(Q, B) = \{0\}$ for all $p \geq 1$ because Q is flat.

(iv) If $\mathcal{B} = \mathbf{Sh}(X)$ and $\Gamma \colon \mathbf{Sh}(X) \to \mathbf{Ab}$ is the global sections functor, then every flabby sheaf $\mathcal{F}$ is right Γ-acyclic.   ◀

There are four Grothendieck spectral sequences, depending on the variances of the functors involved; two are first quadrant (homology) and two are third quadrant (cohomology). We will prove one of the third quadrant versions, but we will merely state the other versions.

The following elementary fact is used in the next proof. If $\alpha \colon U \to W$ is a homomorphism of abelian groups, and if $V \subseteq U$ is a subgroup, then $\alpha_* \colon U/V \to \alpha(U)/\alpha(V)$, given by $u + V \mapsto \alpha u + \alpha(V)$, is a surjection. Moreover, if α is an injection, then α_* is an isomorphism. By the Full Imbedding Theorem, this is also true in any abelian category.

---

[5]There is an earlier discussion of derived functors of composites in Cartan–Eilenberg, *Homological Algebra*, pp. 376–377, using *hyperhomology*.

Theorem 10.47 (Grothendieck). *Let $\mathcal{A} \xrightarrow{G} \mathcal{B} \xrightarrow{F} \mathcal{C}$ be covariant additive functors, where $\mathcal{A}, \mathcal{B},$ and $\mathcal{C}$ are abelian categories with enough injectives. Assume that F is left exact and that GE is right F-acyclic for every injective object E in $\mathcal{A}$. Then, for every object A in $\mathcal{A}$, there is a third quadrant spectral sequence with*

$$E_2^{p,q} = (R^p F)(R^q G)A \underset{p}{\Rightarrow} R^n(FG)A.$$

Proof. Choose an injective resolution $\mathbf{E} = 0 \to A \to E^0 \to E^1 \to$, and apply G to its deletion $\mathbf{E}^A$ to obtain the complex

$$G\mathbf{E}^A = 0 \to GE^0 \to GE^1 \to GE^2 \to .$$

By Theorem 10.45, there exists a Cartan–Eilenberg injective resolution of $G\mathbf{E}^A$: a third quadrant bicomplex M whose rows are complexes and whose columns are deleted injective resolutions (the definition also gives injective resolutions of cycles, boundaries, and homology). Here is the picture of M after raising indices.

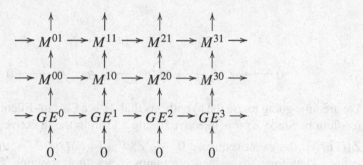

Consider the bicomplex FM and its total complex $\operatorname{Tot}(FM)$. It is very easy to compute its first iterated homology. For fixed p, the pth column $FM^{p,*}$ is a deleted injective resolution of GE^p, and so $FM^{p,*}$ is a complex whose qth homology is $(RF)^q(GE^p)$:

$$H^q(FM^{p,*}) = (R^q F)(GE^p).$$

Now E^p is injective, so that GE^p is right F-acyclic; that is, $(R^q F)(GE^p) = \{0\}$ for all $q \geq 1$. Hence,

$$H^q(FM^{p,*}) = \begin{cases} (R^0 F)(GE^p) & \text{if } q = 0, \\ \{0\} & \text{if } q > 0. \end{cases}$$

But F is assumed to be left exact, so that $R^0 F = F$, by Theorem 6.45. All that survives is on the p-axis,

$$0 \to FG(E^0) \to FG(E^1) \to FG(E^2) \to,$$

and this is FG applied to the deleted injective resolution $\mathbf{E}^A$. Hence, its pth homology is $R^p(FG)A$:

$$^I E_2^{p,q} = \begin{cases} R^p(FG)A & \text{if } q = 0, \\ \{0\} & \text{if } q > 0. \end{cases}$$

Thus, the first spectral sequence of FM collapses on the p-axis, and we have

$$H^n(\text{Tot}(FM)) \cong R^n(FG)A.$$

To compute the second iterated homology, we first transpose the indices p, q in the bicomplex FM, noting that

$$H^q(FM^{*,p}) = \frac{\ker Fd^{q,p}}{\operatorname{im} Fd^{q-1,p}}.$$

Apply F to the commutative diagram in which $j \colon B \to Z$ and $i \colon Z \to M$ are inclusions, and $\delta \colon M \to B$ is the surjection arising from d by changing its target; note that $d = ij\delta$.

$$
\begin{array}{ccccccccc}
 & & & & & & M^{q+1,p} & & \\
 & & & & \nearrow^{d} & & \uparrow^{ij} & & \\
0 & \longrightarrow & Z^{q,p} & \xrightarrow{\ i\ } & M^{q,p} & \xrightarrow{\ \delta\ } & B^{q+1,p} & \longrightarrow & 0
\end{array}
$$

We are now going to use the hypothesis that M is a Cartan–Eilenberg injective resolution. Since $Z^{q,p}$ is injective [being a term in the injective resolution of $\mathbf{Z}(GE^p)$], the exact sequence $0 \to Z^{q,p} \xrightarrow{i} M^{q,p} \xrightarrow{\delta} B^{q+1,p} \to 0$ splits. Therefore, the sequence remains exact after applying F, so that Fi is monic, $\ker F\delta = \operatorname{im} Fi$, and $F\delta$ is epic. Similarly, the exact sequence $0 \to B^{q,p} \xrightarrow{j} Z^{q,p} \to H^{q,p} \to 0$ splits, because $B^{q,p}$ is injective, so that it, too, remains exact after applying F. Hence, Fj is monic.

It is clearer to give the next argument in **Ab** (as usual, this is no loss in generality, thanks to the Full Imbedding Theorem). We compute $\ker Fd / \operatorname{im} Fd$. Now $Fd = F(ij\delta) = (Fi)(Fj)(F\delta)$. Since both Fi and Fj are injections, the numerator

$$\ker Fd = \ker F\delta = \operatorname{im} Fi = (Fi)(FZ).$$

The denominator

$$\operatorname{im} Fd = (Fd)(FM) = (Fi)[(Fj)(F\delta)(FM)] = (Fi)[(Fj)(FB)],$$

because $F\delta \colon FM \to FB$ is a surjection. Invoke the elementary fact: the homomorphism $Fi \colon FZ \to FM$ and the subgroup $(Fj)(FB) \subseteq FZ$ give a surjection $(Fi)(FZ)/(Fi)[(Fj)(FB)] \to FZ/(Fj)(FB)$; moreover, this is an isomorphism because Fi is an injection. Therefore,

$$\frac{\ker Fd}{\operatorname{im} Fd} = \frac{(Fi)(FZ)}{(Fi)[(Fj)(FB)]} \cong \frac{FZ}{(Fj)(FB)}.$$

But $FZ/(Fj)(FB) = \operatorname{coker} Fj \cong FH$, because $0 \to FB \xrightarrow{Fj} FZ \to FH \to 0$ is exact. Restoring indices, we conclude that

$$H^q(FM^{*,p}) = \frac{\ker Fd^{q,p}}{\operatorname{im} Fd^{q-1,p}} \cong FH^{q,p};$$

that is, F commutes with H^q. By hypothesis, each $0 \to H^q(G\mathbf{E}^A) \to H^{q,0} \to H^{q,1} \to \cdots \to H^{q,p} \to$ is an injective resolution of $H^q(M^{*,p})$. By definition, $H^q(G\mathbf{E}^A) = (R^qG)A$, so that the modules $H^q(M^{*,p})$ form an injective resolution of $(R^qG)A$. Hence,

$$^{\mathrm{II}}E_2^{p,q} = H^pH^q(FM) = (R^pF)(R^qG)A,$$

for F commutes with H^q, and so $(R^pF)(R^qG)A \underset{p}{\Rightarrow} R^n(FG)A$, because both spectral sequences have the same limit, namely, $R^n(FG)A$. $\quad\bullet$

Here are several variations of the Grothendieck spectral sequence, whose proofs are routine modifications of the proof just given.

Theorem 10.48. *Let $\mathcal{A} \xrightarrow{G} \mathcal{B} \xrightarrow{F} \mathcal{C}$ be covariant additive functors, where $\mathcal{A}, \mathcal{B},$ and $\mathcal{C}$ are abelian categories with enough projectives. Assume that F is right exact and that GP is left F-acyclic for every projective P in $\mathcal{A}$. Then, for every object A in $\mathcal{A}$, there is a first quadrant spectral sequence with*

$$E_{p,q}^2 = (L_pF)(L_qG)A \underset{p}{\Rightarrow} L_n(FG)A.$$

Theorem 10.49. *Let $\mathcal{A} \xrightarrow{G} \mathcal{B} \xrightarrow{F} \mathcal{C}$ be additive functors, where $\mathcal{A}$ is an abelian category with enough projectives, $\mathcal{B}$ is an abelian category with enough injectives, and $\mathcal{C}$ is an abelian category with enough projectives. Assume that F is contravariant left exact, G is covariant, and GP is right F-acyclic for every projective P in $\mathcal{A}$. Then, for every object A in $\mathcal{A}$, there is a third quadrant spectral sequence with*

$$E_2^{p,q} = (R^pF)(L_qG)A \underset{p}{\Rightarrow} R^n(FG)A.$$

Theorem 10.50. Let $\mathcal{A} \xrightarrow{G} \mathcal{B} \xrightarrow{F} \mathcal{C}$ be additive contravariant functors, where $\mathcal{A}$ is an abelian category with enough injectives, and $\mathcal{B}, \mathcal{C}$ are abelian categories with enough projectives. Assume that F is left exact and that GP is right F-acyclic for every projective P in $\mathcal{A}$. Then, for every object A in $\mathcal{A}$, there is a first quadrant spectral sequence with

$$E_{p,q}^2 = (R^p F)(R^q G)A \underset{p}{\Rightarrow} L_n(FG)A.$$

10.7 Groups

The most common use of spectral sequences in Group Theory involves the Lyndon–Hochschild–Serre spectral sequence. Lyndon discovered spectral sequences in his dissertation: "The cohomology theory of group extensions," *Duke Math J.* 15 (1948), 271–292 (it is interesting to watch spectral sequences arising naturally), and his results were extended by Hochschild and Serre, "Cohomology of Group Extensions," *Trans. AMS* 74 (1953), pp. 110–134; as they developed their results, Hochschild and Serre rewrote Lyndon's spectral sequence in the the present language. Here is one of Lyndon's main results.

Notation. Let k be a positive integer and let A be an abelian group. Denote the direct sum of A with itself k times by kA, and define $0A = \{0\}$. Regard $-kA$ as a symbol such that $nA \oplus -kA = (n-k)A$ when $n \geq k$, while $-(-kA) = kA$. If $s < 0$, let $\binom{s}{j}$ denote the coefficient of x^j in the power series expansion of $1/(1+x)^s$.

Theorem (Lyndon). Let $\pi = \mathbb{I}_{q_1} \oplus \cdots \oplus \mathbb{I}_{q_m}$ be a finite abelian group, in which $q_m \mid q_{m-1} \mid \cdots \mid q_1$ (this is the reverse of the usual notation). Then, for all $n \geq 2$,

$$H^n(\pi, \mathbb{Z}) = (-1)^n \bigoplus_{k=1}^{m} \bigoplus_{j=0}^{n-2} \binom{-k}{j} \mathbb{I}_{q_k}.$$

Proof. The cohomology theory of group extensions, Theorem 6. •

We now specialize the Grothendieck spectral sequence to groups.

Lemma 10.51. *Let π be a group with normal subgroup N, and let $_\pi \mathbf{Mod}$ and $_{\pi/N}\mathbf{Mod}$ be, respectively, the categories of left π-modules and left (π/N)-modules.*

(i) *Let* $_\pi\mathbf{Mod} \xrightarrow{G} {}_{\pi/N}\mathbf{Mod} \xrightarrow{F} \mathbf{Ab}$ *be defined by* $G = \mathrm{Hom}_N(\mathbb{Z}, \square)$ *and* $F = \mathrm{Hom}_{\pi/N}(\mathbb{Z}, \square)$. *If* E *is an injective* π*-module, then* GE *is an injective* π/N*-module and, hence, is right* F*-acyclic.*

(ii) *Let* $_\pi\mathbf{Mod} \xrightarrow{G} {}_{\pi/N}\mathbf{Mod} \xrightarrow{F} \mathbf{Ab}$ *be defined by* $G = \mathbb{Z} \otimes_N \square$ *and* $F = \mathbb{Z} \otimes_{\pi/N} \square$. *If* P *is a projective* π*-module, then* GP *is a projective* π/N*-module and, hence, is left* F*-acyclic.*

Proof.

(i) By change of rings, the ring map $\varphi\colon \mathbb{Z}\pi \to \mathbb{Z}(\pi/N)$ allows us to re-gard every (π/N)-module as a π-module and every (π/N)-map as a π-map. Recall that $GE = \mathrm{Hom}_N(\mathbb{Z}, E) = E^N$; we claim that E^N is an injective (π/N)-module. Consider the diagram in $_\pi\mathbf{Mod}$

$$
\begin{array}{ccc}
0 \longrightarrow E^N & \xrightarrow{\ j\ } & E \\
{\scriptstyle f}\big\uparrow & & \big\uparrow{\scriptstyle \widetilde{f}} \\
0 \longrightarrow M' & \xrightarrow{\ i\ } & M,
\end{array}
$$

where $M' \subseteq M$ and i and j are inclusions. Given $f\colon M' \to E^N$, injectivity of E as a π-module gives a π-map $\widetilde{f}\colon M \to E$ making the diagram commute. We claim that $\mathrm{im}\, \widetilde{f} \subseteq E^N$: if $n \in N$ and $m \in M$, then $n\widetilde{f}(m) = \widetilde{f}(nm) = \widetilde{f}(m)$, for M is a (π/N)-module and hence $nm = m$. Therefore, $\widetilde{f}$ is a (π/N)-map, and so E^N is (π/N)-injective.

(ii) The proof in part (i) dualizes. •

Theorem 10.52 (Lyndon–Hochschild–Serre). [6] *Let* N *be a normal subgroup of a group* π.

(i) [7] *For each* π*-module* A, *there is a third quadrant spectral sequence with*

$$
H^p(\pi/N, H^q(N, A)) \underset{p}{\Rightarrow} H^n(\pi, A).
$$

(ii) *For each* π*-module* A, *there is a first quadrant spectral sequence with*

$$
H_p(\pi/N, H_q(N, A)) \underset{p}{\Rightarrow} H_n(\pi, A).
$$

---

[6]There is an analog of this spectral sequence for Lie algebras; see Weibel, *An Introduction to Homological Algebra*, p. 232.

[7]Lemma 9.82 says that $H^q(N, A)$ is a (π/N)-module. Similarly, the map in homology induced by conjugation makes $H_q(N, A)$ into a (π/N)-module.

Proof.

(i) Define covariant functors $_\pi\mathbf{Mod} \xrightarrow{G} {}_{\pi/N}\mathbf{Mod} \xrightarrow{F} \mathbf{Ab}$ by

$$G = \mathrm{Hom}_N(\mathbb{Z}, \square) \quad \text{and} \quad F = \mathrm{Hom}_{\pi/N}(\mathbb{Z}, \square).$$

By Lemma 10.51(i), the ordered pair of functors (F, G) satisfies the hypotheses of Theorem 10.47: F is left exact (it is a Hom functor) and, if E is an injective π-module, then GE is right F-acyclic. The Grothendieck spectral sequence is

$$E_2^{p,q} = (R^p F)(R^q G)A \underset{p}{\Rightarrow} R^n(FG)A.$$

It is easy to see that $E_2^{p,q} = H^p(\pi/N, H^q(N, A))$, and $R^n(FG)(A) = R^n(\mathrm{Hom}_{\pi/N}(\mathbb{Z}, \mathrm{Hom}_N(\mathbb{Z}, A))) = R^n(\mathrm{Hom}_\pi(\mathbb{Z}, A) = H^n(\pi, A)$.

(ii) Define covariant functors $_\pi\mathbf{Mod} \xrightarrow{G} {}_{\pi/N}\mathbf{Mod} \xrightarrow{F} \mathbf{Ab}$ by

$$G = \mathbb{Z} \otimes_N \square \quad \text{and} \quad F = \mathbb{Z} \otimes_{\pi/N} \square.$$

By Lemma 10.51(ii), the ordered pair of functors (F, G) satisfies the hypotheses of Theorem 10.48: F is right exact (it is a tensor functor) and, if P is a projective π-module, then GE is left F-acyclic. It is easy to check that $E_{p,q}^2 = H_p(\pi/N, H_q(N, A))$ and $L_n(FG)(A) = L_n(\mathbb{Z} \otimes_N (\mathbb{Z} \otimes_{\pi/N} A)) \cong L_n(\mathbb{Z} \otimes_\pi A) = L_n(FG)A = H_n(\pi, A)$. •

Here is a new proof of Theorem 9.84 that lengthens the exact sequence there.

Theorem 10.53 (Five-Term Sequences). *If $N \lhd \pi$ and A is a π-module, there are exact sequences of abelian groups:*

(i)

$$0 \to H^1(\pi/N, A^N) \to H^1(\pi, A) \to H^1(N, A)^{\pi/N}$$
$$\to H^2(\pi/N, A^N) \to H^2(\pi, A);$$

(ii)

$$H_2(\pi, A) \to H_2(\pi/N, A_N) \to H_1(N, A)_{\pi/N}$$
$$\to H_1(\pi, A) \to H_1(\pi/N, A_N) \to 0.$$

Proof. By Theorem 10.33, there is an exact sequence

$$0 \to E_2^{1,0} \to H^1(\mathrm{Tot}(M)) \to E_2^{0,1} \xrightarrow{d_2} E_2^{2,0} \to H^2(\mathrm{Tot}(M)).$$

Now substitute the values for $E_2^{p,q}$ from Theorem 10.52.

The proof for homology groups is dual, and it is left to the reader. •

Recall Lemma 9.82: if $N \lhd \pi$ and A is a π-module, then $H^p(N, A)$ and $H_p(N, A)$ are (π/N)-modules; moreover, if A is a trivial π-module, then both $H^p(N, A)$ and $H_p(N, A)$ are trivial (π/N)-modules. We use this result in the next proof.

Theorem 10.54 (Green).   *If ℓ is a prime and G is a group of order ℓ^n, then*

$$\mathrm{card}(H_2(G, \mathbb{Z}) \le \ell^{n(n-1)/2}.$$

Proof.   If $n = 1$, then π is cyclic and $H_2(\pi, \mathbb{Z}) = \{0\}$, by Corollary 9.49. We proceed by induction on n.

As every finite ℓ-group has a nontrivial center, there is a (necessarily) normal subgroup N of order ℓ with $N \subseteq Z(\pi)$, and

$$H_p(\pi/N, H_q(N, A)) \underset{p}{\Rightarrow} H_n(\pi, A),$$

by Theorem 10.52. Thus, there is a filtration

$$\{0\} = \Phi^{-1}H_2 \subseteq \Phi^0 H_2 \subseteq \Phi^1 H_2 \subseteq \Phi^2 H_2 = H_2(G, \mathbb{Z})$$

with

$$\Phi^p H_2/\Phi^{p-1}H_2 \cong E_{p,q}^{\infty}, \quad \text{where } p + q = 2.$$

Writing $|X|$ for $\mathrm{card}(X)$ when X is a set,

$$|H_2(\pi, \mathbb{Z})| = |E_{2,0}^{\infty}||E_{1,1}^{\infty}||E_{0,2}^{\infty}| \le |E_{2,0}^2||E_{1,1}^2||E_{0,2}^2|,$$

for $E_{p,q}^{\infty}$ is a subquotient of $E_{p,q}^2$.

We use the fact that $H_q(\pi/N, \mathbb{Z})$ is (π/N)-trivial to compute $E_{p,q}^2$. Now $E_{2,0}^2 = H_2(\pi, H_0(N, \mathbb{Z})) \cong H_2(\pi/N, \mathbb{Z})$, so that induction gives $|E_{2,0}^2| \le \ell^{(n-1)(n-2)/2}$. The term $E_{1,1}^2 = H_1(\pi/N, H_1(N, \mathbb{Z})) \cong H_1(\pi/N, \mathbb{I}_\ell)$, for $H_1(N, \mathbb{Z}) \cong N/N' = N \cong \mathbb{I}_\ell$. Exercise 9.24 on page 558 now applies: if we denote π/N by Q, then $H_1(Q, \mathbb{I}_\ell) \cong Q/Q'Q^\ell$, which is a quotient of Q. Hence, $|E_{1,1}^2| \le |Q| = |\pi/N| = \ell^{n-1}$. Now $E_{0,2}^2 = H_0(\pi/N, H_2(N, \mathbb{Z}))$; but $H_2(N, \mathbb{Z}) = \{0\}$, by the base step $n = 1$, so that $|E_{0,2}^2| = |\{0\}| = 1$. We conclude that

$$|H_2(\pi, \mathbb{Z})| \le \ell^{(n-1)(n-2)/2} \cdot \ell^{n-1} \cdot 1 = \ell^{n(n-1)/2}. \quad \bullet$$

The next result shows that Green's inequality is sharp (we have already sketched a proof using exterior algebra in Example 9.65). Recall that if ℓ is a prime, then an *elementary abelian group* of order ℓ^n is the direct product of n copies of $\mathbb{I}_\ell$. As we remarked at the beginning of this section, the next theorem was proved in Lyndon's thesis.

Theorem 10.55. *If ℓ is a prime and π is an elementary abelian group of order ℓ^n, then $H_2(\pi, \mathbb{Z})$ is elementary abelian of order $\ell^{n(n-1)/2}$.*

Proof. First, we show that $\ell H_2(\pi, \mathbb{Z}) = \{0\}$. A presentation of π is

$$(x_1, \ldots, x_n \mid x_1^\ell, \ldots, x_n^\ell, \ [x_i, x_j] \text{ for } i < j).$$

Let F be a free group with basis $\{x_1, \ldots, x_n\}$ and let R be the normal sub-group of F generated by the relations. We proceed as we did in Example 9.64 when we computed the Schur multiplier of the four-group. Define $K = F/[R, F]$; since $F' \subseteq R$, Hopf's Formula, Theorem 9.63, gives $K' = F'/[R, F] \cong H_2(\pi, \mathbb{Z})$. Define $a_i = x_i[R, F]$, and observe, as in the four-group example, that

$$K' = \langle [a_i, a_j] : i < j \rangle, \ [a_i, a_j] \in Z(K), \ a_j^\ell \in Z(K).$$

Hence,

$$a_j^\ell = a_i a_j^\ell a_i^{-1} = (a_i a_j a_i^{-1})^\ell = ([a_i, a_j]a_j)^\ell = [a_i, a_j]^\ell a_j^\ell.$$

Therefore, $[a_i, a_j]^\ell = 1$ for all $i < j$, as desired.

The theorem is proved by induction on n, the base step $n = 1$ being Theorem 10.54. Choose a subgroup N of π of index ℓ so that π/N is cyclic of order ℓ (note that $N \lhd \pi$ because π is abelian). Theorem 10.52 gives a spectral sequence with

$$E_{p,q}^2 = H_p(\pi/N, H_q(N, \mathbb{Z})) \underset{p}{\Rightarrow} H_n(\pi, \mathbb{Z}).$$

Now $E_{2,0}^2 = H_2(\pi/N, H_0(N, \mathbb{Z}))$; but Corollary 9.49 gives $H_0(N, \mathbb{Z}) = \mathbb{Z}$ and $H_2(N, \mathbb{Z}) = \{0\}$, so that $E_{2,0}^2 = \{0\}$. Hence, $E_{2,0}^\infty = \{0\}$, and the filtration of $H_2(\pi, \mathbb{Z})$ has only two steps. The usual bidegree argument (see the proof of Theorem 10.31, for example) shows that $E_{0,2}^\infty = E_{0,2}^2$ and $E_{1,1}^\infty = E_{1,1}^2$. Thus, there is an exact sequence

$$0 \to E_{0,2}^2 \to H_2(\pi, \mathbb{Z}) \to E_{1,1}^2 \to 0.$$

As $\ell H_2(\pi, \mathbb{Z}) = \{0\}$, the middle term $H_2(\pi, \mathbb{Z})$ is a vector space over $\mathbb{F}_\ell$; thus, the flanking terms are also vector spaces and the sequence splits. It remains to compute dimensions. Now

$$H_1(\pi, \mathbb{Z}) \cong N/N' = N \cong \bigoplus^{n-1} \mathbb{I}_\ell,$$

so that Exercise 9.24 on page 558 gives

$$E_{1,1}^2 = H_1(\pi/N, H_1(N, \mathbb{Z})) \cong \bigoplus^{n-1} H_1(\pi/N, \mathbb{I}_\ell) \cong \bigoplus^{n-1} \mathbb{I}_\ell,$$

for N is elementary abelian of order ℓ^{n-1}. The term

$$E_{0,2}^2 = H_0(\pi/N, H_2(N, \mathbb{Z})) \cong H_2(N, \mathbb{Z});$$

by induction, $E_{0,2}^2$ is elementary abelian of order $\ell^{(n-1)(n-2)/2}$. Therefore, $H_2(\pi, \mathbb{Z})$ has dimension $(n-1) + \frac{1}{2}(n-1)(n-2) = \frac{1}{2}n(n-1)$ and order $\ell^{n(n-1)/2}$. •

Our final application computes the cohomological dimension of a free abelian group.

Lemma 10.56.   *If π is a free abelian group of finite rank r, then $cd(\pi) \leq r$.*

Proof.   If $\pi = \pi' \times \pi''$ is a direct product of groups, then the reader may prove that there is a ring isomorphism $\mathbb{Z}\pi \cong \mathbb{Z}\pi' \otimes_{\mathbb{Z}} \mathbb{Z}\pi''$. It follows that if $\mathbf{P}'$ is a deleted π'-projective resolution of $\mathbb{Z}$ and $\mathbf{P}''$ is a deleted π''-projective resolution of $\mathbb{Z}$, then $\mathbf{P}' \otimes_{\mathbb{Z}} \mathbf{P}''$ is a deleted π-projective resolution of $\mathbb{Z} \otimes_{\mathbb{Z}} \mathbb{Z} = \mathbb{Z}$. (Of course, if π', π'' are abelian, then $\pi' \times \pi'' = \pi' \oplus \pi''$.)

We prove the lemma by induction on r. The base step $r = 1$ is in Example 9.110(iii): if $\pi = \mathbb{Z}$, then the augmentation ideal $\mathcal{P}$ is projective and

$$0 \to \mathcal{P} \to \mathbb{Z}\pi \to \mathbb{Z} \to 0 \tag{1}$$

is a projective resolution. For the inductive step, let $\pi = \mathbb{Z}^{r+1} = \mathbb{Z}^r \oplus \mathbb{Z}$. By induction, there is a projective $\mathbb{Z}^r$-projective resolution $\mathbf{P}$ of $\mathbb{Z}$ of length r, and the tensor product of $\mathbf{P}$ and the resolution (1) is a projective π-resolution of $\mathbb{Z}$ of length $\leq r + 1$. •

Theorem 10.57.   *If π is a free abelian group of finite rank r, then $H^n(\pi, \mathbb{Z})$ is free abelian of rank $\binom{r}{n}$ (we interpret the binomial coefficient $\binom{r}{n}$ to be 0 if $n > r$).*

Proof.   First of all, Lemma 10.56 shows that $H^n(\pi, \mathbb{Z}) = \{0\}$ for $n > r$. We proceed by induction on r.

If $r = 1$, then $\pi = \mathbb{Z}$. Now $H^0(\pi, \mathbb{Z}) = \mathbb{Z}^\pi = \mathbb{Z}$ and $H^1(\pi, \mathbb{Z}) = \operatorname{Hom}_\pi(\mathbb{Z}, \mathbb{Z}) = \mathbb{Z}$, so the induction begins.

For the inductive step, choose a subgroup N with $\pi/N \cong \mathbb{Z}$ (N is necessarily free abelian of rank $r - 1$). The Lyndon-Hochschild-Serre spectral sequence satisfies

$$E_2^{p,q} = H^p(\pi/N, H^q(N, \mathbb{Z})) \underset{p}{\Rightarrow} H^n(\pi, \mathbb{Z}).$$

Since $\pi/N \cong \mathbb{Z}$, we know that $E_2^{p,q} = \{0\}$ for $p \neq 0, 1$. Therefore, the two-adjacent column hypothesis of Corollary 10.29 (rather, the third quadrant version of that corollary) holds, so there are exact sequences

$$0 \to E_2^{1,n-1} \to H^n(\pi, \mathbb{Z}) \to E_2^{0,n} \to 0. \tag{2}$$

Now $E_2^{1,n-1} = H^1(\pi/N, H^{n-1}(N, \mathbb{Z}))$. Hence, $H_p(N, A)$ is (π/N)-trivial for all p, by Lemma 9.82, and

$$H^1(\pi/N, H^{n-1}(N, \mathbb{Z})) \cong \text{Hom}_{\pi/N}(\mathbb{Z}, H^{n-1}(N, \mathbb{Z})) \cong H^{n-1}(N, \mathbb{Z}).$$

By induction, this group is free abelian of rank $\binom{r-1}{n-1}$. The term $E_2^{0,n} = H^0(\pi/N, H^n(N, \mathbb{Z})) \cong H^n(N, \mathbb{Z})$, using Lemma 9.82 once again. By induction, this group is free abelian of rank $\binom{r}{n-1}$. It follows that exact sequence (2) of abelian groups splits, whence $H^n(\pi, \mathbb{Z})$ is free abelian of rank $\binom{r-1}{n-1} + \binom{r-1}{n} = \binom{r}{n}$. •

Corollary 10.58. *If π is a free abelian group of finite rank r, then*

$$\text{cd}(\pi) = r.$$

Proof.   Lemma 10.56 shows that $\text{cd}(\pi) \leq k$, while Theorem 10.57 shows that this inequality is not strict. •

10.8   Rings

We now examine composites of suitable pairs of functors and the corresponding Grothendieck spectral sequences. Let R and S be rings, and consider the situation $(A_R, {}_RB_S, {}_SC)$; the associative law for tensor product gives a natural isomorphism

$$A \otimes_R (B \otimes_S C) \cong (A \otimes_R B) \otimes_S C.$$

Thus, if ${}_S\textbf{Mod} \xrightarrow{G} {}_R\textbf{Mod} \xrightarrow{F} \textbf{Ab}$ are defined by $G = B \otimes_S \square$ and $F = A \otimes_R \square$, then $FG = (A \otimes_R B) \otimes_S \square : {}_S\textbf{Mod} \to \textbf{Ab}$. Clearly, F is right exact, so that Theorem 10.48 applies when there is acyclicity.

Theorem 10.59.   *Assume that A_R and ${}_RB_S$ satisfy $\text{Tor}_i^R(A, B \otimes_S P) = \{0\}$ for all $i \geq 1$ whenever ${}_SP$ is projective. Then there is a first quadrant spectral sequence for every ${}_SC$:*

$$\text{Tor}_p^R(A, \text{Tor}_q^S(B, C)) \underset{p}{\Rightarrow} \text{Tor}_n^S(A \otimes_R B, C).$$

Let us write the five-term exact sequence in this case so that the reader can see that it is not something he or she would have invented otherwise.

$$\text{Tor}_2^S(A \otimes_R B, C) \to \text{Tor}_2^R(A, B \otimes_S C) \to A \otimes_R \text{Tor}_1^S(B, C)$$
$$\to \text{Tor}_1^S(A \otimes_R B, C) \to \text{Tor}_1^R(A, B \otimes_S C) \to 0.$$

Assume that A_R is flat (which ensures the hypothesis of Theorem 10.59). The spectral sequence now collapses on the q-axis, and Proposition 10.21 gives

$$A \otimes_R \operatorname{Tor}_n^S(B, C) \cong \operatorname{Tor}_n^S(A \otimes_R B, C).$$

There is a second way to look at the associative law for tensor product. Let $\mathbf{Mod}_R \xrightarrow{G} \mathbf{Mod}_S \xrightarrow{F} \mathbf{Ab}$ be defined by $G = \square \otimes_R B$ and $F = \square \otimes_S C$; then $FG = \square \otimes_R (B \otimes_S C)$. Here is the corresponding Grothendieck spectral sequence.

Theorem 10.60. *Assume that $_R B_S$ and $_S C$ satisfy $\operatorname{Tor}_i^S(Q \otimes_R B, C) = \{0\}$ for all $i \geq 1$ whenever Q_R is projective. Then there is a first quadrant spectral sequence for each A_R:*

$$\operatorname{Tor}_p^S(\operatorname{Tor}_q^R(A, B), C) \underset{p}{\Rightarrow} \operatorname{Tor}_n^R(A, B \otimes_S C).$$

Here is another way to force these spectral sequences to collapse.

Corollary 10.61. *If $_R B_S$ is flat on either side, then*

$$\operatorname{Tor}_n^S(A \otimes_R B, C) \cong \operatorname{Tor}_n^R(A, B \otimes_S C).$$

Proof. If $_R B$ is flat and Q_R is projective, hence flat, then $Q \otimes_R B$ is a flat right S-module, and the hypothesis of Theorem 10.60 holds. Moreover, the spectral sequence collapses, giving the desired isomorphism. A similar argument, using Theorem 10.59, works when B_S is flat.   •

We can do something more interesting: we present various "mixed identities" involving Hom, tensor, and their derived functors Ext and Tor. For example, consider the adjoint isomorphism in the situation $(_R A, _S B_R, _S C)$:

$$\operatorname{Hom}_S(B \otimes_R A, C) \cong \operatorname{Hom}_S(A, \operatorname{Hom}_R(B, C)).$$

If $G = B \otimes_S \square$ and $F = \operatorname{Hom}_R(\square, C)$, then $FG = \operatorname{Hom}_S(\square, \operatorname{Hom}_R(B, C))$. Now F is left exact, being a Hom functor, and Theorem 10.49 applies.

Theorem 10.62. *Assume that $_S B_R$ and $_S C$ satisfy $\operatorname{Ext}_S^i(B \otimes_R P, C) = \{0\}$ for all $i \geq 1$ whenever $_R P$ is projective. Then there is a third quadrant spectral sequence for every $_R A$:*

$$\operatorname{Ext}_R^p(\operatorname{Tor}_q^S(B, A), C) \underset{p}{\Rightarrow} \operatorname{Ext}_S^n(A, \operatorname{Hom}_R(B, C)).$$

Corollary 10.63. *If $_S C$ is injective, then*

$$\operatorname{Hom}_R(\operatorname{Tor}_q^S(B, A), C) \cong \operatorname{Ext}_S^n(A, \operatorname{Hom}_R(B, C)).$$

Proof. In this case, the spectral sequence collapses on the q-axis. •

We may view the adjoint isomorphism differently: if $G = \operatorname{Hom}_R(B, \square)$ and $F = \operatorname{Hom}_S(A, \square)$, then $FG = \operatorname{Hom}_R(B \otimes_R A, \square)$.

Theorem 10.64. *Assume that $_R A$ and $_S B_R$ satisfy $\operatorname{Ext}_S^i(A, \operatorname{Hom}_R(B, E)) = \{0\}$ for all $i \geq 1$ whenever $_S E$ is injective. Then there is a third quadrant spectral sequence*

$$\operatorname{Ext}_S^p(A, \operatorname{Ext}_R^q(B, C)) \underset{p}{\Rightarrow} \operatorname{Ext}_R^n(B \otimes_S A, C).$$

Corollary 10.65. *If $_S B_R$ is projective on either side, then*

$$\operatorname{Ext}_S^n(A, \operatorname{Hom}_R(B, C)) \cong \operatorname{Ext}_R^n(B \otimes_S A, C).$$

Proof. If B is projective on either side, then the hypothesis of either Theorem 10.62 or 10.64 holds, and the corresponding spectral sequence collapses on the p-axis. •

If we assume that $_R A$ is projective (instead of assuming projectivity of B), then we obtain

$$\operatorname{Ext}_S^n(B \otimes_R A, C) \cong \operatorname{Hom}_R(A, \operatorname{Ext}_S^n(B, C)).$$

Yet another isomorphism arises if we choose $_S C$ injective. The hypothesis of Theorem 10.62 holds and the spectral sequence collapses on the p-axis. Hence, there are isomorphisms

$$\operatorname{Hom}_R(\operatorname{Tor}_n^S(B, A), C) \cong \operatorname{Ext}_S^n(A, \operatorname{Hom}_R(B, C)).$$

Theorem 10.66. *Let R and S be rings with R left noetherian, and consider the situation $(_R A, {}_R B_S, C_S)$ in which A is finitely generated and C is injective. Then there are isomorphisms for all $n \geq 0$,*

$$\operatorname{Hom}_S(B, C) \otimes_R A \cong \operatorname{Tor}_n^R(\operatorname{Hom}_S(B, C), A).$$

Proof. Since R is left noetherian, A finitely generated implies A is finitely presented. The hypothesis of Lemma 3.55 is satisfied, and there is thus a natural isomorphism

$$\operatorname{Hom}_S(B, C) \otimes_R A \cong \operatorname{Hom}_S(\operatorname{Hom}_R(A, B), C).$$

Define $G = \text{Hom}_R(\square, B)\colon \mathcal{A} \to {}_S\text{Mod}$, where $\mathcal{A}$ is the category of all finitely generated left R-modules. (The reason we assume that R is left noetherian is that $\mathcal{A}$ has enough projectives.) If P is a finitely generated projective left R-module, then $\text{Ext}_S^i(\text{Hom}_R(P, B), C) = \{0\}$ for all $i \geq 1$, because C is injective. Having verified acyclicity, Theorem 10.50 gives

$$\text{Ext}_S^p(\text{Ext}_R^q(A, B), C) \underset{p}{\Rightarrow} \text{Tor}_n^R(\text{Hom}_S(B, C), A).$$

Since C is injective, this spectral sequence collapses on the q-axis, yielding isomorphisms

$$\text{Hom}_S(\text{Ext}_R^n(A, B), C) \cong \text{Tor}_n^R(\text{Hom}_S(B, C), A). \quad \bullet$$

The mixed identities above arise from the collapsing of a Grothendieck spectral sequence. Here is a variation of this technique in which we use two composite functors. The next theorem generalizes Proposition 7.39.

Theorem 10.67. *Let S be a multiplicative subset of a commutative noetherian ring R. If N is a finitely generated R-module, then there are isomorphisms*

$$S^{-1}\text{Ext}_R^n(N, M) \cong \text{Ext}_{S^{-1}R}^n(S^{-1}N, S^{-1}M)$$

for every R-module M.

Proof. Consider covariant functors ${}_R\text{Mod} \xrightarrow{G} {}_R\text{Mod} \xrightarrow{F} {}_{S^{-1}R}\text{Mod}$ defined by $G = \text{Hom}_R(N, \square)$ and $F = S^{-1} = S^{-1}R \otimes_R \square$. Now F is left exact; in fact, F is an exact functor (since $S^{-1}R$ is flat), and so

$$(R^q F)M = \begin{cases} \{0\} & \text{if } q \neq 0, \\ S^{-1}M & \text{if } q = 0. \end{cases}$$

Thus, the acyclicity condition holds: if E is injective, then GE is right F-acyclic:

$$(R^i F)(GE) = \{0\}$$

for all $i \geq 1$. Hence, there is a spectral sequence given by Theorem 10.47:

$$(R^p F)(R^q G)M = (R^p F)(\text{Ext}_R^q(N, M)) \underset{p}{\Rightarrow} R^n(FG)M.$$

This spectral sequence collapses on the q-axis, and

$$S^{-1}\text{Ext}_R^n(N, M) \cong R^n(FG)M.$$

Now consider covariant functors ${}_R\text{Mod} \xrightarrow{\Gamma} {}_{S^{-1}R}\text{Mod} \xrightarrow{\Phi} {}_{S^{-1}R}\text{Mod}$ defined by $\Gamma = S^{-1}$ and $\Phi = \text{Hom}_{S^{-1}R}(S^{-1}N, \square)$. Clearly, Φ is left exact, being a Hom, and the acyclicity condition holds: if E is an injective R-module, then ΓE is right Φ-acyclic:

$$\text{Ext}_{S^{-1}R}^i(S^{-1}N, S^{-1}E) = \{0\}$$

for all $i \geq 1$ [because R is noetherian and $S^{-1}E$ is $(S^{-1}R)$-injective, by Theorem 4.88]. Thus, Theorem 10.47 gives a spectral sequence

$$\operatorname{Ext}^p_{S^{-1}R}(S^{-1}N, (R^q\Gamma)M) \underset{p}{\Rightarrow} R^n(\Phi\Gamma)M.$$

Since Γ is exact, $(R^q\Gamma)M = \{0\}$ for $q \neq 0$, and this spectral sequence collapses on the p-axis, yielding isomorphisms

$$\operatorname{Ext}^n_{S^{-1}R}(S^{-1}N, S^{-1}M) \cong R^n(\Phi\Gamma)M.$$

But FG and $\Phi\Gamma$ are naturally isomorphic, by Lemma 4.87, and hence their derived functors are naturally isomorphic: $R^n(FG) \cong R^n(\Phi\Gamma)$. Therefore,

$$S^{-1}\operatorname{Ext}^n_R(N, M) \cong \operatorname{Ext}^n_{S^{-1}R}(S^{-1}N, S^{-1}M). \quad \bullet$$

Our final topic is a continuation of our discussion in Chapter 8 of *change of rings*, which will lead to change of rings spectral sequences. Recall that a ring map $\varphi: R \to T$ defines an exact functor $U: {}_T\mathbf{Mod} \to {}_R\mathbf{Mod}$ (in essence, every left T-module may be viewed as a left R-module and every T-map may be viewed as an R-map) by defining

$$rm = \varphi(r)m$$

for every $r \in R$ and every m in a left T-module M. At our pleasure, therefore, a module M may be regarded as either an R-module or a T-module; we will not provide different notations.

The ring T plays a special role: it is an (R, T)-bimodule, for the associative law in T gives

$$r(t_1t_2) = \varphi(r)(t_1t_2) = (\varphi(r)t_1))t_2 = (rt_1)t_2.$$

Similarly, one may equip T with a right R-module structure, namely, $tr = t\varphi(r)$, making T a (T, R)-bimodule.

Since a T-module may also be regarded as an R-module, we ask what is the relation between "homological properties" of modules over T and modules over R.

It is not necessary to be fancy, but it helps to put this in proper context.

Lemma 10.68. *If $\varphi: R \to T$ is a ring map and $U: {}_T\mathbf{Mod} \to {}_R\mathbf{Mod}$ is the corresponding change of rings functor, then both*

$$(U, \operatorname{Hom}_R(T, \square)) \quad and \quad (T \otimes_R \square, U)$$

are adjoint pairs.

Remark.   There is also a change of rings functor for right modules, $\mathbf{Mod}_T \to$ $\mathbf{Mod}_R$, and a similar lemma holds for it.   ◄

Proof.   A routine exercise.   •

Nature is telling us to look at $\mathrm{Hom}_R(T, \square)$ and $T \otimes_R \square$. Actually, there are more cases to consider, because of the various right/left possibilities due to the noncommutativity of the rings R and T. Let us assume that $A = A_R$ and $B = {}_R B$. Here is the elaborate notation. Denote

$$_\varphi B = T \otimes_R B \in \mathrm{obj}({}_T\mathbf{Mod}), \qquad T = {}_T T_R,$$

and

$$A_\varphi = A \otimes_R T \in \mathrm{obj}(\mathbf{Mod}_T), \qquad T = {}_R T_T.$$

There are also left and right modules resulting from $\mathrm{Hom}_R(T, \square)$:

$$B^\varphi = \mathrm{Hom}_R(T, B) \in \mathrm{obj}(\mathbf{Mod}_T), \qquad T = {}_R T_T,$$

and

$$^\varphi A = \mathrm{Hom}_R(T, A) \in \mathrm{obj}({}_T\mathbf{Mod}) \qquad T = {}_T T_R.$$

In short, when the symbol φ is a subscript, it denotes tensor; when φ is a superscript, it denotes Hom; it is set right or left depending on whether the resulting T-module is a right or a left T-module.

We need two more technical facts before we can state change of rings spectral sequences.

Lemma 10.69.   *Let $\varphi\colon R \to T$ be a ring map, and let $A = A_R$ and $B = {}_R B$.*

(i) *If A is R-projective, then A_φ is T-projective.*

(ii) *If B is R-projective, then $_\varphi B$ is T-projective.*

(iii) *If A is R-injective, then $^\varphi A$ is T-injective.*

(iv) *If B is R-injective, then B^φ is T-injective.*

Proof.   Straightforward.   •

Remember that a left T-module L can be viewed as a left R-module and that a right T-module M can be viewed as a right R-module.

Lemma 10.70. *Let $\varphi\colon R \to T$ be a ring map, and let $A = A_R$, $B = {}_R B$, $L = {}_T L$, and $M = M_T$. Then there are natural isomorphisms in the following cases:*

(i) $(A_R, {}_T L)$: $A \otimes_R L \cong A_\varphi \otimes_T L$;

(ii) $(M_T, {}_R B)$: $M \otimes_R B \cong M \otimes_T ({}_\varphi B)$;

(iii) $({}_R B, {}_T L)$: $\operatorname{Hom}_R(B, L) \cong \operatorname{Hom}_T({}_\varphi B, L)$;

(iv) $({}_T L, {}_R B)$: $\operatorname{Hom}_R(L, B) \cong \operatorname{Hom}_T(L, {}^\varphi B)$.

Proof.   (i) and (ii) follow from associativity of tensor product. For example,

$$A_\varphi \otimes_T L = (A \otimes_R T) \otimes_T L \cong A \otimes_R (T \otimes_T L) \cong A \otimes_R L.$$

(iii) and (iv) follow from adjoint isomorphisms:

$$\operatorname{Hom}_T({}_\varphi B, L) = \operatorname{Hom}_T(T \otimes_R B, L)$$
$$\cong \operatorname{Hom}_R(B, \operatorname{Hom}_T(T, L)) \cong \operatorname{Hom}_R(B, L).$$
$$\operatorname{Hom}_T(L, {}^\varphi B) = \operatorname{Hom}_T(L, \operatorname{Hom}_R(T, B))$$
$$\cong \operatorname{Hom}_R(T \otimes_T L, B) \cong \operatorname{Hom}_R(L, B). \quad \bullet$$

Theorem 10.71 (Change of Rings).   *Let $\varphi\colon R \to T$ be a ring map, and let $A = A_R$ and $B = {}_R B$. There is a spectral sequence*

$$\operatorname{Tor}_p^T(\operatorname{Tor}_q^R(A, T), L) \underset{p}{\Rightarrow} \operatorname{Tor}_n^R(A, L).$$

Proof.   Let $G\colon A \mapsto A_\varphi$ [that is, $G = \square \otimes_R T$], and let $F = \square \otimes_T L$. By Lemma 10.70, $FG = \square \otimes_R L$. Now F is right exact (being a tensor functor), and left acyclicity holds: if P_R is projective, then P_φ is T-projective and $\operatorname{Tor}_i^T(P_\varphi, L) = \{0\}$ for all $i \geq 1$. Now apply Theorem 10.48. $\bullet$

The next corollary generalizes Proposition 7.17 when R is commutative and $T = S^{-1}R$.

Corollary 10.72.   *Let $\varphi\colon R \to T$ be a ring map. If ${}_R T$ is flat, then there are isomorphisms for all $n \geq 0$,*

$$\operatorname{Tor}_n^T(A_\varphi, L) \cong \operatorname{Tor}_n^R(A, L)$$

for all $A = A_R$ and $K = {}_T L$.

Proof.   Flatness of T forces the spectral sequence to collapse. $\bullet$

The other isomorphisms in Lemma 10.70 also give spectral sequences; we merely state the results.

Theorem 10.73 (Change of Rings). *If $\varphi \colon R \to T$ is a ring map, then there is a spectral sequence*

$$\mathrm{Tor}_p^T(M, \mathrm{Tor}_q^R(T, B)) \underset{p}{\Rightarrow} \mathrm{Tor}_n^R(M, B)$$

for $M = M_T$ and $B = {}_R B$. Moreover, if T_R is flat, then there are isomorphisms for all $n \geq 0$:

$$\mathrm{Tor}_n^T(M, {}_\varphi B) \cong \mathrm{Tor}_n^R(M, B).$$

Theorem 10.74 (Change of Rings). *If $\varphi \colon R \to T$ is a ring map, then there is a spectral sequence*

$$\mathrm{Ext}_T^p(\mathrm{Tor}_q^R(T, B), L) \underset{p}{\Rightarrow} \mathrm{Ext}_R^n(B, L)$$

for $L = {}_T L$ and $B = {}_R B$. Moreover, if T_R is flat, then there are isomorphisms for all $n \geq 0$:

$$\mathrm{Ext}_T^n({}_\varphi B, L) \cong \mathrm{Ext}_R^n(B, L).$$

Theorem 10.75 (Change of Rings). *If $\varphi \colon R \to T$ is a ring map, then there is a spectral sequence*

$$\mathrm{Ext}_T^p(L, \mathrm{Ext}_R^q(T, B)) \underset{p}{\Rightarrow} \mathrm{Ext}_R^n(L, B)$$

for $L = {}_T L$ and $B = {}_R B$. Moreover, if ${}_R T$ is projective, then there are isomorphisms for all $n \geq 0$:

$$\mathrm{Ext}_T^n(L, {}^\varphi B) \cong \mathrm{Ext}_R^n(L, B).$$

We give three applications of these change of rings theorems. The first is another proof of the Eckmann–Shapiro Lemma.

Corollary 10.76. *If S is a subgroup of a group G and B is a G-module, then*

$$H^n(S, B) \cong H^n(G, \mathrm{Hom}_S(\mathbb{Z}G, B)).$$

Proof. The inclusion $S \to G$ induces a ring map $\varphi \colon \mathbb{Z}S \to \mathbb{Z}G$. Since $H^n(G, \square) = \mathrm{Ext}_{\mathbb{Z}G}^n(\mathbb{Z}, \square)$, we can apply Theorem 10.75 with $L = \mathbb{Z}$. Since $\mathbb{Z}G$ is a projective S-module (it is even free) and ${}^\varphi B = \mathrm{Hom}_S(\mathbb{Z}G, B)$, we are merely asserting the isomorphism of Theorem 10.75. •

Here is the spectral sequence proof of Theorem 8.34 of Rees.

Theorem 10.77. *Let R be a ring, let $x \in Z(R)$ be a nonzero element that is neither a unit nor a zero-divisor, and let $T = R/xR$. If L is a left R-module and x is regular on L, then there is an isomorphism*

$$\operatorname{Ext}^n_T(A, L/xL) \cong \operatorname{Ext}^{n+1}_R(A, L)$$

for every left T-module A and every $n \geq 0$.

Proof.   Since x is central, multiplication by x is an R-map; since x is not a zero-divisor, there is an R-exact sequence

$$0 \to R \xrightarrow{x} R \to T \to 0.$$

Applying $\operatorname{Hom}_R(\square, B)$ gives exactness of

$$0 \to \operatorname{Hom}_R(T, B) \to \operatorname{Hom}_R(R, B) \xrightarrow{x} \operatorname{Hom}_R(R, B)$$
$$\to \operatorname{Ext}^1_R(T, B) \to \operatorname{Ext}^1_R(R, B) \to$$

and

$$\operatorname{Ext}^q_R(R, B) \to \operatorname{Ext}^{q+1}_R(T, B) \to \operatorname{Ext}^{q+1}_R(R, B)$$

for all $q \geq 1$. Since $\operatorname{Hom}_R(R, B) = B$ and multiplication by x is an injection, we have

$$\operatorname{Hom}_R(T, B) = \{0\}.$$

Since $\operatorname{Ext}^1_R(R, B) = \{0\}$, we also have

$$\operatorname{Ext}^1_R(T, B) \cong B/xB.$$

Finally, the other exact sequences give

$$\operatorname{Ext}^q_R(T, B) = \{0\} \quad \text{for all } q \geq 2.$$

Consider the change of rings spectral sequence, Theorem 10.75:

$$E_2^{p,q} = \operatorname{Ext}^p_T(L, \operatorname{Ext}^q_R(T, B)) \underset{p}{\Rightarrow} \operatorname{Ext}^n_R(L, B).$$

We have just seen that $E_2^{p,q} = \{0\}$ for all $q \neq 1$, so there is collapsing on the line $q = 1$. Thus, there are isomorphisms for all $n \geq 1$:

$$\operatorname{Ext}^{n-1}_T(L, \operatorname{Ext}^1_R(T, B)) \cong \operatorname{Ext}^n_R(L, B),$$

and replacing n by $n + 1$ gives

$$\operatorname{Ext}^n_T(L, B/xB) \cong \operatorname{Ext}^{n+1}_R(L, B). \quad \bullet$$

Our last application is to projective dimension.

Theorem 10.78. *If $\varphi: R \to T$ is a ring map and L is a left T-module, then*

$$\mathrm{pd}_R(L) \leq \mathrm{pd}_T(L) + \mathrm{pd}_R(T).$$

Proof. Clearly, we may assume that both $\mathrm{pd}_T(L) = \ell$ and $\mathrm{pd}_R(T) = t$ are finite. By Theorem 10.75, there is a spectral sequence

$$E_2^{p,q} = \mathrm{Ext}_T^p(L, \mathrm{Ext}_R^q(T, B)) \underset{p}{\Rightarrow} \mathrm{Ext}_R^n(L, B)$$

for every left R-module B. If $m = p + q > \ell + t$, then either $p > \ell$ or $q > t$. If $p > \ell$, then $E_2^{p,q} = \{0\}$ because we have exceeded $\mathrm{pd}_T(L)$; if $q > t$, then $E_2^{p,q} = \{0\}$ because we have exceeded $\mathrm{pd}_R(T)$. It follows that $E_\infty^{p,q} = \{0\}$ for all (p, q) with $m = p + q > \ell + t$, whence $\mathrm{Ext}_R^m(L, B) = \{0\}$. As B is arbitrary, $\mathrm{pd}(R(L) \leq \ell + t$. •

10.9 Sheaves

We now discuss the relation between sheaf cohomology and Čech cohomology. Since our goal is merely to see how spectral sequences are used in this context, we do not give complete proofs. Recall that *sheaf cohomology* is defined as right derived functors of global sections Γ: if $\mathcal{F}$ is a sheaf of abelian groups over a space X, then

$$H^q(\mathcal{F}) = H^q(\Gamma \mathbf{E}^{\mathcal{F}}),$$

where $\mathbf{E}^{\mathcal{F}}$ is a deleted injective resolution of $\mathcal{F}$ in the category of sheaves.

Čech cohomology is introduced because sheaf cohomology is very difficult to compute. Let $\mathcal{U} = (U_i)_{i \in I}$ be an open cover of a space X, and let $N(\mathcal{U})$ be its nerve. Thus, $N(\mathcal{U})$ is an abstract simplicial complex with q-simplexes all $(q + 1)$-tuples $\sigma = [U_{i_0}, \ldots, U_{i_q}]$ of distinct open sets in $\mathcal{U}$ for which $U_\sigma \neq \varnothing$, where $U_\sigma = U_{i_0} \cap \cdots \cap U_{i_q}$. Denote the set of all q-simplexes in $N(\mathcal{U})$ by $\Sigma_q(\mathcal{U})$ or, more simply, by Σ_q.

Recall that Čech cohomology $\check{H}^q(\mathcal{F})$ is defined (see §6.3.1) in terms of the *Čech complex* $C^\bullet(\mathcal{U}, \mathcal{F})$, where

$$C^q(\mathcal{U}, \mathcal{F}) = \prod_{\sigma \in \Sigma_q} \mathcal{F}(U_\sigma).$$

The differentials are defined as follows: if $\alpha = (\alpha_\sigma) \in C^q(\mathcal{U}, \mathcal{F})$, then $\delta\alpha = \beta \in C^{q+1}(\mathcal{U}, \mathcal{F}) = \prod_{\tau \in \Sigma_{q+1}} \mathcal{F}(U_\tau)$; that is, $\beta = (\beta_\tau)$, where, if $\tau = [U_{i_0}, \ldots, U_{i_{q+1}}]$ is a $(q + 1)$-simplex, then

$$\beta_\tau = \beta_{[U_{i_0}, \ldots, U_{i_{q+1}}]} = \sum_{j=0}^{q+1} (-1)^j \alpha_{[U_{i_0}, \ldots, \widehat{U}_{i_j}, \ldots, U_{i_{q+1}}]}.$$

Define
$$\check{H}^q(\mathcal{U}, \mathcal{F}) = H^q(C^{\bullet}(\mathcal{U}, \mathcal{F})).$$

Čech cohomology $\check{H}^q(\mathcal{F})$ is then defined as a certain direct limit of $\check{H}^q(\mathcal{U}, \mathcal{F})$ over all $\mathcal{U}$.

Here are two comparison theorems.

Theorem 10.79 (Leray).   *Let $\mathcal{F}$ be a sheaf of abelian groups over a space X, and let $\mathcal{U}$ be an open cover of X such that $\check{H}^q(\mathcal{F}|Y) = \{0\}$ for every intersection Y of finitely many terms of $\mathcal{U}$ and all $q \geq 1$. Then there are isomorphisms, for all $q \geq 0$,*

$$\check{H}^q(\mathcal{U}, \mathcal{F}) \cong H^q(\mathcal{F}).$$

Proof.   Godement, *Topologie Algébrique et Théorie des Faisceaux*, p. 209.
●

Theorem 10.80 (Cartan).   *Let $\mathcal{F}$ be a sheaf over a space X. Assume that $\mathcal{U}$ is an open cover of X that contains arbitrarily small open sets and that is closed under finite intersections. If $\check{H}^q(U, \mathcal{F}) = \{0\}$ for all $U \in \mathcal{U}$ and all $q \geq 1$, then there are isomorphisms for all $q \geq 0$*

$$\check{H}^q(\mathcal{F}) \cong H^q(\mathcal{F}).$$

Proof.   Godement, *Topologie Algébrique et TThéorie des Faisceaux*, p. 227.
●

In each case, one constructs a third quadrant bicomplex involving a resolution of $\mathcal{F}$ and an open cover $\mathcal{U}$. The desired isomorphism is obtained because the hypothesis on $\mathcal{U}$ forces the first or second spectral sequence to collapse.

In Chapter 6, we gave a sheaf version of the Čech complex $C^{\bullet}(\mathcal{U}, \mathcal{F})$ of an open cover $\mathcal{U}$ of X and a sheaf $\mathcal{F}$ of abelian groups over X. In fancy language, define

$$\mathfrak{C}^q(\mathcal{U}, \mathcal{F}) = \prod_{\sigma \in \Sigma_q(\mathcal{U})} (j_\sigma)_*(\mathcal{F}|U_\sigma),$$

where $j_\sigma : U_\sigma \to X$ is the inclusion, where $\mathcal{F}|U_\sigma$ is the restriction sheaf over U_σ [defined on open sets $W \subseteq U_\sigma$ by $(\mathcal{F}|U_\sigma)(W) = \mathcal{F}(W)$], and where $(j_\sigma)_*(\mathcal{F}|U_\sigma)$ is the *direct image* sheaf. A less fancy, but longer, definition of $\mathfrak{C}^q(\mathcal{U}, \mathcal{F})$ is given in Godement, *Topologie Algébrique et TThèorie des Faisceaux*, §5.2.1. If U is an open set in X, then

$$\mathfrak{C}^q(\mathcal{U}, \mathcal{F}): U \mapsto \prod_{\sigma \in \Sigma_q} \mathcal{F}(U_\sigma \cap U).$$

In particular, if $U = X$, then global sections are

$$\Gamma(X) = \mathfrak{C}^q(\mathcal{U}, \mathcal{F})(X) = \prod_{\sigma \in \Sigma_q} \mathcal{F}(U_\sigma \cap X) = C^q(\mathcal{U}, \mathcal{F}).$$

Recall the following result from Chapter 6.

Lemma 6.79. *Let $\mathcal{U}$ be an open cover of a space X, and let $\mathcal{F}$ be a sheaf of abelian groups over X.*

(i) *There is an exact sequence of sheaves*

$$0 \to \mathcal{F} \to \mathfrak{C}^0(\mathcal{U}, \mathcal{F}) \to \mathfrak{C}^1(\mathcal{U}, \mathcal{F}) \to \mathfrak{C}^2(\mathcal{U}, \mathcal{F}) \to \qquad (1)$$

with $\Gamma(\mathfrak{C}^\bullet(\mathcal{U}, \mathcal{F})) = C^\bullet(\mathcal{U}, \mathcal{F})$.

(ii) *If $\mathcal{F}$ is flabby, then $\mathfrak{C}^q(\mathcal{U}, \mathcal{F})$ is flabby for all $q \geq 0$, and so (1) is a flabby resolution.*

For Leray's Theorem, let $\mathbf{E} = 0 \to \mathcal{F} \to \mathcal{E}^0 \to \mathcal{E}^1 \to$ be a flabby resolution of $\mathcal{F}$, and define a bicomplex M with

$$M^{p,q} = \Gamma(\mathcal{G}^p \mathcal{E}^q),$$

where $\mathcal{G}^q \mathcal{E}^q$ is the pth term of the Godement resolution of the sheaf $\mathcal{E}^q$. Let $\mathcal{H}^q(\mathbf{E}^{\mathcal{F}})$ be the qth cohomology; it is a sheaf because the deleted resolution $\mathbf{E}^{\mathcal{F}}$ is a complex of sheaves. Godement computes the E_2-term of the first spectral sequence:

$${}^{\mathrm{I}}E_2^{p,q} = \check{H}^p(\mathcal{U}, \mathcal{H}^q(\mathcal{F})),$$

and he then proves

$$\check{H}^p(\mathcal{U}, \mathcal{H}^q(\mathcal{F})) \underset{p}{\Rightarrow} H^n(\mathcal{F}).$$

The hypothesis that $\check{H}^q(\mathcal{F}|Y) = \{0\}$ for every intersection Y of finitely many terms of $\mathcal{U}$ and all $q \geq 1$ forces the spectral sequence to collapse, yielding $\check{H}^q(\mathcal{U}, \mathcal{F}) \cong H^q(\mathcal{F})$ for all q.

Cartan's Theorem is a bit trickier. Given a sheaf $\mathcal{F}$ of abelian groups over a space X, define a complex of abelian groups

$$\check{C}^\bullet(X, \mathcal{F}) = \varinjlim_{\mathcal{U}} \Gamma C^\bullet(\mathcal{U}, \mathcal{F})$$

(we discussed such direct limits when we introduced Čech cohomology in Chapter 6). Now define a third quadrant bicomplex K with

$$K^{p,q} = \check{C}^p(X, \mathcal{G}^q \mathcal{F}).$$

Godement proves that the second spectral sequence collapses on the p-axis, while

$$^{\mathrm{I}}E_2^{p,q} = \check{H}^p(\mathcal{H}^q(\mathcal{F})).$$

Hence,

$$\check{H}^p(\mathcal{H}^q(\mathcal{F})) \underset{p}{\Rightarrow} H^n(\mathcal{F}).$$

The Cartan hypothesis forces the first spectral sequence to collapse as well, giving the desired isomorphism.

10.10   Künneth Theorems

The Eilenberg–Zilber Theorem says that the singular complex $\mathbf{S}_\bullet(X \times Y)$ of the cartesian product $X \times Y$ of topological spaces X and Y is isomorphic to the tensor product $\mathbf{S}_\bullet(X) \otimes_{\mathbb{Z}} \mathbf{S}_\bullet(Y)$. This theorem poses the problem: given complexes of R-modules, $\mathbf{A}$ and $\mathbf{C}$, find $H_\bullet(\mathbf{A} \otimes_R \mathbf{C})$ in terms of $H_\bullet(\mathbf{A})$ and $H_\bullet(\mathbf{C})$. The Künneth Formula does exactly this.

The problem is not simple even in the special case when $\mathbf{C}$ is a complex with $C_0 = C$ and $C_q = \{0\}$ for $q \neq 0$. If we further assume that R is hereditary and each term A_p of $\mathbf{A}$ is projective, then the problem is solved by the Universal Coefficient Theorem, Theorem 7.55. To indicate why the general problem involving two complexes is difficult, even when $\mathbf{A}$ has all terms projective, let us see what happens when we try the obvious idea of using the bicomplex M having $M_{p,q} = A_p \otimes C_q$; after all, $\mathrm{Tot}(M) = \mathbf{A} \otimes \mathbf{C}$. The first iterated homology is $^{\mathrm{I}}E_{p,q}^2 = H'_p H''_q(M)$. The pth column of M is $A_p \otimes \mathbf{C}$; since A_p is projective, hence flat, $H_q(A_p \otimes \mathbf{C}) \cong A_p \otimes H_q(\mathbf{C})$, for homology commutes with exact functors. Hence, $H_q(M_{p,*}) = A_p \otimes H_q(\mathbf{C})$. The Universal Coefficient Theorem now gives

$$^{\mathrm{I}}E_{p,q}^2 = H_p(\mathbf{A} \otimes H_q(\mathbf{C})) = H_p(\mathbf{A}) \otimes H_q(\mathbf{C}) \oplus \mathrm{Tor}_1^R(H_{p-1}(\mathbf{A}), H_q(\mathbf{C})).$$

Though it is true that $E^2 \Rightarrow H(\mathbf{A} \otimes \mathbf{C})$, we are essentially helpless because we cannot compute E^∞. The spectral sequence arising from $^{\mathrm{II}}E_{p,q}^2$ is worse; the Universal Coefficient Theorem may not apply, and so we may not even be able to compute its E^2-term. Indeed, there seems to be only one meager result that can be salvaged: if each A_p is projective and if either $\mathbf{A}$ or $\mathbf{C}$ is acyclic, then $\mathbf{A} \otimes \mathbf{C}$ is acyclic (the proof is left to the reader).

We give two solutions to the problem of computing $H_\bullet(\mathbf{A} \otimes \mathbf{C})$. The first is strong enough for the Eilenberg–Zilber Theorem, and its proof does not use spectral sequences (we have deferred the proof to this chapter because it uses bicomplexes and graded modules). In essence, the Künneth Formula is the Universal Coefficient Theorem in a suitable category of complexes.

Theorem 10.81 (Künneth Formula for Homology, I). *Let R be a ring and let $(\mathbf{A}, \delta)$ be a complex of right R-modules such that all the terms of $\mathbf{A}$ and its subcomplex $\mathbf{B}$ of boundaries are flat. Given a complex $(\mathbf{C}, d)$ of left R-modules, there is an exact sequence for each n:*

$$\bigoplus_{p+q=n} H_p(\mathbf{A}) \otimes_R H_q(\mathbf{C}) \xrightarrow{\alpha} H_n(\mathbf{A} \otimes \mathbf{C}) \xrightarrow{\beta} \bigoplus_{p+q=n-1} \mathrm{Tor}_1^R(H_p(\mathbf{A}), H_q(\mathbf{C})),$$

where $\alpha_n \colon \sum_p \mathrm{cls}(b_p) \otimes c_{n-p} \mapsto \sum_p \mathrm{cls}(b_p \otimes c_{n-p})$. Moreover, α and β are natural in $\mathbf{A}$ and $\mathbf{C}$.

Proof. (**Heller**) We adapt the proof of Theorem 7.55 to the abelian category **Comp**. There is a short exact sequence of complexes,

$$0 \to \mathbf{Z} \xrightarrow{i} \mathbf{A} \to \mathbf{B}[-1] \to 0,$$

where $\mathbf{Z}$ is the subcomplex of cycles, i is the inclusion, and $\mathbf{B}[-1]$ is the complex obtained from $\mathbf{B}$ by setting $\mathbf{B}[-1]_n = \mathbf{B}_{n-1}$. By Exercise 10.10 on page 647, there is an exact sequence of complexes

$$0 \to \mathbf{Z} \otimes_R \mathbf{C} \xrightarrow{i \otimes 1} \mathbf{A} \otimes_R \mathbf{C} \to \mathbf{B}[-1] \otimes_R \mathbf{C} \to 0.$$

The corresponding long exact sequence is an exact sequence of complexes

$$H_{n+1}(\mathbf{B}[-1] \otimes_R \mathbf{C}) \xrightarrow{\partial_{n+1}} H_n(\mathbf{Z} \otimes_R \mathbf{C}) \to H_n(\mathbf{A} \otimes_R \mathbf{C})$$
$$\to H_n(\mathbf{B}[-1] \otimes_R \mathbf{C}) \xrightarrow{\partial_n} H_{n-1}(\mathbf{Z} \otimes_R \mathbf{C}).$$

Now $H_{n+1}(\mathbf{B}[-1] \otimes_R \mathbf{C}) = (\mathbf{B} \otimes_R \mathbf{C})_n$ and $H_n(\mathbf{Z} \otimes_R \mathbf{C}) = (\mathbf{Z} \otimes_R \mathbf{C})_n$, because $\mathbf{Z}$ and $\mathbf{B}[-1]$ have zero differentials. Thus, we may rewrite the long exact sequence as

$$(\mathbf{B} \otimes_R \mathbf{C})_n \xrightarrow{\partial_{n+1}} (\mathbf{Z} \otimes_R \mathbf{C})_n \to H_n(\mathbf{A} \otimes_R \mathbf{C})$$
$$\to (\mathbf{B} \otimes_R \mathbf{C})_{n-1} \xrightarrow{\partial_n} (\mathbf{Z} \otimes_R \mathbf{C})_{n-1}.$$

By Exercise 2.17 (which holds in any abelian category), there are short exact sequences

$$0 \to (\mathrm{coker}\, \partial_{n+1})_n \xrightarrow{\alpha_n} H_n(\mathbf{A} \otimes_R \mathbf{C}) \xrightarrow{\beta_n} (\ker \partial_n)_n \to 0, \qquad (1)$$

where $\alpha_n \colon \sum_p \mathrm{cls}(b_p) \otimes c_{n-p} \mapsto \sum_p \mathrm{cls}(i_p b_p \otimes c_{n-p})$. The reader may prove that both α_n and β_n are natural.

We claim that the definition of the connecting homomorphism gives $\partial_n = (j_n \otimes 1)_*$, as in the proof of Theorem 7.55, where $j_n \colon B_n \to Z_n$ is the inclusion. The connecting homomorphism arises from the diagram.

$$
\begin{array}{ccc}
(\mathbf{A} \otimes_R \mathbf{C})_n & \xrightarrow{\ \delta \otimes 1\ } & (\mathbf{B} \otimes_R \mathbf{C})_{n-1} \\
{\scriptstyle D}\downarrow & & \\
(\mathbf{Z} \otimes_R \mathbf{C})_{n-1} & \xrightarrow{\ i \otimes 1\ } & (\mathbf{A} \otimes_R \mathbf{C})_{n-1},
\end{array}
$$

where D is the usual differential of a total complex. If $z \in (\mathbf{B} \otimes_R \mathbf{C})_{n-1}$ is a cycle, then $z = \sum_{p+q=n-1} b_p \otimes c_q$, where $b_p \in B_p$ and $c_q \in C_q$. Since each b_p is a boundary, there are $a_{p+1} \in A_{p+1}$ with $b_p = \delta a_{p+1}$. Hence [abusing notation by writing $\partial_n(z)$ instead of $\partial_n \operatorname{cls}(z)$],

$$
\begin{aligned}
\partial_n(z) &= (i \otimes 1)^{-1} D(\delta \otimes 1)^{-1}\Big[\sum_{p+q=n-1} b_p \otimes c_q \Big] \\
&= (i \otimes 1)^{-1} D\Big[\sum a_{p+1} \otimes c_q \Big] \\
&= (i \otimes 1)^{-1}\Big[\sum \delta a_{p+1} \otimes c_q + (-1)^{p+1} a_{p+1} \otimes dc_q \Big] \\
&= (i \otimes 1)^{-1}\Big[\sum b_p \otimes c_q + (-1)^{p+1} a_{p+1} \otimes dc_q \Big] \\
&= (i \otimes 1)^{-1}(z) + (i \otimes 1)^{-1}\Big[\sum (-1)^{p+1} a_{p+1} \otimes dc_q \Big].
\end{aligned}
$$

Applying $(i \otimes 1)^{-1}$ tells us to regard elements as lying in $(\mathbf{Z} \otimes_R \mathbf{C})_{n-1}$. Thus, $\partial_n(z) - z \in (\mathbf{Z} \otimes_R \mathbf{C})_{n-1}$, and so $\sum (-1)^{p+1} a_{p+1} \otimes dc_q \in (\mathbf{Z} \otimes_R \mathbf{C})_{n-1}$. Since $\mathbf{Z}$ has all differentials zero, the defining formula for the differential in $\mathbf{Z} \otimes \mathbf{C}$ says that $\sum (-1)^{p+1} a_{p+1} \otimes dc_q$ is a boundary; that is, $\partial_n(z)$ is homologous to z. Thus, if z is a cycle, then $\partial_n \operatorname{cls}(z) = \operatorname{cls}((j_n \otimes 1)z)$; that is, $\partial_n = (j_n \otimes 1)_*$. Hence, exact sequence (1) is

$$
0 \to (\operatorname{coker}(j_n \otimes 1))_n \xrightarrow{\ \alpha_n\ } H_n(\mathbf{A} \otimes_R \mathbf{C}) \xrightarrow{\ \beta_n\ } (\ker(j_{n-1} \otimes 1))_n \to 0. \quad (2)
$$

Consider the flanking terms in (2). Since $\mathbf{B}$ and $\mathbf{Z}$ are flat, the exact sequence of complexes

$$
0 \to \mathbf{B} \xrightarrow{\ j\ } \mathbf{Z} \to H_\bullet(\mathbf{A}) \to 0
$$

is a flat resolution of $H_\bullet(\mathbf{A})$, where $H_\bullet(\mathbf{A})$ is viewed as a complex with zero differentials. Thus, $0 \to \mathbf{B} \xrightarrow{\ j\ } \mathbf{Z} \to 0$ is a deleted flat resolution of $H_\bullet(\mathbf{A})$; after tensoring by $\mathbf{C}$, its homology in degree n is given by

$$
(H_1)_{n-1} = (\ker(j \otimes 1))_{n-1} = \mathbf{Tor}_1^R(H_\bullet(\mathbf{A}), \mathbf{C})_{n-1}
$$

and

$$
(H_0)_n = (\operatorname{coker}(j \otimes 1))_n = \mathbf{Tor}_0^R(H_\bullet(\mathbf{A}), \mathbf{C})_n.
$$

Thus, exact sequence (2) is

$$0 \to \operatorname{Tor}_0^R(H_\bullet(\mathbf{A}), \mathbf{C})_n \to H_n(\mathbf{A} \otimes_R \mathbf{C}) \to \operatorname{Tor}_1^R(H_\bullet(\mathbf{A}), \mathbf{C})_{n-1} \to 0.$$

But

$$\operatorname{Tor}_0^R(H_\bullet(\mathbf{A}), \mathbf{C})_n \cong (H_\bullet(\mathbf{A}) \otimes_R \mathbf{C})_n = \bigoplus_{p+q=n} H_p(\mathbf{A}) \otimes_R H_q(\mathbf{C})$$

and

$$\operatorname{Tor}_1^R(H_\bullet(\mathbf{A}), \mathbf{C})_{n-1} = \bigoplus_{p+q=n-1} \operatorname{Tor}_1^R(H_p(\mathbf{A}), H_q(\mathbf{C})). \quad \bullet$$

Corollary 10.82 (Künneth Formula for Homology, II). *Let R be a right hereditary ring, let $(\mathbf{A}, \delta)$ be a complex of projective right R-modules, and let $\mathbf{C}$ be a complex of left R-modules.*

(i) *For all $n \geq 0$, there is an exact sequence*

$$\bigoplus_{p+q=n} H_p(\mathbf{A}) \otimes_R H_q(\mathbf{C}) \xrightarrow{\alpha} H_n(\mathbf{A} \otimes \mathbf{C}) \xrightarrow{\beta} \bigoplus_{p+q=n-1} \operatorname{Tor}_1^R(H_p(\mathbf{A}), H_q(\mathbf{C})),$$

where $\alpha_n \colon \sum_p \operatorname{cls}(b_p) \otimes c_{n-p} \mapsto \sum_p \operatorname{cls}(b_p \otimes c_{n-p})$, and both λ_n and μ_n are natural.

(ii) *For all $n \geq 0$, the exact sequence splits:*[8]

$$H_n(\mathbf{A} \otimes_R \mathbf{C}) \cong \left[\bigoplus_{p+q=n} H_p(\mathbf{A}) \otimes_R H_q(\mathbf{C}) \right] \oplus \bigoplus_{p+q=n-1} \operatorname{Tor}_1^R(H_p(\mathbf{A}), H_q(\mathbf{C})).$$

Proof. The first statement is true because R right hereditary implies that every submodule of a projective module is projective, and hence flat. To prove that the sequence splits, there are inclusions of subcomplexes

$$\operatorname{im}(\delta \otimes 1) \subseteq \operatorname{im}(i \otimes 1) \subseteq \ker(\delta \otimes 1) \subseteq \mathbf{A} \otimes \mathbf{C}.$$

The proof is completed by adapting the proof of Theorem 7.56. $\quad \bullet$

Corollary 10.83. *For every pair of topological spaces X and Y, there are isomorphisms for every $n \geq 0$,*

$$H_n(X \times Y) \cong \left[\bigoplus_{p+q=n} H_p(X) \otimes_{\mathbb{Z}} H_q(Y) \right] \oplus \left[\bigoplus_{p+q=n-1} \operatorname{Tor}_1^{\mathbb{Z}}(H_p(X), H_q(Y)) \right].$$

Proof. This follows from the Eilenberg–Zilber Theorem and the Künneth Formula, for $\mathbb{Z}$ is a hereditary ring and the terms of the singular complex are free abelian groups. $\quad \bullet$

The following corollary will be used in our discussion of the Künneth spectral sequence.

---

[8]The splitting need not be natural.

Corollary 10.84. *Let R be a ring, let $\mathbf{A}$ be a complex of right R-modules, and let $\mathbf{C}$ be a complex of left R-modules. If all the terms of $\mathbf{A}$, its subcomplex $\mathbf{B}$ of boundaries, and its homology $H_n(\mathbf{A})$ are flat, then*

$$\bigoplus_{p+q=n} H_p(\mathbf{A}) \otimes_R H_q(\mathbf{C}) \cong H_n(\mathbf{A} \otimes_R \mathbf{C}).$$

Proof. By Theorem 10.81, there is an exact sequence

$$\bigoplus_{p+q=n} H_p(\mathbf{A}) \otimes_R H_q(\mathbf{C}) \rightarrowtail H_n(\mathbf{A} \otimes \mathbf{C}) \twoheadrightarrow \bigoplus_{p+q=n-1} \operatorname{Tor}_1^R(H_p(\mathbf{A}), H_q(\mathbf{C})).$$

But flatness of every $H_p(\mathbf{A})$ forces all the Tor terms to vanish. $\quad\bullet$

We merely state the dual of Corollary 10.82; the proof is left to the reader.

Theorem 10.85 ([Künneth Formula for Cohomology). *Let R be a ring and let $\mathbf{A}$ be a complex of left R-modules such that all the terms of $\mathbf{A}$ and its subcomplex $\mathbf{B}$ of boundaries are projective.*

(i) *For all $n \geq 0$ and every complex $\mathbf{C}$ of left R-modules, there is an exact sequence*

$$\prod_{p-q=n-1} \operatorname{Ext}_R^1(H_p, H^q) \xrightarrow{\alpha_n} H^n(\operatorname{Hom}_R(\mathbf{A}, \mathbf{C})) \xrightarrow{\beta_n} \prod_{p-q=n} \operatorname{Hom}_R(H_p, H^q),$$

where H_p abbreviates $H_p(\mathbf{A})$, H^q abbreviates $H^q(\mathbf{C}) = H_{-q}(\mathbf{C})$, and both α_n and β_n are natural.

(ii) *If R is left hereditary, then the exact sequence splits for all $n \geq 0$.*[9]

Consider the special case of Theorem 10.85 when $\mathbf{C}$ is the complex with a module C concentrated in degree 0; that is, $H^0 = C$ and $H^q = \{0\}$ for all $q > 0$. We obtain exactness of

$$\operatorname{Ext}_R^1(H_{n-1}(\mathbf{A}), C) \xrightarrow{\alpha_n} H^n(\operatorname{Hom}_R(\mathbf{A}, C)) \xrightarrow{\beta_n} \operatorname{Hom}_R(H_n(\mathbf{A}), C),$$

which is Theorem 7.59, the Universal Coefficient Theorem for Cohomology.

The Künneth Formulas give some curious identities, which we have used in our study of cotorsion groups.

---

[9]The splitting need not be natural.

Proposition 10.86. *Given a commutative hereditary ring R and R-modules A, B, and C, there is an isomorphism*

$$\operatorname{Ext}_R^1(\operatorname{Tor}_1^R(A, B), C) \cong \operatorname{Ext}_R^1(A, \operatorname{Ext}_R^1(B, C)).$$

Proof. Consider the case $n = 2$ of Theorem 10.85 with $\mathbf{A} = \mathbf{P}_A \otimes_R B$, where $\mathbf{P}_A$ is a deleted projective resolution of A and $\mathbf{C} = \mathbf{E}^C$ is a deleted injective resolution of C. Thus, $H_p(\mathbf{A}) = H_p(\mathbf{P}_A \otimes_R B) = \operatorname{Tor}_p^R(A, B)$, and, since $\mathbf{C} = \mathbf{E}^C$ is a deleted resolution, $H^0(\mathbf{C}) = C$ and $H^q(\mathbf{C}) = \{0\}$ for all $q > 0$. Now the exact sequence of Theorem 10.85 splits (because R is hereditary), and so the only possible nonzero terms involved in $H^2(\operatorname{Hom}_R(\mathbf{P}_A \otimes_R B, \mathbf{E}^C))$ are

$$\bigoplus_{p-q=1} \operatorname{Ext}_R^1(H_p, H^q) \quad \text{and} \quad \bigoplus_{p-q=2} \operatorname{Hom}_R(H_p, H^q).$$

The only possible nonzero Ext term is $\operatorname{Ext}_R^1(H_1, H^0) = \operatorname{Ext}_R^1(\operatorname{Tor}_1^R(A, B), C)$. The only possible nonzero Hom term is $\operatorname{Hom}_R(H_2, H^0)$; but, since R is hereditary, $H_2 = \{0\}$ (for $\operatorname{Tor}_2^R = \{0\}$ because $\operatorname{wD}(R) \le \ell\operatorname{D}(R) = 1$). We conclude that

$$H^2(\operatorname{Hom}_R(\mathbf{P}_A \otimes_R B, \mathbf{E}^C)) \cong \operatorname{Ext}_R^1(\operatorname{Tor}_1^R(A, B), C).$$

A similar argument with the complexes $\mathbf{A}' = \mathbf{P}_A$ and $\mathbf{C}' = \operatorname{Hom}(B, \mathbf{E}^C)$ gives

$$H^2(\operatorname{Hom}_R(\mathbf{P}_A, \operatorname{Hom}_R(B, \mathbf{E}^C))) \cong \operatorname{Ext}_R^1(A, \operatorname{Ext}_R^1(B, C)).$$

But naturality of the adjoint isomorphism in Theorem 2.75 gives an isomorphism of the complexes $\operatorname{Hom}_R(\mathbf{P}_A \otimes_R B, \mathbf{E}^C) \cong \operatorname{Hom}_R(\mathbf{P}_A, \operatorname{Hom}_R(B, \mathbf{E}^C))$, so that these complexes have the same homology. We conclude that

$$\operatorname{Ext}_R^1(\operatorname{Tor}_1^R(A, B), C) \cong \operatorname{Ext}_R^1(A, \operatorname{Ext}_R^1(B, C)). \quad \bullet$$

Here is a variant of Proposition 10.86.

Proposition 10.87. *If R is a commutative hereditary ring and A, B, C are R-modules, then*

$$\operatorname{Ext}_R^1(A \otimes_R B, C) \oplus \operatorname{Hom}_R(\operatorname{Tor}_1^R(A, B), C))$$
$$\cong \operatorname{Ext}_R^1(A, \operatorname{Hom}_R(B, C)) \oplus \operatorname{Hom}_R(A, \operatorname{Ext}_R^1(B, C)).$$

Proof. Consider the case $n = 1$ of Theorem 10.85 with $\mathbf{A} = \mathbf{P}_A \otimes_R B$, where $\mathbf{P}_A$ is a deleted projective resolution of A and $\mathbf{C} = \mathbf{E}^C$ is a deleted injective resolution of C. Thus, $H_p(\mathbf{A}) = H_p(\mathbf{P}_A \otimes_R B) = \operatorname{Tor}_p^R(A, B)$ and, since $\mathbf{C} = \mathbf{E}^C$ is a deleted resolution, $H^0(\mathbf{C}) = C$ and $H^q(\mathbf{C}) = \{0\}$ for $q > 0$.

Now the exact sequence of Theorem 10.85 splits (because R is hereditary), and so the only possible nonzero terms are

$$H^1(\operatorname{Hom}_R(\mathbf{P}_A \otimes_R B, \mathbf{E}^C)) \cong \operatorname{Ext}_R^1(H_0, H^0) \oplus \operatorname{Hom}_R(H_1, H^0)$$
$$= \operatorname{Ext}_R^1(A \otimes_R B, C) \oplus \operatorname{Hom}_R(\operatorname{Tor}_1^R(A, B), C).$$

Similarly, the Künneth Formula for the complexes $\mathbf{A}' = \mathbf{P}_A$ and $\mathbf{C}' = B \otimes_R \mathbf{E}^C$ gives an isomorphism

$$H^1(\operatorname{Hom}_R(\mathbf{A}', \mathbf{C}')) \cong \operatorname{Ext}_R^1(A, \operatorname{Hom}_R(B, C)) \oplus \operatorname{Hom}_R(A, \operatorname{Ext}_R^1(B, C)).$$

Naturality of the adjoint isomorphism gives an isomorphism of complexes

$$\operatorname{Hom}_R(\mathbf{P}_A \otimes_R B, \mathbf{E}^C) \cong \operatorname{Hom}_R(\mathbf{P}_A, \operatorname{Hom}_R(B, \mathbf{E}^C)).$$

Therefore, $H^1(\operatorname{Hom}_R(\mathbf{A}, \mathbf{C})) \cong H^1(\operatorname{Hom}_R(\mathbf{A}', \mathbf{C}'))$, which gives the desired isomorphism. •

Proposition 10.88. *Given a right hereditary ring R, a left hereditary ring S, and modules A_R, $_R B_S$, and $_S C$, there is a natural isomorphism*

$$\operatorname{Tor}_1^R(A, \operatorname{Tor}_1^S(B, C)) \cong \operatorname{Tor}_1^S(\operatorname{Tor}_1^R(A, B), C).$$

Proof. The proof is similar to that of Proposition 10.86, and we leave it to the reader. It uses the fact that $\operatorname{wD}(R) \le \ell D(R) = 1$; that is, $\operatorname{Tor}_2^R = \{0\}$. •

Proposition 10.89. *If R is a commutative hereditary ring and A, B, and C are R-modules, then there is an isomorphism*

$$\operatorname{Tor}_1^R(A, B) \otimes_R C \oplus \operatorname{Tor}_1^R(A \otimes_R B, C)$$
$$\cong A \otimes_R \operatorname{Tor}_1^R(B, C) \oplus \operatorname{Tor}_1^R(A, B \otimes_R C).$$

Proof. The proof is modeled on the proof of Proposition 10.87. Consider the case $n = 1$ of Theorem 10.82:

$$\bigoplus_{p+q=n} H_p(\mathbf{A}) \otimes_R H_q(\mathbf{C}) \xrightarrow{\alpha} H_n(\mathbf{A} \otimes \mathbf{C}) \xrightarrow{\beta} \bigoplus_{p+q=n-1} \operatorname{Tor}_1^R(H_p(\mathbf{A}), H_q(\mathbf{C})),$$

with $\mathbf{A} = \mathbf{P}_A \otimes_R B$ and $\mathbf{C} = \mathbf{Q}_C$, where $\mathbf{P}_A$, $\mathbf{Q}_C$ are deleted projective resolutions of A and C, respectively. Thus, $H_p(\mathbf{A}) = H_p(\mathbf{P}_A \otimes_R B) = \operatorname{Tor}_p^R(A, B)$ and, since $\mathbf{C} = \mathbf{Q}_C$ is a deleted resolution, $H_1(\mathbf{C}) = C$ and $H_q(\mathbf{C}) = \{0\}$ for $q > 0$. Now the exact sequence of Theorem 10.82 splits (because R is hereditary), and so the only possible nonzero terms are

$$H_1((\mathbf{P}_A \otimes_R B) \otimes_R \mathbf{Q}_C) \cong \operatorname{Tor}_1^R(A, B) \otimes_R C \oplus \operatorname{Tor}_1^R(A \otimes_R B, C).$$

Similarly,

$$H_1(\mathbf{P}_A \otimes_R (B \otimes_R \mathbf{Q}_C)) \cong A \otimes_R \operatorname{Tor}_1^R(B, C) \oplus \operatorname{Tor}_1^R(A, B \otimes_R C).$$

Finally, naturality of associativity of tensor product gives

$$H_1((\mathbf{P}_A \otimes_R B) \otimes_R \mathbf{Q}_C) \cong H_1(\mathbf{P}_A \otimes_R (B \otimes_R \mathbf{Q}_C)),$$

as desired. •

We are going to use spectral sequences to prove a more general version of the Künneth Formula. A constant theme in Homological Algebra is the replacement of a module by a deleted resolution of it. Thus, it is natural for us to replace a complex $\mathbf{C}$ by a deleted projective resolution of it in **Comp**; an exact sequence of projective complexes $M_{\bullet,q}$

$$0 \leftarrow \mathbf{C} \leftarrow \mathbf{M}_{\bullet,0} \leftarrow \mathbf{M}_{\bullet,1} \leftarrow \cdots \leftarrow \mathbf{M}_{\bullet,q} \leftarrow .$$

Using the sign trick of Example 10.4, M is a bicomplex in $\mathcal{A}$ whose rows are projective complexes and whose columns $M_{p,*}$ are projective resolutions.

$$
\begin{array}{ccccccc}
& \downarrow & & \downarrow & & \downarrow & \\
\leftarrow & M_{01} & \leftarrow & M_{11} & \leftarrow & M_{21} & \leftarrow \\
& \downarrow & & \downarrow & & \downarrow & \\
\leftarrow & M_{00} & \leftarrow & M_{10} & \leftarrow & M_{20} & \leftarrow \\
& \downarrow & & \downarrow & & \downarrow & \\
\leftarrow & C_0 & \leftarrow & C_1 & \leftarrow & C_2 & \leftarrow \\
& \downarrow & & \downarrow & & \downarrow & \\
& 0 & & 0 & & 0 &
\end{array}
$$

How can we compute homology by replacing a complex by a Cartan–Eilenberg resolution of it? A **tricomplex** is a 4-tuple (L, d', d'', d'''), where L is triply graded and the three differentials pairwise anticommute. For example, given a complex $\mathbf{A}$ and a bicomplex M, Mac Lane (*Homology*, p. 403) constructs a tricomplex $\mathbf{A} \otimes M$ with $L_{p,q,r} = A_p \otimes M_{q,r}$, and he defines its **total complex** to have terms

$$\operatorname{Tot}(\mathbf{A} \otimes M)_n = \bigoplus_{p+q+r=n} A_p \otimes M_{q,r}.$$

The **hyperhomology** of $\mathbf{A}$ and M is defined as

$$\mathcal{H}_n(\mathbf{A}, M) = H_n(\operatorname{Tot}(\mathbf{A} \otimes M)).$$

A bicomplex $(K, d'+d''', d'')$ can be constructed from the tricomplex $\mathbf{A} \otimes M$, where

$$K_{p,q} = \bigoplus_{s+t=p} A_s \otimes M_{q,t}.$$

This construction will appear in the next proof.

Theorem 10.90 (Künneth Homology Spectral Sequence).   *Let R be a ring, let $\mathbf{A}$ be a positive complex of flat right R-modules, and let $\mathbf{C}$ be a positive complex of left R-modules. There is a first quadrant spectral sequence*

$$E^2_{p,q} = \bigoplus_{s+t=q} \operatorname{Tor}^R_p(H_s(\mathbf{A}), H_t(\mathbf{C})) \underset{p}{\Rightarrow} H_n(\mathbf{A} \otimes_R \mathbf{C}).$$

Proof. We have already seen, at the beginning of this section, that the homology of the obvious bicomplex [having (p, q) term $A_p \otimes_R C_q$] is too complicated; we cannot even compute its E^2-term. Let us replace $\mathbf{C}$ by a Cartan–Eilenberg projective resolution M of $\mathbf{C}$. Note that the bicomplex M is first quadrant because $\mathbf{C}$ is a positive complex. Define the bicomplex (K, d', d''), where

$$K_{p,q} = \bigoplus_{s+t=p} A_s \otimes_R M_{q,t},$$

the horizontal d' is $1_{A_s} \otimes \delta$ (here, δ is the horizontal differential of M), and the vertical d'' is built from the differential of $\mathbf{A}$ and the vertical differential of M.

The first filtration begins with the bicomplex $H_q(K_{p,*})$. The pth column $K_{p,*}$ is

$$\rightarrow \bigoplus_{s+t=p} A_s \otimes_R M_{2,t} \rightarrow \bigoplus_{s+t=p} A_s \otimes_R M_{1,t} \rightarrow \bigoplus_{s+t=p} A_s \otimes_R M_{0,t} \rightarrow 0.$$

For fixed (s, t) with $s + t = p$, we have the complex

$$\rightarrow A_s \otimes_R M_{2,t} \rightarrow A_s \otimes_R M_{1,t} \rightarrow A_s \otimes_R M_{0,t} \rightarrow 0, \tag{3}$$

which is just $A_s \otimes_R \square$ applied to $\rightarrow M_{2,t} \rightarrow M_{1,t} \rightarrow M_{0,t} \rightarrow 0$, a deleted projective resolution of C_t. Since A_s is flat, the complex (3) is a deleted flat resolution of $A_s \otimes_R C_t$. Since Tor commutes with direct sums,

$$H''_{p,q}(K) = \bigoplus_{s+t=p} \operatorname{Tor}^R_q(A_s, C_t).$$

Using flatness of A_s once again, we see that all these terms vanish for $q > 0$. We conclude that

$$^{\mathrm{I}}E^2_{p,q} = H'_p H''_q(K) = \begin{cases} \bigoplus_{s+t=p} H_p\left(A_s \otimes_R C_t\right) & \text{if } q = 0; \\ \{0\} & \text{if } q > 0. \end{cases}$$

Hence, the first spectral sequence collapses, and Proposition 10.21 gives

$$H_n(\operatorname{Tot}(K)) = {}^{\mathrm{I}}E^2_{n,0} = H_n(\mathbf{A} \otimes_R \mathbf{C}).$$

Since the limit of the second spectral sequence is the same as the limit of the first, namely, $H_n(\mathbf{A} \otimes_R \mathbf{C})$, it is only necesssary to show that

$$^{\mathrm{II}}E_{p,q}^2 = H_p'' H_q'(K) = \bigoplus_{s+t=q} \mathrm{Tor}_p^R(H_s(\mathbf{A}), H_t(\mathbf{C})).$$

To compute the second iterated homology of K, first transpose the indices:

$$K_{q,p} = \bigoplus_{s+t=q} A_s \otimes_R M_{p,t}.$$

Fixing p gives the complex with qth term $\bigoplus_{s+t=q} A_s \otimes_R M_{p,t}$; this complex is just $\mathrm{Tot}(\mathbf{A} \otimes_R M_{p,*})$. Since M is a Cartan–Eilenberg projective resolution of $\mathbf{C}$, the cycles, boundaries, and homology of $M_{p,*}$ are projective, hence flat. Hence, Corollary 10.84 applies to give

$$H_q'(K_{p,*}) = H_q(\mathrm{Tot}(\mathbf{A} \otimes_R M_{p,*})) = \bigoplus_{s+t=q} H_s(\mathbf{A}) \otimes_R H_t(M_{p,*}).$$

The definition of Cartan–Eilenberg projective resolution says that

$$H_t(M_{p,*}) = \to H_t(M_{p,2}) \to H_t(M_{p,1}) \to H_t(M_{p,0}) \to H_t(\mathbf{C}) \to 0$$

is a projective resolution of $H_t(\mathbf{C})$. Now

$$H_p\big(H_s(\mathbf{A}) \otimes H_t(M_{q,*})\big) = \mathrm{Tor}_p^R(H_s(\mathbf{A}), H_t(\mathbf{C})),$$

for this is the definition of Tor. Therefore,

$$^{\mathrm{II}}E_{p,q}^2 = H_p'' H_q'(K) = H_p\Big(\bigoplus_{s+t=q} H_s(\mathbf{A}) \otimes_R H_t(\mathbf{C})\Big)$$

$$= \bigoplus_{s+t=q} H_p(H_s(\mathbf{A}) \otimes_R H_t(\mathbf{C}))$$

$$= \bigoplus_{s+t=q} \mathrm{Tor}_p^R(H_s(\mathbf{A}) \otimes_R H_t(\mathbf{C})). \quad \bullet$$

Let us now see that the Künneth spectral sequence gives the Künneth Formula, Theorem 10.81 (but only for positive complexes).

Corollary 10.91 (Künneth Formula Again). *Let* $\mathbf{A}$ *be a positive complex of right R-modules whose subcomplexes* $\mathbf{Z}$ *of cycles and* $\mathbf{B}$ *of boundaries have each term flat. Given a positive complex* $\mathbf{C}$ *of left R-modules, there is an exact sequence for each $n \geq 0$*

$$\bigoplus_{p+q=n} H_p(\mathbf{A}) \otimes_R H_q(\mathbf{C}) \xrightarrow{\alpha} H_n(\mathbf{A} \otimes \mathbf{C}) \xrightarrow{\beta} \bigoplus_{p+q=n-1} \mathrm{Tor}_1^R(H_P(\mathbf{A}), H_q(\mathbf{C})).$$

Remark. One can prove α and β are natural in $\mathbf{A}$ and $\mathbf{C}$ if one is more careful. ◄

Proof. Since $\mathbf{A}$ and $\mathbf{C}$ are positive complexes, Theorem 10.90 applies to give

$$E_{p,q}^2 = \bigoplus_{s+t=q} \operatorname{Tor}_p^R(H_s(\mathbf{A}), H_t(\mathbf{C})) \underset{p}{\Rightarrow} H_n(\mathbf{A} \otimes_R \mathbf{C}).$$

Now exactness of $0 \to B_s \to Z_s \to H_s(\mathbf{A}) \to 0$ gives exactness of

$$\operatorname{Tor}_p^R(Z_s, X) \to \operatorname{Tor}_p^R(H_s(\mathbf{A}), X) \to \operatorname{Tor}_{p-1}^R(B_s, X)$$

for every module X. If $p - 1 \geq 1$, then flatness of Z_s and B_s forces the middle term to vanish:

$$\operatorname{Tor}_p^R(H_s(\mathbf{A}), X) = \{0\}$$

for all X and all $p \geq 2$. Thus, $E_{p,q}^2 = \{0\}$ for $p \neq 0$ and $p \neq 1$, which is the adjacent column hypothesis of Corollary 10.29. We conclude that there are exact sequences

$$0 \to E_{0,n}^2 \to H_n(\mathbf{A} \otimes_R \mathbf{C}) \to E_{1,n-1}^2 \to 0,$$

and this is what is sought once we replace the terms $E_{0,n}^2$ and $E_{1,n-1}^2$ by their values. •

The reader should now be convinced that using spectral sequences can prove interesting theorems. Moreover, even if there are "elementary" proofs of these results (i.e., avoiding spectral sequences), these more "sophisticated" proofs offer a systematic approach in place of sporadic success.

References

Adem, A., and Milgram, R. J., *Cohomology of Finite Groups*, 2d ed., Springer, New York, 2004.

Anderson, F., and Fuller, K., *Rings and Categories of Modules*, Springer-Verlag, New York, 1974.

Babakhanian, A., *Cohomological Methods in Group Theory*, Monographs and Textbooks in Pure and Applied Mathematics 11, Marcel Dekker, New York, 1972.

Benson, D. J., *Representations and Cohomology* I: *Basic Representation Theory of Finite Groups and Associative Algebras*, 2d ed., Cambridge University Press, Cambridge, 2000.

Blyth, T. S., *Module Theory; An Approach to Linear Algebra*, 2d ed., Oxford University Press, New York, 1990.

Bott, R., and Tu, I., *Differential Forms in Algebraic Topology*, Springer-Verlag, New York, 1982.

Brown, K. S., *Cohomology of Groups*, Springer-Verlag, New York, 1982.

Cartan, H., and Eilenberg, S., *Homological Algebra*, Princeton University Press, Princeton, NJ, 1956.

Cassels, J. W. S., and Fröhlich, A. Eds., *Algebraic Number Theory*, Academic Press, London, 1967.

Chase, S. U., Harrison, D. K., and Rosenberg, A., *Galois Theory and Cohomology of Commutative Rings*, Memoir American Mathematical Society, Providence, RI, 1965.

Cohen, D. E., *Groups of Cohomological Dimension One*, Lecture Notes 245, Springer-Verlag, New York, 1972.

Cohn, P. M., *Free Rings and Their Relations*, Academic Press, New York, 1971.

Curtis, C., and Reiner, I., *Representation Theory of Finite Groups and Associative Algebras*, Wiley, New York, 1962; reprinted, American Mathematical Society, Chelsea, Providence, RI, 2006.

De Meyer, F., and Ingraham, E., *Separable Algebras over Commutative Rings*, Lecture Notes 181, Springer-Verlag, New York, 1971.

Dixon, J. D., du Sautoy, M. P. F., Mann, A., and Segal, D., *Analytic pro-p groups*, 2d ed., Cambridge Studies in Advanced Mathematics 61, Cambridge University Press, Cambridge, 1999.

Douady, R., and Douady, A., *Algèbre et Théories Galoisiennes* I, CEDIC, Paris, 1977; reprinted, Cassini, Paris, 2005.

Dowker, C. H., *Lectures on Sheaf Theory*, Tata Institute, Bombay, 1956.

Dugundji, J., *Topology*, Allyn & Bacon, Boston, 1966.

Evens, L., *The Cohomology of Groups*, Oxford University Press, New York, 1991.

Fontana, M., Huckaba, J. A., and Papick, I. J., *Prüfer Domains*, Monographs and Textbooks in Pure and Applied Mathematics 203, Marcel Dekker, New York, 1997.

Fossum, R. M., Griffith, P., and Reiten, I., *Trivial Extensions of Abelian Categories*, Lecture Notes 456, Springer-Verlag, New York, 1975.

Freyd, P., *Abelian Categories; An Introduction to the Theory of Functors*, Harper & Row, New York, 1964.

Fuchs, L., *Infinite Abelian Groups* I, Academic Press, New York, 1970.

Fuchs, L., *Infinite Abelian Groups* II, Academic Press, New York, 1973.

Fulton, W., *Algebraic Topology: A First Course*, Springer-Verlag, New York, 1995.

Gelfand, S. I., and Manin, Y. I., *Methods of Homological Algebra*, 2d ed., Springer, Berlin, 2003.

Gille, P., and Szamuely, T., *Central Simple Algebras and Galois Cohomology*, Cambridge University Press, Cambridge, 2006.

Godement, R., *Topologie Algébrique et Théorie des Faisceaux*, Actualités Scientifiques et Industrielles 1252, Hermann, Paris, 1964.

Grothendieck, A., "Sur quelques points d'algèbre homologique," *Tohoku Math J.* 9 (1957), 119–221.

Grothendieck, A., *Éléments de Géométrie Algèbrique* III; *Étude Cohomologique des Faisceaux Cohérents*, IHES, Paris, 1961.

Gruenberg, K. W., *Cohomological Topics in Group Theory*, Lecture Notes 143, Springer-Verlag, New York, 1970.

Gunning, R. C., *Introduction to Holomorphic Functions of Several Variables* III, *Homological Theory*, Wadsworth & Brooks/Cole, Monterey, CA, 1990.

Gunning, R. C., *Lectures on Riemann Surfaces*, Princeton University Press, Princeton, NJ, 1966.

Hartshorne, R., *Algebraic Geometry*, Springer-Verlag, New York, 1977.

Herrlich, H., and Strecker, G. E., *Category Theory*, Allyn and Bacon, Boston, 1973.

Herstein, I. N., *Noncommutative Rings*, Carus Mathematical Monograph, American Mathematical Society, Providence, RI, 1968.

Hewitt, E., and Ross, K. A., *Abstract Harmonic Analysis* I, Springer-Verlag, New York, 1963.

Hilton, P., and Stammbach, U., *A Course in Homological Algebra*, Springer-Verlag, New York, 1971.

Hilton, P., and Wylie, S., *Homology Theory*, Cambridge University Press, Cambridge, 1960.

Humphreys, J., *Introduction to Lie Algebras and Representation Theory*, Springer-Verlag, New York, 1972.

Hurewicz, W., and Wallman, H., *Dimension Theory*, Princeton University Press, Princeton, NJ, 1948.

Jacobson, N., *Basic Algebra* I, Freeman, San Francisco, 1974.

Jacobson, N., *Basic Algebra* II, Freeman, San Francisco, 1980.

James, I. M., *History of Topology*, North-Holland, Amsterdam, 1999.

Jans, J., *Rings and Homology*, Holt, New York, 1964.

Janusz, G. J., *Algebraic Number Theory*, Academic Press, New York, 1973.

Kaplansky, I., *Infinite Abelian Groups*, University of Michigan Press, Ann Arbor, MI, 1954.

Kaplansky, I., *Commutative Rings*, Allyn and Bacon, Boston, 1970.

Kaplansky, I., *Set Theory and Metric Spaces*, 2d ed., Chelsea, New York, 1977.

Kashiwara, M., and Schapira, P., *Categories and Sheaves*, Springer, New York, 2006.

Kendig, K., *Elementary Algebraic Geometry*, Springer-Verlag, New York, 1977.

Lam, T. Y., *A First Course in Noncommutative Rings*, Springer-Verlag, New York, 1991.

Lam, T. Y., *Lectures on Modules and Rings*, Springer-Verlag, New York, 1999.

Lam, T. Y., *Serre's Problem on Projective Modules*, Springer, New York, 2006.

Lambek, J., *Lectures on Rings and Modules*, Ginn (Blaisdell), Boston, 1966.

Lang, S., *Algebra*, Addison-Wesley, Reading, MA, 1965.

Lyndon, R. C., "The cohomology theory of group extensions," *Duke Math J.* 15 (1948), 271–292.

Mac Lane, S., *Categories for the Working Mathematician*, Springer-Verlag, New York, 1971.

Mac Lane, S., *Homology*, Springer-Verlag, New York, 1975.

Macdonald, I. G., *Algebraic Geometry: Introduction to Schemes*, W. A. Benjamin, Inc., New York, 1968.

Matsumura, H., *Commutative Ring Theory*, Cambridge University Press, Cambridge, 1986.

May, J. P., *Simplicial Objects in Algebraic Topology*, van Nostrand Reinhold, New York, 1967.

McCleary, J., *User's Guide to Spectral Sequences*, Publish or Perish, Wilmington, DE, 1985.

Milnor, J., *Introduction to Algebraic K-Theory*, Princeton University Press, Princeton, NJ, 1971.

Mitchell, B., *Theory of Categories*, Academic Press, New York, 1965.

Munkres, J. R., *Elements of Algebraic Topology*, Addison-Wesley, Menlo Park, CA, 1984.

Osborne, M. S., *Basic Homological Algebra*, Springer-Verlag, New York, 2000.

Osofsky, B. L., *Homological Dimension of Modules*, CBMS Regional Conf. Ser. Math. 12, American Mathematical Society, Providence, RI, 1973.

Pareigis, B., *Categories and Functors*, Academic Press, New York, 1970.

Reiner, I., *Maximal Orders*, Academic Press, New York, 1975; reprinted, Oxford University Press, New York, 2003.

Robinson, D. J. S., *A Course in the Theory of Groups*, 2d ed., Springer-Verlag, New York, 1996.

Rosen, M., *Number Theory in Function Fields*, Springer-Verlag, New York, 2002.

Rosenberg, J., *Algebraic K-theory and Its Applications*, Springer-Verlag, New York, 1994.

Rotman, J. J., *Advanced Modern Algebra*, Prentice-Hall, Upper Saddle River, NJ, 2002.

Rotman, J. J., *An Introduction to Homological Algebra*, Academic Press, New York, 1979.

Rotman, J. J., *An Introduction to the Theory of Groups*, 4th ed., Springer-Verlag, New York, 1995.

Rotman, J. J., *Notes on Homological Algebra*, van Nostrand Reinhold, New York, 1970.

Rowen, L. H., *Ring Theory* I, Academic Press, New York, 1988.

Serre, J.-P., *Algebraic Groups and Class Fields*, Springer-Verlag, New York, 1988.

Serre, J.-P., *Local Fields*, Springer-Verlag, New York, 1979.

Serre, J.-P., "Faisceaux algèbriques cohérents," *Annals Math.* 61 (1955), 197–278.

Spanier, E. H., *Algebraic Topology*, Springer-Verlag, New York, 1981.

Suzuki, M., *Group Theory* I, Springer-Verlag, New York, 1982.

Swan, R. G., *The Theory of Sheaves*, University of Chicago Press, Chicago, 1964.

Tennison, B. R., *Sheaf Theory*, London Mathematical Society Lecture Note Series 20, Cambridge University Press, Cambridge, 1975.

Verschoren, A., *Relative Invariants of Sheaves*, Monographs and Textbooks in Pure and Applied Mathematics 11, Marcel Dekker, New York, 1987.

Weibel, C. A., *An Introduction to Homological Algebra*, Cambridge University Press, New York, 1994.

Weiss, E., *Cohomology of Groups*, Academic Press, New York, 1969.

Wells, R. O., Jr., *Differential Analysis on Complex Manifolds*, 2d ed., Graduate Texts in Mathematics, 65, Springer-Verlag, New York, 1980.

Zariski, O., and Samuel, P. *Commutative Algebra*, I, 2d ed., van Nostrand Reinhold, Princeton, NJ, 1975.

Special Notation

$\mathbb{N}$	natural numbers		$\mathbb{Z}$	integers	
$\mathbb{Q}$	rationals		$\mathbb{R}$	real numbers	
$\mathbb{C}$	complex numbers		$\mathbb{I}_n$	integers mod n	
$A \subseteq B$	A a subset of B		$A \subsetneq B$	A a proper subset of B	
$|X|$	cardinal number of a set X		$X \ni x$	$x \in X$	
$X \rightarrowtail Y$	injection, monomorphism		$X \twoheadrightarrow Y$	surjection, epimorphism	

$$v_0, \ldots, \widehat{v_i}, \ldots, v_n \quad \text{list } v_0, \ldots, v_n \text{ with } v_i \text{ omitted}$$

∂_n	5	R^{op}	15	$\bigoplus_{i \in I} A_i$	53
Z_n	6	$\mathbf{Mod}_R$	16	$\prod_{i \in I} A_i$	53
B_n	6	$\mathrm{Hom}_R(A, \square)$	17	A^X	53
H_n	6	f_*	17	$(A_i)_{i \in I}$	53
$\mathrm{Hom}_{\mathcal{C}}(A, B)$	8	$\mathrm{Hom}_R(\square, B)$	20	$A \otimes_R B$	71
$\mathrm{obj}(\mathcal{C})$	8	f^*	20	$A \otimes_R \square, \square \otimes_R B$	75
Sets	9	V^*	20	$_R M_S$	75
Groups	9	$\mathcal{C}^{\mathrm{op}}$	22	A^e	83
Top	10	$\mathrm{Nat}(F, G)$	25	$\mathrm{Frac}(R)$	94
Asc	11	$\mathcal{B}^{\mathcal{A}}$	27	$\mathrm{Env}(M)$	127
$N(\mathcal{U})$	11	$\mathbf{pSh}(X, \mathbf{Ab})$	28	$\mathbb{Z}(p^{\infty})$	131
Htp	12	Δ^n	29	tM	134
Ab	12	$\mathbf{S}_{\bullet}(X)$	33	$R^{\times}$	189
ComRings	12	kG	38	R_{p}	194
$_R M$	13	$\langle X \rangle$	42	$S^{-1}R$	195
$Z(R)$	13	$\ker f$	43	$A \sqcup B$	214
M_R	14	$\mathrm{im}\, f$	43	$A \sqcap B$	217
$\mathrm{Hom}_R(A, B)$	15	$\mathrm{coker}\, f$	43	$X \vee Y$	227
$_R \mathbf{Mod}$	15	$S \oplus T$	49	$\{M_i, \psi_i^j\}$	230
				$\varprojlim M_i$	231

Index